CALCULUS
SECOND EDITION

STUDENT'S SOLUTIONS MANUAL

Ross L. Finney

Franklin D. Demana
The Ohio State University

Bert K. Waits
The Ohio State University

Daniel Kennedy
Baylor School

An imprint of Addison Wesley Longman, Inc.

Reading, Massachusetts • Menlo Park, California • New York • Harlow, England
Don Mills, Ontario • Sydney • Mexico City • Madrid • Amsterdam

Introduction

This publication provides complete, worked out solutions for Quick Review, odd numbered Exercises, and lesson Explorations, including Cumulative Review and Appendices exercises.
      Many of the analytic solutions could alternatively be done graphically. Teachers are encouraged to accept creative, graphical solutions that show understanding of the graphing utilities. In general, the solutions in this manual provide one possible method of solution.

Reproduced by Addison Wesley Longman from camera copy.

Copyright © 1999 Addison Wesley Longman.

All rights reserved. No part of this publication may be reproduced, stored in a retrieval system, or transmitted, in any form or by any means, electronic, mechanical, photocopying, recording, or otherwise, without the prior written permission of the publisher. Printed in the United States of America.

ISBN 0-201-44139-X

3 4 5 6 7 8 9 10 VG 00

CONTENTS

Chapter 1 Prerequisites for Calculus — 1
  1.1 Lines — 1
  1.2 Functions and Graphs — 3
  1.3 Exponential Functions — 7
  1.4 Parametric Equations — 9
  1.5 Functions and Logarithms — 14
  1.6 Trigonometric Functions — 19
  Chapter Review — 21

Chapter 2 Limits and Continuity — 27
  2.1 Rates of Change and Limits — 27
  2.2 Limits Involving Infinity — 31
  2.3 Continuity — 35
  2.4 Rates of Change and Tangent Lines — 37
  Chapter Review — 42

Chapter 3 Derivatives — 45
  3.1 Derivative of a Function — 45
  3.2 Differentiability — 48
  3.3 Rules for Differentiation — 50
  3.4 Velocity and Other Rates of Change — 52
  3.5 Derivatives of Trigonometric Functions — 56
  3.6 Chain Rule — 59
  3.7 Implicit Differentiation — 64
  3.8 Derivatives of Inverse Trigonometric Functions — 69
  3.9 Derivatives of Exponential and Logarithmic Functions — 72
  Chapter Review — 74

Chapter 4 Applications of Derivities — 83
  4.1 Extreme Values of Functions — 83
  4.2 Mean Value Theorem — 87
  4.3 Connecting f' and f'' with the Graph of f — 90
  4.4 Modeling and Optimization — 98
  4.5 Linearization and Newton's Method — 108
  4.6 Related Rates — 112
  Chapter Review — 118

Chapter 5 The Definite Integral — 137
  5.1 Estimating with Finite Sums — 137
  5.2 Definite Integrals — 140
  5.3 Definite Integrals and Antiderivatives — 143
  5.4 Fundamental Theorem of Calculus — 146
  5.5 Trapezoidal Rule — 150
  Chapter Review — 151

Chapter 6	**Differential Equations and Mathematical Modeling**	**156**
6.1	Antiderivatives and Slope Fields	156
6.2	Integration by Substitution	161
6.3	Integration by Parts	166
6.4	Exponential Growth and Decay	171
6.5	Population Growth	175
6.6	Numerical Methods	180
	Chapter Review	184

Chapter 7	**Applications of Definite Integrals**	**192**
7.1	Integral as Net Change	192
7.2	Areas in the Plane	196
7.3	Volumes	199
7.4	Lengths of Curves	205
7.5	Applications from Science and Statistics	206
	Chapter Review	209

Chapter 8	**L'Hôpital's Rule, Improper Integrals, and Partial Fractions**	**214**
8.1	L'Hôpital's Rule	214
8.2	Relative Rates of Growth	219
8.3	Improper Integrals	221
8.4	Partial Fractions and Integral Tables	227
	Chapter Review	233

Chapter 9	**Infinite Series**	**242**
9.1	Power Series	242
9.2	Taylor Series	247
9.3	Taylor's Theorem	250
9.4	Radius of Convergence	253
9.5	Testing Convergence at Endpoints	256
	Chapter Review	259

Chapter 10	**Vectors**	**267**
10.1	Parametric Functions	267
10.2	Vectors in the Plane	270
10.3	Vector-valued Functions	273
10.4	Modeling Projectile Motion	276
10.5	Polar Coordinates and Polar Graphs	279
10.6	Calculus of Polar Curves	282
	Chapter Review	287

	Appendix	**294**
	Cumulative Review	294
A2	Mathematical Induction	304
A3	Using the Limit Definition	305
A5.1	Conic Sections and Quadratic Equations	306
A5.2	Classifying Conic Sections by Eccentricity	309
A5.3	Quadratic Equations and Rotations	310
A6	Hyperbolic Functions	312

Chapter 1
Prerequisites for Calculus

■ Section 1.1 Lines (pp. 1–9)

Quick Review 1.1

1. $y = -2 + 4(3 - 3) = -2 + 4(0) = -2 + 0 = -2$
2. $3 = 3 - 2(x + 1)$
   $3 = 3 - 2x - 2$
   $2x = -2$
   $x = -1$
3. $m = \dfrac{2 - 3}{5 - 4} = \dfrac{-1}{1} = -1$
4. $m = \dfrac{2 - (-3)}{3 - (-1)} = \dfrac{5}{4}$
5. (a) $3(2) - 4\left(\dfrac{1}{4}\right) \stackrel{?}{=} 5$
   $6 - 1 = 5$ Yes
   (b) $3(3) - 4(-1) \stackrel{?}{=} 5$
   $13 \neq 5$ No
6. (a) $7 \stackrel{?}{=} -2(-1) + 5$
   $7 = 2 + 5$ Yes
   (b) $1 = -2(-2) + 5$
   $1 \neq 9$ No
7. $d = \sqrt{(x_2 - x_1)^2 + (y_2 - y_1)^2}$
   $= \sqrt{(0 - 1)^2 + (1 - 0)^2}$
   $= \sqrt{2}$
8. $d = \sqrt{(x_2 - x_1)^2 + (y_2 - y_1)^2}$
   $= \sqrt{(1 - 2)^2 + \left(-\dfrac{1}{3} - 1\right)^2}$
   $= \sqrt{(-1)^2 + \left(-\dfrac{4}{3}\right)^2}$
   $= \sqrt{1 + \dfrac{16}{9}}$
   $= \sqrt{\dfrac{25}{9}}$
   $= \dfrac{5}{3}$
9. $4x - 3y = 7$
   $-3y = -4x + 7$
   $y = \dfrac{4}{3}x - \dfrac{7}{3}$
10. $-2x + 5y = -3$
    $5y = 2x - 3$
    $y = \dfrac{2}{5}x - \dfrac{3}{5}$

Section 1.1 Exercises

1. $\Delta x = -1 - 1 = -2$
   $\Delta y = -1 - 2 = -3$
3. $\Delta x = -8 - (-3) = -5$
   $\Delta y = 1 - 1 = 0$
5. (a, c)

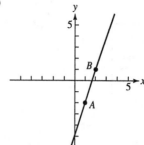

(b) $m = \dfrac{1 - (-2)}{2 - 1} = \dfrac{3}{1} = 3$

7. (a, c)

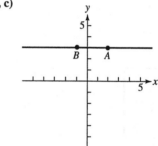

(b) $m = \dfrac{3 - 3}{-1 - 2} = \dfrac{0}{-3} = 0$

9. (a) $x = 2$
   (b) $y = 3$
11. (a) $x = 0$
    (b) $y = -\sqrt{2}$
13. $y = 1(x - 1) + 1$
15. $y = 2(x - 0) + 3$
17. $m = \dfrac{3 - 0}{2 - 0} = \dfrac{3}{2}$
    $y = \dfrac{3}{2}(x - 0) + 0$
    $y = \dfrac{3}{2}x$
    $2y = 3x$
    $3x - 2y = 0$
19. $m = \dfrac{-2 - 0}{-2 - (-2)} = \dfrac{-2}{0}$ (undefined)
    Vertical line: $x = -2$
21. $y = 3x - 2$
23. $y = -\dfrac{1}{2}x - 3$
25. The line contains $(0, 0)$ and $(10, 25)$.
    $m = \dfrac{25 - 0}{10 - 0} = \dfrac{25}{10} = \dfrac{5}{2}$
    $y = \dfrac{5}{2}x$

2 Section 1.1

27. $3x + 4y = 12$

$4y = -3x + 12$

$y = -\frac{3}{4}x + 3$

(a) Slope: $-\frac{3}{4}$

(b) y-intercept: 3

(c)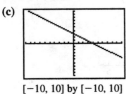
[−10, 10] by [−10, 10]

29. $\frac{x}{3} + \frac{y}{4} = 1$

$\frac{y}{4} = -\frac{x}{3} + 1$

$y = -\frac{4}{3}x + 4$

(a) Slope: $-\frac{4}{3}$

(b) y-intercept: 4

(c)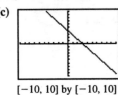
[−10, 10] by [−10, 10]

31. (a) The desired line has slope -1 and passes through $(0, 0)$:
$y = -1(x - 0) + 0$ or $y = -x$.

(b) The desired line has slope $\frac{-1}{-1} = 1$ and passes through $(0, 0)$:
$y = 1(x - 0) + 0$ or $y = x$.

33. (a) The given line is vertical, so we seek a vertical line through $(-2, 4)$: $x = -2$.

(b) We seek a horizontal line through $(-2, 4)$: $y = 4$.

35. $m = \frac{9 - 2}{3 - 1} = \frac{7}{2}$

$f(x) = \frac{7}{2}(x - 1) + 2 = \frac{7}{2}x - \frac{3}{2}$

Check: $f(5) = \frac{7}{2}(5) - \frac{3}{2} = 16$, as expected.

Since $f(x) = \frac{7}{2}x - \frac{3}{2}$, we have $m = \frac{7}{2}$ and $b = -\frac{3}{2}$.

37. $-\frac{2}{3} = \frac{y - 3}{4 - (-2)}$

$-\frac{2}{3}(6) = y - 3$

$-4 = y - 3$

$-1 = y$

39. (a) $y = 0.680x + 9.013$

(b) The slope is 0.68. It represents the approximate average weight gain in pounds per month.

(c)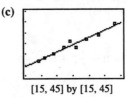
[15, 45] by [15, 45]

(d) When $x = 30$, $y \approx 0.680(30) + 9.013 = 29.413$. She weighs about 29 pounds.

41. $y = 1 \cdot (x - 3) + 4$

$y = x - 3 + 4$

$y = x + 1$

This is the same as the equation obtained in Example 5.

43. (a) The given equations are equivalent to $y = -\frac{2}{k}x + \frac{3}{k}$ and $y = -x + 1$, respectively, so the slopes are $-\frac{2}{k}$ and -1. The lines are parallel when $-\frac{2}{k} = -1$, so $k = 2$.

(b) The lines are perpendicular when $-\frac{2}{k} = \frac{-1}{-1}$, so $k = -2$.

45. Slope: $k = \frac{\Delta p}{\Delta d} = \frac{10.94 - 1}{100 - 0} = \frac{9.94}{100}$
$= 0.0994$ atmospheres per meter

At 50 meters, the pressure is

$p = 0.0994(50) + 1 = 5.97$ atmospheres.

47. (a) $y = 5632x - 11,080,280$

(b) The rate at which the median price is increasing in dollars per year

(c) $y = 2732x - 5,362,360$

(d) The median price is increasing at a rate of about $5632 per year in the Northeast, and about $2732 per year in the Midwest. It is increasing more rapidly in the Northeast.

49. The coordinates of the three missing vertices are $(5, 2)$, $(-1, 4)$ and $(-1, -2)$, as shown below.

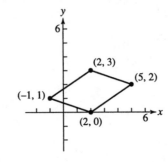

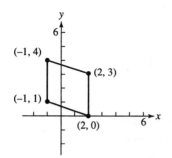

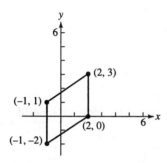

51. The radius through (3, 4) has slope $\frac{4-0}{3-0} = \frac{4}{3}$.

The tangent line is tangent to this radius, so its slope is $\frac{-1}{4/3} = -\frac{3}{4}$. We seek the line of slope $-\frac{3}{4}$ that passes through (3, 4).

$y = -\frac{3}{4}(x - 3) + 4$

$y = -\frac{3}{4}x + \frac{9}{4} + 4$

$y = -\frac{3}{4}x + \frac{25}{4}$

■ Section 1.2 Functions and Graphs

(pp. 9–19)

Exploration 1 Composing Functions

1. $y_3 = g \circ f$, $y_4 = f \circ g$
2. Domain of y_3: [−2, 2] Range of y_3: [0, 2]

y_1:
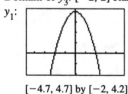
[−4.7, 4.7] by [−2, 4.2]

y_2:
[−4.7, 4.7] by [−2, 4.2]

y_3:
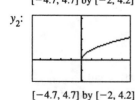
[−4.7, 4.7] by [−2, 4.2]

3. Domain of y_4: [0, ∞); Range of y_4: (−∞, 4]

y_4:
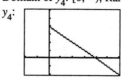
[−2, 6] by [−2, 6]

4. $y_3 = y_2(y_1(x)) = \sqrt{y_1(x)} = \sqrt{4 - x^2}$
   $y_4 = y_1(y_2(x)) = 4 - (y_2(x))^2 = 4 - (\sqrt{x})^2 = 4 - x$, $x \geq 0$

Quick Review 1.2

1. $3x - 1 \leq 5x + 3$
   $-2x \leq 4$
   $x \geq -2$
   Solution: $[-2, \infty)$

2. $x(x - 2) > 0$
   Solutions to $x(x - 2) = 0$: $x = 0$, $x = 2$
   Test $x = -1$: $-1(-1 - 2) = 3 > 0$
   $x(x - 2) > 0$ is true when $x < 0$.
   Test $x = 1$: $1(1 - 2) = -1 < 0$
   $x(x - 2) > 0$ is false when $0 < x < 2$.
   Test $x = 3$: $3(3 - 2) = 3 > 0$
   $x(x - 2) > 0$ is true when $x > 2$.
   Solution set: $(-\infty, 0) \cup (2, \infty)$

3. $|x - 3| \leq 4$
   $-4 \leq x - 3 \leq 4$
   $-1 \leq x \leq 7$
   Solution set: $[-1, 7]$

4. $|x - 2| \geq 5$
   $x - 2 \leq -5$ or $x - 2 \geq 5$
   $x \leq -3$ or $x \geq 7$
   Solution set: $(-\infty, -3] \cup [7, \infty)$

5. $x^2 < 16$
   Solutions to $x^2 = 16$: $x = -4$, $x = 4$
   Test $x = -6$ $(-6)^2 = 36 > 16$
   $x^2 < 16$ is false when $x < -4$
   Test $x = 0$: $0^2 = 0 < 16$
   $x^2 < 16$ is true when $-4 < x < 4$
   Test $x = 6$: $6^2 = 36 > 16$
   $x^2 < 16$ is false when $x > 4$.
   Solution set: $(-4, 4)$

6. $9 - x^2 \geq 0$
   Solutions to $9 - x^2 = 0$: $x = -3$, $x = 3$
   Test $x = -4$: $9 - (-4)^2 = 9 - 16 = -7 < 0$
   $9 - x^2 \geq 0$ is false when $x < -3$.
   Test $x = 0$: $9 - 0^2 = 9 > 0$
   $9 - x^2 \geq 0$ is true when $-3 < x < 3$.
   Test $x = 4$: $9 - 4^2 = 9 - 16 = -7 < 0$
   $9 - x^2 \geq 0$ is false when $x > 3$.
   Solution set: $[-3, 3]$

7. Translate the graph of f 2 units left and 3 units downward.
8. Translate the graph of f 5 units right and 2 units upward.
9. (a) $f(x) = 4$
   $x^2 - 5 = 4$
   $x^2 - 9 = 0$
   $(x + 3)(x - 3) = 0$
   $x = -3$ or $x = 3$

   (b) $f(x) = -6$
   $x^2 - 5 = -6$
   $x^2 = -1$
   No real solution

4 Section 1.2

10. (a) $f(x) = -5$
$$\frac{1}{x} = -5$$
$$x = -\frac{1}{5}$$

(b) $f(x) = 0$
$$\frac{1}{x} = 0$$
No solution

11. (a) $f(x) = 4$
$$\sqrt{x+7} = 4$$
$$x + 7 = 16$$
$$x = 9$$
Check: $\sqrt{9+7} = \sqrt{16} = 4$; it checks.

(b) $f(x) = 1$
$$\sqrt{x+7} = 1$$
$$x + 7 = 1$$
$$x = -6$$
Check: $\sqrt{-6+7} = 1$; it checks.

12. (a) $f(x) = -2$
$$\sqrt[3]{x-1} = -2$$
$$x - 1 = -8$$
$$x = -7$$

(b) $f(x) = 3$
$$\sqrt[3]{x-1} = 3$$
$$x - 1 = 27$$
$$x = 28$$

Section 1.2 Exercises

1. Since $A = \pi r^2 = \pi\left(\frac{d}{2}\right)^2$, the formula is $A = \frac{\pi d^2}{4}$, where A represents area and d represents diameter.

3. $S = 6e^2$, where S represents surface area and e represents edge length.

5. (a) $(-\infty, \infty)$ or all real numbers

(b) $(-\infty, 4]$

(c)

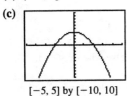

$[-5, 5]$ by $[-10, 10]$

(d) Symmetric about y-axis (even)

7. (a) Since we require $x - 1 \geq 0$, the domain is $[1, \infty)$.

(b) $[2, \infty)$

(c)

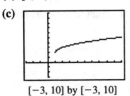

$[-3, 10]$ by $[-3, 10]$

(d) None

9. (a) Since we require $3 - x \geq 0$, the domain is $(-\infty, 3]$.

(b) $[0, \infty)$

(c)

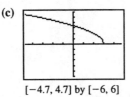

$[-4.7, 4.7]$ by $[-6, 6]$

(d) None

11. (a) $(-\infty, \infty)$ or all real numbers

(b) $(-\infty, \infty)$ or all real numbers

(c)

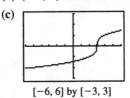

$[-6, 6]$ by $[-3, 3]$

(d) None

13. (a) Since we require $-x \geq 0$, the domain is $(-\infty, 0]$.

(b) $[0, \infty)$

(c)

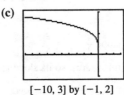

$[-10, 3]$ by $[-1, 2]$

(d) None

15. (a) Since we require $4 - x^2 \geq 0$, the domain is $[-2, 2]$.

(b) Since $4 - x^2$ will be between 0 and 4, inclusive (for x in the domain), its square root is between 0 and 2, inclusive. The range is $[0, 2]$.

(c)

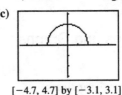

$[-4.7, 4.7]$ by $[-3.1, 3.1]$

(d) Symmetric about the y-axis (even)

17. (a) Since we require $x^2 \neq 0$, the domain is $(-\infty, 0) \cup (0, \infty)$.

(b) Since $\frac{1}{x^2} > 0$ for all x, the range is $(1, \infty)$.

(c)

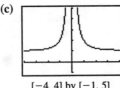

$[-4, 4]$ by $[-1, 5]$

(d) Symmetric about the y-axis (even)

19. Even, since the function is an even power of x.

21. Neither, since the function is a sum of even and odd powers of x ($x^1 + 2x^0$).

23. Even, since the function involves only even powers of x.

25. Odd, since the function is a quotient of an odd function (x^3) and an even function ($x^2 - 1$).

27. Neither, since, (for example), $y(-1)$ is defined and $y(1)$ is undefined.

29. (a)

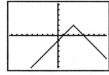

[−9.4, 9.4] by [−6.2, 6.2]

Note that $f(x) = -|x - 3| + 2$, so its graph is the graph of the absolute value function reflected across the x-axis and then shifted 3 units right and 2 units upward.

(b) $(-\infty, \infty)$

(c) $(-\infty, 2]$

31. (a)

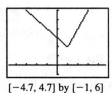

[−4.7, 4.7] by [−1, 6]

(b) $(-\infty, \infty)$ or all real numbers

(c) $[2, \infty)$

33. (a)

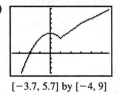

[−3.7, 5.7] by [−4, 9]

(b) $(-\infty, \infty)$ or all real numbers

(c) $(-\infty, \infty)$ or all real numbers

35. Because if the vertical line test holds, then for each x-coordinate, there is at most one y-coordinate giving a point on the curve. This y-coordinate would correspond to the value assigned to the x-coordinate. Since there is only one y-coordinate, the assignment would be unique.

37. No

39. Yes

41. Line through (0, 0) and (1, 1): $y = x$
Line through (1, 1) and (2, 0): $y = -x + 2$

$$f(x) = \begin{cases} x, & 0 \le x \le 1 \\ -x + 2, & 1 < x \le 2 \end{cases}$$

43. Line through (0, 2) and (2, 0): $y = -x + 2$
Line through (2, 1) and (5, 0): $m = \dfrac{0-1}{5-2} = \dfrac{-1}{3} = -\dfrac{1}{3}$,
so $y = -\dfrac{1}{3}(x - 2) + 1 = -\dfrac{1}{3}x + \dfrac{5}{3}$

$$f(x) = \begin{cases} -x + 2, & 0 < x \le 2 \\ -\dfrac{1}{3}x + \dfrac{5}{3}, & 2 < x \le 5 \end{cases}$$

45. Line through (−1, 1) and (0, 0): $y = -x$

Line through (0, 1) and (1, 1): $y = 1$

Line through (1, 1) and (3, 0):
$m = \dfrac{0-1}{3-1} = \dfrac{-1}{2} = -\dfrac{1}{2}$,
so $y = -\dfrac{1}{2}(x - 1) + 1 = -\dfrac{1}{2}x + \dfrac{3}{2}$

$$f(x) = \begin{cases} -x, & -1 \le x < 0 \\ 1, & 0 < x \le 1 \\ -\dfrac{1}{2}x + \dfrac{3}{2}, & 1 < x < 3 \end{cases}$$

47. Line through $\left(\dfrac{T}{2}, 0\right)$ and $(T, 1)$:
$m = \dfrac{1-0}{T - (T/2)} = \dfrac{2}{T}$, so $y = \dfrac{2}{T}\left(x - \dfrac{T}{2}\right) + 0 = \dfrac{2}{T}x - 1$

$$f(x) = \begin{cases} 0, & 0 \le x \le \dfrac{T}{2} \\ \dfrac{2}{T}x - 1, & \dfrac{T}{2} < x \le T \end{cases}$$

49. (a) $f(g(x)) = (x^2 - 3) + 5 = x^2 + 2$

(b) $g(f(x)) = (x + 5)^2 - 3$
$= (x^2 + 10x + 25) - 3$
$= x^2 + 10x + 22$

(c) $f(g(0)) = 0^2 + 2 = 2$

(d) $g(f(0)) = 0^2 + 10 \cdot 0 + 22 = 22$

(e) $g(g(-2)) = [(-2)^2 - 3]^2 - 3 = 1^2 - 3 = -2$

(f) $f(f(x)) = (x + 5) + 5 = x + 10$

51. (a) Enter $y_1 = f(x) = x - 7$, $y_2 = g(x) = \sqrt{x}$,
$y_3 = (f \circ g)(x) = y_1(y_2(x))$,
and $y_4 = (g \circ f)(x) = y_2(y_1(x))$

$f \circ g$:

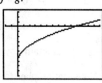

[−10, 70] by [−10, 3]
Domain: $[0, \infty)$
Range: $[-7, \infty)$

(b) $(f \circ g)(x) = \sqrt{x} - 7$

$g \circ f$:

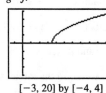

[−3, 20] by [−4, 4]
Domain: $[7, \infty)$
Range: $[0, \infty)$

$(g \circ f)(x) = \sqrt{x - 7}$

6 Section 1.2

53. (a) Enter $y_1 = f(x) = x^2 - 3$, $y_2 = g(x) = \sqrt{x+2}$,
$y_3 = (f \circ g)(x) = y_1(y_2(x))$,
and $y_4 = (g \circ f)(x) = y_2(y_1(x))$.
$f \circ g$:

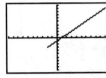

[−10, 10] by [−10, 10]
Domain: $[-2, \infty)$
Range: $[-3, \infty)$
$g \circ f$:

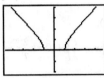

[−4.7, 4.7] by [−2, 4]
Domain: $(-\infty, -1] \cup [1, \infty)$
Range: $[0, \infty)$

(b) $(f \circ g)(x) = (\sqrt{x+2})^2 - 3$
$= (x+2) - 3, x \geq -2$
$= x - 1, x \geq -2$
$(g \circ f)(x) = \sqrt{(x^2 - 3) + 2} = \sqrt{x^2 - 1}$

55.

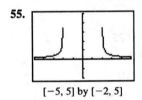

[−5, 5] by [−2, 5]

We require $x^2 - 4 \geq 0$ (so that the square root is defined) and $x^2 - 4 \neq 0$ (to avoid division by zero), so the domain is $(-\infty, -2) \cup (2, \infty)$. For values of x in the domain, $x^2 - 4$ (and hence $\sqrt{x^2 - 4}$ and $\dfrac{1}{\sqrt{x^2 - 4}}$) can attain any positive value, so the range is $(0, \infty)$. (Note that grapher failure may cause the range to appear as a finite interval on a grapher.

57.

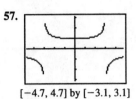

[−4.7, 4.7] by [−3.1, 3.1]

We require $9 - x^2 \neq 0$, so the domain is $(-\infty, -3) \cup (-3, 3) \cup (3, \infty)$. For values of x in the domain, $9 - x^2$ can attain any value in $(-\infty, 0) \cup (0, 9]$, so $\sqrt[3]{9 - x^2}$ can attain any value in $(-\infty, 0) \cup (0, \sqrt[3]{9}]$. Therefore, $\dfrac{2}{\sqrt[3]{9 - x^2}}$ can attain any value in $(-\infty, 0) \cup \left[\dfrac{2}{\sqrt[3]{9}}, \infty\right)$. The range is $(-\infty, 0) \cup \left[\dfrac{2}{\sqrt[3]{9}}, \infty\right)$ or approximately $(-\infty, 0) \cup [0.96, \infty)$.
(Note that grapher failure can cause the intervals in the range to appear as finite intervals on a grapher.)

59. (a)

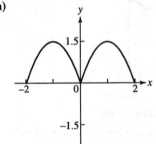

(b)

61. (a)

(b)
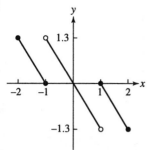

63. (a) Since $(f \circ g)(x) = \sqrt{g(x) - 5} = \sqrt{x^2 - 5}$, $g(x) = x^2$.

(b) Since $(f \circ g)(x) = 1 + \dfrac{1}{g(x)} = x$, we know that $\dfrac{1}{g(x)} = x - 1$, so $g(x) = \dfrac{1}{x - 1}$.

(c) Since $(f \circ g)(x) = f\left(\dfrac{1}{x}\right) = x$, $f(x) = \dfrac{1}{x}$.

(d) Since $(f \circ g)(x) = f(\sqrt{x}) = |x|$, $f(x) = x^2$.
The completed table is shown. Note that the absolute value sign in part (d) is optional.

$g(x)$	$f(x)$	$(f \circ g)(x)$
x^2	$\sqrt{x - 5}$	$\sqrt{x^2 - 5}$
$\dfrac{1}{x - 1}$	$1 + \dfrac{1}{x}$	$x, x \neq -1$
$\dfrac{1}{x}$	$\dfrac{1}{x}$	$x, x \neq 0$
$\sqrt{x}$	x^2	$|x|, x \geq 0$

65. (a) Because the circumference of the original circle was 8π and a piece of length x was removed.

(b) $r = \dfrac{8\pi - x}{2\pi} = 4 - \dfrac{x}{2\pi}$

(c) $h = \sqrt{16 - r^2}$

$= \sqrt{16 - \left(4 - \dfrac{x}{2\pi}\right)^2}$

$= \sqrt{16 - \left(16 - \dfrac{4x}{\pi} + \dfrac{x^2}{4\pi^2}\right)}$

$= \sqrt{\dfrac{4x}{\pi} - \dfrac{x^2}{4\pi^2}}$

$= \sqrt{\dfrac{16\pi x}{4\pi^2} - \dfrac{x^2}{4\pi^2}}$

$= \dfrac{\sqrt{16\pi x - x^2}}{2\pi}$

(d) $V = \dfrac{1}{3}\pi r^2 h$

$= \dfrac{1}{3}\pi\left(\dfrac{8\pi - x}{2\pi}\right)^2 \cdot \dfrac{\sqrt{16\pi x - x^2}}{2\pi}$

$= \dfrac{(8\pi - x)^2 \sqrt{16\pi x - x^2}}{24\pi^2}$

67. (a)

$[-3, 3]$ by $[-1, 3]$

(b) Domain of y_1: $[0, \infty)$
Domain of y_2: $(-\infty, 1]$
Domain of y_3: $[0, 1]$

(c) The functions $y_1 - y_2$, $y_2 - y_1$, and $y_1 \cdot y_2$ all have domain $[0, 1]$, the same as the domain of $y_1 + y_2$ found in part (b).

Domain of $\dfrac{y_1}{y_2}$: $[0, 1)$

Domain of $\dfrac{y_2}{y_1}$: $(0, 1]$

(d) The domain of a sum, difference, or product of two functions is the intersection of their domains.

The domain of a quotient of two functions is the intersection of their domains with any zeros of the denominator removed.

■ Section 1.3 Exponential Functions
(pp. 20–26)

Exploration 1   Exponential Functions

1.

$[-5, 5]$ by $[-2, 5]$

2. $x > 0$

3. $x < 0$

4. $x = 0$

5.
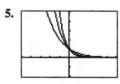
$[-5, 5]$ by $[-2, 5]$

6. $2^{-x} < 3^{-x} < 5^{-x}$ for $x < 0$; $2^{-x} > 3^{-x} > 5^{-x}$ for $x > 0$; $2^{-x} = 3^{-x} = 5^{-x}$ for $x = 0$.

Quick Review 1.3

1. Using a calculator, $5^{2/3} \approx 2.924$.

2. Using a calculator, $3^{\sqrt{2}} \approx 4.729$.

3. Using a calculator, $3^{-1.5} \approx 0.192$.

4. $x^3 = 17$
$x = \sqrt[3]{17}$
$x \approx 2.5713$

5. $x^5 = 24$
$x = \sqrt[5]{24}$
$x \approx 1.8882$

8 Section 1.3

6. $x^{10} = 1.4567$
$x = \pm \sqrt[10]{1.4567}$
$x \approx \pm 1.0383$

7. $500(1.0475)^5 \approx \$630.58$

8. $1000(1.063)^3 \approx \$1201.16$

9. $\dfrac{(x^{-3}y^2)^2}{(x^4y^3)^3} = \dfrac{x^{-6}y^4}{x^{12}y^9}$
$= x^{-6-12}y^{4-9}$
$= x^{-18}y^{-5}$
$= \dfrac{1}{x^{18}y^5}$

10. $\left(\dfrac{a^3b^{-2}}{c^4}\right)^2 \left(\dfrac{a^4c^{-2}}{b^3}\right)^{-1} = \dfrac{a^6b^{-4}}{c^8} \cdot \dfrac{b^3}{a^4c^{-2}}$
$= \dfrac{a^6}{b^4c^8} \cdot \dfrac{b^3c^2}{a^4}$
$= a^{6-4}b^{-4+3}c^{-8+2}$
$= a^2b^{-1}c^{-6} = \dfrac{a^2}{bc^6}$

Section 1.3 Exercises

1. The graph of $y = 2^x$ is increasing from left to right and has the negative x-axis as an asymptote. (a)

3. The graph of $y = -3^{-x}$ is the reflection about the x-axis of the graph in Exercise 2. (e)

5. The graph of $y = 2^{-x} - 2$ is decreasing from left to right and has the line $y = -2$ as an asymptote. (b)

7.
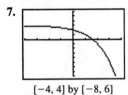
[−4, 4] by [−8, 6]
Domain: $(-\infty, \infty)$
Range: $(-\infty, 3)$
x-intercept: ≈ 1.585
y-intercept: 2

9.
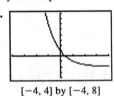
[−4, 4] by [−4, 8]
Domain: $(-\infty, \infty)$
Range: $(-2, \infty)$
x-intercept: ≈ 0.405
y-intercept: 1

11. $9^{2x} = (3^2)^{2x} = 3^{4x}$

13. $\left(\dfrac{1}{8}\right)^{2x} = (2^{-3})^{2x} = 2^{-6x}$

15.
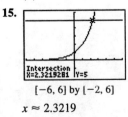
[−6, 6] by [−2, 6]
$x \approx 2.3219$

17.
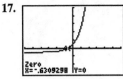
[−6, 6] by [−3, 5]
$x \approx -0.6309$

19.

x	y	Δy
1	−1	
		2
2	1	
		2
3	3	
		2
4	5	

21.

x	y	Δy
1	1	
		3
2	4	
		5
3	9	
		7
4	16	

23. Let t be the number of years. Solving $500{,}000(1.0375)^t = 1{,}000{,}000$ graphically, we find that $t \approx 18.828$. The population will reach 1 million in about 19 years.

25. (a) $A(t) = 6.6\left(\dfrac{1}{2}\right)^{t/14}$

(b) Solving $A(t) = 1$ graphically, we find that $t \approx 38.1145$. There will be 1 gram remaining after about 38.1145 days.

27. Let A be the amount of the initial investment, and let t be the number of years. We wish to solve $A(1.0625)^t = 2A$, which is equivalent to $1.0625^t = 2$. Solving graphically, we find that $t \approx 11.433$. It will take about 11.433 years. (If the interest is credited at the end of each year, it will take 12 years.)

29. Let A be the amount of the initial investment, and let t be the number of years. We wish to solve $Ae^{0.0625t} = 2A$, which is equivalent to $e^{0.0625t} = 2$. Solving graphically, we find that $t \approx 11.090$. It will take about 11.090 years.

31. Let A be the amount of the initial investment, and let t be the number of years. We wish to solve

$A\left(1 + \dfrac{0.0575}{365}\right)^{365t} = 3A$, which is equivalent to

$\left(1 + \dfrac{0.0575}{365}\right)^{365t} = 3$. Solving graphically, we find that

$t \approx 19.108$. It will take about 19.108 years.

33. After t hours, the population is $P(t) = 2^{t/0.5}$ or, equivalently, $P(t) = 2^{2t}$. After 24 hours, the population is $P(24) = 2^{48} \approx 2.815 \times 10^{14}$ bacteria.

35. Since $\Delta x = 1$, the corresponding value of Δy is equal to the slope of the line. If the changes in x are constant for a linear function, then the corresponding changes in y are constant as well.

37. (a) Let $x = 0$ represent 1900, $x = 1$ represent 1901, and so on. The regression equation is $P(x) = 6.033(1.030)^x$.

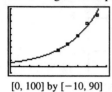

[0, 100] by [−10, 90]

(b) The regression equation gives an estimate of $P(0) \approx 6.03$ million, which is not very close to the actual population.

(c) Since the equation is of the form $P(x) = P(0) \cdot 1.030^x$, the annual rate of growth is about 3%.

39. $5422(1.018)^{19} \approx 7609.7$ million

41. Since $f(1) = 4.5$ we have $ka = 4.5$, and since $f(-1) = 0.5$ we have $ka^{-1} = 0.5$.
Dividing, we have
$$\frac{ka}{ka^{-1}} = \frac{4.5}{0.5}$$
$$a^2 = 9$$
$$a = \pm 3$$

Since $f(x) = k \cdot a^x$ is an exponential function, we require $a > 0$, so $a = 3$. Then $ka = 4.5$ gives $3k = 4.5$, so $k = 1.5$. The values are $a = 3$ and $k = 1.5$.

■ Section 1.4 Parametric Equations
(pp. 26–31)

Exploration 1 Parametrizing Circles

1. Each is a circle with radius $|a|$. As $|a|$ increases, the radius of the circle increases.

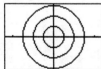

[−4.7, 4.7] by [−3.1, 3.1]

2. $0 \leq t \leq \frac{\pi}{2}$:

[−4.7, 4.7] by [−3.1, 3.1]

$0 \leq t \leq \pi$:

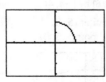

[−4.7, 4.7] by [−3.1, 3.1]

$0 \leq t \leq \frac{3\pi}{2}$:

[−4.7, 4.7] by [−3.1, 3.1]

$2\pi \leq t \leq 4\pi$:

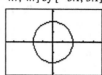

[−4.7, 4.7] by [−3.1, 3.1]

$0 \leq t \leq 4\pi$:

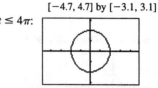

[−4.7, 4.7] by [−3.1, 3.1]

Let d be the length of the parametric interval. If $d < 2\pi$, you get $\frac{d}{2\pi}$ of a complete circle. If $d = 2\pi$, you get the complete circle. If $d > 2\pi$, you get the complete circle but portions of the circle will be traced out more than once. For example, if $d = 4\pi$ the entire circle is traced twice.

3.

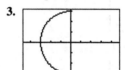

$\frac{\pi}{2} \leq t \leq \frac{3\pi}{2}$
initial point: (0, 3)
terminal point: (0, −3)

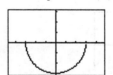

$\pi \leq t \leq 2\pi$
initial point: (−3, 0)
terminal point: (3, 0)

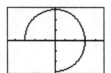

$\frac{3\pi}{2} \leq t \leq 3\pi$
initial point: (0, −3)
terminal point: (−3, 0)

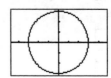

$\pi \leq t \leq 5\pi$
initial point: (−3, 0)
terminal point: (−3, 0)

10 Section 1.4

4. For $0 \le t \le 2\pi$ the complete circle is traced once clockwise beginning and ending at (2, 0).

For $\pi \le t \le 3\pi$ the complete circle is traced once clockwise beginning and ending at (−2, 0).

For $\dfrac{\pi}{2} \le t \le \dfrac{3\pi}{2}$ the half circle below is traced clockwise starting at (0, −2) and ending at (0, 2).

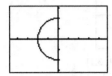

Exploration 2 Parametrizing Ellipses

1. $a = 2, b = 3$:

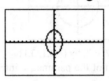

[−12, 12] by [−8, 8]

$a = 2, b = 4$:

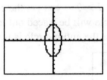

[−12, 12] by [−8, 8]

$a = 2, b = 5$:

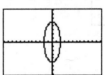

[−12, 12] by [−8, 8]

$a = 2, b = 6$:

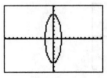

[−12, 12] by [−8, 8]

2. $a = 3, b = 4$:

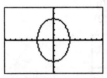

[−9, 9] by [−6, 6]

$a = 5, b = 4$:

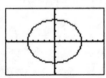

[−9, 9] by [−6, 6]

$a = 6, b = 4$:

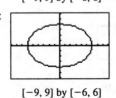

[−9, 9] by [−6, 6]

$a = 7, b = 4$:

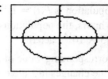

[−9, 9] by [−6, 6]

3. If $|a| > |b|$, then the major axis is on the x-axis and the minor on the y-axis. If $|a| < |b|$, then the major axis is on the y-axis and the minor on the x-axis.

4. $0 \le t \le \dfrac{\pi}{2}$:

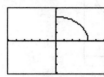

[−6, 6] by [−4, 4]

$0 \le t \le \pi$:

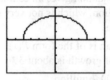

[−6, 6] by [−4, 4]

$0 \le t \le \dfrac{3\pi}{2}$:

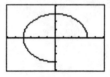

[−6, 6] by [−4, 4]

$0 \le t \le 4\pi$:

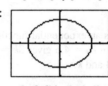

[−6, 6] by [−4, 4]

Let d be the length of the parametric interval. If $d < 2\pi$, you get $\dfrac{d}{2\pi}$ of a complete ellipse. If $d = 2\pi$, you get the complete ellipse. If $d > 2\pi$, you get the complete ellipse but portions of the ellipse will be traced out more than once. For example, if $d = 4\pi$ the entire ellipse is traced twice.

5. $0 \le t \le 2\pi$:

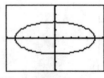

[−6, 6] by [−4, 4]

initial point: (5, 0)
terminal point: (5, 0)

$\pi \le t \le 3\pi$:

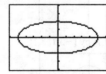

[−6, 6] by [−4, 4]

initial point: (−5, 0)
terminal point: (−5, 0)

$\frac{\pi}{2} \le t \le \frac{3\pi}{2}$:

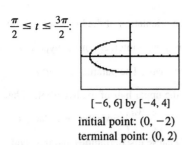

[−6, 6] by [−4, 4]

initial point: (0, −2)
terminal point: (0, 2)

Each curve is traced clockwise from the initial point to the terminal point.

Exploration 3 Graphing the Witch of Agnesi

1. We used the parameter interval $[0, \pi]$ because our graphing calculator ignored the fact that the curve is not defined when $t = 0$ or π. The curve is traced from right to left across the screen. x ranges from $-\infty$ to ∞.

2. $-\frac{\pi}{2} \le t \le \frac{\pi}{2}$:

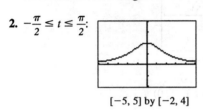

[−5, 5] by [−2, 4]

$0 < t \le \frac{\pi}{2}$:

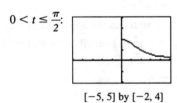

[−5, 5] by [−2, 4]

$\frac{\pi}{2} \le t < \pi$:

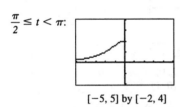

[−5, 5] by [−2, 4]

For $-\frac{\pi}{2} \le t \le \frac{\pi}{2}$, the entire graph described in part 1 is drawn. The left branch is drawn from right to left across the screen starting at the point (0, 2). Then the right branch is drawn from right to left across the screen stopping at the point (0, 2). If you leave out $-\frac{\pi}{2}$ and $\frac{\pi}{2}$, then the point (0, 2) is not drawn.

For $0 < t \le \frac{\pi}{2}$, the right branch is drawn from right to left across the screen stopping at the point (0, 2). If you leave out $\frac{\pi}{2}$, then the point (0, 2) is not drawn.

For $\frac{\pi}{2} \le t < \pi$, the left branch is drawn from right to left across the screen starting at the point (0, 2). If you leave out $\frac{\pi}{2}$, then the point (0, 2) is not drawn.

3. If you replace $x = 2 \cot t$ by $x = -2 \cot t$, the same graph is drawn except it is traced from left to right across the screen. If you replace $x = 2 \cot t$ by $x = 2 \cot (\pi - t)$, the same graph is drawn except it is traced from left to right across the screen.

Quick Review 1.4

1. $m = \frac{3 - 8}{4 - 1} = \frac{-5}{3} = -\frac{5}{3}$

   $y = -\frac{5}{3}(x - 1) + 8$

   $y = -\frac{5}{3}x + \frac{29}{3}$

2. $y = -4$

3. $x = 2$

4. When $y = 0$, we have $\frac{x^2}{9} = 1$, so the x-intercepts are -3 and 3. When $x = 0$, we have $\frac{y^2}{16} = 1$, so the y-intercepts are -4 and 4.

5. When $y = 0$, we have $\frac{x^2}{16} = 1$, so the x-intercepts are -4 and 4. When $x = 0$, we have $-\frac{y^2}{9} = 1$, which has no real solution, so there are no y-intercepts.

6. When $y = 0$, we have $0 = x + 1$, so the x-intercept is -1. When $x = 0$, we have $2y^2 = 1$, so the y-intercepts are $-\frac{1}{\sqrt{2}}$ and $\frac{1}{\sqrt{2}}$.

7. (a) $2(1)^2(1) + 1^2 \stackrel{?}{=} 3$
   $3 = 3$ Yes

   (b) $2(-1)^2(-1) + (-1)^2 \stackrel{?}{=} 3$
   $-2 + 1 \stackrel{?}{=} 3$
   $-1 \ne 3$ No

   (c) $2\left(\frac{1}{2}\right)^2(-2) + (-2)^2 \stackrel{?}{=} 3$
   $-1 + 4 \stackrel{?}{=} 3$
   $3 = 3$ Yes

8. (a) $9(1)^2 - 18(1) + 4(3)^2 \stackrel{?}{=} 27$
   $9 - 18 + 36 \stackrel{?}{=} 27$
   $27 = 27$ Yes

   (b) $9(1)^2 - 18(1) + 4(-3)^2 \stackrel{?}{=} 27$
   $9 - 18 + 36 \stackrel{?}{=} 27$
   $27 = 27$ Yes

   (c) $9(-1)^2 - 18(-1) + 4(3)^2 \stackrel{?}{=} 27$
   $9 + 18 + 36 \stackrel{?}{=} 27$
   $63 \ne 27$ No

9. (a) $2x + 3t = -5$
   $3t = -2x - 5$
   $t = \frac{-2x - 5}{3}$

   (b) $3y - 2t = -1$
   $-2t = -3y - 1$
   $2t = 3y + 1$
   $t = \frac{3y + 1}{2}$

12 Section 1.4

10. (a) The equation is true for $a \geq 0$.

  (b) The equation is equivalent to "$\sqrt{a^2} = a$ or $\sqrt{a^2} = -a$." Since $\sqrt{a^2} = a$ is true for $a \geq 0$ and $\sqrt{a^2} = -a$ is true for $a \leq 0$, at least one of the two equations is true for all real values of a. Therefore, the given equation $\sqrt{a^2} = \pm a$ is true for all real values of a.

  (c) The equation is true for all real values of a.

Section 1.4 Exercises

1. Graph (c). Window: $[-4, 4]$ by $[-3, 3]$, $0 \leq t \leq 2\pi$

3. Graph (d). Window: $[-10, 10]$ by $[-10, 10]$, $0 \leq t \leq 2\pi$

5. (a) The resulting graph appears to be the right half of a hyperbola in the first and fourth quadrants. The parameter a determines the x-intercept. The parameter b determines the shape of the hyperbola. If b is smaller, the graph has less steep slopes and appears "sharper." If b is larger, the slopes are steeper and the graph appears more "blunt." The graphs for $a = 2$ and $b = 1, 2,$ and 3 are shown.

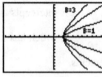
$[-10, 10]$ by $[-10, 10]$

  (b)
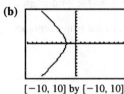
$[-10, 10]$ by $[-10, 10]$

  This appears to be the left half of the same hyberbola.

  (c)
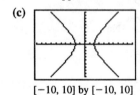
$[-10, 10]$ by $[-10, 10]$

  One must be careful because both sec t and tan t are discontinuous at these points. This might cause the grapher to include extraneous lines (the asymptotes of the hyperbola) in its graph. The extraneous lines can be avoided by using the grapher's dot mode instead of connected mode.

  (d) Note that $\sec^2 t - \tan^2 t = 1$ by a standard trigonometric identity. Substituting $\dfrac{x}{a}$ for sec t and $\dfrac{y}{b}$ for tan t gives $\left(\dfrac{x}{a}\right)^2 - \left(\dfrac{y}{b}\right)^2 = 1$.

  (e) This changes the orientation of the hyperbola. In this case, b determines the y-intercept of the hyperbola, and a determines the shape. The parameter interval $\left(-\dfrac{\pi}{2}, \dfrac{\pi}{2}\right)$ gives the upper half of the hyperbola. The parameter interval $\left(\dfrac{\pi}{2}, \dfrac{3\pi}{2}\right)$ gives the lower half. The same values of t cause discontinuities and may add extraneous lines to the graph. Substituting $\dfrac{y}{b}$ for sec t and $\dfrac{x}{a}$ for tan t in the identity $\sec^2 t - \tan^2 t = 1$ gives $\left(\dfrac{y}{b}\right)^2 - \left(\dfrac{x}{a}\right)^2 = 1$.

7. (a)
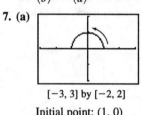
$[-3, 3]$ by $[-2, 2]$

Initial point: $(1, 0)$
Terminal point: $(-1, 0)$

  (b) $x^2 + y^2 = \cos^2 t + \sin^2 t = 1$

  The parametrized curve traces the upper half of the circle defined by $x^2 + y^2 = 1$ (or all of the semicircle defined by $y = \sqrt{1 - x^2}$).

9. (a)
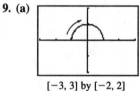
$[-3, 3]$ by $[-2, 2]$

Initial point: $(-1, 0)$
Terminal point: $(0, 1)$

  (b) $x^2 + y^2 = \cos^2(\pi - t) + \sin^2(\pi - t) = 1$

  The parametrized curve traces the upper half of the circle defined by $x^2 + y^2 = 1$ (or all of the semicircle defined by $y = \sqrt{1 - x^2}$).

11. (a)
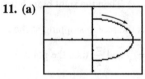
$[-4.7, 4.7]$ by $[-3.1, 3.1]$

Initial point: $(0, 2)$
Terminal point: $(0, -2)$

  (b) $\left(\dfrac{x}{4}\right)^2 + \left(\dfrac{y}{2}\right)^2 = \sin^2 t + \cos^2 t = 1$

  The parametrized curve traces the right half of the ellipse defined by $\left(\dfrac{x}{4}\right)^2 + \left(\dfrac{y}{2}\right)^2 = 1$ (or all of the curve defined by $x = 2\sqrt{4 - y^2}$).

13. (a)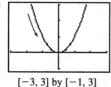

[−3, 3] by [−1, 3]

No initial or terminal point.

(b) $y = 9t^2 = (3t)^2 = x^2$

The parametrized curve traces all of the parabola defined by $y = x^2$.

15. (a)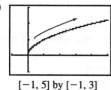

[−1, 5] by [−1, 3]

Initial point: (0, 0)
Terminal point: None

(b) $y = \sqrt{t} = \sqrt{x}$

The parametrized curve traces all of the curve defined by $y = \sqrt{x}$ (or the upper half of the parabola defined by $x = y^2$).

17. (a)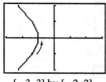

[−3, 3] by [−2, 2]

No initial or terminal point. Note that it may be necessary to use a t-interval such as [−1.57, 1.57] or use dot mode in order to avoid "asymptotes" showing on the calculator screen.

(b) $x^2 − y^2 = \sec^2 t − \tan^2 t = 1$

The parametrized curve traces the left branch of the hyperbola defined by $x^2 − y^2 = 1$ (or all of the curve defined by $x = −\sqrt{y^2 + 1}$).

19. (a)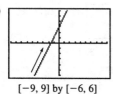

[−9, 9] by [−6, 6]

No initial or terminal point.

(b) $y = 4t − 7 = 2(2t − 5) + 3 = 2x + 3$

The parametrized curve traces all of the line defined by $y = 2x + 3$.

21. (a)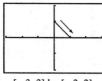

[−3, 3] by [−2, 2]

Initial point: (0, 1)
Terminal point: (1, 0)

(b) $y = 1 − t = 1 − x = −x + 1$

The Cartesian equation is $y = −x + 1$. The portion traced by the parametrized curve is the segment from (0, 1) to (1, 0).

23. (a)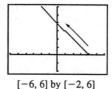

[−6, 6] by [−2, 6]

Initial point: (4, 0)
Terminal point: None

(b) $y = \sqrt{t} = 4 − (4 − \sqrt{t}) = 4 − x = −x + 4$

The parametrized curve traces the portion of the line defined by $y = −x + 4$ to the left of (4, 0), that is, for $x \leq 4$.

25. (a)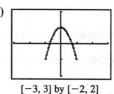

[−3, 3] by [−2, 2]

No initial or terminal point, since the t-interval has no beginning or end. The curve is traced and retraced in both directions.

(b) $y = \cos 2t$
$= \cos^2 t − \sin^2 t$
$= 1 − 2\sin^2 t$
$= 1 − 2x^2$
$= −2x^2 + 1$

The parametrized curve traces the portion of the parabola defined by $y = −2x^2 + 1$ corresponding to $−1 \leq x \leq 1$.

27. Using (−1, −3) we create the parametric equations $x = −1 + at$ and $y = −3 + bt$, representing a line which goes through (−1, −3) at $t = 0$. We determine a and b so that the line goes through (4, 1) when $t = 1$.
Since $4 = −1 + a$, $a = 5$.
Since $1 = −3 + b$, $b = 4$.
Therefore, one possible parametrization is $x = −1 + 5t$, $y = −3 + 4t$, $0 \leq t \leq 1$.

29. The lower half of the parabola is given by $x = y^2 + 1$ for $y \leq 0$. Substituting t for y, we obtain one possible parametrization: $x = t^2 + 1$, $y = t$, $t \leq 0$.

31. For simplicity, we assume that x and y are linear functions of t and that the point (x, y) starts at $(2, 3)$ for $t = 0$ and passes through $(−1, −1)$ at $t = 1$. Then $x = f(t)$, where $f(0) = 2$ and $f(1) = −1$.

Since slope $= \dfrac{\Delta x}{\Delta t} = \dfrac{-1 - 2}{1 - 0} = -3$,

$x = f(t) = −3t + 2 = 2 − 3t$. Also, $y = g(t)$, where $g(0) = 3$ and $g(1) = −1$.

Since slope $= \dfrac{\Delta y}{\Delta t} = \dfrac{-1 - 3}{1 - 0} = -4$,

$y = g(t) = −4t + 3 = 3 − 4t$.

One possible parametrization is:

$x = 2 − 3t$, $y = 3 − 4t$, $t \geq 0$.

33. The graph is in Quadrant I when $0 < y < 2$, which corresponds to $1 < t < 3$. To confirm, note that $x(1) = 2$ and $x(3) = 0$.

14 Section 1.5

35. The graph is in Quadrant III when $-6 \leq y < -4$, which corresponds to $-5 \leq t < -3$. To confirm, note that $x(-5) = -2$ and $x(-3) = 0$.

37. The graph of $y = x^2 + 2x + 2$ lies in Quadrant I for all $x > 0$. Substituting t for x, we obtain one possible parametrization:
$x = t, y = t^2 + 2t + 2, t > 0$.

39. Possible answers:
 (a) $x = a \cos t, y = -a \sin t, 0 \leq t \leq 2\pi$
 (b) $x = a \cos t, y = a \sin t, 0 \leq t \leq 2\pi$
 (c) $x = a \cos t, y = -a \sin t, 0 \leq t \leq 4\pi$
 (d) $x = a \cos t, y = a \sin t, 0 \leq t \leq 4\pi$

41. Note that $m\angle OAQ = t$, since alternate interior angles formed by a transversal of parallel lines are congruent.

Therefore, $\tan t = \dfrac{OQ}{AQ} = \dfrac{2}{x}$, so $x = \dfrac{2}{\tan t} = 2 \cot t$.

Now, by equation (iii), we know that
$AB = \dfrac{(AQ)^2}{AO}$
$= \left(\dfrac{AQ}{AO}\right)(AQ)$
$= (\cos t)(x)$
$= (\cos t)(2 \cot t)$
$= \dfrac{2 \cos^2 t}{\sin t}$.

Then equation (ii) gives
$y = 2 - AB \sin t = 2 - \dfrac{2 \cos^2 t}{\sin t} \cdot \sin t = 2 - 2 \cos^2 t$
$= 2 \sin^2 t$.

The parametric equations are:
$x = 2 \cot t, y = 2 \sin^2 t, 0 < t < \pi$

Note: Equation (iii) may not be immediately obvious, but it may be justified as follows. Sketch segment QB. Then $\angle OBQ$ is a right angle, so $\triangle ABQ \sim \triangle AQO$, which gives
$\dfrac{AB}{AQ} = \dfrac{AQ}{AO}$.

Section 1.5 Functions and Logarithms
(pp. 32–40)

Exploration 1 Testing for Inverses Graphically

1. It appears that $(f \circ g)(x) = (g \circ f)(x) = x$, suggesting that f and g may be inverses of each other.
 (a) f and g:

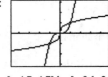

[−4.7, 4.7] by [−3.1, 3.1]

 (b) $f \circ g$:

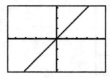

[−4.7, 4.7] by [−3.1, 3.1]

 (c) $g \circ f$:

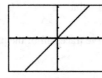

[−4.7, 4.7] by [−3.1, 3.1]

2. It appears that $f \circ g = g \circ f = g$, suggesting that f may be the identity function.
 (a) f and g:

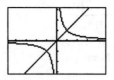

[−4.7, 4.7] by [−3.1, 3.1]

 (b) $f \circ g$:

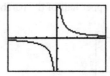

[−4.7, 4.7] by [−3.1, 3.1]

 (c) $g \circ f$:

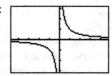

[−4.7, 4.7] by [−3.1, 3.1]

3. It appears that $(f \circ g)(x) = (g \circ f)(x) = x$, suggesting that f and g may be inverses of each other.
 (a) f and g:

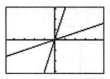

[−4.7, 4.7] by [−3.1, 3.1]

 (b) $f \circ g$:

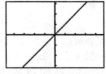

[−4.7, 4.7] by [−3.1, 3.1]

 (c) $g \circ f$:

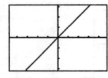

[−4.7, 4.7] by [−3.1, 3.1]

4. It appears that $(f \circ g)(x) = (g \circ f)(x) = x$, suggesting that f and g may be inverse of each other. (Notice that the domain of $f \circ g$ is $(0, \infty)$ and the domain of $g \circ f$ is $(-\infty, \infty)$.)

(a) f and g:

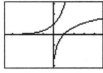

[−4.7, 4.7] by [−3.1, 3.1]

(b) $f \circ g$:

[−4.7, 4.7] by [−3.1, 3.1]

(c) $g \circ f$:

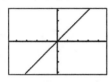

[−4.7, 4.7] by [−3.1, 3.1]

Exploration 2 Supporting the Product Rule

1. They appear to be vertical translations of each other.

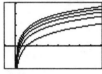

[−1, 8] by [−2, 4]

2. This graph suggests that each difference $(y_3 = y_1 - y_2)$ is a constant.

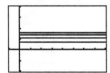

[−1, 8] by [−2, 4]

3. $y_3 = y_1 - y_2 = \ln(ax) - \ln x = \ln a + \ln x - \ln x = \ln a$
Thus, the difference $y_3 = y_1 - y_2$ is the constant $\ln a$.

Quick Review 1.5

1. $(f \circ g)(1) = f(g(1)) = f(2) = 1$
2. $(g \circ f)(-7) = g(f(-7)) = g(-2) = 5$
3. $(f \circ g)(x) = f(g(x)) = f(x^2 + 1) = \sqrt[3]{(x^2 + 1) - 1}$
$= \sqrt[3]{x^2} = x^{2/3}$
4. $(g \circ f)(x) = g(f(x)) = g(\sqrt[3]{x - 1})$
$= (\sqrt[3]{x - 1})^2 + 1$
$= (x - 1)^{2/3} + 1$
5. Substituting t for x, one possible answer is:
$x = t, y = \dfrac{1}{t - 1}, t \geq 2$.
6. Substituting t for x, one possible answer is:
$x = t, y = t, t < -3$.

7.

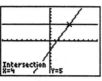

[−10, 10] by [−10, 10]

(4, 5)

8.

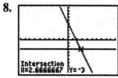

[−10, 10] by [−10, 10]

$\left(\dfrac{8}{3}, -3\right) \approx (2.67, -3)$

9. (a)

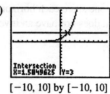

[−10, 10] by [−10, 10]

(1.58, 3)

(b) No points of intersection, since $2^x > 0$ for all values of x.

10. (a)

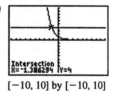

[−10, 10] by [−10, 10]

(−1.39, 4)

(b) No points of intersection, since $e^{-x} > 0$ for all values of x.

Section 1.5 Exercises

1. No, since (for example) the horizontal line $y = 2$ intersects the graph twice.
3. Yes, since each horizontal line intersects the graph at most once.
5. Yes, since each horizontal line intersects the graph only once.
7.

[−10, 10] by [−10, 10]

Yes, the function is one-to-one since each horizontal line intersects the graph at most once, so it has an inverse function.

16   Section 1.5

9.

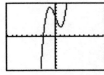

[−10, 10] by [−10, 10]

No, the function is not one-to-one since (for example) the horizontal line $y = 5$ intersects the graph more than once, so it does not have an inverse function.

11.

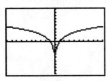

[−10, 10] by [−10, 10]

No, the function is not one-to-one since each horizontal line intersects the graph twice, so it does not have an inverse function.

13. $y = 2x + 3$
$y - 3 = 2x$
$\dfrac{y-3}{2} = x$
Interchange x and y.
$\dfrac{x-3}{2} = y$
$f^{-1}(x) = \dfrac{x-3}{2}$.
Verify.
$(f \circ f^{-1})(x) = f\left(\dfrac{x-3}{2}\right)$
$= 2\left(\dfrac{x-3}{2}\right) + 3$
$= (x - 3) + 3$
$= x$
$(f^{-1} \circ f)(x) = f^{-1}(2x + 3)$
$= \dfrac{(2x+3) - 3}{2}$
$= \dfrac{2x}{2}$
$= x$

15. $y = x^3 - 1$
$y + 1 = x^3$
$(y + 1)^{1/3} = x$
Interchange x and y.
$(x + 1)^{1/3} = y$
$f^{-1}(x) = (x + 1)^{1/3}$ or $\sqrt[3]{x+1}$
Verify.
$(f \circ f^{-1})(x) = f(\sqrt[3]{x+1})$
$= (\sqrt[3]{x+1})^3 - 1 = (x + 1) - 1 = x$
$(f^{-1} \circ f)(x) = f^{-1}(x^3 - 1)$
$= \sqrt[3]{(x^3 - 1) + 1} = \sqrt[3]{x^3} = x$

17. $y = x^2, x \le 0$
$x = -\sqrt{y}$
Interchange x and y.
$y = -\sqrt{x}$
$f^{-1}(x) = -\sqrt{x}$ or $-x^{1/2}$
Verify.
For $x \ge 0$ (the domain of f^{-1}),
$(f \circ f^{-1})(x) = f(-\sqrt{x}) = (-\sqrt{x})^2 = x$
For $x \le 0$, (the domain of f),
$(f^{-1} \circ f)(x) = f^{-1}(x^2) = -\sqrt{x^2} = -|x| = x$

19. $y = -(x - 2)^2, x \le 2$
$(x - 2)^2 = -y, x \le 2$
$x - 2 = -\sqrt{-y}$
$x = 2 - \sqrt{-y}$
Interchange x and y.
$y = 2 - \sqrt{-x}$
$f^{-1}(x) = 2 - \sqrt{-x}$ or $2 - (-x)^{1/2}$
Verify.
For $x \le 0$ (the domain of f^{-1})
$(f \circ f^{-1})(x) = f(2 - \sqrt{-x})$
$= -[(2 - \sqrt{-x}) - 2]^2$
$= -(-\sqrt{-x})^2 = -|x| = x$
For $x \le 2$ (the domain of f),
$(f^{-1} \circ f)(x) = f^{-1}(-(x - 2)^2)$
$= 2 - \sqrt{(x - 2)^2}$
$= 2 - |x - 2| = 2 + (x - 2) = x$

21. $y = \dfrac{1}{x^2}, x > 0$
$x^2 = \dfrac{1}{y}, x > 0$
$x = \sqrt{\dfrac{1}{y}} = \dfrac{1}{\sqrt{y}}$
Interchange x and y.
$y = \dfrac{1}{\sqrt{x}}$
$f^{-1}(x) = \dfrac{1}{\sqrt{x}}$ or $\dfrac{1}{x^{1/2}}$
Verify.
For $x > 0$ (the domain of f^{-1}),
$(f \circ f^{-1})(x) = f\left(\dfrac{1}{\sqrt{x}}\right) = \dfrac{1}{(1/\sqrt{x})^2} = x$
For $x > 0$ (the domain of f),
$(f^{-1} \circ f)(x) = f^{-1}\left(\dfrac{1}{x^2}\right) = \dfrac{1}{\sqrt{1/x^2}} = \sqrt{x^2} = |x| = x$

23. $y = \dfrac{2x+1}{x+3}$

$xy + 3y = 2x + 1$

$xy - 2x = 1 - 3y$

$(y-2)x = 1 - 3y$

$x = \dfrac{1-3y}{y-2}$

Interchange x and y.

$y = \dfrac{1-3x}{x-2}$

$f^{-1}(x) = \dfrac{1-3x}{x-2}$

Verify.

$(f \circ f^{-1})(x) = f\left(\dfrac{1-3x}{x-2}\right)$

$= \dfrac{2\left(\dfrac{1-3x}{x-2}\right)+1}{\dfrac{1-3x}{x-2}+3}$

$= \dfrac{2(1-3x)+(x-2)}{(1-3x)+3(x-2)}$

$= \dfrac{-5x}{-5} = x$

$(f^{-1} \circ f)(x) = f^{-1}\left(\dfrac{2x+1}{x+3}\right)$

$= \dfrac{1 - 3\left(\dfrac{2x+1}{x+3}\right)}{\dfrac{2x+1}{x+3} - 2}$

$= \dfrac{(x+3) - 3(2x+1)}{(2x+1) - 2(x+3)}$

$= \dfrac{-5x}{-5} = x$

25. Graph of f: $x_1 = t$, $y_1 = e^t$
Graph of f^{-1}: $x_2 = e^t$, $y_2 = t$
Graph of $y = x$: $x_3 = t$, $y_3 = t$

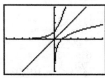

$[-6, 6]$ by $[-4, 4]$

27. Graph of f: $x_1 = t$, $y_1 = 2^{-t}$
Graph of f^{-1}: $x_2 = 2^{-t}$, $y_2 = t$
Graph of $y = x$: $x_3 = t$, $y_3 = t$

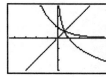

$[-4.5, 4.5]$ by $[-3, 3]$

29. Graph of f: $x_1 = t$, $y_1 = \ln t$
Graph of f^{-1}: $x_2 = \ln t$, $y_2 = t$
Graph of $y = x$: $x_3 = t$, $y_3 = t$

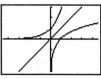

$[-4.5, 4.5]$ by $[-3, 3]$

31. Graph of f: $x_1 = t$, $y_1 = \sin^{-1} t$
Graph of f^{-1}: $x_2 = \sin^{-1} t$, $y_2 = t$
Graph of $y = x$: $x_3 = t$, $y_3 = t$

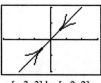

$[-3, 3]$ by $[-2, 2]$

33.

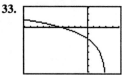

$[-10, 5]$ by $[-7, 3]$

Domain: $(\infty, 3)$
Range: $(-\infty, \infty)$

35.

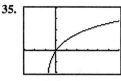

$[-3, 6]$ by $[-2, 4]$

Domain: $(-1, \infty)$
Range: $(-\infty, \infty)$

37. $(1.045)^t = 2$

$\ln(1.045)^t = \ln 2$

$t \ln 1.045 = \ln 2$

$t = \dfrac{\ln 2}{\ln 1.045} \approx 15.75$

Graphical support:

$[-2, 18]$ by $[-1, 3]$

18 Section 1.5

39. $e^x + e^{-x} = 3$
$e^x - 3 + e^{-x} = 0$
$e^x(e^x - 3 + e^{-x}) = e^x(0)$
$(e^x)^2 - 3e^x + 1 = 0$

$e^x = \dfrac{3 \pm \sqrt{(-3)^2 - 4(1)(1)}}{2(1)}$

$e^x = \dfrac{3 \pm \sqrt{5}}{2}$

$x = \ln\left(\dfrac{3 \pm \sqrt{5}}{2}\right) \approx -0.96 \text{ or } 0.96$

Graphical support:

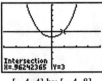

$[-4, 4]$ by $[-4, 8]$

41. $\ln y = 2t + 4$
$e^{\ln y} = e^{2t + 4}$
$y = e^{2t + 4}$

43. $y = \dfrac{100}{1 + 2^{-x}}$

$1 + 2^{-x} = \dfrac{100}{y}$

$2^{-x} = \dfrac{100}{y} - 1$

$\log_2(2^{-x}) = \log_2\left(\dfrac{100}{y} - 1\right)$

$-x = \log_2\left(\dfrac{100}{y} - 1\right)$

$x = -\log_2\left(\dfrac{100}{y} - 1\right)$

$= -\log_2\left(\dfrac{100 - y}{y}\right)$

$= \log_2\left(\dfrac{y}{100 - y}\right)$

Interchange x and y.

$y = \log_2\left(\dfrac{x}{100 - x}\right)$

$f^{-1}(x) = \log_2\left(\dfrac{x}{100 - x}\right)$

Verify.

$(f \circ f^{-1})(x) = f\left(\log_2 \dfrac{x}{100 - x}\right)$

$= \dfrac{100}{1 + 2^{-\log_2\left(\frac{x}{100 - x}\right)}}$

$= \dfrac{100}{1 + 2^{\log_2\left(\frac{100 - x}{x}\right)}}$

$= \dfrac{100}{1 + \dfrac{100 - x}{x}}$

$= \dfrac{100x}{x + (100 - x)} = \dfrac{100x}{100} = x$

$(f^{-1} \circ f)(x) = f^{-1}\left(\dfrac{100}{1 + 2^{-x}}\right)$

$= \log_2\left(\dfrac{\dfrac{100}{1 + 2^{-x}}}{100 - \dfrac{100}{1 + 2^{-x}}}\right)$

$= \log_2\left(\dfrac{100}{100(1 + 2^{-x}) - 100}\right)$

$= \log_2\left(\dfrac{1}{2^{-x}}\right) = \log_2(2^x) = x$

45. (a) $f(f(x)) = \sqrt{1 - (f(x))^2}$

$= \sqrt{1 - (1 - x^2)}$

$= \sqrt{x^2} = |x| = x$, since $x \geq 0$

(b) $f(f(x)) = f\left(\dfrac{1}{x}\right) = \dfrac{1}{1/x} = x$ for all $x \neq 0$

47. $500(1.0475)^t = 1000$

$1.0475^t = 2$

$\ln(1.0475^t) = \ln 2$

$t \ln 1.0475 = \ln 2$

$t = \dfrac{\ln 2}{\ln 1.0475} \approx 14.936$

It will take about 14.936 years. (If the interest is paid at the end of each year, it will take 15 years.)

49. (a) $y = -2539.852 + 636.896 \ln x$

(b) When $x = 75$, $y \approx 209.94$. About 209.94 million metric tons were produced.

(c) $-2539.852 + 636.896 \ln x = 400$

$636.896 \ln x = 2939.852$

$\ln x = \dfrac{2939.852}{636.896}$

$x = e^{\frac{2939.852}{636.896}} \approx 101.08$

According to the regression equation, Saudi Arabian oil production will reach 400 million metric tons when $x \approx 101.08$, in about 2001.

51. (a) Suppose that $f(x_1) = f(x_2)$. Then $mx_1 + b = mx_2 + b$ so $mx_1 = mx_2$. Since $m \neq 0$, this gives $x_1 = x_2$.

(b) $y = mx + b$

$y - b = mx$

$\dfrac{y - b}{m} = x$

Interchange x and y.

$\dfrac{x - b}{m} = y$

$f^{-1}(x) = \dfrac{x - b}{m}$

The slopes are reciprocals.

(c) If the original functions both have slope m, each of the inverse functions will have slope $\dfrac{1}{m}$. The graphs of the inverses will be parallel lines with nonzero slope.

(d) If the original functions have slopes m and $-\dfrac{1}{m}$, respectively, then the inverse functions will have slopes $\dfrac{1}{m}$ and $-m$, respectively. Since each of $\dfrac{1}{m}$ and $-m$ is the negative reciprocal of the other, the graphs of the inverses will be perpendicular lines with nonzero slopes.

53. If the graph of $f(x)$ passes the horizontal line test, so will the graph of $g(x) = -f(x)$ since it's the same graph reflected about the x-axis.
Alternate answer: If $g(x_1) = g(x_2)$ then $-f(x_1) = -f(x_2)$, $f(x_1) = f(x_2)$, and $x_1 = x_2$ since f is one-to-one.

55. (a) The expression $a(b^{c-x}) + d$ is defined for all values of x, so the domain is $(-\infty, \infty)$. Since b^{c-x} attains all positive values, the range is (d, ∞) if $a > 0$ and the range is $(-\infty, d)$ if $a < 0$.

(b) The expression $a \log_b(x - c) + d$ is defined when $x - c > 0$, so the domain is (c, ∞).
Since $a \log_b(x - c) + d$ attains every real value for some value of x, the range is $(-\infty, \infty)$.

■ Section 1.6 Trigonometric Functions (pp. 41–51)

Exploration 1 Unwrapping Trigonometric Functions

1. (x_1, y_1) is the circle of radius 1 centered at the origin (unit circle). (x_2, y_2) is one period of the graph of the sine function.

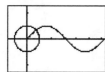

2. The y-values are the same in the interval $0 \leq t \leq 2\pi$.

3. The y-values are the same in the interval $0 \leq t \leq 4\pi$.

4. The x_1-values and the y_2-values are the same in each interval.

5. $y_2 = \tan t$:

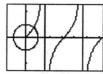

$y_2 = \csc t$:

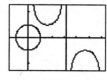

$y_2 = \sec t$:

$y_2 = \cot t$:

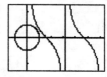

For each value of t, the value of $y_2 = \tan t$ is equal to the ratio $\dfrac{y_1}{x_1}$.

For each value of t, the value of $y_2 = \csc t$ is equal to the ratio $\dfrac{1}{y_1}$.

For each value of t, the value of $y_2 = \sec t$ is equal to the ratio $\dfrac{1}{x_1}$.

For each value of t, the value of $y_2 = \cot t$ is equal to the ratio $\dfrac{x_1}{y_1}$.

Exploration 2 Finding Sines and Cosines

1. The decimal viewing window $[-4.7, 4.7]$ by $[-3.1, 3.1]$ is square on the TI-82/83 and many other calculators. There are many other possibilities.

$[-4.7, 4.7]$ by $[-3.1, 3.1]$

2. Using the Ask table setting for the independent variable on the TI-83 we obtain

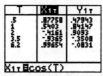

3.

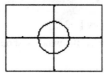

$[-3, 3]$ by $[-2, 2]$

Using trace, cos t and sin t are being computed for 0, 15, 30, ... , 360 degrees.

20   Section 1.6

Quick Review 1.6

1. $\dfrac{\pi}{3} \cdot \dfrac{180°}{\pi} = 60°$

2. $-2.5 \cdot \dfrac{180°}{\pi} = \left(-\dfrac{450}{\pi}\right)° \approx -143.24°$

3. $-40° \cdot \dfrac{\pi}{180°} = -\dfrac{2\pi}{9}$

4. $45° \cdot \dfrac{\pi}{180°} = \dfrac{\pi}{4}$

5.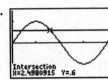
   $[0, 2\pi]$ by $[-1.5, 1.5]$
   $x \approx 0.6435, x \approx 2.4981$

6.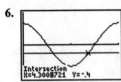
   $[0, 2\pi]$ by $[-1.5, 1.5]$
   $x \approx 1.9823, x \approx 4.3009$

7.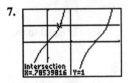
   $\left[-\dfrac{\pi}{2}, \dfrac{3\pi}{2}\right]$ by $[-2, 2]$
   $x \approx 0.7854 \left(\text{or } \dfrac{\pi}{4}\right), x \approx 3.9270 \left(\text{or } \dfrac{5\pi}{4}\right)$

8. $f(-x) = 2(-x)^2 - 3 = 2x^2 - 3 = f(x)$
   The graph is symmetric about the y-axis because if a point (a, b) is on the graph, then so is the point $(-a, b)$.

9. $f(-x) = (-x)^3 - 3(-x)$
   $= -x^3 + 3x$
   $= -(x^3 - 3x) = -f(x)$
   The graph is symmetric about the origin because if a point (a, b) is on the graph, then so is the point $(-a, -b)$.

10. $x \geq 0$

Section 1.6 Exercises

1. Arc length $= \left(\dfrac{5\pi}{8}\right)(2) = \dfrac{5\pi}{4}$

3. Angle $= \dfrac{7}{14} = \dfrac{1}{2}$ radian or about $28.65°$

5. (a) The period of $y = \sec x$ is 2π, so the window should have length 4π.
   One possible answer: $[0, 4\pi]$ by $[-3, 3]$

   (b) The period of $y = \csc x$ is 2π, so the window should have length 4π.
   One possible answer: $[0, 4\pi]$ by $[-3, 3]$

   (c) The period of $y = \cot x$ is π, so the window should have length 2π.
   One possible answer: $[0, 2\pi]$ by $[-3, 3]$

7. Since $\dfrac{\pi}{6}$ is in the range $\left[-\dfrac{\pi}{2}, \dfrac{\pi}{2}\right]$ of $y = \sin^{-1} x$ and $\sin \dfrac{\pi}{6} = 0.5$, $\sin^{-1}(0.5) = \dfrac{\pi}{6}$ radian or $\dfrac{\pi}{6} \cdot \dfrac{180°}{\pi} = 30°$.

9. Using a calculator, $\tan^{-1}(-5) \approx -1.3734$ radians or $-78.6901°$.

11. (a) Period $= \dfrac{2\pi}{2} = \pi$
    (b) Amplitude $= 1.5$
    (c) $[-2\pi, 2\pi]$ by $[-2, 2]$

13. (a) Period $= \dfrac{2\pi}{2} = \pi$
    (b) Amplitude $= 3$
    (c) $[-2\pi, 2\pi]$ by $[-4, 4]$

15. (a) Period $= \dfrac{2\pi}{\pi/3} = 6$
    (b) Amplitude $= 4$
    (c) $[-3, 3]$ by $[-5, 5]$

17. (a) Period $= \dfrac{2\pi}{3}$

    (b) Domain: Since $\csc(3x + \pi) = \dfrac{1}{\sin(3x + \pi)}$, we require $3x + \pi \neq k\pi$, or $x \neq \dfrac{(k-1)\pi}{3}$. This requirement is equivalent to $x \neq \dfrac{k\pi}{3}$ for integers k.

    (c) Since $|\csc(3x + \pi)| \geq 1$, the range excludes numbers between $-3 - 2 = -5$ and $3 - 2 = 1$. The range is $(-\infty, -5] \cup [1, \infty)$.

    (d)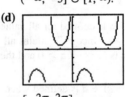
    $\left[-\dfrac{2\pi}{3}, \dfrac{2\pi}{3}\right]$ by $[-8, 8]$

19. (a) Period $= \dfrac{\pi}{3}$

    (b) Domain: We require $3x + \pi \neq \dfrac{k\pi}{2}$ for odd integers k. Therefore, $x \neq \dfrac{(k-2)\pi}{6}$ for odd integers k. This requirement is equivalent to $x \neq \dfrac{k\pi}{6}$ for odd integers k.

    (c) Since the tangent function attains all real values, the range is $(-\infty, \infty)$.

    (d)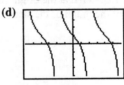
    $\left[-\dfrac{\pi}{2}, \dfrac{\pi}{2}\right]$ by $[-8, 8]$

21. Note that $\sqrt{8^2 + 15^2} = 17$.

Since $\sin\theta = \frac{8}{17}$ and $-\frac{\pi}{2} \le \theta \le \frac{\pi}{2}$,
$\cos\theta = \sqrt{1 - \sin^2\theta} = \sqrt{1 - \left(\frac{8}{17}\right)^2} = \frac{15}{17}$.

Therefore: $\sin\theta = \frac{8}{17}$, $\cos\theta = \frac{15}{17}$, $\tan\theta = \frac{\sin\theta}{\cos\theta} = \frac{8}{15}$,
$\cot\theta = \frac{1}{\tan\theta} = \frac{15}{8}$, $\sec\theta = \frac{1}{\cos\theta} = \frac{17}{15}$, $\csc\theta = \frac{1}{\sin\theta} = \frac{17}{8}$

23. Note that $r = \sqrt{(-3)^2 + 4^2} = 5$. Then:
$\sin\theta = \frac{y}{r} = \frac{4}{5}$, $\cos\theta = \frac{x}{r} = -\frac{3}{5}$, $\tan\theta = \frac{y}{x} = -\frac{4}{3}$,
$\cot\theta = \frac{x}{y} = -\frac{3}{4}$, $\sec\theta = \frac{r}{x} = -\frac{5}{3}$, $\csc\theta = \frac{r}{y} = \frac{5}{4}$

25. The angle $\tan^{-1}(2.5) \approx 1.190$ is the solution to this equation in the interval $-\frac{\pi}{2} < x < \frac{\pi}{2}$. Another solution in $0 \le x < 2\pi$ is $\tan^{-1}(2.5) + \pi \approx 4.332$. The solutions are $x \approx 1.190$ and $x \approx 4.332$.

27. This equation is equivalent to $\sin x = \frac{1}{2}$, so the solutions in the interval $0 \le x < 2\pi$ are $x = \frac{\pi}{6}$ and $x = \frac{5\pi}{6}$.

29. The solutions in the interval $0 \le x < 2\pi$ are $x = \frac{7\pi}{6}$ and $x = \frac{11\pi}{6}$. Since $y = \sin x$ has period 2π, the solutions are all of the form $x = \frac{7\pi}{6} + 2k\pi$ or $x = \frac{11\pi}{6} + 2k\pi$, where k is any integer.

31. Let $\theta = \cos^{-1}\left(\frac{7}{11}\right)$. Then $0 \le \theta \le \pi$ and $\cos\theta = \frac{7}{11}$, so
$\sin\left(\cos^{-1}\left(\frac{7}{11}\right)\right) = \sin\theta = \sqrt{1 - \cos^2\theta} = \sqrt{1 - \left(\frac{7}{11}\right)^2}$
$= \frac{\sqrt{72}}{11} = \frac{6\sqrt{2}}{11} \approx 0.771$.

33. (a) Using a graphing calculator with the sinusoidal regression feature, the equation is
$y = 1.543 \sin(2468.635x - 0.494) + 0.438$.

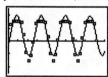

[0, 0.01] by [−2.5, 2.5]

(b) The frequency is 2468.635 radians per second, which is equivalent to $\frac{2468.635}{2\pi} \approx 392.9$ cycles per second (Hz). The note is a "G."

35. (a) Amplitude = 37

(b) Period $= \frac{2\pi}{(2\pi/365)} = 365$

(c) Horizontal shift = 101

(d) Vertical shift = 25

37. (a) $\cot(-x) = \frac{\cos(-x)}{\sin(-x)} = \frac{\cos(x)}{-\sin(x)} = -\cot(x)$

(b) Assume that f is even and g is odd.
Then $\frac{f(-x)}{g(-x)} = \frac{f(x)}{-g(x)} = -\frac{f(x)}{g(x)}$ so $\frac{f}{g}$ is odd. The situation is similar for $\frac{g}{f}$.

39. Assume that f is even and g is odd.
Then $f(-x)g(-x) = (f(x))(-g(x)) = -f(x)g(x)$ so (fg) is odd.

41. (a) Using a graphing calculator with the sinusoidal regression feature, the equation is
$y = 3.0014 \sin(0.9996x + 2.0012) + 2.9999$.

(b) $y = 3 \sin(x + 2) + 3$

43. (a) $\sqrt{2} \sin\left(ax + \frac{\pi}{4}\right)$

(b) See part (a).

(c) It works.

(d) $\sin\left(ax + \frac{\pi}{4}\right) = (\sin ax)\left(\cos\frac{\pi}{4}\right) + (\cos ax)\left(\sin\frac{\pi}{4}\right)$
$= (\sin ax)\left(\frac{1}{\sqrt{2}}\right) + (\cos ax)\left(\frac{1}{\sqrt{2}}\right)$
$= \frac{1}{\sqrt{2}}(\sin ax + \cos ax)$
So, $\sin(ax) + \cos(ax) = \sqrt{2}\sin\left(ax + \frac{\pi}{4}\right)$.

45. Since $\sin x$ has period 2π, $\sin^3(x + 2\pi) = \sin^3(x)$. This function has period 2π. A graph shows that no smaller number works for the period.

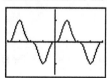

[−2π, 2π] by [−1.5, 1.5]

47. The period is $\frac{2\pi}{60} = \frac{\pi}{30}$. One possible graph:

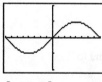

$\left[-\frac{\pi}{60}, \frac{\pi}{60}\right]$ by [−2, 2]

Chapter 1 Review Exercises
(pp. 52–53)

1. $y = 3(x - 1) + (-6)$
$y = 3x - 9$

2. $y = -\frac{1}{2}(x + 1) + 2$
$y = -\frac{1}{2}x + \frac{3}{2}$

3. $x = 0$

4. $m = \frac{-2 - 6}{1 - (-3)} = \frac{-8}{4} = -2$
$y = -2(x + 3) + 6$
$y = -2x$

22 Chapter Review

5. $y = 2$

6. $m = \dfrac{5-3}{-2-3} = \dfrac{2}{-5} = -\dfrac{2}{5}$

$y = -\dfrac{2}{5}(x - 3) + 3$

$y = -\dfrac{2}{5} + \dfrac{21}{5}$

7. $y = -3x + 3$

8. Since $2x - y = -2$ is equivalent to $y = 2x + 2$, the slope of the given line (and hence the slope of the desired line) is 2.
$y = 2(x - 3) + 1$
$y = 2x - 5$

9. Since $4x + 3y = 12$ is equivalent to $y = -\dfrac{4}{3}x + 4$, the slope of the given line (and hence the slope of the desired line) is $-\dfrac{4}{3}$.

$y = -\dfrac{4}{3}(x - 4) - 12$

$y = -\dfrac{4}{3}x - \dfrac{20}{3}$

10. Since $3x - 5y = 1$ is equivalent to $y = \dfrac{3}{5}x - \dfrac{1}{5}$, the slope of the given line is $\dfrac{3}{5}$ and the slope of the perpendicular line is $-\dfrac{5}{3}$.

$y = -\dfrac{5}{3}(x + 2) - 3$

$y = -\dfrac{5}{3}x - \dfrac{19}{3}$

11. Since $\dfrac{1}{2}x + \dfrac{1}{3}y = 1$ is equivalent to $y = -\dfrac{3}{2}x + 3$, the slope of the given line is $-\dfrac{3}{2}$ and the slope of the perpendicular line is $\dfrac{2}{3}$.

$y = \dfrac{2}{3}(x + 1) + 2$

$y = \dfrac{2}{3}x + \dfrac{8}{3}$

12. The line passes through $(0, -5)$ and $(3, 0)$

$m = \dfrac{0 - (-5)}{3 - 0} = \dfrac{5}{3}$

$y = \dfrac{5}{3}x - 5$

13. $m = \dfrac{2 - 4}{2 - (-2)} = \dfrac{-2}{4} = -\dfrac{1}{2}$

$f(x) = -\dfrac{1}{2}(x + 2) + 4$

$f(x) = -\dfrac{1}{2}x + 3$

Check: $f(4) = -\dfrac{1}{2}(4) + 3 = 1$, as expected.

14. The line passes through $(4, -2)$ and $(-3, 0)$.

$m = \dfrac{0 - (-2)}{-3 - 4} = \dfrac{2}{-7} = -\dfrac{2}{7}$

$y = -\dfrac{2}{7}(x - 4) - 2$

$y = -\dfrac{2}{7}x - \dfrac{6}{7}$

15.

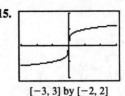

$[-3, 3]$ by $[-2, 2]$

Symmetric about the origin.

16.

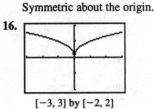

$[-3, 3]$ by $[-2, 2]$

Symmetric about the y-axis.

17.

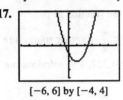

$[-6, 6]$ by $[-4, 4]$

Neither

18.

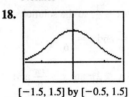

$[-1.5, 1.5]$ by $[-0.5, 1.5]$

Symmetric about the y-axis.

19. $y(-x) = (-x)^2 + 1 = x^2 + 1 = y(x)$
Even

20. $y(-x) = (-x)^5 - (-x)^3 - (-x) = -x^5 + x^3 + x = -y(x)$
Odd

21. $y(-x) = 1 - \cos(-x) = 1 - \cos x = y(x)$
Even

22. $y(-x) = \sec(-x) \tan(-x)$

$= \dfrac{\sin(-x)}{\cos^2(-x)} = \dfrac{-\sin x}{\cos^2 x}$

$= -\sec x \tan x = -y(x)$

Odd

23. $y(-x) = \dfrac{(-x)^4 + 1}{(-x)^3 - 2(-x)} = \dfrac{x^4 + 1}{-x^3 + 2x} = -\dfrac{x^4 + 1}{x^3 - 2x} = -y(x)$
Odd

24. $y(-x) = 1 - \sin(-x) = 1 + \sin x$
Neither even nor odd

25. $y(-x) = -x + \cos(-x) = -x + \cos x$
Neither even nor odd

26. $y(-x) = \sqrt{(-x)^4 - 1} = \sqrt{x^4 - 1}$
Even

27. (a) The function is defined for all values of x, so the domain is $(-\infty, \infty)$.
   (b) Since $|x|$ attains all nonnegative values, the range is $[-2, \infty)$.
   (c)
   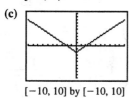
   $[-10, 10]$ by $[-10, 10]$

28. (a) Since the square root requires $1 - x \geq 0$, the domain is $(-\infty, 1]$.
   (b) Since $\sqrt{1-x}$ attains all nonnegative values, the range is $[-2, \infty)$.
   (c)
   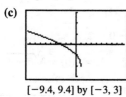
   $[-9.4, 9.4]$ by $[-3, 3]$

29. (a) Since the square root requires $16 - x^2 \geq 0$, the domain is $[-4, 4]$.
   (b) For values of x in the domain, $0 \leq 16 - x^2 \leq 16$, so $0 \leq \sqrt{16-x^2} \leq 4$. The range is $[0, 4]$.
   (c)
   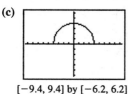
   $[-9.4, 9.4]$ by $[-6.2, 6.2]$

30. (a) The function is defined for all values of x, so the domain is $(-\infty, \infty)$.
   (b) Since 3^{2-x} attains all positive values, the range is $(1, \infty)$.
   (c)
   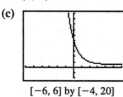
   $[-6, 6]$ by $[-4, 20]$

31. (a) The function is defined for all values of x, so the domain is $(-\infty, \infty)$.
   (b) Since $2e^{-x}$ attains all positive values, the range is $(-3, \infty)$.
   (c)
   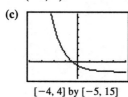
   $[-4, 4]$ by $[-5, 15]$

32. (a) The function is equivalent to $y = \tan 2x$, so we require $2x \neq \frac{k\pi}{2}$ for odd integers k. The domain is given by $x \neq \frac{k\pi}{4}$ for odd integers k.
   (b) Since the tangent function attains all values, the range is $(-\infty, \infty)$.
   (c)
   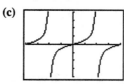
   $\left[-\frac{\pi}{2}, \frac{\pi}{2}\right]$ by $[-8, 8]$

33. (a) The function is defined for all values of x, so the domain is $(-\infty, \infty)$.
   (b) The sine function attains values from -1 to 1, so $-2 \leq 2\sin(3x + \pi) \leq 2$, and hence $-3 \leq 2\sin(3x + \pi) - 1 \leq 1$. The range is $[-3, 1]$.
   (c)
   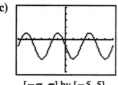
   $[-\pi, \pi]$ by $[-5, 5]$

34. (a) The function is defined for all values of x, so the domain is $(-\infty, \infty)$.
   (b) The function is equivalent to $y = \sqrt[5]{x^2}$, which attains all nonnegative values. The range is $[0, \infty)$.
   (c)
   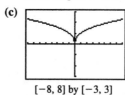
   $[-8, 8]$ by $[-3, 3]$

35. (a) The logarithm requires $x - 3 > 0$, so the domain is $(3, \infty)$.
   (b) The logarithm attains all real values, so the range is $(-\infty, \infty)$.
   (c)
   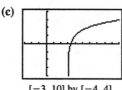
   $[-3, 10]$ by $[-4, 4]$

36. (a) The function is defined for all values of x, so the domain is $(-\infty, \infty)$.
   (b) The cube root attains all real values, so the range is $(-\infty, \infty)$.
   (c)
   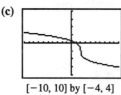
   $[-10, 10]$ by $[-4, 4]$

24 Chapter Review

37. (a) The function is defined for $-4 \le x \le 4$, so the domain is $[-4, 4]$.

(b) The function is equivalent to $y = \sqrt{|x|}$, $-4 \le x \le 4$, which attains values from 0 to 2 for x in the domain. The range is $[0, 2]$.

(c)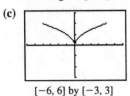

$[-6, 6]$ by $[-3, 3]$

38. (a) The function is defined for $-2 \le x \le 2$, so the domain is $[-2, 2]$.

(b) See the graph in part (c). The range is $[-1, 1]$.

(c)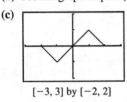

$[-3, 3]$ by $[-2, 2]$

39. First piece: Line through $(0, 1)$ and $(1, 0)$

$m = \dfrac{0-1}{1-0} = \dfrac{-1}{1} = -1$

$y = -x + 1$ or $1 - x$

Second piece:
Line through $(1, 1)$ and $(2, 0)$

$m = \dfrac{0-1}{2-1} = \dfrac{-1}{1} = -1$

$y = -(x - 1) + 1$

$y = -x + 2$ or $2 - x$

$f(x) = \begin{cases} 1 - x, & 0 \le x < 1 \\ 2 - x, & 1 \le x \le 2 \end{cases}$

40. First piece: Line through $(0, 0)$ and $(2, 5)$

$m = \dfrac{5-0}{2-0} = \dfrac{5}{2}$

$y = \dfrac{5}{2}x$

Second piece: Line through $(2, 5)$ and $(4, 0)$

$m = \dfrac{0-5}{4-2} = \dfrac{-5}{2} = -\dfrac{5}{2}$

$y = -\dfrac{5}{2}(x - 2) + 5$

$y = -\dfrac{5}{2}x + 10$ or $10 - \dfrac{5x}{2}$

$f(x) = \begin{cases} \dfrac{5x}{2}, & 0 \le x < 2 \\ 10 - \dfrac{5x}{2}, & 2 \le x \le 4 \end{cases}$

(Note: $x = 2$ can be included on either piece.)

41. (a) $(f \circ g)(-1) = f(g(-1)) = f\left(\dfrac{1}{\sqrt{-1+2}}\right) = f(1) = \dfrac{1}{1} = 1$

(b) $(g \circ f)(2) = g(f(2)) = g\left(\dfrac{1}{2}\right) = \dfrac{1}{\sqrt{1/2 + 2}} = \dfrac{1}{\sqrt{2.5}}$ or $\sqrt{\dfrac{2}{5}}$

(c) $(f \circ f)(x) = f(f(x)) = f\left(\dfrac{1}{x}\right) = \dfrac{1}{1/x} = x, x \ne 0$

(d) $(g \circ g)(x) = g(g(x)) = g\left(\dfrac{1}{\sqrt{x+2}}\right) = \dfrac{1}{\sqrt{1/\sqrt{x+2} + 2}}$

$= \dfrac{\sqrt[4]{x+2}}{\sqrt{1 + 2\sqrt{x+2}}}$

42. (a) $(f \circ g)(-1) = f(g(-1))$

$= f(\sqrt[3]{-1} + 1)$

$= f(0) = 2 - 0 = 2$

(b) $(g \circ f)(2) = g(f(2)) = g(2 - 2) = g(0) = \sqrt[3]{0} + 1 = 1$

(c) $(f \circ f)(x) = f(f(x)) = f(2 - x) = 2 - (2 - x) = x$

(d) $(g \circ g)(x) = g(g(x)) = g(\sqrt[3]{x} + 1) = \sqrt[3]{\sqrt[3]{x} + 1} + 1$

43. (a) $(f \circ g)(x) = f(g(x))$

$= f(\sqrt{x + 2})$

$= 2 - (\sqrt{x+2})^2$

$= -x, x \ge -2$

$(g \circ f)(x) = g(f(x))$

$= g(2 - x^2)$

$= \sqrt{(2 - x^2) + 2} = \sqrt{4 - x^2}$

(b) Domain of $f \circ g$: $[-2, \infty)$
Domain of $g \circ f$: $[-2, 2]$

(c) Range of $f \circ g$: $(-\infty, 2]$
Range of $g \circ f$: $[0, 2]$

44. (a) $(f \circ g)(x) = f(g(x))$

$= f(\sqrt{1 - x})$

$= \sqrt{\sqrt{1 - x}}$

$= \sqrt[4]{1 - x}$

$(g \circ f)(x) = g(f(x)) = g(\sqrt{x}) = \sqrt{1 - \sqrt{x}}$

(b) Domain of $f \circ g$: $(-\infty, 1]$
Domain of $g \circ f$: $[0, 1]$

(c) Range of $f \circ g$: $[0, \infty)$
Range of $g \circ f$: $[0, 1]$

45. (a)

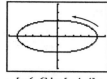

[−6, 6] by [−4, 4]

Initial point: (5, 0)
Terminal point: (5, 0)
The ellipse is traced exactly once in a counterclockwise direction starting and ending at the point (5, 0).

(b) Substituting $\cos t = \frac{x}{5}$ and $\sin t = \frac{y}{2}$ in the identity $\cos^2 t + \sin^2 t = 1$ gives the Cartesian equation $\left(\frac{x}{5}\right)^2 + \left(\frac{y}{2}\right)^2 = 1$.
The entire ellipse is traced by the curve.

46. (a)

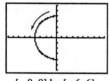

[−9, 9] by [−6, 6]

Initial point: (0, 4)
Terminal point: None (since the endpoint $\frac{3\pi}{2}$ is not included in the t-interval)
The semicircle is traced in a counterclockwise direction starting at (0, 4) and extending to, but not including, (0, −4).

(b) Substituting $\cos t = \frac{x}{4}$ and $\sin t = \frac{y}{4}$ in the identity $\cos^2 t + \sin^2 t = 1$ gives the Cartesian equation $\left(\frac{x}{4}\right)^2 + \left(\frac{y}{4}\right)^2 = 1$, or $x^2 + y^2 = 16$. The left half of the circle is traced by the parametrized curve.

47. (a)

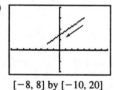

[−8, 8] by [−10, 20]

Initial point: (4, 15)
Terminal point: (−2, 3)
The line segment is traced from right to left starting at (4, 15) and ending at (−2, 3).

(b) Substituting $t = 2 - x$ into $y = 11 - 2t$ gives the Cartesian equation $y = 11 - 2(2 - x)$, or $y = 2x + 7$. The part of the line from (4, 15) to (−2, 3) is traced by the parametrized curve.

48. (a)

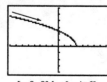

[−8, 8] by [−4, 6]

Initial point: None
Terminal point: (3, 0)
The curve is traced from left to right ending at the point (3, 0).

(b) Substituting $t = x - 1$ into $y = \sqrt{4 - 2t}$ gives the Cartesian equation $y = \sqrt{4 - 2(x - 1)}$, or $y = \sqrt{6 - 2x}$. The entire curve is traced by the parametrized curve.

49. (a) For simplicity, we assume that x and y are linear functions of t, and that the point (x, y) starts at (−2, 5) for $t = 0$ and ends at (4, 3) for $t = 1$. Then $x = f(t)$, where $f(0) = -2$ and $f(1) = 4$. Since
$$\text{slope} = \frac{\Delta x}{\Delta t} = \frac{4 - (-2)}{1 - 0} = 6,$$
$x = f(t) = 6t - 2 = -2 + 6t$.
Also, $y = g(t)$, where $g(0) = 5$ and $g(1) = 3$. Since
$$\text{slope} = \frac{\Delta y}{\Delta t} = \frac{3 - 5}{1 - 0} = -2,$$
$y = g(t) = -2t + 5 = 5 - 2t$.
One possible parametrization is:
$x = -2 + 6t, y = 5 - 2t, 0 \leq t \leq 1$

50. For simplicity, we assume that x and y are linear functions of t and that the point (x, y) passes through (−3, −2) for $t = 0$ and (4, −1) for $t = 1$. Then $x = f(t)$, where $f(0) = -3$ and $f(1) = 4$. Since
$$\text{slope} = \frac{\Delta x}{\Delta t} = \frac{4 - (-3)}{1 - 0} = 7,$$
$x = f(t) = 7t - 3 = -3 + 7t$.
Also, $y = g(t)$, where $g(0) = -2$ and $g(1) = -1$. Since
$$\text{slope} = \frac{\Delta y}{\Delta t} = \frac{-1 - (-2)}{1 - 0} = 1$$
$y = g(t) = t - 2 = -2 + t$.
One possible parametrization is:
$x = -3 + 7t, y = -2 + t, -\infty < t < \infty$.

51. For simplicity, we assume that x and y are linear functions of t and that the point (x, y) starts at (2, 5) for $t = 0$ and passes through (−1, 0) for $t = 1$. Then $x = f(t)$, where $f(0) = 2$ and $f(1) = -1$. Since
$$\text{slope} = \frac{\Delta x}{\Delta t} = \frac{-1 - 2}{1 - 0} = -3, x = f(t) = -3t + 2 = 2 - 3t.$$
Also, $y = g(t)$, where $g(0) = 5$ and $g(1) = 0$. Since
$$\text{slope} = \frac{\Delta y}{\Delta t} = \frac{0 - 5}{1 - 0} = -5, y = g(t) = -5t + 5 = 5 - 5t.$$
One possible parametrization is:
$x = 2 - 3t, y = 5 - 5t, t \geq 0$.

52. One possible parametrization is:
$x = t, y = t(t - 4), t \leq 2$.

26 Chapter Review

53. (a) $y = 2 - 3x$
$3x = 2 - y$
$x = \dfrac{2-y}{3}$
Interchange x and y.
$y = \dfrac{2-x}{3}$
$f^{-1}(x) = \dfrac{2-x}{3}$
Verify.
$(f \circ f^{-1})(x) = f(f^{-1}(x))$
$= f\left(\dfrac{2-x}{3}\right)$
$= 2 - 3\left(\dfrac{2-x}{3}\right)$
$= 2 - (2 - x) = x$

$(f^{-1} \circ f)(x) = f^{-1}(f(x))$
$= f^{-1}(2 - 3x)$
$= \dfrac{2 - (2 - 3x)}{3}$
$= \dfrac{3x}{3} = x$

(b)
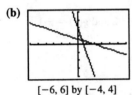
$[-6, 6]$ by $[-4, 4]$

54. (a) $y = (x + 2)^2,\ x \geq -2$
$\sqrt{y} = x + 2$
$x = \sqrt{y} - 2$
Interchange x and y.
$y = \sqrt{x} - 2$
$f^{-1}(x) = \sqrt{x} - 2$
Verify.
For $x \geq 0$ (the domain of f^{-1})
$(f \circ f^{-1})(x) = f(f^{-1}(x))$
$= f(\sqrt{x} - 2)$
$= [(\sqrt{x} - 2) + 2]^2$
$= (\sqrt{x})^2 = x$
For $x \geq -2$ (the domain of f),
$(f^{-1} \circ f)(x) = f^{-1}(f(x))$
$= f^{-1}((x + 2)^2)$
$= \sqrt{(x+2)^2} - 2$
$= |x + 2| - 2$
$= (x + 2) - 2 = x$

(b)
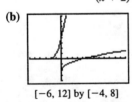
$[-6, 12]$ by $[-4, 8]$

55. Using a calculator, $\sin^{-1}(0.6) \approx 0.6435$ radians or $36.8699°$.

56. Using a calculator, $\tan^{-1}(-2.3) \approx -1.1607$ radians or $-66.5014°$.

57. Since $\cos\theta = \dfrac{3}{7}$ and $0 \leq \theta \leq \pi$,
$\sin\theta = \sqrt{1 - \cos^2\theta} = \sqrt{1 - \left(\dfrac{3}{7}\right)^2} = \sqrt{\dfrac{40}{49}} = \dfrac{\sqrt{40}}{7}$.
Therefore,
$\sin\theta = \dfrac{\sqrt{40}}{7},\ \cos\theta = \dfrac{3}{7},\ \tan\theta = \dfrac{\sin\theta}{\cos\theta} = \dfrac{\sqrt{40}}{3}$,
$\cot\theta = \dfrac{1}{\tan\theta} = \dfrac{3}{\sqrt{40}},\ \sec\theta = \dfrac{1}{\cos\theta} = \dfrac{7}{3}$,
$\csc\theta = \dfrac{1}{\sin\theta} = \dfrac{7}{\sqrt{40}}$

58. (a) Note that $\sin^{-1}(-0.2) \approx -0.2014$. In $[0, 2\pi)$, the solutions are $x = \pi - \sin^{-1}(-0.2) \approx 3.3430$ and $x = \sin^{-1}(-0.2) + 2\pi \approx 6.0818$.

(b) Since the period of $\sin x$ is 2π, the solutions are $x \approx 3.3430 + 2k\pi$ and $x \approx 6.0818 + 2k\pi$, k any integer.

59. $e^{-0.2x} = 4$
$\ln e^{-0.2x} = \ln 4$
$-0.2x = \ln 4$
$x = \dfrac{\ln 4}{-0.2} = -5 \ln 4$

60. (a) The given graph is reflected about the y-axis.

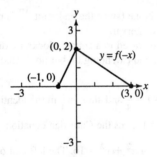

(b) The given graph is reflected about the x-axis.

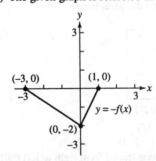

(c) The given graph is shifted left 1 unit, stretched vertically by a factor of 2, reflected about the x-axis, and then shifted upward 1 unit.

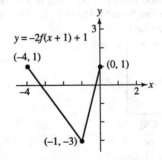

(d) The given graph is shifted right 2 units, stretched vertically by a factor of 3, and then shifted downward 2 units.

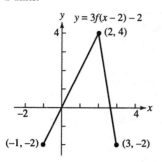

61. (a)

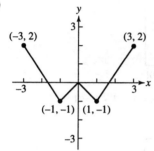

(b)

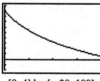

62. (a) $V = 100{,}000 - 10{,}000x$, $0 \le x \le 10$

(b) $V = 55{,}000$
$100{,}000 - 10{,}000x = 55{,}000$
$-10{,}000x = -45{,}000$
$x = 4.5$
The value is \$55,000 after 4.5 years.

63. (a) $f(0) = 90$ units

(b) $f(2) = 90 - 52 \ln 3 \approx 32.8722$ units

(c)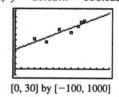

[0, 4] by [−20, 100]

64. $1500(1.08)^t = 5000$

$1.08^t = \dfrac{5000}{1500} = \dfrac{10}{3}$

$\ln(1.08)^t = \ln \dfrac{10}{3}$

$t \ln 1.08 = \ln \dfrac{10}{3}$

$t = \dfrac{\ln(10/3)}{\ln 1.08}$

$t \approx 15.6439$

It will take about 15.6439 years. (If the bank only pays interest at the end of the year, it will take 16 years.)

65. (a) $N(t) = 4 \cdot 2^t$

(b) 4 days: $4 \cdot 2^4 = 64$ guppies
1 week: $4 \cdot 2^7 = 512$ guppies

(c) $N(t) = 2000$
$4 \cdot 2^t = 2000$
$2^t = 500$
$\ln 2^t = \ln 500$
$t \ln 2 = \ln 500$
$t = \dfrac{\ln 500}{\ln 2} \approx 8.9658$

There will be 2000 guppies after 8.9658 days, or after nearly 9 days.

(d) Because it suggests the number of guppies will continue to double indefinitely and become arbitrarily large, which is impossible due to the finite size of the tank and the oxygen supply in the water.

66. (a) $y = 20.627x + 338.622$

[0, 30] by [−100, 1000]

(b) When $x = 30$, $y \approx 957.445$. According to the regression equation, about 957 degrees will be earned.

(c) The slope is 20.627. It represents the approximate annual increase in the number of doctorates earned by Hispanic Americans per year.

67. (a) $y = 14.60175 \cdot 1.00232^x$

(b) Solving $y = 25$ graphically, we obtain $x \approx 232$. According to the regression equation, the population will reach 25 million in the year 2132.

(c) 0.232%

Chapter 2
Limits and Continuity

■ Section 2.1 Rates of Change and Limits
(pp. 55–65)

Quick Review 2.1

1. $f(2) = 2(2^3) - 5(2)^2 + 4 = 0$

2. $f(2) = \dfrac{4(2)^2 - 5}{2^3 + 4} = \dfrac{11}{12}$

3. $f(2) = \sin\left(\pi \cdot \dfrac{2}{2}\right) = \sin \pi = 0$

4. $f(2) = \dfrac{1}{2^2 - 1} = \dfrac{1}{3}$

5. $|x| < 4$
$-4 < x < 4$

6. $|x| < c^2$
$-c^2 < x < c^2$

7. $|x - 2| < 3$
$-3 < x - 2 < 3$
$-1 < x < 5$

28   Section 2.1

8. $|x - c| < d^2$
   $-d^2 < x - c < d^2$
   $-d^2 + c < x < d^2 + c$

9. $\dfrac{x^2 - 3x - 18}{x + 3} = \dfrac{(x + 3)(x - 6)}{x + 3} = x - 6, x \neq -3$

10. $\dfrac{2x^2 - x}{2x^2 + x - 1} = \dfrac{x(2x - 1)}{(2x - 1)(x + 1)} = \dfrac{x}{x + 1}, x \neq \dfrac{1}{2}$

Section 2.1 Exercises

1. (a) $\lim\limits_{x \to 3^-} f(x) = 3$

   (b) $\lim\limits_{x \to 3^+} f(x) = -2$

   (c) $\lim\limits_{x \to 3} f(x)$ does not exist, because the left- and right-hand limits are not equal.

   (d) $f(3) = 1$

3. (a) $\lim\limits_{h \to 0^-} f(h) = -4$

   (b) $\lim\limits_{h \to 0^+} f(h) = -4$

   (c) $\lim\limits_{h \to 0} f(h) = -4$

   (d) $f(0) = -4$

5. (a) $\lim\limits_{x \to 0^-} F(x) = 4$

   (b) $\lim\limits_{x \to 0^+} F(x) = -3$

   (c) $\lim\limits_{x \to 0} F(x)$ does not exist, because the left- and right-hand limits are not equal.

   (d) $F(0) = 4$

7. $\lim\limits_{x \to -1/2} 3x^2(2x - 1) = 3\left(-\dfrac{1}{2}\right)^2 \left[2\left(-\dfrac{1}{2}\right) - 1\right] = 3\left(\dfrac{1}{4}\right)(-2)$
   $= -\dfrac{3}{2}$

   Graphical support:
   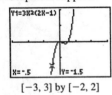
   $[-3, 3]$ by $[-2, 2]$

9. $\lim\limits_{x \to 1} (x^3 + 3x^2 - 2x - 17) = (1)^3 + 3(1)^2 - 2(1) - 17$
   $= 1 + 3 - 2 - 17 = -15$

   Graphical support:
   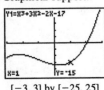
   $[-3, 3]$ by $[-25, 25]$

11. $\lim\limits_{y \to -3} \dfrac{y^2 + 4y + 3}{y^2 - 3} = \dfrac{(-3)^2 + 4(-3) + 3}{(-3)^2 - 3} = \dfrac{0}{6} = 0$

    Graphical support:
    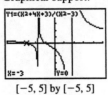
    $[-5, 5]$ by $[-5, 5]$

13. $\lim\limits_{x \to -2} (x - 6)^{2/3} = (-2 - 6)^{2/3} = \sqrt[3]{(-8)^2} = \sqrt[3]{64} = 4$

    Graphical support:
    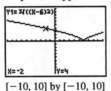
    $[-10, 10]$ by $[-10, 10]$

15. $\lim\limits_{x \to 0} (e^x \cos x) = e^0 \cos 0 = 1 \cdot 1 = 1$

    Graphical support:
    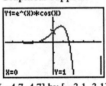
    $[-4.7, 4.7]$ by $[-3.1, 3.1]$

17. You cannot use substitution because the expression $\sqrt{x - 2}$ is not defined at $x = -2$. Since the expression is not defined at points near $x = -2$, the limit does not exist.

19. You cannot use substitution because the expression $\dfrac{|x|}{x}$ is not defined at $x = 0$. Since $\lim\limits_{x \to 0^-} \dfrac{|x|}{x} = -1$ and $\lim\limits_{x \to 0^+} \dfrac{|x|}{x} = 1$, the left- and right-hand limits are not equal and so the limit does not exist.

21.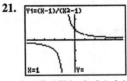
    $[-4.7, 4.7]$ by $[-3.1, 3.1]$

    $\lim\limits_{x \to 1} \dfrac{x - 1}{x^2 - 1} = \dfrac{1}{2}$

    Algebraic confirmation:

    $\lim\limits_{x \to 1} \dfrac{x - 1}{x^2 - 1} = \lim\limits_{x \to 1} \dfrac{x - 1}{(x + 1)(x - 1)} = \lim\limits_{x \to 1} \dfrac{1}{x + 1} = \dfrac{1}{1 + 1} = \dfrac{1}{2}$

23.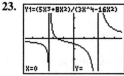
[−4.7, 4.7] by [−3.1, 3.1]

$$\lim_{x \to 0} \frac{5x^3 + 8x^2}{3x^4 - 16x^2} = -\frac{1}{2}$$

Algebraic confirmation:

$$\lim_{x \to 0} \frac{5x^3 + 8x^2}{3x^4 - 16x^2} = \lim_{x \to 0} \frac{x^2(5x + 8)}{x^2(3x^2 - 16)}$$
$$= \lim_{x \to 0} \frac{5x + 8}{3x^2 - 16}$$
$$= \frac{5(0) + 8}{3(0)^2 - 16}$$
$$= \frac{8}{-16} = -\frac{1}{2}$$

25.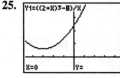
[−4.7, 4.7] by [−5, 20]

$$\lim_{x \to 0} \frac{(2 + x)^3 - 8}{x} = 12$$

Algebraic confirmation:

$$\lim_{x \to 0} \frac{(2 + x)^3 - 8}{x} = \lim_{x \to 0} \frac{12x + 6x^2 + x^3}{x}$$
$$= \lim_{x \to 0} (12 + 6x + x^2)$$
$$= 12 + 6(0) + (0)^2 = 12$$

27.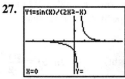
[−4.7, 4.7] by [−3.1, 3.1]

$$\lim_{x \to 0} \frac{\sin x}{2x^2 - x} = -1$$

Algebraic confirmation:

$$\lim_{x \to 0} \frac{\sin x}{2x^2 - x} = \lim_{x \to 0} \left(\frac{\sin x}{x} \cdot \frac{1}{2x - 1} \right)$$
$$= \left(\lim_{x \to 0} \frac{\sin x}{x} \right)\left(\lim_{x \to 0} \frac{1}{2x - 1} \right) = (1)\left(\frac{1}{2(0) - 1} \right) = -1$$

29.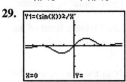
[−4.7, 4.7] by [−3.1, 3.1]

$$\lim_{x \to 0} \frac{\sin^2 x}{x} = 0$$

Algebraic confirmation:

$$\lim_{x \to 0} \frac{\sin^2 x}{x} = \lim_{x \to 0} \left(\sin x \cdot \frac{\sin x}{x} \right)$$
$$= \left(\lim_{x \to 0} \sin x \right) \cdot \left(\lim_{x \to 0} \frac{\sin x}{x} \right)$$
$$= (\sin 0)(1) = 0$$

31. (a) True
(b) True
(c) False, since $\lim_{x \to 0^-} f(x) = 0$.
(d) True, since both are equal to 0.
(e) True, since (d) is true.
(f) True
(g) False, since $\lim_{x \to 0} f(x) = 0$.
(h) False, $\lim_{x \to 1^-} f(x) = 1$, but $\lim_{x \to 1} f(x)$ is undefined.
(i) False, $\lim_{x \to 1^+} f(x) = 0$, but $\lim_{x \to 1} f(x)$ is undefined.
(j) False, since $\lim_{x \to 2^-} f(x) = 0$.

33. $y_1 = \frac{x^2 + x - 2}{x - 1} = \frac{(x - 1)(x + 2)}{x - 1} = x + 2, x \neq 1$
(c)

35. $y_1 = \frac{x^2 - 2x + 1}{x - 1} = \frac{(x - 1)^2}{x - 1} = x - 1, x \neq 1$
(d)

37. Since int $x = 0$ for x in $(0, 1)$, $\lim_{x \to 0^+}$ int $x = 0$.

39. Since int $x = 0$ for x in $(0, 1)$, $\lim_{x \to 0.01}$ int $x = 0$.

41. Since $\frac{x}{|x|} = 1$ for $x > 0$, $\lim_{x \to 0^+} \frac{x}{|x|} = 1$.

43. (a) $\lim_{x \to 4} (g(x) + 3) = \left(\lim_{x \to 4} g(x) \right) + \left(\lim_{x \to 4} 3 \right) = 3 + 3 = 6$
(b) $\lim_{x \to 4} xf(x) = \left(\lim_{x \to 4} x \right)\left(\lim_{x \to 4} f(x) \right) = 4 \cdot 0 = 0$
(c) $\lim_{x \to 4} g^2(x) = \left(\lim_{x \to 4} g(x) \right)^2 = 3^2 = 9$
(d) $\lim_{x \to 4} \frac{g(x)}{f(x) - 1} = \frac{\lim_{x \to 4} g(x)}{\left(\lim_{x \to 4} f(x) \right) - \left(\lim_{x \to 4} 1 \right)} = \frac{3}{0 - 1} = -3$

45. (a)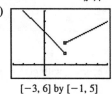
[−3, 6] by [−1, 5]
(b) $\lim_{x \to 2^+} f(x) = 2$; $\lim_{x \to 2^-} f(x) = 1$
(c) No, because the two one-sided limits are different.

47. (a)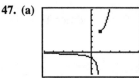
[−5, 5] by [−4, 8]
(b) $\lim_{x \to 1^+} f(x) = 4$; $\lim_{x \to 1^-} f(x)$ does not exist.
(c) No, because the left-hand limit does not exist.

30 Section 2.1

49. (a)

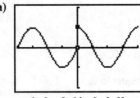

$[-2\pi, 2\pi]$ by $[-2, 2]$

(b) $(-2\pi, 0) \cup (0, 2\pi)$

(c) $c = 2\pi$

(d) $c = -2\pi$

51. (a)

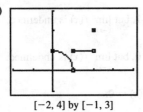

$[-2, 4]$ by $[-1, 3]$

(b) $(0, 1) \cup (1, 2)$

(c) $c = 2$ **(d)** $c = 0$

53.

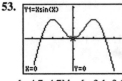

$[-4.7, 4.7]$ by $[-3.1, 3.1]$

$\lim_{x \to 0} (x \sin x) = 0$

Confirm using the Sandwich Theorem, with $g(x) = -|x|$ and $h(x) = |x|$.

$|x \sin x| = |x| \cdot |\sin x| \leq |x| \cdot 1 = |x|$

$-|x| \leq x \sin x \leq |x|$

Because $\lim_{x \to 0} (-|x|) = \lim_{x \to 0} |x| = 0$, the Sandwich Theorem gives $\lim_{x \to 0} (x \sin x) = 0$.

55.

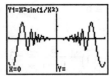

$[-0.5, 0.5]$ by $[-0.25, 0.25]$

$\lim_{x \to 0} \left(x^2 \sin \frac{1}{x^2} \right) = 0$

Confirm using the Sandwich Theorem, with $g(x) = -x^2$ and $h(x) = x^2$.

$\left| x^2 \sin \frac{1}{x^2} \right| = |x^2| \cdot \left| \sin \frac{1}{x^2} \right| \leq |x^2| \cdot 1 = x^2$.

$-x^2 \leq x^2 \sin \frac{1}{x^2} \leq x^2$

Because $\lim_{x \to 0} (-x^2) = \lim_{x \to 0} x^2 = 0$, the Sandwich Theorem give $\lim_{x \to 0} \left(x^2 \sin \frac{1}{x^2} \right) = 0$.

57. (a) In three seconds, the ball falls $4.9(3)^2 = 44.1$ m, so its average speed is $\dfrac{44.1}{3} = 14.7$ m/sec.

(b) The average speed over the interval from time $t = 3$ to time $3 + h$ is

$\dfrac{\Delta y}{\Delta t} = \dfrac{4.9(3 + h)^2 - 4.9(3)^2}{(3 + h) - 3} = \dfrac{4.9(6h + h^2)}{h}$

$= 29.4 + 4.9h$

Since $\lim_{h \to 0} (29.4 + 4.9h) = 29.4$, the instantaneous speed is 29.4 m/sec.

59. (a)

x	-0.1	-0.01	-0.001	-0.0001
$f(x)$	-0.054402	-0.005064	-0.000827	-0.000031

(b)

x	0.1	0.01	0.001	0.0001
$f(x)$	-0.054402	-0.005064	-0.000827	-0.000031

The limit appears to be 0.

61. (a)

x	-0.1	-0.01	-0.001	-0.0001
$f(x)$	2.0567	2.2763	2.2999	2.3023

(b)

x	0.1	0.01	0.001	0.0001
$f(x)$	2.5893	2.3293	2.3052	2.3029

The limit appears to be approximately 2.3.

63. (a) Because the right-hand limit at zero depends only on the values of the function for positive x-values near zero.

(b) Area of $\triangle OAP = \dfrac{1}{2}(\text{base})(\text{height}) = \dfrac{1}{2}(1)(\sin \theta) = \dfrac{\sin \theta}{2}$

Area of sector $OAP = \dfrac{(\text{angle})(\text{radius})^2}{2} = \dfrac{\theta(1)^2}{2} = \dfrac{\theta}{2}$

Area of $\triangle OAT = \dfrac{1}{2}(\text{base})(\text{height}) = \dfrac{1}{2}(1)(\tan \theta) = \dfrac{\tan \theta}{2}$

(c) This is how the areas of the three regions compare.

(d) Multiply by 2 and divide by $\sin \theta$.

(e) Take reciprocals, remembering that all of the values involved are positive.

(f) The limits for $\cos \theta$ and 1 are both equal to 1. Since $\dfrac{\sin \theta}{\theta}$ is between them, it must also have a limit of 1.

(g) $\dfrac{\sin(-\theta)}{-\theta} = \dfrac{-\sin \theta}{-\theta} = \dfrac{\sin(\theta)}{\theta}$

(h) If the function is symmetric about the y-axis, and the right-hand limit at zero is 1, then the left-hand limit at zero must also be 1.

(i) The two one-sided limits both exist and are equal to 1.

65. (a) $f\left(\dfrac{\pi}{6}\right) = \sin\dfrac{\pi}{6} = \dfrac{1}{2}$

(b) The graphs of $y_1 = f(x)$, $y_2 = 0.3$, and $y_3 = 0.7$ are shown.

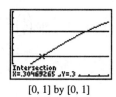

[0, 1] by [0, 1]

The intersections of y_1 with y_2 and y_3 are at $x \approx 0.3047$ and $x \approx 0.7754$, respectively, so we may choose any value of a in $\left[0.3047, \dfrac{\pi}{6}\right)$ and any value of b in $\left(\dfrac{\pi}{6}, 0.7754\right]$, where the interval endpoints are approximate.

One possible answer: $a = 0.305$, $b = 0.775$

(c) The graphs of $y_1 = f(x)$, $y_2 = 0.49$, and $y_3 = 0.51$ are shown.

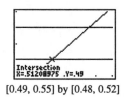

[0.49, 0.55] by [0.48, 0.52]

The intersections of y_1 with y_2 and y_3 are at $x \approx 0.5121$ and $x \approx 0.5352$, respectively, so we may choose any value of a in $\left[0.5121, \dfrac{\pi}{6}\right)$ and any value of b in $\left(\dfrac{\pi}{6}, 0.5352\right]$, where the interval endpoints are approximate.

One possible answer: $a = 0.513$, $b = 0.535$

■ Section 2.2 Limits Involving Infinity
(pp. 65–73)

Exploration 1 Exploring Theorem 5

1. Neither $\lim\limits_{x\to\infty} f(x)$ or $\lim\limits_{x\to\infty} g(x)$ exist. In this case, we can describe the behavior of f and g as $x \to \infty$ by writing $\lim\limits_{x\to\infty} f(x) = \infty$ and $\lim\limits_{x\to\infty} g(x) = \infty$. We cannot apply the quotient rule because both limits must exist. However, from Example 5,

$$\lim_{x\to\infty} \dfrac{5x + \sin x}{x} = \lim_{x\to\infty}\left(5 + \dfrac{\sin x}{x}\right) = 5 + 0 = 5,$$

so the limit of the quotient exists.

2. Both f and g oscillate between 0 and 1 as $x \to \infty$, taking on each value infinitely often. We cannot apply the sum rule because neither limit exists. However,

$$\lim_{x\to\infty}(\sin^2 x + \cos^2 x) = \lim_{x\to\infty}(1) = 1,$$

so the limit of the sum exists.

3. The limit of f and g as $x \to \infty$ do not exist, so we cannot apply the difference rule to $f - g$. We can say that $\lim\limits_{x\to\infty} f(x) = \lim\limits_{x\to\infty} g(x) = \infty$. We can write the difference as

$$f(x) - g(x) = \ln(2x) - \ln(x+1) = \ln\dfrac{2x}{x+1}.$$

We can use graphs or tables to convince ourselves that this limit is equal to $\ln 2$.

4. The fact that the limits of f and g as $x \to \infty$ do not exist does not necessarily mean that the limits of $f + g$, $f - g$ or $\dfrac{f}{g}$ do not exist, just that Theorem 5 cannot be applied.

Quick Review 2.2

1. $y = 2x - 3$

$y + 3 = 2x$

$\dfrac{y+3}{2} = x$

Interchange x and y.

$\dfrac{x+3}{2} = y$

$f^{-1}(x) = \dfrac{x+3}{2}$

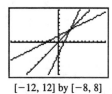

[−12, 12] by [−8, 8]

2. $y = e^x$

$\ln y = x$

Interchange x and y.

$\ln x = y$

$f^{-1}(x) = \ln x$

[−6, 6] by [−4, 4]

32 Section 2.2

3. $y = \tan^{-1} x$

$\tan y = x, -\dfrac{\pi}{2} < y < \dfrac{\pi}{2}$

Interchange x and y.

$\tan x = y, -\dfrac{\pi}{2} < x < \dfrac{\pi}{2}$

$f^{-1}(x) = \tan x, -\dfrac{\pi}{2} < x < \dfrac{\pi}{2}$

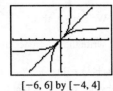

$[-6, 6]$ by $[-4, 4]$

4. $y = \cot^{-1} x$

$\cot y = x, 0 < x < \pi$

Interchange x and y.

$\cot x = y, 0 < y < \pi$

$f^{-1}(x) = \cot x, 0 < x < \pi$

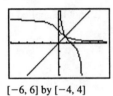

$[-6, 6]$ by $[-4, 4]$

5.
$$3x^3 + 4x - 5 \overline{)\, 2x^3 - 3x^2 + x - 1\,}^{\tfrac{2}{3}}$$
$$2x^3 + 0x^2 + \tfrac{8}{3}x - \tfrac{10}{3}$$
$$-3x^2 - \tfrac{5}{3}x + \tfrac{7}{3}$$

$q(x) = \dfrac{2}{3}$

$r(x) = -3x^2 - \dfrac{5}{3}x + \dfrac{7}{3}$

6.
$$x^3 - x^2 + 1 \overline{)\, 2x^5 + 0x^4 - x^3 + 0x^2 + x - 1\,}^{2x^2 + 2x + 1}$$
$$2x^5 - 2x^4 + 0x^3 + 2x^2$$
$$2x^4 - x^3 - 2x^2 + x - 1$$
$$2x^4 - 2x^3 + 0x^2 + 2x$$
$$x^3 - 2x^2 - x - 1$$
$$x^3 - x^2 + 0x + 1$$
$$-x^2 - x - 2$$

$q(x) = 2x^2 + 2x + 1$
$r(x) = -x^2 - x - 2$

7. (a) $f(-x) = \cos(-x) = \cos x$

(b) $f\!\left(\dfrac{1}{x}\right) = \cos\!\left(\dfrac{1}{x}\right)$

8. (a) $f(-x) = e^{-(-x)} = e^x$

(b) $f\!\left(\dfrac{1}{x}\right) = e^{-1/x}$

9. (a) $f(-x) = \dfrac{\ln(-x)}{-x} = -\dfrac{\ln(-x)}{x}$

(b) $f\!\left(\dfrac{1}{x}\right) = \dfrac{\ln 1/x}{1/x} = x \ln x^{-1} = -x \ln x$

10. (a) $f(-x) = \left(-x + \dfrac{1}{-x}\right) \sin(-x) = -\left(x + \dfrac{1}{x}\right)(-\sin x)$
$= \left(x + \dfrac{1}{x}\right) \sin x$

(b) $f\!\left(\dfrac{1}{x}\right) = \left(\dfrac{1}{x} + \dfrac{1}{1/x}\right) \sin\!\left(\dfrac{1}{x}\right) = \left(\dfrac{1}{x} + x\right) \sin\!\left(\dfrac{1}{x}\right)$

Section 2.2 Exercises

1.

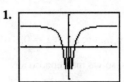

$[-5, 5]$ by $[-1.5, 1.5]$

X	Y1
100	.99995
200	.99999
300	.99999
400	1
-100	.99995
-200	.99999
-300	.99999

Y1■cos(1/X)

(a) $\lim\limits_{x \to \infty} f(x) = 1$

(b) $\lim\limits_{x \to -\infty} f(x) = 1$

(c) $y = 1$

3.

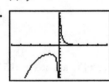

$[-5, 5]$ by $[-10, 10]$

X	Y1
100	4E-46
200	7E-90
300	0
400	0
-100	-3E41
-200	-4E84
-300	ERROR

Y1■e^(-X)/X

(a) $\lim\limits_{x \to \infty} f(x) = 0$

(b) $\lim\limits_{x \to -\infty} f(x) = -\infty$

(c) $y = 0$

5.

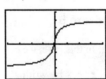

$[-20, 20]$ by $[-4, 4]$

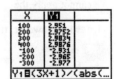

X	Y1
100	2.951
200	2.9752
300	2.9834
400	2.9876
-100	-2.931
-200	-2.965
-300	-2.977

Y1■(3X+1)/(abs(...

(a) $\lim\limits_{x \to \infty} f(x) = 3$

(b) $\lim\limits_{x \to -\infty} f(x) = -3$

(c) $y = 3, y = -3$

7.

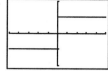

[−5, 5] by [−2, 2]

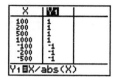

(a) $\lim_{x \to \infty} f(x) = 1$

(b) $\lim_{x \to -\infty} f(x) = -1$

(c) $y = 1, y = -1$

9.

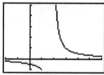

[−2, 6] by [−1, 5]

$\lim_{x \to 2^+} \dfrac{1}{x-2} = \infty$

11.

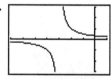

[−7, 1] by [−3, 3]

$\lim_{x \to -3^-} \dfrac{1}{x+3} = -\infty$

13.

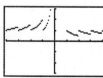

[−4, 4] by [−3, 3]

$\lim_{x \to 0^+} \dfrac{\text{int } x}{x} = 0$

15.

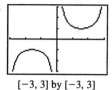

[−3, 3] by [−3, 3]

$\lim_{x \to 0^+} \csc x = \infty$

17.

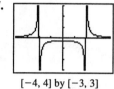

[−4, 4] by [−3, 3]

(a) $x = -2, x = 2$

(b) Left-hand limit at −2 is ∞.
Right-hand limit at −2 is −∞.
Left-hand limit at 2 is −∞.
Right-hand limit at 2 is ∞.

19.

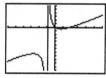

[−6, 6] by [−12, 6]

(a) $x = -1$

(b) Left-hand limit at −1 is −∞.
Right-hand limit at −1 is ∞.

21.

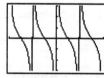

[−2π, 2π] by [−3, 3]

(a) $x = k\pi$, k any integer

(b) at each vertical asymptote:
Left-hand limit is −∞.
Right-hand limit is ∞.

23. $y = \left(2 - \dfrac{x}{x+1}\right)\left(\dfrac{x^2}{5+x^2}\right) = \left(\dfrac{2(x+1) - x}{x+1}\right)\left(\dfrac{x^2}{5+x^2}\right)$

$= \left(\dfrac{x+2}{x+1}\right)\left(\dfrac{x^2}{5+x^2}\right) = \dfrac{x^3 + 2x^2}{x^3 + x^2 + 5x + 5}$

An end behavior model for y is $\dfrac{x^3}{x^3} = 1$.

$\lim_{x \to \infty} y = \lim_{x \to \infty} 1 = 1$

$\lim_{x \to -\infty} y = \lim_{x \to -\infty} 1 = 1$

34 Section 2.2

25. Use the method of Example 10 in the text.

$$\lim_{x\to\infty} \frac{\cos\left(\frac{1}{x}\right)}{1+\frac{1}{x}} = \lim_{x\to 0^+} \frac{\cos x}{1+x} = \frac{\cos(0)}{1+0} = \frac{1}{1} = 1$$

$$\lim_{x\to-\infty} \frac{\cos\left(\frac{1}{x}\right)}{1+\frac{1}{x}} = \lim_{x\to 0^-} \frac{\cos x}{1+x} = \frac{\cos(0)}{1+0} = \frac{1}{1} = 1$$

27. Use $y = \frac{\sin x}{2x^2 + x} = \frac{\sin x}{x} \cdot \frac{1}{2x+1}$

$\lim_{x\to\pm\infty} \frac{\sin x}{x} = 0$

$\lim_{x\to\pm\infty} \frac{1}{2x+1} = 0$

So, $\lim_{x\to\infty} y = 0$ and $\lim_{x\to-\infty} y = 0$.

29. An end behavior model is $\frac{2x^3}{x} = 2x^2$. (a)

31. An end behavior model is $\frac{2x^4}{-x} = -2x^3$. (d)

33. (a) $3x^2$
(b) None

35. (a) $\frac{x}{2x^2} = \frac{1}{2x}$
(b) $y = 0$

37. (a) $\frac{4x^3}{x} = 4x^2$
(b) None

39. (a) The function $y = e^x$ is a right end behavior model because $\lim_{x\to\infty} \frac{e^x - 2x}{e^x} = \lim_{x\to\infty}\left(1 - \frac{2x}{e^x}\right) = 1 - 0 = 1$.

(b) The function $y = -2x$ is a left end behavior model because $\lim_{x\to-\infty} \frac{e^x - 2x}{-2x} = \lim_{x\to-\infty}\left(-\frac{e^x}{2x} + 1\right) = 0 + 1 = 1$.

41. (a, b) The function $y = x$ is both a right end behavior model and a left end behavior model because

$\lim_{x\to\pm\infty} \left(\frac{x + \ln|x|}{x}\right) = \lim_{x\to\pm\infty}\left(1 + \frac{\ln|x|}{x}\right) = 1 + 0 = 1$.

43.

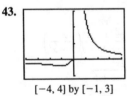

[-4, 4] by [-1, 3]

The graph of $y = f\left(\frac{1}{x}\right) = \frac{1}{x}e^{1/x}$ is shown.

$\lim_{x\to\infty} f(x) = \lim_{x\to 0^+} f\left(\frac{1}{x}\right) = \infty$

$\lim_{x\to\infty^-} f(x) = \lim_{x\to 0^-} f\left(\frac{1}{x}\right) = 0$

45.

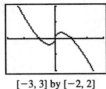

[-3, 3] by [-2, 2]

The graph of $y = f\left(\frac{1}{x}\right) = x \ln\left|\frac{1}{x}\right|$ is shown.

$\lim_{x\to\infty} f(x) = \lim_{x\to 0^+} f\left(\frac{1}{x}\right) = 0$

$\lim_{x\to-\infty} f(x) = \lim_{x\to 0^-} f\left(\frac{1}{x}\right) = 0$

47. (a) $\lim_{x\to-\infty} f(x) = \lim_{x\to-\infty}\left(\frac{1}{x}\right) = 0$

(b) $\lim_{x\to\infty} f(x) = \lim_{x\to\infty}(-1) = -1$

(c) $\lim_{x\to 0^-} f(x) = \lim_{x\to 0^-} \frac{1}{x} = -\infty$

(d) $\lim_{x\to 0^+} f(x) = \lim_{x\to 0^+}(-1) = -1$

49. One possible answer:

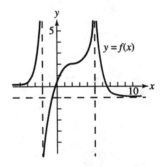

51. Note that $\frac{f_1(x)/f_2(x)}{g_1(x)/g_2(x)} = \frac{f_1(x)g_2(x)}{g_1(x)f_2(x)} = \frac{f_1(x)/g_1(x)}{f_2(x)/g_2(x)}$.

As x becomes large, $\frac{f_1}{g_1}$ and $\frac{f_2}{g_2}$ both approach 1. Therefore, using the above equation, $\frac{f_1/f_2}{g_1/g_2}$ must also approach 1.

53. (a) Using 1980 as $x = 0$:
$y = -2.2316x^3 + 54.7134x^2 - 351.0933x + 733.2224$

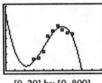

[0, 20] by [0, 800]

(b) Again using 1980 as $x = 0$:
$y = 1.458561x^4 - 60.5740x^3 + 905.8877x^2 - 5706.0943x + 12967.6288$

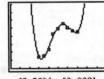

[0, 20] by [0, 800]

(c) Cubic: approximately -2256 dollars
Quartic: approximately 9979 dollars

(d) Cubic: End behavior model is $-2.2316x^3$.
This model predicts that the grants will become negative by 1996.
Quartic: End behavior model is $1.458561x^4$.
This model predicts that the size of the grants will grow very rapidly after 1995.
Neither of these seems reasonable. There is no reason to expect the grants to disappear (become negative) based on the data. Similarly, the data give no indication that a period of rapid growth is about to occur.

55. (a) This follow from $x - 1 < \text{int } x \leq x$, which is true for all x. Dividing by x gives the result.

(b, c) Since $\lim_{x \to \pm\infty} \frac{x-1}{x} = \lim_{x \to \pm\infty} 1 = 1$, the Sandwich Theorem gives $\lim_{x \to \infty} \frac{\text{int } x}{x} = \lim_{x \to -\infty} \frac{\text{int } x}{x} = 1$.

57. This is because as x approaches infinity, $\sin x$ continues to oscillate between 1 and -1 and doesn't approach any given real number.

59. $\lim_{x \to \infty} \frac{\ln x}{\log x} = \ln (10)$, since $\frac{\ln x}{\log x} = \frac{\ln x}{(\ln x)/(\ln 10)}$
$= \ln 10$.

■ Section 2.3 Continuity
(pp. 73–81)

Exploration 1 Removing a Discontinuity

1. $x^2 - 9 = (x - 3)(x + 3)$. The domain of f is $(-\infty, -3) \cup (-3, 3) \cup (3, \infty)$ or all $x \neq \pm 3$.

2. It appears that the limit of f as $x \to 3$ exists and is a little more than 3.

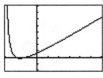

[−3, 6] by [−2, 8]

3. $f(3)$ should be defined as $\frac{10}{3}$.

4. $x^3 - 7x - 6 = (x - 3)(x + 1)(x + 2)$, $x^2 - 9$
$= (x - 3)(x + 3)$, so $f(x) = \frac{(x+1)(x+2)}{x+3}$ for $x \neq 3$.
Thus, $\lim_{x \to 3} \frac{(x+1)(x+2)}{x+3} = \frac{20}{6} = \frac{10}{3}$.

5. $\lim_{x \to 3} g(x) = \frac{10}{3} = g(3)$, so g is continuous at $x = 3$.

Quick Review 2.3

1. $\lim_{x \to -1} \frac{3x^2 - 2x + 1}{x^3 + 4} = \frac{3(-1)^2 - 2(-1) + 1}{(-1)^3 + 4} = \frac{6}{3} = 2$

2. (a) $\lim_{x \to -1^-} f(x) = \lim_{x \to -1^-} \text{int } (x) = -2$

(b) $\lim_{x \to -1^+} f(x) = \lim_{x \to -1^+} f(x) = -1$

(c) $\lim_{x \to -1} f(x)$ does not exist, because the left- and right-hand limits are not equal.

(d) $f(-1) = \text{int } (-1) = -1$

3. (a) $\lim_{x \to 2^-} f(x) = \lim_{x \to 2^-} (x^2 - 4x + 5) = 2^2 - 4(2) + 5 = 1$

(b) $\lim_{x \to 2^+} f(x) = \lim_{x \to 2^+} (4 - x) = 4 - 2 = 2$

(c) $\lim_{x \to 2} f(x)$ does not exist, because the left- and right-hand limits are not equal.

(d) $f(2) = 4 - 2 = 2$

4. $(f \circ g)(x) = f(g(x)) = f\left(\frac{1}{x} + 1\right) = \frac{2\left(\frac{1}{x} + 1\right) - 1}{\left(\frac{1}{x} + 1\right) + 5}$
$= \frac{2(1+x) - x}{(1+x) + 5x} = \frac{x+2}{6x+1}, x \neq 0$

$(g \circ f)(x) = g(f(x)) = g\left(\frac{2x-1}{x+5}\right) = \frac{1}{\frac{2x-1}{x+5}} + 1$
$= \frac{x+5}{2x-1} + \frac{2x-1}{2x-1} = \frac{3x+4}{2x-1}, x \neq -5$

5. Note that $\sin x^2 = (g \circ f)(x) = g(f(x)) = g(x^2)$.
Therefore: $g(x) = \sin x, x \geq 0$
$(f \circ g)(x) = f(g(x)) = f(\sin x) = (\sin x)^2$ or $\sin^2 x, x \geq 0$

6. Note that $\frac{1}{x} = (g \circ f)(x) = g(f(x)) = \sqrt{f(x) - 1}$.
Therefore, $\sqrt{f(x) - 1} = \frac{1}{x}$ for $x > 0$. Squaring both sides gives $f(x) - 1 = \frac{1}{x^2}$. Therefore, $f(x) = \frac{1}{x^2} + 1, x > 0$.
$(f \circ g)(x) = f(g(x)) = \frac{1}{(\sqrt{x-1})^2} + 1 = \frac{1}{x-1} + 1$
$= \frac{1 + x - 1}{x - 1} = \frac{x}{x-1}, x > 1$

7. $2x^2 + 9x - 5 = 0$
$(2x - 1)(x + 5) = 0$
Solutions: $x = \frac{1}{2}, x = -5$

8.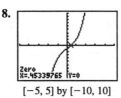

[−5, 5] by [−10, 10]
Solution: $x \approx 0.453$

9. For $x \leq 3, f(x) = 4$ when $5 - x = 4$, which gives $x = 1$. (Note that this value is, in fact, ≤ 3.)

For $x > 3, f(x) = 4$ when $-x^2 + 6x - 8 = 4$, which gives $x^2 - 6x + 12 = 0$. The discriminant of this equation is $b^2 - 4ac = (-6)^2 - 4(1)(12) = -12$. Since the discriminant is negative, the quadratic equation has no solution.

The only solution to the original equation is $x = 1$.

Section 2.3

10.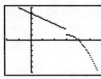
[−2.7, 6.7] by [−6, 6]

A graph of $f(x)$ is shown. The range of $f(x)$ is $(-\infty, 1) \cup [2, \infty)$. The values of c for which $f(x) = c$ has no solution are the values that are excluded from the range. Therefore, c can be any value in $[1, 2)$.

Section 2.3 Exercises

1. The function $y = \dfrac{1}{(x + 2)^2}$ is continuous because it is a quotient of polynomials, which are continuous. Its only point of discontinuity occurs where it is undefined. There is an infinite discontinuity at $x = -2$.

3. The function $y = \dfrac{1}{x^2 + 1}$ is continuous because it is a quotient of polynomials, which are continuous. Furthermore, the domain is all real numbers because the denominator, $x^2 + 1$, is never zero. Since the function is continuous and has domain $(-\infty, \infty)$, there are no points of discontinuity.

5. The function $y = \sqrt{2x + 3}$ is a composition $(f \circ g)(x)$ of the continuous functions $f(x) = \sqrt{x}$ and $g(x) = 2x + 3$, so it is continuous. Its points of discontinuity are the points not in the domain, i.e., all $x < -\dfrac{3}{2}$.

7. The function $y = \dfrac{|x|}{x}$ is equivalent to
$$y = \begin{cases} -1, & x < 0 \\ 1, & x > 0. \end{cases}$$
It has a jump discontinuity at $x = 0$.

9. The function $y = e^{1/x}$ is a composition $(f \circ g)(x)$ of the continuous functions $f(x) = e^x$ and $g(x) = \dfrac{1}{x}$, so it is continuous. Its only point of discontinuity occurs at $x = 0$, where it is undefined. Since $\lim\limits_{x \to 0^+} e^{1/x} = \infty$, this may be considered an infinite discontinuity.

11. (a) Yes, $f(-1) = 0$.

    (b) Yes, $\lim\limits_{x \to -1^+} = 0$.

    (c) Yes

    (d) Yes, since -1 is a left endpoint of the domain of f and $\lim\limits_{x \to -1^+} f(x) = f(-1)$, f is continuous at $x = -1$.

13. (a) No

    (b) No, since $x = 2$ is not in the domain.

15. Since $\lim\limits_{x \to 2} f(x) = 0$, we should assign $f(2) = 0$.

17. No, because the right-hand and left-hand limits are not the same at zero.

19.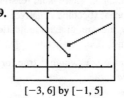
[−3, 6] by [−1, 5]

    (a) $x = 2$

    (b) Not removable, the one-sided limits are different.

21.
[−5, 5] by [−4, 8]

    (a) $x = 1$

    (b) Not removable, it's an infinite discontinuity.

23. (a) All points not in the domain along with $x = 0, 1$

    (b) $x = 0$ is a removable discontinuity, assign $f(0) = 0$.
    $x = 1$ is not removable, the one-sided limits are different.

25. For $x \neq -3$, $f(x) = \dfrac{x^2 - 9}{x + 3} = \dfrac{(x + 3)(x - 3)}{x + 3} = x - 3$.
The extended function is $y = x - 3$.

27. Since $\lim\limits_{x \to 0} \dfrac{\sin x}{x} = 1$, the extended function is
$$y = \begin{cases} \dfrac{\sin x}{x}, & x \neq 0 \\ 1, & x = 0. \end{cases}$$

29. For $x \neq 4$ (and $x > 0$),
$$f(x) = \dfrac{x - 4}{\sqrt{x} - 2} = \dfrac{(\sqrt{x} + 2)(\sqrt{x} - 2)}{\sqrt{x} - 2} = \sqrt{x} + 2.$$

The extended function is $y = \sqrt{x} + 2$.

31. One possible answer:
Assume $y = x$, constant functions, and the square root function are continuous.

By the sum theorem, $y = x + 2$ is continuous.

By the composite theorem, $y = \sqrt{x + 2}$ is continuous.

By the quotient theorem, $y = \dfrac{1}{\sqrt{x + 2}}$ is continuous.

Domain: $(-2, \infty)$

33. Possible answer:
Assume $y = x$ and $y = |x|$ are continuous.
By the product theorem, $y = x^2 = x \cdot x$ is continuous.
By the constant multiple theorem, $y = 4x$ is continuous.
By the difference theorem, $y = x^2 - 4x$ is continuous.
By the composite theorem, $y = |x^2 - 4x|$ is continuous.
Domain: $(-\infty, \infty)$

35. One possible answer:

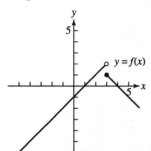

37. One possible answer:

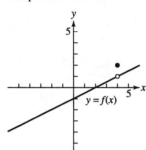

39.

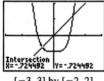

[−3, 3] by [−2, 2]

Solving $x = x^4 - 1$, we obtain the solutions $x \approx -0.724$ and $x \approx 1.221$.

41. We require that $\lim\limits_{x \to 3^+} 2ax = \lim\limits_{x \to 3^-} (x^2 - 1)$:

$2a(3) = 3^2 - 1$

$6a = 8$

$a = \dfrac{4}{3}$

43. (a) Sarah's salary is $36,500 = \$36,500(1.035)^0$ for the first year ($0 \le t < 1$), $36,500(1.035)$ for the second year ($1 \le t < 2$), $36,500(1.035)^2$ for the third year ($2 \le t < 3$), and so on. This corresponds to $y = 36,500(1.035)^{\text{int } t}$.

(b)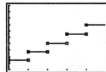

[0, 4.98] by [35,000, 45,000]

The function is continuous at all points in the domain [0, 5) except at $t = 1, 2, 3, 4$.

45. (a) The function is defined when $1 + \dfrac{1}{x} > 0$, that is, on $(-\infty, -1) \cup (0, \infty)$. (It can be argued that the domain should also include certain values in the interval $(-1, 0)$, namely, those rational numbers that have odd denominators when expressed in lowest terms.)

(b)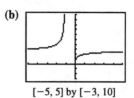

[−5, 5] by [−3, 10]

(c) If we attempt to evaluate $f(x)$ at these values, we obtain
$f(-1) = \left(1 + \dfrac{1}{-1}\right)^{-1} = 0^{-1} = \dfrac{1}{0}$ (undefined) and
$f(0) = \left(1 + \dfrac{1}{0}\right)^0$ (undefined). Since f is undefined at these values due to division by zero, both values are points of discontinuity.

(d) The discontinuity at $x = 0$ is removable because the right-hand limit is 0. The discontinuity at $x = -1$ is not removable because it is an infinite discontinuity.

(e)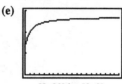

[0, 20] by [0, 3]

X	Y1
10	2.5937
100	2.7048
1000	2.7169
10000	2.7181
100000	2.7183
1E6	2.7183
1E7	2.7183

Y1 ◻ (1+1/X)^X

The limit is about 2.718, or e.

47. Suppose not. Then f would be negative somewhere in the interval and positive somewhere else in the interval. So, by the Intermediate Value Theorem, it would have to be zero somewhere in the interval, which contradicts the hypothesis.

49. For any real number a, the limit of this function as x approaches a cannot exist. This is because as x approaches a, the values of the function will continually oscillate between 0 and 1.

■ Section 2.4 Rates of Change and Tangent Lines (pp. 82–90)

Quick Review 2.4

1. $\Delta x = 3 - (-5) = 8$
$\Delta y = 5 - 2 = 3$

2. $\Delta x = a - 1$
$\Delta y = b - 3$

3. $m = \dfrac{-1 - 3}{5 - (-2)} = \dfrac{-4}{7} = -\dfrac{4}{7}$

38 Section 2.4

4. $m = \dfrac{3-(-1)}{3-(-3)} = \dfrac{4}{6} = \dfrac{2}{3}$

5. $y = \dfrac{3}{2}[x-(-2)] + 3$
   $y = \dfrac{3}{2}x + 6$

6. $m = \dfrac{-1-6}{4-1} = \dfrac{-7}{3} = -\dfrac{7}{3}$
   $y = -\dfrac{7}{3}(x-1) + 6$
   $y = -\dfrac{7}{3}x + \dfrac{25}{3}$

7. $y = -\dfrac{3}{4}(x-1) + 4$
   $y = -\dfrac{3}{4}x + \dfrac{19}{4}$

8. $m = -\dfrac{1}{-3/4} = \dfrac{4}{3}$
   $y = \dfrac{4}{3}(x-1) + 4$
   $y = \dfrac{4}{3}x + \dfrac{8}{3}$

9. Since $2x + 3y = 5$ is equivalent to $y = -\dfrac{2}{3}x + \dfrac{5}{3}$, we use $m = -\dfrac{2}{3}$.
   $y = -\dfrac{2}{3}[x-(-1)] + 3$
   $y = -\dfrac{2}{3}x + \dfrac{7}{3}$

10. $\dfrac{b-3}{4-2} = \dfrac{5}{3}$
    $b - 3 = \dfrac{10}{3}$
    $b = \dfrac{19}{3}$

Section 2.4 Exercises

1. (a) $\dfrac{\Delta f}{\Delta x} = \dfrac{f(3) - f(2)}{3 - 2} = \dfrac{28 - 9}{1} = 19$

   (b) $\dfrac{\Delta f}{\Delta x} = \dfrac{f(1) - f(-1)}{1-(-1)} = \dfrac{2 - 0}{2} = 1$

3. (a) $\dfrac{\Delta f}{\Delta x} = \dfrac{f(0) - f(-2)}{0-(-2)} = \dfrac{1 - e^{-2}}{2} \approx 0.432$

   (b) $\dfrac{\Delta f}{\Delta x} = \dfrac{f(3) - f(1)}{3 - 1} = \dfrac{e^3 - e}{2} \approx 8.684$

5. (a) $\dfrac{\Delta f}{\Delta x} = \dfrac{f(3\pi/4) - f(\pi/4)}{(3\pi/4) - (\pi/4)} = \dfrac{-1-1}{\pi/2} = -\dfrac{4}{\pi} \approx -1.273$

   (b) $\dfrac{\Delta f}{\Delta x} = \dfrac{f(\pi/2) - f(\pi/6)}{(\pi/2) - (\pi/6)} = \dfrac{0 - \sqrt{3}}{\pi/3} = -\dfrac{3\sqrt{3}}{\pi} \approx -1.654$

7. We use $Q_1 = (10, 225)$, $Q_2 = (14, 375)$, $Q_3 = (16.5, 475)$, $Q_4 = (18, 550)$, and $P = (20, 650)$.

   (a) Slope of PQ_1: $\dfrac{650 - 225}{20 - 10} \approx 43$

   Slope of PQ_2: $\dfrac{650 - 375}{20 - 14} \approx 46$

   slope of PQ_3: $\dfrac{650 - 475}{20 - 16.5} = 50$

   Slope of PQ_4: $\dfrac{650 - 550}{20 - 18} = 50$

Secant	Slope
PQ_1	43
PQ_2	46
PQ_3	50
PQ_4	50

   The appropriate units are meters per second.

   (b) Approximately 50 m/sec

9. (a) $\lim\limits_{h \to 0} \dfrac{y(-2+h) - y(-2)}{h} = \lim\limits_{h \to 0} \dfrac{(-2+h)^2 - (-2)^2}{h}$
   $= \lim\limits_{h \to 0} \dfrac{4 - 4h + h^2 - 4}{h}$
   $= \lim\limits_{h \to 0} \dfrac{-4h + h^2}{h}$
   $= \lim\limits_{h \to 0} (-4 + h)$
   $= -4$

   (b) The tangent line has slope -4 and passes through $(-2, y(-2)) = (-2, 4)$.
   $y = -4[x-(-2)] + 4$
   $y = -4x - 4$

   (c) The normal line has slope $-\dfrac{1}{-4} = \dfrac{1}{4}$ and passes through $(-2, y(-2)) = (-2, 4)$.
   $y = \dfrac{1}{4}[x-(-2)] + 4$
   $y = \dfrac{1}{4}x + \dfrac{9}{2}$

   (d)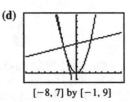
   $[-8, 7]$ by $[-1, 9]$

11. (a) $\lim_{h \to 0} \dfrac{y(2+h) - y(2)}{h} = \lim_{h \to 0} \dfrac{\dfrac{1}{(2+h)-1} - \dfrac{1}{2-1}}{h}$

$= \lim_{h \to 0} \dfrac{\dfrac{1}{h+1} - 1}{h}$

$= \lim_{h \to 0} \dfrac{1 - (h+1)}{h(h+1)}$

$= \lim_{h \to 0} \left(-\dfrac{1}{h+1}\right)$

$= -1$

(b) The tangent line has slope -1 and passes through $(2, y(2)) = (2, 1)$.
$y = -(x - 2) + 1$
$y = -x + 3$

(c) The normal line has slope $-\dfrac{1}{-1} = 1$ and passes through $(2, y(2)) = (2, 1)$.

$y = 1(x - 2) + 1$
$y = x - 1$

(d)

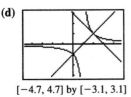

$[-4.7, 4.7]$ by $[-3.1, 3.1]$

13. (a) Near $x = 2$, $f(x) = |x| = x$.

$\lim_{h \to 0} \dfrac{f(2+h) - f(2)}{h} = \lim_{h \to 0} \dfrac{(2+h) - 2}{h} = \lim_{h \to 0} 1 = 1$

(b) Near $x = -3$, $f(x) = |x| = -x$.

$\lim_{h \to 0} \dfrac{f(-3+h) - f(-3)}{h} = \lim_{h \to 0} \dfrac{(3-h) - 3}{h}$

$= \lim_{h \to 0} -1 = -1$

15. First, note that $f(0) = 2$.

$\lim_{h \to 0^-} \dfrac{f(0+h) - f(0)}{h} = \lim_{h \to 0^-} \dfrac{(2 - 2h - h^2) - 2}{h}$

$= \lim_{h \to 0^-} \dfrac{-2h - h^2}{h}$

$= \lim_{h \to 0^-} (-2 - h)$

$= -2$

$\lim_{h \to 0^+} \dfrac{f(0+h) - f(0)}{h} = \lim_{h \to 0^+} \dfrac{(2h + 2) - 2}{h}$

$= \lim_{h \to 0^+} 2$

$= 2$

No, the slope from the left is -2 and the slope from the right is 2. The two-sided limit of the difference quotient does not exist.

17. First, note that $f(2) = \dfrac{1}{2}$.

$\lim_{h \to 0^-} \dfrac{f(2+h) - f(2)}{h} = \lim_{h \to 0^-} \dfrac{\dfrac{1}{2+h} - \dfrac{1}{2}}{h}$

$= \lim_{h \to 0^-} \dfrac{2 - (2+h)}{2h(2+h)}$

$= \lim_{h \to 0^-} \dfrac{-h}{2h(2+h)}$

$= \lim_{h \to 0^-} -\dfrac{1}{2(2+h)}$

$= -\dfrac{1}{4}$

$\lim_{h \to 0^+} \dfrac{f(2+h) - f(2)}{h} = \lim_{h \to 0^+} \dfrac{\dfrac{4-(2+h)}{4} - \dfrac{1}{2}}{h}$

$= \lim_{h \to 0^+} \dfrac{[4 - (2+h)] - 2}{4h}$

$= \lim_{h \to 0^+} \dfrac{-h}{4h}$

$= -\dfrac{1}{4}$

Yes. The slope is $-\dfrac{1}{4}$.

19. (a) $\lim_{h \to 0} \dfrac{f(a+h) - f(a)}{h} = \lim_{h \to 0} \dfrac{[(a+h)^2 + 2] - (a^2 + 2)}{h}$

$= \lim_{h \to 0} \dfrac{a^2 + 2ah + h^2 + 2 - a^2 - 2}{h}$

$= \lim_{h \to 0} \dfrac{2ah + h^2}{h}$

$= \lim_{h \to 0} (2a + h)$

$= 2a$

(b) The slope of the tangent steadily increases as a increases.

21. (a) $\lim_{h \to 0} \dfrac{f(a+h) - f(a)}{h} = \lim_{h \to 0} \dfrac{\dfrac{1}{a+h-1} - \dfrac{1}{a-1}}{h}$

$= \lim_{h \to 0} \dfrac{(a-1) - (a+h-1)}{h(a-1)(a+h-1)}$

$= \lim_{h \to 0} -\dfrac{1}{(a-1)(a+h-1)}$

$= -\dfrac{1}{(a-1)^2}$

(b) The slope of the tangent is always negative. The tangents are very steep near $x = 1$ and nearly horizontal as a moves away from the origin.

23. Let $f(t) = 100 - 4.9t^2$.

$\lim_{h \to 0} \dfrac{f(2+h) - f(2)}{h}$

$= \lim_{h \to 0} \dfrac{[100 - 4.9(2+h)^2] - [100 - 4.9(2)^2]}{h}$

$= \lim_{h \to 0} \dfrac{100 - 19.6 - 19.6h - 4.9h^2 - 100 + 19.6}{h}$

$= \lim_{h \to 0} (-19.6 - 4.9h)$

$= -19.6$

The object is falling at a speed of 19.6 m/sec.

40   Section 2.4

25. Let $f(r) = \pi r^2$, the area of a circle of radius r.

$$\lim_{h \to 0} \frac{f(3+h) - f(3)}{h} = \lim_{h \to 0} \frac{\pi(3+h)^2 - \pi(3)^2}{h}$$
$$= \lim_{h \to 0} \frac{9\pi + 6\pi h + \pi h^2 - 9\pi}{h}$$
$$= \lim_{h \to 0} (6\pi + \pi h)$$
$$= 6\pi$$

The area is changing at a rate of 6π in²/in., that is, 6π square inches of area per inch of radius.

27. $\lim_{h \to 0} \dfrac{s(1+h) - s(1)}{h} = \lim_{h \to 0} \dfrac{1.86(1+h)^2 - 1.86(1)^2}{h}$
$= \lim_{h \to 0} \dfrac{1.86 + 3.72h + 1.86h^2 - 1.86}{h}$
$= \lim_{h \to 0} (3.72 + 1.86h)$
$= 3.72$

The speed of the rock is 3.72 m/sec.

29. First, find the slope of the tangent at $x = a$.

$\lim_{h \to 0} \dfrac{f(a+h) - f(a)}{h}$
$= \lim_{h \to 0} \dfrac{[(a+h)^2 + 4(a+h) - 1] - (a^2 + 4a - 1)}{h}$
$= \lim_{h \to 0} \dfrac{a^2 + 2ah + h^2 + 4a + 4h - 1 - a^2 - 4a + 1}{h}$
$= \lim_{h \to 0} \dfrac{2ah + h^2 + 4h}{h}$
$= \lim_{h \to 0} (2a + h + 4)$
$= 2a + 4$

The tangent at $x = a$ is horizontal when $2a + 4 = 0$, or $a = -2$. The tangent line is horizontal at $(-2, f(-2)) = (-2, -5)$.

31. (a) From Exercise 21, the slope of the curve at $x = a$, is $-\dfrac{1}{(a-1)^2}$. The tangent has slope -1 when $-\dfrac{1}{(a-1)^2} = -1$, which gives $(a-1)^2 = 1$, so $a = 0$ or $a = 2$. Note that $y(0) = \dfrac{1}{0-1} = -1$ and $y(2) = \dfrac{1}{2-1} = 1$, so we need to find the equations of lines of slope -1 passing through $(0, -1)$ and $(2, 1)$, respectively.

At $x = 0$: $y = -1(x - 0) - 1$
$\qquad\qquad y = -x - 1$
At $x = 2$: $y = -1(x - 2) + 1$
$\qquad\qquad y = -x + 3$

(b) The normal has slope 1 when the tangent has slope $\dfrac{-1}{1} = -1$, so we again need to find lines through $(0, -1)$ and $(2, 1)$, this time using slope 1.

At $x = 0$: $y = 1(x - 0) - 1$
$\qquad\qquad y = x - 1$
At $x = 2$: $y = 1(x - 2) + 1$
$\qquad\qquad y = x - 1$

There is only one such line. It is normal to the curve at two points and its equation is $y = x - 1$.

33. (a) $\dfrac{2.1 - 1.5}{1995 - 1993} = 0.3$

The rate of change was 0.3 billion dollars per year.

(b) $\dfrac{3.1 - 2.1}{1997 - 1995} = 0.5$

The rate of change was 0.5 billion dollars per year.

(c) $y = 0.0571x^2 - 0.1514x + 1.3943$

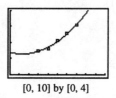

[0, 10] by [0, 4]

(d) $\dfrac{y(5) - y(3)}{5 - 3} \approx 0.31$

$\dfrac{y(7) - y(5)}{7 - 5} \approx 0.53$

According to the regression equation, the rates were 0.31 billion dollars per year and 0.53 billion dollars per year.

(e) $\lim_{h \to 0} \dfrac{y(7+h) - y(7)}{h} = \lim_{h \to 0} \dfrac{[0.0571(7+h)^2 - 0.1514(7+h) + 1.3943] - [0.0571(7)^2 - 0.1514(7) + 13943]}{h}$

$= \lim_{h \to 0} \dfrac{0.0571(14h + h^2) - 0.1514h}{h}$

$= \lim_{h \to 0} [0.0571(14) - 0.1514 + 0.0571h]$

≈ 0.65

The funding was growing at a rate of about 0.65 billion dollars per year.

35. (a) $\dfrac{f(1+h) - f(1)}{h} = \dfrac{e^{1+h} - e}{h}$

(b)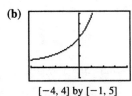

[−4, 4] by [−1, 5]

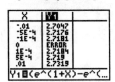

Limit ≈ 2.718

(c) They're about the same.

(d) Yes, it has a tangent whose slope is about e.

37. Let $f(x) = x^{2/5}$. The graph of $y = \dfrac{f(0+h) - f(0)}{h} = \dfrac{f(h)}{h}$ is shown.

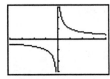

[−4, 4] by [−3, 3]

The left- and right-hand limits are $-\infty$ and ∞, respectively. Since they are not the same, the curve does not have a vertical tangent at $x = 0$. No.

39. Let $f(x) = x^{1/3}$. The graph of $y = \dfrac{f(0+h) - f(0)}{h} = \dfrac{f(h)}{h}$ is shown.

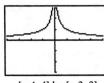

[−4, 4] by [−3, 3]

Yes, the curve has a vertical tangent at $x = 0$ because

$\lim_{h \to 0} \dfrac{f(0+h) - f(0)}{h} = \infty$.

41. This function has a tangent with slope zero at the origin. It is sandwiched between two functions, $y = x^2$ and $y = -x^2$, both of which have slope zero at the origin.

Looking at the difference quotient,

$-h \leq \dfrac{f(0+h) - f(0)}{h} \leq h$,

so the Sandwich Theorem tells us the limit is 0.

43. Let $f(x) = \sin x$. The difference quotient is
$$\frac{f(1+h) - f(1)}{h} = \frac{\sin(1+h) - \sin(1)}{h}.$$
A graph and table for the difference quotient are shown.

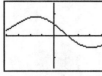

[−4, 4] by [−1.5, 1.5]

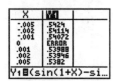

Since the limit as $h \to 0$ is about 0.540, the slope of $y = \sin x$ at $x = 1$ is about 0.540.

■ Chapter 2 Review Exercises
(pp. 91–93)

1. $\lim\limits_{x \to -2} (x^3 - 2x^2 + 1) = (-2)^3 - 2(-2)^2 + 1 = -15$

2. $\lim\limits_{x \to -2} \dfrac{x^2 + 1}{3x^2 - 2x + 5} = \dfrac{(-2)^2 + 1}{3(-2)^2 - 2(-2) + 5} = \dfrac{5}{21}$

3. No limit, because the expression $\sqrt{1 - 2x}$ is undefined for values of x near 4.

4. No limit, because the expression $\sqrt[4]{9 - x^2}$ is undefined for values of x near 5.

5. $\lim\limits_{x \to 0} \dfrac{\frac{1}{2+x} - \frac{1}{2}}{x} = \lim\limits_{x \to 0} \dfrac{2 - (2+x)}{2x(2+x)} = \lim\limits_{x \to 0} \dfrac{-x}{2x(2+x)}$
$= \lim\limits_{x \to 0} \left(-\dfrac{1}{2(2+x)}\right) = -\dfrac{1}{2(2+0)} = -\dfrac{1}{4}$

6. $\lim\limits_{x \to \pm\infty} \dfrac{2x^2 + 3}{5x^2 + 7} = \lim\limits_{x \to \pm\infty} \dfrac{2x^2}{5x^2} = \dfrac{2}{5}$

7. An end behavior model for $\dfrac{x^4 + x^3}{12x^3 + 128}$ is $\dfrac{x^4}{12x^3} = \dfrac{1}{12}x$.

Therefore:
$\lim\limits_{x \to \infty} \dfrac{x^4 + x^3}{12x^3 + 128} = \lim\limits_{x \to \infty} \dfrac{1}{12}x = \infty$
$\lim\limits_{x \to -\infty} \dfrac{x^4 + x^3}{12x^3 + 128} = \lim\limits_{x \to -\infty} \dfrac{1}{12}x = -\infty$

8. $\lim\limits_{x \to 0} \dfrac{\sin 2x}{4x} = \dfrac{1}{2} \lim\limits_{x \to 0} \dfrac{\sin 2x}{2x} = \dfrac{1}{2}(1) = \dfrac{1}{2}$

9. Multiply the numerator and denominator by $\sin x$.
$\lim\limits_{x \to 0} \dfrac{x \csc x + 1}{x \csc x} = \lim\limits_{x \to 0} \dfrac{x + \sin x}{x} = \lim\limits_{x \to 0} \left(1 + \dfrac{\sin x}{x}\right)$
$= \left(\lim\limits_{x \to 0} 1\right) + \left(\lim\limits_{x \to 0} \dfrac{\sin x}{x}\right) = 1 + 1 = 2$

10. $\lim\limits_{x \to 0} e^x \sin x = e^0 \sin 0 = 1 \cdot 0 = 0$

11. Let $x = \dfrac{7}{2} + h$, where h is in $\left(0, \dfrac{1}{2}\right)$. Then
$\operatorname{int}(2x - 1) = \operatorname{int}\left[2\left(\dfrac{7}{2}\right) + 2h - 1\right] = \operatorname{int}(6 + 2h) = 6$,
because $6 + 2h$ is in (6, 7).

Therefore, $\lim\limits_{x \to 7/2^+} \operatorname{int}(2x - 1) = \lim\limits_{x \to 7/2^+} 6 = 6$.

12. Let $x = \dfrac{7}{2} + h$, where h is in $\left(-\dfrac{1}{2}, 0\right)$. Then
$\operatorname{int}(2x - 1) = \operatorname{int}\left[2\left(\dfrac{7}{2}\right) + 2h - 1\right] = \operatorname{int}(6 + 2h) = 5$,
because $6 + 2h$ is in (5, 6).

Therefore, $\lim\limits_{x \to 7/2^-} \operatorname{int}(2x - 1) = \lim\limits_{x \to 7/2^-} 5 = 5$

13. Since $\lim\limits_{x \to \infty}(-e^{-x}) = \lim\limits_{x \to \infty} e^{-x} = 0$, and
$-e^{-x} \le e^{-x} \cos x \le e^{-x}$ for all x, the Sandwich Theorem gives $\lim\limits_{x \to \infty} e^{-x} \cos x = 0$.

14. Since the expression x is an end behavior model for both $x + \sin x$ and $x + \cos x$, $\lim\limits_{x \to \infty} \dfrac{x + \sin x}{x + \cos x} = \lim\limits_{x \to \infty} \dfrac{x}{x} = 1$.

15. Limit exists. **16.** Limit exists.
17. Limit exists. **18.** Limit does not exist.
19. Limit exists. **20.** Limit exists.
21. Yes **22.** No
23. No **24.** Yes

25. (a) $\lim\limits_{x \to 3^-} g(x) = 1$

(b) $g(3) = 1.5$

(c) No, since $\lim\limits_{x \to 3^-} g(x) \ne g(3)$.

(d) g is discontinuous at $x = 3$ (and at points not in the domain).

(e) Yes, the discontinuity at $x = 3$ can be removed by assigning the value 1 to $g(3)$.

26. (a) $\lim\limits_{x \to 1^-} k(x) = 1.5$

(b) $\lim\limits_{x \to 1^+} k(x) = 0$

(c) $k(1) = 0$

(d) No, since $\lim\limits_{x \to 1^-} k(x) \ne k(1)$

(e) k is discontinuous at $x = 1$ (and at points not in the domain).

(f) No, the discontinuity at $x = 1$ is not removable because the one-sided limits are different.

27.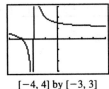

[−4, 4] by [−3, 3]

(a) Vertical asymptote: $x = -2$

(b) Left-hand limit $= \lim_{x \to -2^-} \dfrac{x+3}{x+2} = -\infty$

Right-hand limit: $\lim_{x \to -2^+} \dfrac{x+3}{x+2} = \infty$

28.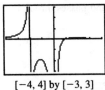

[−4, 4] by [−3, 3]

(a) Vertical asymptotes: $x = 0,\ x = -2$

(b) At $x = 0$:

Left-hand limit $= \lim_{x \to 0^-} \dfrac{x-1}{x^2(x+2)} = -\infty$

Right-hand limit $= \lim_{x \to 0^+} \dfrac{x-1}{x^2(x+2)} = -\infty$

At $x = -2$:

Left-hand limit $= \lim_{x \to -2^-} \dfrac{x-1}{x^2(x+2)} = \infty$

Right-hand limit $= \lim_{x \to -2^+} \dfrac{x-1}{x^2(x+2)} = -\infty$

29. (a) At $x = -1$:

Left-hand limit $= \lim_{x \to -1^-} f(x) = \lim_{x \to -1^-} (1) = 1$

Right-hand limit $= \lim_{x \to -1^+} f(x) = \lim_{x \to -1^+} (-x) = 1$

At $x = 0$:

Left-hand limit $= \lim_{x \to 0^-} f(x) = \lim_{x \to 0^-} (-x) = 0$

Right-hand limit $= \lim_{x \to 0^+} f(x) = \lim_{x \to 0^+} (-x) = 0$

At $x = 1$:

Left-hand limit $= \lim_{x \to 1^-} f(x) = \lim_{x \to 1^-} (-x) = -1$

Right-hand limit $= \lim_{x \to 1^+} f(x) = \lim_{x \to 1^+} (1) = 1$

(b) At $x = -1$: Yes, the limit is 1.
At $x = 0$: Yes, the limit is 0.
At $x = 1$: No, the limit doesn't exist because the two one-sided limits are different.

(c) At $x = -1$: Continuous because $f(-1) =$ the limit.
At $x = 0$: Discontinuous because $f(0) \ne$ the limit.
At $x = 1$: Discontinuous because the limit does not exist.

30. (a) Left-hand limit $= \lim_{x \to 1^-} f(x) = \lim_{x \to 1^-} |x^3 - 4x|$
$= |(1)^3 - 4(1)| = |-3| = 3$

Right-hand limit $= \lim_{x \to 1^+} f(x) = \lim_{x \to 1^+} (x^2 - 2x - 2)$
$= (1)^2 - 2(1) - 2 = -3$

(b) No, because the two one-sided limits are different.

(c) Every place except for $x = 1$

(d) At $x = 1$

31. Since $f(x)$ is a quotient of polynomials, it is continuous and its points of discontinuity are the points where it is undefined, namely $x = -2$ and $x = 2$.

32. There are no points of discontinuity, since $g(x)$ is continuous and defined for all real numbers.

33. (a) End behavior model: $\dfrac{2x}{x^2}$, or $\dfrac{2}{x}$

(b) Horizontal asymptote: $y = 0$ (the x-axis)

34. (a) End behavior model: $\dfrac{2x^2}{x^2}$, or 2

(b) Horizontal asymptote: $y = 2$

35. (a) End behavior model: $\dfrac{x^3}{x}$, or x^2

(b) Since the end behavior model is quadratic, there are no horizontal asymptotes.

36. (a) End behavior model: $\dfrac{x^4}{x^3}$, or x

(b) Since the end behavior model represents a nonhorizontal line, there are no horizontal asymptotes.

37. (a) Since $\lim_{x \to \infty} \dfrac{x + e^x}{e^x} = \lim_{x \to \infty} \left(\dfrac{x}{e^x} + 1\right) = 1$, a right end behavior model is e^x.

(b) Since $\lim_{x \to -\infty} \dfrac{x + e^x}{x} = \lim_{x \to -\infty} \left(1 + \dfrac{e^x}{x}\right) = 1$, a left end behavior model is x.

38. (a, b) Note that $\lim_{x \to \pm\infty} \left(-\dfrac{1}{\ln|x|}\right) = \lim_{x \to \pm\infty} \left(\dfrac{1}{\ln|x|}\right) = 0$ and
$-\dfrac{1}{\ln|x|} < \dfrac{\sin x}{\ln|x|} < \dfrac{1}{\ln|x|}$ for all $x \ne 0$.

Therefore, the Sandwich Theorem gives

$\lim_{x \to \pm\infty} \dfrac{\sin x}{\ln|x|} = 0$. Hence

$\lim_{x \to \pm\infty} \dfrac{\ln|x| + \sin x}{\ln|x|} = \lim_{x \to \pm\infty} \left(1 + \dfrac{\sin x}{\ln|x|}\right) = 1 + 0 = 1$,

so $\ln|x|$ is both a right end behavior model and a left end behavior model.

39. $\lim_{x \to 3} f(x) = \lim_{x \to 3} \dfrac{x^2 + 2x - 15}{x - 3} = \lim_{x \to 3} \dfrac{(x-3)(x+5)}{x-3}$
$= \lim_{x \to 3} (x + 5) = 3 + 5 = 8$.

Assign the value $k = 8$.

44 Chapter 2 Review

40. $\lim\limits_{x\to 0} f(x) = \lim\limits_{x\to 0} \dfrac{\sin x}{2x} = \dfrac{1}{2}\lim\limits_{x\to 0} \dfrac{\sin x}{x} = \dfrac{1}{2}(1) = \dfrac{1}{2}$

Assign the value $k = \dfrac{1}{2}$.

41. One possible answer:

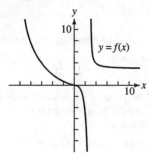

42. One possible answer:

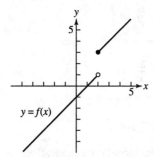

43. $\dfrac{f(\pi/2) - f(0)}{\pi/2 - 0} = \dfrac{2-1}{\pi/2} = \dfrac{2}{\pi}$

44. $\lim\limits_{h\to 0} \dfrac{V(a+h) - V(a)}{h} = \lim\limits_{h\to 0} \dfrac{\frac{1}{3}\pi(a+h)^2 H - \frac{1}{3}\pi a^2 H}{h}$

$= \dfrac{1}{3}\pi H \lim\limits_{h\to 0} \dfrac{a^2 + 2ah + h^2 - a^2}{h}$

$= \dfrac{1}{3}\pi H \lim\limits_{h\to 0} (2a + h)$

$= \dfrac{1}{3}\pi H(2a)$

$= \dfrac{2}{3}\pi a H$

45. $\lim\limits_{h\to 0} \dfrac{S(a+h) - S(a)}{h} = \lim\limits_{h\to 0} \dfrac{6(a+h)^2 - 6a^2}{h}$

$= \lim\limits_{h\to 0} \dfrac{6a^2 + 12ah + 6h^2 - 6a^2}{h}$

$= \lim\limits_{h\to 0} (12a + 6h)$

$= 12a$

46. $\lim\limits_{h\to 0} \dfrac{y(a+h) - y(a)}{h}$

$= \lim\limits_{h\to 0} \dfrac{[(a+h)^2 - (a+h) - 2] - (a^2 - a - 2)}{h}$

$= \lim\limits_{h\to 0} \dfrac{a^2 + 2ah + h^2 - a - h - 2 - a^2 + a + 2}{h}$

$= \lim\limits_{h\to 0} \dfrac{2ah + h^2 - h}{h}$

$= \lim\limits_{h\to 0} (2a + h - 1)$

$= 2a - 1$

47. (a) $\lim\limits_{h\to 0} \dfrac{f(1+h) - f(1)}{h} = \lim\limits_{h\to 0} \dfrac{[(1+h)^2 - 3(1+h)] - (-2)}{h}$

$= \lim\limits_{h\to 0} \dfrac{1 + 2h + h^2 - 3 - 3h + 2}{h}$

$= \lim\limits_{h\to 0} (-1 + h)$

$= -1$

(b) The tangent at P has slope -1 and passes through $(1, -2)$.

$y = -1(x - 1) - 2$

$y = -x - 1$

(c) The normal at P has slope 1 and passes through $(1, -2)$.

$y = 1(x - 1) - 2$

$y = x - 3$

48. At $x = a$, the slope of the curve is

$\lim\limits_{h\to 0} \dfrac{f(a+h) - f(a)}{h} = \lim\limits_{h\to 0} \dfrac{[(a+h)^2 - 3(a+h)] - (a^2 - 3a)}{h}$

$= \lim\limits_{h\to 0} \dfrac{a^2 + 2ah + h^2 - 3a - 3h - a^2 + 3a}{h}$

$= \lim\limits_{h\to 0} \dfrac{2ah - 3h + h^2}{h}$

$= \lim\limits_{h\to 0} (2a - 3 + h)$

$= 2a - 3$

The tangent is horizontal when $2a - 3 = 0$, at $a = \dfrac{3}{2}$

$\left(\text{or } x = \dfrac{3}{2}\right)$. Since $f\left(\dfrac{3}{2}\right) = -\dfrac{9}{4}$, the point where this occurs is $\left(\dfrac{3}{2}, -\dfrac{9}{4}\right)$.

49. (a) $p(0) = \dfrac{200}{1 + 7e^{-0.1(0)}} = \dfrac{200}{8} = 25$

Perhaps this is the number of bears placed in the reserve when it was established.

(b) $\lim\limits_{t\to\infty} p(t) = \lim\limits_{t\to\infty} \dfrac{200}{1 + 7e^{-0.1t}} = \dfrac{200}{1} = 200$

(c) Perhaps this is the maximum number of bears which the reserve can support due to limitations of food, space, or other resources. Or, perhaps the number is capped at 200 and excess bears are moved to other locations.

50. (a) $f(x) = \begin{cases} 3.20 - 1.35 \text{ int } (-x+1), & 0 < x \le 20 \\ 0, & x = 0 \end{cases}$

(Note that we cannot use the formula $f(x) = 3.20 + 1.35 \text{ int } x$, because it gives incorrect results when x is an integer.)

(b)

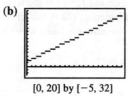

[0, 20] by [−5, 32]

f is discontinuous at integer values of x: 0, 1, 2, ..., 19.

51. (a) Cubic: $y = -1.644x^3 + 42.981x^2 - 254.369x + 300.232$
    Quartic: $y = 2.009x^4 - 102.081x^3 + 1884.997x^2 - 14918.180x + 43004.464$

    (b) Cubic: $-1.644x^3$, predicts spending will go to 0
    Quartic: $2.009x^4$, predicts spending will go to ∞

52. Let $A = \lim_{x \to c} f(x)$ and $B = \lim_{x \to c} g(x)$. Then $A + B = 2$ and $A - B = 1$. Adding, we have $2A = 3$, so $A = \frac{3}{2}$, whence $\frac{3}{2} + B = 2$, which gives $B = \frac{1}{2}$. Therefore, $\lim_{x \to c} f(x) = \frac{3}{2}$ and $\lim_{x \to c} g(x) = \frac{1}{2}$.

53. (a) [3, 12] by [−2, 24]

    (b)

Year of Q	Slope of PQ
1995	$\frac{20.1 - 2.7}{2000 - 1995} = 3.48$
1996	$\frac{20.1 - 4.8}{2000 - 1996} = 3.825$
1997	$\frac{20.1 - 7.8}{2000 - 1997} = 4.1$
1998	$\frac{20.1 - 11.2}{2000 - 1998} = 4.45$
1999	$\frac{20.1 - 15.2}{2000 - 1999} = 4.9$

    (c) Approximately 5 billion dollars per year.

    (d) $y = 0.3214x^2 - 1.3471x + 1.3857$

    $$\lim_{h \to 0} \frac{y(10+h) - y(10)}{h} = \lim_{h \to 0} \frac{[0.3214(10+h)^2 - 1.3471(10+h) + 1.3857] - [0.3214(10)^2 - 1.3471(10) + 1.3857]}{h}$$

    $$= \lim_{h \to 0} \frac{0.3214(20h + h^2) - 1.3471h}{h}$$

    $$= 0.3214(20) - 1.3471$$

    $$\approx 5.081$$

    The predicted rate of change in 2000 is about 5.081 billion dollars per year.

Chapter 3
Derivatives

■ Section 3.1 Derivative of a Function (pp. 95–104)

Exploration 1   Reading the Graphs

1. The graph in Figure 3.3b represents the rate of change of the depth of the water in the puddle with respect to time. Since y is measured in inches and x is measured in days, the derivative $\frac{dy}{dx}$ would be measured in inches per day. Those are the units that should be used along the y-axis in Figure 3.3b.

2. The water in the ditch is 1 inch deep at the start of the first day and rising rapidly. It continues to rise, at a gradually decreasing rate, until the end of the second day, when it achieves a maximum depth of 5 inches. During days 3, 4, 5, and 6, the water level goes down, until it reaches a depth of 1 inch at the end of day 6. During the seventh day it rises again, almost to a depth of 2 inches.

3. The weather appears to have been wettest at the beginning of day 1 (when the water level was rising fastest) and driest at the end of day 4 (when the water level was declining the fastest).

4. The highest point on the graph of the derivative shows where the water is rising the fastest, while the lowest point (most negative) on the graph of the derivative shows where the water is declining the fastest.

5. The y-coordinate of point C gives the maximum depth of the water level in the ditch over the 7-day period, while the x-coordinate of C gives the time during the 7-day period that the maximum depth occurred. The derivative of the function changes sign from positive to negative at C', indicating that this is when the water level stops rising and begins falling.

6. Water continues to run down sides of hills and through underground streams long after the rain has stopped falling. Depending on how much high ground is located near the ditch, water from the first day's rain could still be flowing into the ditch several days later. Engineers responsible for flood control of major rivers must take this into consideration when they predict when floodwaters will "crest," and at what levels.

Quick Review 3.1

1. $\lim_{h \to 0} \frac{(2+h)^2 - 4}{4} = \lim_{h \to 0} \frac{(4 + 4h + h^2) - 4}{h}$
$= \lim_{h \to 0} 4 + h$
$= 4 + 0 = 4$

2. $\lim_{x \to 2^+} \frac{x+3}{2} = \frac{2+3}{2} = \frac{5}{2}$

3. Since $\frac{|y|}{y} = -1$ for $y < 0$, $\lim_{y \to 0^-} \frac{|y|}{y} = -1$.

4. $\lim_{x \to 4} \frac{2x - 8}{\sqrt{x} - 2} = \lim_{x \to 4} \frac{2(\sqrt{x} + 2)(\sqrt{x} - 2)}{\sqrt{x} - 2}$
$= \lim_{h \to 4} 2(\sqrt{x} + 2) = 2(\sqrt{4} + 2) = 8$

5. The vertex of the parabola is at $(0, 1)$. The slope of the line through $(0, 1)$ and another point $(h, h^2 + 1)$ on the parabola is $\frac{(h^2 + 1) - 1}{h - 0} = h$. Since $\lim_{h \to 0} h = 0$, the slope of the line tangent to the parabola at its vertex is 0.

6. Use the graph of f in the window $[-6, 6]$ by $[-4, 4]$ to find that $(0, 2)$ is the coordinate of the high point and $(2, -2)$ is the coordinate of the low point. Therefore, f is increasing on $(-\infty, 0]$ and $[2, \infty)$.

7. $\lim_{x \to 1^+} f(x) = \lim_{x \to 1^+} (x - 1)^2 = (1 - 1)^2 = 0$
$\lim_{x \to 1^-} f(x) = \lim_{x \to 1^-} (x + 2) = 1 + 2 = 3$

8. $\lim_{h \to 0^+} f(1 + h) = \lim_{x \to 1^+} f(x) = 0$

9. No, the two one-sided limits are different (see Exercise 7).

10. No, f is discontinuous at $x = 1$ because $\lim_{x \to 1} f(x)$ does not exist.

Section 3.1 Exercises

1. (a) The tangent line has slope 5 and passes through $(2, 3)$.
$y = 5(x - 2) + 3$
$y = 5x - 7$

(b) The normal line has slope $-\frac{1}{5}$ and passes through $(2, 3)$.
$y = -\frac{1}{5}(x - 2) + 3$
$y = -\frac{1}{5}x + \frac{17}{5}$

3. $f'(3) = \lim_{x \to 3} \frac{f(x) - f(3)}{x - 3}$
$= \lim_{x \to 3} \frac{\frac{1}{x} - \frac{1}{3}}{x - 3}$
$= \lim_{x \to 3} \frac{3 - x}{(x - 3)(x)(3)}$
$= \lim_{x \to 3} -\frac{1}{3x} = -\frac{1}{9}$

5. $\frac{dy}{dx} = \lim_{h \to 0} \frac{y(x + h) - y(x)}{h}$
$= \lim_{h \to 0} \frac{7(x + h) - 7x}{h}$
$= \lim_{h \to 0} \frac{7h}{h} = \lim_{h \to 0} 7 = 7$

7. The graph of $y = f_1(x)$ is decreasing for $x < 0$ and increasing for $x > 0$, so its derivative is negative for $x < 0$ and positive for $x > 0$. (b)

9. The graph of $y = f_3(x)$ oscillates between increasing and decreasing, so its derivative oscillates between positive and negative. (d)

11. $\frac{dy}{dx} = \lim_{h \to 0} \frac{y(x + h) - y(x)}{h}$
$= \lim_{h \to 0} \frac{[2(x + h)^2 - 13(x + h) + 5] - (2x^2 - 13x + 5)}{h}$
$= \lim_{h \to 0} \frac{2x^2 + 4xh + 2h^2 - 13x - 13h + 5 - 2x^2 + 13x - 5}{h}$
$= \lim_{h \to 0} \frac{4xh + 2h^2 - 13h}{h}$
$= \lim_{h \to 0} (4x + 2h - 13) = 4x - 13$

At $x = 3$, $\frac{dy}{dx} = 4(3) - 13 = -1$, so the tangent line has slope -1 and passes through $(3, y(3)) = (3, -16)$.

$y = -1(x - 3) - 16$

$y = -x - 13$

13. Since the graph of $y = x \ln x - x$ is decreasing for $0 < x < 1$ and increasing for $x > 1$, its derivative is negative for $0 < x < 1$ and positive for $x > 1$. The only one of the given functions with this property is $y = \ln x$. Note also that $y = \ln x$ is undefined for $x < 0$, which further agrees with the given graph. (ii)

15. (a) The amount of daylight is increasing at the fastest rate when the slope of the graph is largest. This occurs about one-fourth of the way through the year, sometime around April 1. The rate at this time is approximately $\frac{4 \text{ hours}}{24 \text{ days}}$ or $\frac{1}{6}$ hour per day.

(b) Yes, the rate of change is zero when the tangent to the graph is horizontal. This occurs near the beginning of the year and halfway through the year, around January 1 and July 1.

(c) Positive: January 1 through July 1
Negative: July 1 through December 31

17. (a) Using Figure 3.10a, the number of rabbits is largest after 40 days and smallest from about 130 to 200 days. Using Figure 3.10b, the derivative is 0 at these times.

(b) Using Figure 3.10b, the derivative is largest after 20 days and smallest after about 63 days. Using Figure 3.10a, there were 1700 and about 1300 rabbits, respectively, at these times.

19.

Midpoint of Interval (x)	Slope $\left(\frac{\Delta y}{\Delta x}\right)$
0.5	$\frac{3.3 - 0}{1 - 0} = 3.3$
1.5	$\frac{13.3 - 3.3}{2 - 1} = 10.0$
2.5	$\frac{29.9 - 13.3}{3 - 2} = 16.6$
3.5	$\frac{53.2 - 29.9}{4 - 3} = 23.3$
4.5	$\frac{83.2 - 53.2}{5 - 4} = 30.0$
5.5	$\frac{119.8 - 83.2}{6 - 5} = 36.6$
6.5	$\frac{163.0 - 119.8}{7 - 6} = 43.2$
7.5	$\frac{212.9 - 163.0}{8 - 7} = 49.9$
8.5	$\frac{269.5 - 212.9}{9 - 8} = 56.6$
9.5	$\frac{332.7 - 269.5}{10 - 9} = 63.2$

A graph of the derivative data is shown.

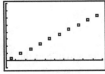

[0, 10] by [−10, 80]

(a) The derivative represents the speed of the skier.

(b) Since the distances are given in feet and the times are given in seconds, the units are feet per second.

(c) The graph appears to be approximately linear and passes through (0, 0) and (9.5, 63.2), so the slope is $\frac{63.2 - 0}{9.5 - 0} \approx 6.65$. The equation of the derivative is approximately $D = 6.65t$.

21.

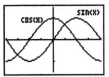

$[-\pi, \pi]$ by $[-1.5, 1.5]$

The cosine function could be the derivative of the sine function. The values of the cosine are positive where the sine is increasing, zero where the sine has horizontal tangents, and negative where sine is decreasing.

23. $\lim\limits_{h \to 0^+} \frac{f(0 + h) - f(0)}{h} = \lim\limits_{h \to 0^+} \frac{\sqrt{h} - \sqrt{0}}{h} = \lim\limits_{h \to 0^+} \frac{\sqrt{h}}{h}$
$= \lim\limits_{h \to 0^+} \frac{1}{\sqrt{h}} = \infty$

Thus, the right-hand derivative at 0 does not exist.

25. For $x > -1$, the graph of $y = f(x)$ must lie on a line of slope -2 that passes through $(0, -1)$: $y = -2x - 1$. Then $y(-1) = -2(-1) - 1 = 1$, so for $x < -1$, the graph of $y = f(x)$ must lie on a line of slope 1 that passes through $(-1, 1)$: $y = 1(x + 1) + 1$ or $y = x + 2$.

Thus $f(x) = \begin{cases} x + 2, & x < -1 \\ -2x - 1, & x \geq -1 \end{cases}$

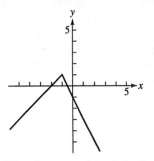

27. The y-intercept of the derivative is $b - a$.

29. (a) $1 \cdot \frac{364}{365} \cdot \frac{363}{365} \approx 0.992$

Alternate method: $\frac{365P_3}{365^3} \approx 0.992$

(b) Using the answer to part (a), the probability is about $1 - 0.992 = 0.008$.

(c) Let P represent the answer to part (b), $P \approx 0.008$. Then the probability that three people all have different birthdays is $1 - P$. Adding a fourth person, the probability that all have different birthdays is $(1 - P)\left(\frac{362}{365}\right)$, so the probability of a shared birthday is $1 - (1 - P)\left(\frac{362}{365}\right) \approx 0.016$.

(d) No. Clearly February 29 is a much less likely birth date. Furthermore, census data do not support the assumption that the other 365 birth dates are equally likely. However, this simplifying assumption may still give us some insight into this problem even if the calculated probabilities aren't completely accurate.

Section 3.2 Differentiability (pp. 105–112)

Exploration 1 Zooming in to "See" Differentiability

1. Zooming in on the graph of f at the point (0, 1) always produces a graph exactly like the one shown below, provided that a square window is used. The corner shows no sign of straightening out.

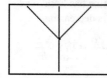

$[-0.25, 0.25]$ by $[0.836, 1.164]$

2. Zooming in on the graph of g at the point (0, 1) begins to reveal a smooth turning point. This graph shows the result of three zooms, each by a factor of 4 horizontally and vertically, starting with the window $[-4, 4]$ by $[-1.624, 3.624]$.

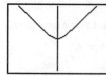

$[-0.0625, 0.0625]$ by $[0.959, 1.041]$

3. On our grapher, the graph became horizontal after 8 zooms. Results can vary on different machines.

4. As we zoom in on the graphs of f and g together, the differentiable function gradually straightens out to resemble its tangent line, while the nondifferentiable function stubbornly retains its same shape.

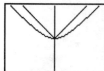

$[-0.03125, 0.03125]$ by $[0.9795, 1.0205]$

Exploration 2 Looking at the Symmetric Difference Quotient Analytically

1. $\dfrac{f(10+h) - f(10)}{h} = \dfrac{(10.01)^2 - 10^2}{0.01} = 20.01$

   $f'(10) = 2 \cdot 10 = 20$

   The difference quotient is 0.01 away from $f'(10)$.

2. $\dfrac{f(10+h) - f(10-h)}{2h} = \dfrac{(10.01)^2 - (9.99)^2}{0.02} = 20$

   The symmetric difference quotient exactly equals $f'(10)$.

3. $\dfrac{f(10+h) - f(10)}{h} = \dfrac{(10.01)^3 - 10^3}{0.01} = 300.3001$

   $f'(10) = 3 \cdot 10^2 = 300$

   The difference quotient is 0.3001 away from $f'(10)$.

   $\dfrac{f(10+h) - f(10-h)}{2h} = \dfrac{(10.01)^3 - (9.99)^3}{0.02} = 300.0001$.

   The symmetric difference quotient is 0.0001 away from $f'(10)$.

Quick Review 3.2

1. Yes
2. No (The $f(h)$ term in the numerator is incorrect.)
3. Yes
4. Yes
5. No (The denominator for this expression should be $2h$).
6. All reals
7. $[0, \infty)$
8. $[3, \infty)$
9. The equation is equivalent to $y = 3.2x + (3.2\pi + 5)$, so the slope is 3.2.
10. $\dfrac{f(3 + 0.001) - f(3 - 0.001)}{0.002} = \dfrac{5(3 + 0.001) - 5(3 - 0.001)}{0.002}$

    $= \dfrac{5(0.002)}{0.002} = 5$

Section 3.2 Exercises

1. Left-hand derivative:

   $\lim\limits_{h \to 0^-} \dfrac{f(0+h) - f(0)}{h} = \lim\limits_{h \to 0^-} \dfrac{h^2 - 0}{h} = \lim\limits_{h \to 0^-} h = 0$

   Right-hand derivative:

   $\lim\limits_{h \to 0^+} \dfrac{f(0+h) - f(0)}{h} = \lim\limits_{h \to 0^+} \dfrac{h - 0}{h} = \lim\limits_{h \to 0^+} 1 = 1$

   Since $0 \ne 1$, the function is not differentiable at the point P.

3. Left-hand derivative:

   $\lim\limits_{h \to 0^-} \dfrac{f(1+h) - f(1)}{h} = \lim\limits_{h \to 0^-} \dfrac{\sqrt{1+h} - 1}{h}$

   $= \lim\limits_{h \to 0^-} \dfrac{(\sqrt{1+h} - 1)(\sqrt{1+h} + 1)}{h(\sqrt{1+h} + 1)}$

   $= \lim\limits_{h \to 0^-} \dfrac{(1+h) - 1}{h(\sqrt{1+h} + 1)}$

   $= \lim\limits_{h \to 0^-} \dfrac{1}{\sqrt{1+h} + 1} = \dfrac{1}{2}$

   Right-hand derivative:

   $\lim\limits_{h \to 0^+} \dfrac{f(1+h) - f(1)}{h} = \lim\limits_{h \to 0^+} \dfrac{[2(1+h) - 1] - 1}{h}$

   $= \lim\limits_{h \to 0^+} \dfrac{2h}{h}$

   $= \lim\limits_{h \to 0^+} 2 = 2$

   Since $\dfrac{1}{2} \ne 2$, the function is not differentiable at the point P.

5. (a) All points in $[-3, 2]$
   (b) None
   (c) None

7. (a) All points in $[-3, 3]$ except $x = 0$
   (b) None
   (c) $x = 0$

9. (a) All points in $[-1, 2]$ except $x = 0$
   (b) $x = 0$
   (c) None

11. Since $\lim\limits_{x \to 0} \tan^{-1} x = \tan^{-1} 0 = 0 \neq y(0)$, the problem is a discontinuity.

13. Note that $y = x + \sqrt{x^2} + 2 = x + |x| + 2$
    $= \begin{cases} 2, & x \leq 0 \\ 2x + 2, & x > 0. \end{cases}$
    $\lim\limits_{h \to 0^-} \dfrac{y(0+h) - y(0)}{h} = \lim\limits_{h \to 0^-} \dfrac{2-2}{h} = \lim\limits_{h \to 0^-} 0 = 0$
    $\lim\limits_{h \to 0^+} \dfrac{y(0+h) - y(0)}{h} = \lim\limits_{h \to 0^+} \dfrac{(2h+2) - 2}{h} = \lim\limits_{h \to 0^+} 2 = 2$
    The problem is a corner.

15. Note that $y = 3x - 2|x| - 1 = \begin{cases} 5x - 1, & x \leq 0 \\ x - 1, & x > 0 \end{cases}$
    $\lim\limits_{h \to 0^-} \dfrac{y(0+h) - y(0)}{h} = \lim\limits_{h \to 0^-} \dfrac{(5h-1) - (-1)}{h} = \lim\limits_{h \to 0^-} 5 = 5$
    $\lim\limits_{h \to 0^+} \dfrac{y(0+h) - y(0)}{h} = \lim\limits_{h \to 0^+} \dfrac{(h-1) - (-1)}{h} = \lim\limits_{h \to 0^+} 1 = 1$
    The problem is a corner.

17. Find the zeros of the denominator.
    $x^2 - 4x - 5 = 0$
    $(x+1)(x-5) = 0$
    $x = -1$ or $x = 5$
    The function is a rational function, so it is differentiable for all x in its domain: all reals except $x = -1, 5$.

19. Note that the sine function is odd, so
    $P(x) = \sin(|x|) - 1 = \begin{cases} -\sin x - 1, & x < 0 \\ \sin x - 1, & x \geq 0. \end{cases}$
    The graph of $P(x)$ has a corner at $x = 0$. The function is differentiable for all reals except $x = 0$.

21. The function is piecewise-defined in terms of polynomials, so it is differentiable everywhere except possibly at $x = 0$ and at $x = 3$. Check $x = 0$:
    $\lim\limits_{h \to 0^-} \dfrac{g(0+h) - g(0)}{h} = \lim\limits_{h \to 0^-} \dfrac{(h+1)^2 - 1}{h} = \lim\limits_{h \to 0^-} \dfrac{h^2 + 2h}{h}$
    $= \lim\limits_{h \to 0^-} (h+2) = 2$
    $\lim\limits_{h \to 0^+} \dfrac{g(0+h) - g(0)}{h} = \lim\limits_{h \to 0^+} \dfrac{(2h+1) - 1}{h} = \lim\limits_{h \to 0^+} 2 = 2$
    The function is differentiable at $x = 0$.

    Check $x = 3$:
    Since $g(3) = (4-3)^2 = 1$ and
    $\lim\limits_{x \to 3^-} g(x) = \lim\limits_{x \to 3^-} (2x+1) = 2(3) + 1 = 7$, the function is not continuous (and hence not differentiable) at $x = 3$.

    The function is differentiable for all reals except $x = 3$.

23. (a) $x = 0$ is not in their domains, or, they are both discontinuous at $x = 0$.

    (b) For $\dfrac{1}{x}$: NDER $\left(\dfrac{1}{x}, 0\right) = 1{,}000{,}000$
    For $\dfrac{1}{x^2}$: NDER $\left(\dfrac{1}{x^2}, 0\right) = 0$

    (c) It returns an incorrect response because even though these functions are not defined at $x = 0$, they are defined at $x = \pm 0.001$. The responses differ from each other because $\dfrac{1}{x^2}$ is even (which automatically makes NDER $\left(\dfrac{1}{x^2}, 0\right) = 0$) and $\dfrac{1}{x}$ is odd.

25.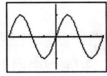
    $[-2\pi, 2\pi]$ by $[-1.5, 1.5]$
    $\dfrac{dy}{dx} = \sin x$

27.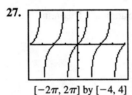
    $[-2\pi, 2\pi]$ by $[-4, 4]$
    $\dfrac{dy}{dx} = \tan x$

    Note: Due to the way NDER is defined, the graph of $y = \text{NDER}(x)$ actually has two asymptotes for each asymptote of $y = \tan x$. The asymptotes of $y = \text{NDER}(x)$ occur at $x = \dfrac{\pi}{2} + k\pi \pm 0.001$, where k is an integer. A good window for viewing this behavior is $[1.566, 1.576]$ by $[-1000, 1000]$.

29. (a) $\lim\limits_{x \to 1^-} f(x) = f(1)$
    $\lim\limits_{x \to 1^-} (3 - x) = a(1)^2 + b(1)$
    $2 = a + b$

    The relationship is $a + b = 2$.

50 Section 3.3

(b) Since the function needs to be continuous, we may assume that $a + b = 2$ and $f(1) = 2$.

$$\lim_{h \to 0^-} \frac{f(1+h) - f(1)}{h} = \lim_{h \to 0^-} \frac{3 - (1+h) - 2}{h}$$
$$= \lim_{h \to 0^-} (-1) = -1$$

$$\lim_{h \to 0^+} \frac{f(1+h) - f(1)}{h} = \lim_{h \to 0^-} \frac{a(1+h)^2 + b(1+h) - 2}{h}$$
$$= \lim_{h \to 0^-} \frac{a + 2ah + ah^2 + b + bh - 2}{h}$$
$$= \lim_{h \to 0^-} \frac{2ah + ah^2 + bh + (a + b - 2)}{h}$$
$$= \lim_{h \to 0^-} (2a + ah + b)$$
$$= 2a + b$$

Therefore, $2a + b = -1$. Substituting $2 - a$ for b gives $2a + (2 - a) = -1$, so $a = -3$.
Then $b = 2 - a = 2 - (-3) = 5$. The values are $a = -3$ and $b = 5$.

31. (a) Note that $-x \le \sin \frac{1}{x} \le x$, for all x,
so $\lim_{x \to 0} \left(x \sin \frac{1}{x} \right) = 0$ by the Sandwich Theorem.
Therefore, f is continuous at $x = 0$.

(b) $\dfrac{f(0+h) - f(0)}{h} = \dfrac{h \sin \frac{1}{h} - 0}{h} = \sin \frac{1}{h}$

(c) The limit does not exist because $\sin \frac{1}{h}$ oscillates between -1 and 1 an infinite number of times arbitrarily close to $h = 0$ (that is, for h in any open interval containing 0).

(d) No, because the limit in part (c) does not exist.

(e) $\dfrac{g(0+h) - g(0)}{h} = \dfrac{h^2 \sin\left(\frac{1}{h}\right) - 0}{h} = h \sin \frac{1}{h}$

As noted in part (a), the limit of this as x approaches zero is 0, so $g'(0) = 0$.

■ Section 3.3 Rules for Differentiation
(pp. 112–121)

Quick Review 3.3

1. $(x^2 - 2)(x^{-1} + 1) = x^2 x^{-1} + x^2 \cdot 1 - 2x^{-1} - 2 \cdot 1$
$= x + x^2 - 2x^{-1} - 2$

2. $\left(\dfrac{x}{x^2 + 1}\right)^{-1} = \dfrac{x^2 + 1}{x} = \dfrac{x^2}{x} + \dfrac{1}{x} = x + x^{-1}$

3. $3x^2 - \dfrac{2}{x} + \dfrac{5}{x^2} = 3x^2 - 2x^{-1} + 5x^{-2}$

4. $\dfrac{3x^4 - 2x^3 + 4}{2x^2} = \dfrac{3x^4}{2x^2} - \dfrac{2x^3}{2x^2} + \dfrac{4}{2x^2} = \dfrac{3}{2}x^2 - x + 2x^{-2}$

5. $(x^{-1} + 2)(x^{-2} + 1) = x^{-1}x^{-2} + x^{-1} \cdot 1 + 2x^{-2} + 2 \cdot 1$
$= x^{-3} + x^{-1} + 2x^{-2} + 2$

6. $\dfrac{x^{-1} + x^{-2}}{x^{-3}} = x^3(x^{-1} + x^{-2}) = x^2 + x$

7.

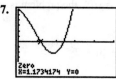

[0, 5] by [−6, 6]

At $x \approx 1.173$, $500x^6 \approx 1305$.
At $x \approx 2.394$, $500x^6 \approx 94{,}212$.
After rounding, we have:
At $x \approx 1$, $500x^6 \approx 1305$.
At $x \approx 2$, $500x^6 \approx 94{,}212$.

8. (a) $f(10) = 7$

(b) $f(0) = 7$

(c) $f(x + h) = 7$

(d) $\lim_{x \to a} \dfrac{f(x) - f(a)}{x - a} = \lim_{x \to a} \dfrac{7 - 7}{x - a} = \lim_{x \to a} 0 = 0$

9. These are all constant functions, so the graph of each function is a horizontal line and the derivative of each function is 0.

10. (a) $f'(x) = \lim_{h \to 0} \dfrac{f(x+h) - f(x)}{h} = \lim_{h \to 0} \dfrac{\frac{x+h}{\pi} - \frac{x}{\pi}}{h}$
$= \lim_{h \to 0} \dfrac{x + h - x}{\pi h} = \lim_{h \to 0} \dfrac{1}{\pi} = \dfrac{1}{\pi}$

(b) $f'(x) = \lim_{h \to 0} \dfrac{f(x+h) - f(x)}{h} = \lim_{h \to 0} \dfrac{\frac{\pi}{x+h} - \frac{\pi}{x}}{h}$
$= \lim_{h \to 0} \dfrac{\pi x - \pi(x+h)}{hx(x+h)} = \lim_{h \to 0} \dfrac{-\pi h}{hx(x+h)}$
$= \lim_{h \to 0} -\dfrac{\pi}{x(x+h)} = -\dfrac{\pi}{x^2} = -\pi x^{-2}$

Section 3.3 Exercises

1. $\dfrac{dy}{dx} = \dfrac{d}{dx}(-x^2) + \dfrac{d}{dx}(3) = -2x + 0 = -2x$

$\dfrac{d^2 y}{dx^2} = \dfrac{d}{dx}(-2x) = -2$

3. $\dfrac{dy}{dx} = \dfrac{d}{dx}(2x) + \dfrac{d}{dx}(1) = 2 + 0 = 2$

$\dfrac{d^2 y}{dx^2} = \dfrac{d}{dx}(2) = 0$

5. $\dfrac{dy}{dx} = \dfrac{d}{dx}\left(\dfrac{1}{3}x^3\right) + \dfrac{d}{dx}\left(\dfrac{1}{2}x^2\right) + \dfrac{d}{dx}(x)$
$= x^2 + x + 1$

$\dfrac{d^2 y}{dx^2} = \dfrac{d}{dx}(x^2) + \dfrac{d}{dx}(x) + \dfrac{d}{dx}(1) = 2x + 1 + 0 = 2x + 1$

7. $\dfrac{dy}{dx} = \dfrac{d}{dx}(x^4) - \dfrac{d}{dx}(7x^3) + \dfrac{d}{dx}(2x^2) + \dfrac{d}{dx}(15)$
$= 4x^3 - 21x^2 + 4x + 0 = 4x^3 - 21x^2 + 4x$

$\dfrac{d^2 y}{dx^2} = \dfrac{d}{dx}(4x^3) - \dfrac{d}{dx}(21x^2) + \dfrac{d}{dx}(4x) = 12x^2 - 42x + 4$

9. $\dfrac{dy}{dx} = \dfrac{d}{dx}(4x^{-2}) - \dfrac{d}{dx}(8x) + \dfrac{d}{dx}(1)$
$= -8x^{-3} - 8 + 0 = -8x^{-3} - 8$
$\dfrac{d^2y}{dx^2} = \dfrac{d}{dx}(-8x^{-3}) - \dfrac{d}{dx}(8) = 24x^{-4} - 0 = 24x^{-4}$

11. (a) $\dfrac{dy}{dx} = \dfrac{d}{dx}[(x+1)(x^2+1)]$
$= (x+1)\dfrac{d}{dx}(x^2+1) + (x^2+1)\dfrac{d}{dx}(x+1)$
$= (x+1)(2x) + (x^2+1)(1)$
$= 2x^2 + 2x + x^2 + 1$
$= 3x^2 + 2x + 1$

(b) $\dfrac{dy}{dx} = \dfrac{d}{dx}[(x+1)(x^2+1)]$
$= \dfrac{d}{dx}(x^3 + x^2 + x + 1)$
$= 3x^2 + 2x + 1$

13. $\dfrac{dy}{dx} = \dfrac{d}{dx}\dfrac{2x+5}{3x-2} = \dfrac{(3x-2)(2) - (2x+5)(3)}{(3x-2)^2} = -\dfrac{19}{(3x-2)^2}$

15. $\dfrac{dy}{dx} = \dfrac{d}{dx}\left(\dfrac{(x-1)(x^2+x+1)}{x^3}\right) = \dfrac{d}{dx}\left(\dfrac{x^3-1}{x^3}\right)$
$= \dfrac{d}{dx}(1 - x^{-3}) = 0 + 3x^{-4} = \dfrac{3}{x^4}$

17. $\dfrac{dy}{dx} = \dfrac{d}{dx}\left(\dfrac{x^2}{1-x^3}\right) = \dfrac{(1-x^3)(2x) - x^2(-3x^2)}{(1-x^3)^2} = \dfrac{x^4 + 2x}{(1-x^3)^2}$

19. $\dfrac{dy}{dx} = \dfrac{d}{dx}\left(\dfrac{(x+1)(x+2)}{(x-1)(x-2)}\right) = \dfrac{d}{dx}\left(\dfrac{x^2+3x+2}{x^2-3x+2}\right)$
$= \dfrac{(x^2-3x+2)(2x+3) - (x^2+3x+2)(2x-3)}{(x^2-3x+2)^2}$
$= \dfrac{(2x^3 - 3x^2 - 5x + 6) - (2x^3 + 3x^2 - 5x - 6)}{(x^2-3x+2)^2}$
$= \dfrac{12 - 6x^2}{(x^2-3x+2)^2}$

21. $\dfrac{d}{dx}(c \cdot f(x)) = c \cdot \dfrac{d}{dx}f(x) + f(x) \cdot \dfrac{d}{dx}(c)$
$= c \cdot \dfrac{d}{dx}f(x) + 0 = c \cdot \dfrac{d}{dx}f(x)$

23. (a) At $x = 0$, $\dfrac{d}{dx}(uv) = u(0)v'(0) + v(0)u'(0)$
$= (5)(2) + (-1)(-3) = 13$

(b) At $x = 0$, $\dfrac{d}{dx}\left(\dfrac{u}{v}\right) = \dfrac{v(0)u'(0) - u(0)v'(0)}{[v(0)]^2}$
$= \dfrac{(-1)(-3) - (5)(2)}{(-1)^2} = -7$

(c) At $x = 0$, $\dfrac{d}{dx}\left(\dfrac{v}{u}\right) = \dfrac{u(0)v'(0) - v(0)u'(0)}{[u(0)]^2}$
$= \dfrac{(5)(2) - (-1)(-3)}{(5)^2} = \dfrac{7}{25}$

(d) At $x = 0$, $\dfrac{d}{dx}(7v - 2u) = 7v'(0) - 2u'(0)$
$= 7(2) - 2(-3) = 20$

25. $y'(x) = 2x + 5$
$y'(3) = 2(3) + 5 = 11$
The slope is 11. (iii)

27. $y'(x) = 3x^2 - 3$
$y'(2) = 3(2)^2 - 3 = 9$

The tangent line has slope 9, so the perpendicular line has slope $-\dfrac{1}{9}$ and passes through (2, 3).
$y = -\dfrac{1}{9}(x - 2) + 3$
$y = -\dfrac{1}{9}x + \dfrac{29}{9}$

Graphical support:

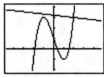

[-4.7, 4.7] by [-2.1, 4.1]

29. $y'(x) = 6x^2 - 6x - 12$
$= 6(x^2 - x - 2)$
$= 6(x + 1)(x - 2)$

The tangent is parallel to the x-axis when $y' = 0$, at $x = -1$ and at $x = 2$. Since $y(-1) = 27$ and $y(2) = 0$, the two points where this occurs are $(-1, 27)$ and $(2, 0)$.

Graphical support:

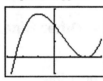

[-3, 3] by [-10, 30]

31. $y'(x) = \dfrac{(x^2+1)(4) - 4x(2x)}{(x^2+1)^2} = \dfrac{-4x^2 + 4}{(x^2+1)^2}$

At the origin: $y'(0) = 4$

The tangent is $y = 4x$.

At (1, 2): $y'(1) = 0$

The tangent is $y = 2$.

Graphical support:

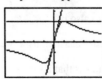

[-4.7, 4.7] by [-3.1, 3.1]

33. $\dfrac{dP}{dV} = \dfrac{d}{dV}\left(\dfrac{nRT}{V-nb} - \dfrac{an^2}{V^2}\right)$
$= \dfrac{(V-nb)\dfrac{d}{dV}(nRT) - (nRT)\dfrac{d}{dV}(V-nb)}{(V-nb)^2} - \dfrac{d}{dV}(an^2V^{-2})$
$= \dfrac{0 - nRT}{(V-nb)^2} + 2an^2V^{-3}$
$= -\dfrac{nRT}{(V-nb)^2} + \dfrac{2an^2}{V^3}$

52 Section 3.4

35. $\dfrac{dR}{dM} = \dfrac{d}{dM}\left[M^2\left(\dfrac{C}{2} - \dfrac{M}{3}\right)\right]$

$= \dfrac{d}{dM}\left(\dfrac{C}{2}M^2 - \dfrac{1}{3}M^3\right)$

$= CM - M^2$

37. If the radius of a sphere is changed by a very small amount Δr, the change in the volume can be thought of as a very thin layer with an area given by the surface area, $4\pi r^2$, and a thickness given by Δr. Therefore, the change in the volume can be thought of as $(4\pi r^2)(\Delta r)$, which means that the change in the volume divided by the change in the radius is just $4\pi r^2$.

39. Let $m(x)$ be the number of members and $c(x)$ be the pavillion cost x years from now. Then $m(0) = 65$, $c(0) = 250$, $m'(0) = 6$, and $c'(0) = 10$. The rate of change of each member's share is $\dfrac{d}{dx}\left(\dfrac{c}{m}\right) = \dfrac{m(0)c'(0) - c(0)m'(0)}{[m(0)]^2}$

$= \dfrac{(65)(10) - (250)(6)}{(65)^2} \approx -0.201$ dollars per year. Each member's share of the cost is decreasing by approximately 20 cents per year.

■ Section 3.4 Velocity and Other Rates of Change (pp. 122–133)

Exploration 1 Growth Rings on a Tree

1. Figure 3.22 is a better model, as it shows rings of equal *area* as opposed to rings of equal *width*. It is not likely that a tree could sustain increased growth year after year, although climate conditions do produce some years of greater growth than others.

2. Rings of equal area suggest that the tree adds approximately the same amount of wood to its girth each year. With access to approximately the same raw materials from which to make the wood each year, this is how most trees actually grow.

3. Since change in area is constant, so also is $\dfrac{\text{change in area}}{2\pi}$. If we denote this latter constant by k, we have $\dfrac{k}{\text{change in radius}} = r$, which means that r varies inversely as the change in the radius. In other words, the change in radius must get smaller when r gets bigger, and vice-versa.

Exploration 2 Modeling Horizontal Motion

1. The particle reverses direction at about $t = 0.61$ and $t = 2.06$.

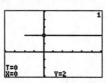

2. When the trace cursor is moving to the right the particle is moving to the right, and when the cursor is moving to the left the particle is moving to the left. Again we find the particle reverses direction at about $t = 0.61$ and $t = 2.06$.

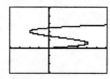

3. When the trace cursor is moving upward the particle is moving to the right, and when the cursor is moving downward the particle is moving to the left. Again we find the same values of t for when the particle reverses direction.

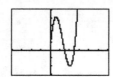

4. We can represent the velocity by graphing the parametric equations
$x_4(t) = x_1'(t) = 12t^2 - 32t + 15$, $y_4(t) = 2$ (part 1),
$x_5(t) = x_1'(t) = 12t^2 - 32t + 15$, $y_5(t) = t$ (part 2),
$x_6(t) = t$, $y_6(t) = x_1'(t) = 12t^2 - 32t + 15$ (part 3)

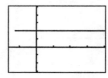

$[-8, 20]$ by $[-3, 5]$
(x_4, y_4)

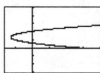

$[-8, 20]$ by $[-3, 5]$
(x_5, y_5)

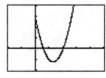

$[-2, 5]$ by $[-10, 20]$
(x_6, y_6)

For (x_4, y_4) and (x_5, y_5), the particle is moving to the right when the x-coordinate of the graph (velocity) is positive, moving to the left when the x-coordinate of the graph (velocity) is negative, and is stopped when the x-coordinate of the graph (velocity) is 0. For (x_6, y_6), the particle is moving to the right when the y-coordinate of the graph (velocity) is positive, moving to the left when the y-coordinate of the graph (velocity) is negative, and is stopped when the y-coordinate of the graph (velocity) is 0.

Exploration 3 Seeing Motion on a Graphing Calculator

1. Let tMin $= 0$ and tMax $= 10$.
2. Since the rock achieves a maximum height of 400 feet, set yMax to be slightly greater than 400, for example yMax $= 420$.
4. The grapher proceeds with constant increments of t (time), so pixels appear on the screen at regular time intervals. When the rock is moving more slowly, the pixels appear closer together. When the rock is moving faster, the pixels appear farther apart. We observe faster motion when the pixels are farther apart.

Quick Review 3.4

1. The coefficient of x^2 is negative, so the parabola opens downward.

   Graphical support:

   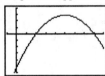

   $[-1, 9]$ by $[-300, 200]$

2. The y-intercept is $f(0) = -256$.
   See the solution to Exercise 1 for graphical support.

3. The x-intercepts occur when $f(x) = 0$.
   $-16x^2 + 160x - 256 = 0$
   $-16(x^2 - 10x + 16) = 0$
   $-16(x - 2)(x - 8) = 0$
   $x = 2$ or $x = 8$
   The x-intercepts are 2 and 8. See the solution to Exercise 1 for graphical support.

4. Since $f(x) = -16(x^2 - 10x + 16)$
   $= -16(x^2 - 10x + 25 - 9) = -16(x - 5)^2 + 144$,
   the range is $(-\infty, 144]$.
   See the solution to Exercise 1 for graphical support.

5. Since $f(x) = -16(x^2 - 10x + 16)$
   $= -16(x^2 - 10x + 25 - 9) = -16(x - 5)^2 + 144$,
   the vertex is at $(5, 144)$. See the solution to Exercise 1 for graphical support.

6. $$f(x) = 80$$
   $-16x^2 + 160x - 256 = 80$
   $-16x^2 + 160x - 336 = 0$
   $-16(x^2 - 10x + 21) = 0$
   $-16(x - 3)(x - 7) = 0$
   $x = 3$ or $x = 7$
   $f(x) = 80$ at $x = 3$ and at $x = 7$.
   See the solution to Exercise 1 for graphical support.

7. $\dfrac{dy}{dx} = 100$
   $-32x + 160 = 100$
   $60 = 32x$
   $x = \dfrac{15}{8}$
   $\dfrac{dy}{dx} = 100$ at $x = \dfrac{15}{8}$

   Graphical support: the graph of NDER $f(x)$ is shown.

   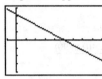

   $[-1, 9]$ by $[-200, 200]$

8. $\dfrac{dy}{dx} > 0$
   $-32x + 160 > 0$
   $-32x > -160$
   $x < 5$

   $\dfrac{dy}{dx} > 0$ when $x < 5$.
   See the solution to Exercise 7 for graphical support.

9. Note that $f'(x) = -32x + 160$.
   $\lim\limits_{h \to 0} \dfrac{f(3 + h) - f(3)}{h} = f'(3) = -32(3) + 160 = 64$

   For graphical support, use the graph shown in the solution to Exercise 7 and observe that NDER $(f(x), 3) \approx 64$.

10. $f'(x) = -32x + 160$
    $f''(x) = -32$
    At $x = 7$ (and, in fact, at any other of x),
    $\dfrac{d^2y}{dx^2} = -32$.

    Graphical support: the graph of NDER(NDER $f(x)$) is shown.

    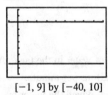

    $[-1, 9]$ by $[-40, 10]$

Section 3.4 Exercises

1. Since $V = s^3$, the instantaneous rate of change is $\dfrac{dV}{ds} = 3s^2$.

54 Section 3.4

3. (a) Velocity: $v(t) = \dfrac{ds}{dt} = \dfrac{d}{dt}(24t - 0.8t^2) = 24 - 1.6t$ m/sec

   Acceleration: $a(t) = \dfrac{dv}{dt} = \dfrac{d}{dt}(24 - 1.6t) = -1.6$ m/sec^2

   (b) The rock reaches its highest point when
   $v(t) = 24 - 1.6t = 0$, at $t = 15$. It took 15 seconds.

   (c) The maximum height was $s(15) = 180$ meters.

   (d) $\quad s(t) = \dfrac{1}{2}(180)$

   $24t - 0.8t^2 = 90$

   $0 = 0.8t^2 - 24t + 90$

   $t = \dfrac{24 \pm \sqrt{(-24)^2 - 4(0.8)(90)}}{2(0.8)}$

   $\approx 4.393, 25.607$

   It took about 4.393 seconds to reach half its maximum height.

   (e) $\quad s(t) = 0$
   $24t - 0.8t^2 = 0$
   $0.8t(30 - t) = 0$
   $t = 0$ or $t = 30$
   The rock was aloft from $t = 0$ to $t = 30$, so it was aloft for 30 seconds.

5. The rock reaches its maximum height when the velocity $s'(t) = 24 - 9.8t = 0$, at $t \approx 2.449$. Its maximum height is about $s(2.449) \approx 29.388$ meters.

7. The following is one way to simulate the problem situation.
   For the moon:
   $x_1(t) = 3(t < 160) + 3.1(t \geq 160)$
   $y_1(t) = 832t - 2.6t^2$
   t-values: 0 to 320
   window: [0, 6] by [−10,000, 70,000]
   For the earth:
   $x_1(t) = 3(t < 26) + 3.1(t \geq 26)$
   $y_1(t) = 832t - 16t^2$
   t-values: 0 to 52
   window: [0, 6] by [−1000, 11,000]

9. $Q(t) = 200(30 - t)^2 = 200(900 - 60t + t^2)$
   $= 180{,}000 - 12{,}000t + 200t^2$
   $Q'(t) = -12{,}000 + 400t$
   The rate of change of the amount of water in the tank after 10 minutes is $Q'(10) = -8000$ gallons per minute.
   Note that $Q'(10) < 0$, so the rate at which the water is running *out* is positive. The water is running out at the rate of 8000 gallons per minute.

   The average rate for the first 10 minutes is
   $\dfrac{Q(10) - Q(0)}{10 - 0} = \dfrac{80{,}000 - 180{,}000}{10} = -10{,}000$ gallons per minute.

   The water is flowing out at an average rate of 10,000 gallons per minute over the first 10 min.

11. (a)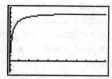

    [0, 50] by [−500, 2200]
    The values of x which make sense are the whole numbers, $x \geq 0$.

    (b) Marginal revenue $= r'(x) = \dfrac{d}{dx}\left[2000\left(1 - \dfrac{1}{x+1}\right)\right]$
    $= \dfrac{d}{dx}\left(2000 - \dfrac{2000}{x+1}\right)$
    $= 0 - \dfrac{(x+1)(0) - (2000)(1)}{(x+1)^2} = \dfrac{2000}{(x+1)^2}$

    (c) $r'(5) = \dfrac{2000}{(5+1)^2} = \dfrac{2000}{36} \approx 55.56$
    The increase in revenue is approximately $55.56.

    (d) The limit is 0. This means that as x gets large, one reaches a point where very little extra revenue can be expected from selling more desks.

13. $a(t) = v'(t) = 6t^2 - 18t + 12$
    Find when acceleration is zero.
    $6t^2 - 18t + 12 = 0$
    $6(t^2 - 3t + 2) = 0$
    $6(t-1)(t-2) = 0$
    $t = 1$ or $t = 2$
    At $t = 1$, the speed is $|v(1)| = |0| = 0$ m/sec.
    At $t = 2$, the speed is $|v(2)| = |-1| = 1$ m/sec.

15. (a)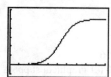

    [0, 200] by [−2, 12]

    (b) The values of x which make sense are the whole numbers, $x \geq 0$.

    (c)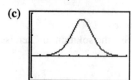

    [0, 200] by [−0.1, 0.2]
    P is most sensitive to changes in x when $|P'(x)|$ is largest. It is relatively sensitive to changes in x between approximately $x = 60$ and $x = 160$.

    (d) The marginal profit, $P'(x)$, is greatest at $x \approx 106.44$.
    Since x must be an integer,
    $P(106) \approx 4.924$ thousand dollars or $4924.

    (e) $P'(50) \approx 0.013$, or $13 per package sold
    $P'(100) \approx 0.165$, or $165 per package sold
    $P'(125) \approx 0.118$, or $118 per package sold
    $P'(150) \approx 0.031$, or $31 per package sold
    $P'(175) \approx 0.006$, or $6 per package sold
    $P'(300) \approx 10^{-6}$, or $0.001 per package sold

    (f) The limit is 10. The maximum possible profit is $10,000 monthly.

    (g) Yes. In order to sell more and more packages, the company might need to lower the price to a point where they won't make any additional profit.

17. Note that "downward velocity" is positive when McCarthy is falling downward. His downward velocity increases steadily until the parachute opens, and then decreases to a constant downward velocity. One possible sketch:

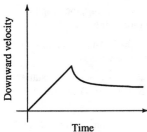

19. The particle is at (5, 2) when $4t^3 - 16t^2 + 15t = 5$, which occurs at $t \approx 2.83$.

21. (a)

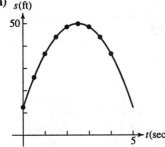

(b) $s'(1) = 18$, $s'(2.5) = 0$, $s'(3.5) = -12$

23. (a) The body reverses direction when v changes sign, at $t = 2$ and at $t = 7$.

(b) The body is moving at a constant speed, $|v| = 3$ m/sec, between $t = 3$ and $t = 6$.

(c) The speed graph is obtained by reflecting the negative portion of the velocity graph, $2 < t < 7$, about the x-axis.

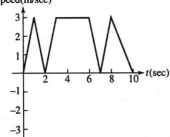

(d) For $0 \le t < 1$: $a = \dfrac{3-0}{1-0} = 3$ m/sec^2

For $1 < t < 3$: $a = \dfrac{-3-3}{3-1} = -3$ m/sec^2

For $3 < t < 6$: $a = \dfrac{-3-(-3)}{6-3} = 0$ m/sec^2

For $6 < t < 8$: $a = \dfrac{3-(-3)}{8-6} = 3$ m/sec^2

For $8 < t \le 10$: $a = \dfrac{0-3}{10-8} = -1.5$ m/sec^2

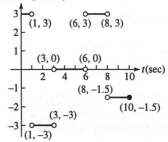

25. (a) The particle moves forward when $v > 0$, for $0 \le t < 1$ and for $5 < t < 7$.
The particle moves backward when $v < 0$, for $1 < t < 5$.
The particle speeds up when v is negative and decreasing, for $1 < t < 2$, and when v is positive and increasing, for $5 < t < 6$.
The particle slows down when v is positive and decreasing, for $0 \le t < 1$ and for $6 < t < 7$, and when v is negative and increasing, for $3 < t < 5$.

(b) Note that the acceleration $a = \dfrac{dv}{dt}$ is undefined at $t = 2$, $t = 3$, and $t = 6$.
The acceleration is positive when v is increasing, for $3 < t < 6$.
The acceleration is negative when v is decreasing, for $0 \le t < 2$ and for $6 < t < 7$.
The acceleration is zero when v is constant, for $2 < t < 3$ and for $7 < t \le 9$.

(c) The particle moves at its greatest speed when $|v|$ is maximized, at $t = 0$ and for $2 < t < 3$.

(d) The particle stands still for more than an instant when v stays at zero, for $7 < t \le 9$.

27. (a) Solving $160 = 490t^2$ gives $t = \pm\dfrac{4}{7}$.

It took $\dfrac{4}{7}$ of a second. The average velocity was $\dfrac{160 \text{ cm}}{\left(\frac{4}{7}\right) \text{ sec}} = 280$ cm/sec.

(b) $v = s'(t) = 980t$

$a = s''(t) = 980$

The velocity was $s'\left(\dfrac{4}{7}\right) = 560$ cm/sec.

The acceleration was $s''\left(\dfrac{4}{7}\right) = 980$ cm/sec^2.

(c) Since there were about 16 flashes during $\dfrac{4}{7}$ of a second, the light was flashing at a rate of about 28 flashes per second.

29. Graph C is position, graph B is velocity, and graph A is acceleration.
B is the derivative of C because it is negative and zero where C is decreasing and has horizontal tangents, respectively.
A is the derivative of B because it is positive, negative, and zero where B is increasing, decreasing, and has horizontal tangents, respectively.

31. (a) $\dfrac{dV}{dr} = \dfrac{d}{dr}\left(\dfrac{4}{3}\pi r^3\right) = 4\pi r^2$

When $r = 2$, $\dfrac{dV}{dr} = 4\pi(2)^2 = 16\pi$ cubic feet of volume per foot of radius.

(b) The increase in the volume is
$\dfrac{4}{3}\pi(2.2)^3 - \dfrac{4}{3}\pi(2)^3 \approx 11.092$ cubic feet.

33. Let v_0 be the exit velocity of a particle of lava. Then
$s(t) = v_0 t - 16t^2$ feet, so the velocity is $\dfrac{ds}{dt} = v_0 - 32t$.
Solving $\dfrac{ds}{dt} = 0$ gives $t = \dfrac{v_0}{32}$. Then the maximum height, in feet, is $s\left(\dfrac{v_0}{32}\right) = v_0\left(\dfrac{v_0}{32}\right) - 16\left(\dfrac{v_0}{32}\right)^2 = \dfrac{v_0^2}{64}$. Solving $\dfrac{v_0^2}{64} = 1900$ gives $v_0 \approx \pm 348.712$. The exit velocity was about 348.712 ft/sec. Multiplying by $\dfrac{3600 \text{ sec}}{1 \text{ h}} \cdot \dfrac{1 \text{ mi}}{5280 \text{ ft}}$, we find that this is equivalent to about 237.758 mi/h.

35. Since profit = revenue − cost, the Sum and Difference Rule gives $\dfrac{d}{dx}(\text{profit}) = \dfrac{d}{dx}(\text{revenue}) - \dfrac{d}{dx}(\text{cost})$, where x is the number of units produced. This means that marginal profit = marginal revenue − marginal cost.

37. (a) Assume that f is even. Then,

$f'(-x) = \lim_{h\to 0} \dfrac{f(-x+h) - f(-x)}{h}$

$= \lim_{h\to 0} \dfrac{f(x-h) - f(x)}{h}$, and substituting $k = -h$,

$= \lim_{k\to 0} \dfrac{f(x+k) - f(x)}{-k}$

$= -\lim_{k\to 0} \dfrac{f(x+k) - f(x)}{k} = -f'(x)$

So, f' is an odd function.

(b) Assume that f is odd. Then,

$f'(-x) = \lim_{h\to 0} \dfrac{f(-x+h) - f(-x)}{h}$

$= \lim_{h\to 0} \dfrac{-f(x-h) + f(x)}{h}$,

and substituting $k = -h$,

$= \lim_{k\to 0} \dfrac{-f(x+k) + f(x)}{-k}$

$= \lim_{k\to 0} \dfrac{f(x+k) - f(x)}{k} = f'(x)$

So, f' is an even function.

■ Section 3.5 Derivatives of Trigonometric Functions (pp. 134–141)

Exploration 1 Making a Conjecture with NDER

1. When the graph of sin x is increasing, the graph of NDER (sin x) is positive (above the x-axis).

2. When the graph of sin x is decreasing, the graph of NDER (sin x) is negative (below the x-axis).

3. When the graph of sin x stops increasing and starts decreasing, the graph of NDER (sin x) crosses the x-axis from above to below.

4. The slope of the graph of sin x matches the value of NDER (sin x) at these points.

5. We conjecture that NDER (sin x) = cos x. The graphs coincide, supporting our conjecture.

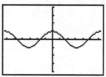

$[-2\pi, 2\pi]$ by $[-4, 4]$

6. When the graph of cos x is increasing, the graph of NDER (cos x) is positive (above the x-axis).
When the graph of cos x is decreasing, the graph of NDER (cos x) is negative (below the x-axis).
When the graph of cos x stops increasing and starts decreasing, the graph of NDER (cos x) crosses the x-axis from above to below.
The slope of the graph of cos x matches the value of NDER (cos x) at these points.
We conjecture that NDER (cos x) = −sin x. The graphs coincide, supporting our conjecture.

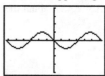

$[-2\pi, 2\pi]$ by $[-4, 4]$

Quick Review 3.5

1. $135° \cdot \dfrac{\pi}{180°} = \dfrac{3\pi}{4} \approx 2.356$

2. $1.7 \cdot \dfrac{180°}{\pi} = \left(\dfrac{306}{\pi}\right)° \approx 97.403°$

3. $\sin \dfrac{\pi}{3} = \dfrac{\sqrt{3}}{2}$

4. Domain: All reals
Range: $[-1, 1]$

5. Domain: $x \neq \dfrac{k\pi}{2}$ for odd integers k
Range: All reals

6. $\cos a = \pm\sqrt{1 - \sin^2 a} = \pm\sqrt{1 - (-1)^2} = \pm\sqrt{0} = 0$

7. If $\tan a = -1$, then $a = \dfrac{3\pi}{4} + k\pi$ for some integer k, so $\sin a = \pm\dfrac{1}{\sqrt{2}}$.

8. $\dfrac{1 - \cos h}{h} = \dfrac{(1 - \cos h)(1 + \cos h)}{h(1 + \cos h)} = \dfrac{1 - \cos^2 h}{h(1 + \cos h)}$
$= \dfrac{\sin^2 h}{h(1 + \cos h)}$

9. $y'(x) = 6x^2 - 14x$
$y'(3) = 12$
The tangent line has slope 12 and passes through $(3, 1)$, so its equation is $y = 12(x - 3) + 1$, or $y = 12x - 35$.

10. $a(t) = v'(t) = 6t^2 - 14t$
$a(3) = 12$

Section 3.5 Exercises

1. $\dfrac{d}{dx}(1 + x - \cos x) = 0 + 1 - (-\sin x) = 1 + \sin x$

3. $\dfrac{d}{dx}\left(\dfrac{1}{x} + 5\sin x\right) = -\dfrac{1}{x^2} + 5\cos x$

5. $\dfrac{d}{dx}(4 - x^2 \sin x) = \dfrac{d}{dx}(4) - \left[x^2\dfrac{d}{dx}(\sin x) + (\sin x)\dfrac{d}{dx}(x^2)\right]$
$= 0 - [x^2 \cos x + (\sin x)(2x)]$
$= -x^2 \cos x - 2x \sin x$

7. $\dfrac{d}{dx}\left(\dfrac{4}{\cos x}\right) = \dfrac{d}{dx}(4 \sec x) = 4 \sec x \tan x$

9. $\dfrac{d}{dx}\dfrac{\cot x}{1 + \cot x} = \dfrac{(1 + \cot x)\dfrac{d}{dx}(\cot x) - (\cot x)\dfrac{d}{dx}(1 + \cot x)}{(1 + \cot x)^2}$
$= \dfrac{(1 + \cot x)(-\csc^2 x) - (\cot x)(-\csc^2 x)}{(1 + \cot x)^2}$
$= -\dfrac{\csc^2 x}{(1 + \cot x)^2} = -\dfrac{\csc^2 x \sin^2 x}{(1 + \cot x)^2 \sin^2 x} = -\dfrac{1}{(\sin x + \cos x)^2}$

11. $y'(x) = \dfrac{d}{dx}(\sin x + 3) = \cos x$
$y'(\pi) = \cos \pi = -1$

The tangent line has slope -1 and passes through
$(\pi, \sin \pi + 3) = (\pi, 3)$.
Its equation is $y = -1(x - \pi) + 3$, or $y = -x + \pi + 3$.

13. $y'(x) = \dfrac{d}{dx}(x^2 \sin x) = x^2\dfrac{d}{dx}(\sin x) + (\sin x)\dfrac{d}{dx}(x^2)$
$= x^2 \cos x + 2x \sin x$
$y'(3) = 9\cos 3 + 6\sin 3$

The tangent line has slope $9\cos 3 + 6\sin 3$ and passes through $(3, 9\sin 3)$. Its equation is
$y = (9\cos 3 + 6\sin 3)(x - 3) + 9\sin 3$, or
$y = (9\cos 3 + 6\sin 3)x - 27\cos 3 - 9\sin 3$. Using decimals, this equation is approximately
$y = -8.063x + 25.460$.

15. (a) $\dfrac{d}{dx}\tan x = \dfrac{d}{dx}\dfrac{\sin x}{\cos x} = \dfrac{(\cos x)\dfrac{d}{dx}(\sin x) - (\sin x)\dfrac{d}{dx}(\cos x)}{(\cos x)^2}$
$= \dfrac{(\cos x)(\cos x) - (\sin x)(-\sin x)}{\cos^2 x}$
$= \dfrac{\cos^2 x + \sin^2 x}{\cos^2 x} = \dfrac{1}{\cos^2 x} = \sec^2 x$

(b) $\dfrac{d}{dx}\sec x = \dfrac{d}{dx}\dfrac{1}{\cos x} = \dfrac{(\cos x)\dfrac{d}{dx}(1) - (1)\dfrac{d}{dx}(\cos x)}{(\cos x)^2}$
$= \dfrac{(\cos x)(0) - (1)(-\sin x)}{\cos^2 x}$
$= \dfrac{\sin x}{\cos^2 x} = \sec x \tan x$

17. $\dfrac{d}{dx}\sec x = \sec x \tan x$ which is 0 at $x = 0$, so the slope of the tangent line is 0. $\dfrac{d}{dx}\cos x = -\sin x$ which is 0 at $x = 0$, so the slope of the tangent line is 0.

19. $y'(x) = \dfrac{d}{dx}(\sqrt{2}\cos x) = -\sqrt{2}\sin x$
$y'\left(\dfrac{\pi}{4}\right) = -\sqrt{2}\sin\dfrac{\pi}{4} = -\sqrt{2}\left(\dfrac{1}{\sqrt{2}}\right) = -1$
The tangent line has slope -1 and passes through
$\left(\dfrac{\pi}{4}, \sqrt{2}\cos\dfrac{\pi}{4}\right) = \left(\dfrac{\pi}{4}, 1\right)$, so its equation is
$y = -1\left(x - \dfrac{\pi}{4}\right) + 1$, or $y = -x + \dfrac{\pi}{4} + 1$.
The normal line has slope 1 and passes through $\left(\dfrac{\pi}{4}, 1\right)$, so its equation is $y = 1\left(x - \dfrac{\pi}{4}\right) + 1$, or $y = x + 1 - \dfrac{\pi}{4}$.

21. $y'(x) = \dfrac{d}{dx}(4 + \cot x - 2\csc x)$
$= 0 - \csc^2 x + 2\csc x \cot x$
$= -\csc^2 x + 2\csc x \cot x$

(a) $y'\left(\dfrac{\pi}{2}\right) = -\csc^2\dfrac{\pi}{2} + 2\csc\dfrac{\pi}{2}\cot\dfrac{\pi}{2}$
$= -1^2 + 2(1)(0) = -1$

The tangent line has slope -1 and passes through $P\left(\dfrac{\pi}{2}, 2\right)$. Its equation is $y = -1\left(x - \dfrac{\pi}{2}\right) + 2$, or
$y = -x + \dfrac{\pi}{2} + 2$.

21. continued

(b)
$$y'(x) = 0$$
$$-\csc^2 x + 2\csc x \cot x = 0$$
$$-\frac{1}{\sin^2 x} + \frac{2\cos x}{\sin^2 x} = 0$$
$$\frac{1}{\sin^2 x}(2\cos x - 1) = 0$$
$$\cos x = \frac{1}{2}$$
$$x = \frac{\pi}{3} \text{ at point } Q$$

$$y\left(\frac{\pi}{3}\right) = 4 + \cot\frac{\pi}{3} - 2\csc\frac{\pi}{3}$$
$$= 4 + \frac{1}{\sqrt{3}} - 2\left(\frac{2}{\sqrt{3}}\right)$$
$$= 4 - \frac{3}{\sqrt{3}} = 4 - \sqrt{3}$$

The coordinates of Q are $\left(\frac{\pi}{3}, 4 - \sqrt{3}\right)$.

The equation of the horizontal line is $y = 4 - \sqrt{3}$.

23.
(a) Velocity: $s'(t) = -2\cos t$ m/sec
Speed: $|s'(t)| = |2\cos t|$ m/sec
Acceleration: $s''(t) = 2\sin t$ m/sec^2
Jerk: $s'''(t) = 2\cos t$ m/sec^3

(b) Velocity: $-2\cos\frac{\pi}{4} = -\sqrt{2}$ m/sec
Speed: $|-\sqrt{2}| = \sqrt{2}$ m/sec
Acceleration: $2\sin\frac{\pi}{4} = \sqrt{2}$ m/sec^2
Jerk: $2\cos\frac{\pi}{4} = \sqrt{2}$ m/sec^3

(c) The body starts at 2, goes to 0 and then oscillates between 0 and 4.

Speed:

Greatest when $\cos t = \pm 1$ (or $t = k\pi$), at the center of the interval of motion.

Zero when $\cos t = 0$ (or $t = \frac{k\pi}{2}$, k odd), at the endpoints of the interval of motion.

Acceleration:

Greatest (in magnitude) when $\sin t = \pm 1$ (or $t = \frac{k\pi}{2}$, k odd)

Zero when $\sin t = 0$ (or $t = k\pi$)

Jerk:

Greatest (in magnitude) when $\cos t = \pm 1$ (or $t = k\pi$)

Zero when $\cos t = 0$ $\left(\text{or } t = \frac{k\pi}{2}, k \text{ odd}\right)$

25.
(a)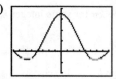

$[-360, 360]$ by $[-0.01, 0.02]$

The limit is $\frac{\pi}{180}$ because this is the conversion factor for changing from degrees to radians.

(b)

$[-360, 360]$ by $[-0.02, 0.02]$

This limit is still 0.

(c) $\frac{d}{dx}\sin x = \lim_{h \to 0} \frac{\sin(x+h) - \sin x}{h}$
$$= \lim_{h \to 0} \frac{\sin x \cos h + \cos x \sin h - \sin x}{h}$$
$$= \lim_{h \to 0} \frac{\sin x(\cos h - 1) + \cos x \sin h}{h}$$
$$= \left(\lim_{h \to 0} \sin x\right)\left(\lim_{h \to 0} \frac{\cos h - 1}{h}\right) + \left(\lim_{h \to 0} \cos x\right)\left(\lim_{h \to 0} \frac{\sin h}{h}\right)$$
$$= (\sin x)(0) + (\cos x)\left(\frac{\pi}{180}\right)$$
$$= \frac{\pi}{180}\cos x$$

(d) $\frac{d}{dx}\cos x = \lim_{h \to 0} \frac{\cos(x+h) - \cos x}{h}$
$$= \lim_{h \to 0} \frac{\cos x \cos h - \sin x \sin h - \cos x}{h}$$
$$= \lim_{h \to 0} \frac{(\cos x)(\cos h - 1) - \sin x \sin h}{h}$$
$$= \left(\lim_{h \to 0} \cos x\right)\left(\lim_{h \to 0} \frac{\cos h - 1}{h}\right) - \left(\lim_{h \to 0} \sin x\right)\left(\lim_{h \to 0} \frac{\sin h}{h}\right)$$
$$= (\cos x)(0) - (\sin x)\left(\frac{\pi}{180}\right)$$
$$= -\frac{\pi}{180}\sin x$$

(e) $\frac{d^2}{dx^2}\sin x = \frac{d}{dx}\frac{\pi}{180}\cos x = \frac{\pi}{180}\left(-\frac{\pi}{180}\sin x\right)$
$$= -\frac{\pi^2}{180^2}\sin x$$
$\frac{d^3}{dx^3}\sin x = \frac{d}{dx}\left(-\frac{\pi^2}{180^2}\sin x\right) = -\frac{\pi^2}{180^2}\left(\frac{\pi}{180}\cos x\right)$
$$= -\frac{\pi^3}{180^3}\cos x$$
$\frac{d^2}{dx^2}\cos x = \frac{d}{dx}\left(-\frac{\pi}{180}\sin x\right) = -\frac{\pi}{180}\left(\frac{\pi}{180}\cos x\right)$
$$= -\frac{\pi^2}{180^2}\cos x$$
$\frac{d^3}{dx^3}\cos x = \frac{d}{dx}\left(-\frac{\pi^2}{180^2}\cos x\right) = -\frac{\pi^2}{180^2}\left(-\frac{\pi}{180}\sin x\right)$
$$= \frac{\pi^3}{180^3}\sin x$$

27. $y' = \dfrac{d}{d\theta}(\theta \tan \theta)$

$= \theta \dfrac{d}{d\theta}(\tan \theta) + (\tan \theta)\dfrac{d}{d\theta}(\theta)$

$= \theta \sec^2 \theta + \tan \theta$

$y'' = \dfrac{d}{d\theta}(\theta \sec^2 \theta + \tan \theta)$

$= \theta \dfrac{d}{d\theta}[(\sec \theta)(\sec \theta)] + (\sec^2 \theta)\dfrac{d}{d\theta}(\theta) + \dfrac{d}{d\theta}(\tan \theta)$

$= \theta\left[(\sec \theta)\dfrac{d}{d\theta}(\sec \theta) + (\sec \theta)\dfrac{d}{d\theta}(\sec \theta)\right]$
$\quad + \sec^2 \theta + \sec^2 \theta$

$= 2\theta \sec^2 \theta \tan \theta + 2 \sec^2 \theta$

$= (2\theta \tan \theta + 2)(\sec^2 \theta)$

or, writing in terms of sines and cosines,

$= \dfrac{2 + 2\theta \tan \theta}{\cos^2 \theta}$

$= \dfrac{2 \cos \theta + 2\theta \sin \theta}{\cos^3 \theta}$

29. Observe the pattern:

$\dfrac{d}{dx}\cos x = -\sin x \qquad \dfrac{d^5}{dx^5}\cos x = -\sin x$

$\dfrac{d^2}{dx^2}\cos x = -\cos x \qquad \dfrac{d^6}{dx^6}\cos x = -\cos x$

$\dfrac{d^3}{dx^3}\cos x = \sin x \qquad \dfrac{d^7}{dx^7}\cos x = \sin x$

$\dfrac{d^4}{dx^4}\cos x = \cos x \qquad \dfrac{d^8}{dx^8}\cos x = \cos x$

Continuing the pattern, we see that

$\dfrac{d^n}{dx^n}\cos x = \sin x$ when $n = 4k + 3$ for any whole number k.

Since $999 = 4(249) + 3$, $\dfrac{d^{999}}{dx^{999}}\cos x = \sin x$.

31. The line is tangent to the graph of $y = \sin x$ at $(0, 0)$. Since $y'(0) = \cos(0) = 1$, the line has slope 1 and its equation is $y = x$.

33. $\dfrac{d}{dx}\sin 2x = \dfrac{d}{dx}(2 \sin x \cos x)$

$= 2\dfrac{d}{dx}(\sin x \cos x)$

$= 2\left[(\sin x)\dfrac{d}{dx}(\cos x) + (\cos x)\dfrac{d}{dx}(\sin x)\right]$

$= 2[(\sin x)(-\sin x) + (\cos x)(\cos x)]$

$= 2(\cos^2 x - \sin^2 x)$

$= 2\cos 2x$

35. $\displaystyle\lim_{h\to 0}\dfrac{(\cos h - 1)}{h} = \lim_{h\to 0}\dfrac{(\cos h - 1)(\cos h + 1)}{h(\cos h + 1)}$

$= \displaystyle\lim_{h\to 0}\dfrac{\cos^2 h - 1}{h(\cos h + 1)}$

$= \displaystyle\lim_{h\to 0}\dfrac{-\sin^2 h}{h(\cos h + 1)}$

$= -\left(\displaystyle\lim_{h\to 0}\dfrac{\sin h}{h}\right)\left(\displaystyle\lim_{h\to 0}\dfrac{\sin h}{\cos h + 1}\right)$

$= -(1)\left(\dfrac{0}{2}\right) = 0$

Section 3.6 Chain Rule (pp. 141–149)

Quick Review 3.6

1. $f(g(x)) = f(x^2 + 1) = \sin(x^2 + 1)$
2. $f(g(h(x))) = f(g(7x)) = f((7x)^2 + 1)$
   $= \sin[(7x)^2 + 1] = \sin(49x^2 + 1)$
3. $(g \circ h)(x) = g(h(x)) = g(7x) = (7x)^2 + 1 = 49x^2 + 1$
4. $(h \circ g)(x) = h(g(x)) = h(x^2 + 1) = 7(x^2 + 1) = 7x^2 + 7$
5. $f\left(\dfrac{g(x)}{h(x)}\right) = f\left(\dfrac{x^2 + 1}{7x}\right) = \sin\dfrac{x^2 + 1}{7x}$
6. $\sqrt{\cos x + 2} = g(\cos x) = g(f(x))$
7. $\sqrt{3\cos^2 x + 2} = g(3\cos^2 x) = g(h(\cos x)) = g(h(f(x)))$
8. $3\cos x + 6 = 3(\cos x + 2) = 3(\sqrt{\cos x + 2})^2$
   $= h(\sqrt{\cos x + 2}) = h(g(\cos x)) = h(g(f(x)))$
9. $\cos 27x^4 = f(27x^4) = f(3(3x^2)^2) = f(h(3x^2)) = f(h(h(x)))$
10. $\cos\sqrt{2 + 3x^2} = \cos\sqrt{3x^2 + 2} = f(\sqrt{3x^2 + 2})$
    $= f(g(3x^2)) = f(g(h(x)))$

Section 3.6 Exercises

1. $\dfrac{dy}{dx} = \dfrac{d}{dx}\sin(3x + 1) = [\cos(3x+1)]\dfrac{d}{dx}(3x + 1)$

   $= [\cos(3x + 1)](3) = 3\cos(3x + 1)$

3. $\dfrac{dy}{dx} = \dfrac{d}{dx}\cos(\sqrt{3}x) = [-\sin(\sqrt{3}x)]\dfrac{d}{dx}(\sqrt{3}x)$

   $= [-\sin(\sqrt{3}x)](\sqrt{3}) = -\sqrt{3}\sin(\sqrt{3}x)$

5. $\dfrac{dy}{dx} = \dfrac{d}{dx}\left[5\cot\left(\dfrac{2}{x}\right)\right] = \left[-5\csc^2\left(\dfrac{2}{x}\right)\right]\dfrac{d}{dx}(2x^{-1})$

   $= \left[-5\csc^2\left(\dfrac{2}{x}\right)\right](-2x^{-2}) = \dfrac{10}{x^2}\csc^2\left(\dfrac{2}{x}\right)$

7. $\dfrac{dy}{dx} = \dfrac{d}{dx}\cos(\sin x) = [-\sin(\sin x)]\dfrac{d}{dx}(\sin x)$

   $= -\sin(\sin x)\cos x$

9. $\dfrac{dy}{dx} = \dfrac{d}{dx}(x + \sqrt{x})^{-2} = -2(x + \sqrt{x})^{-3}\dfrac{d}{dx}(x + \sqrt{x})$

   $= -2(x + \sqrt{x})^{-3}\left(1 + \dfrac{1}{2\sqrt{x}}\right)$

11. $\dfrac{dy}{dx} = \dfrac{d}{dx}(\sin^{-5} x - \cos^3 x)$

    $= (-5\sin^{-6} x)\dfrac{d}{dx}(\sin x) - (3\cos^2 x)\dfrac{d}{dx}(\cos x)$

    $= -5\sin^{-6} x \cos x + 3\cos^2 x \sin x$

13. $\dfrac{dy}{dx} = \dfrac{d}{dx}(\sin^3 x \tan 4x)$

    $= (\sin^3 x)\dfrac{d}{dx}(\tan 4x) + (\tan 4x)\dfrac{d}{dx}(\sin^3 x)$

    $= (\sin^3 x)(\sec^2 4x)\dfrac{d}{dx}(4x) + (\tan 4x)(3\sin^2 x)\dfrac{d}{dx}(\sin x)$

    $= (\sin^3 x)(\sec^2 4x)(4) + (\tan 4x)(3\sin^2 x)(\cos x)$

    $= 4\sin^3 x \sec^2 4x + 3\sin^2 x \cos x \tan 4x$

15. $\dfrac{dy}{dx} = \dfrac{d}{dx}\left(\dfrac{3}{\sqrt{2x+1}}\right)$

$= \dfrac{(\sqrt{2x+1})\dfrac{d}{dx}(3) - 3\dfrac{d}{dx}(\sqrt{2x+1})}{(\sqrt{2x+1})^2}$

$= \dfrac{(\sqrt{2x+1})(0) - 3\left(\dfrac{1}{2\sqrt{2x+1}}\right)\dfrac{d}{dx}(2x+1)}{2x+1}$

$= \dfrac{-3\left(\dfrac{1}{2\sqrt{2x+1}}\right)(2)}{2x+1}$

$= -\dfrac{3}{(2x+1)\sqrt{2x+1}}$

$= -3(2x+1)^{-3/2}$

17. The last step here uses the identity $2\sin a \cos a = \sin 2a$.

$\dfrac{dy}{dx} = \dfrac{d}{dx}\sin^2(3x-2)$

$= 2\sin(3x-2)\dfrac{d}{dx}\sin(3x-2)$

$= 2\sin(3x-2)\cos(3x-2)\dfrac{d}{dx}(3x-2)$

$= 2\sin(3x-2)\cos(3x-2)(3)$

$= 6\sin(3x-2)\cos(3x-2)$

$= 3\sin(6x-4)$

19. $\dfrac{dy}{dx} = \dfrac{d}{dx}(1+\cos^2 7x)^3$

$= 3(1+\cos^2 7x)^2\dfrac{d}{dx}(1+\cos^2 7x)$

$= 3(1+\cos^2 7x)^2(2\cos 7x)\dfrac{d}{dx}(\cos 7x)$

$= 3(1+\cos^2 7x)^2(2\cos 7x)(-\sin 7x)\dfrac{d}{dx}(7x)$

$= 3(1+\cos^2 7x)^2(2\cos 7x)(-\sin 7x)(7)$

$= -42(1+\cos^2 7x)^2 \cos 7x \sin 7x$

21. $\dfrac{ds}{dt} = \dfrac{d}{dt}\cos\left(\dfrac{\pi}{2} - 3t\right)$

$= \left[-\sin\left(\dfrac{\pi}{2} - 3t\right)\right]\dfrac{d}{dt}\left(\dfrac{\pi}{2} - 3t\right)$

$= \left[-\sin\left(\dfrac{\pi}{2} - 3t\right)\right](-3)$

$= 3\sin\left(\dfrac{\pi}{2} - 3t\right)$

23. $\dfrac{ds}{dt} = \dfrac{d}{dt}\left(\dfrac{4}{3\pi}\sin 3t + \dfrac{4}{5\pi}\cos 5t\right)$

$= \dfrac{4}{3\pi}(\cos 3t)\dfrac{d}{dt}(3t) + \dfrac{4}{5\pi}(-\sin 5t)\dfrac{d}{dt}(5t)$

$= \dfrac{4}{3\pi}(\cos 3t)(3) + \dfrac{4}{5\pi}(-\sin 5t)(5)$

$= \dfrac{4}{\pi}\cos 3t - \dfrac{4}{\pi}\sin 5t$

25. $\dfrac{dr}{d\theta} = \dfrac{d}{d\theta}\tan(2-\theta) = \sec^2(2-\theta)\dfrac{d}{d\theta}(2-\theta)$

$= \sec^2(2-\theta)(-1) = -\sec^2(2-\theta)$

27. $\dfrac{dr}{d\theta} = \dfrac{d}{d\theta}\sqrt{\theta \sin\theta} = \dfrac{1}{2\sqrt{\theta\sin\theta}}\dfrac{d}{d\theta}(\theta\sin\theta)$

$= \dfrac{1}{2\sqrt{\theta\sin\theta}}\left[\theta\dfrac{d}{d\theta}(\sin\theta) + (\sin\theta)\dfrac{d}{d\theta}(\theta)\right]$

$= \dfrac{1}{2\sqrt{\theta\sin\theta}}(\theta\cos\theta + \sin\theta)$

$= \dfrac{\theta\cos\theta + \sin\theta}{2\sqrt{\theta\sin\theta}}$

29. $y' = \dfrac{d}{dx}\tan x = \sec^2 x$

$y'' = \dfrac{d}{dx}\sec^2 x = (2\sec x)\dfrac{d}{dx}(\sec x)$

$= (2\sec x)(\sec x \tan x)$

$= 2\sec^2 x \tan x$

31. $y' = \dfrac{d}{dx}\cot(3x-1) = -\csc^2(3x-1)\dfrac{d}{dx}(3x-1)$

$= -3\csc^2(3x-1)$

$y'' = \dfrac{d}{dx}[-3\csc^2(3x-1)]$

$= -3[2\csc(3x-1)]\dfrac{d}{dx}\csc(3x-1)$

$= -3[2\csc(3x-1)] \cdot$

$\quad [-\csc(3x-1)\cot(3x-1)]\dfrac{d}{dx}(3x-1)$

$= -3[2\csc(3x-1)][-\csc(3x-1)\cot(3x-1)](3)$

$= 18\csc^2(3x-1)\cot(3x-1)$

33. $f'(u) = \dfrac{d}{du}(u^5+1) = 5u^4$

$g'(x) = \dfrac{d}{dx}(\sqrt{x}) = \dfrac{1}{2\sqrt{x}}$

$(f \circ g)'(1) = f'(g(1))g'(1) = f'(1)g'(1) = (5)\left(\dfrac{1}{2}\right) = \dfrac{5}{2}$

35. $f'(u) = \dfrac{d}{du}\left(\cot\dfrac{\pi u}{10}\right) = -\csc^2\left(\dfrac{\pi u}{10}\right)\dfrac{d}{du}\left(\dfrac{\pi u}{10}\right)$

$= -\dfrac{\pi}{10}\csc^2\left(\dfrac{\pi u}{10}\right)$

$g'(x) = \dfrac{d}{dx}(5\sqrt{x}) = \dfrac{5}{2\sqrt{x}}$

$(f \circ g)'(1) = f'(g(1))g'(1) = f'(5)g'(1)$

$= -\dfrac{\pi}{10}\left[\csc^2\left(\dfrac{\pi}{2}\right)\right]\left(\dfrac{5}{2}\right)$

$= -\dfrac{\pi}{10}(1)\left(\dfrac{5}{2}\right) = -\dfrac{\pi}{4}$

37. $f'(u) = \dfrac{d}{du}\dfrac{2u}{u^2+1} = \dfrac{(u^2+1)\dfrac{d}{du}(2u) - (2u)\dfrac{d}{du}(u^2+1)}{(u^2+1)^2}$

$= \dfrac{(u^2+1)(2) - (2u)(2u)}{(u^2+1)^2} = \dfrac{-2u^2+2}{(u^2+1)^2}$

$g'(x) = \dfrac{d}{dx}(10x^2+x+1) = 20x+1$

$(f \circ g)'(0) = f'(g(0))g'(0) = f'(1)g'(0) = (0)(1) = 0$

39. (a) $\dfrac{dy}{dx} = \dfrac{dy}{du}\dfrac{du}{dx}$

$= \dfrac{d}{du}(\cos u)\dfrac{d}{dx}(6x+2)$

$= (-\sin u)(6)$

$= -6\sin u$

$= -6\sin(6x+2)$

(b) $\dfrac{dy}{dx} = \dfrac{dy}{du}\dfrac{du}{dx}$

$= \dfrac{d}{du}(\cos 2u)\dfrac{d}{dx}(3x+1)$

$= (-\sin 2u)(2)\cdot(3)$

$= -6\sin 2u$

$= -6\sin(6x+2)$

41. $\dfrac{dx}{dt} = \dfrac{d}{dt}(2\cos t) = -2\sin t$

$\dfrac{dy}{dt} = \dfrac{d}{dt}(2\sin t) = 2\cos t$

$\dfrac{dy}{dx} = \dfrac{\frac{dy}{dt}}{\frac{dx}{dt}} = \dfrac{2\cos t}{-2\sin t} = -\cot t$

The line passes through $\left(2\cos\dfrac{\pi}{4},\, 2\sin\dfrac{\pi}{4}\right) = (\sqrt{2},\sqrt{2})$

and has slope $-\cot\dfrac{\pi}{4} = -1$. Its equation is

$y = -(x-\sqrt{2}) + \sqrt{2}$, or $y = -x + 2\sqrt{2}$.

43. $\dfrac{dx}{dt} = \dfrac{d}{dt}(\sec^2 t - 1) = (2\sec t)\dfrac{d}{dt}(\sec t)$

$= (2\sec t)(\sec t\tan t)$

$= 2\sec^2 t\tan t$

$\dfrac{dy}{dt} = \dfrac{d}{dt}\tan t = \sec^2 t$

$\dfrac{dy}{dx} = \dfrac{\frac{dy}{dt}}{\frac{dx}{dt}} = \dfrac{\sec^2 t}{2\sec^2 t\tan t} = \dfrac{1}{2}\cot t.$

The line passes through

$\left(\sec^2\left(-\dfrac{\pi}{4}\right) - 1,\, \tan\left(-\dfrac{\pi}{4}\right)\right) = (1,-1)$ and has

slope $\dfrac{1}{2}\cot\left(-\dfrac{\pi}{4}\right) = -\dfrac{1}{2}$. Its equation is

$y = -\dfrac{1}{2}(x-1) - 1$, or $y = -\dfrac{1}{2}x - \dfrac{1}{2}$.

45. $\dfrac{dx}{dt} = \dfrac{d}{dt}t = 1$

$\dfrac{dy}{dt} = \dfrac{d}{dt}\sqrt{t} = \dfrac{1}{2\sqrt{t}}$

$\dfrac{dy}{dx} = \dfrac{\frac{dy}{dt}}{\frac{dx}{dt}} = \dfrac{1/(2\sqrt{t})}{1} = \dfrac{1}{2\sqrt{t}}$

The line passes through $\left(\dfrac{1}{4},\sqrt{\dfrac{1}{4}}\right) = \left(\dfrac{1}{4},\dfrac{1}{2}\right)$ and has slope

$\dfrac{1}{2\sqrt{\frac{1}{4}}} = 1$. Its equation is $y = 1\left(x - \dfrac{1}{4}\right) + \dfrac{1}{2}$, or

$y = x + \dfrac{1}{4}$.

47. $\dfrac{dx}{dt} = \dfrac{d}{dt}(t - \sin t) = 1 - \cos t$

$\dfrac{dy}{dt} = \dfrac{d}{dt}(1 - \cos t) = \sin t$

$\dfrac{dy}{dx} = \dfrac{\frac{dy}{dt}}{\frac{dx}{dt}} = \dfrac{\sin t}{1 - \cos t}$

The line passes through

$\left(\dfrac{\pi}{3} - \sin\dfrac{\pi}{3},\, 1 - \cos\dfrac{\pi}{3}\right) = \left(\dfrac{\pi}{3} - \dfrac{\sqrt{3}}{2},\dfrac{1}{2}\right)$ and has slope

$\dfrac{\sin\left(\frac{\pi}{3}\right)}{1 - \cos\left(\frac{\pi}{3}\right)} = \sqrt{3}$. Its equation is

$y = \sqrt{3}\left(x - \dfrac{\pi}{3} + \dfrac{\sqrt{3}}{2}\right) + \dfrac{1}{2}$, or

$y = \sqrt{3}x + 2 - \dfrac{\pi}{\sqrt{3}}$.

49. (a) $\dfrac{dx}{dt} = \dfrac{d}{dt}(t^2 + t) = 2t + 1$

$\dfrac{dy}{dt} = \dfrac{d}{dt}\sin t = \cos t$

$\dfrac{dy}{dx} = \dfrac{\frac{dy}{dt}}{\frac{dx}{dt}} = \dfrac{\cos t}{2t + 1}$

(b) $\dfrac{d}{dt}\left(\dfrac{dy}{dx}\right) = \dfrac{d}{dt}\dfrac{\cos t}{2t + 1}$

$= \dfrac{(2t+1)\dfrac{d}{dt}(\cos t) - (\cos t)\dfrac{d}{dt}(2t+1)}{(2t+1)^2}$

$= \dfrac{(2t+1)(-\sin t) - (\cos t)(2)}{(2t+1)^2}$

$= -\dfrac{(2t+1)(\sin t) + 2\cos t}{(2t+1)^2}$

49. continued

(c) Let $u = \dfrac{dy}{dx}$.

Then $\dfrac{du}{dt} = \dfrac{du}{dx}\dfrac{dx}{dt}$, so $\dfrac{du}{dx} = \dfrac{du}{dt} \div \dfrac{dx}{dt}$. Therefore,

$$\dfrac{d}{dx}\left(\dfrac{dy}{dx}\right) = \dfrac{d}{dt}\left(\dfrac{dy}{dx}\right) \div \dfrac{dx}{dt}$$

$$= -\dfrac{(2t+1)(\sin t) + 2\cos t}{(2t+1)^2} \div (2t+1)$$

$$= -\dfrac{(2t+1)(\sin t) + 2\cos t}{(2t+1)^3}$$

(d) The expression in part (c).

51.
$\dfrac{ds}{dt} = \dfrac{ds}{d\theta}\dfrac{d\theta}{dt} = \dfrac{d}{d\theta}(\cos\theta)\dfrac{d\theta}{dt}$

$= (-\sin\theta)\left(\dfrac{d\theta}{dt}\right)$

When $\theta = \dfrac{3\pi}{2}$ and $\dfrac{d\theta}{dt} = 5$, $\dfrac{ds}{dt} = \left(-\sin\dfrac{3\pi}{2}\right)(5) = 5$.

53.
$\dfrac{dy}{dx} = \dfrac{d}{dx}\sin\dfrac{x}{2} = \left(\cos\dfrac{x}{2}\right)\dfrac{d}{dx}\left(\dfrac{x}{2}\right) = \dfrac{1}{2}\cos\dfrac{x}{2}$

Since the range of the function $f(x) = \dfrac{1}{2}\cos\dfrac{x}{2}$ is $\left[-\dfrac{1}{2}, \dfrac{1}{2}\right]$, the largest possible value of $\dfrac{dy}{dx}$ is $\dfrac{1}{2}$.

55.
$\dfrac{dy}{dx} = \dfrac{d}{dx} 2\tan\dfrac{\pi x}{4} = \left(2\sec^2\dfrac{\pi x}{4}\right)\dfrac{d}{dx}\left(\dfrac{\pi x}{4}\right)$

$= \dfrac{\pi}{2}\sec^2\left(\dfrac{\pi x}{4}\right)$

$y'(1) = \dfrac{\pi}{2}\sec^2\left(\dfrac{\pi}{4}\right) = \dfrac{\pi}{2}(\sqrt{2})^2 = \pi$.

The tangent line has slope π and passes through $\left(1, 2\tan\dfrac{\pi}{4}\right) = (1, 2)$. Its equation is $y = \pi(x-1) + 2$, or $y = \pi x - \pi + 2$.

The normal line has slope $-\dfrac{1}{\pi}$ and passes through $(1, 2)$. Its equation is $y = -\dfrac{1}{\pi}(x-1) + 2$, or $y = -\dfrac{1}{\pi}x + \dfrac{1}{\pi} + 2$.

Graphical support:

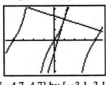

$[-4.7, 4.7]$ by $[-3.1, 3.1]$

57.
(a) $\dfrac{d}{dx}[5f(x) - g(x)] = 5f'(x) - g'(x)$

At $x = 1$, the derivative is

$5f'(1) - g'(1) = 5\left(-\dfrac{1}{3}\right) - \left(-\dfrac{8}{3}\right) = 1$.

(b) $\dfrac{d}{dx}f(x)g^3(x) = f(x)\dfrac{d}{dx}g^3(x) + g^3(x)\dfrac{d}{dx}f(x)$

$= f(x)[3g^2(x)]\dfrac{d}{dx}g(x) + g^3(x)f'(x)$

$= 3f(x)g^2(x)g'(x) + g^3(x)f'(x)$

At $x = 0$, the derivative is $3f(0)g^2(0)g'(0) + g^3(0)f'(0)$

$= 3(1)(1)^2\left(\dfrac{1}{3}\right) + (1)^3(5) = 6$.

(c) $\dfrac{d}{dx}\dfrac{f(x)}{g(x)+1} = \dfrac{[g(x)+1]\dfrac{d}{dx}f(x) - f(x)\dfrac{d}{dx}[g(x)+1]}{[g(x)+1]^2}$

$= \dfrac{[g(x)+1]f'(x) - f(x)g'(x)}{[g(x)+1]^2}$

At $x = 1$, the derivative is

$\dfrac{[g(1)+1]f'(1) - f(1)g'(1)}{[g(1)+1]^2} = \dfrac{(-4+1)\left(-\dfrac{1}{3}\right) - (3)\left(-\dfrac{8}{3}\right)}{(-4+1)^2}$

$= \dfrac{9}{9} = 1$.

(d) $\dfrac{d}{dx}f(g(x)) = f'(g(x))g'(x)$

At $x = 0$, the derivative is

$f'(g(0))g'(0) = f'(1)g'(0) = \left(-\dfrac{1}{3}\right)\left(\dfrac{1}{3}\right) = -\dfrac{1}{9}$.

(e) $\dfrac{d}{dx}g(f(x)) = g'(f(x))f'(x)$

At $x = 0$, the derivative is

$g'(f(0))f'(0) = g'(1)f'(0) = \left(-\dfrac{8}{3}\right)(5) = -\dfrac{40}{3}$

(f) $\dfrac{d}{dx}[g(x)+f(x)]^{-2} = -2[g(x)+f(x)]^{-3}\dfrac{d}{dx}[g(x)+f(x)]$

$= -\dfrac{2[g'(x)+f'(x)]}{[g(x)+f(x)]^3}$

At $x = 1$, the derivative is

$-\dfrac{2[g'(1)+f'(1)]}{[g(1)+f(1)]^3} = -\dfrac{2\left(-\dfrac{8}{3}-\dfrac{1}{3}\right)}{(-4+3)^3} = -\dfrac{-6}{-1} = -6$.

(g) $\frac{d}{dx}[f(x + g(x))] = f'(x + g(x))\frac{d}{dx}[x + g(x)]$

$= f'(x + g(x))(1 + g'(x))$

At $x = 0$, the derivative is

$f'(0 + g(0))(1 + g'(0)) = f'(0 + 1)\left(1 + \frac{1}{3}\right)$

$= f'(1)\left(\frac{4}{3}\right)$

$= \left(-\frac{1}{3}\right)\left(\frac{4}{3}\right) = -\frac{4}{9}$.

59. Because the symbols $\frac{dy}{dx}, \frac{dy}{du}$, and $\frac{du}{dx}$ are not fractions. The individual symbols dy, dx, and du do not have numerical values.

61. (a) $y'(t) = \frac{d}{dt} 37 \sin\left[\frac{2\pi}{365}(x - 101)\right] + \frac{d}{dt}(25)$

$= 37 \cos\left[\frac{2\pi}{365}(x - 101)\right] \cdot \frac{d}{dx}\left[\frac{2\pi}{365}(x - 101)\right] + 0$

$= 37 \cos\left[\frac{2\pi}{365}(x - 101)\right] \cdot \frac{2\pi}{365}$

$= \frac{74\pi}{365} \cos\left[\frac{2\pi}{365}(x - 101)\right]$

Since $\cos u$ is greatest when $u = 0, \pm 2\pi$, and so on, $y'(t)$ is greatest when $\frac{2\pi}{365}(x - 101) = 0$, or $x = 101$. The temperature is increasing the fastest on day 101 (April 11).

(b) The rate of increase is

$y'(101) = \frac{74\pi}{365} \approx 0.637$ degrees per day.

63. Acceleration $= \frac{dv}{dt} = \frac{dv}{ds}\frac{ds}{dt} = \left(\frac{dv}{ds}\right)(v) = \left[\frac{d}{ds}(k\sqrt{s})\right](k\sqrt{s})$

$= \left(\frac{k}{2\sqrt{s}}\right)(k\sqrt{s}) = \frac{k^2}{2}$, a constant.

65. Acceleration $= \frac{dv}{dt} = \frac{df(x)}{dt} = \frac{df(x)}{dx}\frac{dx}{dt} = f'(x)f(x)$

67. No, this does not contradict the Chain Rule. The Chain Rule states that if two functions are differentiable at the appropriate points, then their composite must also be differentiable. It does not say: If a composite is differentiable, then the functions which make up the composite must all be differentiable.

69. For $h = 1$:

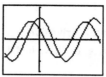

[−2, 3.5] by [−3, 3]

For $h = 0.5$:

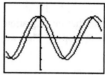

[−2, 3.5] by [−3, 3]

For $h = 0.2$:

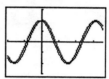

[−2, 3.5] by [−3, 3]

As $h \to 0$, the second curve (the difference quotient) approaches the first ($y = 2 \cos 2x$). This is because $2 \cos 2x$ is the derivative of $\sin 2x$, and the second curve is the difference quotient used to define the derivative of $\sin 2x$. As $h \to 0$, the difference quotient expression should be approaching the derivative.

71. (a) Let $f(x) = |x|$.

Then

$\frac{d}{dx}|u| = \frac{d}{dx}f(u) = f'(u)\frac{du}{dx} = \left(\frac{d}{du}|u|\right)\left(\frac{du}{dx}\right) = \frac{u}{|u|}u'$.

The derivative of the absolute value function is $+1$ for positive values, -1 for negative values, and undefined at 0. So $f'(u) = \begin{cases} -1, & u < 0 \\ 1, & u > 0. \end{cases}$

But this is exactly how the expression $\frac{u}{|u|}$ evaluates.

(b) $f'(x) = \left[\frac{d}{dx}(x^2 - 9)\right] \cdot \frac{x^2 - 9}{|x^2 - 9|} = \frac{(2x)(x^2 - 9)}{|x^2 - 9|}$

$g'(x) = \frac{d}{dx}(|x| \sin x)$

$= |x|\frac{d}{dx}(\sin x) + (\sin x)\frac{d}{dx}|x|$

$= |x| \cos x + \frac{x \sin x}{|x|}$

Note: The expression for $g'(x)$ above is undefined at $x = 0$, but actually

$g'(0) = \lim_{h \to 0} \frac{g(0 + h) - g(0)}{h} = \lim_{h \to 0} \frac{|h| \sin h}{h} = 0$.

Therefore, we may express the derivative as

$g'(x) = \begin{cases} |x| \cos x + \frac{x \sin x}{|x|}, & x \neq 0 \\ 0, & x = 0. \end{cases}$

Section 3.7 Implicit Differentiation
(pp. 149–157)

Exploration 1 An Unexpected Derivative

1. $2x - 2y - 2xy' + 2yy' = 0$. Solving for y', we find that $\dfrac{dy}{dx} = 1$ (provided $y \ne x$).

2. With a constant derivative of 1, the graph would seem to be a line with slope 1.

3. Letting $x = 0$ in the original equation, we find that $y = \pm 2$. This would seem to indicate that this equation defines two lines implicitly, both with slope 1. The two lines are $y = x + 2$ and $y = x - 2$.

4. Factoring the original equation, we have
$[(x - y) - 2][(x - y) + 2] = 0$
$\therefore x - y - 2 = 0$ or $x - y + 2 = 0$
$\therefore y = x - 2$ or $y = x + 2$.
The graph is shown below.

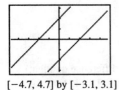

[−4.7, 4.7] by [−3.1, 3.1]

5. At each point (x, y) on either line, $\dfrac{dy}{dx} = 1$. The condition $y \ne x$ is true because both lines are parallel to the line $y = x$. The derivative is surprising because it does not depend on x or y, but there are no inconsistencies.

Quick Review 3.7

1. $x - y^2 = 0$
$x = y^2$
$\pm\sqrt{x} = y$
$y_1 = \sqrt{x}, y_2 = -\sqrt{x}$

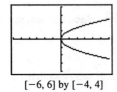

[−6, 6] by [−4, 4]

2. $4x^2 + 9y^2 = 36$
$9y^2 = 36 - 4x^2$
$y^2 = \dfrac{36 - 4x^2}{9} = \dfrac{4}{9}(9 - x^2)$
$y = \pm\dfrac{2}{3}\sqrt{9 - x^2}$
$y_1 = \dfrac{2}{3}\sqrt{9 - x^2}, y_2 = -\dfrac{2}{3}\sqrt{9 - x^2}$

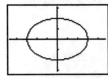

[−4.7, 4.7] by [−3.1, 3.1]

3. $x^2 - 4y^2 = 0$
$(x + 2y)(x - 2y) = 0$
$y = \pm\dfrac{x}{2}$
$y_1 = \dfrac{x}{2}, y_2 = -\dfrac{x}{2}$

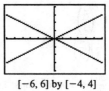

[−6, 6] by [−4, 4]

4. $x^2 + y^2 = 9$
$y^2 = 9 - x^2$
$y = \pm\sqrt{9 - x^2}$
$y_1 = \sqrt{9 - x^2}, y_2 = -\sqrt{9 - y^2}$

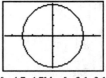

[−4.7, 4.7] by [−3.1, 3.1]

5. $x^2 + y^2 = 2x + 3$
$y^2 = 2x + 3 - x^2$
$y = \pm\sqrt{2x + 3 - x^2}$
$y_1 = \sqrt{2x + 3 - x^2}, y_2 = -\sqrt{2x + 3 - x^2}$

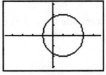

[−4.7, 4.7] by [−3.1, 3.1]

6. $x^2 y' - 2xy = 4x - y$
$x^2 y' = 4x - y + 2xy$
$y' = \dfrac{4x - y + 2xy}{x^2}$

7. $y' \sin x - x \cos x = xy' + y$
$y' \sin x - xy' = y + x \cos x$
$(\sin x - x)y' = y + x \cos x$
$y' = \dfrac{y + x \cos x}{\sin x - x}$

8. $x(y^2 - y') = y'(x^2 - y)$
$xy^2 = y'(x^2 - y + x)$
$y' = \dfrac{xy^2}{x^2 - y + x}$

9. $\sqrt{x}(x - \sqrt[3]{x}) = x^{1/2}(x - x^{1/3})$
$= x^{1/2} x - x^{1/2} x^{1/3}$
$= x^{3/2} - x^{5/6}$

10. $\dfrac{x + \sqrt[3]{x^2}}{\sqrt{x^3}} = \dfrac{x + x^{2/3}}{x^{3/2}}$

$= \dfrac{x}{x^{3/2}} + \dfrac{x^{2/3}}{x^{3/2}}$

$= x^{-1/2} + x^{-5/6}$

Section 3.7 Exercises

1. $\dfrac{dy}{dx} = \dfrac{d}{dx}x^{9/4} = \dfrac{9}{4}x^{(9/4)-1} = \dfrac{9}{4}x^{5/4}$

3. $\dfrac{dy}{dx} = \dfrac{d}{dx}\sqrt[3]{x} = \dfrac{d}{dx}x^{1/3} = \dfrac{1}{3}x^{(1/3)-1} = \dfrac{1}{3}x^{-2/3}$

5. $\dfrac{dy}{dx} = \dfrac{d}{dx}(2x+5)^{-1/2} = -\dfrac{1}{2}(2x+5)^{(-1/2)-1}\dfrac{d}{dx}(2x+5)$

$= -\dfrac{1}{2}(2x+5)^{-3/2}(2) = -(2x+5)^{-3/2}$

7. $\dfrac{dy}{dx} = \dfrac{d}{dx}(x\sqrt{x^2+1})$

$= x\dfrac{d}{dx}\sqrt{x^2+1} + \sqrt{x^2+1}\,\dfrac{d}{dx}(x)$

$= x\dfrac{d}{dx}(x^2+1)^{1/2} + (x^2+1)^{1/2}$

$= x \cdot \dfrac{1}{2}(x^2+1)^{-1/2}(2x) + (x^2+1)^{1/2}$

$= x^2(x^2+1)^{-1/2} + (x^2+1)^{1/2}$

Note: This answer is equivalent to $\dfrac{2x^2+1}{\sqrt{x^2+1}}$.

9. $\qquad x^2 y + xy^2 = 6$

$\dfrac{d}{dx}(x^2 y) + \dfrac{d}{dx}(xy^2) = \dfrac{d}{dx}(6)$

$x^2\dfrac{dy}{dx} + y(2x) + x(2y)\dfrac{dy}{dx} + y^2(1) = 0$

$x^2\dfrac{dy}{dx} + 2xy\dfrac{dy}{dx} = -(2xy + y^2)$

$(2xy + x^2)\dfrac{dy}{dx} = -(2xy + y^2)$

$\dfrac{dy}{dx} = -\dfrac{2xy + y^2}{2xy + x^2}$

11. $\qquad y^2 = \dfrac{x-1}{x+1}$

$\dfrac{d}{dx}y^2 = \dfrac{d}{dx}\dfrac{x-1}{x+1}$

$2y\dfrac{dy}{dx} = \dfrac{(x+1)(1)-(x-1)(1)}{(x+1)^2}$

$2y\dfrac{dy}{dx} = \dfrac{2}{(x+1)^2}$

$\dfrac{dy}{dx} = \dfrac{1}{y(x+1)^2}$

13. $\dfrac{dy}{dx} = \dfrac{d}{dx}(1-x^{1/2})^{1/2}$

$= \dfrac{1}{2}(1-x^{1/2})^{-1/2}\dfrac{d}{dx}(1-x^{1/2})$

$= \dfrac{1}{2}(1-x^{1/2})^{-1/2}\left(-\dfrac{1}{2}x^{-1/2}\right)$

$= -\dfrac{1}{4}(1-x^{1/2})^{-1/2}x^{-1/2}$

15. $\dfrac{dy}{dx} = \dfrac{d}{dx}3(\csc x)^{3/2}$

$= \dfrac{9}{2}(\csc x)^{1/2}\dfrac{d}{dx}(\csc x)$

$= \dfrac{9}{2}(\csc x)^{1/2}(-\csc x \cot x)$

$= -\dfrac{9}{2}(\csc x)^{3/2}\cot x$

17. $\qquad x = \tan y$

$\dfrac{d}{dx}(x) = \dfrac{d}{dx}(\tan y)$

$1 = \sec^2 y\dfrac{dy}{dx}$

$\dfrac{dy}{dx} = \dfrac{1}{\sec^2 y} = \cos^2 y$

19. $\qquad x + \tan xy = 0$

$\dfrac{d}{dx}(x) + \dfrac{d}{dx}(\tan xy) = 0$

$1 + \sec^2(xy)\dfrac{d}{dx}(xy) = 0$

$1 + (\sec^2 xy)\left[x\dfrac{dy}{dx} + (y)(1)\right] = 0$

$(\sec^2 xy)(x)\dfrac{dy}{dx} = -1 - (\sec^2 xy)(y)$

$\dfrac{dy}{dx} = \dfrac{-1 - y\sec^2 xy}{x\sec^2 xy}$

$\dfrac{dy}{dx} = -\dfrac{1}{x}\cos^2 xy - \dfrac{y}{x}$

21. (a) If $f(x) = \dfrac{3}{2}x^{2/3} - 3$, then

$f'(x) = x^{-1/3}$ and $f''(x) = -\dfrac{1}{3}x^{-4/3}$

which contradicts the given equation $f''(x) = x^{-1/3}$.

(b) If $f(x) = \dfrac{9}{10}x^{5/3} - 7$, then

$f'(x) = \dfrac{3}{2}x^{2/3}$ and $f''(x) = x^{-1/3}$,

which matches the given equation.

(c) Differentiating both sides of the given equation

$f''(x) = x^{-1/3}$ gives $f'''(x) = -\dfrac{1}{3}x^{-4/3}$, so it *must* be true

that $f'''(x) = -\dfrac{1}{3}x^{-4/3}$.

(d) If $f'(x) = \dfrac{3}{2}x^{2/3} + 6$, then $f''(x) = x^{-1/3}$, which matches

the given equation.

Conclusion: (b), (c), and (d) could be true.

23. $x^2 + y^2 = 1$

$\frac{d}{dx}(x^2) + \frac{d}{dx}(y^2) = \frac{d}{dx}(1)$

$2x + 2yy' = 0$

$2yy' = -2x$

$y' = -\frac{x}{y}$

$y'' = \frac{d}{dx}\left(-\frac{x}{y}\right)$

$= -\frac{(y)(1) - (x)(y')}{y^2}$

$= -\frac{y - x\left(-\frac{x}{y}\right)}{y^2}$

$= -\frac{x^2 + y^2}{y^3}$

Since our original equation was $x^2 + y^2 = 1$, we may substitute 1 for $x^2 + y^2$, giving $y'' = -\frac{1}{y^3}$.

25. $y^2 = x^2 + 2x$

$\frac{d}{dy}(y^2) = \frac{d}{dy}(x^2) + \frac{d}{dy}(2x)$

$2yy' = 2x + 2$

$y' = \frac{2x + 2}{2y} = \frac{x + 1}{y}$

$y'' = \frac{d}{dx}\frac{x + 1}{y}$

$= \frac{(y)(1) - (x + 1)y'}{y^2}$

$= \frac{y - (x + 1)\left(\frac{x + 1}{y}\right)}{y^2}$

$= \frac{y^2 - (x + 1)^2}{y^3}$

Since our original equation was $y^2 = x^2 + 2x$, we may write $y^2 - (x + 1)^2 = (x^2 + 2x) - (x^2 + 2x + 1) = -1$, which gives $y'' = -\frac{1}{y^3}$.

27. $x^2 + xy - y^2 = 1$

$\frac{d}{dx}(x^2) + \frac{d}{dx}(xy) - \frac{d}{dx}(y^2) = \frac{d}{dx}(1)$

$2x + x\frac{dy}{dx} + (y)(1) - 2y\frac{dy}{dx} = 0$

$(x - 2y)\frac{dy}{dx} = -2x - y$

$\frac{dy}{dx} = \frac{-2x - y}{x - 2y} = \frac{2x + y}{2y - x}$

Slope at $(2, 3)$: $\frac{2(2) + 3}{2(3) - 2} = \frac{7}{4}$

(a) Tangent: $y = \frac{7}{4}(x - 2) + 3$ or $y = \frac{7}{4}x - \frac{1}{2}$

(b) Normal: $y = -\frac{4}{7}(x - 2) + 3$ or $y = -\frac{4}{7}x + \frac{29}{7}$

29. $x^2y^2 = 9$

$\frac{d}{dx}(x^2y^2) = \frac{d}{dx}(9)$

$(x^2)(2y)\frac{dy}{dx} + (y^2)(2x) = 0$

$2x^2y\frac{dy}{dx} = -2xy^2$

$\frac{dy}{dx} = -\frac{2xy^2}{2x^2y} = -\frac{y}{x}$

Slope at $(-1, 3)$: $-\frac{3}{-1} = 3$

(a) Tangent: $y = 3(x + 1) + 3$ or $y = 3x + 6$

(b) Normal: $y = -\frac{1}{3}(x + 1) + 3$ or $y = -\frac{1}{3}x + \frac{8}{3}$

31. $6x^2 + 3xy + 2y^2 + 17y - 6 = 0$

$\frac{d}{dx}(6x^2) + \frac{d}{dx}(3xy) + \frac{d}{dx}(2y^2) + \frac{d}{dx}(17y) - \frac{d}{dx}(6) = \frac{d}{dx}(0)$

$12x + 3x\frac{dy}{dx} + (3y)(1) + 4y\frac{dy}{dx} + 17\frac{dy}{dx} - 0 = 0$

$3x\frac{dy}{dx} + 4y\frac{dy}{dx} + 17\frac{dy}{dx} = -12x - 3y$

$(3x + 4y + 17)\frac{dy}{dx} = -12x - 3y$

$\frac{dy}{dx} = \frac{-12x - 3y}{3x + 4y + 17}$

Slope at $(-1, 0)$: $\frac{-12(-1) - 3(0)}{3(-1) + 4(0) + 17} = \frac{12}{14} = \frac{6}{7}$

(a) Tangent: $y = \frac{6}{7}(x + 1) + 0$ or $y = \frac{6}{7}x + \frac{6}{7}$

(b) Normal: $y = -\frac{7}{6}(x + 1) + 0$ or $y = -\frac{7}{6}x - \frac{7}{6}$

33. $2xy + \pi \sin y = 2\pi$

$2\frac{d}{dx}(xy) + \pi\frac{d}{dx}(\sin y) = \frac{d}{dx}(2\pi)$

$2x\frac{dy}{dx} + 2y(1) + \pi \cos y \frac{dy}{dx} = 0$

$(2x + \pi \cos y)\frac{dy}{dx} = -2y$

$\frac{dy}{dx} = -\frac{2y}{2x + \pi \cos y}$

Slope at $\left(1, \frac{\pi}{2}\right)$: $-\frac{2(\pi/2)}{2(1) + \pi \cos(\pi/2)} = -\frac{\pi}{2}$

(a) Tangent: $y = -\frac{\pi}{2}(x - 1) + \frac{\pi}{2}$ or $y = -\frac{\pi}{2}x + \pi$

(b) Normal: $y = \frac{2}{\pi}(x - 1) + \frac{\pi}{2}$ or $y = \frac{2}{\pi}x - \frac{2}{\pi} + \frac{\pi}{2}$

35.
$$y = 2\sin(\pi x - y)$$
$$\frac{dy}{dx} = \frac{d}{dx} 2\sin(\pi x - y)$$
$$\frac{dy}{dx} = 2\cos(\pi x - y)\left(\pi - \frac{dy}{dx}\right)$$
$$[1 + 2\cos(\pi x - y)]\frac{dy}{dx} = 2\pi\cos(\pi x - y)$$
$$\frac{dy}{dx} = \frac{2\pi\cos(\pi x - y)}{1 + 2\cos(\pi x - y)}$$

Slope at $(1, 0)$: $\dfrac{2\pi\cos\pi}{1 + 2\cos\pi} = \dfrac{2\pi(-1)}{1 + 2(-1)} = 2\pi$

(a) Tangent: $y = 2\pi(x - 1) + 0$ or $y = 2\pi x - 2\pi$

(b) Normal: $y = -\dfrac{1}{2\pi}(x - 1) + 0$ or $y = -\dfrac{x}{2\pi} + \dfrac{1}{2\pi}$

37. (a)
$$y^4 = y^2 - x^2$$
$$\frac{d}{dx}(y^4) = \frac{d}{dx}(y^2) - \frac{d}{dx}x^2$$
$$4y^3\frac{dy}{dx} = 2y\frac{dy}{dx} - 2x$$
$$(4y^3 - 2y)\frac{dy}{dx} = -2x$$
$$\frac{dy}{dx} = \frac{-2x}{4y^3 - 2y} = \frac{x}{y - 2y^3}$$

At $\left(\dfrac{\sqrt{3}}{4}, \dfrac{\sqrt{3}}{2}\right)$:

$$\text{Slope} = \frac{\frac{\sqrt{3}}{4}}{\frac{\sqrt{3}}{2} - 2\left(\frac{\sqrt{3}}{2}\right)^3}$$
$$= \frac{\frac{\sqrt{3}}{4}}{\frac{\sqrt{3}}{2} - \frac{3\sqrt{3}}{4}} \cdot \frac{\frac{4}{\sqrt{3}}}{\frac{4}{\sqrt{3}}} = \frac{1}{2 - 3} = -1$$

At $\left(\dfrac{\sqrt{3}}{4}, \dfrac{1}{2}\right)$:

$$\text{Slope} = \frac{\frac{\sqrt{3}}{4}}{\frac{1}{2} - 2\left(\frac{1}{2}\right)^3} = \frac{\frac{\sqrt{3}}{4}}{\frac{1}{2} - \frac{1}{4}} \cdot \frac{4}{4} = \frac{\sqrt{3}}{1} = \sqrt{3}$$

(b)

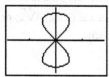

$[-1.8, 1.8]$ by $[-1.2, 1.2]$
Parameter interval: $-1 \le t \le 1$

39. (a) $(-1)^3(1)^2 = \cos(\pi)$ is true since both sides equal -1.

(b)
$$x^3 y^2 = \cos(\pi y)$$
$$\frac{d}{dx}(x^3 y^2) = \frac{d}{dx}\cos(\pi y)$$
$$(x^3)(2y)\frac{dy}{dx} + (y^2)(3x^2) = (-\sin\pi y)(\pi)\frac{dy}{dx}$$
$$(2x^3 y + \pi\sin\pi y)\frac{dy}{dx} = -3x^2 y^2$$
$$\frac{dy}{dx} = -\frac{3x^2 y^2}{2x^3 y + \pi\sin\pi y}$$

Slope at $(-1, 1)$: $-\dfrac{3(-1)^2(1)}{2(-1)^3(1) + \pi\sin\pi} = \dfrac{-3}{-2} = \dfrac{3}{2}$

The slope of the tangent line is $\dfrac{3}{2}$.

41. Find the two points:
The curve crosses the x-axis when $y = 0$, so the equation becomes $x^2 + 0x + 0 = 7$, or $x^2 = 7$. The solutions are $x = \pm\sqrt{7}$, so the points are $(\pm\sqrt{7}, 0)$.
Show tangents are parallel:
$$x^2 + xy + y^2 = 7$$
$$\frac{d}{dx}(x^2) + \frac{d}{dx}(xy) + \frac{d}{dx}(y^2) = \frac{d}{dx}(7)$$
$$2x + x\frac{dy}{dx} + (y)(1) + 2y\frac{dy}{dx} = 0$$
$$(x + 2y)\frac{dy}{dx} = -(2x + y)$$
$$\frac{dy}{dx} = -\frac{2x + y}{x + 2y}$$

Slope at $(\sqrt{7}, 0)$: $-\dfrac{2\sqrt{7} + 0}{\sqrt{7} + 2(0)} = -2$

Slope at $(-\sqrt{7}, 0)$: $-\dfrac{2(-\sqrt{7}) + 0}{-\sqrt{7} + 2(0)} = -2$

The tangents at these points are parallel because they have the same slope. The common slope is -2.

68 Section 3.7

43. First curve:
$$2x^2 + 3y^2 = 5$$
$$\frac{d}{dx}(2x^2) + \frac{d}{dx}(3y^2) = \frac{d}{dx}(5)$$
$$4x + 6y\frac{dy}{dx} = 0$$
$$\frac{dy}{dx} = -\frac{4x}{6y} = -\frac{2x}{3y}$$

Second curve:
$$y^2 = x^3$$
$$\frac{d}{dx}y^2 = \frac{d}{dx}x^3$$
$$2y\frac{dy}{dx} = 3x^2$$
$$\frac{dy}{dx} = \frac{3x^2}{2y}$$

At $(1, 1)$, the slopes are $-\frac{2}{3}$ and $\frac{3}{2}$ respectively. At $(1, -1)$, the slopes are $\frac{2}{3}$ and $-\frac{3}{2}$ respectively. In both cases, the tangents are perpendicular. To graph the curves and normal lines, we may use the following parametric equations for $-\pi \le t \le \pi$:

First curve: $x = \sqrt{\frac{5}{2}} \cos t, \; y = \sqrt{\frac{5}{3}} \sin t$
Second curve: $x = \sqrt[3]{t^2}, \; y = t$

Tangents at $(1, 1)$: $\quad x = 1 + 3t, \; y = 1 - 2t$
$\qquad\qquad\qquad\quad x = 1 + 2t, \; y = 1 + 3t$
Tangents at $(1, -1)$: $\; x = 1 + 3t, \; y = -1 + 2t$
$\qquad\qquad\qquad\quad x = 1 + 2t, \; y = -1 - 3t$

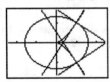

$[-2.4, 2.4]$ by $[-1.6, 1.6]$

45. Acceleration $= \frac{dv}{dt} = \frac{d}{dt}[8(s - t)^{1/2} + 1]$
$= 4(s - t)^{-1/2}\left(\frac{ds}{dt} - 1\right)$
$= 4(s - t)^{-1/2}(v - 1)$
$= 4(s - t)^{-1/2}[(8(s - t)^{1/2} + 1) - 1]$
$= 32(s - t)^{-1/2}(s - t)^{1/2}$
$= 32$ ft/sec^2

47. (a) $\qquad\qquad x^3 + y^3 - 9xy = 0$
$$\frac{d}{dx}(x^3) + \frac{d}{dx}(y^3) - 9\frac{d}{dx}(xy) = \frac{d}{dx}(0)$$
$$3x^2 + 3y^2\frac{dy}{dx} - 9x\frac{dy}{dx} - 9(y)(1) = 0$$
$$(3y^2 - 9x)\frac{dy}{dx} = 9y - 3x^2$$
$$\frac{dy}{dx} = \frac{9y - 3x^2}{3y^2 - 9x} = \frac{3y - x^2}{y^2 - 3x}$$

Slope at $(4, 2)$: $\frac{3(2) - (4)^2}{(2)^2 - 3(4)} = \frac{-10}{-8} = \frac{5}{4}$

Slope at $(2, 4)$: $\frac{3(4) - (2)^2}{(4)^2 - 3(2)} = \frac{8}{10} = \frac{4}{5}$

(b) The tangent is horizontal when
$$\frac{dy}{dx} = \frac{3y - x^2}{y^2 - 3x} = 0, \text{ or } y = \frac{x^2}{3}.$$
Substituting $\frac{x^2}{3}$ for y in the original equation, we have:
$$x^3 + y^3 - 9xy = 0$$
$$x^3 + \left(\frac{x^2}{3}\right)^3 - 9x\left(\frac{x^2}{3}\right) = 0$$
$$x^3 + \frac{x^6}{27} - 3x^3 = 0$$
$$\frac{x^3}{27}(x^3 - 54) = 0$$
$$x = 0 \text{ or } x = \sqrt[3]{54} = 3\sqrt[3]{2}$$

At $x = 0$, we have $y = \frac{0^2}{3} = 0$, which gives the point $(0, 0)$, which is the origin. At $x = 3\sqrt[3]{2}$, we have $y = \frac{1}{3}(3\sqrt[3]{2})^2 = \frac{1}{3}(9\sqrt[3]{4}) = 3\sqrt[3]{4}$, so the point other than the origin is $(3\sqrt[3]{2}, 3\sqrt[3]{4})$ or approximately $(3.780, 4.762)$.

(c) The equation $x^3 + y^3 - 9xy$ is not affected by interchanging x and y, so its graph is symmetric about the line $y = x$ and we may find the desired point by interchanging the x-value and the y-value in the answer to part (b). The desired point is $(3\sqrt[3]{4}, 3\sqrt[3]{2})$ or approximately $(4.762, 3.780)$.

49.
$$xy + 2x - y = 0$$
$$\frac{d}{dx}(xy) + \frac{d}{dx}(2x) - \frac{d}{dx}(y) = 0$$
$$x\frac{dy}{dx} + (y)(1) + 2 - \frac{dy}{dx} = 0$$
$$(x - 1)\frac{dy}{dx} = -2 - y$$
$$\frac{dy}{dx} = \frac{-2 - y}{x - 1} = \frac{2 + y}{1 - x}$$

Since the slope of the line $2x + y = 0$ is -2, we wish to find points where the normal has slope -2, that is, where the tangent has slope $\frac{1}{2}$. Thus, we have

$$\frac{2 + y}{1 - x} = \frac{1}{2}$$
$$2(2 + y) = 1 - x$$
$$4 + 2y = 1 - x$$
$$x = -2y - 3$$

Substituting $-2y - 3$ in the original equation, we have:
$$xy + 2x - y = 0$$
$$(-2y - 3)y + 2(-2y - 3) - y = 0$$
$$-2y^2 - 8y - 6 = 0$$
$$-2(y + 1)(y + 3) = 0$$
$$y = -1 \text{ or } y = -3$$
At $y = -1$, $x = -2y - 3 = 2 - 3 = -1$.
At $y = -3$: $x = -2y - 3 = 6 - 3 = 3$.
The desired points are $(-1, -1)$ and $(3, -3)$.
Finally, we find the desired normals to the curve, which are the lines of slope -2 passing through each of these points. At $(-1, -1)$, the normal line is $y = -2(x + 1) - 1$ or $y = -2x - 3$. At $(3, -3)$, the normal line is $y = -2(x - 3) - 3$ or $y = -2x + 3$.

51. (a)
$$\frac{x^2}{a^2} + \frac{y^2}{b^2} = 1$$
$$b^2 x^2 + a^2 y^2 = a^2 b^2$$
$$\frac{d}{dx}(b^2 x^2) + \frac{d}{dx}(a^2 y^2) = \frac{d}{dx}(a^2 b^2)$$
$$2b^2 x + 2a^2 y \frac{dy}{dx} = 0$$
$$\frac{dy}{dx} = -\frac{2b^2 x}{2a^2 y} = -\frac{b^2 x}{a^2 y}$$

The slope at (x_1, y_1) is $-\frac{b^2 x_1}{a^2 y_1}$.

The tangent line is $y - y_1 = -\frac{b^2 x_1}{a^2 y_1}(x - x_1)$. This gives:

$$a^2 y_1 y - a^2 y_1^2 = -b^2 x_1 x + b^2 x_1^2$$
$$a^2 y_1 y + b^2 x_1 x = a^2 y_1^2 + b^2 x_1^2.$$

But $a^2 y_1^2 + b^2 x_1^2 = a^2 b^2$ since (x_1, y_1) is on the ellipse. Therefore, $a^2 y_1 y + b^2 x_1 x = a^2 b^2$, and dividing by $a^2 b^2$ gives $\frac{x_1 x}{a^2} + \frac{y_1 y}{b^2} = 1$.

(b)
$$\frac{x^2}{a^2} - \frac{y^2}{b^2} = 1$$
$$b^2 x^2 - a^2 y^2 = a^2 b^2$$
$$\frac{d}{dx}(b^2 x^2) - \frac{d}{dx}(a^2 y^2) = \frac{d}{dx}(a^2 b^2)$$
$$2b^2 x - 2a^2 y \frac{dy}{dx} = 0$$
$$\frac{dy}{dx} = \frac{-2b^2 x}{-2a^2 y} = \frac{b^2 x}{a^2 y}$$

The slope at (x_1, y_1) is $\frac{b^2 x_1}{a^2 y_1}$.

The tangent line is $y - y_1 = \frac{b^2 x_1}{a^2 y_1}(x - x_1)$.

This gives:
$$a^2 y_1 y - a^2 y_1^2 = b^2 x_1 x - b^2 x_1^2$$
$$b^2 x_1^2 - a^2 y_1^2 = b^2 x_1 x - a^2 y_1 y$$

But $b^2 x_1^2 - a^2 y_1^2 = a^2 b^2$ since (x_1, y_1) is on the hyperbola. Therefore, $b^2 x_1 x - a^2 y_1 y = a^2 b^2$, and dividing by $a^2 b^2$ gives $\frac{x_1 x}{a^2} - \frac{y_1 y}{b^2} = 1$.

■ Section 3.8 Derivatives of Inverse Trigonometric Functions (pp. 157–163)

Exploration 1 Finding a Derivative on an Inverse Graph Geometrically

1. The graph is shown at the right. It appears to be a one-to-one function

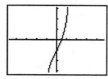

[−4.7, 4.7] by [−3.1, 3.1]

2. $f'(x) = 5x^4 + 2$. The fact that this function is always positive enables us to conclude that f is everywhere increasing, and hence one-to-one.

3. The graph of f^{-1} is shown to the right, along with the graph of f. The graph of f^{-1} is obtained from the graph of f by reflecting it in the line $y = x$.

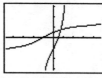

[−4.7, 4.7] by [−3.1, 3.1]

4. The line L is tangent to the graph of f^{-1} at the point $(2, 1)$.

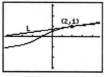

[−4.7, 4.7] by [−3.1, 3.1]

70   Section 3.8

5. The reflection of line L is tangent to the graph of f at the point $(1, 2)$.

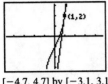

$[-4.7, 4.7]$ by $[-3.1, 3.1]$

6. The reflection of line L is the tangent line to the graph of $y = x^5 + 2x - 1$ at the point $(1, 2)$. The slope is $\frac{dy}{dx}$ at $x = 1$, which is 7.

7. The slope of L is the reciprocal of the slope of its reflection $\left(\text{since } \frac{\Delta y}{\Delta x} \text{ gets reflected to become } \frac{\Delta x}{\Delta y}\right)$. It is $\frac{1}{7}$.

8. $\frac{1}{7}$

Quick Review 3.8

1. Domain: $[-1, 1]$
Range: $\left[-\frac{\pi}{2}, \frac{\pi}{2}\right]$
At 1: $\frac{\pi}{2}$

2. Domain: $[-1, 1]$
Range: $[0, \pi]$
At 1: 0

3. Domain: all reals
Range: $\left(-\frac{\pi}{2}, \frac{\pi}{2}\right)$
At 1: $\frac{\pi}{4}$

4. Domain: $(-\infty, -1] \cup [1, \infty)$
Range: $\left[0, \frac{\pi}{2}\right) \cup \left(\frac{\pi}{2}, \pi\right]$
At 1: 0

5. Domain: all reals
Range: all reals
At 1: 1

6. $f(x) = y = 3x - 8$
$y + 8 = 3x$
$x = \frac{y + 8}{3}$
Interchange x and y:
$y = \frac{x + 8}{3}$
$f^{-1}(x) = \frac{x + 8}{3}$

7. $f(x) = y = \sqrt[3]{x + 5}$
$y^3 = x + 5$
$x = y^3 - 5$
Interchange x and y:
$y = x^3 - 5$
$f^{-1}(x) = x^3 - 5$

8. $f(x) = y = \frac{8}{x}$
$x = \frac{8}{y}$
Interchange x and y:
$y = \frac{8}{x}$
$f^{-1}(x) = \frac{8}{x}$

9. $f(x) = y = \frac{3x - 2}{x}$
$xy = 3x - 2$
$(y - 3)x = -2$
$x = \frac{-2}{y - 3} = \frac{2}{3 - y}$
Interchange x and y:
$y = \frac{2}{3 - x}$
$f^{-1}(x) = \frac{2}{3 - x}$

10. $f(x) = y = \arctan \frac{x}{3}$
$\tan y = \frac{x}{3}, -\frac{\pi}{2} < y < \frac{\pi}{2}$
$x = 3 \tan y, -\frac{\pi}{2} < y < \frac{\pi}{2}$
Interchange x and y:
$y = 3 \tan x, -\frac{\pi}{2} < x < \frac{\pi}{2}$
$f^{-1}(x) = 3 \tan x, -\frac{\pi}{2} < x < \frac{\pi}{2}$

Section 3.8 Exercises

1. $\frac{dy}{dx} = \frac{d}{dx} \cos^{-1}(x^2) = -\frac{1}{\sqrt{1 - (x^2)^2}} \frac{d}{dx}(x^2)$
$= -\frac{1}{\sqrt{1 - x^4}} (2x) = -\frac{2x}{\sqrt{1 - x^4}}$

3. $\frac{dy}{dt} = \frac{d}{dt} \sin^{-1} \sqrt{2}t = \frac{1}{\sqrt{1 - (\sqrt{2}t)^2}} \frac{d}{dt}(\sqrt{2}t) = \frac{\sqrt{2}}{\sqrt{1 - 2t^2}}$

5. $\frac{dy}{ds} = \frac{d}{ds} \sec^{-1}(2s + 1)$
$= \frac{1}{|2s + 1| \sqrt{(2s + 1)^2 - 1}} \frac{d}{ds}(2s + 1)$
$= \frac{1}{|2s + 1| \sqrt{4s^2 + 4s}} (2) = \frac{1}{|2s + 1| \sqrt{s^2 + s}}$

7. $\dfrac{dy}{dx} = \dfrac{d}{dx} \csc^{-1}(x^2 + 1)$

$= -\dfrac{1}{|x^2+1|\sqrt{(x^2+1)^2 - 1}} \dfrac{d}{dx}(x^2+1)$

$= -\dfrac{2x}{(x^2+1)\sqrt{x^4 + 2x^2}} = -\dfrac{2}{(x^2+1)\sqrt{x^2+2}}$

Note that the condition $x > 0$ is required in the last step.

9. $\dfrac{dy}{dt} = \dfrac{d}{dt} \sec^{-1}\left(\dfrac{1}{t}\right) = \dfrac{1}{\left|\dfrac{1}{t}\right|\sqrt{\left(\dfrac{1}{t}\right)^2 - 1}} \dfrac{d}{dt}\left(\dfrac{1}{t}\right)$

$= \dfrac{1}{\left|\dfrac{1}{t}\right|\sqrt{\left(\dfrac{1}{t}\right)^2 - 1}} \left(-\dfrac{1}{t^2}\right) = -\dfrac{1}{\sqrt{1 - t^2}}$

Note that the condition $t > 0$ is required in the last step.

11. $\dfrac{dy}{dt} = \dfrac{d}{dt} \cot^{-1}\sqrt{t} = -\dfrac{1}{1 + (\sqrt{t})^2} \dfrac{d}{dt}\sqrt{t}$

$= -\dfrac{1}{2\sqrt{t}(t+1)}$

13. $\dfrac{dy}{ds} = \dfrac{d}{ds}(s\sqrt{1-s^2}) + \dfrac{d}{ds}(\cos^{-1}s)$

$= (s)\left(\dfrac{1}{2\sqrt{1-s^2}}\right)(-2s) + (\sqrt{1-s^2})(1) - \dfrac{1}{\sqrt{1-s^2}}$

$= -\dfrac{s^2}{\sqrt{1-s^2}} + \sqrt{1-s^2} - \dfrac{1}{\sqrt{1-s^2}}$

$= \dfrac{-s^2 + (1 - s^2) - 1}{\sqrt{1-s^2}}$

$= -\dfrac{2s^2}{\sqrt{1-s^2}}$

15. $\dfrac{dy}{dx} = \dfrac{d}{dx}(\tan^{-1}\sqrt{x^2 - 1}) + \dfrac{d}{dx}(\csc^{-1}x)$

$= \dfrac{1}{1 + (\sqrt{x^2-1})^2} \dfrac{d}{dx}(\sqrt{x^2-1}) - \dfrac{1}{|x|\sqrt{x^2-1}}$

$= \dfrac{1}{x^2} \dfrac{1}{2\sqrt{x^2-1}}(2x) - \dfrac{1}{|x|\sqrt{x^2-1}}$

$= \dfrac{1}{x\sqrt{x^2-1}} - \dfrac{1}{|x|\sqrt{x^2-1}}$

$= 0$

Note that the condition $x > 1$ is required in the last step.

17. $\dfrac{dy}{dx} = \dfrac{d}{dx}(x \sin^{-1}x) + \dfrac{d}{dx}(\sqrt{1-x^2})$

$= (x)\left(\dfrac{1}{\sqrt{1-x^2}}\right) + (\sin^{-1}x)(1) + \dfrac{1}{2\sqrt{1-x^2}}(-2x)$

$= \sin^{-1}x$

19. (a) Since $\dfrac{dy}{dx} = \sec^2 x$, the slope at $\left(\dfrac{\pi}{4}, 1\right)$ is $\sec^2\left(\dfrac{\pi}{4}\right) = 2$.
The tangent line is given by $y = 2\left(x - \dfrac{\pi}{4}\right) + 1$, or
$y = 2x - \dfrac{\pi}{2} + 1$.

(b) Since $\dfrac{dy}{dx} = \dfrac{1}{1+x^2}$, the slope at $\left(1, \dfrac{\pi}{4}\right)$ is $\dfrac{1}{1+1^2} = \dfrac{1}{2}$.
The tangent line is given by $y = \dfrac{1}{2}(x - 1) + \dfrac{\pi}{4}$, or
$y = \dfrac{1}{2}x - \dfrac{1}{2} + \dfrac{\pi}{4}$.

21. (a) Note that $f'(x) = -\sin x + 3$, which is always between 2 and 4. Thus f is differentiable at every point on the interval $(-\infty, \infty)$ and $f'(x)$ is never zero on this interval, so f has a differentiable inverse by Theorem 3.

(b) $f(0) = \cos 0 + 3(0) = 1$;
$f'(0) = -\sin 0 + 3 = 3$

(c) Since the graph of $y = f(x)$ includes the point $(0, 1)$ and the slope of the graph is 3 at this point, the graph of $y = f^{-1}(x)$ will include $(1, 0)$ and the slope will be $\dfrac{1}{3}$. Thus, $f^{-1}(1) = 0$ and $(f^{-1})'(1) = \dfrac{1}{3}$.

23. (a) $v(t) = \dfrac{dx}{dt} = \dfrac{1}{1+t^2}$ which is always positive.

(b) $a(t) = \dfrac{dv}{dt} = -\dfrac{2t}{(1+t^2)^2}$ which is always negative.

(c) $\dfrac{\pi}{2}$

25. $\dfrac{d}{dx} \cot^{-1} x = \dfrac{d}{dx}\left(\dfrac{\pi}{2} - \tan^{-1}(x)\right)$

$= 0 - \dfrac{d}{dx} \tan^{-1}(x)$

$= -\dfrac{1}{1+x^2}$

27. (a) $y = \dfrac{\pi}{2}$

(b) $y = -\dfrac{\pi}{2}$

(c) None, since $\dfrac{d}{dx} \tan^{-1} x = \dfrac{1}{1+x^2} \ne 0$.

29. (a) $y = \dfrac{\pi}{2}$

(b) $y = \dfrac{\pi}{2}$

(c) None, since $\dfrac{d}{dx} \sec^{-1} x = \dfrac{1}{|x|\sqrt{x^2-1}} \ne 0$.

31. (a) None, since $\sin^{-1} x$ is undefined for $x > 1$.

(b) None, since $\sin^{-1} x$ is undefined for $x < -1$.

(c) None, since $\dfrac{d}{dx} \sin^{-1} x = \dfrac{1}{\sqrt{1-x^2}} \ne 0$.

72   Section 3.9

33. (a)

$\alpha = \cos^{-1} x, \beta = \sin^{-1} x$
So $\cos^{-1} x + \sin^{-1} x = \alpha + \beta = \dfrac{\pi}{2}$.

(b)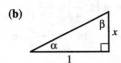

$\alpha = \tan^{-1} x, \beta = \cot^{-1} x$
So $\tan^{-1} x + \cot^{-1} x = \alpha + \beta = \dfrac{\pi}{2}$.

(c)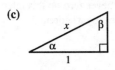

$\alpha = \sec^{-1} x, \beta = \csc^{-1} x$
So $\sec^{-1} x + \csc^{-1} x = \alpha + \beta = \dfrac{\pi}{2}$.

35.

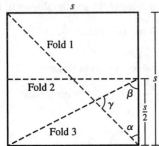

If s is the length of a side of the square, then

$\tan \alpha = \dfrac{s}{s} = 1$, so $\alpha = \tan^{-1} 1$ and

$\tan \beta = \dfrac{s}{\frac{s}{2}} = 2$, so $\beta = \tan^{-1} 2$.

From Exercise 34, we have

$\gamma = \pi - \alpha - \beta = \pi - \tan^{-1} 1 - \tan^{-1} 2 = \tan^{-1} 3$.

■ Section 3.9 Derivatives of Exponential and Logarithmic Functions (pp. 163–171)

Exploration 1   Leaving Milk on the Counter

1. The temperature of the refrigerator is 42°F, the temperature of the milk at time $t = 0$.

2. The temperature of the room is 72°F, the limit to which y tends as t increases.

3. The milk is warming up the fastest at $t = 0$. The second derivative $y'' = -30(\ln(0.98))^2(0.98)^t$ is negative, so y' (the rate at which the milk is warming) is maximized at the lowest value of t.

4. We set $y = 55$ and solve:
$72 - 30(0.98)^t = 55$
$(0.98)^t = \dfrac{17}{30}$
$t \ln (0.98) = \ln \left(\dfrac{17}{30}\right)$
$t = \dfrac{\ln\left(\dfrac{17}{30}\right)}{\ln (0.98)} \approx 28.114$

The milk reaches a temperature of 55°F after about 28 minutes.

5. $\dfrac{dy}{dt} = -30 \ln (0.98) \cdot (0.98)^t$. At $t = \dfrac{\ln\left(\dfrac{17}{30}\right)}{\ln (0.98)}$,
$\dfrac{dy}{dt} \approx 0.343$ degrees/minute.

Quick Review 3.9

1. $\log_5 8 = \dfrac{\ln 8}{\ln 5}$

2. $7^x = e^{\ln 7^x} = e^{x \ln 7}$

3. $\ln (e^{\tan x}) = \tan x$

4. $\ln (x^2 - 4) - \ln (x + 2) = \ln \dfrac{x^2 - 4}{x + 2} = \ln \dfrac{(x + 2)(x - 2)}{x + 2}$
$= \ln (x - 2)$

5. $\log_2 (8^{x-5}) = \log_2 (2^3)^{x-5} = \log_2 2^{3x-15} = 3x - 15$

6. $\dfrac{\log_4 x^{15}}{\log_4 x^{12}} = \dfrac{15 \log_4 x}{12 \log_4 x} = \dfrac{15}{12} = \dfrac{5}{4}, x > 0$

7. $3 \ln x - \ln 3x + \ln (12x^2) = \ln x^3 - \ln 3x + \ln (12x^2)$
$= \ln \dfrac{(x^3)(12x^2)}{3x} = \ln (4x^4)$

8. $\quad 3^x = 19$
$\ln 3^x = \ln 19$
$x \ln 3 = \ln 19$
$x = \dfrac{\ln 19}{\ln 3} \approx 2.68$

9. $5^t \ln 5 = 18$
$5^t = \dfrac{18}{\ln 5}$
$\ln 5^t = \ln \dfrac{18}{\ln 5}$
$t \ln 5 = \ln 18 - \ln (\ln 5)$
$t = \dfrac{\ln 18 - \ln (\ln 5)}{\ln 5} \approx 1.50$

10. $\quad 3^{x+1} = 2^x$
$\ln 3^{x+1} = \ln 2^x$
$(x + 1) \ln 3 = x \ln 2$
$x(\ln 3 - \ln 2) = -\ln 3$
$x = \dfrac{\ln 3}{\ln 2 - \ln 3} \approx -2.71$

Section 3.9 Exercises

1. $\dfrac{dy}{dx} = \dfrac{d}{dx}(2e^x) = 2e^x$

3. $\dfrac{dy}{dx} = \dfrac{d}{dx}e^{-x} = e^{-x}\dfrac{d}{dx}(-x) = -e^{-x}$

5. $\dfrac{dy}{dx} = \dfrac{d}{dx}e^{2x/3} = e^{2x/3}\dfrac{d}{dx}\left(\dfrac{2x}{3}\right) = \dfrac{2}{3}e^{2x/3}$

7. $\dfrac{dy}{dx} = \dfrac{d}{dx}(xe^2) - \dfrac{d}{dx}(e^x) = e^2 - e^x$

9. $\dfrac{dy}{dx} = \dfrac{d}{dx}e^{\sqrt{x}} = e^{\sqrt{x}}\dfrac{d}{dx}(\sqrt{x}) = \dfrac{e^{\sqrt{x}}}{2\sqrt{x}}$

11. $\dfrac{dy}{dx} = \dfrac{d}{dx}(x^\pi) = \pi x^{\pi-1}$

13. $\dfrac{dy}{dx} = \dfrac{d}{dx}x^{-\sqrt{2}} = -\sqrt{2}x^{-\sqrt{2}-1}$

15. $\dfrac{dy}{dx} = \dfrac{d}{dx}8^x = 8^x \ln 8$

17. $\dfrac{dy}{dx} = \dfrac{d}{dx}3^{\csc x} = 3^{\csc x}(\ln 3)\dfrac{d}{dx}(\csc x)$
$= 3^{\csc x}(\ln 3)(-\csc x \cot x)$
$= -3^{\csc x}(\ln 3)(\csc x \cot x)$

19. Use logarithmic differentiation.
$y = x^{\ln x}$
$\ln y = \ln x^{\ln x}$
$\ln y = \ln x \ln x$
$\dfrac{d}{dx}(\ln y) = \dfrac{d}{dx}(\ln x)^2$
$\dfrac{1}{y}\dfrac{dy}{dx} = (2\ln x)\left(\dfrac{1}{x}\right)$
$\dfrac{dy}{dx} = \dfrac{2y \ln x}{x}$
$\dfrac{dy}{dx} = \dfrac{2x^{\ln x} \ln x}{x}$

21. $\dfrac{dy}{dx} = \dfrac{d}{dx}\ln(x^2) = \dfrac{1}{x^2}\dfrac{d}{dx}(x^2) = \dfrac{1}{x^2}(2x) = \dfrac{2}{x}$

23. $\dfrac{dy}{dx} = \dfrac{d}{dx}\ln(x^{-1}) = \dfrac{d}{dx}(-\ln x) = -\dfrac{1}{x}, x > 0$

25. $\dfrac{dy}{dx} = \dfrac{d}{dx}\ln(x+2) = \dfrac{1}{x+2}\dfrac{d}{dx}(x+2) = \dfrac{1}{x+2}, x > -2$

27. $\dfrac{dy}{dx} = \dfrac{d}{dx}\ln(2 - \cos x) = \dfrac{1}{2 - \cos x}\dfrac{d}{dx}(2 - \cos x)$
$= \dfrac{\sin x}{2 - \cos x}$

29. $\dfrac{d}{dx}\ln(\ln x) = \dfrac{1}{\ln x}\dfrac{d}{dx}\ln x = \dfrac{1}{\ln x}\cdot\dfrac{1}{x} = \dfrac{1}{x \ln x}$

31. $\dfrac{dy}{dx} = \dfrac{d}{dx}(\log_4 x^2) = \dfrac{d}{dx}\dfrac{\ln x^2}{\ln 4} = \dfrac{d}{dx}\left[\left(\dfrac{2}{\ln 4}\right)(\ln x)\right]$
$= \dfrac{2}{\ln 4}\cdot\dfrac{1}{x} = \dfrac{2}{x \ln 4} = \dfrac{1}{x \ln 2}$

33. $\dfrac{dy}{dx} = \dfrac{d}{dx}\log_2(3x+1) = \dfrac{1}{(3x+1)\ln 2}\dfrac{d}{dx}(3x+1)$
$= \dfrac{3}{(3x+1)\ln 2}, x > -\dfrac{1}{3}$

35. $\dfrac{dy}{dx} = \dfrac{d}{dx}\log_2\left(\dfrac{1}{x}\right) = \dfrac{d}{dx}(-\log_2 x) = -\dfrac{1}{x \ln 2}, x > 0$

37. $\dfrac{dy}{dx} = \dfrac{d}{dx}(\ln 2 \cdot \log_2 x) = (\ln 2)\dfrac{d}{dx}(\log_2 x)$
$= (\ln 2)\left(\dfrac{1}{x \ln 2}\right) = \dfrac{1}{x}, x > 0$

39. $\dfrac{dy}{dx} = \dfrac{d}{dx}(\log_{10} e^x) = \dfrac{d}{dx}(x \log_{10} e) = \log_{10} e = \dfrac{\ln e}{\ln 10}$
$= \dfrac{1}{\ln 10}$

41. The line passes through (a, e^a) for some value of a and has slope $m = e^a$. Since the line also passes through the origin, the slope is also given by $m = \dfrac{e^a - 0}{a - 0}$ and we have
$e^a = \dfrac{e^a}{a}$, so $a = 1$. Hence, the slope is e and the equation is
$y = ex$.

43.
$y = (\sin x)^x$
$\ln y = \ln (\sin x)^x$
$\ln y = x \ln (\sin x)$
$\dfrac{d}{dx}\ln y = \dfrac{d}{dx}[x \ln (\sin x)]$
$\dfrac{1}{y}\dfrac{dy}{dx} = (x)\left(\dfrac{1}{\sin x}\right)(\cos x) + \ln (\sin x)(1)$
$\dfrac{dy}{dx} = y[x \cot x + \ln (\sin x)]$
$\dfrac{dy}{dx} = (\sin x)^x[x \cot x + \ln (\sin x)]$

45.
$y = \sqrt[5]{\dfrac{(x-3)^4(x^2+1)}{(2x+5)^3}} = \left(\dfrac{(x-3)^4(x^2+1)}{(2x+5)^3}\right)^{1/5}$
$\ln y = \ln\left(\dfrac{(x-3)^4(x^2+1)}{(2x+5)^3}\right)^{1/5}$
$\ln y = \dfrac{1}{5}\ln\dfrac{(x-3)^4(x^2+1)}{(2x+5)^3}$
$\ln y = \dfrac{1}{5}[4 \ln (x-3) + \ln (x^2+1) - 3 \ln (2x+5)]$
$\dfrac{d}{dx}(\ln y) = \dfrac{4}{5}\dfrac{d}{dx}\ln(x-3) +$
$\dfrac{1}{5}\dfrac{d}{dx}\ln(x^2+1) - \dfrac{3}{5}\dfrac{d}{dx}\ln(2x+5)$
$\dfrac{1}{y}\dfrac{dy}{dx} = \dfrac{4}{5}\dfrac{1}{x-3} + \dfrac{1}{5}\dfrac{1}{x^2+1}(2x) - \dfrac{3}{5}\dfrac{1}{2x+5}(2)$
$\dfrac{dy}{dx} = y\left(\dfrac{4}{5(x-3)} + \dfrac{2x}{5(x^2+1)} - \dfrac{6}{5(2x+5)}\right)$
$\dfrac{dy}{dx} = \left(\dfrac{(x-3)^4(x^2+1)}{(2x+5)^3}\right)^{1/5}\cdot$
$\left(\dfrac{4}{5(x-3)} + \dfrac{2x}{5(x^2+1)} - \dfrac{6}{5(2x+5)}\right)$

47. $\dfrac{dA}{dt} = 20\dfrac{d}{dt}\left(\dfrac{1}{2}\right)^{t/140}$

$= 20\dfrac{d}{dt}2^{-t/140}$

$= 20\,(2^{-t/140})(\ln 2)\dfrac{d}{dt}\left(-\dfrac{t}{140}\right)$

$= 20(2^{-t/140})(\ln 2)\left(-\dfrac{1}{140}\right)$

$= -\dfrac{(2^{-t/140})(\ln 2)}{7}$

At $t = 2$ days, we have

$\dfrac{dA}{dt} = -\dfrac{(2^{-1/70})(\ln 2)}{7} \approx -0.098$ grams/day.

This means that the rate of *decay* is the positive rate of approximately 0.098 grams/day.

49. (a) Since $f'(x) = 2^x \ln 2, f'(0) = 2^0 \ln 2 = \ln 2$.

(b) $f'(0) = \lim\limits_{h \to 0}\dfrac{f(h) - f(0)}{h} = \lim\limits_{h \to 0}\dfrac{2^h - 2^0}{h} = \lim\limits_{h \to 0}\dfrac{2^h - 1}{h}$

(c) Since quantities in parts (a) and (b) are equal,

$\lim\limits_{h \to 0}\dfrac{2^h - 1}{h} = \ln 2.$

(d) By following the same procedure as above using $g(x) = 7^x$, we may see that $\lim\limits_{h \to 0}\dfrac{7^h - 1}{h} = \ln 7.$

51. (a) The graph y_4 is a horizontal line at $y = a$.

(b) The graph of y_3 is always a horizontal line.

a	2	3	4	5
y_3	0.693147	1.098613	1.386295	1.609439
ln a	0.693147	1.098612	1.386294	1.609438

We conclude that the graph of y_3 is a horizontal line at $y = \ln a$.

(c) $\dfrac{d}{dx}a^x = a^x$ if and only if $y_3 = \dfrac{y_2}{y_1} = 1$.

So if $y_3 = \ln a$, then $\dfrac{d}{dx}a^x$ will equal a^x if and only if $\ln a = 1$, or $a = e$.

(d) $y_2 = \dfrac{d}{dx}a^x = a^x \ln a$. This will equal $y_1 = a^x$ if and only if $\ln a = 1$, or $a = e$.

53. (a) Since the line passes through the origin and has slope $\dfrac{1}{e}$, its equation is $y = \dfrac{x}{e}$.

(b) The graph of $y = \ln x$ lies below the graph of the line $y = \dfrac{x}{e}$ for all positive $x \ne e$. Therefore, $\ln x < \dfrac{x}{e}$ for all positive $x \ne e$.

(c) Multiplying by e, $e \ln x < x$ or $\ln x^e < x$.

(d) Exponentiating both sides of $\ln x^e < x$, we have $e^{\ln x^e} < e^x$, or $x^e < e^x$ for all positive $x \ne e$.

(e) Let $x = \pi$ to see that $\pi^e < e^\pi$. Therefore, e^π is bigger.

■ Chapter 3 Review Exercises
(pp. 172–175)

1. $\dfrac{dy}{dx} = \dfrac{d}{dx}\left(x^5 - \dfrac{1}{8}x^2 + \dfrac{1}{4}x\right) = 5x^4 - \dfrac{1}{4}x + \dfrac{1}{4}$

2. $\dfrac{dy}{dx} = \dfrac{d}{dx}(3 - 7x^3 + 3x^7) = -21x^2 + 21x^6$

3. $\dfrac{dy}{dx} = \dfrac{d}{dx}(2 \sin x \cos x)$

$= 2(\sin x)\dfrac{d}{dx}(\cos x) + 2(\cos x)\dfrac{d}{dx}(\sin x)$

$= -2\sin^2 x + 2\cos^2 x$

Alternate solution:

$\dfrac{dy}{dx} = \dfrac{d}{dx}(2 \sin x \cos x) = \dfrac{d}{dx}\sin 2x = (\cos 2x)(2)$

$= 2\cos 2x$

4. $\dfrac{dy}{dx} = \dfrac{d}{dx}\dfrac{2x+1}{2x-1} = \dfrac{(2x-1)(2) - (2x+1)(2)}{(2x-1)^2} = -\dfrac{4}{(2x-1)^2}$

5. $\dfrac{ds}{dt} = \dfrac{d}{dt}\cos(1 - 2t) = -\sin(1 - 2t)(-2) = 2\sin(1 - 2t)$

6. $\dfrac{ds}{dt} = \dfrac{d}{dt}\cot\left(\dfrac{2}{t}\right) = -\csc^2\left(\dfrac{2}{t}\right)\dfrac{d}{dt}\left(\dfrac{2}{t}\right) = -\csc^2\left(\dfrac{2}{t}\right)\left(-\dfrac{2}{t^2}\right)$

$= \dfrac{2}{t^2}\csc^2\left(\dfrac{2}{t}\right)$

7. $\dfrac{dy}{dx} = \dfrac{d}{dx}\left(\sqrt{x} + 1 + \dfrac{1}{\sqrt{x}}\right) = \dfrac{d}{dx}(x^{1/2} + 1 + x^{-1/2})$

$= \dfrac{1}{2}x^{-1/2} - \dfrac{1}{2}x^{-3/2} = \dfrac{1}{2\sqrt{x}} - \dfrac{1}{2x^{3/2}}$

8. $\dfrac{dy}{dx} = \dfrac{d}{dx}(x\sqrt{2x+1}) = (x)\left(\dfrac{1}{2\sqrt{2x+1}}\right)(2) + (\sqrt{2x+1})(1)$

$= \dfrac{x + (2x+1)}{\sqrt{2x+1}} = \dfrac{3x+1}{\sqrt{2x+1}}$

9. $\dfrac{dr}{d\theta} = \dfrac{d}{d\theta}\sec(1 + 3\theta) = \sec(1 + 3\theta)\tan(1 + 3\theta)(3)$

$= 3\sec(1 + 3\theta)\tan(1 + 3\theta)$

10. $\dfrac{dr}{d\theta} = \dfrac{d}{d\theta} \tan^2 (3 - \theta^2)$
$= 2 \tan (3 - \theta^2) \dfrac{d}{d\theta} \tan (3 - \theta^2)$
$= 2 \tan (3 - \theta^2) \sec^2 (3 - \theta^2)(-2\theta)$
$= -4\theta \tan (3 - \theta^2) \sec^2 (3 - \theta^2)$

11. $\dfrac{dy}{dx} = \dfrac{d}{dx}(x^2 \csc 5x)$
$= (x^2)(-\csc 5x \cot 5x)(5) + (\csc 5x)(2x)$
$= -5x^2 \csc 5x \cot 5x + 2x \csc 5x$

12. $\dfrac{dy}{dx} = \dfrac{d}{dx} \ln \sqrt{x} = \dfrac{1}{\sqrt{x}} \dfrac{d}{dx} \sqrt{x} = \dfrac{1}{\sqrt{x}} \cdot \dfrac{1}{2\sqrt{x}} = \dfrac{1}{2x},\; x > 0$

13. $\dfrac{dy}{dx} = \dfrac{d}{dx} \ln (1 + e^x) = \dfrac{1}{1 + e^x} \dfrac{d}{dx}(1 + e^x) = \dfrac{e^x}{1 + e^x}$

14. $\dfrac{dy}{dx} = \dfrac{d}{dx}(xe^{-x}) = (x)(e^{-x})(-1) + (e^{-x})(1) = -xe^{-x} + e^{-x}$

15. $\dfrac{dy}{dx} = \dfrac{d}{dx}(e^{1+\ln x}) = \dfrac{d}{dx}(e^1 e^{\ln x}) = \dfrac{d}{dx}(ex) = e$

16. $\dfrac{dy}{dx} = \dfrac{d}{dx} \ln (\sin x) = \dfrac{1}{\sin x} \dfrac{d}{dx}(\sin x) = \dfrac{\cos x}{\sin x} = \cot x$, for values of x in the intervals $(k\pi, (k+1)\pi)$, where k is even.

17. $\dfrac{dr}{dx} = \dfrac{d}{dx} \ln (\cos^{-1} x) = \dfrac{1}{\cos^{-1} x} \dfrac{d}{dx} \cos^{-1} x$
$= \dfrac{1}{\cos^{-1} x} \left(-\dfrac{1}{\sqrt{1-x^2}}\right) = -\dfrac{1}{\cos^{-1} x \sqrt{1-x^2}}$

18. $\dfrac{dr}{d\theta} = \dfrac{d}{d\theta} \log_2 (\theta^2) = \dfrac{1}{\theta^2 \ln 2} \dfrac{d}{d\theta}(\theta^2) = \dfrac{2\theta}{\theta^2 \ln 2} = \dfrac{2}{\theta \ln 2}$

19. $\dfrac{ds}{dt} = \dfrac{d}{dt} \log_5 (t - 7) = \dfrac{1}{(t-7) \ln 5} \dfrac{d}{dt}(t - 7) = \dfrac{1}{(t-7) \ln 5}$, $t > 7$

20. $\dfrac{ds}{dt} = \dfrac{d}{dt}(8^{-t}) = 8^{-t}(\ln 8)\dfrac{d}{dt}(-t) = -8^{-t} \ln 8$

21. Use logarithmic differentiation.
$y = x^{\ln x}$
$\ln y = \ln (x^{\ln x})$
$\ln y = (\ln x)(\ln x)$
$\dfrac{d}{dx} \ln y = \dfrac{d}{dx}(\ln x)^2$
$\dfrac{1}{y} \dfrac{dy}{dx} = 2 \ln x \dfrac{d}{dx} \ln x$
$\dfrac{dy}{dx} = \dfrac{2y \ln x}{x}$
$\dfrac{dy}{dx} = \dfrac{2x^{\ln x} \ln x}{x}$

22. $\dfrac{dy}{dx} = \dfrac{d}{dx} \dfrac{(2x)2^x}{\sqrt{x^2+1}}$
$= \dfrac{\sqrt{x^2+1}\dfrac{d}{dx}[(2x)2^x] - (2x)(2^x)\dfrac{d}{dx}\sqrt{x^2+1}}{x^2+1}$
$= \dfrac{\sqrt{x^2+1}[(2x)(2^x)(\ln 2) + (2^x)(2)] - (2x)(2^x)\dfrac{1}{2\sqrt{x^2+1}}(2x)}{x^2+1}$
$= \dfrac{(x^2+1)(2^x)(2x \ln 2 + 2) - 2x^2(2^x)}{(x^2+1)^{3/2}}$
$= \dfrac{(2 \cdot 2^x)[(x^2+1)(x \ln 2 + 1) - x^2]}{(x^2+1)^{3/2}}$
$= \dfrac{(2 \cdot 2^x)(x^3 \ln 2 + x^2 + x \ln 2 + 1 - x^2)}{(x^2+1)^{3/2}}$
$= \dfrac{(2 \cdot 2^x)(x^3 \ln 2 + x \ln 2 + 1)}{(x^2+1)^{3/2}}$

Alternate solution, using logarithmic differentiation:
$y = \dfrac{(2x)2^x}{\sqrt{x^2+1}}$
$\ln y = \ln (2x) + \ln (2^x) - \ln \sqrt{x^2+1}$
$\ln y = \ln 2 + \ln x + x \ln 2 - \dfrac{1}{2} \ln (x^2+1)$
$\dfrac{d}{dx} \ln y = \dfrac{d}{dx}[\ln 2 + \ln x + x \ln 2 - \dfrac{1}{2} \ln (x^2+1)]$
$\dfrac{1}{y} \dfrac{dy}{dx} = 0 + \dfrac{1}{x} + \ln 2 - \dfrac{1}{2} \dfrac{1}{x^2+1}(2x)$
$\dfrac{dy}{dx} = y\left(\dfrac{1}{x} + \ln 2 - \dfrac{x}{x^2+1}\right)$
$\dfrac{dy}{dx} = \dfrac{(2x)2^x}{\sqrt{x^2+1}}\left(\dfrac{1}{x} + \ln 2 - \dfrac{x}{x^2+1}\right)$

23. $\dfrac{dy}{dx} = \dfrac{d}{dx} e^{\tan^{-1}x} = e^{\tan^{-1}x} \dfrac{d}{dx} \tan^{-1} x = \dfrac{e^{\tan^{-1}x}}{1+x^2}$

24. $\dfrac{dy}{du} = \dfrac{d}{du} \sin^{-1} \sqrt{1-u^2}$
$= \dfrac{1}{\sqrt{1-(\sqrt{1-u^2})^2}} \dfrac{d}{du} \sqrt{1-u^2}$
$= \dfrac{1}{\sqrt{u^2}} \dfrac{1}{2\sqrt{1-u^2}}(-2u) = -\dfrac{u}{|u|\sqrt{1-u^2}}$

25. $\dfrac{dy}{dt} = \dfrac{d}{dt}(t \sec^{-1} t - \dfrac{1}{2} \ln t)$
$= (t)\left(\dfrac{1}{|t|\sqrt{t^2-1}}\right) + (\sec^{-1} t)(1) - \dfrac{1}{2t}$
$= \dfrac{t}{|t|\sqrt{t^2-1}} + \sec^{-1} t - \dfrac{1}{2t}$

26. $\dfrac{dy}{dt} = \dfrac{d}{dt}[(1+t^2)\cot^{-1} 2t]$
$= (1+t^2)\left(-\dfrac{1}{1+(2t)^2}\right)(2) + (\cot^{-1} 2t)(2t)$
$= -\dfrac{2+2t^2}{1+4t^2} + 2t \cot^{-1} 2t$

27. $\dfrac{dy}{dz} = \dfrac{d}{dz}(z\cos^{-1} z - \sqrt{1-z^2})$

$= (z)\left(-\dfrac{1}{\sqrt{1-z^2}}\right) + (\cos^{-1} z)(1) - \dfrac{1}{2\sqrt{1-z^2}}(-2z)$

$= -\dfrac{z}{\sqrt{1-z^2}} + \cos^{-1} z + \dfrac{z}{\sqrt{1-z^2}}$

$= \cos^{-1} z$

28. $\dfrac{dy}{dx} = \dfrac{d}{dx}(2\sqrt{x-1}\,\csc^{-1}\sqrt{x})$

$= (2\sqrt{x-1})\left(-\dfrac{1}{|\sqrt{x}|\sqrt{(\sqrt{x})^2 - 1}}\right)\left(\dfrac{1}{2\sqrt{x}}\right)$

$\quad + (2\csc^{-1}\sqrt{x})\left(\dfrac{1}{2\sqrt{x-1}}\right)$

$= -\dfrac{\sqrt{x-1}}{(\sqrt{x})^2\sqrt{x-1}} + \dfrac{\csc^{-1}\sqrt{x}}{\sqrt{x-1}}$

$= -\dfrac{1}{x} + \dfrac{\csc^{-1}\sqrt{x}}{\sqrt{x-1}}$

29. $\dfrac{dy}{dx} = \dfrac{d}{dx}\csc^{-1}(\sec x)$

$= \left(-\dfrac{1}{|\sec x|\sqrt{\sec^2 x - 1}}\right)\dfrac{d}{dx}(\sec x)$

$= -\dfrac{1}{|\sec x|\sqrt{\tan^2 x}}\sec x \tan x$

$= -\dfrac{\sec x \tan x}{|\sec x \tan x|}$

$= -\dfrac{\frac{1}{\cos x}\frac{\sin x}{\cos x}}{\left|\frac{1}{\cos x}\frac{\sin x}{\cos x}\right|} = -\dfrac{\sin x}{|\sin x|}$

$= -\operatorname{sign}(\sin x),\ x \ne \dfrac{\pi}{2}, \pi, \dfrac{3\pi}{2}$

Alternate method:

On the domain $0 \le x \le 2\pi$, $x \ne \dfrac{\pi}{2}$, $x \ne \dfrac{3\pi}{2}$, we may rewrite the function as follows:

$y = \csc^{-1}(\sec x)$

$= \dfrac{\pi}{2} - \sec^{-1}(\sec x)$

$= \dfrac{\pi}{2} - \cos^{-1}(\cos x)$

$= \begin{cases} \dfrac{\pi}{2} - x, & 0 \le x \le \pi,\ x \ne \dfrac{\pi}{2} \\ \dfrac{\pi}{2} - (\pi - x), & \pi < x \le 2\pi,\ x \ne \dfrac{3\pi}{2} \end{cases}$

$= \begin{cases} \dfrac{\pi}{2} - x, & 0 \le x \le \pi,\ x \ne \dfrac{\pi}{2} \\ -\dfrac{\pi}{2} + x, & \pi < x \le 2\pi,\ x \ne \dfrac{3\pi}{2} \end{cases}$

Therefore, $\dfrac{dy}{dx} = \begin{cases} -1, & 0 \le x < \pi,\ x \ne \dfrac{\pi}{2} \\ 1, & \pi < x \le 2\pi,\ x \ne \dfrac{\pi}{2} \end{cases}$

Note that the derivative exists at 0 and 2π only because these are the endpoints of the given domain; the two-sided derivative of $y = \csc^{-1}(\sec x)$ does not exist at these points.

30. $\dfrac{dr}{d\theta} = \dfrac{d}{d\theta}\left(\dfrac{1 + \sin\theta}{1 - \cos\theta}\right)^2$

$= 2\left(\dfrac{1 + \sin\theta}{1 - \cos\theta}\right)\left(\dfrac{(1 - \cos\theta)(\cos\theta) - (1 + \sin\theta)(\sin\theta)}{(1 - \cos\theta)^2}\right)$

$= 2\left(\dfrac{1 + \sin\theta}{1 - \cos\theta}\right)\left(\dfrac{\cos\theta - \cos^2\theta - \sin\theta - \sin^2\theta}{(1 - \cos\theta)^2}\right)$

$= 2\left(\dfrac{1 + \sin\theta}{1 - \cos\theta}\right)\left(\dfrac{\cos\theta - \sin\theta - 1}{(1 - \cos\theta)^2}\right)$

31. Since $y = \ln x^2$ is defined for all $x \ne 0$ and

$\dfrac{dy}{dx} = \dfrac{1}{x^2}\dfrac{d}{dx}(x^2) = \dfrac{2x}{x^2} = \dfrac{2}{x}$, the function is differentiable for all $x \ne 0$.

32. Since $y = \sin x - x\cos x$ is defined for all real x and

$\dfrac{dy}{dx} = \cos x - (x)(-\sin x) - (\cos x)(1) = x\sin x$, the function is differentiable for all real x.

33. Since $y = \sqrt{\dfrac{1-x}{1+x^2}}$ is defined for all $x < 1$ and

$\dfrac{dy}{dx} = \dfrac{1}{2\sqrt{\dfrac{1-x}{1+x^2}}} \cdot \dfrac{(1 + x^2)(-1) - (1 - x)(2x)}{(1 + x^2)^2}$

$= \dfrac{x^2 - 2x - 1}{2\sqrt{1-x}\,(1 + x^2)^{3/2}}$, which is defined only for $x < 1$,

the function is differentiable for all $x < 1$.

34. Since $y = (2x - 7)^{-1}(x + 5) = \dfrac{x + 5}{2x - 7}$ is defined for all

$x \ne \dfrac{7}{2}$ and $\dfrac{dy}{dx} = \dfrac{(2x - 7)(1) - (x + 5)(2)}{(2x - 7)^2} = -\dfrac{17}{(2x - 7)^2}$, the

function is differentiable for all $x \ne \dfrac{7}{2}$.

35. Use implicit differentiation.
$$xy + 2x + 3y = 1$$
$$\frac{d}{dx}(xy) + \frac{d}{dx}(2x) + \frac{d}{dx}(3y) = \frac{d}{dx}(1)$$
$$x\frac{dy}{dx} + (y)(1) + 2 + 3\frac{dy}{dx} = 0$$
$$(x+3)\frac{dy}{dx} = -(y+2)$$
$$\frac{dy}{dx} = -\frac{y+2}{x+3}$$

36. Use implicit differentiation.
$$5x^{4/5} + 10y^{6/5} = 15$$
$$\frac{d}{dx}(5x^{4/5}) + \frac{d}{dx}(10y^{6/5}) = \frac{d}{dx}(15)$$
$$4x^{-1/5} + 12y^{1/5}\frac{dy}{dx} = 0$$
$$\frac{dy}{dx} = -\frac{4x^{-1/5}}{12y^{1/5}} = -\frac{1}{3(xy)^{1/5}}$$

37. Use implicit differentiation.
$$\sqrt{xy} = 1$$
$$\frac{d}{dx}\sqrt{xy} = \frac{d}{dx}(1)$$
$$\frac{1}{2\sqrt{xy}}\left[x\frac{dy}{dx} + (y)(1)\right] = 0$$
$$x\frac{dy}{dx} + y = 0$$
$$\frac{dy}{dx} = -\frac{y}{x}$$

Alternate method:

Since $\sqrt{xy} = 1$, we have $xy = 1$ and $y = \frac{1}{x}$.

Therefore, $\frac{dy}{dx} = -\frac{1}{x^2}$.

38. Use implicit differentiation.
$$y^2 = \frac{x}{x+1}$$
$$\frac{d}{dx}y^2 = \frac{d}{dx}\frac{x}{x+1}$$
$$2y\frac{dy}{dx} = \frac{(x+1)(1) - (x)(1)}{(x+1)^2}$$
$$\frac{dy}{dx} = \frac{1}{2y(x+1)^2}$$

39.
$$x^3 + y^3 = 1$$
$$\frac{d}{dx}(x^3) + \frac{d}{dx}(y^3) = \frac{d}{dx}(1)$$
$$3x^2 + 3y^2 y' = 0$$
$$y' = -\frac{x^2}{y^2}$$
$$y'' = \frac{d}{dx}\left(-\frac{x^2}{y^2}\right)$$
$$= -\frac{(y^2)(2x) - (x^2)(2y)(y')}{y^4}$$
$$= -\frac{(y^2)(2x) - (x^2)(2y)\left(-\frac{x^2}{y^2}\right)}{y^4}$$
$$= -\frac{2xy^3 + 2x^4}{y^5}$$
$$= -\frac{2x(x^3 + y^3)}{y^5}$$
$$= -\frac{2x}{y^5}$$

since $x^3 + y^3 = 1$

40.
$$y^2 = 1 - \frac{2}{x}$$
$$\frac{d}{dx}(y^2) = \frac{d}{dx}(1) - \frac{d}{dx}\left(\frac{2}{x}\right)$$
$$2yy' = \frac{2}{x^2}$$
$$y' = \frac{2}{x^2(2y)} = \frac{1}{x^2 y}$$
$$y'' = \frac{d}{dx}\left(\frac{1}{x^2 y}\right)$$
$$= -\frac{1}{(x^2 y)^2}\frac{d}{dx}(x^2 y)$$
$$= -\frac{1}{(x^2 y)^2}[(x^2)(y') + (y)(2x)]$$
$$= -\frac{1}{(x^2 y)^2}\left[(x^2)\left(\frac{1}{x^2 y}\right) + 2xy\right]$$
$$= -\frac{1}{x^4 y^2}\left(\frac{1}{y} + 2xy\right)$$
$$= -\frac{1 + 2xy^2}{x^4 y^3}$$

Chapter 3 Review

41.
$$y^3 + y = 2\cos x$$

$$\frac{d}{dx}(y^3) + \frac{d}{dx}(y) = \frac{d}{dx}(2\cos x)$$

$$3y^2 y' + y' = -2\sin x$$

$$(3y^2 + 1)y' = -2\sin x$$

$$y' = -\frac{2\sin x}{3y^2 + 1}$$

$$y'' = \frac{d}{dx}\left(-\frac{2\sin x}{3y^2 + 1}\right)$$

$$= -\frac{(3y^2 + 1)(2\cos x) - (2\sin x)(6yy')}{(3y^2 + 1)^2}$$

$$= -\frac{(3y^2 + 1)(2\cos x) - (12y\sin x)\left(-\frac{2\sin x}{3y^2 + 1}\right)}{(3y^2 + 1)^2}$$

$$= -2\frac{(3y^2 + 1)^2 \cos x + 12y\sin^2 x}{(3y^2 + 1)^3}$$

42.
$$x^{1/3} + y^{1/3} = 4$$

$$\frac{d}{dx}(x^{1/3}) + \frac{d}{dx}(y^{1/3}) = \frac{d}{dx}(4)$$

$$\frac{1}{3}x^{-2/3} + \frac{1}{3}y^{-2/3} y' = 0$$

$$y' = -\frac{x^{-2/3}}{y^{-2/3}} = -\left(\frac{y}{x}\right)^{2/3}$$

$$y'' = \frac{d}{dx}\left[-\left(\frac{y}{x}\right)^{2/3}\right]$$

$$= -\frac{2}{3}\left(\frac{y}{x}\right)^{-1/3}\left(\frac{xy' - (y)(1)}{x^2}\right)$$

$$= -\frac{2}{3}\left(\frac{y}{x}\right)^{-1/3}\left(\frac{(x)\left[-\left(\frac{y}{x}\right)^{2/3}\right] - y}{x^2}\right)$$

$$= -\frac{2}{3}x^{1/3}y^{-1/3}(-x^{-5/3}y^{2/3} - x^{-2}y)$$

$$= \frac{2}{3}x^{-4/3}y^{1/3} + \frac{2}{3}x^{-5/3}y^{2/3}$$

43. $y' = 2x^3 - 3x - 1$,
$y'' = 6x^2 - 3$,
$y''' = 12x$,
$y^{(4)} = 12$, and the rest are all zero.

44. $y' = \dfrac{x^4}{24}$,
$y'' = \dfrac{x^3}{6}$,
$y''' = \dfrac{x^2}{2}$,
$y^{(4)} = x$,
$y^{(5)} = 1$, and the rest are all zero.

45. $\dfrac{dy}{dx} = \dfrac{d}{dx}\sqrt{x^2 - 2x} = \dfrac{1}{2\sqrt{x^2 - 2x}}(2x - 2) = \dfrac{x - 1}{\sqrt{x^2 - 2x}}$

At $x = 3$, we have $y = \sqrt{3^2 - 2(3)} = \sqrt{3}$

and $\dfrac{dy}{dx} = \dfrac{3 - 1}{\sqrt{3^2 - 2(3)}} = \dfrac{2}{\sqrt{3}}$.

(a) Tangent: $y = \dfrac{2}{\sqrt{3}}(x - 3) + \sqrt{3}$ or $y = \dfrac{2}{\sqrt{3}}x - \sqrt{3}$

(b) Normal: $y = -\dfrac{\sqrt{3}}{2}(x - 3) + \sqrt{3}$

or $y = -\dfrac{\sqrt{3}}{2}x + \dfrac{5\sqrt{3}}{2}$

46. $\dfrac{dy}{dx} = \dfrac{d}{dx}(4 + \cot x - 2\csc x)$

$= -\csc^2 x + 2\csc x \cot x$

At $x = \dfrac{\pi}{2}$, we have

$y = 4 + \cot \dfrac{\pi}{2} - 2\csc \dfrac{\pi}{2} = 4 + 0 - 2 = 2$ and

$\dfrac{dy}{dx} = -\csc^2 \dfrac{\pi}{2} + 2\csc \dfrac{\pi}{2}\cot \dfrac{\pi}{2} = -1 + 2(1)(0) = -1$.

(a) Tangent: $y = -1\left(x - \dfrac{\pi}{2}\right) + 2$ or $y = -x + \dfrac{\pi}{2} + 2$

(b) Normal: $y = 1\left(x - \dfrac{\pi}{2}\right) + 2$ or $y = x - \dfrac{\pi}{2} + 2$

47. Use implicit differentiation.

$$x^2 + 2y^2 = 9$$

$$\frac{d}{dx}(x^2) + \frac{d}{dx}(2y^2) = \frac{d}{dx}(9)$$

$$2x + 4y\frac{dy}{dx} = 0$$

$$\frac{dy}{dx} = -\frac{2x}{4y} = -\frac{x}{2y}$$

Slope at $(1, 2)$: $-\dfrac{1}{2(2)} = -\dfrac{1}{4}$

(a) Tangent: $y = -\dfrac{1}{4}(x - 1) + 2$ or $y = -\dfrac{1}{4}x + \dfrac{9}{4}$

(b) Normal: $y = 4(x - 1) + 2$ or $y = 4x - 2$

48. Use implicit differentiation.

$$x + \sqrt{xy} = 6$$

$$\frac{d}{dx}(x) + \frac{d}{dx}(\sqrt{xy}) = \frac{d}{dx}(6)$$

$$1 + \frac{1}{2\sqrt{xy}}\left[(x)\left(\frac{dy}{dx}\right) + (y)(1)\right] = 0$$

$$\frac{x}{2\sqrt{xy}}\frac{dy}{dx} = -1 - \frac{y}{2\sqrt{xy}}$$

$$\frac{dy}{dx} = \frac{2\sqrt{xy}}{x}\left(-1 - \frac{y}{2\sqrt{xy}}\right)$$

$$= -2\sqrt{\frac{y}{x}} - \frac{y}{x}$$

Slope at (4, 1): $-2\sqrt{\frac{1}{4}} - \frac{1}{4} = -\frac{2}{2} - \frac{1}{4} = -\frac{5}{4}$

(a) Tangent: $y = -\frac{5}{4}(x - 4) + 1$ or $y = -\frac{5}{4}x + 6$

(b) Normal: $y = \frac{4}{5}(x - 4) + 1$ or $y = \frac{4}{5}x - \frac{11}{5}$

49. $\frac{dy}{dx} = \frac{\frac{dy}{dt}}{\frac{dx}{dt}} = \frac{-2\sin t}{2\cos t} = -\tan t$

At $t = \frac{3\pi}{4}$, we have $x = 2\sin\frac{3\pi}{4} = \sqrt{2}$,

$y = 2\cos\frac{3\pi}{4} = -\sqrt{2}$, and $\frac{dy}{dx} = -\tan\frac{3\pi}{4} = 1$.

The equation of the tangent line is

$y = 1(x - \sqrt{2}) + (-\sqrt{2})$, or $y = x - 2\sqrt{2}$.

50. $\frac{dy}{dx} = \frac{\frac{dy}{dt}}{\frac{dx}{dt}} = \frac{4\cos t}{-3\sin t} = -\frac{4}{3}\cot t$

At $t = \frac{3\pi}{4}$, we have $x = 3\cos\frac{3\pi}{4} = -\frac{3\sqrt{2}}{2}$,

$y = 4\sin\frac{3\pi}{4} = 2\sqrt{2}$, and $\frac{dy}{dx} = -\frac{4}{3}\cot\frac{3\pi}{4} = \frac{4}{3}$.

The equation of the tangent line is

$y = \frac{4}{3}\left(x + \frac{3\sqrt{2}}{2}\right) + 2\sqrt{2}$, or $y = \frac{4}{3}x + 4\sqrt{2}$.

51. $\frac{dy}{dx} = \frac{\frac{dy}{dt}}{\frac{dx}{dt}} = \frac{5\sec^2 t}{3\sec t \tan t} = \frac{5\sec t}{3\tan t} = \frac{5}{3\sin t}$

At $t = \frac{\pi}{6}$, we have $x = 3\sec\frac{\pi}{6} = 2\sqrt{3}$,

$y = 5\tan\frac{\pi}{6} = \frac{5\sqrt{3}}{3}$, and $\frac{dy}{dx} = \frac{5}{3\sin\left(\frac{\pi}{6}\right)} = \frac{10}{3}$.

The equation of the tangent line is

$y = \frac{10}{3}(x - 2\sqrt{3}) + \frac{5\sqrt{3}}{3}$, or $y = \frac{10}{3}x - 5\sqrt{3}$.

52. $\frac{dy}{dx} = \frac{\frac{dy}{dt}}{\frac{dx}{dt}} = \frac{1 + \cos t}{-\sin t}$

At $t = -\frac{\pi}{4}$, we have $x = \cos\left(-\frac{\pi}{4}\right) = \frac{\sqrt{2}}{2}$,

$y = -\frac{\pi}{4} + \sin\left(-\frac{\pi}{4}\right) = -\frac{\pi}{4} - \frac{\sqrt{2}}{2}$, and

$\frac{dy}{dx} = \frac{1 + \cos\left(-\frac{\pi}{4}\right)}{-\sin\left(-\frac{\pi}{4}\right)} = \frac{1 + \frac{\sqrt{2}}{2}}{\frac{\sqrt{2}}{2}} = \sqrt{2} + 1$.

The equation of the tangent line is

$y = (\sqrt{2} + 1)\left(x - \frac{\sqrt{2}}{2}\right) - \frac{\pi}{4} - \frac{\sqrt{2}}{2}$, or

$y = (1 + \sqrt{2})x - \sqrt{2} - 1 - \frac{\pi}{4}$.

This is approximately $y = 2.414x - 3.200$.

53. (a)

[−1, 3] by $\left[-1, \frac{5}{3}\right]$

(b) Yes, because both of the one-sided limits as $x \to 1$ are equal to $f(1) = 1$.

(c) No, because the left-hand derivative at $x = 1$ is $+1$ and the right-hand derivative at $x = 1$ is -1.

54. (a) The function is continuous for all values of m, because the right-hand limit as $x \to 0$ is equal to $f(0) = 0$ for any value of m.

(b) The left-hand derivative at $x = 0$ is $2\cos(2 \cdot 0) = 2$, and the right-hand derivative at $x = 0$ is m, so in order for the function to be differentiable at $x = 0$, m must be 2.

55. (a) For all $x \neq 0$ (b) At $x = 0$
(c) Nowhere

56. (a) For all x (b) Nowhere
(c) Nowhere

57. Note that $\lim_{x \to 0^-} f(x) = \lim_{x \to 0^-}(2x - 3) = -3$ and

$\lim_{x \to 0^+} f(x) = \lim_{x \to 0^+}(x - 3) = -3$. Since these values agree with $f(0)$, the function is continuous at $x = 0$. On the other hand,

$f'(x) = \begin{cases} 2, & -1 \leq x < 0 \\ 1, & 0 < x \leq 4 \end{cases}$, so the derivative is undefined at $x = 0$.

(a) $[-1, 0) \cup (0, 4]$ (b) At $x = 0$
(c) Nowhere in its domain

80 Chapter 3 Review

58. Note that the function is undefined at $x = 0$.
(a) $[-2, 0) \cup (0, 2]$ (b) Nowhere
(c) Nowhere in its domain

59.
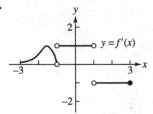

60.
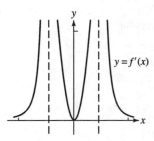

61. (a) iii (b) i
(c) ii

62. The graph passes through $(0, 5)$ and has slope -2 for $x < 2$ and slope -0.5 for $x > 2$.

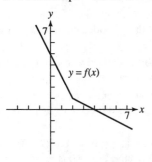

63. The graph passes through $(-1, 2)$ and has slope -2 for $x < 1$, slope 1 for $1 < x < 4$, and slope -1 for $4 < x < 6$.

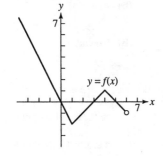

64. i. If $f(x) = \frac{9}{28}x^{7/3} + 9$, then $f'(x) = \frac{3}{4}x^{4/3}$ and $f''(x) = x^{1/3}$, which matches the given equation.

ii. If $f'(x) = \frac{9}{28}x^{7/3} - 2$, then $f''(x) = \frac{3}{4}x^{4/3}$, which contradicts the given equation $f''(x) = x^{1/3}$.

iii. If $f'(x) = \frac{3}{4}x^{4/3} + 6$, then $f''(x) = x^{1/3}$, which matches the given equation.

iv. If $f(x) = \frac{3}{4}x^{4/3} - 4$, then $f'(x) = x^{1/3}$ and $f''(x) = \frac{1}{3}x^{-2/3}$, which contradicts the given equation $f''(x) = x^{1/3}$.

Answer is **D**: **i** and **iii** only could be true. Note, however that **i** and **iii** could not simultaneously be true.

65. (a)
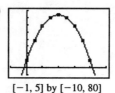
$[-1, 5]$ by $[-10, 80]$

(b)

t interval	avg. vel.
$[0, 0.5]$	$\frac{38 - 10}{0.5 - 0} = 56$
$[0.5, 1]$	$\frac{58 - 38}{1 - 0.5} = 40$
$[1, 1.5]$	$\frac{70 - 58}{1.5 - 1} = 24$
$[1.5, 2]$	$\frac{74 - 70}{2 - 1.5} = 8$
$[2, 2.5]$	$\frac{70 - 74}{2.5 - 2} = -8$
$[2.5, 3]$	$\frac{58 - 70}{3 - 2.5} = -24$
$[3, 3.5]$	$\frac{38 - 58}{3.5 - 3} = -40$
$[3.5, 4]$	$\frac{10 - 38}{4 - 3.5} = -56$

(c)

$[-1, 5]$ by $[-80, 80]$

(d) Average velocity is a good approximation to velocity.

66. (a) $\frac{d}{dx}[\sqrt{x}f(x)] = \sqrt{x}f'(x) + \frac{1}{2\sqrt{x}}f(x)$

At $x = 1$, the derivative is

$\sqrt{1}f'(1) + \frac{1}{2\sqrt{1}}f(1) = 1\left(\frac{1}{5}\right) + \left(\frac{1}{2}\right)(-3) = -\frac{13}{10}$.

(b) $\frac{d}{dx}\sqrt{f(x)} = \frac{1}{2\sqrt{f(x)}}f'(x) = \frac{f'(x)}{2\sqrt{f(x)}}$

At $x = 0$, the derivative is $\frac{f'(0)}{2\sqrt{f(0)}} = -\frac{2}{2\sqrt{9}} = -\frac{1}{3}$.

(c) $\frac{d}{dx}f(\sqrt{x}) = f'(\sqrt{x})\frac{d}{dx}\sqrt{x} = \frac{f'(\sqrt{x})}{2\sqrt{x}}$

At $x = 1$, the derivative is $\frac{f'(\sqrt{1})}{2\sqrt{1}} = \frac{f'(1)}{2} = \frac{\frac{1}{5}}{2} = \frac{1}{10}$.

(d) $\frac{d}{dx}f(1 - 5\tan x) = f'(1 - 5\tan x)(-5\sec^2 x)$

At $x = 0$, the derivative is

$f'(1 - 5\tan 0)(-5\sec^2 0) = f'(1)(-5) = \left(\frac{1}{5}\right)(-5)$

$= -1$.

(e) $\frac{d}{dx}\frac{f(x)}{2 + \cos x} = \frac{(2 + \cos x)(f'(x)) - (f(x))(-\sin x)}{(2 + \cos x)^2}$

At $x = 0$, the derivative is

$\frac{(2 + \cos 0)(f'(0)) - (f(0))(-\sin 0)}{(2 + \cos 0)^2} = \frac{3f'(0)}{3^2} = -\frac{2}{3}$.

(f) $\frac{d}{dx}\left[10\sin\left(\frac{\pi x}{2}\right)f^2(x)\right]$

$= 10\left(\sin\frac{\pi x}{2}\right)(2f(x)f'(x)) + 10f^2(x)\left(\cos\frac{\pi x}{2}\right)\left(\frac{\pi}{2}\right)$

$= 20f(x)f'(x)\sin\frac{\pi x}{2} + 5\pi f^2(x)\cos\frac{\pi x}{2}$

At $x = 1$, the derivative is

$20f(1)f'(1)\sin\frac{\pi}{2} + 5\pi f^2(1)\cos\frac{\pi}{2}$

$= 20(-3)\left(\frac{1}{5}\right)(1) + 5\pi(-3)^2(0)$

$= -12$.

67. (a) $\frac{d}{dx}[3f(x) - g(x)] = 3f'(x) - g'(x)$

At $x = -1$, the derivative is

$3f'(-1) - g'(-1) = 3(2) - 1 = 5$.

(b) $\frac{d}{dx}[f^2(x)g^3(x)]$

$= f^2(x) \cdot 3g^2(x)g'(x) + g^3(x) \cdot 2f(x)f'(x)$

$= f(x)g^2(x)[3f(x)g'(x) + 2g(x)f'(x)]$

At $x = 0$, the derivative is

$f(0)g^2(0)[3f(0)g'(0) + 2g(0)f'(0)]$

$= (-1)(-3)^2[3(-1)(4) + 2(-3)(-2)]$

$= -9[-12 + 12] = 0$.

(c) $\frac{d}{dx}g(f(x)) = g'(f(x))f'(x)$

At $x = -1$, the derivative is

$g'(f(-1))f'(-1) = g'(0)f'(-1) = (4)(2) = 8$.

(d) $\frac{d}{dx}f(g(x)) = f'(g(x))g'(x)$

At $x = -1$, the derivative is

$f'(g(-1))g'(-1) = f'(-1)g'(-1) = (2)(1) = 2$.

(e) $\frac{d}{dx}\frac{f(x)}{g(x) + 2} = \frac{(g(x) + 2)f'(x) - f(x)g'(x)}{(g(x) + 2)^2}$

At $x = 0$, the derivative is

$\frac{(g(0) + 2)f'(0) - f(0)g'(0)}{(g(0) + 2)^2} = \frac{(-3 + 2)(-2) - (-1)(4)}{(-3 + 2)^2}$

$= 6$.

(f) $\frac{d}{dx}g(x + f(x)) = g'(x + f(x))\frac{d}{dx}(x + f(x))$

$= g'(x + f(x))(1 + f'(x))$

At $x = 0$, the derivative is $g'(0 + f(0))[1 + f'(0)]$

$= g'(0 - 1)[1 + (-2)] = (1)(-1) = -1$

68. $\frac{dw}{ds} = \frac{dw}{dr}\frac{dr}{ds} = \frac{d}{dr}[\sin(\sqrt{r} - 2)]\frac{d}{ds}\left[8\sin\left(s + \frac{\pi}{6}\right)\right]$

$= \left[\cos(\sqrt{r} - 2)\frac{1}{2\sqrt{r}}\right]\left[8\cos\left(s + \frac{\pi}{6}\right)\right]$

At $s = 0$, we have $r = 8\sin\left(0 + \frac{\pi}{6}\right) = 4$ and so

$\frac{dw}{ds} = \left[\cos(\sqrt{4} - 2)\frac{1}{2\sqrt{4}}\right]\left[8\cos\left(0 + \frac{\pi}{6}\right)\right]$

$= \left(\frac{\cos 0}{4}\right)\left(8\cos\frac{\pi}{6}\right) = \left(\frac{1}{4}\right)\left(\frac{8\sqrt{3}}{2}\right) = \sqrt{3}$

82 Chapter 3 Review

69. Solving $\theta^2 t + \theta = 1$ for t, we have

$t = \dfrac{1-\theta}{\theta^2} = \theta^{-2} - \theta^{-1}$, and we may write:

$$\dfrac{dr}{d\theta} = \dfrac{dr}{dt} \dfrac{dt}{d\theta}$$

$$\dfrac{d}{d\theta}(\theta^2 + 7)^{1/3} = \dfrac{dr}{dt} \dfrac{d}{d\theta}(\theta^{-2} - \theta^{-1})$$

$$\dfrac{1}{3}(\theta^2 + 7)^{-2/3}(2\theta) = \left(\dfrac{dr}{dt}\right)(-2\theta^{-3} + \theta^{-2})$$

$$\dfrac{dr}{dt} = \dfrac{2\theta(\theta^2 + 7)^{-2/3}}{3(-2\theta^{-3} + \theta^{-2})} = \dfrac{2\theta^4(\theta^2 + 7)^{-2/3}}{3(\theta - 2)}$$

At $t = 0$, we may solve $\theta^2 t + \theta = 1$ to obtain $\theta = 1$,

and so $\dfrac{dr}{dt} = \dfrac{2(1)^4(1^2 + 7)^{-2/3}}{3(1-2)} = \dfrac{2(8)^{-2/3}}{-3} = -\dfrac{1}{6}$.

70. (a) One possible answer:

$$x(t) = 10 \cos\left(t + \dfrac{\pi}{4}\right),\ y(t) = 1$$

(b) $s(0) = 10 \cos \dfrac{\pi}{4} = 5\sqrt{2}$

(c) Farthest left:

When $\cos\left(t + \dfrac{\pi}{4}\right) = -1$, we have $s(t) = -10$.

Farthest right:

When $\cos\left(t + \dfrac{\pi}{4}\right) = 1$, we have $s(t) = 10$.

(d) Since $\cos \dfrac{\pi}{2} = 0$, the particle first reaches the origin at $t = \dfrac{\pi}{4}$. The velocity is given by

$v(t) = -10 \sin\left(t + \dfrac{\pi}{4}\right)$, so the velocity at $t = \dfrac{\pi}{4}$ is $-10 \sin \dfrac{\pi}{2} = -10$, and the

speed at $t = \dfrac{\pi}{4}$ is $|-10| = 10$. The acceleration is given by $a(t) = -10 \cos\left(t + \dfrac{\pi}{4}\right)$, so the acceleration at $t = \dfrac{\pi}{4}$ is $-10 \cos \dfrac{\pi}{2} = 0$.

71. (a) $\dfrac{ds}{dt} = \dfrac{d}{dt}(64t - 16t^2) = 64 - 32t$

$\dfrac{d^2s}{dt^2} = \dfrac{d}{dt}(64 - 32t) = -32$

(b) The maximum height is reached when $\dfrac{ds}{dt} = 0$, which occurs at $t = 2$ sec.

(c) When $t = 0$, $\dfrac{ds}{dt} = 64$, so the velocity is 64 ft/sec.

(d) Since $\dfrac{ds}{dt} = \dfrac{d}{dt}(64t - 2.6t^2) = 64 - 5.2t$, the maximum height is reached at $t = \dfrac{64}{5.2} \approx 12.3$ sec.

The maximum height is $s\left(\dfrac{64}{5.2}\right) \approx 393.8$ ft.

72. (a) Solving $160 = 490t^2$, it takes $\dfrac{4}{7}$ sec.

The average velocity is $\dfrac{160}{\frac{4}{7}} = 280$ cm/sec.

(b) Since $v(t) = \dfrac{ds}{dt} = 980t$, the velocity is

$(980)\left(\dfrac{4}{7}\right) = 560$ cm/sec. Since $a(t) = \dfrac{dv}{dt} = 980$, the acceleration is 980 cm/sec^2.

73. $\dfrac{dV}{dx} = \dfrac{d}{dx}\left[\pi\left(10 - \dfrac{x}{3}\right)x^2\right] = \dfrac{d}{dx}\left[\pi\left(10x^2 - \dfrac{1}{3}x^3\right)\right]$
$= \pi(20x - x^2)$

74. (a) $r(x) = \left(3 - \dfrac{x}{40}\right)^2 x = 9x - \dfrac{3}{20}x^2 + \dfrac{1}{1600}x^3$

(b) The marginal revenue is

$r'(x) = 9 - \dfrac{3}{10}x + \dfrac{3}{1600}x^2$

$= \dfrac{3}{1600}(x^2 - 160x + 4800)$

$= \dfrac{3}{1600}(x - 40)(x - 120),$

which is zero when $x = 40$ or $x = 120$. Since the bus holds only 60 people, we require $0 \le x \le 60$. The marginal revenue is 0 when there are 40 people, and the corresponding fare is $p(40) = \left(3 - \dfrac{40}{40}\right)^2 = \$4.00$.

(c) One possible answer:
If the current ridership is less than 40, then the proposed plan may be good. If the current ridership is greater than or equal to 40, then the plan is not a good idea. Look at the graph of $y = r(x)$.

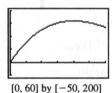

[0, 60] by [−50, 200]

75. (a) Since $x = \tan \theta$, we have

$\dfrac{dx}{dt} = (\sec^2 \theta)\dfrac{d\theta}{dt} = -0.6 \sec^2 \theta$. At point A, we have

$\theta = 0$ and $\dfrac{dx}{dt} = -0.6 \sec^2 0 = -0.6$ km/sec.

(b) $0.6 \dfrac{\text{rad}}{\text{sec}} \cdot \dfrac{1 \text{ revolution}}{2\pi \text{ rad}} \cdot \dfrac{60 \text{ sec}}{1 \text{ min}} = \dfrac{18}{\pi}$ revolutions per minute or approximately 5.73 revolutions per minute.

76. Let $f(x) = \sin(x - \sin x)$. Then

$$f'(x) = \cos(x - \sin x)\frac{d}{dx}(x - \sin x)$$
$$= \cos(x - \sin x)(1 - \cos x).$$ This derivative is zero when $\cos(x - \sin x) = 0$ (which we need not solve) or when $\cos x = 1$, which occurs at $x = 2k\pi$ for integers k. For each of these values, $f(x) = f(2k\pi) = \sin(2k\pi - \sin 2k\pi)$ $= \sin(2k\pi - 0) = 0$. Thus, $f(x) = f'(x) = 0$ for $x = 2k\pi$, which means that the graph has a horizontal tangent at each of these values of x.

77. $y'(r) = \frac{d}{dr}\left(\frac{1}{2rl}\sqrt{\frac{T}{\pi d}}\right) = \left(\frac{1}{2l}\sqrt{\frac{T}{\pi d}}\right)\frac{d}{dr}\left(\frac{1}{r}\right) = -\frac{1}{2r^2l}\sqrt{\frac{T}{\pi d}}$

$y'(l) = \frac{d}{dl}\left(\frac{1}{2rl}\sqrt{\frac{T}{\pi d}}\right) = \left(\frac{1}{2r}\sqrt{\frac{T}{\pi d}}\right)\frac{d}{dl}\left(\frac{1}{l}\right) = -\frac{1}{2rl^2}\sqrt{\frac{T}{\pi d}}$

$y'(d) = \frac{d}{dd}\left(\frac{1}{2rl}\sqrt{\frac{T}{\pi d}}\right) = \left(\frac{1}{2rl}\sqrt{\frac{T}{\pi}}\right)\frac{d}{dd}(d^{-1/2})$

$= \frac{1}{2rl}\sqrt{\frac{T}{\pi}}\left(-\frac{1}{2}d^{-3/2}\right) = -\frac{1}{4rl}\sqrt{\frac{T}{\pi d^3}}$

$y'(T) = \frac{d}{dT}\left(\frac{1}{2rl}\sqrt{\frac{T}{\pi d}}\right) = \left(\frac{1}{2rl}\sqrt{\frac{1}{\pi d}}\right)\frac{d}{dT}(\sqrt{T})$

$= \frac{1}{2rl}\sqrt{\frac{1}{\pi d}}\left(\frac{1}{2\sqrt{T}}\right) = \frac{1}{4rl\sqrt{\pi dT}}$

Since $y'(r) < 0$, $y'(l) < 0$, and $y'(d) < 0$, increasing r, l, or d would decrease the frequency. Since $y'(T) > 0$, increasing T would increase the frequency.

78. (a) $P(0) = \frac{200}{1 + e^5} \approx 1$ student

(b) $\lim_{t \to \infty} P(t) = \lim_{t \to \infty} \frac{200}{1 + e^{5-t}} = \frac{200}{1} = 200$ students

(c) $P'(t) = \frac{d}{dt} 200(1 + e^{5-t})^{-1}$

$= -200(1 + e^{5-t})^{-2}(e^{5-t})(-1)$

$= \frac{200e^{5-t}}{(1 + e^{5-t})^2}$

$P''(t) = \frac{(1 + e^{5-t})^2(200e^{5-t})(-1) - (200e^{5-t})(2)(1 + e^{5-t})(e^{5-t})(-1)}{(1 + e^{5-t})^4}$

$= \frac{(1 + e^{5-t})(-200e^{5-t}) + 400(e^{5-t})^2}{(1 + e^{5-t})^3}$

$= \frac{(200e^{5-t})(e^{5-t} - 1)}{(1 + e^{5-t})^3}$

Since $P'' = 0$ when $t = 5$, the critical point of $y = P'(t)$ occurs at $t = 5$. To confirm that this corresponds to the maximum value of $P'(t)$, note that $P''(t) > 0$ for $t < 5$ and $P''(t) < 0$ for $t > 5$. The maximum rate occurs at $t = 5$, and this rate is

$P'(5) = \frac{200e^0}{(1 + e^0)^2} = \frac{200}{2^2} = 50$ students per day.

Note: This problem can also be solved graphically.

79.

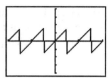

$[-\pi, \pi]$ by $[-4, 4]$

(a) $x \neq k\frac{\pi}{4}$, where k is an odd integer

(b) $\left(-\frac{\pi}{2}, \frac{\pi}{2}\right)$

(c) Where it's not defined, at $x = k\frac{\pi}{4}$, k an odd integer

(d) It has period $\frac{\pi}{2}$ and continues to repeat the pattern seen in this window.

80. Use implicit differentiation.

$$x^2 - y^2 = 1$$

$$\frac{d}{dx}(x^2) - \frac{d}{dx}(y^2) = \frac{d}{dx}(1)$$

$$2x - 2yy' = 0$$

$$y' = \frac{2x}{2y} = \frac{x}{y}$$

$$y'' = \frac{d}{dx}\frac{x}{y}$$

$$= \frac{(y)(1) - (x)(y')}{y^2}$$

$$= \frac{y - x\left(\frac{x}{y}\right)}{y^2}$$

$$= \frac{y^2 - x^2}{y^3}$$

$$= -\frac{1}{y^3}$$

(since the given equation is $x^2 - y^2 = 1$)

At $(2, \sqrt{3})$, $\frac{d^2y}{dx^2} = -\frac{1}{y^3} = -\frac{1}{(\sqrt{3})^3} = -\frac{1}{3\sqrt{3}}$.

Chapter 4
Applications of Derivatives

Section 4.1 Extreme Values of Functions
(pp. 177–185)

Exploration 1   Finding Extreme Values

1. From the graph we can see that there are three critical points: $x = -1, 0, 1$.
Critical point values: $f(-1) = 0.5, f(0) = 0, f(1) = 0.5$
Endpoint values: $f(-2) = 0.4, f(2) = 0.4$
Thus f has absolute maximum value of 0.5 at $x = -1$ and $x = 1$, absolute minimum value of 0 at $x = 0$, and local minimum value of 0.4 at $x = -2$ and $x = 2$.

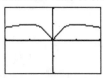

$[-2, 2]$ by $[-1, 1]$

Section 4.1

2. The graph of f' has zeros at $x = -1$ and $x = 1$ where the graph of f has local extreme values. The graph of f' is not defined at $x = 0$, another extreme value of the graph of f.

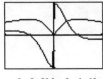

[−2, 2] by [−1, 1]

3. Using the chain rule and $\frac{d}{dx}(|x|) = \frac{|x|}{x}$, we find

$\frac{df}{dx} = \frac{|x|}{x} \cdot \frac{1 - x^2}{(x^2 + 1)^2}$.

Quick Review 4.1

1. $f'(x) = \frac{1}{2\sqrt{4 - x}} \cdot \frac{d}{dx}(4 - x) = \frac{-1}{2\sqrt{4 - x}}$

2. $f'(x) = \frac{3}{4}x^{-1/4}$

3. $f'(x) = \frac{d}{dx}2(9 - x^2)^{-1/2} = -(9 - x^2)^{-3/2} \cdot \frac{d}{dx}(9 - x^2)$
$= -(9 - x^2)^{-3/2}(-2x) = \frac{2x}{(9 - x^2)^{3/2}}$

4. $f'(x) = \frac{d}{dx}(x^2 - 1)^{-1/3} = -\frac{1}{3}(x^2 - 1)^{-4/3} \cdot \frac{d}{dx}(x^2 - 1)$
$= -\frac{1}{3}(x^2 - 1)^{-4/3}(2x) = \frac{-2x}{3(x^2 - 1)^{4/3}}$

5. $g'(x) = \frac{1}{x^2 + 1} \cdot \frac{d}{dx}(x^2 + 1) = \frac{2x}{x^2 + 1}$

6. $g'(x) = -\sin(\ln x) \cdot \frac{d}{dx} \ln x = -\frac{\sin(\ln x)}{x}$

7. $h'(x) = e^{2x} \cdot \frac{d}{dx}2x = 2e^{2x}$

8. $h'(x) = \frac{d}{dx}e^{\ln x} = \frac{d}{dx}x = 1$

9. As $x \to 3^-$, $\sqrt{9 - x^2} \to 0^+$. Therefore, $\lim_{x \to 3^-} f(x) = \infty$.

10. As $x \to -3^+$, $\sqrt{9 - x^2} \to 0^+$. Therefore, $\lim_{x \to -3^+} f(x) = \infty$.

11. (a) $\frac{d}{dx}(x^3 - 2x) = 3x^2 - 2$

$f'(1) = 3(1)^2 - 2 = 1$

(b) $\frac{d}{dx}(x + 2) = 1$

$f'(3) = 1$

(c) Left-hand derivative:

$\lim_{h \to 0^-} \frac{f(2 + h) - f(2)}{h} = \lim_{h \to 0^-} \frac{[(2 + h)^3 - 2(2 + h)] - 4}{h}$

$= \lim_{h \to 0^-} \frac{h^3 + 6h^2 + 10h}{h}$

$= \lim_{h \to 0^-} (h^2 + 6h + 10)$

$= 10$

Right-hand derivative:

$\lim_{h \to 0^+} \frac{f(2 + h) - f(2)}{h} = \lim_{h \to 0^+} \frac{[(2 + h) + 2] - 4}{h}$

$= \lim_{h \to 0^+} \frac{h}{h}$

$= \lim_{h \to 0^+} 1$

$= 1$

Since the left- and right-hand derivatives are not equal, $f'(2)$ is undefined.

12. (a) The domain is $x \neq 2$. (See the solution for 11.(c)).

(b) $f'(x) = \begin{cases} 3x^2 - 2, & x < 2 \\ 1, & x > 2 \end{cases}$

Section 4.1 Exercises

1. Maximum at $x = b$, minimum at $x = c_2$;
The Extreme Value Theorem applies because f is continuous on $[a, b]$, so both the maximum and minimum exist.

3. Maximum at $x = c$, no minimum;
The Extreme Value Theorem does not apply, because the function is not defined on a closed interval.

5. Maximum at $x = c$, minimum at $x = a$;
The Extreme Value Theorem does not apply, because the function is not continuous.

7. Local minimum at $(-1, 0)$, local maximum at $(1, 0)$

9. Maximum at $(0, 5)$ Note that there is no minimum since the endpoint $(2, 0)$ is excluded from the graph.

11. The first derivative $f'(x) = -\dfrac{1}{x^2} + \dfrac{1}{x}$ has a zero at $x = 1$.

Critical point value: $f(1) = 1 + \ln 1 = 1$

Endpoint values: $f(0.5) = 2 + \ln 0.5 \approx 1.307$
$$f(4) = \dfrac{1}{4} + \ln 4 \approx 1.636$$
Maximum value is $\dfrac{1}{4} + \ln 4$ at $x = 4$;

minimum value is 1 at $x = 1$;

local maximum at $\left(\dfrac{1}{2}, 2 - \ln 2\right)$

13. The first derivative $h'(x) = \dfrac{1}{x+1}$ has no zeros, so we need only consider the endpoints.
$$h(0) = \ln 1 = 0 \qquad h(3) = \ln 4$$
Maximum value is $\ln 4$ at $x = 3$;
minimum value is 0 at $x = 0$.

15. The first derivative $f'(x) = \cos\left(x + \dfrac{\pi}{4}\right)$, has zeros at

$x = \dfrac{\pi}{4}, x = \dfrac{5\pi}{4}$.

Critical point values: $x = \dfrac{\pi}{4} \quad f(x) = 1$
$\qquad\qquad\qquad\qquad x = \dfrac{5\pi}{4} \quad f(x) = -1$

Endpoint values: $\quad x = 0 \quad f(x) = \dfrac{1}{\sqrt{2}}$
$\qquad\qquad\qquad\quad x = \dfrac{7\pi}{4} \quad f(x) = 0$

Maximum value is 1 at $x = \dfrac{\pi}{4}$;

minimum value is -1 at $x = \dfrac{5\pi}{4}$;

local minimum at $\left(0, \dfrac{1}{\sqrt{2}}\right)$;

local maximum at $\left(\dfrac{7\pi}{4}, 0\right)$

17. The first derivative $f'(x) = \dfrac{2}{5}x^{-3/5}$ is never zero but is

undefined at $x = 0$.

Critical point value: $\quad x = 0 \quad f(x) = 0$
Endpoint value: $\qquad\quad x = -3 \quad f(x) = (-3)^{2/5}$
$\qquad\qquad\qquad\qquad\qquad\qquad\qquad = 3^{2/5} \approx 1.552$

Since $f(x) > 0$ for $x \neq 0$, the critical point at $x = 0$ is a local minimum, and since $f(x) \leq (-3)^{2/5}$ for $-3 \leq x < 1$, the endpoint value at $x = -3$ is a global maximum.
Maximum value is $3^{2/5}$ at $x = -3$;
minimum value is 0 at $x = 0$.

19.

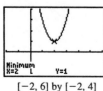

$[-2, 6]$ by $[-2, 4]$

Minimum value is 1 at $x = 2$.

21.

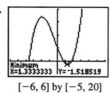

$[-6, 6]$ by $[-5, 20]$

To find the exact values, note that

$y' = 3x^2 + 2x - 8 = (3x - 4)(x + 2)$, which is zero when

$x = -2$ or $x = \dfrac{4}{3}$. Local maximum at $(-2, 17)$;

local minimum at $\left(\dfrac{4}{3}, -\dfrac{41}{27}\right)$

23.

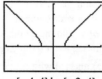

$[-4, 4]$ by $[-2, 4]$

Minimum value is 0 at $x = -1$ and at $x = 1$.

25.

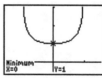

$[-1.5, 1.5]$ by $[-0.5, 3]$

The minimum value is 1 at $x = 0$.

27.

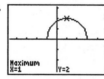

$[-4.7, 4.7]$ by $[-3.1, 3.1]$

Maximum value is 2 at $x = 1$;
minimum value is 0 at $x = -1$ and at $x = 3$.

29.

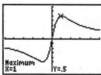

$[-5, 5]$ by $[-0.7, 0.7]$

Maximum value is $\dfrac{1}{2}$ at $x = 1$;

minimum value is $-\dfrac{1}{2}$ at $x = -1$.

31.

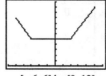

$[-6, 6]$ by $[0, 12]$

Maximum value is 11 at $x = 5$;
minimum value is 5 on the interval $[-3, 2]$;
local maximum at $(-5, 9)$

33.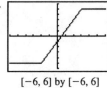
[−6, 6] by [−6, 6]

Maximum value is 5 on the interval [3, ∞); minimum value is −5 on the interval (−∞, −2].

35. (a) No, since $f'(x) = \frac{2}{3}(x-2)^{-1/3}$, which is undefined at $x = 2$.

(b) The derivative is defined and nonzero for all $x \neq 2$. Also, $f(2) = 0$ and $f(x) > 0$ for all $x \neq 2$.

(c) No, $f(x)$ need not not have a global maximum because its domain is all real numbers. Any restriction of f to a closed interval of the form $[a, b]$ would have both a maximum value and a minimum value on the interval.

(d) The answers are the same as (a) and (b) with 2 replaced by a.

37.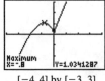
[−4, 4] by [−3, 3]

$y' = x^{2/3}(1) + \frac{2}{3}x^{-1/3}(x+2) = \frac{5x+4}{3\sqrt[3]{x}}$

crit. pt.	derivative	extremum	value
$x = -\frac{4}{5}$	0	local max	$\frac{12}{25}10^{1/3} \approx 1.034$
$x = 0$	undefined	local min	0

39.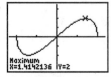
[−2.35, 2.35] by [−3.5, 3.5]

$y' = x \cdot \frac{1}{2\sqrt{4-x^2}}(-2x) + (1)\sqrt{4-x^2}$

$= \frac{-x^2 + (4-x^2)}{\sqrt{4-x^2}} = \frac{4-2x^2}{\sqrt{4-x^2}}$

crit. pt.	derivative	extremum	value
$x = -2$	undefined	local max	0
$x = -\sqrt{2}$	0	minimum	−2
$x = \sqrt{2}$	0	maximum	2
$x = 2$	undefined	local min	0

41.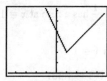
[−4.7, 4.7] by [0, 6.2]

$y' = \begin{cases} -2, & x < 1 \\ 1, & x > 1 \end{cases}$

crit. pt.	derivative	extremum	value
$x = 1$	undefined	minimum	2

43.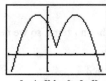
[−4, 6] by [−2, 6]

$y' = \begin{cases} -2x - 2, & x < 1 \\ -2x + 6, & x > 1 \end{cases}$

crit. pt.	derivative	extremum	value
$x = -1$	0	maximum	5
$x = 1$	undefined	local min	1
$x = 3$	0	maximum	5

45. Graph (c), since this is the only graph that has positive slope at c.

47. Graph (d), since this is the only graph representing a function that is differentiable at b but not at a.

49. (a) $V(x) = 160x - 52x^2 + 4x^3$
$V'(x) = 160 - 104x + 12x^2 = 4(x-2)(3x-20)$
The only critical point in the interval (0, 5) is at $x = 2$. The maximum value of $V(x)$ is 144 at $x = 2$.

(b) The largest possible volume of the box is 144 cubic units, and it occurs when $x = 2$.

51. (a) $f'(x) = 3ax^2 + 2bx + c$ is a quadratic, so it can have 0, 1, or 2 zeros, which would be the critical points of f.
Examples:

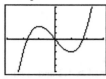

[−3, 3] by [−5, 5]

The function $f(x) = x^3 - 3x$ has two critical points at $x = -1$ and $x = 1$.

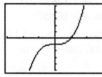

[−3, 3] by [−5, 5]

The function $f(x) = x^3 - 1$ has one critical point at $x = 0$.

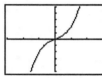

[−3, 3] by [−5, 5]

The function $f(x) = x^3 + x$ has no critical points.

(b) The function can have either two local extreme values or no extreme values. (If there is only one critical point, the cubic function has no extreme values.)

53. (a)

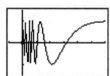

[−0.1, 0.6] by [−1.5, 1.5]

$f(0) = 0$ is not a local extreme value because in any open interval containing $x = 0$, there are infinitely many points where $f(x) = 1$ and where $f(x) = -1$.

(b) One possible answer, on the interval [0, 1]:

$$f(x) = \begin{cases} (1-x)\cos\dfrac{1}{1-x}, & 0 \le x < 1 \\ 0, & x = 1 \end{cases}$$

This function has no local extreme value at $x = 1$. Note that it is continuous on [0, 1].

■ Section 4.2 Mean Value Theorem
(pp. 186–194)

Quick Review 4.2

1. $2x^2 - 6 < 0$
$2x^2 < 6$
$x^2 < 3$
$-\sqrt{3} < x < \sqrt{3}$
Interval: $(-\sqrt{3}, \sqrt{3})$

2. $3x^2 - 6 > 0$
$3x^2 > 6$
$x^2 > 2$
$x < -\sqrt{2}$ or $x > \sqrt{2}$
Intervals: $(-\infty, -\sqrt{2}) \cup (\sqrt{2}, \infty)$

3. Domain: $8 - 2x^2 \ge 0$
$8 \ge 2x^2$
$4 \ge x^2$
$-2 \le x \le 2$
The domain is $[-2, 2]$.

4. f is continuous for all x in the domain, or, in the interval $[-2, 2]$.

5. f is differentiable for all x in the interior of its domain, or, in the interval $(-2, 2)$.

6. We require $x^2 - 1 \ne 0$, so the domain is $x \ne \pm 1$.

7. f is continuous for all x in the domain, or, for all $x \ne \pm 1$.

8. f is differentiable for all x in the domain, or, for all $x \ne \pm 1$.

9. $7 = -2(-2) + C$
$7 = 4 + C$
$C = 3$

10. $-1 = (1)^2 + 2(1) + C$
$-1 = 3 + C$
$C = -4$

Section 4.2 Exercises

1. (a) $f'(x) = 5 - 2x$

Since $f'(x) > 0$ on $\left(-\infty, \dfrac{5}{2}\right)$, $f'(x) = 0$ at $x = \dfrac{5}{2}$, and $f'(x) < 0$ on $\left(\dfrac{5}{2}, \infty\right)$, we know that $f(x)$ has a local maximum at $x = \dfrac{5}{2}$. Since $f\left(\dfrac{5}{2}\right) = \dfrac{25}{4}$, the local maximum occurs at the point $\left(\dfrac{5}{2}, \dfrac{25}{4}\right)$. (This is also a global maximum.)

(b) Since $f'(x) > 0$ on $\left(-\infty, \dfrac{5}{2}\right)$, $f(x)$ is increasing on $\left(-\infty, \dfrac{5}{2}\right]$.

(c) Since $f'(x) < 0$ on $\left(\dfrac{5}{2}, \infty\right)$, $f(x)$ is decreasing on $\left[\dfrac{5}{2}, \infty\right)$.

3. (a) $h'(x) = -\dfrac{2}{x^2}$

Since $h'(x)$ is never zero and is undefined only where $h(x)$ is undefined, there are no critical points. Also, the domain $(-\infty, 0) \cup (0, \infty)$ has no endpoints. Therefore, $h(x)$ has no local extrema.

(b) Since $h'(x)$ is never positive, $h(x)$ is not increasing on any interval.

(c) Since $h'(x) < 0$ on $(-\infty, 0) \cup (0, \infty)$, $h(x)$ is decreasing on $(-\infty, 0)$ and on $(0, \infty)$.

5. (a) $f'(x) = 2e^{2x}$

Since $f'(x)$ is never zero or undefined, and the domain of $f(x)$ has no endpoints, $f(x)$ has no extrema.

(b) Since $f'(x)$ is always positive, $f(x)$ is increasing on $(-\infty, \infty)$.

(c) Since $f'(x)$ is never negative, $f(x)$ is not decreasing on any interval.

88 Section 4.2

7. (a) $y' = -\dfrac{1}{2\sqrt{x+2}}$

In the domain $[-2, \infty)$, y' is never zero and is undefined only at the endpoint $x = -2$. The function y has a local maximum at $(-2, 4)$. (This is also a global maximum.)

(b) Since y' is never positive, y is not increasing on any interval.

(c) Since y' is negative on $(-2, \infty)$, y is decreasing on $[-2, \infty)$.

9.

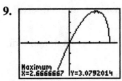

[−4.7, 4.7] by [−3.1, 3.1]

(a) $f'(x) = x \cdot \dfrac{1}{2\sqrt{4-x}}(-1) + \sqrt{4-x}$

$= \dfrac{-3x + 8}{2\sqrt{4-x}}$

The local extrema occur at the critical point $x = \dfrac{8}{3}$ and at the endpoint $x = 4$. There is a local (and absolute) maximum at $\left(\dfrac{8}{3}, \dfrac{16}{3\sqrt{3}}\right)$ or approximately $(2.67, 3.08)$, and a local minimum at $(4, 0)$.

(b) Since $f'(x) > 0$ on $\left(-\infty, \dfrac{8}{3}\right)$, $f(x)$ is increasing on $\left(-\infty, \dfrac{8}{3}\right]$.

(c) Since $f'(x) < 0$ on $\left(\dfrac{8}{3}, 4\right)$, $f(x)$ is decreasing on $\left[\dfrac{8}{3}, 4\right]$.

11.

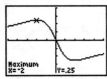

[−5, 5] by [−0.4, 0.4]

(a) $h'(x) = \dfrac{(x^2 + 4)(-1) - (-x)(2x)}{(x^2 + 4)^2} = \dfrac{x^2 - 4}{(x^2 + 4)^2}$

$= \dfrac{(x + 2)(x - 2)}{(x^2 + 4)^2}$

The local extrema occur at the critical points, $x = \pm 2$.

There is a local (and absolute) maximum at $\left(-2, \dfrac{1}{4}\right)$ and a local (and absolute) minimum at $\left(2, -\dfrac{1}{4}\right)$.

(b) Since $h'(x) > 0$ on $(-\infty, -2)$ and $(2, \infty)$, $h(x)$ is increasing on $(-\infty, -2]$ and $[2, \infty)$.

(c) Since $h'(x) < 0$ on $(-2, 2)$, $h(x)$ is decreasing on $[-2, 2]$.

13.

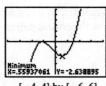

[−4, 4] by [−6, 6]

(a) $f'(x) = 3x^2 - 2 + 2 \sin x$

Note that $3x^2 - 2 > 2$ for $|x| \geq 1.2$ and $|2 \sin x| \leq 2$ for all x, so $f'(x) > 0$ for $|x| \geq 1.2$. Therefore, all critical points occur in the interval $(-1.2, 1.2)$, as suggested by the graph. Using grapher techniques, there is a local maximum at approximately $(-1.126, -0.036)$, and a local minimum at approximately $(0.559, -2.639)$.

(b) $f(x)$ is increasing on the intervals $(-\infty, -1.126]$ and $[0.559, \infty)$, where the interval endpoints are approximate.

(c) $f(x)$ is decreasing on the interval $[-1.126, 0.559]$, where the interval endpoints are approximate.

15. (a) f is continuous on $[0, 1]$ and differentiable on $(0, 1)$.

(b) $f'(c) = \dfrac{f(1) - f(0)}{1 - 0}$

$2c + 2 = \dfrac{2 - (-1)}{1}$

$2c = 1$

$c = \dfrac{1}{2}$

17. (a) f is continuous on $[-1, 1]$ and differentiable on $(-1, 1)$.

(b) $f'(c) = \dfrac{f(1) - f(-1)}{1 - (-1)}$

$\dfrac{1}{\sqrt{1 - c^2}} = \dfrac{\dfrac{\pi}{2} - \left(-\dfrac{\pi}{2}\right)}{2}$

$\sqrt{1 - c^2} = \dfrac{2}{\pi}$

$1 - c^2 = \dfrac{4}{\pi^2}$

$c^2 = 1 - \dfrac{4}{\pi^2}$

$c = \pm\sqrt{1 - \dfrac{4}{\pi^2}} \approx \pm 0.771$

19. (a) The secant line passes through $(0.5, f(0.5)) = (0.5, 2.5)$ and $(2, f(2)) = (2, 2.5)$, so its equation is $y = 2.5$.

(b) The slope of the secant line is 0, so we need to find c such that $f'(c) = 0$.

$1 - c^{-2} = 0$

$c^{-2} = 1$

$c = 1$

$f(c) = f(1) = 2$

The tangent line has slope 0 and passes through $(1, 2)$, so its equation is $y = 2$.

21. (a) Since $f'(x) = \frac{1}{3}x^{-2/3}$, f is not differentiable at $x = 0$.

(b)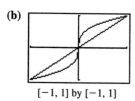

[−1, 1] by [−1, 1]

(c) $f'(c) = \dfrac{f(1) - f(-1)}{1 - (-1)}$

$\dfrac{1}{3}c^{-2/3} = \dfrac{1 - (-1)}{2}$

$\dfrac{1}{3}c^{-2/3} = 1$

$c^{-2/3} = 3$

$c = \pm 3^{-3/2} \approx 0.192$

23. (a) Since $f'(x) = \begin{cases} 1, & -1 \le x < 0 \\ -1, & 1 \ge x > 0 \end{cases}$,

f is not differentiable at $x = 0$. (If f were differentiable at $x = 1$, it would violate the Intermediate Value Theorem for Derivatives.)

(b)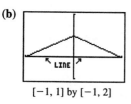

[−1, 1] by [−1, 2]

(c) We require $f'(c) = \dfrac{f(1) - f(-1)}{1 - (-1)} = \dfrac{0 - 0}{2} = 0$, but $f'(x) = \pm 1$ for all x where $f'(x)$ is defined. Therefore, there is no such value of c.

25. $f(x) = \dfrac{x^2}{2} + C$

27. $f(x) = x^3 - x^2 + x + C$

29. $f(x) = e^x + C$

31. $f(x) = \dfrac{1}{x} + C, x > 0$

$f(2) = 1$

$\dfrac{1}{2} + C = 1$

$C = \dfrac{1}{2}$

$f(x) = \dfrac{1}{x} + \dfrac{1}{2}, x > 0$

33. $f(x) = \ln(x + 2) + C$

$f(-1) = 3$

$\ln(-1 + 2) + C = 3$

$0 + C = 3$

$C = 3$

$f(x) = \ln(x + 2) + 3$

35. Possible answers:

(a)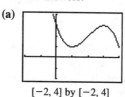

[−2, 4] by [−2, 4]

(b)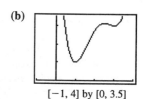

[−1, 4] by [0, 3.5]

(c)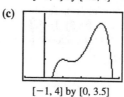

[−1, 4] by [0, 3.5]

37. One possible answer:

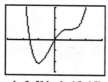

[−3, 3] by [−15, 15]

39. Because the trucker's average speed was 79.5 mph, and by the Mean Value Theorem, the trucker must have been going that speed at least once during the trip.

41. Because its average speed was approximately 7.667 knots, and by the Mean Value Theorem, it must have been going that speed at least once during the trip.

43. (a) Since $v'(t) = 1.6$, $v(t) = 1.6t + C$. But $v(0) = 0$, so $C = 0$ and $v(t) = 1.6t$. Therefore, $v(30) = 1.6(30) = 48$. The rock will be going 48 m/sec.

(b) Let $s(t)$ represent position. Since $s'(t) = v(t) = 1.6t$, $s(t) = 0.8t^2 + D$. But $s(0) = 0$, so $D = 0$ and $s(t) = 0.8t^2$. Therefore, $s(30) = 0.8(30)^2 = 720$. The rock travels 720 meters in the 30 seconds it takes to hit bottom, so the bottom of the crevasse is 720 meters below the point of release.

(c) The velocity is now given by $v(t) = 1.6t + C$, where $v(0) = 4$. (Note that the sign of the initial velocity is the same as the sign used for the acceleration, since both act in a downward direction.) Therefore, $v(t) = 1.6t + 4$, and $s(t) = 0.8t^2 + 4t + D$, where $s(0) = 0$ and so $D = 0$. Using $s(t) = 0.8t^2 + 4t$ and the known crevasse depth of 720 meters, we solve $s(t) = 720$ to obtain the positive solution $t \approx 27.604$, and so $v(t) = v(27.604) = 1.6(27.604) + 4 \approx 48.166$. The rock will hit bottom after about 27.604 seconds, and it will be going about 48.166 m/sec.

45. Because the function is not continuous on [0, 1]. The function does not satisfy the hypotheses of the Mean Value Theorem, and so it need not satisfy the conclusion of the Mean Value Theorem.

47. $f(x)$ must be zero at least once between a and b by the Intermediate Value Theorem. Now suppose that $f(x)$ is zero twice between a and b. Then by the Mean Value Theorem, $f'(x)$ would have to be zero at least once between the two zeros of $f(x)$, but this can't be true since we are given that $f'(x) \ne 0$ on this interval. Therefore, $f(x)$ is zero once and only once between a and b.

49. Let $f(x) = x + \ln(x + 1)$. Then $f(x)$ is continuous and differentiable everywhere on $[0, 3]$. $f'(x) = 1 + \dfrac{1}{x+1}$, which is never zero on $[0, 3]$. Now $f(0) = 0$, so $x = 0$ is one solution of the equation. If there were a second solution, $f(x)$ would be zero twice in $[0, 3]$, and by the Mean Value Theorem, $f'(x)$ would have to be zero somewhere between the two zeros of $f(x)$. But this can't happen, since $f'(x)$ is never zero on $[0, 3]$. Therefore, $f(x) = 0$ has exactly one solution in the interval $[0, 3]$.

51. (a) Increasing: $[-2, -1.3]$ and $[1.3, 2]$;
decreasing: $[-1.3, 1.3]$;
local max: $x \approx -1.3$
local min: $x \approx 1.3$

(b) Regression equation: $y = 3x^2 - 5$

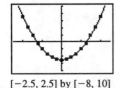

$[-2.5, 2.5]$ by $[-8, 10]$

(c) Since $f'(x) = 3x^2 - 5$, we have $f(x) = x^3 - 5x + C$. But $f(0) = 0$, so $C = 0$. Then $f(x) = x^3 - 5x$.

53. $\dfrac{f(b) - f(a)}{b - a} = \dfrac{\frac{1}{b} - \frac{1}{a}}{b - a} = -\dfrac{1}{ab}$

$f'(c) = -\dfrac{1}{c^2}$, so $-\dfrac{1}{c^2} = -\dfrac{1}{ab}$ and $c^2 = ab$.
Thus, $c = \sqrt{ab}$.

55. By the Mean Value Theorem,
$\sin b - \sin a = (\cos c)(b - a)$ for some c between a and b. Taking the absolute value of both sides and using $|\cos c| \leq 1$ gives the result.

57. Let $f(x)$ be a monotonic function defined on an interval D. For any two values in D, we may let x_1 be the smaller value and let x_2 be the larger value, so $x_1 < x_2$. Then either $f(x_1) < f(x_2)$ (if f is increasing), or $f(x_1) > f(x_2)$ (if f is decreasing), which means $f(x_1) \neq f(x_2)$. Therefore, f is one-to-one.

■ Section 4.3 Connecting f' and f'' with the Graph of f (pp. 194–206)

Exploration 1 Finding f from f'

1. Any function $f(x) = x^4 - 4x^3 + C$ where C is a real number. For example, let $C = 0, 1, 2$. Their graphs are all vertical shifts of each other.

2. Their behavior is the same as the behavior of the function f of Example 8.

Exploration 2 Finding f from f' and f''

1. f has an absolute maximum at $x = 0$ and an absolute minimum of 1 at $x = 4$. We are not given enough information to determine $f(0)$.

2. f has a point of inflection at $x = 2$.

3.

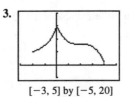

$[-3, 5]$ by $[-5, 20]$

Quick Review 4.3

1. $x^2 - 9 < 0$
$(x + 3)(x - 3) < 0$

Intervals	$x < -3$	$-3 < x < 3$	$3 < x$
Sign of $(x+3)(x-3)$	+	−	+

Solution set: $(-3, 3)$

2. $x^3 - 4x > 0$
$x(x + 2)(x - 2) > 0$

Intervals	$x < -2$	$-2 < x < 0$	$0 < x < 2$	$2 < x$
Sign of $x(x+2)(x-2)$	−	+	−	+

Solution set: $(-2, 0) \cup (2, \infty)$

3. f: all reals
f': all reals, since $f'(x) = xe^x + e^x$

4. f: all reals
f': $x \neq 0$, since $f'(x) = \dfrac{3}{5}x^{-2/5}$

5. f: $x \neq 2$
f': $x \neq 2$, since $f'(x) = \dfrac{(x-2)(1) - (x)(1)}{(x-2)^2} = \dfrac{-2}{(x-2)^2}$

6. f: all reals
f': $x \neq 0$, since $f'(x) = \dfrac{2}{5}x^{-3/5}$

7. Left end behavior model: 0
Right end behavior model: $-x^2 e^x$
Horizontal asymptote: $y = 0$

8. Left end behavior model: $x^2 e^{-x}$
Right end behavior model: 0
Horizontal asymptote: $y = 0$

9. Left end behavior model: 0
Right end behavior model: 200
Horizontal asymptotes: $y = 0, y = 200$

10. Left end behavior model: 0
Right end behavior model: 375
Horizontal asymptotes: $y = 0, y = 375$

Section 4.3 Exercises

1. (a) Zero: $x = \pm 1$;
   positive: $(-\infty, -1)$ and $(1, \infty)$;
   negative: $(-1, 1)$

   (b) Zero: $x = 0$;
   positive: $(0, \infty)$;
   negative: $(-\infty, 0)$

3. (a) $(-\infty, -2]$ and $[0, 2]$
   (b) $[-2, 0]$ and $[2, \infty)$
   (c) Local maxima: $x = -2$ and $x = 2$;
   local minimum: $x = 0$

5. (a) $[0, 1]$, $[3, 4]$, and $[5.5, 6]$
   (b) $[1, 3]$ and $[4, 5.5]$
   (c) Local maxima: $x = 1$, $x = 4$
   (if f is continuous at $x = 4$), and $x = 6$;
   local minima: $x = 0$, $x = 3$, and $x = 5.5$

7. $y' = 2x - 1$

Intervals	$x < \frac{1}{2}$	$x > \frac{1}{2}$
Sign of y'	$-$	$+$
Behavior of y	Decreasing	Increasing

$y'' = 2$ (always positive: concave up)

Graphical support:

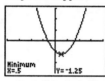

$[-4, 4]$ by $[-3, 3]$

(a) $\left[\frac{1}{2}, \infty\right)$

(b) $\left(-\infty, \frac{1}{2}\right]$

(c) $(-\infty, \infty)$

(d) Nowhere

(e) Local (and absolute) minimum at $\left(\frac{1}{2}, -\frac{5}{4}\right)$

(f) None

9. $y' = 8x^3 - 8x = 8x(x - 1)(x + 1)$

Intervals	$x < -1$	$-1 < x < 0$	$0 < x < 1$	$1 < x$
Sign of y'	$-$	$+$	$-$	$+$
Behavior of y	Decreasing	Increasing	Decreasing	Increasing

$y'' = 24x^2 - 8 = 8(\sqrt{3}x - 1)(\sqrt{3}x + 1)$

Intervals	$x < -\frac{1}{\sqrt{3}}$	$-\frac{1}{\sqrt{3}} < x < \frac{1}{\sqrt{3}}$	$\frac{1}{\sqrt{3}} < x$
Sign of y''	$+$	$-$	$+$
Behavior of y	Concave up	Concave down	Concave up

Graphical support:

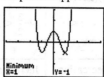

$[-4, 4]$ by $[-3, 3]$

(a) $[-1, 0]$ and $[1, \infty)$
(b) $(-\infty, -1]$ and $[0, 1]$
(c) $\left(-\infty, -\frac{1}{\sqrt{3}}\right)$ and $\left(\frac{1}{\sqrt{3}}, \infty\right)$
(d) $\left(-\frac{1}{\sqrt{3}}, \frac{1}{\sqrt{3}}\right)$
(e) Local maximum: $(0, 1)$;
local (and absolute) minima: $(-1, -1)$ and $(1, -1)$
(f) $\left(\pm\frac{1}{\sqrt{3}}, -\frac{1}{9}\right)$

11. $y' = x \cdot \dfrac{1}{2\sqrt{8-x^2}}(-2x) + (\sqrt{8-x^2})(1) = \dfrac{8 - 2x^2}{\sqrt{8-x^2}}$

Intervals	$-\sqrt{8} < x < -2$	$-2 < x < 2$	$2 < x < \sqrt{8}$
Sign of y'	$-$	$+$	$-$
Behavior of y	Decreasing	Increasing	Decreasing

$$y'' = \dfrac{(\sqrt{8-x^2})(-4x) - (8 - 2x^2)\dfrac{1}{2\sqrt{8-x^2}}(-2x)}{(\sqrt{8-x^2})^2}$$

$$= \dfrac{2x^3 - 24x}{(8-x^2)^{3/2}} = \dfrac{2x(x^2 - 12)}{(8-x^2)^{3/2}}$$

Intervals	$-\sqrt{8} < x < 0$	$0 < x < \sqrt{8}$
Sign of y''	$+$	$-$
Behavior of y	Concave up	Concave down

Graphical support:

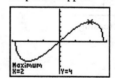

$[-3.02, 3.02]$ by $[-6.5, 6.5]$

(a) $[-2, 2]$
(b) $[-\sqrt{8}, -2]$ and $[2, \sqrt{8}]$
(c) $(-\sqrt{8}, 0)$
(d) $(0, \sqrt{8})$
(e) Local maxima: $(-\sqrt{8}, 0)$ and $(2, 4)$;
local minima: $(-2, -4)$ and $(\sqrt{8}, 0)$
Note that the local extrema at $x = \pm 2$ are also absolute extrema.
(f) $(0, 0)$

13. $y' = 12x^2 + 42x + 36 = 6(x+2)(2x+3)$

Intervals	$x < -2$	$-2 < x < -\frac{3}{2}$	$-\frac{3}{2} < x$
Sign of y'	+	−	+
Behavior of y	Increasing	Decreasing	Increasing

$y'' = 24x + 42 = 6(4x+7)$

Intervals	$x < -\frac{7}{4}$	$-\frac{7}{4} < x$
Sign of y''	−	+
Behavior of y	Concave down	Concave up

Graphical support:

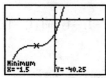

[−4, 4] by [−80, 20]

(a) $(-\infty, -2]$ and $\left[-\frac{3}{2}, \infty\right)$

(b) $\left[-2, -\frac{3}{2}\right]$

(c) $\left(-\frac{7}{4}, \infty\right)$

(d) $\left(-\infty, -\frac{7}{4}\right)$

(e) Local maximum: $(-2, -40)$; local minimum: $\left(-\frac{3}{2}, -\frac{161}{4}\right)$

(f) $\left(-\frac{7}{4}, -\frac{321}{8}\right)$

15. $y' = \frac{2}{5}x^{-4/5}$

Intervals	$x < 0$	$0 < x$
Sign of y'	+	+
Behavior of y	Increasing	Increasing

$y'' = -\frac{8}{25}x^{-9/5}$

Intervals	$x < 0$	$0 < x$
Sign of y''	+	−
Behavior of y	Concave up	Concave down

Graphical support:

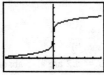

[−6, 6] by [−1.5, 7.5]

(a) $(-\infty, \infty)$

(b) None

(c) $(-\infty, 0)$

(d) $(0, \infty)$

(e) None

(f) $(0, 3)$

17. This problem can be solved using either graphical or analytic methods. For a graphical solution, use NDER to obtain the graphs shown.

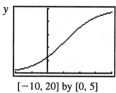
$[-10, 20]$ by $[0, 5]$

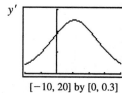
$[-10, 20]$ by $[0, 0.3]$

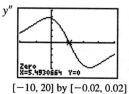
$[-10, 20]$ by $[-0.02, 0.02]$

An analytic solution follows.

$$y' = \frac{(e^x + 3e^{0.8x})(5e^x) - 5e^x(e^x + 2.4e^{0.8x})}{(e^x + 3e^{0.8x})^2}$$

$$= \frac{5e^{2x} + 15e^{1.8x} - 5e^{2x} - 12e^{1.8x}}{(e^x + 3e^{0.8x})^2}$$

$$= \frac{3e^{1.8x}}{(e^x + 3e^{0.8x})^2}$$

Since $y' > 0$ for all x, y is increasing for all x.

$$y'' = \frac{(e^x + 3e^{0.8x})^2(5.4e^{1.8x}) - (3e^{1.8x})(2)(e^x + 3e^{0.8x})(e^x + 2.4e^{0.8x})}{(e^x + 3e^{0.8x})^4}$$

$$= \frac{(e^x + 3e^{0.8x})(5.4e^{1.8x}) - (6e^{1.8x})(e^x + 2.4e^{0.8x})}{(e^x + 3e^{0.8x})^3}$$

$$= \frac{(-0.6e^x + 1.8e^{0.8x})e^{1.8x}}{(e^x + 3e^{0.8x})^3}$$

$$= \frac{0.6(3 - e^{0.2x})e^{2.6x}}{(e^x + 3e^{0.8x})^3}$$

Solve $y'' = 0$: $3 - e^{0.2x} = 0$

$$0.2x = \ln 3$$

$$x = 5 \ln 3$$

Intervals	$x < 5 \ln 3$	$5 \ln 3 < x$
Sign of y''	+	−
Behavior of y	Concave up	Concave down

(a) $(-\infty, \infty)$

(b) None

(c) $(-\infty, 5 \ln 3) \approx (-\infty, 5.49)$

(d) $(5 \ln 3, \infty) \approx (5.49, \infty)$

(e) None

(f) $\left(5 \ln 3, \frac{5}{2}\right) \approx (5.49, 2.50)$

19. $y' = \begin{cases} 2, & x < 1 \\ -2x, & x > 1 \end{cases}$

Intervals	$x < 1$	$1 < x$
Sign of y'	+	−
Behavior of y	Increasing	Decreasing

$y'' = \begin{cases} 0, & x < 1 \\ -2, & x > 1 \end{cases}$

Intervals	$x < 1$	$1 < x$
Sign of y''	0	−
Behavior of y	Linear	Concave down

Graphical support:

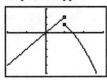

[−2, 3] by [−5, 3]

(a) $(-\infty, 1)$

(b) $[1, \infty)$

(c) None

(d) $(1, \infty)$

(e) None

(f) None

21. $y' = xe^{1/x^2}(-2x^{-3}) + (e^{1/x^2})(1)$

$\quad = e^{1/x^2}(1 - 2x^{-2}) = e^{1/x^2}\left(\dfrac{x^2 - 2}{x^2}\right)$

Intervals	$x < -\sqrt{2}$	$-\sqrt{2} < x < 0$	$0 < x < \sqrt{2}$	$\sqrt{2} < x$
Sign of y'	+	−	−	+
Behavior of y	Increasing	Decreasing	Decreasing	Increasing

$y'' = (e^{1/x^2})(4x^{-3}) + (1 - 2x^{-2})(e^{1/x^2})(-2x^{-3})$

$\quad = (e^{1/x^2})(2x^{-3} + 4x^{-5})$

$\quad = 2e^{1/x^2}\left(\dfrac{x^2 + 2}{x^5}\right)$

Intervals	$x < 0$	$0 < x$
Sign of y''	−	+
Behavior of y	Concave down	Concave up

Graphical support:

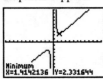

[−12, 12] by [−9, 9]

(a) $(-\infty, -\sqrt{2}]$ and $[\sqrt{2}, \infty)$

(b) $[-\sqrt{2}, 0)$ and $(0, \sqrt{2}]$

(c) $(0, \infty)$

(d) $(-\infty, 0)$

(e) Local maximum: $(-\sqrt{2}, -\sqrt{2e}) \approx (-1.41, -2.33)$; local minimum: $(\sqrt{2}, \sqrt{2e}) \approx (1.41, 2.33)$

(f) None

23. $y' = \dfrac{1}{1+x^2}$

Since $y' > 0$ for all x, y is always increasing.

$y'' = \dfrac{d}{dx}(1+x^2)^{-1} = -(1+x^2)^{-2}(2x) = \dfrac{-2x}{(1+x^2)^2}$

Intervals	$x < 0$	$0 < x$
Sign of y''	+	−
Behavior of y	Concave up	Concave down

Graphical support:

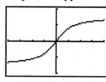

$[-4, 4]$ by $[-2, 2]$

(a) $(-\infty, \infty)$

(b) None

(c) $(-\infty, 0)$

(d) $(0, \infty)$

(e) None

(f) $(0, 0)$

25. $y = x^{1/3}(x - 4) = x^{4/3} - 4x^{1/3}$

$y' = \dfrac{4}{3}x^{1/3} - \dfrac{4}{3}x^{-2/3} = \dfrac{4x - 4}{3x^{2/3}}$

Intervals	$x < 0$	$0 < x < 1$	$1 < x$
Sign of y'	−	−	+
Behavior of y	Decreasing	Decreasing	Increasing

$y'' = \dfrac{4}{9}x^{-2/3} + \dfrac{8}{9}x^{-5/3} = \dfrac{4x + 8}{9x^{5/3}}$

Intervals	$x < -2$	$-2 < x < 0$	$0 < x$
Sign of y''	+	−	+
Behavior of y	Concave up	Concave down	Concave up

Graphical support:

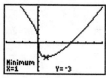

$[-4, 8]$ by $[-6, 8]$

(a) $[1, \infty)$

(b) $(-\infty, 1]$

(c) $(-\infty, -2)$ and $(0, \infty)$

(d) $(-2, 0)$

(e) Local minimum: $(1, -3)$

(f) $(-2, 6\sqrt[3]{2}) \approx (-2, 7.56)$ and $(0, 0)$

96 Section 4.3

27. We use a combination of analytic and grapher techniques to solve this problem. Depending on the viewing window chosen, graphs obtained using NDER may exhibit strange behavior near $x = 2$ because, for example, NDER $(y, 2) \approx 1{,}000{,}000$ while y' is actually undefined at $x = 2$. The graph of $y = \dfrac{x^3 - 2x^2 + x - 1}{x - 2}$ is shown below.

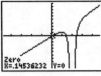

$[-4.7, 4.7]$ by $[-5, 15]$

$$y' = \frac{(x - 2)(3x^2 - 4x + 1) - (x^3 - 2x^2 + x - 1)(1)}{(x - 2)^2}$$

$$= \frac{2x^3 - 8x^2 + 8x - 1}{(x - 2)^2}$$

The graph of y' is shown below.

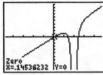

$[-4.7, 4.7]$ by $[-10, 10]$

The zeros of y' are $x \approx 0.15$, $x \approx 1.40$, and $x \approx 2.45$.

Intervals	$x < 0.15$	$0.15 < x < 1.40$	$1.40 < x < 2$	$2 < x < 2.45$	$2.45 < x$
Sign of y'	$-$	$+$	$-$	$-$	$+$
Behavior of y	Decreasing	Increasing	Decreasing	Decreasing	Increasing

$$y'' = \frac{(x-2)^2(6x^2 - 16x + 8) - (2x^3 - 8x^2 + 8x - 1)(2)(x - 2)}{(x - 2)^4}$$

$$= \frac{(x - 2)(6x^2 - 16x + 8) - 2(2x^3 - 8x^2 + 8x - 1)}{(x - 2)^3}$$

$$= \frac{2x^3 - 12x^2 + 24x - 14}{(x - 2)^3}$$

$$= \frac{2(x - 1)(x^2 - 5x + 7)}{(x - 2)^3}$$

The graph of y'' is shown below.

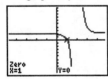

$[-4.7, 4.7]$ by $[-10, 10]$

Note that the discriminant of $x^2 - 5x + 7$ is $(-5)^2 - 4(1)(7) = -3$, so the only solution of $y'' = 0$ is $x = 1$.

Intervals	$x < 1$	$1 < x < 2$	$2 < x$
Sign of y''	$+$	$-$	$+$
Behavior of y	Concave up	Concave down	Concave up

(a) Approximately $[0.15, 1.40]$ and $[2.45, \infty)$

(b) Approximately $(-\infty, 0.15]$, $[1.40, 2)$, and $(2, 2.45]$

(c) $(-\infty, 1)$ and $(2, \infty)$

(d) $(1, 2)$

(e) Local maximum: $\approx (1.40, 1.29)$; local minima: $\approx (0.15, 0.48)$ and $(2.45, 9.22)$

(f) $(1, 1)$

Section 4.3   97

29. $y' = (x-1)^2(x-2)$

Intervals	$x < 1$	$1 < x < 2$	$2 < x$
Sign of y'	−	−	+
Behavior of y	Decreasing	Decreasing	Increasing

$y'' = (x-1)^2(1) + (x-2)(2)(x-1)$
$= (x-1)[(x-1) + 2(x-2)]$
$= (x-1)(3x-5)$

Intervals	$x < 1$	$1 < x < \frac{5}{3}$	$\frac{5}{3} < x$
Sign of y''	+	−	+
Behavior of y	Concave up	Concave down	Concave up

(a) There are no local maxima.

(b) There is a local (and absolute) minimum at $x = 2$.

(c) There are points of inflection at $x = 1$ and at $x = \frac{5}{3}$.

31.

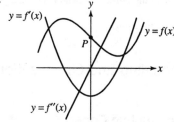

33. (a) Absolute maximum at $(1, 2)$; absolute minimum at $(3, -2)$

(b) None

(c) One possible answer:

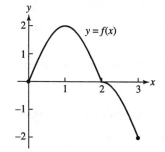

35.

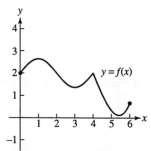

37. (a) $v(t) = s'(t) = 2t - 4$

(b) $a(t) = v'(t) = 2$

(c) It begins at position 3 moving in a negative direction. It moves to position -1 when $t = 2$, and then changes direction, moving in a positive direction thereafter.

98 Section 4.4

39. (a) $v(t) = s'(t) = 3t^2 - 3$
  (b) $a(t) = v'(t) = 6t$
  (c) It begins at position 3 moving in a negative direction. It moves to position 1 when $t = 1$, and then changes direction, moving in a positive direction thereafter.

41. (a) The velocity is zero when the tangent line is horizontal, at approximately $t = 2.2$, $t = 6$, and $t = 9.8$.
  (b) The acceleration is zero at the inflection points, approximately $t = 4$, $t = 8$, and $t = 11$.

43. No. f must have a horizontal tangent at that point, but f could be increasing (or decreasing), and there would be no local extremum. For example, if $f(x) = x^3$, $f'(0) = 0$ but there is no local extremum at $x = 0$.

45. One possible answer:

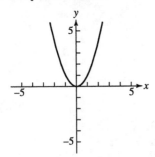

47. One possible answer:

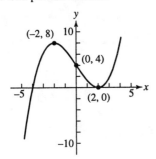

49. (a) Regression equation:
$$y = \frac{2161.4541}{1 + 28.1336e^{-0.8627x}}$$

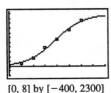

[0, 8] by [−400, 2300]

  (b) At approximately $x = 3.868$ (late in 1996), when the sales are about 1081 million dollars/year
  (c) 2161.45 million dollars/year

51. (a) $f'(x) = \dfrac{(1 + ae^{-bx})(0) - (c)(-abe^{-bx})}{(1 + ae^{-bx})^2}$

$= \dfrac{abce^{-bx}}{(1 + ae^{-bx})^2}$

$= \dfrac{abce^{bx}}{(e^{bx} + a)^2}$,

so the sign of $f'(x)$ is the same as the sign of abc.

  (b) $f''(x) = \dfrac{(e^{bx} + a)^2(ab^2ce^{bx}) - (abce^{bx})2(e^{bx} + a)(be^{bx})}{(e^{bx} + a)^4}$

$= \dfrac{(e^{bx} + a)(ab^2ce^{bx}) - (abce^{bx})(2be^{bx})}{(e^{bx} + a)^3}$

$= -\dfrac{ab^2ce^{bx}(e^{bx} - a)}{(e^{bx} + a)^3}$

Since $a > 0$, this changes sign when $x = \dfrac{\ln a}{b}$ due to the $e^{bx} - a$ factor in the numerator, and $f(x)$ has a point of inflection at that location.

■ Section 4.4 Modeling and Optimization
(pp. 206–220)

Exploration 1 Constructing Cones

1. The circumference of the base of the cone is the circumference of the circle of radius 4 minus x, or $8\pi - x$. Thus, $r = \dfrac{8\pi - x}{2\pi}$. Use the Pythagorean Theorem to find h, and the formula for the volume of a cone to find V.

2. The expression under the radical must be nonnegative, that is, $16 - \left(\dfrac{8\pi - x}{2\pi}\right)^2 \geq 0$.

Solving this inequality for x gives: $0 \leq x \leq 16\pi$.

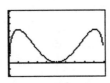

[0, 16π] by [−10, 40]

3. The circumference of the original circle of radius 4 is 8π. Thus, $0 \leq x \leq 8\pi$.

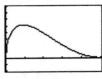

[0, 8π] by [−10, 40]

4. The maximum occurs at about $x = 4.61$. The maximum volume is about $V = 25.80$.

5. Start with $\frac{dV}{dx} = \frac{2\pi}{3}rh\frac{dr}{dx} + \frac{\pi}{3}r^2\frac{dh}{dx}$. Compute $\frac{dr}{dx}$ and $\frac{dh}{dx}$, substitute these values in $\frac{dV}{dx}$, set $\frac{dV}{dx} = 0$, and solve for x to obtain $x = \frac{8(3-\sqrt{6})\pi}{3} \approx 4.61$.

Then $V = \frac{128\pi\sqrt{3}}{27} \approx 25.80$.

Quick Review 4.4

1. $y' = 3x^2 - 12x + 12 = 3(x-2)^2$
Since $y' \geq 0$ for all x (and $y' > 0$ for $x \neq 2$), y is increasing on $(-\infty, \infty)$ and there are no local extrema.

2. $y' = 6x^2 + 6x - 12 = 6(x+2)(x-1)$
$y'' = 12x + 6$
The critical points occur at $x = -2$ or $x = 1$, since $y' = 0$ at these points. Since $y''(-2) = -18 < 0$, the graph has a local maximum at $x = -2$. Since $y''(1) = 18 > 0$, the graph has a local minimum at $x = 1$. In summary, there is a local maximum at $(-2, 17)$ and a local minimum at $(1, -10)$.

3. $V = \frac{1}{3}\pi r^2 h = \frac{1}{3}\pi(5)^2(8) = \frac{200\pi}{3}$ cm^3

4. $V = \pi r^2 h = 1000$
$SA = 2\pi rh + 2\pi r^2 = 600$
Solving the volume equation for h gives $h = \frac{1000}{\pi r^2}$.
Substituting into the surface area equation gives $\frac{2000}{r} + 2\pi r^2 = 600$. Solving graphically, we have $r \approx -11.14$, $r \approx 4.01$, or $r \approx 7.13$. Discarding the negative value and using $h = \frac{1000}{\pi r^2}$ to find the corresponding values of h, the two possibilities for the dimensions of the cylinder are:

$r \approx 4.01$ cm and $h \approx 19.82$ cm, or,

$r \approx 7.13$ cm and $h \approx 6.26$ cm.

5. Since $y = \sin x$ is an odd function, $\sin(-\alpha) = -\sin\alpha$.
6. Since $y = \cos x$ is an even function, $\cos(-\alpha) = \cos\alpha$.
7. $\sin(\pi - \alpha) = \sin\pi\cos\alpha - \cos\pi\sin\alpha$
$= 0\cos\alpha - (-1)\sin\alpha$
$= \sin\alpha$
8. $\cos(\pi - \alpha) = \cos\pi\cos\alpha + \sin\pi\sin\alpha$
$= (-1)\cos\alpha + 0\sin\alpha$
$= -\cos\alpha$
9. $x^2 + y^2 = 4$ and $y = \sqrt{3}x$
$x^2 + (\sqrt{3}x)^2 = 4$
$x^2 + 3x^2 = 4$
$4x^2 = 4$
$x = \pm 1$
Since $y = \sqrt{3}x$, the solutions are:
$x = 1$ and $y = \sqrt{3}$, or, $x = -1$ and $y = -\sqrt{3}$.
In ordered pair notation, the solutions are $(1, \sqrt{3})$ and $(-1, -\sqrt{3})$.

10. $\frac{x^2}{4} + \frac{y^2}{9} = 1$ and $y = x + 3$
$\frac{x^2}{4} + \frac{(x+3)^2}{9} = 1$
$9x^2 + 4(x+3)^2 = 36$
$9x^2 + 4x^2 + 24x + 36 = 36$
$13x^2 + 24x = 0$
$x(13x + 24) = 0$
$x = 0$ or $x = -\frac{24}{13}$

Since $y = x + 3$, the solutions are:
$x = 0$ and $y = 3$, or, $x = -\frac{24}{13}$ and $y = \frac{15}{13}$.

In ordered pair notation, the solutions are $(0, 3)$ and $\left(-\frac{24}{13}, \frac{15}{13}\right)$.

Section 4.4 Exercises

1. Represent the numbers by x and $20 - x$, where $0 \leq x \leq 20$.

(a) The sum of the squares is given by
$f(x) = x^2 + (20-x)^2 = 2x^2 - 40x + 400$. Then $f'(x) = 4x - 40$. The critical point and endpoints occur at $x = 0$, $x = 10$, and $x = 20$. Then $f(0) = 400$, $f(10) = 200$, and $f(20) = 400$. The sum of the squares is as large as possible for the numbers 0 and 20, and is as small as possible for the numbers 10 and 10.

Graphical support:

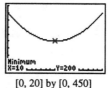

[0, 20] by [0, 450]

(b) The sum of one number plus the square root of the other is given by $g(x) = x + \sqrt{20 - x}$. Then
$g'(x) = 1 - \frac{1}{2\sqrt{20-x}}$. The critical point occurs when $2\sqrt{20-x} = 1$, so $20 - x = \frac{1}{4}$ and $x = \frac{79}{4}$. Testing the endpoints and critical point, we find
$g(0) = \sqrt{20} \approx 4.47$, $g\left(\frac{79}{4}\right) = \frac{81}{4} = 20.25$, and $g(20) = 20$.

1. continued

The sum is as large as possible when the numbers are $\frac{79}{4}$ and $\frac{1}{4}$ (summing $\frac{79}{4} + \sqrt{\frac{1}{4}}$), and is as small as possible when the numbers are 0 and 20 (summing $0 + \sqrt{20}$).

Graphical support:

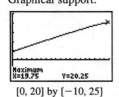

[0, 20] by [−10, 25]

3.
Let x represent the length of the rectangle in inches ($x > 0$). Then the width is $\frac{16}{x}$ and the perimeter is
$P(x) = 2\left(x + \frac{16}{x}\right) = 2x + \frac{32}{x}$.
Since $P'(x) = 2 - 32x^{-2} = \frac{2(x^2 - 16)}{x^2}$ the critical point occurs at $x = 4$. Since $P'(x) < 0$ for $0 < x < 4$ and $P'(x) > 0$ for $x > 0$, this critical point corresponds to the minimum perimeter. The smallest possible perimeter is $P(4) = 16$ in., and the rectangle's dimensions are 4 in. by 4 in.

Graphical support:

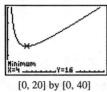

[0, 20] by [0, 40]

5.
(a) The equation of line AB is $y = -x + 1$, so the y-coordinate of P is $-x + 1$.

(b) $A(x) = 2x(1 - x)$

(c) Since $A'(x) = \frac{d}{dx}(2x - 2x^2) = 2 - 4x$, the critical point occurs at $x = \frac{1}{2}$. Since $A'(x) > 0$ for $0 < x < \frac{1}{2}$ and $A'(x) < 0$ for $\frac{1}{2} < x < 1$, this critical point corresponds to the maximum area. The largest possible area is $A\left(\frac{1}{2}\right) = \frac{1}{2}$ square unit, and the dimensions of the rectangle are $\frac{1}{2}$ unit by 1 unit.

Graphical support:

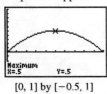

[0, 1] by [−0.5, 1]

7.
Let x be the side length of the cut-out square ($0 < x < 4$). Then the base measures $8 - 2x$ in. by $15 - 2x$ in., and the volume is

$V(x) = x(8 - 2x)(15 - 2x) = 4x^3 - 46x^2 + 120x$. Then
$V'(x) = 12x^2 - 92x + 120 = 4(3x - 5)(x - 6)$. Then the critical point (in $0 < x < 4$) occurs at $x = \frac{5}{3}$. Since $V'(x) > 0$ for $0 < x < \frac{5}{3}$ and $V'(x) < 0$ for $\frac{5}{3} < x < 4$, the critical point corresponds to the maximum volume. The maximum volume is $V\left(\frac{5}{3}\right) = \frac{2450}{27} \approx 90.74$ in^3, and the dimensions are $\frac{5}{3}$ in. by $\frac{14}{3}$ in. by $\frac{35}{3}$ in.

Graphical support:

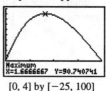

[0, 4] by [−25, 100]

9.
Let x be the length in meters of each side that adjoins the river. Then the side parallel to the river measures $800 - 2x$ meters and the area is
$A(x) = x(800 - 2x) = 800x - 2x^2$ for $0 < x < 400$.
Therefore, $A'(x) = 800 - 4x$ and the critical point occurs at $x = 200$. Since $A'(x) > 0$ for $0 < x < 200$ and $A'(x) < 0$ for $200 < x < 400$, the critical point corresponds to the maximum area. The largest possible area is $A(200) = 80,000$ m^2 and the dimensions are 200 m (perpendicular to the river) by 400 m (parallel to the river).

Graphical support:

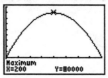

[0, 400] by [−25,000, 90,000]

11.
(a) Let x be the length in feet of each side of the square base. Then the height is $\frac{500}{x^2}$ ft and the surface area (not including the open top) is
$S(x) = x^2 + 4x\left(\frac{500}{x^2}\right) = x^2 + 2000x^{-1}$. Therefore,
$S'(x) = 2x - 2000x^{-2} = \frac{2(x^3 - 1000)}{x^2}$ and the critical point occurs at $x = 10$. Since $S'(x) < 0$ for $0 < x < 10$ and $S'(x) > 0$ for $x > 10$, the critical point corresponds to the minimum amount of steel used. The dimensions should be 10 ft by 10 ft by 5 ft, where the height is 5 ft.

(b) Assume that the weight is minimized when the total area of the bottom and the four sides is minimized.

13. Let x be the height in inches of the printed area. Then the width of the printed area is $\frac{50}{x}$ in. and the overall dimensions are $x + 8$ in. by $\frac{50}{x} + 4$ in. The amount of paper used is

$A(x) = (x + 8)\left(\frac{50}{x} + 4\right) = 4x + 82 + \frac{400}{x}$ in^2. Then $A'(x) = 4 - 400x^{-2} = \frac{4(x^2 - 100)}{x^2}$ and the critical point (for $x > 0$) occurs at $x = 10$. Since $A'(x) < 0$ for $0 < x < 10$ and $A'(x) > 0$ for $x > 10$, the critical point corresponds to the minimum amount of paper. Using $x + 8$ and $\frac{50}{x} + 4$ for $x = 10$, the overall dimensions are 18 in. high by 9 in. wide.

15. We assume that a and b are held constant. Then $A(\theta) = \frac{1}{2}ab \sin \theta$ and $A'(\theta) = \frac{1}{2}ab \cos \theta$. The critical point (for $0 < \theta < \pi$) occurs at $\theta = \frac{\pi}{2}$. Since $A'(\theta) > 0$ for $0 < \theta < \frac{\pi}{2}$ and $A'(\theta) < 0$ for $\frac{\pi}{2} < \theta < \pi$, the critical point corresponds to the maximum area. The angle that maximizes the triangle's area is $\theta = \frac{\pi}{2}$ (or 90°).

17. Note that $\pi r^2 h = 1000$, so $h = \frac{1000}{\pi r^2}$. Then $A = 8r^2 + 2\pi rh = 8r^2 + \frac{2000}{r}$, so $\frac{dA}{dr} = 16r - 2000r^{-2} = \frac{16(r^3 - 125)}{r^2}$. The critical point occurs at $r = \sqrt[3]{125} = 5$ cm. Since $\frac{dA}{dr} < 0$ for $0 < r < 5$ and $\frac{dA}{dr} > 0$ for $r > 5$, the critical point corresponds to the least amount of aluminum used or wasted and hence the most economical can. The dimensions are $r = 5$ cm and $h = \frac{40}{\pi}$, so the ratio of h to r is $\frac{8}{\pi}$ to 1.

19. (a) The "sides" of the suitcase will measure $24 - 2x$ in. by $18 - 2x$ in. and will be $2x$ in. apart, so the volume formula is
$V(x) = 2x(24 - 2x)(18 - 2x) = 8x^3 - 168x^2 + 864x$.

(b) We require $x > 0$, $2x < 18$, and $2x < 24$. Combining these requirements, the domain is the interval (0, 9).

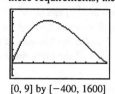

[0, 9] by [−400, 1600]

(c)

[0, 9] by [−400, 1600]

The maximum volume is approximately 1309.95 in^3 when $x \approx 3.39$ in.

(d) $V'(x) = 24x^2 - 336x + 864 = 24(x^2 - 14x + 36)$
The critical point is at
$x = \frac{14 \pm \sqrt{(-14)^2 - 4(1)(36)}}{2(1)} = \frac{14 \pm \sqrt{52}}{2} = 7 \pm \sqrt{13}$,
that is, $x \approx 3.39$ or $x \approx 10.61$. We discard the larger value because it is not in the domain. Since $V''(x) = 24(2x - 14)$, which is negative when $x \approx 3.39$, the critical point corresponds to the maximum volume. The maximum value occurs at $x = 7 - \sqrt{13} \approx 3.39$, which confirms the results in (c).

(e) $8x^3 - 168x^2 + 864x = 1120$
$8(x^3 - 21x^2 + 108x - 140) = 0$
$8(x - 2)(x - 5)(x - 14) = 0$
Since 14 is not in the domain, the possible values of x are $x = 2$ in. or $x = 5$ in.

(f) The dimensions of the resulting box are $2x$ in., $(24 - 2x)$ in., and $(18 - 2x)$ in. Each of these measurements must be positive, so that gives the domain of (0, 9).

21. If the upper right vertex of the rectangle is located at $(x, 4 \cos 0.5x)$ for $0 < x < \pi$, then the rectangle has width $2x$ and height $4 \cos 0.5x$, so the area is $A(x) = 8x \cos 0.5x$. Then $A'(x) = 8x(-0.5 \sin 0.5x) + 8(\cos 0.5x)(1)$
$= -4x \sin 0.5x + 8 \cos 0.5x$.
Solving $A'(x)$ graphically for $0 < x < \pi$, we find that $x \approx 1.72$. Evaluating $2x$ and $4 \cos 0.5x$ for $x \approx 1.72$, the dimensions of the rectangle are approximately 3.44 (width) by 2.61 (height), and the maximum area is approximately 8.98.

23. (a) $f'(x) = x(-e^{-x}) + e^{-x}(1) = e^{-x}(1-x)$
The critical point occurs at $x = 1$. Since $f'(x) > 0$ for $0 \le x < 1$ and $f'(x) < 0$ for $x > 1$, the critical point corresponds to the maximum value of f. The absolute maximum of f occurs at $x = 1$.

(b) To find the values of b, use grapher techniques to solve $xe^{-x} = 0.1e^{-0.1}$, $xe^{-x} = 0.2e^{-0.2}$, and so on. To find the values of A, calculate $(b - a)ae^{-a}$, using the unrounded values of b. (Use the *list* features of the grapher in order to keep track of the unrounded values for part (d).)

a	b	A
0.1	3.71	0.33
0.2	2.86	0.44
0.3	2.36	0.46
0.4	2.02	0.43
0.5	1.76	0.38
0.6	1.55	0.31
0.7	1.38	0.23
0.8	1.23	0.15
0.9	1.11	0.08
1.0	1.00	0.00

(c)

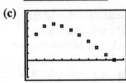

[0, 1.1] by [−0.2, 0.6]

(d) Quadratic:
$A \approx -0.91a^2 + 0.54a + 0.34$

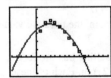

[−0.5, 1.5] by [−0.2, 0.6]

Cubic:
$A \approx 1.74\, a^3 - 3.78a^2 + 1.86a + 0.19$

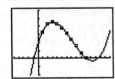

[−0.5, 1.5] by [−0.2, 0.6]

Quartic:
$A \approx -1.92a^4 + 5.96a^3 - 6.87a^2 + 2.71a + 0.12$

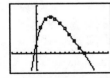

[−0.5, 1.5] by [−0.2, 0.6]

(e) Quadratic:

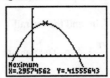

[−0.5, 1.5] by [−0.2, 0.6]

According to the quadratic regression equation, the maximum area occurs at $a \approx 0.30$ and is approximately 0.42.

Cubic:

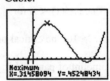

[−0.5, 1.5] by [−0.2, 0.6]

According to the cubic regression equation, the maximum area occurs at $a \approx 0.31$ and is approximately 0.45.

Quartic:

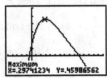

[−0.5, 1.5] by [−0.2, 0.6]

According to the quartic regression equation the maximum area occurs at $a \approx 0.30$ and is approximately 0.46.

25. Let x be the length in inches of each edge of the square end, and let y be the length of the box. Then we require $4x + y \le 108$. Since our goal is to maximize volume, we assume $4x + y = 108$ and so $y = 108 - 4x$. The volume is $V(x) = x^2(108 - 4x) = 108x^2 - 4x^3$, where $0 < x < 27$. Then $V' = 216x - 12x^2 = -12x(x - 18)$, so the critical point occurs at $x = 18$ in. Since $V'(x) > 0$ for $0 < x < 18$ and $V'(x) < 0$ for $18 < x < 27$, the critical point corresponds to the maximum volume. The dimensions of the box with the largest possible volume are 18 in. by 18 in. by 36 in.

27. Note that $h^2 + r^2 = 3$ and so $r = \sqrt{3 - h^2}$. Then the volume is given by $V = \frac{\pi}{3}r^2 h = \frac{\pi}{3}(3 - h^2)h = \pi h - \frac{\pi}{3}h^3$ for $0 < h < \sqrt{3}$, and so $\frac{dV}{dh} = \pi - \pi h^2 = \pi(1 - h^2)$. The critical point (for $h > 0$) occurs at $h = 1$. Since $\frac{dV}{dh} > 0$ for $0 < h < 1$ and $\frac{dV}{dh} < 0$ for $1 < h < \sqrt{3}$, the critical point corresponds to the maximum volume. The cone of greatest volume has radius $\sqrt{2}$ m, height 1 m, and volume $\frac{2\pi}{3}$ m^3.

29. $f'(x) = 2x - ax^{-2} = \dfrac{2x^3 - a}{x^2}$, so the only sign change in $f'(x)$ occurs at $x = \left(\dfrac{a}{2}\right)^{1/3}$, where the sign changes from negative to positive. This means there is a local minimum at that point, and there are no local maxima.

31. Refer to the illustration in the problem statement. Since $x^2 + y^2 = 9$, we have $x = \sqrt{9 - y^2}$. Then the volume of the cone is given by

$$V = \tfrac{1}{3}\pi r^2 h = \tfrac{1}{3}\pi x^2 (y + 3)$$
$$= \tfrac{1}{3}\pi (9 - y^2)(y + 3)$$
$$= \tfrac{\pi}{3}(-y^3 - 3y^2 + 9y + 27),$$

for $-3 < y < 3$.

Thus $\dfrac{dV}{dy} = \tfrac{\pi}{3}(-3y^2 - 6y + 9) = -\pi(y^2 + 2y - 3)$
$= -\pi(y + 3)(y - 1)$, so the critical point in the interval $(-3, 3)$ is $y = 1$. Since $\dfrac{dV}{dy} > 0$ for $-3 < y < 1$ and $\dfrac{dV}{dy} < 0$ for $1 < y < 3$, the critical point does correspond to the maximum value, which is $V(1) = \dfrac{32\pi}{3}$ cubic units.

33. (a) Note that $w^2 + d^2 = 12^2$, so $d = \sqrt{144 - w^2}$. Then we may write $S = kwd^3 = kw(144 - w^2)^{3/2}$, so

$\dfrac{dS}{dw} = kw \cdot \tfrac{3}{2}(144 - w^2)^{1/2}(-2w) + k(144 - w^2)^{3/2}(1)$
$= (k\sqrt{144 - w^2})(-3w^2 + 144 - w^2)$
$= (-4k\sqrt{144 - w^2})(w^2 - 36)$

The critical point (for $0 < w < 12$) occurs at $w = 6$. Since $\dfrac{dS}{dw} > 0$ for $0 < w < 6$ and $\dfrac{dS}{dw} < 0$ for $6 < w < 12$, the critical point corresponds to the maximum stiffness. The dimensions are 6 in. wide by $6\sqrt{3}$ in. deep.

(b)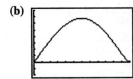

[0, 12] by [−2000, 8000]

The graph of $S = w(144 - w^2)^{3/2}$ is shown. The maximum stiffness shown in the graph occurs at $w = 6$, which agrees with the answer to part (a).

(c)

[0, 12] by [−2000, 8000]

The graph of $S = d^3\sqrt{144 - d^2}$ is shown. The maximum stiffness shown in the graph occurs at $d = 6\sqrt{3} \approx 10.4$ agrees with the answer to part (a), and its value is the same as the maximum value found in part (b), as expected.

Changing the value of k changes the maximum stiffness, but not the dimensions of the stiffest beam. The graphs for different values of k look the same except that the vertical scale is different.

35. Since $\dfrac{di}{dt} = -2 \sin t + 2 \cos t$, the largest magnitude of the current occurs when $-2 \sin t + 2 \cos t = 0$, or $\sin t = \cos t$. Squaring both sides gives $\sin^2 t = \cos^2 t$, and we know that $\sin^2 t + \cos^2 t = 1$, so $\sin^2 t = \cos^2 t = \dfrac{1}{2}$. Thus the possible values of t are $\dfrac{\pi}{4}, \dfrac{3\pi}{4}, \dfrac{5\pi}{4}$, and so on. Eliminating extraneous solutions, the solutions of $\sin t = \cos t$ are $t = \dfrac{\pi}{4} + k\pi$ for integers k, and at these times $|i| = |2 \cos t + 2 \sin t| = 2\sqrt{2}$. The peak current is $2\sqrt{2}$ amps.

37. Calculus method:

The square of the distance from the point $(1, \sqrt{3})$ to $(x, \sqrt{16 - x^2})$ is given by

$D(x) = (x - 1)^2 + (\sqrt{16 - x^2} - \sqrt{3})^2$
$= x^2 - 2x + 1 + 16 - x^2 - 2\sqrt{48 - 3x^2} + 3$
$= -2x + 20 - 2\sqrt{48 - 3x^2}$. Then

$D'(x) = -2 - \dfrac{2}{2\sqrt{48 - 3x^2}}(-6x) = -2 + \dfrac{6x}{\sqrt{48 - 3x^2}}$.

Solving $D'(x) = 0$, we have:
$6x = 2\sqrt{48 - 3x^2}$
$36x^2 = 4(48 - 3x^2)$
$9x^2 = 48 - 3x^2$
$12x^2 = 48$
$x = \pm 2$

We discard $x = -2$ as an extraneous solution, leaving $x = 2$. Since $D'(x) < 0$ for $-4 < x < 2$ and $D'(x) > 0$ for $2 < x < 4$, the critical point corresponds to the minimum distance. The minimum distance is $\sqrt{D(2)} = 2$.

37. continued

Geometry method:

The semicircle is centered at the origin and has radius 4.

The distance from the origin to $(1, \sqrt{3})$ is $\sqrt{1^2 + (\sqrt{3})^2} = 2$. The shortest distance from the point to the semicircle is the distance along the radius containing the point $(1, \sqrt{3})$. That distance is $4 - 2 = 2$.

39. (a) Because $f(x)$ is periodic with period 2π.

(b) No. Since $f(x)$ is continuous on $[0, 2\pi]$, its absolute minimum occurs at a critical point or endpoint.
Find the critical points in $[0, 2\pi]$:
$$f'(x) = -4\sin x - 2\sin 2x = 0$$
$$-4\sin x - 4\sin x \cos x = 0$$
$$-4(\sin x)(1 + \cos x) = 0$$
$$\sin x = 0 \text{ or } \cos x = -1$$
$$x = 0, \pi, 2\pi$$
The critical points (and endpoints) are $(0, 8)$, $(\pi, 0)$, and $(2\pi, 8)$. Thus, $f(x)$ has an absolute minimum at $(\pi, 0)$ and it is never negative.

41. (a)
$$\sin t = \sin\left(t + \frac{\pi}{3}\right)$$
$$\sin t = \sin t \cos \frac{\pi}{3} + \cos t \sin \frac{\pi}{3}$$
$$\sin t = \frac{1}{2}\sin t + \frac{\sqrt{3}}{2}\cos t$$
$$\frac{1}{2}\sin t = \frac{\sqrt{3}}{2}\cos t$$
$$\tan t = \sqrt{3}$$

Solving for t, the particles meet at $t = \frac{\pi}{3}$ sec and at $t = \frac{4\pi}{3}$ sec.

(b) The distance between the particles is the absolute value of $f(t) = \sin\left(t + \frac{\pi}{3}\right) - \sin t = \frac{\sqrt{3}}{2}\cos t - \frac{1}{2}\sin t$. Find the critical points in $[0, 2\pi]$:
$$f'(t) = -\frac{\sqrt{3}}{2}\sin t - \frac{1}{2}\cos t = 0$$
$$-\frac{\sqrt{3}}{2}\sin t = \frac{1}{2}\cos t$$
$$\tan t = -\frac{1}{\sqrt{3}}$$

The solutions are $t = \frac{5\pi}{6}$ and $t = \frac{11\pi}{6}$, so the critical points are at $\left(\frac{5\pi}{6}, -1\right)$ and $\left(\frac{11\pi}{6}, 1\right)$, and the interval endpoints are at $\left(0, \frac{\sqrt{3}}{2}\right)$ and $\left(2\pi, \frac{\sqrt{3}}{2}\right)$. The particles are farthest apart at $t = \frac{5\pi}{6}$ sec and at $t = \frac{11\pi}{6}$ sec, and the maximum distance between the particles is 1 m.

(c) We need to maximize $f'(t)$, so we solve $f''(t) = 0$.
$$f''(t) = -\frac{\sqrt{3}}{2}\cos t + \frac{1}{2}\sin t = 0$$
$$\frac{1}{2}\sin t = \frac{\sqrt{3}}{2}\cos t$$

This is the same equation we solved in part (a), so the solutions are $t = \frac{\pi}{3}$ sec and $t = \frac{4\pi}{3}$ sec.

For the function $y = f'(t)$, the critical points occur at $\left(\frac{\pi}{3}, -1\right)$ and $\left(\frac{4\pi}{3}, 1\right)$, and the interval endpoints are at $\left(0, -\frac{1}{2}\right)$ and $\left(2\pi, -\frac{1}{2}\right)$.

Thus, $|f'(t)|$ is maximized at $t = \frac{\pi}{3}$ and $t = \frac{4\pi}{3}$. But these are the instants when the particles pass each other, so the graph of $y = |f(t)|$ has corners at these points and $\frac{d}{dt}|f(t)|$ is undefined at these instants. We cannot say that the distance is changing the fastest at any particular instant, but we can say that near $t = \frac{\pi}{3}$ or $t = \frac{4\pi}{3}$ the distance is changing faster than at any other time in the interval.

43. (a)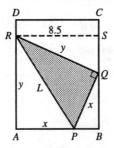

Sketch segment RS as shown, and let y be the length of segment QR. Note that $PB = 8.5 - x$, and so $QB = \sqrt{x^2 - (8.5 - x)^2} = \sqrt{8.5(2x - 8.5)}$. Also note that triangles QRS and PQB are similar.

$$\frac{QR}{RS} = \frac{PQ}{QB}$$
$$\frac{y}{8.5} = \frac{x}{\sqrt{8.5(2x - 8.5)}}$$
$$\frac{y^2}{8.5^2} = \frac{x^2}{8.5(2x - 8.5)}$$
$$y^2 = \frac{8.5x^2}{2x - 8.5}$$
$$L^2 = x^2 + y^2$$
$$L^2 = x^2 + \frac{8.5x^2}{2x - 8.5}$$
$$L^2 = \frac{x^2(2x - 8.5) + 8.5x^2}{2x - 8.5}$$
$$L^2 = \frac{2x^3}{2x - 8.5}$$

(b) Note that $x > 4.25$, and let $f(x) = L^2 = \dfrac{2x^3}{2x - 8.5}$. Since $y \le 11$, the approximate domain of f is $5.20 \le x \le 8.5$.

Then

$$f'(x) = \dfrac{(2x - 8.5)(6x^2) - (2x^3)(2)}{(2x - 8.5)^2} = \dfrac{x^2(8x - 51)}{(2x - 8.5)^2}$$

For $x > 5.20$, the critical point occurs at

$x = \dfrac{51}{8} = 6.375$ in., and this corresponds to a minimum value of $f(x)$ because $f'(x) < 0$ for $5.20 < x < 6.375$ and $f'(x) > 0$ for $x > 6.375$. Therefore, the value of x that minimizes L^2 is $x = 6.375$ in.

(c) The minimum value of L is

$$\sqrt{\dfrac{2(6.375)^3}{2(6.375) - 8.5}} \approx 11.04 \text{ in.}$$

45. The profit is given by

$$P(x) = (n)(x - c) = a + b(100 - x)(x - c)$$
$$= -bx^2 + (100 + c)bx + (a - 100bc).$$

Then $P'(x) = -2bx + (100 + c)b$
$$= b(100 + c - 2x).$$

The critical point occurs at $x = \dfrac{100 + c}{2} = 50 + \dfrac{c}{2}$, and this value corresponds to the maximum profit because $P'(x) > 0$ for $x < 50 + \dfrac{c}{2}$ and $P'(x) < 0$ for $x > 50 + \dfrac{c}{2}$.

A selling price of $50 + \dfrac{c}{2}$ will bring the maximum profit.

47. $\dfrac{dv}{dx} = ka - 2kx$

The critical point occurs at $x = \dfrac{ka}{2k} = \dfrac{a}{2}$, which represents a maximum value because $\dfrac{d^2v}{dx^2} = -2k$, which is negative for all x. The maximum value of v is

$$kax - kx^2 = ka\left(\dfrac{a}{2}\right) - k\left(\dfrac{a}{2}\right)^2 = \dfrac{ka^2}{4}.$$

49. Revenue: $r(x) = [200 - 2(x - 50)]x = -2x^2 + 300x$
Cost: $c(x) = 6000 + 32x$
Profit: $p(x) = r(x) - c(x)$
$\qquad = -2x^2 + 268x - 6000,\ 50 \le x \le 80$

Since $p'(x) = -4x + 268 = -4(x - 67)$, the critical point occurs at $x = 67$. This value represents the maximum because $p''(x) = -4$, which is negative for all x in the domain. The maximum profit occurs if 67 people go on the tour.

51. The profit is given by
$$p(x) = r(x) - c(x)$$
$$= 6x - (x^3 - 6x^2 + 15x)$$
$$= -x^3 + 6x^2 - 9x, \text{ for } x \ge 0.$$

Then $p'(x) = -3x^2 + 12x - 9 = -3(x - 1)(x - 3)$, so the critical points occur at $x = 1$ and $x = 3$. Since $p'(x) < 0$ for $0 \le x < 1$, $p'(x) > 0$ for $1 < x < 3$, and $p'(x) < 0$ for $x > 3$, the relative maxima occur at the endpoint $x = 0$ and at the critical point $x = 3$. Since $p(0) = p(3) = 0$, this means that for $x \ge 0$, the function $p(x)$ has its absolute maximum value at the points $(0, 0)$ and $(3, 0)$. This result can also be obtained graphically, as shown.

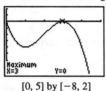

$[0, 5]$ by $[-8, 2]$

53. (a) According to the graph, $y'(0) = 0$.

(b) According to the graph, $y'(-L) = 0$.

(c) $y(0) = 0$, so $d = 0$.

Now $y'(x) = 3ax^2 + 2bx + c$, so $y'(0)$ implies that $c = 0$. Therefore, $y(x) = ax^3 + bx^2$ and $y'(x) = 3ax^2 + 2bx$. Then $y(-L) = -aL^3 + bL^2 = H$ and $y'(-L) = 3aL^2 - 2bL = 0$, so we have two linear equations in the two unknowns a and b. The second equation gives $b = \dfrac{3aL}{2}$. Substituting into the first equation, we have $-aL^3 + \dfrac{3aL^3}{2} = H$, or $\dfrac{aL^3}{2} = H$, so $a = 2\dfrac{H}{L^3}$. Therefore, $b = 3\dfrac{H}{L^2}$ and the equation for y is $y(x) = 2\dfrac{H}{L^3}x^3 + 3\dfrac{H}{L^2}x^2$, or $y(x) = H\left[2\left(\dfrac{x}{L}\right)^3 + 3\left(\dfrac{x}{L}\right)^2\right]$.

106 Section 4.4

55. (a) Let x_0 represent the fixed value of x at point P, so that P has coordinates (x_0, a),
and let $m = f'(x_0)$ be the slope of line RT. Then the equation of line RT is
$y = m(x - x_0) + a$. The y-intercept of this line is $m(0 - x_0) + a = a - mx_0$,
and the x-intercept is the solution of
$m(x - x_0) + a = 0$, or $x = \dfrac{mx_0 - a}{m}$.

Let O designate the origin. Then

(Area of triangle RST)

$= 2$(Area of triangle ORT)

$= 2 \cdot \dfrac{1}{2}(x\text{-intercept of line } RT)(y\text{-intercept of line } RT)$

$= 2 \cdot \dfrac{1}{2}\left(\dfrac{mx_0 - a}{m}\right)(a - mx_0)$

$= -m\left(\dfrac{mx_0 - a}{m}\right)\left(\dfrac{mx_0 - a}{m}\right)$

$= -m\left(\dfrac{mx_0 - a}{m}\right)^2$

$= -m\left(x_0 - \dfrac{a}{m}\right)^2$

Substituting x for x_0, $f'(x)$ for m, and $f(x)$ for a, we have $A(x) = -f'(x)\left[x - \dfrac{f(x)}{f'(x)}\right]^2$.

(b) The domain is the open interval $(0, 10)$.

To graph, let $y_1 = f(x) = 5 + 5\sqrt{1 - \dfrac{x^2}{100}}$,

$y_2 = f'(x) = \text{NDER}(y_1)$, and

$y_3 = A(x) = -y_2\left(x - \dfrac{y_1}{y_2}\right)^2$.

The graph of the area function $y_3 = A(x)$ is shown below.

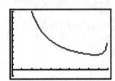

[0, 10] by [−100, 1000]

The vertical asymptotes at $x = 0$ and $x = 10$ correspond to
horizontal or vertical tangent lines, which do not form triangles.

(c) Using our expression for the y-intercept of the tangent line, the height of the triangle is

$a - mx = f(x) - f'(x) \cdot x$

$= 5 + \dfrac{1}{2}\sqrt{100 - x^2} - \dfrac{-x}{2\sqrt{100 - x^2}}x$

$= 5 + \dfrac{1}{2}\sqrt{100 - x^2} + \dfrac{x^2}{2\sqrt{100 - x^2}}$

We may use graphing methods or the analytic method in part (d) to find that the minimum value of $A(x)$ occurs at $x \approx 8.66$.
Substituting this value into the expression above, the height of the triangle is 15. This is 3 times the y-coordinate of the
center of the ellipse.

(d) Part (a) remains unchanged. The domain is $(0, C)$. To graph, note that

$$f(x) = B + B\sqrt{1 - \frac{x^2}{C^2}} = B + \frac{B}{C}\sqrt{C^2 - x^2} \text{ and}$$

$$f'(x) = \frac{B}{C}\frac{1}{2\sqrt{C^2 - x^2}}(-2x) = \frac{-Bx}{C\sqrt{C^2 - x^2}}.$$

Therefore, we have

$$A(x) = -f'(x)\left[x - \frac{f(x)}{f'(x)}\right]^2 = \frac{Bx}{C\sqrt{C^2-x^2}}\left[x - \frac{B + \frac{B}{C}\sqrt{C^2-x^2}}{\frac{-Bx}{C\sqrt{C^2-x^2}}}\right]^2$$

$$= \frac{Bx}{C\sqrt{C^2-x^2}}\left[x - \frac{(BC + B\sqrt{C^2-x^2})\sqrt{C^2-x^2}}{-Bx}\right]^2$$

$$= \frac{1}{BCx\sqrt{C^2-x^2}}[Bx^2 + (BC + B\sqrt{C^2-x^2})(\sqrt{C^2-x^2})]^2$$

$$= \frac{1}{BCx\sqrt{C^2-x^2}}[Bx^2 + BC\sqrt{C^2-x^2} + B(C^2-x^2)]^2$$

$$= \frac{1}{BCx\sqrt{C^2-x^2}}[BC(C + \sqrt{C^2-x^2})]^2$$

$$= \frac{BC(C + \sqrt{C^2-x^2})^2}{x\sqrt{C^2-x^2}}$$

$$A'(x) = BC \cdot \frac{(x\sqrt{C^2-x^2})(2)(C + \sqrt{C^2-x^2})\left(\frac{-x}{\sqrt{C^2-x^2}}\right) - (C + \sqrt{C^2-x^2})^2\left(x\frac{-x}{\sqrt{C^2-x^2}} + \sqrt{C^2-x^2}(1)\right)}{x^2(C^2-x^2)}$$

$$= \frac{BC(C + \sqrt{C^2-x^2})}{x^2(C^2-x^2)}\left[-2x^2 - (C + \sqrt{C^2-x^2})\left(\frac{-x^2}{\sqrt{C^2-x^2}} + \sqrt{C^2-x^2}\right)\right]$$

$$= \frac{BC(C + \sqrt{C^2-x^2})}{x^2(C^2-x^2)}\left[-2x^2 + \frac{Cx^2}{\sqrt{C^2-x^2}} - C\sqrt{C^2-x^2} + x^2 - (C^2-x^2)\right]$$

$$= \frac{BC(C + \sqrt{C^2-x^2})}{x^2(C^2-x^2)}\left(\frac{Cx^2}{\sqrt{C^2-x^2}} - C\sqrt{C^2-x^2} - C^2\right)$$

$$= \frac{BC(C + \sqrt{C^2-x^2})}{x^2(C^2-x^2)^{3/2}}[Cx^2 - C(C^2-x^2) - C^2\sqrt{C^2-x^2}]$$

$$= \frac{BC^2(C + \sqrt{C^2-x^2})}{x^2(C^2-x^2)^{3/2}}(2x^2 - C^2 - C\sqrt{C^2-x^2})$$

To find the critical points for $0 < x < C$, we solve:

$$2x^2 - C^2 = C\sqrt{C^2 - x^2}$$
$$4x^4 - 4C^2x^2 + C^4 = C^4 - C^2x^2$$
$$4x^4 - 3C^2x^2 = 0$$
$$x^2(4x^2 - 3C^2) = 0$$

The minimum value of $A(x)$ for $0 < x < C$ occurs at the critical point $x = \dfrac{C\sqrt{3}}{2}$, or $x^2 = \dfrac{3C^2}{4}$. The corresponding triangle height is
$a - mx = f(x) - f'(x) \cdot x$

$$= B + \dfrac{B}{C}\sqrt{C^2 - x^2} + \dfrac{Bx^2}{C\sqrt{C^2 - x^2}}$$

$$= B + \dfrac{B}{C}\sqrt{C^2 - \dfrac{3C^2}{4}} + \dfrac{B\left(\dfrac{3C^2}{4}\right)}{C\sqrt{C^2 - \dfrac{3C^2}{4}}}$$

$$= B + \dfrac{B}{C}\left(\dfrac{C}{2}\right) + \dfrac{\dfrac{3BC^2}{4}}{\dfrac{C^2}{2}}$$

$$= B + \dfrac{B}{2} + \dfrac{3B}{2}$$

$$= 3B$$

This shows that the triangle has minimum area when its height is $3B$.

■ Section 4.5 Linearization and Newton's Method (pp. 220–232)

Exploration 1 Approximating with Tangent Lines

1. $f'(x) = 2x$, $f'(1) = 2$, so an equation of the tangent line is $y - 1 = 2(x - 1)$ or $y = 2x - 1$.

3. Since $(y_1 - y_2)(1) = y_1(1) - y_2(1) = 1 - 1 = 0$, this view shifts the action from the point $(1, 1)$ to the point $(1, 0)$. Also $(y_1 - y_2)'(1) = y_1'(1) - y_2'(1) = 2 - 2 = 0$. Thus the tangent line to $y_1 - y_2$ at $x = 1$ is horizontal (the x-axis). The measure of how well y_2 fits y_1 at $(1, 1)$ is the same as the measure of how well the x-axis fits $y_1 - y_2$ at $(1, 0)$.

4. These tables show that the values of $y_1 - y_2$ near $x = 1$ are close to 0 so that y_2 is a good approximation to y_1 near $x = 1$. Here are two tables with Δ Table $= 0.0001$.

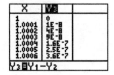

Exploration 2 Using Newton's Method on the Grapher

1–3. Here are the first 11 computations.

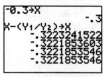

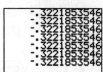

4. Answers will vary. Here is what happens for $x_1 = -2$.

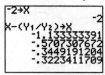

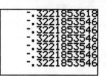

Quick Review 4.5

1. $\dfrac{dy}{dx} = \cos(x^2 + 1) \cdot \dfrac{d}{dx}(x^2 + 1) = 2x \cos(x^2 + 1)$

2. $\dfrac{dy}{dx} = \dfrac{(x+1)(1 - \sin x) - (x + \cos x)(1)}{(x+1)^2}$

   $= \dfrac{x - x \sin x + 1 - \sin x - x - \cos x}{(x+1)^2}$

   $= \dfrac{1 - \cos x - (x+1) \sin x}{(x+1)^2}$

3.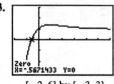

   $[-2, 6]$ by $[-3, 3]$

   $x \approx -0.567$

4.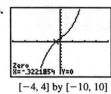

   $[-4, 4]$ by $[-10, 10]$

   $x \approx -0.322$

5. $f'(x) = (x)(-e^{-x}) + (e^{-x})(1) = e^{-x} - xe^{-x}$
   $f'(0) = 1$
   The lines passes through $(0, 1)$ and has slope 1. Its equation is $y = x + 1$.

6. $f'(x) = (x)(-e^{-x}) + (-e^{-x})(1) = e^{-x} - xe^{-x}$
   $f'(-1) = e^1 - (-e^1) = 2e$
   The lines passes through $(-1, -e + 1)$ and has slope $2e$. Its equation is $y = 2e(x + 1) + (-e + 1)$, or $y = 2ex + e + 1$.

7. (a) $x + 1 = 0$
   $x = -1$

   (b) $2ex + e + 1 = 0$
   $2ex = -(e + 1)$
   $x = -\dfrac{e + 1}{2e} \approx -0.684$

8. $f'(x) = 3x^2 - 4$
   $f'(1) = 3(1)^2 - 4 = -1$
   Since $f(1) = -2$ and $f'(1) = -1$, the graph of $g(x)$ passes through $(1, -2)$ and has slope -1. Its equation is
   $g(x) = -1(x - 1) + (-2)$, or $g(x) = -x - 1$.

x	$f(x)$	$g(x)$
0.7	-1.457	-1.7
0.8	-1.688	-1.8
0.9	-1.871	-1.9
1.0	-2	-2
1.1	-2.069	-2.1
1.2	-2.072	-2.2
1.3	-2.003	-2.3

9. $f'(x) = \cos x$
   $f'(1.5) = \cos 1.5$
   Since $f(1.5) = \sin 1.5$ and $f'(1.5) = \cos 1.5$, the tangent line passes through $(1.5, \sin 1.5)$ and has slope $\cos 1.5$. Its equation is $y = (\cos 1.5)(x - 1.5) + \sin 1.5$, or approximately $y = 0.071x + 0.891$

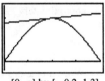

[0, π] by [-0.2, 1.3]

10. For $x > 3$, $f'(x) = \dfrac{1}{2\sqrt{x - 3}}$, and so $f'(4) = \dfrac{1}{2}$. Since $f(4) = 1$ and $f'(4) = \dfrac{1}{2}$, the tangent line passes through $(4, 1)$ and has slope $\dfrac{1}{2}$. Its equation is $y = \dfrac{1}{2}(x - 4) + 1$, or $y = \dfrac{1}{2}x - 1$.

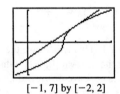

[-1, 7] by [-2, 2]

Section 4.5 Exercises

1. (a) $f'(x) = 3x^2 - 2$
   We have $f(2) = 7$ and $f'(2) = 10$.
   $L(x) = f(2) + f'(2)(x - 2)$
   $= 7 + 10(x - 2)$
   $= 10x - 13$

   (b) Since $f(2.1) = 8.061$ and $L(2.1) = 8$, the approximation differs from the true value in absolute value by less than 10^{-1}.

3. (a) $f'(x) = 1 - x^{-2}$
   We have $f(1) = 2$ and $f'(1) = 0$.
   $L(x) = f(1) + f'(1)(x - 1)$
   $= 2 + 0(x - 1)$
   $= 2$

   (b) Since $f(1.1) \approx 2.009$ and $L(1.1) = 2$, the approximation differs from the true value by less than 10^{-2}.

5. (a) $f'(x) = \sec^2 x$
   We have $f(\pi) = 0$ and $f'(\pi) = 1$.
   $L(x) = f(\pi) + f'(\pi)(x - \pi)$
   $= 0 + 1(x - \pi)$
   $= x - \pi$

   (b) Since $f(\pi + 0.1) \approx 0.10033$ and $L(\pi + 0.1) = 0.1$, the approximation differs from the true value in absolute value by less than 10^{-3}.

7. $f'(x) = k(1 + x)^{k-1}$
   We have $f(0) = 1$ and $f'(0) = k$.
   $L(x) = f(0) + f'(0)(x - 0)$
   $= 1 + k(x - 0)$
   $= 1 + kx$

9. $f'(x) = \dfrac{1}{2\sqrt{x + 1}} + \cos x$
   We have $f(0) = 1$ and $f'(0) = \dfrac{3}{2}$
   $L(x) = f(0) + f'(0)(x - 0)$
   $= 1 + \dfrac{3}{2}x$

   The linearization is the sum of the two individual linearizations, which are x for $\sin x$ and $1 + \dfrac{1}{2}x$ for $\sqrt{x + 1}$.

11. Center $= -1$
    $f'(x) = 4x + 4$
    We have $f(-1) = -5$ and $f'(-1) = 0$
    $L(x) = f(-1) + f'(-1)(x - (-1)) = -5 + 0(x + 1) = -5$

13. Center $= 1$
    $f'(x) = \dfrac{(x + 1)(1) - (x)(1)}{(x + 1)^2} = \dfrac{1}{(x + 1)^2}$
    We have $f(1) = \dfrac{1}{2}$ and $f'(1) = \dfrac{1}{4}$
    $L(x) = f(1) + f'(1)(x - 1) = \dfrac{1}{2} + \dfrac{1}{4}(x - 1) = \dfrac{1}{4}x + \dfrac{1}{4}$

    Alternate solution:
    Using center $= \dfrac{3}{2}$, we have $f\left(\dfrac{3}{2}\right) = \dfrac{3}{5}$ and $f'\left(\dfrac{3}{2}\right) = \dfrac{4}{25}$.
    $L(x) = f\left(\dfrac{3}{2}\right) + f'\left(\dfrac{3}{2}\right)\left(x - \dfrac{3}{2}\right) = \dfrac{3}{5} + \dfrac{4}{25}\left(x - \dfrac{3}{2}\right) = \dfrac{4}{25}x + \dfrac{9}{25}$

Section 4.5

15. Let $f(x) = x^3 + x - 1$. Then $f'(x) = 3x^2 + 1$ and
$$x_{n+1} = x_n - \frac{f(x_n)}{f'(x_n)} = x_n - \frac{x_n^3 + x_n - 1}{3x_n^2 + 1}.$$
Note that f is cubic and f' is always positive, so there is exactly one solution. We choose $x_1 = 0$.
$x_1 = 0$
$x_2 = 1$
$x_3 = 0.75$
$x_4 \approx 0.6860465$
$x_5 \approx 0.6823396$
$x_6 \approx 0.6823278$
$x_7 \approx 0.6823278$
Solution: $x \approx 0.682328$

17. Let $f(x) = x^2 - 2x + 1 - \sin x$.

Then $f'(x) = 2x - 2 - \cos x$ and
$$x_{n+1} = x_n - \frac{f(x_n)}{f'(x_n)} = x_n - \frac{x_n^2 - 2x_n + 1 - \sin x_n}{2x_n - 2 - \cos x_n}.$$
The graph of $y = f(x)$ shows that $f(x) = 0$ has two solutions

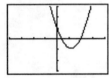

[−4, 4] by [−3, 3]

$x_1 = 0.3$ $x_1 = 2$
$x_2 \approx 0.3825699$ $x_2 \approx 1.9624598$
$x_3 \approx 0.3862295$ $x_3 \approx 1.9615695$
$x_4 \approx 0.3862369$ $x_4 \approx 1.9615690$
$x_5 \approx 0.3862369$ $x_5 \approx 1.9615690$
Solutions: $x \approx 0.386237, 1.961569$

19. (a) Since $\frac{dy}{dx} = 3x^2 - 3$, $dy = (3x^2 - 3)\, dx$.

(b) At the given values,
$dy = (3 \cdot 2^2 - 3)(0.05) = 9(0.05) = 0.45$.

21. (a) Since $\frac{dy}{dx} = (x^2)\left(\frac{1}{x}\right) + (\ln x)(2x) = 2x \ln x + x$,
$dy = (2x \ln x + x)\, dx$.

(b) At the given values,
$dy = [2(1) \ln(1) + 1](0.01) = 1(0.01) = 0.01$

23. (a) Since $\frac{dy}{dx} = e^{\sin x} \cos x$, $dy = (\cos x)e^{\sin x}\, dx$.

(b) At the given values,
$dy = (\cos \pi)(e^{\sin \pi})(-0.1) = (-1)(1)(-0.1) = 0.1$.

25. (a) $y + xy - x = 0$
$y(1 + x) = x$
$y = \frac{x}{x+1}$
Since $\frac{dy}{dx} = \frac{(x+1)(1) - (x)(1)}{(x+1)^2} = \frac{1}{(x+1)^2}$,
$dy = \frac{dx}{(x+1)^2}$.

(b) At the given values,
$dy = \frac{0.01}{(0+1)^2} = 0.01$.

27. (a) $\Delta f = f(0.1) - f(0) = 0.21 - 0 = 0.21$

(b) Since $f'(x) = 2x + 2$, $f'(0) = 2$.
Therefore, $df = 2\, dx = 2(0.1) = 0.2$.

(c) $|\Delta f - df| = |0.21 - 0.2| = 0.01$

29. (a) $\Delta f = f(0.55) - f(0.5) = \frac{20}{11} - 2 = -\frac{2}{11}$

(b) Since $f'(x) = -x^{-2}$, $f'(0.5) = -4$.
Therefore, $df = -4\, dx = -4(0.05) = -0.2 = -\frac{1}{5}$

(c) $|\Delta f - df| = \left|-\frac{2}{11} + \frac{1}{5}\right| = \frac{1}{55}$

31. Note that $\frac{dV}{dr} = 4\pi r^2$, $dV = 4\pi r^2\, dr$. When r changes from a to $a + dr$, the change in volume is approximately $4\pi a^2\, dr$.

33. Note that $\frac{dV}{dx} = 3x^2$, so $dV = 3x^2\, dx$. When x changes from a to $a + dx$, the change in volume is approximately $3a^2\, dx$.

35. Note that $\frac{dV}{dr} = 2\pi rh$, so $dV = 2\pi rh\, dr$. When r changes from a to $a + dr$, the change in volume is approximately $2\pi ah\, dr$.

37. (a) Note that $f'(0) = \cos 0 = 1$.
$L(x) = f(0) + f'(0)(x - 0) = 1 + 1x = x + 1$

(b) $f(0.1) \approx L(0.1) = 1.1$

(c) The actual value is less than 1.1. This is because the derivative is decreasing over the interval [0, 0.1], which means that the graph of $f(x)$ is concave down and lies below its linearization in this interval.

39. Let A = cross section area, C = circumference, and D = diameter. Then $D = \frac{C}{\pi}$, so $\frac{dD}{dC} = \frac{1}{\pi}$ and $dD = \frac{1}{\pi} dC$. Also, $A = \pi \left(\frac{D}{2}\right)^2 = \pi \left(\frac{C}{2\pi}\right)^2 = \frac{C^2}{4\pi}$, so $\frac{dA}{dC} = \frac{C}{2\pi}$ and $dA = \frac{C}{2\pi} dC$. When C increases from 10π in. to $10\pi + 2$ in. the diameter increases by $dD = \frac{1}{\pi}(2) = \frac{2}{\pi} \approx 0.6366$ in. and the area increases by approximately $dA = \frac{10\pi}{2\pi}(2) = 10$ in^2.

41. Let x = side length and A = area. Then $A = x^2$ and $\frac{dA}{dx} = 2x$, so $dA = 2x\, dx$. We want $|dA| \leq 0.02A$, which gives $|2x\, dx| \leq 0.02x^2$, or $|dx| \leq 0.01x$. The side length should be measured with an error of no more than 1%.

43. Let θ = angle of elevation and h = height of building. Then $h = 30 \tan \theta$, so $\frac{dh}{d\theta} = 30 \sec^2 \theta$ and $dh = 30 \sec^2 \theta\, d\theta$. We want $|dh| < 0.04h$, which gives:

$$|30 \sec^2 \theta\, d\theta| < 0.04\, (30 \tan \theta)$$

$$\frac{1}{\cos^2 \theta} |d\theta| < \frac{0.04 \sin \theta}{\cos \theta}$$

$$|d\theta| < 0.04 \sin \theta \cos \theta$$

For $\theta = 75° = \frac{5\pi}{12}$ radians, we have $|d\theta| < 0.04 \sin \frac{5\pi}{12} \cos \frac{5\pi}{12} = 0.01$ radian. The angle should be measured with an error of less than 0.01 radian (or approximately 0.57 degrees), which is a percentage error of approximately 0.76%.

45. (a) Note that $V = \pi r^2 h = 10\pi r^2 = 2.5\pi D^2$, where D is the interior diameter of the tank. Then $\frac{dV}{dD} = 5\pi D$, so $dV = 5\pi D\, dD$. We want $|dV| \leq 0.01V$, which gives $|5\pi D\, dD| \leq 0.01(2.5\pi D^2)$, or $|dD| \leq 0.005D$. The interior diameter should be measured with an error of no more than 0.5%.

(b) Now we let D represent the *exterior* diameter of the tank, and we assume that the paint coverage rate (number of square feet covered per gallon of paint) is known precisely. Then, to determine the amount of paint within 5%, we need to calculate the lateral surface area S with an error of no more than 5%. Note that $S = 2\pi rh = 10\pi D$, so $\frac{dS}{dD} = 10\pi$ and $dS = 10\pi\, dD$. We want $|dS| \leq 0.05S$, which gives $|10\pi\, dD| \leq 0.05(10\pi D)$, or $dD \leq 0.05D$. The exterior diameter should be measured with an error of no more than 5%.

47. We have $\frac{dW}{dg} = -bg^{-2}$, so $dW = -bg^{-2} dg$. Then $\frac{dW_{moon}}{dW_{earth}} = \frac{-b(5.2)^{-2}\, dg}{-b(32)^{-2}\, dg} = \frac{32^2}{5.2^2} \approx 37.87$. The ratio is about 37.87 to 1.

49. If $f'(x) \neq 0$, we have $x_2 = x_1 - \frac{f(x_1)}{f'(x_1)} = x_1 - \frac{0}{f'(x_1)} = x_1$. Therefore $x_2 = x_1$, and all later approximations are also equal to x_1.

51. Note that $f'(x) = \frac{1}{3} x^{-2/3}$ and so $x_{n+1} = x_n - \frac{f(x_n)}{f'(x_n)}$
$= x_n - \frac{x_n^{1/3}}{\frac{x_n^{-2/3}}{3}} = x_n - 3x_n = -2x_n$. For $x_1 = 1$, we have $x_2 = -2, x_3 = 4, x_4 = -8,$ and $x_5 = 16$; $|x_n| = 2^{n-1}$.

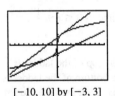

$[-10, 10]$ by $[-3, 3]$

112   Section 4.6

53. Just multiply the corresponding derivative formulas by dx.

(a) Since $\frac{d}{dx}(c) = 0$, $d(c) = 0$.

(b) Since $\frac{d}{dx}(cu) = c\frac{du}{dx}$, $d(cu) = c\,du$.

(c) Since $\frac{d}{dx}(u+v) = \frac{du}{dx} + \frac{dv}{dx}$, $d(u+v) = du + dv$.

(d) Since $\frac{d}{dx}(u \cdot v) = u\frac{dv}{dx} + v\frac{du}{dx}$, $d(u \cdot v) = u\,dv + v\,du$.

(e) Since $\frac{d}{dx}\left(\frac{u}{v}\right) = \frac{v\frac{du}{dx} - u\frac{dv}{dx}}{v^2}$, $d\left(\frac{u}{v}\right) = \frac{v\,du - u\,dv}{v^2}$.

(f) Since $\frac{d}{dx}u^n = nu^{n-1}\frac{du}{dx}$, $d(u^n) = nu^{n-1}\,du$.

55. $g(a) = c$, so if $E(a) = 0$, then $g(a) = f(a)$ and $c = f(a)$.

Then $E(x) = f(x) - g(x) = f(x) - f(a) - m(x-a)$.

Thus, $\frac{E(x)}{x-a} = \frac{f(x)-f(a)}{x-a} - m$.

$\lim_{x \to a} \frac{f(x)-f(a)}{x-a} = f'(a)$, so $\lim_{x \to a} \frac{E(x)}{x-a} = f'(a) - m$.

Therefore, if the limit of $\frac{E(x)}{x-a}$ is zero, then $m = f'(a)$ and

$g(x) = L(x)$.

■ Section 4.6 Related Rates (pp. 232–241)

Exploration 1   Sliding Ladder

1. Here the x-axis represents the ground and the y-axis represents the wall. The curve (x_1, y_1) gives the position of the bottom of the ladder (distance from the wall) at any time t in $0 \le t \le \frac{13}{3}$. The curve (x_2, y_2) gives the position of the top of the ladder at any time in $0 \le t \le \frac{13}{3}$.

2. $0 \le t \le \frac{13}{3}$

3. This is a snapshot at $t \approx 3.1$. The top of the ladder is moving down the y-axis and the bottom of the ladder is moving to the right on the x-axis. Both axes are hidden from view.

[−1, 15] by [−1, 15]

4. $\frac{dy}{dt} = y'(t) = -\frac{9t}{\sqrt{13^2 - 9t^2}}$

$y'(0.5) \approx -0.348$ ft/sec², $y'(1) \approx -0.712$ ft/sec²,

$y'(1.5) \approx -1.107$ ft/sec², $y'(2) \approx -1.561$ ft/sec².

Since $\lim_{t \to (13/3)^-} y'(t) = -\infty$, the speed of the top of the ladder is infinite as it hits the ground.

Quick Review 4.6

1. $D = \sqrt{(7-0)^2 + (0-5)^2} = \sqrt{49+25} = \sqrt{74}$

2. $D = \sqrt{(b-0)^2 + (0-a)^2} = \sqrt{a^2 + b^2}$

3. Use implicit differentiation.

$\frac{d}{dx}(2xy + y^2) = \frac{d}{dx}(x+y)$

$2x\frac{dy}{dx} + 2y(1) + 2y\frac{dy}{dx} = (1) + \frac{dy}{dx}$

$(2x + 2y - 1)\frac{dy}{dx} = 1 - 2y$

$\frac{dy}{dx} = \frac{1-2y}{2x+2y-1}$

4. Use implicit differentiation.

$\frac{d}{dx}(x \sin y) = \frac{d}{dx}(1 - xy)$

$(x)(\cos y)\frac{dy}{dx} + (\sin y)(1) = -x\frac{dy}{dx} - y(1)$

$(x + x \cos y)\frac{dy}{dx} = -y - \sin y$

$\frac{dy}{dx} = \frac{-y - \sin y}{x + x \cos y}$

$\frac{dy}{dx} = -\frac{y + \sin y}{x + x \cos y}$

5. Use implicit differentiation.

$\frac{d}{dx}x^2 = \frac{d}{dx}\tan y$

$2x = \sec^2 y \frac{dy}{dx}$

$\frac{dy}{dx} = \frac{2x}{\sec^2 y}$

$\frac{dy}{dx} = 2x \cos^2 y$

6. Use implicit differentiation.

$\frac{d}{dx}\ln(x+y) = \frac{d}{dx}(2x)$

$\frac{1}{x+y}\left(1 + \frac{dy}{dx}\right) = 2$

$1 + \frac{dy}{dx} = 2(x+y)$

$\frac{dy}{dx} = 2x + 2y - 1$

7. Using $A(-2, 1)$ we create the parametric equations $x = -2 + at$ and $y = 1 + bt$, which determine a line passing through A at $t = 0$. We determine a and b so that the line passes through $B(4, -3)$ at $t = 1$. Since $4 = -2 + a$, we have $a = 6$, and since $-3 = 1 + b$, we have $b = -4$. Thus, one parametrization for the line segment is $x = -2 + 6t$, $y = 1 - 4t$, $0 \le t \le 1$. (Other answers are possible.)

8. Using $A(0, -4)$, we create the parametric equations $x = 0 + at$ and $y = -4 + bt$, which determine a line passing through A at $t = 0$. We now determine a and b so that the line passes through $B(5, 0)$ at $t = 1$. Since $5 = 0 + a$, we have $a = 5$, and since $0 = -4 + b$, we have $b = 4$. Thus, one parametrization for the line segment is $x = 5t$, $y = -4 + 4t$, $0 \le t \le 1$. (Other answers are possible.)

9. One possible answer: $\dfrac{\pi}{2} \le t \le \dfrac{3\pi}{2}$

10. One possible answer: $\dfrac{3\pi}{2} \le t \le 2\pi$

Section 4.6 Exercises

1. Since $\dfrac{dA}{dt} = \dfrac{dA}{dr}\dfrac{dr}{dt}$, we have $\dfrac{dA}{dt} = 2\pi r \dfrac{dr}{dt}$.

3. (a) Since $\dfrac{dV}{dt} = \dfrac{dV}{dh}\dfrac{dh}{dt}$, we have $\dfrac{dV}{dt} = \pi r^2 \dfrac{dh}{dt}$.

   (b) Since $\dfrac{dV}{dt} = \dfrac{dV}{dr}\dfrac{dr}{dt}$, we have $\dfrac{dV}{dt} = 2\pi r h \dfrac{dr}{dt}$.

   (c) $\dfrac{dV}{dt} = \dfrac{d}{dt}\pi r^2 h = \pi \dfrac{d}{dt}(r^2 h)$
   $\dfrac{dV}{dt} = \pi\left(r^2 \dfrac{dh}{dt} + h(2r)\dfrac{dr}{dt}\right)$
   $\dfrac{dV}{dt} = \pi r^2 \dfrac{dh}{dt} + 2\pi r h \dfrac{dr}{dt}$

5. $\dfrac{ds}{dt} = \dfrac{d}{dt}\sqrt{x^2 + y^2 + z^2}$

   $\dfrac{ds}{dt} = \dfrac{1}{2\sqrt{x^2 + y^2 + z^2}} \dfrac{d}{dt}(x^2 + y^2 + z^2)$

   $\dfrac{ds}{dt} = \dfrac{1}{2\sqrt{x^2 + y^2 + z^2}}\left(2x\dfrac{dx}{dt} + 2y\dfrac{dy}{dt} + 2z\dfrac{dz}{dt}\right)$

   $\dfrac{ds}{dt} = \dfrac{x\dfrac{dx}{dt} + y\dfrac{dy}{dt} + z\dfrac{dz}{dt}}{\sqrt{x^2 + y^2 + z^2}}$

7. (a) Since V is increasing at the rate of 1 volt/sec,
   $\dfrac{dV}{dt} = 1$ volt/sec.

   (b) Since I is decreasing at the rate of $\dfrac{1}{3}$ amp/sec,
   $\dfrac{dI}{dt} = -\dfrac{1}{3}$ amp/sec.

   (c) Differentiating both sides of $V = IR$, we have
   $\dfrac{dV}{dt} = I\dfrac{dR}{dt} + R\dfrac{dI}{dt}$.

   (d) Note that $V = IR$ gives $12 = 2R$, so $R = 6$ ohms. Now substitute the known values into the equation in (c).
   $1 = 2\dfrac{dR}{dt} + 6\left(-\dfrac{1}{3}\right)$
   $3 = 2\dfrac{dR}{dt}$
   $\dfrac{dR}{dt} = \dfrac{3}{2}$ ohms/sec

   R is changing at the rate of $\dfrac{3}{2}$ ohms/sec. Since this value is positive, R is increasing.

9. Step 1:
   l = length of rectangle
   w = width of rectangle
   A = area of rectangle
   P = perimeter of rectangle
   D = length of a diagonal of the rectangle

   Step 2:

   At the instant in question,
   $\dfrac{dl}{dt} = -2$ cm/sec, $\dfrac{dw}{dt} = 2$ cm/sec, $l = 12$ cm,
   and $w = 5$ cm.

   Step 3:

   We want to find $\dfrac{dA}{dt}, \dfrac{dP}{dt},$ and $\dfrac{dD}{dt}$.

   Steps 4, 5, and 6:

   (a) $A = lw$
   $\dfrac{dA}{dt} = l\dfrac{dw}{dt} + w\dfrac{dl}{dt}$
   $\dfrac{dA}{dt} = (12)(2) + (5)(-2) = 14$ cm²/sec

   The rate of change of the area is 14 cm²/sec.

   (b) $P = 2l + 2w$
   $\dfrac{dP}{dt} = 2\dfrac{dl}{dt} + 2\dfrac{dw}{dt}$
   $\dfrac{dP}{dt} = 2(-2) + 2(2) = 0$ cm/sec

   The rate of change of the perimeter is 0 cm/sec.

   (c) $D = \sqrt{l^2 + w^2}$
   $\dfrac{dD}{dt} = \dfrac{1}{2\sqrt{l^2 + w^2}}\left(2l\dfrac{dl}{dt} + 2w\dfrac{dw}{dt}\right) = \dfrac{l\dfrac{dl}{dt} + w\dfrac{dw}{dt}}{\sqrt{l^2 + w^2}}$
   $\dfrac{dD}{dt} = \dfrac{(12)(-2) + (5)(2)}{\sqrt{12^2 + 5^2}} = -\dfrac{14}{13}$ cm/sec

   The rate of change of the length of the diameter is $-\dfrac{14}{13}$ cm/sec.

   (d) The area is increasing, because its derivative is positive. The perimeter is not changing, because its derivative is zero. The diagonal length is decreasing, because its derivative is negative.

114 Section 4.6

11. Step 1:
s = (diagonal) distance from antenna to airplane
x = horizontal distance from antenna to airplane

Step 2:
At the instant in question,
$s = 10$ mi and $\dfrac{ds}{dt} = 300$ mph.

Step 3:
We want to find $\dfrac{dx}{dt}$.

Step 4:
$x^2 + 49 = s^2$ or $x = \sqrt{s^2 - 49}$

Step 5:
$\dfrac{dx}{dt} = \dfrac{1}{2\sqrt{s^2 - 49}}\left(2s\dfrac{ds}{dt}\right) = \dfrac{s}{\sqrt{s^2 - 49}}\dfrac{ds}{dt}$

Step 6:
$\dfrac{dx}{dt} = \dfrac{10}{\sqrt{10^2 - 49}}(300) = \dfrac{3000}{\sqrt{51}}$ mph ≈ 420.08 mph

The speed of the airplane is about 420.08 mph.

13. Step 1:
x = distance from wall to base of ladder
y = height of top of ladder
A = area of triangle formed by the ladder, wall, and ground
θ = angle between the ladder and the ground

Step 2:
At the instant in question, $x = 12$ ft and $\dfrac{dx}{dt} = 5$ ft/sec.

Step 3:
We want to find $-\dfrac{dy}{dt}, \dfrac{dA}{dt},$ and $\dfrac{d\theta}{dt}$.

Step 4, 5, and 6:

(a) $x^2 + y^2 = 169$

$2x\dfrac{dx}{dt} + 2y\dfrac{dy}{dt} = 0$

To evaluate, note that, at the instant in question,
$y = \sqrt{169 - x^2} = \sqrt{169 - 12^2} = 5.$

Then $2(12)(5) + 2(5)\dfrac{dy}{dt} = 0$

$\dfrac{dy}{dt} = -12$ ft/sec $\left(\text{or} -\dfrac{dy}{dt} = 12 \text{ ft/sec}\right)$

The top of the ladder is sliding down the wall at the rate of 12 ft/sec. (Note that the *downward* rate of motion is positive.)

(b) $A = \dfrac{1}{2}xy$

$\dfrac{dA}{dt} = \dfrac{1}{2}\left(x\dfrac{dy}{dt} + y\dfrac{dx}{dt}\right)$

Using the results from step 2 and from part (a), we have $\dfrac{dA}{dt} = \dfrac{1}{2}[(12)(-12) + (5)(5)] = -\dfrac{119}{2}$ ft/sec.

The area of the triangle is changing at the rate of -59.5 ft²/sec.

(c) $\tan \theta = \dfrac{y}{x}$

$\sec^2 \theta \dfrac{d\theta}{dt} = \dfrac{x\dfrac{dy}{dt} - y\dfrac{dx}{dt}}{x^2}$

Since $\tan \theta = \dfrac{5}{12}$, we have $\left(\text{for } 0 \leq \theta < \dfrac{\pi}{2}\right)$

$\cos \theta = \dfrac{12}{13}$ and so $\sec^2 \theta = \dfrac{1}{\left(\dfrac{12}{13}\right)^2} = \dfrac{169}{144}.$

Combining this result with the results from step 2 and from part (a), we have $\dfrac{169}{144}\dfrac{d\theta}{dt} = \dfrac{(12)(-12) - (5)(5)}{12^2},$ so

$\dfrac{d\theta}{dt} = -1$ radian/sec. The angle is changing at the rate of -1 radian/sec.

15. Step 1:

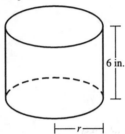

The cylinder shown represents the shape of the hole.
r = radius of cylinder
V = volume of cylinder

Step 2:
At the instant in question, $\dfrac{dr}{dt} = \dfrac{0.001 \text{ in.}}{3 \text{ min}} = \dfrac{1}{3000}$ in./min
and (since the diameter is 3.800 in.), $r = 1.900$ in.

Step 3:
We want to find $\dfrac{dV}{dt}$.

Step 4:
$V = \pi r^2(6) = 6\pi r^2$

Step 5:
$\dfrac{dV}{dt} = 12\pi r \dfrac{dr}{dt}$

Step 6:
$\dfrac{dV}{dt} = 12\pi(1.900)\left(\dfrac{1}{3000}\right) = \dfrac{19\pi}{2500} = 0.0076\pi$
≈ 0.0239 in³/min

The volume is increasing at the rate of approximately 0.0239 in³/min.

17. Step 1:

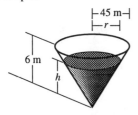

r = radius of top surface of water
h = depth of water in reservoir
V = volume of water in reservoir

Step 2:

At the instant in question, $\frac{dV}{dt} = -50$ m³/min and $h = 5$ m.

Step 3:

We want to find $-\frac{dh}{dt}$ and $\frac{dr}{dt}$.

Step 4:

Note that $\frac{h}{r} = \frac{6}{45}$ by similar cones, so $r = 7.5h$.
Then $V = \frac{1}{3}\pi r^2 h = \frac{1}{3}\pi(7.5h)^2 h = 18.75\pi h^3$

Step 5 and 6:

(a) Since $V = 18.75\pi h^3$, $\frac{dV}{dt} = 56.25\pi h^2 \frac{dh}{dt}$.
Thus $-50 = 56.25\pi(5^2)\frac{dh}{dt}$, and so
$\frac{dh}{dt} = -\frac{8}{225\pi}$ m/min $= -\frac{32}{9\pi}$ cm/min.
The water level is falling by $\frac{32}{9\pi} \approx 1.13$ cm/min.
(Since $\frac{dh}{dt} < 0$, the rate at which the water level is *falling* is positive.)

(b) Since $r = 7.5h$, $\frac{dr}{dt} = 7.5\frac{dh}{dt} = -\frac{80}{3\pi}$ cm/min. The rate of change of the radius of the water's surface is $-\frac{80}{3\pi} \approx -8.49$ cm/min.

19. Step 1:
r = radius of spherical droplet
S = surface area of spherical droplet
V = volume of spherical droplet

Step 2:
No numerical information is given.

Step 3:
We want to show that $\frac{dr}{dt}$ is constant.

Step 4:
$S = 4\pi r^2$, $V = \frac{4}{3}\pi r^3$, $\frac{dV}{dt} = kS$ for some constant k

Steps 5 and 6:
Differentiating $V = \frac{4}{3}\pi r^3$, we have $\frac{dV}{dt} = 4\pi r^2 \frac{dr}{dt}$.
Substituting kS for $\frac{dV}{dt}$ and S for $4\pi r^2$, we have $kS = S\frac{dr}{dt}$, or $\frac{dr}{dt} = k$.

21. Step 1:
l = length of rope
x = horizontal distance from boat to dock
θ = angle between the rope and a vertical line

Step 2:
At the instant in question, $\frac{dl}{dt} = -2$ ft/sec and $l = 10$ ft.

Step 3:
We want to find the values of $-\frac{dx}{dt}$ and $\frac{d\theta}{dt}$.

Step 4, 5, and 6:

(a) $x = \sqrt{l^2 - 36}$
$\frac{dx}{dt} = \frac{l}{\sqrt{l^2 - 36}}\frac{dl}{dt}$
$\frac{dx}{dt} = \frac{10}{\sqrt{10^2 - 36}}(-2) = -2.5$ ft/sec

The boat is approaching the dock at the rate of 2.5 ft/sec.

(b) $\theta = \cos^{-1}\frac{6}{l}$

$\frac{d\theta}{dt} = -\frac{1}{\sqrt{1 - \left(\frac{6}{l}\right)^2}}\left(-\frac{6}{l^2}\right)\frac{dl}{dt}$

$\frac{d\theta}{dt} = -\frac{1}{\sqrt{1 - 0.6^2}}\left(-\frac{6}{10^2}\right)(-2) = -\frac{3}{20}$ radian/sec

The rate of change of angle θ is $-\frac{3}{20}$ radian/sec.

23. (a) $\frac{dc}{dt} = \frac{d}{dt}(x^3 - 6x^2 + 15x)$
$= (3x^2 - 12x + 15)\frac{dx}{dt}$
$= [3(2)^2 - 12(2) + 15](0.1)$
$= 0.3$

$\frac{dr}{dt} = \frac{d}{dt}(9x) = 9\frac{dx}{dt} = 9(0.1) = 0.9$
$\frac{dp}{dt} = \frac{dr}{dt} - \frac{dc}{dt} = 0.9 - 0.3 = 0.6$

(b) $\frac{dc}{dt} = \frac{d}{dt}\left(x^3 - 6x^2 + \frac{45}{x}\right)$
$= \left(3x^2 - 12x - \frac{45}{x^2}\right)\frac{dx}{dt}$
$= \left[3(1.5)^2 - 12(1.5) - \frac{45}{1.5^2}\right](0.05)$
$= -1.5625$

$\frac{dr}{dt} = \frac{d}{dt}(70x) = 70\frac{dx}{dt} = 70(0.05) = 3.5$
$\frac{dp}{dt} = \frac{dr}{dt} - \frac{dc}{dt} = 3.5 - (-1.5625) = 5.0625$

116 Section 4.6

25. Step 1:
Q = rate of CO_2 exhalation (mL/min)
D = difference between CO_2 concentration in blood pumped to the lungs and CO_2 concentration in blood returning from the lungs (mL/L)
y = cardiac output

Step 2:

At the instant in question, $Q = 233$ mL/min, $D = 41$ mL/L, $\frac{dD}{dt} = -2$ (mL/L)/min, and $\frac{dQ}{dt} = 0$ mL/min^2.

Step 3:

We want to find the value of $\frac{dy}{dt}$.

Step 4:

$y = \frac{Q}{D}$

Step 5:

$\frac{dy}{dt} = \frac{D\frac{dQ}{dt} - Q\frac{dD}{dt}}{D^2}$

Step 6:

$\frac{dy}{dt} = \frac{(41)(0) - (233)(-2)}{(41)^2} = \frac{466}{1681} \approx 0.277$ L/min^2

The cardiac output is increasing at the rate of approximately 0.277 L/min^2.

27. Step 1:

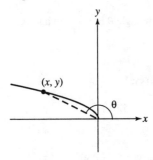

x = x-coordinate of particle's location
y = y-coordinate of particle's location
θ = angle of inclination of line joining the particle to the origin

Step 2:

At the instant in question, $\frac{dx}{dt} = -8$ m/sec and $x = -4$ m.

Step 3:

We want to find $\frac{d\theta}{dt}$.

Step 4:

Since $y = \sqrt{-x}$, we have $\tan\theta = \frac{y}{x} = \frac{\sqrt{-x}}{x} = -(-x)^{-1/2}$, and so, for $x < 0$,

$\theta = \pi + \tan^{-1}[-(-x)^{-1/2}] = \pi - \tan^{-1}(-x)^{-1/2}$.

Step 5:

$\frac{d\theta}{dt} = -\frac{1}{1 + [(-x)^{-1/2}]^2}\left(-\frac{1}{2}(-x)^{-3/2}(-1)\right)\frac{dx}{dt}$

$= -\frac{1}{1 - \left(\frac{1}{x}\right)} \frac{1}{2(-x)^{3/2}} \frac{dx}{dt}$

$= \frac{1}{2\sqrt{-x}(x - 1)} \frac{dx}{dt}$

Step 6:

$\frac{d\theta}{dt} = \frac{1}{2\sqrt{4}(-4-1)}(-8) = \frac{2}{5}$ radian/sec

The angle of inclination is increasing at the rate of $\frac{2}{5}$ radian/sec.

29. Step 1:

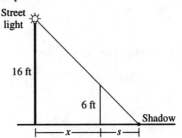

x = distance from streetlight base to man
s = length of shadow

Step 2:

At the instant in question, $\frac{dx}{dt} = -5$ ft/sec and $x = 10$ ft.

Step 3:

We want to find $\frac{ds}{dt}$.

Step 4:

By similar triangles, $\frac{s}{6} = \frac{s + x}{16}$. This is equivalent to $16s = 6s + 6x$, or $s = \frac{3}{5}x$.

Step 5:

$\frac{ds}{dt} = \frac{3}{5}\frac{dx}{dt}$

Step 6:

$\frac{ds}{dt} = \frac{3}{5}(-5) = -3$ ft/sec

The shadow length is changing at the rate of -3 ft/sec.

31. Step 1:
x = position of car ($x = 0$ when car is right in front of you)
θ = camera angle. (We assume θ is negative until the car passes in front of you, and then positive.)

Step 2:
At the first instant in question, $x = 0$ ft and $\frac{dx}{dt} = 264$ ft/sec.
A half second later, $x = \frac{1}{2}(264) = 132$ ft and $\frac{dx}{dt} = 264$ ft/sec.

Step 3:
We want to find $\frac{d\theta}{dt}$ at each of the two instants.

Step 4:
$\theta = \tan^{-1}\left(\frac{x}{132}\right)$

Step 5:
$\frac{d\theta}{dt} = \frac{1}{1 + \left(\frac{x}{132}\right)^2} \cdot \frac{1}{132} \frac{dx}{dt}$

Step 6:
When $x = 0$: $\frac{d\theta}{dt} = \frac{1}{1 + \left(\frac{0}{132}\right)^2}\left(\frac{1}{132}\right)(264) = 2$ radians/sec

When $x = 132$: $\frac{d\theta}{dt} = \frac{1}{1 + \left(\frac{132}{132}\right)^2}\left(\frac{1}{132}\right)(264) = 1$ radian/sec

33. Step 1:
p = x-coordinate of plane's position
x = x-coordinate of car's position
s = distance from plane to car (line-of-sight)

Step 2:
At the instant in question,
$p = 0$, $\frac{dp}{dt} = 120$ mph, $s = 5$ mi, and $\frac{ds}{dt} = -160$ mph.

Step 3:
We want to find $-\frac{dx}{dt}$.

Step 4:
$(x - p)^2 + 3^2 = s^2$

Step 5:
$2(x - p)\left(\frac{dx}{dt} - \frac{dp}{dt}\right) = 2s\frac{ds}{dt}$

Step 6:
Note that, at the instant in question,
$x = \sqrt{5^2 - 3^2} = 4$ mi.

$2(4 - 0)\left(\frac{dx}{dt} - 120\right) = 2(5)(-160)$

$8\left(\frac{dx}{dt} - 120\right) = -1600$

$\frac{dx}{dt} - 120 = -200$

$\frac{dx}{dt} = -80$ mph

The car's speed is 80 mph.

35. Step 1:
a = distance from origin to A
b = distance from origin to B
θ = angle shown in problem statement

Step 2:
At the instant in question, $\frac{da}{dt} = -2$ m/sec, $\frac{db}{dt} = 1$ m/sec, $a = 10$ m, and $b = 20$ m.

Step 3:
We want to find $\frac{d\theta}{dt}$.

Step 4:
$\tan\theta = \frac{a}{b}$ or $\theta = \tan^{-1}\left(\frac{a}{b}\right)$

Step 5:
$\frac{d\theta}{dt} = \frac{1}{1 + \left(\frac{a}{b}\right)^2} \cdot \frac{b\frac{da}{dt} - a\frac{db}{dt}}{b^2} = \frac{b\frac{da}{dt} - a\frac{db}{dt}}{a^2 + b^2}$

Step 6:
$\frac{d\theta}{dt} = \frac{(20)(-2) - (10)(1)}{10^2 + 20^2} = -0.1$ radian/sec
≈ -5.73 degrees/sec

To the nearest degree, the angle is changing at the rate of -6 degrees per second.

37. $\frac{dy}{dt} = \frac{dy}{dx}\frac{dx}{dt} = -10(1 + x^2)^{-2}(2x)\frac{dx}{dt} = -\frac{20x}{(1 + x^2)^2}\frac{dx}{dt}$

Since $\frac{dx}{dt} = 3$ cm/sec, we have
$\frac{dy}{dt} = -\frac{60x}{(1 + x^2)^2}$ cm/sec.

(a) $\frac{dy}{dt} = -\frac{60(-2)}{[1 + (-2)^2]^2} = \frac{120}{5^2} = \frac{24}{5}$ cm/sec

(b) $\frac{dy}{dt} = -\frac{60(0)}{(1 + 0^2)^2} = 0$ cm/sec

(c) $\frac{dy}{dt} = -\frac{60(20)}{(1 + 20^2)^2} \approx -0.00746$ cm/sec

39. (a) The point being plotted would correspond to a point on the edge of the wheel as the wheel turns.

(b) One possible answer is $\theta = 16\pi t$, where t is in seconds. (An arbitrary constant may be added to this expression, and we have assumed counterclockwise motion.)

118 Chapter 4 Review

39. continued

(c) In general, assuming counterclockwise motion:

$$\frac{dx}{dt} = -2\sin\theta\frac{d\theta}{dt} = -2(\sin\theta)(16\pi) = -32\pi\sin\theta$$

$$\frac{dy}{dt} = 2\cos\theta\frac{d\theta}{dt} = 2(\cos\theta)(16\pi) = 32\pi\cos\theta$$

At $\theta = \frac{\pi}{4}$:

$$\frac{dx}{dt} = -32\pi\sin\frac{\pi}{4} = -16\pi(\sqrt{2}) \approx -71.086 \text{ ft/sec}$$

$$\frac{dy}{dt} = 32\pi\cos\frac{\pi}{4} = 16\pi(\sqrt{2}) \approx 71.086 \text{ ft/sec}$$

At $\theta = \frac{\pi}{2}$:

$$\frac{dx}{dt} = -32\pi\sin\frac{\pi}{2} = -32\pi \approx -100.531 \text{ ft/sec}$$

$$\frac{dy}{dt} = 32\pi\cos\frac{\pi}{2} = 0 \text{ ft/sec}$$

At $\theta = \pi$:

$$\frac{dx}{dt} = -32\pi\sin\pi = 0 \text{ ft/sec}$$

$$\frac{dy}{dt} = 32\pi\cos\pi = -32\pi \approx -100.531 \text{ ft/sec}$$

41. (a) $\frac{dy}{dt} = \frac{d}{dt}(uv) = u\frac{dv}{dt} + v\frac{du}{dt}$
$= u(0.05v) + v(0.04u)$
$= 0.09uv$
$= 0.09y$

Since $\frac{dy}{dt} = 0.09y$, the rate of growth of total production is 9% per year.

(b) $\frac{dy}{dt} = \frac{d}{dt}(uv) = u\frac{dv}{dt} + v\frac{du}{dt}$
$= u(0.03v) + v(-0.02u)$
$= 0.01uv$
$= 0.01y$

The total production is increasing at the rate of 1% per year.

■ Chapter 4 Review (pp. 242–245)

1. $y = x\sqrt{2-x}$

$y' = x\left(\frac{1}{2\sqrt{2-x}}\right)(-1) + (\sqrt{2-x})(1)$

$= \frac{-x + 2(2-x)}{2\sqrt{2-x}}$

$= \frac{4 - 3x}{2\sqrt{2-x}}$

The first derivative has a zero at $\frac{4}{3}$.

Critical point value: $x = \frac{4}{3}$ $y = \frac{4\sqrt{6}}{9} \approx 1.09$

Endpoint values: $x = -2$ $y = -4$

$x = 2$ $y = 0$

The global maximum value is $\frac{4\sqrt{6}}{9}$ at $x = \frac{4}{3}$, and the global minimum value is -4 at $x = -2$.

2. Since y is a cubic function with a positive leading coefficient, we have $\lim_{x\to-\infty} y = -\infty$ and $\lim_{x\to\infty} y = \infty$. There are no global extrema.

3. $y' = (x^2)(e^{1/x^2})(-2x^{-3}) + (e^{1/x^2})(2x)$

$= 2e^{1/x^2}\left(-\frac{1}{x} + x\right)$

$= \frac{2e^{1/x^2}(x-1)(x+1)}{x}$

Intervals	$x < -1$	$-1 < x < 0$	$0 < x < 1$	$x > 1$
Sign of y'	−	+	−	+
Behavior of y	Decreasing	Increasing	Decreasing	Increasing

$y'' = \frac{d}{dx}[2e^{1/x^2}(-x^{-1} + x)]$

$= (2e^{1/x^2})(x^{-2} + 1) + (-x^{-1} + x)(2e^{1/x^2})(-2x^{-3})$

$= (2e^{1/x^2})(x^{-2} + 1 + 2x^{-4} - 2x^{-2})$

$= \frac{2e^{1/x^2}(x^4 - x^2 + 2)}{x^4}$

$= \frac{2e^{1/x^2}[(x^2 - 0.5)^2 + 1.75]}{x^4}$

The second derivative is always positive (where defined), so the function is concave up for all $x \neq 0$.

Graphical support:

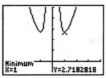

[−4, 4] by [−1, 5]

(a) $[-1, 0)$ and $[1, \infty)$

(b) $(-\infty, -1]$ and $(0, 1]$

(c) $(-\infty, 0)$ and $(0, \infty)$

(d) None

(e) Local (and absolute) minima at $(1, e)$ and $(-1, e)$

(f) None

4. Note that the domain of the function is $[-2, 2]$.

$$y' = x\left(\frac{1}{2\sqrt{4-x^2}}\right)(-2x) + (\sqrt{4-x^2})(1)$$
$$= \frac{-x^2 + (4-x^2)}{\sqrt{4-x^2}}$$
$$= \frac{4 - 2x^2}{\sqrt{4-x^2}}$$

Intervals	$-2 < x < -\sqrt{2}$	$-\sqrt{2} < x < \sqrt{2}$	$\sqrt{2} < x < 2$
Sign of y'	−	+	−
Behavior of y	Decreasing	Increasing	Decreasing

$$y'' = \frac{(\sqrt{4-x^2})(-4x) - (4-2x^2)\left(\frac{1}{2\sqrt{4-x^2}}\right)(-2x)}{4-x^2}$$
$$= \frac{2x(x^2 - 6)}{(4-x^2)^{3/2}}$$

Note that the values $x = \pm\sqrt{6}$ are not zeros of y'' because they fall outside of the domain.

Intervals	$-2 < x < 0$	$0 < x < 2$
Sign of y''	+	−
Behavior of y	Concave up	Concave down

Graphical support:

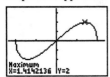

$[-2.35, 2.35]$ by $[-3.5, 3.5]$

(a) $[-\sqrt{2}, \sqrt{2}]$
(b) $[-2, -\sqrt{2}]$ and $[\sqrt{2}, 2]$
(c) $(-2, 0)$
(d) $(0, 2)$
(e) Local maxima: $(-2, 0)$, $(\sqrt{2}, 2)$
   Local minima: $(2, 0)$, $(-\sqrt{2}, -2)$
   Note that the extrema at $x = \pm\sqrt{2}$ are also absolute extrema.
(f) $(0, 0)$

5. $y' = 1 - 2x - 4x^3$
Using grapher techniques, the zero of y' is $x \approx 0.385$.

Intervals	$x < 0.385$	$0.385 < x$
Sign of y'	+	−
Behavior of y	Increasing	Decreasing

$y'' = -2 - 12x^2 = -2(1 + 6x^2)$
The second derivative is always negative so the function is concave down for all x.

Graphical support:

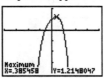

$[-4, 4]$ by $[-4, 2]$

(a) Approximately $(-\infty, 0.385]$
(b) Approximately $[0.385, \infty)$
(c) None
(d) $(-\infty, \infty)$
(e) Local (and absolute) maximum at $\approx (0.385, 1.215)$
(f) None

6. $y' = e^{x-1} - 1$

Intervals	$x < 1$	$1 < x$
Sign of y'	−	+
Behavior of y	Decreasing	Increasing

$y'' = e^{x-1}$

The second derivative is always positive, so the function is concave up for all x.

Graphical support:

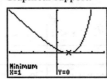

$[-4, 4]$ by $[-2, 4]$

(a) $[1, \infty)$
(b) $(-\infty, 1]$
(c) $(-\infty, \infty)$
(d) None
(e) Local (and absolute) minimum at $(1, 0)$
(f) None

120 Chapter 4 Review

7. Note that the domain is $(-1, 1)$.

$$y = (1 - x^2)^{-1/4}$$

$$y' = -\frac{1}{4}(1 - x^2)^{-5/4}(-2x) = \frac{x}{2(1 - x^2)^{5/4}}$$

Intervals	$-1 < x < 0$	$0 < x < 1$
Sign of y'	−	+
Behavior of y	Decreasing	Increasing

$$y'' = \frac{2(1 - x^2)^{5/4}(1) - (x)(2)\left(\frac{5}{4}\right)(1 - x^2)^{1/4}(-2x)}{4(1 - x^2)^{5/2}}$$

$$= \frac{(1 - x^2)^{1/4}[2 - 2x^2 + 5x^2]}{4(1 - x^2)^{5/2}}$$

$$= \frac{3x^2 + 2}{4(1 - x^2)^{9/4}}$$

The second derivative is always positive, so the function is concave up on its domain $(-1, 1)$.
Graphical support:

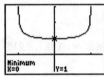

$[-1.3, 1.3]$ by $[-1, 3]$

(a) $[0, 1)$

(b) $(-1, 0]$

(c) $(-1, 1)$

(d) None

(e) Local minimum at $(0, 1)$

(f) None

8. $y' = \frac{(x^3 - 1)(1) - (x)(3x^2)}{(x^3 - 1)^2} = -\frac{2x^3 + 1}{(x^3 - 1)^2}$

Intervals	$x < -2^{-1/3}$	$-2^{-1/3} < x < 1$	$1 < x$
Sign of y'	+	−	−
Behavior of y	Increasing	Decreasing	Decreasing

$$y'' = -\frac{(x^3 - 1)^2(6x^2) - (2x^3 + 1)(2)(x^3 - 1)(3x^2)}{(x^3 - 1)^4}$$

$$= -\frac{(x^3 - 1)(6x^2) - (2x^3 + 1)(6x^2)}{(x^3 - 1)^3}$$

$$= \frac{6x^2(x^3 + 2)}{(x^3 - 1)^3}$$

Intervals	$x < -2^{1/3}$	$-2^{1/3} < x < 0$	$0 < x < 1$	$1 < x$
Sign of y''	+	−	−	+
Behavior of y	Concave up	Concave down	Concave down	Concave up

Graphical support:

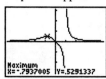

$[-4.7, 4.7]$ by $[-3.1, 3.1]$

(a) $(-\infty, -2^{-1/3}] \approx (-\infty, -0.794]$

(b) $[-2^{-1/3}, 1) \approx [-0.794, 1)$ and $(1, \infty)$

(c) $(-\infty, -2^{1/3}) \approx (-\infty, -1.260)$ and $(1, \infty)$

(d) $(-2^{1/3}, 1) \approx (-1.260, 1)$

(e) Local maximum at
$$\left(-2^{-1/3}, \frac{2}{3} \cdot 2^{-1/3}\right) \approx (-0.794, 0.529)$$

(f) $\left(-2^{1/3}, \frac{1}{3} \cdot 2^{1/3}\right) \approx (-1.260, 0.420)$

9. Note that the domain is $[-1, 1]$.

$$y' = -\frac{1}{\sqrt{1 - x^2}}$$

Since y' is negative on $(-1, 1)$ and y is continuous, y is decreasing on its domain $[-1, 1]$.

$$y'' = \frac{d}{dx}[-(1 - x^2)^{-1/2}]$$
$$= \frac{1}{2}(1 - x^2)^{-3/2}(-2x) = -\frac{x}{(1 - x^2)^{3/2}}$$

Intervals	$-1 < x < 0$	$0 < x < 1$
Sign of y''	+	−
Behavior of y	Concave up	Concave down

Graphical support:

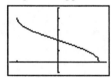

$[-1.175, 1.175]$ by $\left[-\dfrac{\pi}{4}, \dfrac{5\pi}{4}\right]$

(a) None

(b) $[-1, 1]$

(c) $(-1, 0)$

(d) $(0, 1)$

(e) Local (and absolute) maximum at $(-1, \pi)$; local (and absolute) minimum at $(1, 0)$

(f) $\left(0, \dfrac{\pi}{2}\right)$

10. This problem can be solved graphically by using NDER to obtain the graphs shown below.

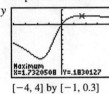
[−4, 4] by [−1, 0.3]

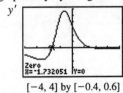
[−4, 4] by [−0.4, 0.6]

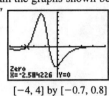
[−4, 4] by [−0.7, 0.8]

An alternative approach using a combination of algebraic and graphical techniques follows.

Note that the denominator of y is always positive because it is equivalent to $(x + 1)^2 + 2$.

$$y' = \frac{(x^2 + 2x + 3)(1) - (x)(2x + 2)}{(x^2 + 2x + 3)^2}$$

$$= \frac{-x^2 + 3}{(x^2 + 2x + 3)^2}$$

Intervals	$x < -\sqrt{3}$	$-\sqrt{3} < x < \sqrt{3}$	$\sqrt{3} < x$
Sign of y'	−	+	−
Behavior of y	Decreasing	Increasing	Decreasing

$$y'' = \frac{(x^2 + 2x + 3)^2(-2x) - (-x^2 + 3)(2)(x^2 + 2x + 3)(2x + 2)}{(x^2 + 2x + 3)^4}$$

$$= \frac{(x^2 + 2x + 3)(-2x) - 2(2x + 2)(-x^2 + 3)}{(x^2 + 2x + 3)^3}$$

$$= \frac{2x^3 - 18x - 12}{(x^2 + 2x + 3)^3}$$

Using graphing techniques, the zeros of $2x^3 - 18x - 12$ (and hence of y'') are at $x \approx -2.584$, $x \approx -0.706$, and $x \approx 3.290$.

Intervals	$(-\infty, -2.584)$	$(-2.584, -0.706)$	$(-0.706, 3.290)$	$(3.290, \infty)$
Sign of y''	−	+	−	+
Behavior of y	Concave down	Concave up	Concave down	Concave up

(a) $[-\sqrt{3}, \sqrt{3}]$

(b) $(-\infty, -\sqrt{3}]$ and $[\sqrt{3}, \infty)$

(c) Approximately $(-2.584, -0.706)$ and $(3.290, \infty)$

(d) Approximately $(-\infty, -2.584)$ and $(-0.706, 3.290)$

(e) Local maximum at $\left(\sqrt{3}, \dfrac{\sqrt{3} - 1}{4}\right)$

$\approx (1.732, 0.183)$;

local minimum at $\left(-\sqrt{3}, \dfrac{-\sqrt{3} - 1}{4}\right)$

$\approx (-1.732, -0.683)$

(f) $\approx(-2.584, -0.573)$, $(-0.706, -0.338)$, and $(3.290, 0.161)$

11. For $x > 0$, $y' = \frac{d}{dx} \ln x = \frac{1}{x}$

For $x < 0$: $y' = \frac{d}{dx} \ln(-x) = \frac{1}{-x}(-1) = \frac{1}{x}$

Thus $y' = \frac{1}{x}$ for all x in the domain.

Intervals	$(-2, 0)$	$(0, 2)$
Sign of y'	−	+
Behavior of y	Decreasing	Increasing

$y'' = -x^{-2}$

The second derivative is always negative, so the function is concave down on each open interval of its domain.

Graphical support:

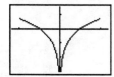

$[-2.35, 2.35]$ by $[-3, 1.5]$

(a) $(0, 2]$

(b) $[-2, 0)$

(c) None

(d) $(-2, 0)$ and $(0, 2)$

(e) Local (and absolute) maxima at $(-2, \ln 2)$ and $(2, \ln 2)$

(f) None

12. $y' = 3 \cos 3x - 4 \sin 4x$

Using graphing techniques, the zeros of y' in the domain $0 \leq x \leq 2\pi$ are $x \approx 0.176$, $x \approx 0.994$, $x = \frac{\pi}{2} \approx 1.57$, $x \approx 2.148$, and $x \approx 2.965$, $x \approx 3.834$, $x = \frac{3\pi}{2}$, $x \approx 5.591$

Intervals	$0 < x < 0.176$	$0.176 < x < 0.994$	$0.994 < x < \frac{\pi}{2}$	$\frac{\pi}{2} < x < 2.148$	$2.148 < x < 2.965$
Sign of y'	+	−	+	−	+
Behavior of y	Increasing	Decreasing	Increasing	Decreasing	Increasing

Intervals	$2.965 < x < 3.834$	$3.834 < x < \frac{3\pi}{2}$	$\frac{3\pi}{2} < x < 5.591$	$5.591 < x < 2\pi$
Sign of y'	−	+	−	+
Behavior of y	Decreasing	Increasing	Decreasing	Increasing

12. continued

$y'' = -9\sin 3x - 16\cos 4x$

Using graphing techniques, the zeros of y'' in the domain $0 \le x \le 2\pi$ are $x \approx 0.542$, $x \approx 1.266$, $x \approx 1.876$, $x \approx 2.600$, $x \approx 3.425$, $x \approx 4.281$, $x \approx 5.144$ and $x \approx 6.000$.

Intervals	$0 < x < 0.542$	$0.542 < x < 1.266$	$1.266 < x < 1.876$	$1.876 < x < 2.600$	$2.600 < x < 3.425$
Sign of y''	−	+	−	+	−
Behavior of y	Concave down	Concave up	Concave down	Concave up	Concave down

Intervals	$3.425 < x < 4.281$	$4.281 < x < 5.144$	$5.144 < x < 6.000$	$6.000 < x < 2\pi$
Sign of y''	+	−	+	−
Behavior of y	Concave up	Concave down	Concave up	Concave down

Graphical support:

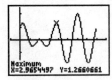

$\left[-\frac{\pi}{4}, \frac{9\pi}{4}\right]$ by $[-2.5, 2.5]$

(a) Approximately $[0, 0.176]$, $\left[0.994, \frac{\pi}{2}\right]$, $[2.148, 2.965]$, $\left[3.834, \frac{3\pi}{2}\right]$, and $[5.591, 2\pi]$

(b) Approximately $[0.176, 0.994]$, $\left[\frac{\pi}{2}, 2.148\right]$, $[2.965, 3.834]$, and $\left[\frac{3\pi}{2}, 5.591\right]$

(c) Approximately $(0.542, 1.266)$, $(1.876, 2.600)$, $(3.425, 4.281)$, and $(5.144, 6.000)$

(d) Approximately $(0, 0.542)$, $(1.266, 1.876)$, $(2.600, 3.425)$, $(4.281, 5.144)$, and $(6.000, 2\pi)$

(e) Local maxima at $\approx(0.176, 1.266)$, $\left(\frac{\pi}{2}, 0\right)$ and $(2.965, 1.266)$, $\left(\frac{3\pi}{2}, 2\right)$, and $(2\pi, 1)$;

local minima at $\approx(0, 1)$, $(0.994, -0.513)$,

$(2.148, -0.513)$, $(3.834, -1.806)$,

and $(5.591, -1.806)$

Note that the local extrema at $x \approx 3.834$, $x = \frac{3\pi}{2}$,

and $x \approx 5.591$ are also absolute extrema.

(f) $\approx(0.542, 0.437)$, $(1.266, -0.267)$, $(1.876, -0.267)$, $(2.600, 0.437)$, $(3.425, -0.329)$, $(4.281, 0.120)$, $(5.144, 0.120)$, and $(6.000, -0.329)$

13. $y' = \begin{cases} -e^{-x}, & x < 0 \\ 4 - 3x^2, & x > 0 \end{cases}$

Intervals	$x < 0$	$0 < x < \dfrac{2}{\sqrt{3}}$	$\dfrac{2}{\sqrt{3}} < x$
Sign of y'	−	+	−
Behavior of y	Decreasing	Increasing	Decreasing

$y'' = \begin{cases} e^{-x}, & x < 0 \\ -6x, & x > 0 \end{cases}$

Intervals	$x < 0$	$0 < x$
Sign of y''	+	−
Behavior of y	Concave up	Concave down

Graphical support:

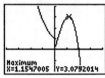

[−4, 4] by [−2, 4]

(a) $\left(0, \dfrac{2}{\sqrt{3}}\right]$

(b) $(-\infty, 0]$ and $\left[\dfrac{2}{\sqrt{3}}, \infty\right)$

(c) $(-\infty, 0)$

(d) $(0, \infty)$

(e) Local maximum at $\left(\dfrac{2}{\sqrt{3}}, \dfrac{16}{3\sqrt{3}}\right) \approx (1.155, 3.079)$

(f) None. Note that there is no point of inflection at $x = 0$ because the derivative is undefined and no tangent line exists at this point.

14. $y' = -5x^4 + 7x^2 + 10x + 4$

Using graphing techniques, the zeros of y' are $x \approx -0.578$ and $x \approx 1.692$.

Intervals	$x < -0.578$	$-0.578 < x < 1.692$	$1.692 < x$
Sign of y'	−	+	−
Behavior of y	Decreasing	Increasing	Decreasing

$y'' = -20x^3 + 14x + 10$
Using graphing techniques, the zero of y'' is $x \approx 1.079$.

Intervals	$x < 1.079$	$1.079 < x$
Sign of y''	+	−
Behavior of y	Concave up	Concave down

Graphical support:

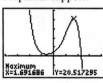

[−4, 4] by [−10, 25]

126 Chapter 4 Review

14. continued

(a) Approximately [−0.578, 1.692]

(b) Approximately (−∞, −0.578] and [1.692, ∞)

(c) Approximately (−∞, 1.079)

(d) Approximately (1.079, ∞)

(e) Local maximum at ≈ (1.692, 20.517); local minimum at ≈ (−0.578, 0.972)

(f) ≈(1.079, 13.601)

15. $y = 2x^{4/5} - x^{9/5}$

$y' = \frac{8}{5}x^{-1/5} - \frac{9}{5}x^{4/5} = \frac{8-9x}{5\sqrt[5]{x}}$

Intervals	$x < 0$	$0 < x < \frac{8}{9}$	$\frac{8}{9} < x$
Sign of y'	−	+	−
Behavior of y	Decreasing	Increasing	Decreasing

$y'' = -\frac{8}{25}x^{-6/5} - \frac{36}{25}x^{-1/5} = -\frac{4(2+9x)}{25x^{6/5}}$

Intervals	$x < -\frac{2}{9}$	$-\frac{2}{9} < x < 0$	$0 < x$
Sign of y''	+	−	−
Behavior of y	Concave up	Concave down	Concave down

Graphical support:

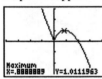

[−4, 4] by [−3, 3]

(a) $\left[0, \frac{8}{9}\right]$

(b) $(-\infty, 0]$ and $\left[\frac{8}{9}, \infty\right)$

(c) $\left(-\infty, -\frac{2}{9}\right)$

(d) $\left(-\frac{2}{9}, 0\right)$ and $(0, \infty)$

(e) Local maximum at $\left(\frac{8}{9}, \frac{10}{9} \cdot \left(\frac{8}{9}\right)^{4/5}\right) \approx (0.889, 1.011)$;
local minimum at (0, 0)

(f) $\left(-\frac{2}{9}, \frac{20}{9} \cdot \left(-\frac{2}{9}\right)^{4/5}\right) \approx \left(-\frac{2}{9}, 0.667\right)$

16. We use a combination of analytic and grapher techniques to solve this problem. Depending on the viewing windows chosen, graphs obtained using NDER may exhibit strange behavior near $x = 2$ because, for example, NDER $(y, 2) \approx 5{,}000{,}000$ while y' is actually undefined at $x = 2$. The graph of $y = \dfrac{5 - 4x + 4x^2 - x^3}{x - 2}$ is shown below.

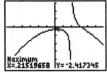

$[-5.875, 5.875]$ by $[-50, 30]$

$$y' = \frac{(x-2)(-4+8x-3x^2)-(5-4x+4x^2-x^3)(1)}{(x-2)^2}$$

$$= \frac{-2x^3+10x^2-16x+3}{(x-2)^2}$$

The graph of y' is shown below.

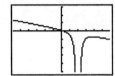

$[-5.875, 5.875]$ by $[-50, 30]$

The zero of y' is $x \approx 0.215$.

Intervals	$x < 0.215$	$0.215 < x < 2$	$2 < x$
Sign of y'	+	−	−
Behavior of y	Increasing	Decreasing	Decreasing

$$y'' = \frac{(x-2)^2(-6x^2+20x-16)-(-2x^3+10x^2-16x+3)(2)(x-2)}{(x-2)^4}$$

$$= \frac{(x-2)(-6x^2+20x-16)-2(-2x^3+10x^2-16x+3)}{(x-2)^3}$$

$$= \frac{-2(x^3-6x^2+12x-13)}{(x-2)^3}$$

The graph of y'' is shown below.

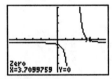

$[-5.875, 5.875]$ by $[-20, 20]$

The zero of $x^3 - 6x^2 + 12x - 13$
(and hence of y'') is $x \approx 3.710$.

Intervals	$x < 2$	$2 < x < 3.710$	$3.710 < x$
Sign of y''	−	+	−
Behavior of y	Concave down	Concave up	Concave down

(a) Approximately $(-\infty, 0.215]$
(b) Approximately $[0.215, 2)$ and $(2, \infty)$
(c) Approximately $(2, 3.710)$
(d) $(-\infty, 2)$ and approximately $(3.710, \infty)$
(e) Local maximum at $\approx (0.215, -2.417)$
(f) $\approx (3.710, -3.420)$

128 Chapter 4 Review

17. $y' = 6(x+1)(x-2)^2$

Intervals	$x < -1$	$-1 < x < 2$	$2 < x$
Sign of y'	−	+	+
Behavior of y	Decreasing	Increasing	Increasing

$y'' = 6(x+1)(2)(x-2) + 6(x-2)^2(1)$
$= 6(x-2)[(2x+2) + (x-2)]$
$= 18x(x-2)$

Intervals	$x < 0$	$0 < x < 2$	$2 < x$
Sign of y''	+	−	+
Behavior of y	Concave up	Concave down	Concave up

(a) There are no local maxima.
(b) There is a local (and absolute) minimum at $x = -1$.
(c) There are points of inflection at $x = 0$ and at $x = 2$.

18. $y' = 6(x+1)(x-2)$

Intervals	$x < -1$	$-1 < x < 2$	$2 < x$
Sign of y'	+	−	+
Behavior of y	Increasing	Decreasing	Increasing

$y'' = \dfrac{d}{dx} 6(x^2 - x - 2) = 6(2x - 1)$

Intervals	$x < \frac{1}{2}$	$\frac{1}{2} < x$
Sign of y''	−	+
Behavior of y	Concave down	Concave up

(a) There is a local maximum at $x = -1$.
(b) There is a local minimum at $x = 2$.
(c) There is a point of inflection at $x = \dfrac{1}{2}$.

19. Since $\dfrac{d}{dx}\left(-\dfrac{1}{4}x^{-4} - e^{-x}\right) = x^{-5} + e^{-x}$,
$f(x) = -\dfrac{1}{4}x^{-4} - e^{-x} + C.$

20. Since $\dfrac{d}{dx} \sec x = \sec x \tan x$, $f(x) = \sec x + C.$

21. Since $\dfrac{d}{dx}\left(2 \ln x + \dfrac{1}{3}x^3 + x\right) = \dfrac{2}{x} + x^2 + 1$,
$f(x) = 2 \ln x + \dfrac{1}{3}x^3 + x + C.$

22. Since $\dfrac{d}{dx}\left(\dfrac{2}{3}x^{3/2} + 2x^{1/2}\right) = \sqrt{x} + \dfrac{1}{\sqrt{x}}$,
$f(x) = \dfrac{2}{3}x^{3/2} + 2x^{1/2} + C.$

23. $f(x) = -\cos x + \sin x + C$
$f(\pi) = 3$
$1 + 0 + C = 3$
$C = 2$
$f(x) = -\cos x + \sin x + 2$

24. $f(x) = \dfrac{3}{4}x^{4/3} + \dfrac{1}{3}x^3 + \dfrac{1}{2}x^2 + x + C$
$f(1) = 0$
$\dfrac{3}{4} + \dfrac{1}{3} + \dfrac{1}{2} + 1 + C = 0$
$C = -\dfrac{31}{12}$
$f(x) = \dfrac{3}{4}x^{4/3} + \dfrac{1}{3}x^3 + \dfrac{1}{2}x^2 + x - \dfrac{31}{12}$

25. $v(t) = s'(t) = 9.8t + 5$
$s(t) = 4.9t^2 + 5t + C$
$s(0) = 10$
$C = 10$
$s(t) = 4.9t^2 + 5t + 10$

26. $a(t) = v'(t) = 32$
$v(t) = 32t + C_1$
$v(0) = 20$
$C_1 = 20$
$v(t) = s'(t) = 32t + 20$
$s(t) = 16t^2 + 20t + C_2$
$s(0) = 5$
$C_2 = 5$
$s(t) = 16t^2 + 20t + 5$

27. $f(x) = \tan x$
$f'(x) = \sec^2 x$
$L(x) = f\left(-\dfrac{\pi}{4}\right) + f'\left(-\dfrac{\pi}{4}\right)\left[x - \left(-\dfrac{\pi}{4}\right)\right]$
$= \tan\left(-\dfrac{\pi}{4}\right) + \sec^2\left(-\dfrac{\pi}{4}\right)\left(x + \dfrac{\pi}{4}\right)$
$= -1 + 2\left(x + \dfrac{\pi}{4}\right)$
$= 2x + \dfrac{\pi}{2} - 1$

28. $f(x) = \sec x$
$f'(x) = \sec x \tan x$
$L(x) = f\left(\dfrac{\pi}{4}\right) + f'\left(\dfrac{\pi}{4}\right)\left(x - \dfrac{\pi}{4}\right)$
$= \sec\left(\dfrac{\pi}{4}\right) + \sec\left(\dfrac{\pi}{4}\right)\tan\left(\dfrac{\pi}{4}\right)\left(x - \dfrac{\pi}{4}\right)$
$= \sqrt{2} + \sqrt{2}(1)\left(x - \dfrac{\pi}{4}\right)$
$= \sqrt{2}x - \dfrac{\pi\sqrt{2}}{4} + \sqrt{2}$

29. $f(x) = \dfrac{1}{1 + \tan x}$
$f'(x) = -(1 + \tan x)^{-2}(\sec^2 x)$
$= -\dfrac{1}{\cos^2 x(1 + \tan x)^2}$
$= -\dfrac{1}{(\cos x + \sin x)^2}$
$L(x) = f(0) + f'(0)(x - 0)$
$= 1 - 1(x - 0)$
$= -x + 1$

30. $f(x) = e^x + \sin x$
$f'(x) = e^x + \cos x$
$L(x) = f(0) + f'(0)(x - 0)$
$ = 1 + 2(x - 0)$
$ = 2x + 1$

31. The global minimum value of $\frac{1}{2}$ occurs at $x = 2$.

32. (a) The values of y' and y'' are both negative where the graph is decreasing and concave down, at T.

(b) The value of y' is negative and the value of y'' is positive where the graph is decreasing and concave up, at P.

33. (a) The function is increasing on the interval $(0, 2]$.

(b) The function is decreasing on the interval $[-3, 0)$.

(c) The local extreme values occur only at the endpoints of the domain. A local maximum value of 1 occurs at $x = -3$, and a local maximum value of 3 occurs at $x = 2$.

34. The 24th day

35.

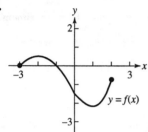

36. (a) We know that f is decreasing on $[0, 1]$ and increasing on $[1, 3]$, the absolute minimum value occurs at $x = 1$ and the absolute maximum value occurs at an endpoint. Since $f(0) = 0, f(1) = -2$, and $f(3) = 3$, the absolute minimum value is -2 at $x = 1$ and the absolute maximum value is 3 at $x = 3$.

(b) The concavity of the graph does not change. There are no points of inflection.

(c)

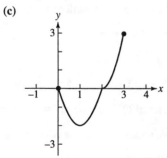

37. (a) $f(x)$ is continuous on $[0.5, 3]$ and differentiable on $(0.5, 3)$.

(b) $f'(x) = (x)\left(\dfrac{1}{x}\right) + (\ln x)(1) = 1 + \ln x$

Using $a = 0.5$ and $b = 3$, we solve as follows.

$$f'(c) = \frac{f(3) - f(0.5)}{3 - 0.5}$$

$$1 + \ln c = \frac{3 \ln 3 - 0.5 \ln 0.5}{2.5}$$

$$\ln c = \frac{\ln\left(\dfrac{3^3}{0.5^{0.5}}\right)}{2.5} - 1$$

$$\ln c = 0.4 \ln(27\sqrt{2}) - 1$$

$$c = e^{-1}(27\sqrt{2})^{0.4}$$

$$c = e^{-1}\sqrt[5]{1458} \approx 1.579$$

(c) The slope of the line is

$$m = \frac{f(b) - f(a)}{b - a} = 0.4 \ln(27\sqrt{2}) = 0.2 \ln 1458,$$ and

the line passes through $(3, 3 \ln 3)$. Its equation is

$y = 0.2(\ln 1458)(x - 3) + 3 \ln 3$, or approximately

$y = 1.457x - 1.075$.

(d) The slope of the line is $m = 0.2 \ln 1458$, and the line passes through
$(c, f(c)) = (e^{-1}\sqrt[5]{1458}, e^{-1}\sqrt[5]{1458}(-1 + 0.2 \ln 1458))$
$\approx (1.579, 0.722)$.
Its equation is
$y = 0.2(\ln 1458)(x - c) + f(c),$
$y = 0.2 \ln 1458(x - e^{-1}\sqrt[5]{1458})$
$ + e^{-1}\sqrt[5]{1458}(-1 + 0.2 \ln 1458),$
$y = 0.2(\ln 1458)x - e^{-1}\sqrt[5]{1458},$
or approximately $y = 1.457x - 1.579$.

38. (a) $v(t) = s'(t) = 4 - 6t - 3t^2$

(b) $a(t) = v'(t) = -6 - 6t$

(c) The particle starts at position 3 moving in the positive direction, but decelerating. At approximately $t = 0.528$, it reaches position 4.128 and changes direction, beginning to move in the negative direction. After that, it continues to accelerate while moving in the negative direction.

39. (a) $L(x) = f(0) + f'(0)(x - 0)$
$ = -1 + 0(x - 0) = -1$

(b) $f(0.1) \approx L(0.1) = -1$

(c) Greater than the approximation in (b), since $f'(x)$ is actually positive over the interval $(0, 0.1)$ and the estimate is based on the derivative being 0.

40. (a) Since $\dfrac{dy}{dx} = (x^2)(-e^{-x}) + (e^{-x})(2x) = (2x - x^2)e^{-x}$,

$dy = (2x - x^2)e^{-x}\,dx$.

(b) $dy = [2(1) - (1)^2](e^{-1})(0.01)$
$ = 0.01e^{-1}$
$ \approx 0.00368$

130 Chapter 4 Review

41. (a) Regression equation $y = \dfrac{2701.73}{1 + 17.28e^{-0.36x}}$

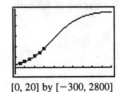

[0, 20] by [−300, 2800]

(b) Note that

$$y' = \frac{d}{dx} 2701.73(1 + 17.28e^{-0.36x})^{-1}$$

$$= -2701.73(1 + 17.28e^{-0.36x})^{-2}(17.28)(-0.36e^{-0.36x})$$

$$\approx \frac{16{,}806.9 e^{-0.36x}}{(1 + 17.28e^{-0.36x})^2}$$

The graph of y' is shown below.

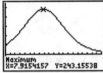

[0, 20] by [−75, 275]

Using graphing techniques, y' has its maximum at $x \approx 7.92$. This corresponds to the year 1998 and represents the inflection point of the logistic curve. The logistic regression equation predicts that the rate of increase in debit card transactions will begin to decrease in 1998, and since $y(7.92) \approx 1351$, there are approximately 1351 million transactions that year.

(c) As x increases, the value of y will increase toward 2701.73. The logistic regression equation predicts a ceiling of approximately 2702 million transactions per year.

42. $f(x) = 2\cos x - \sqrt{1+x}$

$$f'(x) = -2\sin x - \frac{1}{2\sqrt{1+x}}$$

$$x_{n+1} = x_n - \frac{f(x_n)}{f'(x_n)}$$

$$= x_n - \frac{2\cos x_n - \sqrt{1+x_n}}{-2\sin x_n - \dfrac{1}{2\sqrt{1+x_n}}}$$

The graph of $y = f(x)$ shows that $f(x) = 0$ has one solution, near $x = 1$.

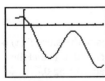

[−2, 10] by [−6, 2]

$x_1 = 1$
$x_2 \approx 0.8361848$
$x_3 \approx 0.8283814$
$x_4 \approx 0.8283608$
$x_5 \approx 0.8283608$
Solution: $x \approx 0.828361$

43. Let t represent time in seconds, where the rocket lifts off at $t = 0$. Since $a(t) = v'(t) = 20$ m/sec^2 and $v(0) = 0$ m/sec, we have $v(t) = 20t$, and so $v(60) = 1200$ m/sec. The speed after 1 minute (60 seconds) will be 1200 m/sec.

44. Let t represent time in seconds, where the rock is blasted upward at $t = 0$. Since $a(t) = v'(t) = -3.72$ m/sec^2 and $v(0) = 93$ m/sec, we have $v(t) = -3.72t + 93$. Since $s'(t) = -3.72t + 93$ and $s(0) = 0$, we have $s(t) = -1.86t^2 + 93t$. Solving $v(t) = 0$, we find that the rock attains its maximum height at $t = 25$ sec and its height at that time is $s(25) = 1162.5$ m.

45. Note that $s = 100 - 2r$ and the sector area is given by

$$A = \pi r^2 \left(\frac{s}{2\pi r}\right) = \frac{1}{2}rs = \frac{1}{2}r(100 - 2r) = 50r - r^2.$$ To find

the domain of $A(r) = 50r - r^2$, note that $r > 0$ and

$0 < s < 2\pi r$, which gives $12.1 \approx \dfrac{50}{\pi + 1} < r < 50$. Since

$A'(r) = 50 - 2r$, the critical point occurs at $r = 25$. This value is in the domain and corresponds to the maximum area because $A''(r) = -2$, which is negative for all r.

The greatest area is attained when $r = 25$ ft and $s = 50$ ft.

46.

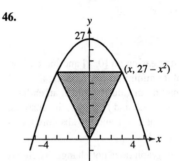

For $0 < x < \sqrt{27}$, the triangle with vertices at (0, 0) and $(\pm x, 27 - x^2)$ has an area given by

$A(x) = \dfrac{1}{2}(2x)(27 - x^2) = 27x - x^3$. Since

$A' = 27 - 3x^2 = 3(3 - x)(3 + x)$ and $A'' = -6x$, the

critical point in the interval $(0, \sqrt{27})$ occurs at $x = 3$ and corresponds to the maximum area because $A''(x)$ is negative in this interval. The largest possible area is

$A(3) = 54$ square units.

47. If the dimensions are x ft by x ft by h ft, then the total amount of steel used is $x^2 + 4xh$ ft^2. Therefore, $x^2 + 4xh = 108$ and so $h = \frac{108 - x^2}{4x}$. The volume is given by $V(x) = x^2 h = \frac{108x - x^3}{4} = 27x - 0.25x^3$. Then $V'(x) = 27 - 0.75x^2 = 0.75(6 + x)(6 - x)$ and $V''(x) = -1.5x$. The critical point occurs at $x = 6$, and it corresponds to the maximum volume because $V''(x) < 0$ for $x > 0$. The corresponding height is $\frac{108 - 6^2}{4(6)} = 3$ ft. The base measures 6 ft by 6 ft, and the height is 3 ft.

48. If the dimensions are x ft by x ft by h ft, then we have $x^2 h = 32$ and so $h = \frac{32}{x^2}$. Neglecting the quarter-inch thickness of the steel, the area of the steel used is $A(x) = x^2 + 4xh = x^2 + \frac{128}{x}$. We can minimize the weight of the vat by minimizing this quantity. Now $A'(x) = 2x - 128x^{-2} = \frac{2}{x^2}(x^3 - 4^3)$ and $A''(x) = 2 + 256x^{-3}$. The critical point occurs at $x = 4$ and corresponds to the minimum possible area because $A''(x) > 0$ for $x > 0$. The corresponding height is $\frac{32}{4^2} = 2$ ft. The base should measure 4 ft by 4 ft, and the height should be 2 ft.

49. We have $r^2 + \left(\frac{h}{2}\right)^2 = 3$, so $r^2 = 3 - \frac{h^2}{4}$. We wish to minimize the cylinder's volume

$V = \pi r^2 h = \pi\left(3 - \frac{h^2}{4}\right)h = 3\pi h - \frac{\pi h^3}{4}$ for $0 < h < 2\sqrt{3}$.

Since $\frac{dV}{dh} = 3\pi - \frac{3\pi h^2}{4} = \frac{3\pi}{4}(2 + h)(2 - h)$ and $\frac{d^2V}{dh^2} = -\frac{3\pi h}{2}$, the critical point occurs at $h = 2$ and it corresponds to the maximum value because $\frac{d^2V}{dh^2} < 0$ for $h > 0$. The corresponding value of r is $\sqrt{3 - \frac{2^2}{4}} = \sqrt{2}$.

The largest possible cylinder has height 2 and radius $\sqrt{2}$.

50. Note that, from similar cones, $\frac{r}{6} = \frac{12 - h}{12}$, so $h = 12 - 2r$. The volume of the smaller cone is given by

$V = \frac{1}{3}\pi r^2 h = \frac{1}{3}\pi r^2(12 - 2r) = 4\pi r^2 - \frac{2\pi}{3}r^3$ for $0 < r < 6$. Then $\frac{dV}{dr} = 8\pi r - 2\pi r^2 = 2\pi r(4 - r)$, so the critical point occurs at $r = 4$. This critical point corresponds to the maximum volume because $\frac{dV}{dr} > 0$ for $0 < r < 4$ and $\frac{dV}{dr} < 0$ for $4 < r < 6$. The smaller cone has the largest possible value when $r = 4$ ft and $h = 4$ ft.

51.

[Figure showing box layout with Lid and Base, 10 in. height, 15 in. width, x dimensions]

(a) $V(x) = x(15 - 2x)(5 - x)$

(b, c) Domain: $0 < x < 5$

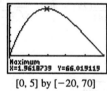

[0, 5] by [−20, 70]

The maximum volume is approximately 66.019 in^3 and it occurs when $x \approx 1.962$ in.

(d) Note that $V(x) = 2x^3 - 25x^2 + 75x$, so $V'(x) = 6x^2 - 50x + 75$. Solving $V'(x) = 0$, we have

$x = \frac{50 \pm \sqrt{(-50)^2 - 4(6)(75)}}{2(6)} = \frac{50 \pm \sqrt{700}}{12}$

$= \frac{50 \pm 10\sqrt{7}}{12} = \frac{25 \pm 5\sqrt{7}}{6}$.

These solutions are approximately $x \approx 1.962$ and $x \approx 6.371$, so the critical point in the appropriate domain occurs at

$x = \frac{25 - 5\sqrt{7}}{6}$.

132 Chapter 4 Review

52.

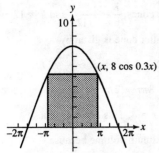

For $0 < x < \dfrac{5\pi}{3}$, the area of the rectangle is given by
$A(x) = (2x)(8 \cos 0.3x) = 16x \cos 0.3x.$

Then $A'(x) = 16x(-0.3 \sin 0.3x) + 16(\cos 0.3x)(1)$
$= 16(\cos 0.3x - 0.3x \sin 0.3x)$

Solving $A'(x) = 0$ graphically, we find that the critical point occurs at $x \approx 2.868$ and the corresponding area is approximately 29.925 square units.

53. The cost (in thousands of dollars) is given by
$C(x) = 40x + 30(20 - y) = 40x + 600 - 30\sqrt{x^2 - 144}.$

Then $C'(x) = 40 - \dfrac{30}{2\sqrt{x^2 - 144}}(2x) = 40 - \dfrac{30x}{\sqrt{x^2 - 144}}.$

Solving $C'(x) = 0$, we have:

$\dfrac{30x}{\sqrt{x^2 - 144}} = 40$

$3x = 4\sqrt{x^2 - 144}$

$9x^2 = 16x^2 - 2304$

$2304 = 7x^2$

Choose the positive solution:

$x = +\dfrac{48}{\sqrt{7}} \approx 18.142 \text{ mi}$

$y = \sqrt{x^2 - 12^2} = \dfrac{36}{\sqrt{7}} \approx 13.607 \text{ mi}$

54. The length of the track is given by $2x + 2\pi r$, so we have
$2x + 2\pi r = 400$ and therefore $x = 200 - \pi r$. Then the area of the rectangle is

$A(r) = 2rx$
$= 2r(200 - \pi r)$
$= 400r - 2\pi r^2$, for $0 < r < \dfrac{200}{\pi}$.

Therefore, $A'(r) = 400 - 4\pi r$ and $A''(r) = -4\pi$, so the critical point occurs at $r = \dfrac{100}{\pi}$ m

and this point corresponds to the maximum rectangle area because $A''(r) < 0$ for all r.

The corresponding value of x is
$x = 200 - \pi\left(\dfrac{100}{\pi}\right) = 100$ m.

The rectangle will have the largest possible area when $x = 100$ m and $r = \dfrac{100}{\pi}$ m.

55. Assume the profit is k dollars per hundred grade B tires and $2k$ dollars per hundred grade A tires.

Then the profit is given by

$P(x) = 2kx + k \cdot \dfrac{40 - 10x}{5 - x}$

$= 2k \cdot \dfrac{(20 - 5x) + x(5 - x)}{5 - x}$

$= 2k \cdot \dfrac{20 - x^2}{5 - x}$

$P'(x) = 2k \cdot \dfrac{(5 - x)(-2x) - (20 - x^2)(-1)}{(5 - x)^2}$

$= 2k \cdot \dfrac{x^2 - 10x + 20}{(5 - x)^2}$

The solutions of $P'(x) = 0$ are

$x = \dfrac{10 \pm \sqrt{(-10)^2 - 4(1)(20)}}{2(1)} = 5 \pm \sqrt{5}$, so the solution in the appropriate domain is $x = 5 - \sqrt{5} \approx 2.76$.

Check the profit for the critical point and endpoints:
Critical point: $x \approx 2.76$ $P(x) \approx 11.06k$
Endpoints: $x = 0$ $P(x) = 8k$
   $x = 4$ $P(x) = 8k$

The highest profit is obtained when $x \approx 2.76$ and $y \approx 5.53$, which corresponds to
276 grade A tires and 553 grade B tires.

56. (a) The distance between the particles is $|f(t)|$ where

$$f(t) = -\cos t + \cos\left(t + \frac{\pi}{4}\right).$$ Then

$$f'(t) = -\sin t + \sin\left(t + \frac{\pi}{4}\right)$$

Solving $f'(t) = 0$ graphically, we obtain $t \approx 1.178$, $t \approx 4.230$, and so on.

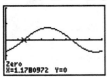

$[0, 2\pi]$ by $[-2, 2]$

Alternatively, $f'(t) = 0$ may be solved analytically as follows.

$$\begin{aligned} f'(t) &= \sin\left[\left(t + \frac{\pi}{8}\right) - \frac{\pi}{8}\right] - \sin\left[\left(t + \frac{\pi}{8}\right) + \frac{\pi}{8}\right] \\ &= \left[\sin\left(t + \frac{\pi}{8}\right)\cos\frac{\pi}{8} - \cos\left(t + \frac{\pi}{8}\right)\sin\frac{\pi}{8}\right] - \left[\sin\left(t + \frac{\pi}{8}\right)\cos\frac{\pi}{8} + \cos\left(t + \frac{\pi}{8}\right)\sin\frac{\pi}{8}\right] \\ &= -2\sin\frac{\pi}{8}\cos\left(t + \frac{\pi}{8}\right), \end{aligned}$$

so the critical points occur when

$\cos\left(t + \frac{\pi}{8}\right) = 0$, or $t = \frac{3\pi}{8} + k\pi$. At each of these values,

$f(t) = \pm 2\cos\frac{3\pi}{8} \approx \pm 0.765$ units, so the maximum distance between

the particles is 0.765 units.

(b) Solving $\cos t = \cos\left(t + \frac{\pi}{4}\right)$ graphically, we obtain $t \approx 2.749$, $t \approx 5.890$, and so on.

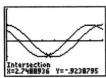

$[0, 2\pi]$ by $[-2, 2]$

Alternatively, this problem may be solved analytically as follows.

$$\begin{aligned} \cos t &= \cos\left(t + \frac{\pi}{4}\right) \\ \cos\left[\left(t + \frac{\pi}{8}\right) - \frac{\pi}{8}\right] &= \cos\left[\left(t + \frac{\pi}{8}\right) + \frac{\pi}{8}\right] \\ \cos\left(t + \frac{\pi}{8}\right)\cos\frac{\pi}{8} + \sin\left(t + \frac{\pi}{8}\right)\sin\frac{\pi}{8} &= \cos\left(t + \frac{\pi}{8}\right)\cos\frac{\pi}{8} - \sin\left(t + \frac{\pi}{8}\right)\cos\frac{\pi}{8} \\ 2\sin\left(t + \frac{\pi}{8}\right)\sin\frac{\pi}{8} &= 0 \\ \sin\left(t + \frac{\pi}{8}\right) &= 0 \\ t &= \frac{7\pi}{8} + k\pi \end{aligned}$$

The particles collide when $t = \frac{7\pi}{8} \approx 2.749$ (plus multiples of π if they keep going.)

57. The dimensions will be x in. by $10 - 2x$ in. by $16 - 2x$ in., so $V(x) = x(10 - 2x)(16 - 2x) = 4x^3 - 52x^2 + 160x$ for $0 < x < 5$.
Then $V'(x) = 12x^2 - 104x + 160 = 4(x - 2)(3x - 20)$, so the critical point in the correct domain is $x = 2$.
This critical point corresponds to the maximum possible volume because $V'(x) > 0$ for $0 < x < 2$ and $V'(x) < 0$ for $2 < x < 5$. The box of largest volume has a height of 2 in. and a base measuring 6 in. by 12 in., and its volume is 144 in³.

Graphical support:

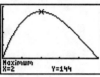

[0, 5] by [−40, 160]

58. Step 1:
r = radius of circle
A = area of circle

Step 2:
At the instant in question, $\dfrac{dr}{dt} = -\dfrac{2}{\pi}$ m/sec and $r = 10$ m.

Step 3:
We want to find $\dfrac{dA}{dt}$.

Step 4:
$A = \pi r^2$

Step 5:
$\dfrac{dA}{dt} = 2\pi r \dfrac{dr}{dt}$

Step 6:
$\dfrac{dA}{dt} = 2\pi(10)\left(-\dfrac{2}{\pi}\right) = -40$

The area is changing at the rate of -40 m²/sec.

59. Step 1:
x = x-coordinate of particle
y = y-coordinate of particle
D = distance from origin to particle

Step 2:
At the instant in question, $x = 5$ m, $y = 12$ m,
$\dfrac{dx}{dt} = -1$ m/sec, and $\dfrac{dy}{dt} = -5$ m/sec.

Step 3:
We want to find $\dfrac{dD}{dt}$.

Step 4:
$D = \sqrt{x^2 + y^2}$

Step 5:
$\dfrac{dD}{dt} = \dfrac{1}{2\sqrt{x^2 + y^2}}\left(2x\dfrac{dx}{dt} + 2y\dfrac{dy}{dt}\right) = \dfrac{x\dfrac{dx}{dt} + y\dfrac{dy}{dt}}{\sqrt{x^2 + y^2}}$

Step 6:
$\dfrac{dD}{dt} = \dfrac{(5)(-1) + (12)(-5)}{\sqrt{5^2 + 12^2}} = -5$ m/sec

Since $\dfrac{dD}{dt}$ is negative, the particle is *approaching* the origin at the *positive* rate of 5 m/sec.

60. Step 1:
x = edge of length of cube
V = volume of cube

Step 2:
At the instant in question,
$\dfrac{dV}{dt} = 1200$ cm³/min and $x = 20$ cm.

Step 3:
We want to find $\dfrac{dx}{dt}$.

Step 4:
$V = x^3$

Step 5:
$\dfrac{dV}{dt} = 3x^2 \dfrac{dx}{dt}$

Step 6:
$1200 = 3(20)^2 \dfrac{dx}{dt}$

$\dfrac{dx}{dt} = 1$ cm/min

The edge length is increasing at the rate of 1 cm/min.

61. Step 1:
$x = $ x-coordinate of point
$y = $ y-coordinate of point
$D = $ distance from origin to point

Step 2:

At the instant in question, $x = 3$ and $\frac{dD}{dt} = 11$ units per sec.

Step 3:

We want to find $\frac{dx}{dt}$.

Step 4:

Since $D^2 = x^2 + y^2$ and $y = x^{3/2}$, we have
$D = \sqrt{x^2 + x^3}$ for $x \geq 0$.

Step 5:

$$\frac{dD}{dt} = \frac{1}{2\sqrt{x^2+x^3}}(2x + 3x^2)\frac{dx}{dt}$$
$$= \frac{2x + 3x^2}{2x\sqrt{1+x}}\frac{dx}{dt} = \frac{3x+2}{2\sqrt{1+x}}\frac{dx}{dt}$$

Step 6:

$11 = \frac{3(3)+2}{2\sqrt{4}}\frac{dx}{dt}$

$\frac{dx}{dt} = 4$ units per sec

62. (a) Since $\frac{h}{r} = \frac{10}{4}$, we may write $h = \frac{5r}{2}$ or $r = \frac{2h}{5}$.

(b) Step 1:
$h = $ depth of water in tank
$r = $ radius of surface of water
$V = $ volume of water in tank

Step 2:

At the instant in question,
$\frac{dV}{dt} = -5$ ft^3/min and $h = 6$ ft.

Step 3:

We want to find $-\frac{dh}{dt}$.

Step 4:

$V = \frac{1}{3}\pi r^2 h = \frac{4}{75}\pi h^3$

Step 5:

$\frac{dV}{dt} = \frac{4}{25}\pi h^2 \frac{dh}{dt}$

Step 6:

$-5 = \frac{4}{25}\pi(6)^2 \frac{dh}{dt}$

$\frac{dh}{dt} = -\frac{125}{144\pi} \approx -0.276$ ft/min

Since $\frac{dh}{dt}$ is negative, the water level is *dropping* at the positive rate of ≈ 0.276 ft/min.

63. Step 1:
$r = $ radius of outer layer of cable on the spool
$\theta = $ clockwise angle turned by spool
$s = $ length of cable that has been unwound

Step 2:

At the instant in question, $\frac{ds}{dt} = 6$ ft/sec and $r = 1.2$ ft

Step 3:

We want to find $\frac{d\theta}{dt}$.

Step 4:

$s = r\theta$

Step 5:

Since r is essentially constant, $\frac{ds}{dt} = r\frac{d\theta}{dt}$.

Step 6:

$6 = 1.2\frac{d\theta}{dt}$

$\frac{d\theta}{dt} = 5$ radians/sec

The spool is turning at the rate of 5 radians per second.

64. $a(t) = v'(t) = -g = -32$ ft/sec^2
Since $v(0) = 32$ ft/sec, $v(t) = s'(t) = -32t + 32$.
Since $s(0) = -17$ ft, $s(t) = -16t^2 + 32t - 17$.
The shovelful of dirt reaches its maximum height when $v(t) = 0$, at $t = 1$ sec. Since $s(1) = -1$, the shovelful of dirt is still below ground level at this time. There was not enough speed to get the dirt out of the hole. Duck!

65. We have $V = \frac{1}{3}\pi r^2 h$, so $\frac{dV}{dr} = \frac{2}{3}\pi rh$ and $dV = \frac{2}{3}\pi rh\, dr$.
When the radius changes from a to $a + dr$, the volume change is approximately $dV = \frac{2}{3}\pi ah\, dr$.

66. (a) Let $x = $ edge of length of cube and $S = $ surface area of cube. Then $S = 6x^2$, which means $\frac{dS}{dx} = 12x$ and $dS = 12x\, dx$. We want $|dS| \leq 0.02S$, which gives $|12x\, dx| \leq 0.02(6x^2)$ or $|dx| \leq 0.01x$. The edge should be measured with an error of no more than 1%.

(b) Let $V = $ volume of cube. Then $V = x^3$, which means $\frac{dV}{dx} = 3x^2$ and $dV = 3x^2\, dx$. We have $|dx| \leq 0.01x$, which means $|3x^2\, dx| \leq 3x^2(0.01x) = 0.03V$, so $|dV| \leq 0.03V$. The volume calculation will be accurate to within approximately 3% of the correct volume.

136 Chapter 4 Review

67. Let $C =$ circumference, $r =$ radius, $S =$ surface area, and $V =$ volume.

(a) Since $C = 2\pi r$, we have $\dfrac{dC}{dr} = 2\pi$ and so $dC = 2\pi\, dr$.

Therefore, $\left|\dfrac{dC}{C}\right| = \left|\dfrac{2\pi\, dr}{2\pi r}\right| = \left|\dfrac{dr}{r}\right| < \dfrac{0.4 \text{ cm}}{10 \text{ cm}} = 0.04$ The calculated radius will be within approximately 4% of the correct radius.

(b) Since $S = 4\pi r^2$, we have $\dfrac{dS}{dr} = 8\pi r$ and so $dS = 8\pi r\, dr$. Therefore, $\left|\dfrac{dS}{S}\right| = \left|\dfrac{8\pi r\, dr}{4\pi r^2}\right|$

$= \left|\dfrac{2\, dr}{r}\right| \leq 2(0.04) = 0.08$. The calculated surface area will be within approximately 8% of the correct surface area.

(c) Since $V = \dfrac{4}{3}\pi r^3$, we have $\dfrac{dV}{dr} = 4\pi r^2$ and so

$dV = 4\pi r^2\, dr$. Therefore $\left|\dfrac{dV}{V}\right| = \left|\dfrac{4\pi r^2\, dr}{\frac{4}{3}\pi r^3}\right| = \left|\dfrac{3\, dr}{r}\right| \leq 3(0.04) = 0.12$.

The calculated volume will be within approximately 12% of the correct volume.

68. By similar triangles, we have $\dfrac{a}{6} = \dfrac{a + 20}{h}$, which gives $ah = 6a + 120$, or $h = 6 + 120a^{-1}$. The height of the lamp post is approximately $6 + 120(15)^{-1} = 14$ ft. The estimated error in measuring a was $|da| \leq 1$ in. $= \dfrac{1}{12}$ ft. Since $\dfrac{dh}{da} = -120a^{-2}$, we have $|dh| = |-120a^{-2}\, da| \leq 120(15)^{-2}\left(\dfrac{1}{12}\right) = \dfrac{2}{45}$ ft, so the estimated possible error is $\pm\dfrac{2}{45}$ ft or $\pm\dfrac{8}{15}$ in.

69. $\dfrac{dy}{dx} = 2\sin x \cos x - 3$. Since $\sin x$ and $\cos x$ are both between 1 and -1, the value of $2\sin x \cos x$ is never greater than 2.

Therefore, $\dfrac{dy}{dx} \leq 2 - 3 = -1$ for all values of x.

Since $\dfrac{dy}{dx}$ is always negative, the function decreases on every interval.

Chapter 5
The Definite Integral

■ Section 5.1 Estimating with Finite Sums
(pp. 247–257)

Exploration 1   Which RAM is the Biggest?

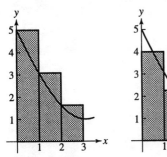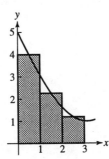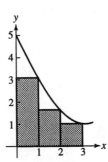

1. LRAM > MRAM > RRAM

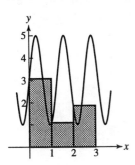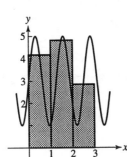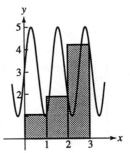

2. MRAM > RRAM > LRAM
3. RRAM > MRAM > LRAM, because the heights of the rectangles increase as you move toward the right under an increasing function.
4. LRAM > MRAM > RRAM, because the heights of the rectangles decrease as you move toward the right under a decreasing function.

Quick Review 5.1

1. 80 mph · 5 hr = 400 mi
2. 48 mph · 3 hr = 144 mi
3. 10 ft/sec^2 · 10 sec = 100 ft/sec

    100 ft/sec · $\frac{1 \text{ mi}}{5280 \text{ ft}}$ · $\frac{3600 \text{ sec}}{1 \text{ h}}$ ≈ 68.18 mph

4. 300,000 km/sec · $\frac{3600 \text{ sec}}{1 \text{ hr}}$ · $\frac{24 \text{ hr}}{1 \text{ day}}$ · $\frac{365 \text{ days}}{1 \text{ yr}}$ · 1 yr ≈ 9.46 × 10^{12} km

5. (6 mph)(3 h) + (5 mph)(2 h) = 18 mi + 10 mi = 28 mi
6. 20 gal/min · 1 h · $\frac{60 \text{ min}}{1 \text{ h}}$ = 1200 gal
7. (−1°C/h)(12 h) + (1.5°C)(6 h) = −3°C
8. 300 ft^3/sec · $\frac{3600 \text{ sec}}{1 \text{ h}}$ · $\frac{24 \text{ h}}{1 \text{ day}}$ · 1 day = 25,920,000 ft^3
9. 350 people/mi^2 · 50 mi^2 = 17,500 people
10. 70 times/sec · $\frac{3600 \text{ sec}}{1 \text{ h}}$ · 1 h · 0.7 = 176,400 times

Section 5.1 Exercises

1. (a)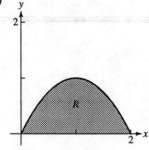

(b)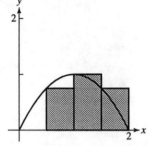

$\Delta x = \frac{1}{2}$

LRAM: $[2(0) - (0)^2](\frac{1}{2}) + [2(\frac{1}{2}) - (\frac{1}{2})^2](\frac{1}{2}) + [2(1) - (1)^2](\frac{1}{2}) + [2(\frac{3}{2}) - (\frac{3}{2})^2](\frac{1}{2}) = \frac{5}{4} = 1.25$

3.

n	$LRAM_n$	$MRAM_n$	$RRAM_n$
10	1.32	1.34	1.32
50	1.3328	1.3336	1.3328
100	1.3332	1.3334	1.3332
500	1.333328	1.333336	1.333328

5.

n	$LRAM_n$	$MRAM_n$	$RRAM_n$
10	12.645	13.4775	14.445
50	13.3218	13.4991	13.6818
100	13.41045	13.499775	13.59045
500	13.482018	13.499991	13.518018

Estimate the area to be 13.5.

7.

n	$LRAM_n$	$MRAM_n$	$RRAM_n$
10	0.98001	0.88220	0.78367
50	0.90171	0.88209	0.86244
100	0.89190	0.88208	0.87226
500	0.88404	0.88208	0.88012
1000	0.88306	0.88208	0.88110

Estimate the area to be 0.8821.

9. LRAM:
   Area
   $\approx f(2) \cdot 2 + f(4) \cdot 2 + f(6) \cdot 2 + \cdots + f(22) \cdot 2$
   $= 2 \cdot (0 + 0.6 + 1.4 + \cdots + 0.5)$
   $= 44.8$ (mg/L) · sec
   RRAM:
   Area
   $\approx f(4) \cdot 2 + f(6) \cdot 2 + f(8) \cdot 2 + \cdots + f(24) \cdot 2$
   $= 2(0.6 + 1.4 + 2.7 + \cdots + 0)$
   $= 44.8$ (mg/L) · sec

   Patient's cardiac output:
   $$\frac{5 \text{ mg}}{44.8 \text{ (mg/L)} \cdot \text{sec}} \cdot \frac{60 \text{ sec}}{1 \text{ min}} \approx 6.7 \text{ L/min}$$
   Note that estimates for the area may vary.

11. 5 min = 300 sec
    (a) LRAM: $300 \cdot (1 + 1.2 + 1.7 + \cdots + 1.2) = 5220$ m
    (b) RRAM: $300 \cdot (1.2 + 1.7 + 2.0 + \cdots + 0) = 4920$ m

13. (a) LRAM: $0.001(0 + 40 + 62 + \cdots + 137) = 0.898$ mi
    RRAM: $0.001(40 + 62 + 82 + \cdots + 142) = 1.04$ mi
    Average $= 0.969$ mi
    (b) The halfway point is 0.4845 mi. The average of LRAM and RRAM is 0.4460 at 0.006 h and 0.5665 at 0.007 h. Estimate that it took 0.006 h = 21.6 sec. The car was going 116 mph.

15. $V = \frac{4}{3}\pi(5)^3 = \frac{500\pi}{3} \approx 523.59878$

n	error	% error
10	2.61799	0.5
20	0.65450	0.125
40	0.16362	0.0312
80	0.04091	0.0078
160	0.01023	0.0020

17. (a) Use RRAM with $\pi(16 - x^2)$.
    $S_8 \approx 120.95132$
    S_8 is an underestimate because each rectangle is below the curve.
    (b) $\frac{|V - S_8|}{V} \approx 0.10 = 10\%$

19. (a) $(5)(6.0 + 8.2 + 9.1 + \cdots + 12.7)(30) \approx 15{,}465$ ft³
    (b) $(5)(8.2 + 9.1 + 9.9 + \cdots + 13.0)(30) \approx 16{,}515$ ft³

21. Use MRAM with πx on the interval $[0, 5]$, $n = 5$.
    $1\left(\frac{1}{2}\pi + \frac{3}{2}\pi + \frac{5}{2}\pi + \frac{7}{2}\pi + \frac{9}{2}\pi\right) = \frac{25}{2}\pi \approx 39.26991$

23. (a) 400 ft/sec $-$ (5 sec)(32 ft/sec²) $=$ 240 ft/sec
    (b) Use RRAM with $400 - 32x$ on $[0, 5]$, $n = 5$.
    $368 + 336 + 304 + 272 + 240 = 1520$ ft

25. (a) Since the release rate of pollutants is increasing, an upper estimate is given by using the data for the end of each month (right rectangles), assuming that new scrubbers were installed before the beginning of January. Upper estimate:
    $30(0.20 + 0.25 + 0.27 + 0.34 + 0.45 + 0.52)$
    ≈ 60.9 tons of pollutants
    A lower estimate is given by using the data for the end of the previous month (left rectangles). We have no data for the beginning of January, but we know that pollutants were released at the new-scrubber rate of 0.05 ton/day, so we may use this value.
    Lower estimate:
    $30(0.05 + 0.20 + 0.25 + 0.27 + 0.34 + 0.45)$
    ≈ 46.8 tons of pollutants

    (b) Using left rectangles, the amount of pollutants released by the end of October is
    $30(0.05 + 0.20 + 0.25 + 0.27 + 0.34 + 0.45 + 0.52 + 0.63 + 0.70 + 0.81) \approx 126.6$ tons.
    Therefore, a total of 125 tons will have been released into the atmosphere by the end of October.

27. (a) The diagonal of the square has length 2, so the side length is $\sqrt{2}$. Area $= (\sqrt{2})^2 = 2$

    (b) Think of the octagon as a collection of 16 right triangles with a hypotenuse of length 1 and an acute angle measuring $\frac{2\pi}{16} = \frac{\pi}{8}$.
    Area $= 16\left(\frac{1}{2}\right)\left(\sin\frac{\pi}{8}\right)\left(\cos\frac{\pi}{8}\right)$
    $= 4 \sin\frac{\pi}{4}$
    $= 2\sqrt{2} \approx 2.828$

    (c) Think of the 16-gon as a collection of 32 right triangles with a hypotenuse of length 1 and an acute angle measuring $\frac{2\pi}{32} = \frac{\pi}{16}$.
    Area $= 32\left(\frac{1}{2}\right)\left(\sin\frac{\pi}{16}\right)\left(\cos\frac{\pi}{16}\right)$
    $= 8 \sin\frac{\pi}{8} \approx 3.061$

    (d) Each area is less than the area of the circle, π. As n increases, the area approaches π.

29. $\text{RRAM}_n f = (\Delta x)[f(x_1) + f(x_2) + \cdots + f(x_{n-1}) + f(x_n)]$
    $= (\Delta x)[f(x_0) + f(x_1) + f(x_2) + \cdots + f(x_{n-1})]$
    $\quad + (\Delta x)[f(x_n) - f(x_0)]$
    $= \text{LRAM}_n f + (\Delta x)[f(x_n) - f(x_0)]$
    But $f(a) = f(b)$ by symmetry, so $f(x_n) - f(x_0) = 0$.
    Therefore, $\text{RRAM}_n f = \text{LRAM}_n f$.

Section 5.2 Definite Integrals
(pp. 258–268)

Exploration 1 Finding Integrals by Signed Areas

1. -2. (This is the same area as $\int_0^\pi \sin x \, dx$, but below the x-axis.)

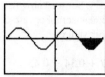
$[-2\pi, 2\pi]$ by $[-3, 3]$

2. 0. (The equal areas above and below the x-axis sum to zero.)

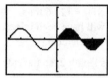
$[-2\pi, 2\pi]$ by $[-3, 3]$

3. 1. (This is half the area of $\int_0^\pi \sin x \, dx$.)

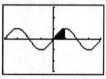
$[-2\pi, 2\pi]$ by $[-3, 3]$

4. $2\pi + 2$. (The same area as $\int_0^\pi \sin x \, dx$ sits above a rectangle of area $\pi \times 2$.)

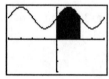
$[-2\pi, 2\pi]$ by $[-3, 3]$

5. 4. (Each rectangle in a typical Riemann sum is twice as tall as in $\int_0^\pi \sin x \, dx$.)

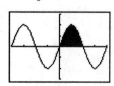
$[-2\pi, 2\pi]$ by $[-3, 3]$

6. 2. (This is the same region as in $\int_0^\pi \sin x \, dx$, translated 2 units to the right.)

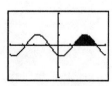
$[-2\pi, 2\pi]$ by $[-3, 3]$

7. 0. (The equal areas above and below the x-axis sum to zero.)

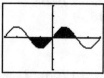
$[-2\pi, 2\pi]$ by $[-3, 3]$

8. 4. (Each rectangle in a typical Riemann sum is twice as wide as in $\int_0^\pi \sin x \, dx$.)

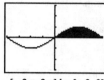
$[-2\pi, 2\pi]$ by $[-3, 3]$

9. 0. (The equal areas above and below the x-axis sum to zero.)

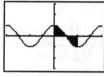
$[-2\pi, 2\pi]$ by $[-3, 3]$

10. 0. (The equal areas above and below the x-axis sum to zero, since $\sin x$ is an odd function.)

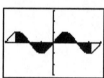
$[-2\pi, 2\pi]$ by $[-3, 3]$

Exploration 2 More Discontinuous Integrands

1. The function has a removable discontinuity at $x = 2$.

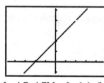
$[-4.7, 4.7]$ by $[-1.1, 5.1]$

2. The thin strip above $x = 2$ has zero area, so the area under the curve is the same as $\int_0^3 (x + 2) \, dx$, which is 10.5.

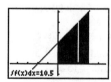
$\int f(x)dx=10.5$
$[-4.7, 4.7]$ by $[-1.1, 5.1]$

Section 5.2 **141**

3. The graph has jump discontinuities at all integer values, but the Riemann sums tend to the area of the shaded region shown. The area is the sum of the areas of 5 rectangles (one of them with height 0):

$$\int_0^5 \text{int}(x)\, dx = 0 + 1 + 2 + 3 + 4 = 10.$$

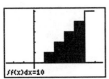

$[-2.7, 6.7]$ by $[-1.1, 5.1]$

Quick Review 5.2

1. $\sum_{n=1}^{5} n^2 = (1)^2 + (2)^2 + (3)^2 + (4)^2 + (5)^2 = 55$

2. $\sum_{k=0}^{4}(3k - 2) = [3(0) - 2] + [3(1) - 2] + [3(2) - 2]$
$+ [3(3) - 2] + [3(4) - 2] = 20$

3. $\sum_{j=0}^{4} 100(j + 1)^2 = 100[(1)^2 + (2)^2 + (3)^2 + (4)^2 + (5)^2]$
$= 5500$

4. $\sum_{k=1}^{99} k$

5. $\sum_{k=0}^{25} 2k$

6. $\sum_{k=1}^{500} 3k^2$

7. $2\sum_{x=1}^{50} x^2 + 3\sum_{x=1}^{50} x = \sum_{x=1}^{50}(2x^2 + 3x)$

8. $\sum_{k=0}^{8} x^k + \sum_{k=9}^{20} x^k = \sum_{k=0}^{20} x^k$

9. $\sum_{k=0}^{n}(-1)^k = 0$ if n is odd.

10. $\sum_{k=0}^{n}(-1)^k = 1$ if n is even.

Section 5.2 Exercises

1. $\lim_{\|P\|\to 0}\sum_{k=1}^{n} c_k^2 \Delta x_k = \int_0^2 x^2\, dx$ where P is any partition of $[0, 2]$.

3. $\lim_{\|P\|\to 0}\sum_{k=1}^{n} \frac{1}{c_k}\Delta x_k = \int_1^4 \frac{1}{x}\, dx$ where P is any partition of $[1, 4]$.

5. $\lim_{\|P\|\to 0}\sum_{k=1}^{n} \sqrt{4 - c_k^2}\,\Delta x_k = \int_0^1 \sqrt{4 - x^2}\, dx$ where P is any partition of $[0, 1]$.

7. $\int_{-2}^{1} 5\, dx = 5[1 - (-2)] = 15$

9. $\int_0^3 (-160)\, dt = (-160)(3 - 0) = -480$

11. $\int_{-2.1}^{3.4} 0.5\, ds = 0.5[3.4 - (-2.1)] = 2.75$

13. Graph the region under $y = \frac{x}{2} + 3$ for $-2 \le x \le 4$.

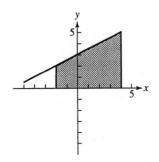

$\int_{-2}^{4}\left(\frac{x}{2} + 3\right) dx = \frac{1}{2}(6)(2 + 5) = 21$

15. Graph the region under $y = \sqrt{9 - x^2}$ for $-3 \le x \le 3$.

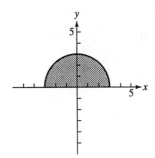

This region is half of a circle of radius 3.

$\int_{-3}^{3} \sqrt{9 - x^2}\, dx = \frac{1}{2}\pi(3)^2 = \frac{9\pi}{2}$

17. Graph the region under $y = |x|$ for $-2 \le x \le 1$.

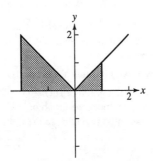

$\int_{-2}^{1} |x|\, dx = \frac{1}{2}(2)(2) + \frac{1}{2}(1)(1) = \frac{5}{2}$

19. Graph the region under $y = 2 - |x|$ for $-1 \le x \le 1$.

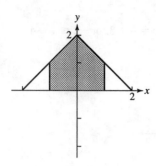

$\int_{-1}^{1} (2 - |x|) \, dx = \frac{1}{2}(1)(1 + 2) + \frac{1}{2}(1)(1 + 2) = 3$

21. Graph the region under $y = \theta$ for $\pi \le \theta \le 2\pi$.

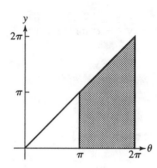

$\int_{\pi}^{2\pi} \theta \, d\theta = \frac{1}{2}(2\pi - \pi)(2\pi + \pi) = \frac{3\pi^2}{2}$

23. $\int_{0}^{b} x \, dx = \frac{1}{2}(b)(b) = \frac{1}{2}b^2$

25. $\int_{a}^{b} 2s \, ds = \frac{1}{2}(b - a)(2b + 2a) = b^2 - a^2$

27. $\int_{a}^{2a} x \, dx = \frac{1}{2}(2a - a)(2a + a) = \frac{3a^2}{2}$

29. Observe that the graph of $f(x) = x^3$ is symmetric with respect to the origin. Hence the area above and below the x-axis is equal for $-1 \le x \le 1$.

$\int_{-1}^{1} x^3 \, dx = -(\text{area below x-axis}) + (\text{area above x-axis}) = 0$

31. Observe that the region under the graph of $f(x) = (x - 2)^3$ for $2 \le x \le 3$ is just the region under the graph of $g(x) = x^3$ for $0 \le x \le 1$ translated two units to the right.

$\int_{2}^{3} (x - 2)^3 \, dx = \int_{0}^{1} x^3 \, dx = \frac{1}{4}$

33. Observe from the graph below that the region under the graph of $f(x) = 1 - x^3$ for $0 \le x \le 1$ cuts out a region R from the square identical to the region under the graph of $g(x) = x^3$ for $0 \le x \le 1$.

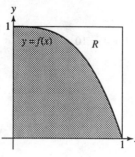

$\int_{0}^{1} (1 - x^3) \, dx = 1 - \int_{0}^{1} x^3 \, dx = 1 - \frac{1}{4} = \frac{3}{4}$

35. Observe that the graph of $f(x) = \left(\frac{x}{2}\right)^3$ for $0 \le x \le 2$ is just a horizontal stretch of the graph of $g(x) = x^3$ for $0 \le x \le 1$ by a factor of 2. Thus the area under $f(x) = \left(\frac{x}{2}\right)^3$ for $0 \le x \le 2$ is twice the area under the graph of $g(x) = x^3$ for $0 \le x \le 1$.

$\int_{0}^{2} \left(\frac{x}{2}\right)^3 dx = 2\int_{0}^{1} x^3 \, dx = \frac{1}{2}$

37. Observe from the graph below that the region between the graph of $f(x) = x^3 - 1$ and the x-axis for $0 \le x \le 1$ cuts out a region R from the square identical to the region under the graph of $g(x) = x^3$ for $0 \le x \le 1$.

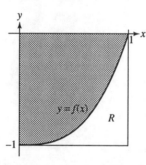

$\int_{0}^{1} (x^3 - 1) \, dx = -1 + \frac{1}{4} = -\frac{3}{4}$

39. NINT$\left(\frac{x}{x^2 + 4}, x, 0, 5\right) \approx 0.9905$

41. NINT$(4 - x^2, x, -2, 2) \approx 10.6667$

43. (a) The function has a discontinuity at $x = 0$.

(b)

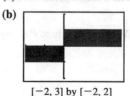

$[-2, 3]$ by $[-2, 2]$

$\int_{-2}^{3} \frac{x}{|x|} \, dx = -2 + 3 = 1$

45. (a) The function has a discontinuity at $x = -1$.

(b)
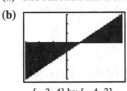
[−3, 4] by [−4, 3]

$$\int_{-3}^{4} \frac{x^2 - 1}{x + 1} \, dx = -\frac{1}{2}(4)(4) + \frac{1}{2}(3)(3) = -\frac{7}{2}$$

47. (a) As x approaches 0 from the right, $f(x)$ goes to ∞.

(b) Using right endpoints we have

$$\int_0^1 \frac{1}{x^2} \, dx = \lim_{n \to \infty} \sum_{k=1}^{n} \left(\frac{1}{\left(\frac{k}{n}\right)^2}\right)\left(\frac{1}{n}\right)$$
$$= \lim_{n \to \infty} \sum_{k=1}^{n} \frac{n}{k^2}$$
$$= \lim_{n \to \infty} n\left(1 + \frac{1}{2^2} + \cdots + \frac{1}{n^2}\right).$$

Note that $n\left(1 + \frac{1}{2^2} + \cdots + \frac{1}{n^2}\right) > n$ and $n \to \infty$, so

$$n\left(1 + \frac{1}{2^2} + \cdots + \frac{1}{n^2}\right) \to \infty.$$

■ Section 5.3 Definite Integrals and Antiderivatives (pp. 268–276)

Exploration 1 How Long is the Average Chord of a Circle?

1. The chord is twice as long as the leg of the right triangle in the first quadrant, which has length $\sqrt{r^2 - x^2}$ by the Pythagorean Theorem.

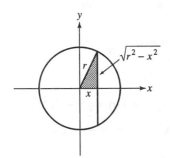

2. Average value $= \dfrac{1}{r - (-r)} \int_{-r}^{r} 2\sqrt{r^2 - x^2} \, dx.$

3. Average value $= \dfrac{2}{2r} \int_{-r}^{r} \sqrt{r^2 - x^2} \, dx$

$= \dfrac{1}{r} \cdot$ (area of semicircle of radius r)

$= \dfrac{1}{r} \cdot \dfrac{\pi r^2}{2}$

$= \dfrac{\pi r}{2}$

4. Although we only computed the average length of chords perpendicular to a particular diameter, the same computation applies to any diameter. The average length of a chord of a circle of radius r is $\dfrac{\pi r}{2}$.

5. The function $y = 2\sqrt{r^2 - x^2}$ is continuous on $[-r, r]$, so the Mean Value Theorem applies and there is a c in $[a, b]$ so that $y(c)$ is the average value $\dfrac{\pi r}{2}$.

Exploration 2 Finding the Derivative of an Integral

Pictures will vary according to the value of x chosen. (Indeed, this is the point of the exploration.) We show a typical solution here.

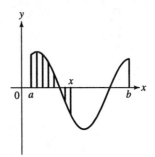

1. We have chosen an arbitrary x between a and b.

2. We have shaded the region using vertical line segments.

3. The shaded region can be written as $\int_a^x f(t) \, dt$ using the definition of the definite integral in Section 5.2. We use t as a dummy variable because x cannot vary between a and itself.

4. The area of the shaded region is our value of $F(x)$.

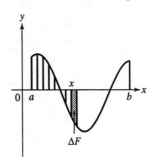

5. We have drawn one more vertical shading segment to represent ΔF.

144 Section 5.3

6. We have moved x a distance of Δx so that it rests above the new shading segment.

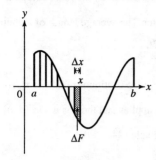

7. Now the (signed) height of the newly-added vertical segment is $f(x)$.

8. The (signed) area of the segment is $\Delta F = \Delta x \cdot f(x)$,

   so $F'(x) = \lim\limits_{\Delta x \to 0} \dfrac{\Delta F}{\Delta x} = f(x)$

Quick Review 5.3

1. $\dfrac{dy}{dx} = \sin x$

2. $\dfrac{dy}{dx} = \cos x$

3. $\dfrac{dy}{dx} = \dfrac{\sec x \tan x}{\sec x} = \tan x$

4. $\dfrac{dy}{dx} = \dfrac{\cos x}{\sin x} = \cot x$

5. $\dfrac{dy}{dx} = \dfrac{\sec x \tan x + \sec^2 x}{\sec x + \tan x} = \sec x$

6. $\dfrac{dy}{dx} = x\left(\dfrac{1}{x}\right) + \ln x - 1 = \ln x$

7. $\dfrac{dy}{dx} = \dfrac{(n+1)x^n}{n+1} = x^n$

8. $\dfrac{dy}{dx} = -\dfrac{1}{(2^x+1)^2} \cdot (\ln 2)2^x = -\dfrac{2^x \ln 2}{(2^x+1)^2}$

9. $\dfrac{dy}{dx} = xe^x + e^x$

10. $\dfrac{dy}{dx} = \dfrac{1}{x^2+1}$

Section 5.3 Exercises

1. (a) $\displaystyle\int_2^2 g(x)\,dx = 0$

   (b) $\displaystyle\int_5^1 g(x)\,dx = -\int_1^5 g(x)\,dx = -8$

   (c) $\displaystyle\int_1^2 3f(x)\,dx = 3\int_1^2 f(x)\,dx = 3(-4) = -12$

   (d) $\displaystyle\int_2^5 f(x)\,dx = \int_2^1 f(x)\,dx + \int_1^5 f(x)\,dx$
   $= -\int_1^2 f(x)\,dx + \int_1^5 f(x)\,dx$
   $= 4 + 6 = 10$

   (e) $\displaystyle\int_1^5 [f(x) - g(x)]\,dx = \int_1^5 f(x)\,dx - \int_1^5 g(x)\,dx$
   $= 6 - 8 = -2$

   (f) $\displaystyle\int_1^5 [4f(x) - g(x)]\,dx = \int_1^5 4f(x)\,dx - \int_1^5 g(x)\,dx$
   $= 4\int_1^5 f(x)\,dx - \int_1^5 g(x)\,dx$
   $= 4(6) - 8 = 16$

3. (a) $\displaystyle\int_1^2 f(u)\,du = 5$

   (b) $\displaystyle\int_1^2 \sqrt{3}f(z)\,dz = \sqrt{3}\int_1^2 f(z)\,dz = 5\sqrt{3}$

   (c) $\displaystyle\int_2^1 f(t)\,dt = -\int_1^2 f(t)\,dt = -5$

   (d) $\displaystyle\int_1^2 [-f(x)]\,dx = -\int_1^2 f(x)\,dx = -5$

5. (a) $\displaystyle\int_3^4 f(z)\,dz = \int_3^0 f(z)\,dz + \int_0^4 f(z)\,dz$
   $= -\int_0^3 f(z)\,dz + \int_0^4 f(z)\,dz$
   $= -3 + 7 = 4$

   (b) $\displaystyle\int_4^3 f(t)\,dt = \int_4^0 f(t)\,dt + \int_0^3 f(t)\,dt$
   $= -\int_0^4 f(t)\,dt + \int_0^3 f(t)\,dt$
   $= -7 + 3 = -4$

7. An antiderivative of 7 is $F(x) = 7x$.

   $\displaystyle\int_3^1 7\,dx = F(1) - F(3) = 7 - 21 = -14$

9. An antiderivative of $\dfrac{x}{8}$ is $F(x) = \dfrac{x^2}{16}$.

   $\displaystyle\int_3^5 \dfrac{x}{8}\,dx = F(5) - F(3) = \dfrac{25}{16} - \dfrac{9}{16} = 1$

11. An antiderivative of $t - \sqrt{2}$ is $F(t) = \dfrac{1}{2}t^2 - t\sqrt{2}$.

    $\displaystyle\int_0^{\sqrt{2}} (t - \sqrt{2})\,dt = F(\sqrt{2}) - F(0) = -1 - 0 = -1$

13. An antiderivative of $\dfrac{1}{1+x^2}$ is $F(x) = \tan^{-1} x$.

    $\displaystyle\int_{-1}^1 \dfrac{1}{1+x^2}\,dx = F(1) - F(-1) = \dfrac{\pi}{4} - \left(-\dfrac{\pi}{4}\right) = \dfrac{\pi}{2}$

15. An antiderivative of e^x is $F(x) = e^x$.

    $\displaystyle\int_0^2 e^x\,dx = F(2) - F(0) = e^2 - 1 \approx 6.389$

17. Divide the shaded area as follows.

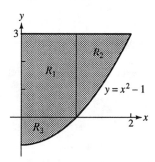

Note that an antiderivative of $x^2 - 1$ is $F(x) = \frac{1}{3}x^3 - x$.

Area of $R_1 = 3(1) = 3$

Area of $R_2 = (3)(1) - \int_1^2 (x^2 - 1)\, dx$

$= 3 - [F(2) - F(1)]$

$= 3 - \left[\left(\frac{2}{3}\right) - \left(-\frac{2}{3}\right)\right] = \frac{5}{3}$

Area of $R_3 = -\int_0^1 (x^2 - 1)\, dx$

$= -[F(1) - F(0)]$

$= -\left[\left(-\frac{2}{3}\right) - 0\right] = \frac{2}{3}$

Total shaded area $= 3 + \frac{5}{3} + \frac{2}{3} = \frac{16}{3}$

19. Divide the shaded area as follows.

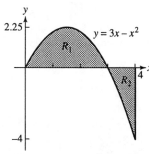

Note that an antiderivative of $3x - x^2$ is $F(x) = \frac{3}{2}x^2 - \frac{1}{3}x^3$.

Area of $R_1 = \int_0^3 (3x - x^2)\, dx = F(3) - F(0) = \frac{9}{2} - 0 = \frac{9}{2}$

Area of $R_2 = -\int_3^4 (3x - x^2)\, dx$

$= -[F(4) - F(3)]$

$= -\left(\frac{8}{3} - \frac{9}{2}\right) = \frac{11}{6}$

Total shaded area $= \frac{9}{2} + \frac{11}{6} = \frac{19}{3}$

21.

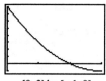

[0, 3] by [−1, 8]

An antiderivative of $x^2 - 6x + 8$ is $F(x) = \frac{1}{3}x^3 - 3x^2 + 8x$.

(a) $\int_0^3 (x^2 - 6x + 8)\, dx = F(3) - F(0) = 6 - 0 = 6$

(b) Area $= \int_0^2 (x^2 - 6x + 8)\, dx - \int_2^3 (x^2 - 6x + 8)\, dx$

$= [F(2) - F(0)] - [F(3) - F(2)]$

$= \left(\frac{20}{3} - 0\right) - \left(6 - \frac{20}{3}\right) = \frac{22}{3}$

23.

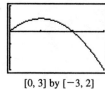

[0, 3] by [−3, 2]

An antiderivative of $2x - x^2$ is $F(x) = x^2 - \frac{1}{3}x^3$.

(a) $\int_0^3 (2x - x^2)\, dx = F(3) - F(0) = 0 - 0 = 0$

(b) Area $= \int_0^2 (2x - x^2)\, dx - \int_2^3 (2x - x^2)\, dx$

$= [F(2) - F(0)] - [F(3) - F(2)]$

$= \left(\frac{4}{3} - 0\right) - \left(0 - \frac{4}{3}\right) = \frac{8}{3}$

25. An antiderivative of $x^2 - 1$ is $F(x) = \frac{1}{3}x^3 - x$.

$av = \frac{1}{\sqrt{3}}\int_0^{\sqrt{3}} (x^2 - 1)\, dx$

$= \frac{1}{\sqrt{3}}\left[F(\sqrt{3}) - F(0)\right]$

$= \frac{1}{\sqrt{3}}(0 - 0) = 0$

Find $x = c$ in $[0, \sqrt{3}]$ such that $c^2 - 1 = 0$

$c^2 = 1$

$c = \pm 1$

Since 1 is in $[0, \sqrt{3}]$, $x = 1$.

27. An antiderivative of $-3x^2 - 1$ is $F(x) = -x^3 - x$.

$av = \frac{1}{1}\int_0^1 (-3x^2 - 1)\, dx = F(1) - F(0) = -2$

Find $x = c$ in $[0, 1]$ such that $-3c^2 - 1 = -2$

$-3c^2 = -1$

$c^2 = \frac{1}{3}$

$c = \pm\frac{1}{\sqrt{3}}$

Since $\frac{1}{\sqrt{3}}$ is in $[0, 1]$, $x = \frac{1}{\sqrt{3}}$.

146 Section 5.4

29. The region between the graph and the x-axis is a triangle of height 3 and base 6, so the area of the region is $\frac{1}{2}(3)(6) = 9$.

$av(f) = \frac{1}{6}\int_{-4}^{2} f(x)\, dx = \frac{9}{6} = \frac{3}{2}$

31. There are equal areas above and below the x-axis.

$av(f) = \frac{1}{2\pi}\int_{0}^{2\pi} f(t)\, dt = \frac{1}{2\pi} \cdot 0 = 0$

33. $\min f = \frac{1}{2}$ and $\max f = 1$

$\frac{1}{2} \leq \int_{0}^{1} \frac{1}{1+x^4}\, dx \leq 1$

35. $\max \sin(x^2) = \sin(1)$ on $[0, 1]$

$\int_{0}^{1} \sin(x^2)\, dx \leq \sin(1) < 1$

37. $(b - a)\min f(x) \geq 0$ on $[a, b]$

$0 \leq (b - a)\min f(x) \leq \int_{a}^{b} f(x)\, dx$

39. Yes, $\int_{a}^{b} av(f)\, dx = \int_{a}^{b} f(x)\, dx$.

This is because $av(f)$ is a constant, so

$\int_{a}^{b} av(f)\, dx = \left[av(f) \cdot x \right]_{a}^{b}$

$= av(f) \cdot b - av(f) \cdot a$

$= (b - a)av(f)$

$= (b - a)\left[\frac{1}{b-a}\int_{a}^{b} f(x)\, dx \right]$

$= \int_{a}^{b} f(x)\, dx$

41. Time for first release $= \dfrac{1000 \text{ m}^3}{10 \text{ m}^3/\text{min}} = 100$ min

Time for second release $= \dfrac{1000 \text{ m}^3}{20 \text{ m}^3/\text{min}} = 50$ min

Average rate $= \dfrac{\text{total released}}{\text{total time}} = \dfrac{2000 \text{ m}^3}{150 \text{ min}} = 13\frac{1}{3}$ m^3/min

43. $\int_{0}^{1} \sec x\, dx \geq \int_{0}^{1}\left(1 + \frac{x^2}{2}\right) dx = \left[x + \frac{x^3}{6}\right]_{0}^{1} = \frac{7}{6}$

45. $av(x^k) = \frac{1}{k}\int_{0}^{k} x^k\, dx = \frac{1}{k}\left[\frac{1}{k+1}x^{k+1}\right]_{0}^{k} = \frac{k^{k+1}}{k(k+1)}$

Graph $y_1 = \dfrac{x^{x+1}}{x(x+1)}$ and $y_2 = x$ on a graphing calculator and find the point of intersection for $x > 1$.

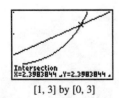

$[1, 3]$ by $[0, 3]$

Thus, $k \approx 2.39838$

■ Section 5.4 Fundamental Theorem of Calculus (pp. 277–288)

Exploration 1 Graphing NINT f

2. The function $y = \tan x$ has vertical asymptotes at all odd multiples of $\frac{\pi}{2}$. There are six of these between -10 and 10.

3. In attempting to find $F(-10) = \int_{3}^{-10} \tan(t)\, dt + 5$, the calculator must find a limit of Riemann sums for the integral, using values of $\tan t$ for t between -10 and 3. The large positive and negative values of $\tan t$ found near the asymptotes cause the sums to fluctuate erratically so that no limit is approached. (We will see in Section 8.3 that the "areas" near the asymptotes are infinite, although NINT is not designed to determine this.)

4. $y = \tan x$

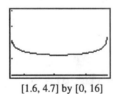

$[1.6, 4.7]$ by $[-2, 2]$

5. The domain of this continuous function is the open interval $\left(\dfrac{\pi}{2}, \dfrac{3\pi}{2}\right)$.

6. The domain of F is the same as the domain of the continuous function in step 4, namely $\left(\dfrac{\pi}{2}, \dfrac{3\pi}{2}\right)$.

7. We need to choose a closed window narrower than $\left(\dfrac{\pi}{2}, \dfrac{3\pi}{2}\right)$ to avoid the asymptotes.

$[1.6, 4.7]$ by $[0, 16]$

8. The graph of F looks like the graph in step 7. It would be decreasing on $\left(\dfrac{\pi}{2}, \pi\right]$ and increasing on $\left[\pi, \dfrac{3\pi}{2}\right)$, with vertical asymptotes at $x = \dfrac{\pi}{2}$ and $x = \dfrac{3\pi}{2}$.

Exploration 2 The Effect of Changing *a* in $\int_a^x f(t)\,dt$

1.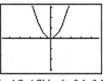
[−4.7, 4.7] by [−3.1, 3.1]

2.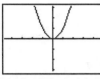
[−4.7, 4.7] by [−3.1, 3.1]

3. Since NINT$(x^2, x, 0, 0) = 0$, the x-intercept is 0.

4. Since NINT$(x^2, x, 5, 5) = 0$, the x-intercept is 5.

5. Changing *a* has no effect on the graph of $y = \dfrac{d}{dx}\int_a^x f(t)\,dt$.

It will always be the same as the graph of $y = f(x)$.

6. Changing *a* shifts the graph of $y = \int_a^x f(t)\,dt$ vertically in such a way that *a* is always the x-intercept. If we change from a_1 to a_2, the distance of the vertical shift is $\int_{a_2}^{a_1} f(t)\,dt$.

Quick Review 5.4

1. $\dfrac{dy}{dx} = \cos(x^2) \cdot 2x = 2x\cos(x^2)$

2. $\dfrac{dy}{dx} = 2(\sin x)(\cos x) = 2\sin x \cos x$

3. $\dfrac{dy}{dx} = 2(\sec x)(\sec x \tan x) - 2(\tan x)(\sec^2 x)$
$= 2\sec^2 x \tan x - 2\tan x \sec^2 x = 0$

4. $\dfrac{dy}{dx} = \dfrac{3}{3x} - \dfrac{7}{7x} = 0$

5. $\dfrac{dy}{dx} = 2^x \ln 2$

6. $\dfrac{dy}{dx} = \dfrac{1}{2}x^{-1/2} = \dfrac{1}{2\sqrt{x}}$

7. $\dfrac{dy}{dx} = \dfrac{(-\sin x)(x) - (\cos x)(1)}{x^2} = -\dfrac{x\sin x + \cos x}{x^2}$

8. $\dfrac{dy}{dt} = \cos t$, $\dfrac{dx}{dt} = -\sin t$
$\dfrac{dy}{dx} = \dfrac{dy/dt}{dx/dt} = \dfrac{\cos t}{-\sin t} = -\cot t$

9. Implicitly differentiate:
$x\dfrac{dy}{dx} + (1)y + 1 = 2y\dfrac{dy}{dx}$
$\dfrac{dy}{dx}(x - 2y) = -(y+1)$
$\dfrac{dy}{dx} = -\dfrac{y+1}{x-2y} = \dfrac{y+1}{2y-x}$

10. $\dfrac{dy}{dx} = \dfrac{1}{dx/dy} = \dfrac{1}{3x}$

Section 5.4 Exercises

1. $\int_{1/2}^{3}\left(2 - \dfrac{1}{x}\right)dx = \left[2x - \ln|x|\right]_{1/2}^{3}$

$= (6 - \ln 3) - \left(1 - \ln\dfrac{1}{2}\right)$

$= 5 - \ln 3 + \ln\dfrac{1}{2}$

$= 5 - \ln 3 - \ln 2$

$= 5 - \ln 6 \approx 3.208$

3. $\int_0^1 (x^2 + \sqrt{x})\,dx = \left[\dfrac{1}{3}x^3 + \dfrac{2}{3}x^{3/2}\right]_0^1 = \left(\dfrac{1}{3} + \dfrac{2}{3}\right) - (0+0) = 1$

5. $\int_1^{32} x^{-6/5}\,dx = \left[-5x^{-1/5}\right]_1^{32} = -5\left(\dfrac{1}{2} - 1\right) = \dfrac{5}{2}$

7. $\int_0^{\pi} \sin x\,dx = \left[-\cos x\right]_0^{\pi} = 1 - (-1) = 2$

9. $\int_0^{\pi/3} 2\sec^2\theta\,d\theta = 2\left[\tan\theta\right]_0^{\pi/3}$
$= 2(\sqrt{3} - 0)$
$= 2\sqrt{3} \approx 3.464$

11. $\int_{\pi/4}^{3\pi/4} \csc x \cot x\,dx = \left[-\csc x\right]_{\pi/4}^{3\pi/4} = (-\sqrt{2}) - (-\sqrt{2}) = 0$

13. $\int_{-1}^{1} (r+1)^2\,dr = \left[\dfrac{1}{3}(r+1)^3\right]_{-1}^{1} = \dfrac{8}{3} - 0 = \dfrac{8}{3}$

15. Graph $y = 2 - x$.

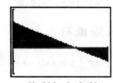
[0, 3] by [−2, 3]

Over [0, 2]: $\int_0^2 (2-x)\,dx = \left[2x - \dfrac{1}{2}x^2\right]_0^2 = 2$

Over [2, 3]: $\int_2^3 (2-x)\,dx = \left[2x - \dfrac{1}{2}x^2\right]_2^3 = \dfrac{3}{2} - 2 = -\dfrac{1}{2}$

Total area $= |2| + \left|-\dfrac{1}{2}\right| = \dfrac{5}{2}$

148 Section 5.4

17. Graph $y = x^3 - 3x^2 + 2x$.

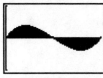

[0, 2] by [−1, 1]

Over [0, 1]:

$$\int_0^1 (x^3 - 3x^2 + 2x)\, dx = \left[\frac{1}{4}x^4 - x^3 + x^2\right]_0^1 = \frac{1}{4} - 0 = \frac{1}{4}$$

Over [1, 2]:

$$\int_1^2 (x^3 - 3x^2 + 2x)\, dx = \left[\frac{1}{4}x^4 - x^3 + x^2\right]_1^2 = 0 - \frac{1}{4} = -\frac{1}{4}$$

Total area $= \left|\frac{1}{4}\right| + \left|-\frac{1}{4}\right| = \frac{1}{2}$

19. (a) No, $f(x) = \dfrac{x^2 - 1}{x + 1}$ is discontinuous at $x = -1$.

(b) $\dfrac{x^2 - 1}{x + 1} = x - 1,\ x \ne -1$

The area between the graph of f and the x-axis over $[-2, 1)$ where f is negative is $\dfrac{1}{2}(3)(3) = \dfrac{9}{2}$. The area between the graph of f and the x-axis over $(1, 3]$ where f is positive is $\dfrac{1}{2}(2)(2) = 2$.

$$\int_{-2}^{3} \dfrac{x^2 - 1}{x + 1}\, dx = -\dfrac{9}{2} + 2 = -\dfrac{5}{2}$$

21. (a) No, $f(x) = \tan x$ is discontinuous at $x = \dfrac{\pi}{2}$ and $x = \dfrac{3\pi}{2}$.

(b) The integral does not have a value. If $0 < b < \dfrac{\pi}{2}$, then

$$\int_0^b \tan x\, dx = \left[-\ln|\cos x|\right]_0^b = -\ln|\cos b|$$ since the Fundamental Theorem applies for $[0, b]$. As $b \to \dfrac{\pi}{2}^-$, $\cos b \to 0^+$ so $-\ln|\cos b| \to \infty$ or $\int_0^b \tan x\, dx \to \infty$.

Hence the integral does not exist over a subinterval of $[0, 2\pi]$, so it doesn't exist over $[0, 2\pi]$.

23. (a) No, $f(x) = \dfrac{\sin x}{x}$ is discontinuous at $x = 0$.

(b) $\text{NINT}\left(\dfrac{\sin x}{x}, x, -1, 2\right) \approx 2.55$. The integral exists since the area is finite because $\dfrac{\sin x}{x}$ is bounded.

25. First, find the area under the graph of $y = x^2$.

$$\int_0^1 x^2\, dx = \left[\dfrac{1}{3}x^3\right]_0^1 = \dfrac{1}{3}$$

Next find the area under the graph of $y = 2 - x$.

$$\int_1^2 (2 - x)\, dx = \left[2x - \dfrac{1}{2}x^2\right]_1^2 = 2 - \dfrac{3}{2} = \dfrac{1}{2}$$

Area of the shaded region $= \dfrac{1}{3} + \dfrac{1}{2} = \dfrac{5}{6}$

27. First find the area under the graph of $y = 1 + \cos x$.

$$\int_0^\pi (1 + \cos x)\, dx = \left[x + \sin x\right]_0^\pi = \pi$$

The area of the rectangle is 2π.

Area of the shaded region $= 2\pi - \pi = \pi$.

29. $\text{NINT}\left(\dfrac{1}{3 + 2\sin x}, x, 0, 10\right) \approx 3.802$

31. $\dfrac{1}{2}\text{NINT}(\sqrt{\cos x}, x, -1, 1) \approx 0.914$

33. Plot $y_1 = \text{NINT}(e^{-t^2}, t, 0, x)$, $y_2 = 0.6$ in a $[0, 1]$ by $[0, 1]$ window, then use the intersect function to find $x \approx 0.699$.

35. $\int_a^x f(t)\, dt + K = \int_b^x f(t)\, dt$

$K = -\int_a^x f(t)\, dt + \int_b^x f(t)\, dt$

$= \int_x^a f(t)\, dt + \int_b^x f(t)\, dt$

$= \int_b^a f(t)\, dt$

$K = \int_2^{-1} (t^2 - 3t + 1)\, dt$

$= \left[\dfrac{1}{3}t^3 - \dfrac{3}{2}t^2 + t\right]_2^{-1}$

$= \left[-\dfrac{1}{3} - \dfrac{3}{2} + (-1)\right] - \left[\dfrac{8}{3} - 6 + 2\right] = -\dfrac{3}{2}$

37. $\dfrac{dy}{dx} = \sqrt{1 + x^2}$

39. $\dfrac{dy}{dx} = \sin\left[(\sqrt{x})^2\right]\dfrac{d}{dx}(\sqrt{x}) = (\sin x) \cdot \dfrac{1}{2}x^{-1/2} = \dfrac{\sin x}{2\sqrt{x}}$

41. $\dfrac{dy}{dx} = \dfrac{d}{dx}\left(\int_0^{x^3} \cos(2t)\, dt - \int_0^{x^2} \cos(2t)\, dt\right)$

$= \cos(2x^3) \cdot 3x^2 - \cos(2x^2) \cdot 2x$

$= 3x^2 \cos(2x^3) - 2x \cos(2x^2)$

43. Choose (d).

$\dfrac{dy}{dx} = \dfrac{d}{dx}\left(\int_\pi^x e^{-t^2}\, dt - 3\right) = e^{-x^2}$

$y(\pi) = \int_\pi^\pi e^{-t^2}\, dt - 3 = 0 - 3 = -3$

45. Choose (b).

$$\frac{dy}{dx} = \frac{d}{dx}\left(\int_0^x \sec t \, dt + 4\right) = \sec x$$

$$y(0) = \int_0^0 \sec t \, dt + 4 = 0 + 4 = 4$$

47. $x = a$ since $\int_a^a f(t) \, dt = 0$

49. $f'(x) = \frac{d}{dx}\left(2 + \int_0^x \frac{10}{1+t} \, dt\right) = \frac{10}{1+x}$

$f'(0) = 10$

$f(0) = 2 + \int_0^0 \frac{10}{1+t} \, dt = 2$

$L(x) = 2 + 10x$

51. One arch of $\sin kx$ is from $x = 0$ to $x = \frac{\pi}{k}$.

Area $= \int_0^{\pi/k} \sin kx \, dx = \left[-\frac{1}{k} \cos kx\right]_0^{\pi/k} = \frac{1}{k} - \left(-\frac{1}{k}\right) = \frac{2}{k}$

53. (a) $H(0) = \int_0^0 f(t) \, dt = 0$

(b) $H'(x) = \frac{d}{dx}\left(\int_0^x f(t) \, dt\right) = f(x)$

$H'(x) > 0$ when $f(x) > 0$.

H is increasing on $[0, 6]$.

(c) H is concave up on the open interval where $H''(x) = f'(x) > 0$.
$f'(x) > 0$ when $9 < x \le 12$.
H is concave up on $(9, 12)$.

(d) $H(12) = \int_0^{12} f(t) \, dt > 0$ because there is more area above the x-axis than below the x-axis.

$H(12)$ is positive.

(e) $H'(x) = f(x) = 0$ at $x = 6$ and $x = 12$. Since $H'(x) = f(x) > 0$ on $[0, 6)$, the values of H are increasing to the left of $x = 6$, and since $H'(x) = f(x) < 0$ on $(6, 12]$, the values of H are decreasing to the right of $x = 6$. H achieves its maximum value at $x = 6$.

(f) $H(x) > 0$ on $(0, 12]$. Since $H(0) = 0$, H achieves its minimum value at $x = 0$.

55. (a) $s'(3) = f(3) = 0$ units/sec

(b) $s''(3) = f'(3) > 0$ so acceleration is positive.

(c) $s(3) = \int_0^3 f(x) \, dx = \frac{1}{2}(-6)(3) = -9$ units

(d) $s(6) = \int_0^6 f(x) \, dx = \frac{1}{2}(-6)(3) + \frac{1}{2}(6)(3) = 0$, so the particle passes through the origin at $t = 6$ sec.

(e) $s''(t) = f'(t) = 0$ at $t = 7$ sec

(f) The particle is moving away from the origin in the negative direction on $(0, 3)$ since $s(0) = 0$ and $s'(t) < 0$ on $(0, 3)$. The particle is moving toward the origin on $(3, 6)$ since $s'(t) > 0$ on $(3, 6)$ and $s(6) = 0$. The particle moves away from the origin in the positive direction for $t > 6$ since $s'(t) > 0$.

(g) The particle is on the positive side since

$s(9) = \int_0^9 f(x) \, dx > 0$ (the area below the x-axis is

smaller than the area above the x-axis).

57. (a) $c(100) - c(1) = \int_1^{100} \left(\frac{dc}{dx}\right) dx$

$$= \int_1^{100} \frac{1}{2\sqrt{x}} \, dx = \left[\sqrt{x}\right]_1^{100}$$

$= 10 - 1 = 9$ or \$9

(b) $c(400) - c(100) = \int_{100}^{400} \left(\frac{dc}{dx}\right) dx$

$$= \int_{100}^{400} \frac{1}{2\sqrt{x}} \, dx = \left[\sqrt{x}\right]_{100}^{400}$$

$= 20 - 10 = 10$ or \$10

59. (a) $\frac{1}{30 - 0}\int_0^{30} \left(450 - \frac{x^2}{2}\right) dx = \frac{1}{30}\left[450x - \frac{x^3}{6}\right]_0^{30}$

$= 300$ drums

(b) $(300 \text{ drums})(\$0.02 \text{ per drum}) = \$6$

61. Since $f(t)$ is odd, $\int_{-x}^0 f(t) \, dt = -\int_0^x f(t) \, dt$ because the area between the curve and the x-axis from 0 to x is the opposite of the area between the curve and the x-axis from $-x$ to 0, but it is on the opposite side of the x-axis.

$$\int_0^{-x} f(t) \, dt = -\int_{-x}^0 f(t) \, dt = -\left[-\int_0^x f(t) \, dt\right] = \int_0^x f(t) \, dt$$

Thus $\int_0^x f(t) \, dt$ is even.

150 Section 5.5

63. If f is an even continuous function, then $\int_0^x f(t)\,dt$ is odd, but $\frac{d}{dx}\int_0^x f(t)\,dt = f(x)$. Therefore, f is the derivative of the odd continuous function $\int_0^x f(t)\,dt$.

Similarly, if f is an odd continuous function, then f is the derivative of the even continuous function $\int_0^x f(t)\,dt$.

■ Section 5.5 Trapezoidal Rule (pp. 289–297)

Exploration 1   Area Under a Parabolic Arc

1. Let $y = f(x) = Ax^2 + Bx + C$
   Then $y_0 = f(-h) = Ah^2 - Bh + C$,
   $y_1 = f(0) = A(0)^2 + B(0) + C = C$, and
   $y_2 = f(h) = Ah^2 + Bh + C$.

2. $y_0 + 4y_1 + y_2 = Ah^2 - Bh + C + 4C + Ah^2 + Bh + C$
   $= 2Ah^2 + 6C$.

3. $A_p = \int_{-h}^{h} (Ax^2 + Bx + C)\,dx$
   $= \left[A\frac{x^3}{3} + B\frac{x^2}{2} + Cx\right]_{-h}^{h}$
   $= A\frac{h^3}{3} + B\frac{h^2}{2} + Ch - \left(-A\frac{h^3}{3} + B\frac{h^2}{2} - Ch\right)$
   $= 2A\frac{h^3}{3} + 2Ch$
   $= \frac{h}{3}(2Ah^2 + 6C)$

4. Substitute the expression in step 2 for the parenthetically enclosed expression in step 3:
   $A_p = \frac{h}{3}(2Ah^2 + 6C)$
   $= \frac{h}{3}(y_0 + 4y_1 + y_2)$.

Quick Review 5.5

1. $y' = -\sin x$
   $y'' = -\cos x$
   $y'' < 0$ on $[-1, 1]$, so the curve is concave down on $[-1, 1]$.

2. $y' = 4x^3 - 12$
   $y'' = 12x^2$
   $y'' > 0$ on $[8, 17]$, so the curve is concave up on $[8, 17]$.

3. $y' = 12x^2 - 6x$
   $y'' = 24x - 6$
   $y'' < 0$ on $[-8, 0]$, so the curve is concave down on $[-8, 0]$.

4. $y' = \frac{1}{2}\cos\frac{x}{2}$
   $y'' = -\frac{1}{4}\sin\frac{x}{2}$
   $y'' \le 0$ on $[48\pi, 50\pi]$, so the curve is concave down on $[48\pi, 50\pi]$.

5. $y' = 2e^{2x}$
   $y'' = 4e^{2x}$
   $y'' > 0$ on $[-5, 5]$, so the curve is concave up on $[-5, 5]$.

6. $y' = \frac{1}{x}$
   $y'' = -\frac{1}{x^2}$
   $y'' < 0$ on $[100, 200]$, so the curve is concave down on $[100, 200]$.

7. $y' = -\frac{1}{x^2}$
   $y'' = \frac{2}{x^3}$
   $y'' > 0$ on $[3, 6]$, so the curve is concave up on $[3, 6]$.

8. $y' = -\csc x \cot x$
   $y'' = (-\csc x)(-\csc^2 x) + (\csc x \cot x)(\cot x)$
   $= \csc^3 x + \csc x \cot^2 x$
   $y'' > 0$ on $[0, \pi]$, so the curve is concave up on $[0, \pi]$.

9. $y' = -100x^9$
   $y'' = -900x^8$
   $y'' < 0$ on $[10, 10^{10}]$, so the curve is concave down on $[10, 10^{10}]$.

10. $y' = \cos x + \sin x$
    $y'' = -\sin x + \cos x$
    $y'' < 0$ on $[1, 2]$, so the curve is concave down.

Section 5.5 Exercises

1. (a) $f(x) = x$, $h = \frac{2-0}{4} = \frac{1}{2}$

x	0	$\frac{1}{2}$	1	$\frac{3}{2}$	2
$f(x)$	0	$\frac{1}{2}$	1	$\frac{3}{2}$	2

   $T = \frac{1}{4}\left(0 + 2\left(\frac{1}{2}\right) + 2(1) + 2\left(\frac{3}{2}\right) + 2\right) = 2$

   (b) $f'(x) = 1$, $f''(x) = 0$
       The approximation is exact.

   (c) $\int_0^2 x\,dx = \left[\frac{1}{2}x^2\right]_0^2 = 2$

3. (a) $f(x) = x^3$, $h = \frac{2-0}{4} = \frac{1}{2}$

x	0	$\frac{1}{2}$	1	$\frac{3}{2}$	2
$f(x)$	0	$\frac{1}{8}$	1	$\frac{27}{8}$	8

   $T = \frac{1}{4}\left(0 + 2\left(\frac{1}{8}\right) + 2(1) + 2\left(\frac{27}{8}\right) + 8\right) = 4.25$

   (b) $f'(x) = 3x^2$, $f''(x) = 6x > 0$ on $[0, 2]$
       The approximation is an overestimate.

   (c) $\int_0^2 x^3\,dx = \left[\frac{1}{4}x^4\right]_0^2 = 4$

5. (a) $f(x) = \sqrt{x}, h = \dfrac{4-0}{4} = 1$

x	0	1	2	3	4
$f(x)$	0	1	$\sqrt{2}$	$\sqrt{3}$	2

$T = \dfrac{1}{2}(0 + 2(1) + 2(\sqrt{2}) + 2(\sqrt{3}) + 2) \approx 5.146$

(b) $f'(x) = \dfrac{1}{2}x^{-1/2}, f''(x) = -\dfrac{1}{4}x^{-3/2} < 0$ on $[0, 4]$

The approximation is an underestimate.

(c) $\displaystyle\int_0^4 \sqrt{x}\, dx = \left[\dfrac{2}{3}x^{3/2}\right]_0^4 = \dfrac{16}{3} \approx 5.333$

7. $\dfrac{5}{2}(6.0 + 2(8.2) + 2(9.1) + \cdots + 2(12.7) + 13.0)(30)$

$= 15{,}990 \text{ ft}^3$

9. Sum the trapezoids and multiply by $\dfrac{1}{3600}$ to change seconds to hours

$\dfrac{1}{2}(2.2(0 + 30) + (3.2 - 2.2)(30 + 40)$
$+ (4.5 - 3.2)(40 + 50) + (5.9 - 4.5)(50 + 60)$
$+ (7.8 - 5.9)(60 + 70) + (10.2 - 7.8)(70 + 80)$
$+ (12.7 - 10.2)(80 + 90) + (16.0 - 12.7)(90 + 100)$
$+ (20.6 - 16.0)(100 + 110)$
$+ (26.2 - 20.6)(110 + 120)$
$+ (37.1 - 26.2)(120 + 130))\left(\dfrac{1}{3600}\right) \approx 0.9785$ miles

11. The average of the 13 discrete temperatures gives equal weight to the low values at the end.

13. $S_{50} \approx 3.13791, S_{100} \approx 3.14029$

15. $S_{50} = 1.37066, S_{100} = 1.37066$ using $a = 0.0001$ as lower limit
$S_{50} = 1.37076, S_{100} = 1.37076$ using $a = 0.000000001$ as lower limit

17. (a) $T_{10} \approx 1.983523538$
$T_{100} \approx 1.999835504$
$T_{1000} \approx 1.999998355$

(b)

n	$|E_T| = 2 - T_n$
10	$0.016476462 = 1.6476462 \times 10^{-2}$
100	1.64496×10^{-4}
1000	1.645×10^{-6}

(c) $|E_{T_{10n}}| \approx 10^{-2}|E_{T_n}|$

(d) $b - a = \pi, h^2 = \dfrac{\pi^2}{n^2}, M = 1$

$|E_{T_n}| \leq \dfrac{\pi}{12}\left(\dfrac{\pi^2}{n^2}\right) = \dfrac{\pi^3}{12n^2}$

$|E_{T_{10n}}| \leq \dfrac{\pi^3}{12(10n)^2} = 10^{-2}|E_{T_n}|$

19. (a) $f'(x) = 2x \cos(x^2)$
$f''(x) = 2x \cdot -2x \sin(x^2) + 2 \cos(x^2)$
$= -4x^2 \sin(x^2) + 2 \cos(x^2)$

(b)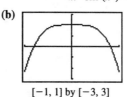

$[-1, 1]$ by $[-3, 3]$

(c) The graph shows that $-3 \leq f''(x) \leq 2$ so $|f''(x)| \leq 3$ for $-1 \leq x \leq 1$.

(d) $|E_T| \leq \dfrac{1 - (-1)}{12}(h^2)(3) = \dfrac{h^2}{2}$

(e) For $0 < h \leq 0.1$, $|E_T| \leq \dfrac{h^2}{2} \leq \dfrac{0.1^2}{2} = 0.005 < 0.01$

(f) $n \geq \dfrac{1 - (-1)}{h} \geq \dfrac{2}{0.1} = 20$

21. $h = \dfrac{24 \text{ in.}}{6} = 4$ in.

Estimate the area to be

$\dfrac{4}{3}[0 + 4(18.75) + 2(24) + 4(26) + 2(24) + 4(18.75) + 0] \approx 466.67 \text{ in}^2$

23. $T_n = \dfrac{h}{2}[y_0 + 2y_1 + 2y_2 + \ldots + 2y_{n-1} + y_n]$

$= \dfrac{h[y_0 + y_1 + \ldots + y_{n-1}] + h[y_1 + y_2 + \ldots + y_n]}{2}$

$= \dfrac{\text{LRAM}_n + \text{RRAM}_n}{2}$

Chapter 5 Review (pp. 298–301)

1.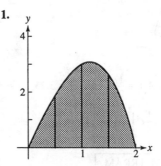

2.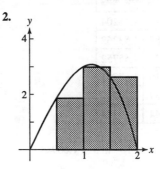

$\text{LRAM}_4: \dfrac{1}{2}\left(0 + \dfrac{15}{8} + 3 + \dfrac{21}{8}\right) = \dfrac{15}{4} = 3.75$

152 Chapter 5 Review

3.

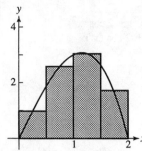

MRAM$_4$: $\frac{1}{2}\left(\frac{63}{64} + \frac{165}{64} + \frac{195}{64} + \frac{105}{64}\right) = 4.125$

4.

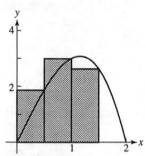

RRAM$_4$: $\frac{1}{2}\left(\frac{15}{8} + 3 + \frac{21}{8} + 0\right) = \frac{15}{4} = 3.75$

5.

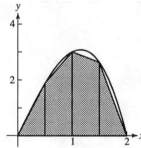

$T_4 = \frac{1}{2}(\text{LRAM}_4 + \text{RRAM}_4) = \frac{1}{2}\left(\frac{15}{4} + \frac{15}{4}\right) = 3.75$

6. $\int_0^2 (4x - x^3)\, dx = \left[2x^2 - \frac{1}{4}x^4\right]_0^2 = 8 - 4 = 4$

7.

n	LRAM$_n$	MRAM$_n$	RRAM$_n$
10	1.78204	1.60321	1.46204
20	1.69262	1.60785	1.53262
30	1.66419	1.60873	1.55752
50	1.64195	1.60918	1.57795
100	1.62557	1.60937	1.59357
1000	1.61104	1.60944	1.60784

8. $\int_1^5 \frac{1}{x}\, dx = \Big[\ln |x|\Big]_1^5 = \ln 5 - \ln 1 = \ln 5 \approx 1.60944$

9. (a) $\int_5^2 f(x)\, dx = -\int_2^5 f(x)\, dx = -3$

The statement is true.

(b) $\int_{-2}^5 [f(x) + g(x)]\, dx$
$= \int_{-2}^5 f(x)\, dx + \int_{-2}^5 g(x)\, dx$
$= \int_{-2}^2 f(x)\, dx + \int_2^5 f(x)\, dx + \int_{-2}^5 g(x)\, dx$
$= 4 + 3 + 2 = 9$

The statement is true.

(c) If $f(x) \leq g(x)$ on $[-2, 5]$, then $\int_{-2}^5 f(x)\, dx \leq \int_{-2}^5 g(x)\, dx$,

but this is not true since

$\int_{-2}^5 f(x)\, dx = \int_{-2}^2 f(x) + \int_2^5 f(x) = 4 + 3 = 7$ and

$\int_{-2}^5 g(x)\, dx = 2$. The statement is false.

10. (a) Volume of one cylinder: $\pi r^2 h = \pi \sin^2(m_i)\, \Delta x$

Total volume: $V = \lim_{n \to \infty} \sum_{i=1}^n \pi \sin^2(m_i)\, \Delta x$

(b) Use $\pi \sin^2 x$ on $[0, \pi]$.

NINT$(\pi \sin^2 x, x, 0, \pi) \approx 4.9348$

11. (a) Approximations may vary. Using Simpson's Rule, the area under the curve is approximately

$\frac{1}{3}[0 + 4(0.5) + 2(1) + 4(2) + 2(3.5) + 4(4.5) +$
$2(4.75) + 4(4.5) + 2(3.5) + 4(2) + 0] = 26.5$

The body traveled about 26.5 m.

(b)

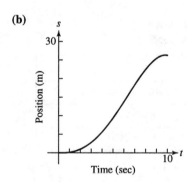

The curve is always increasing because the velocity is always positive, and the graph is steepest when the velocity is highest, at $t = 6$.

12. (a) $\int_0^{10} x^3 \, dx$

(b) $\int_0^{10} x \sin x \, dx$

(c) $\int_0^{10} x(3x-2)^2 \, dx$

(d) $\int_0^{10} \frac{1}{1+x^2} \, dx$

(e) $\int_0^{10} \pi\left(9 - \sin^2 \frac{\pi x}{10}\right) dx$

13. The graph is above the x-axis for $0 \le x < 4$ and below the x-axis for $4 < x \le 6$

Total area $= \int_0^4 (4-x) \, dx - \int_4^6 (4-x) \, dx$
$= \left[4x - \frac{1}{2}x^2\right]_0^4 - \left[4x - \frac{1}{2}x^2\right]_4^6$
$= [8 - 0] - [6 - 8] = 10$

14. The graph is above the x-axis for $0 \le x < \frac{\pi}{2}$ and below the x-axis for $\frac{\pi}{2} < x \le \pi$

Total area $= \int_0^{\pi/2} \cos x \, dx - \int_{\pi/2}^{\pi} \cos x \, dx$
$= \left[\sin x\right]_0^{\pi/2} - \left[\sin x\right]_{\pi/2}^{\pi}$
$= (1 - 0) - (0 - 1) = 2$

15. $\int_{-2}^{2} 5 \, dx = \left[5x\right]_{-2}^{2} = 10 - (-10) = 20$

16. $\int_{2}^{5} 4x \, dx = \left[2x^2\right]_{2}^{5} = 50 - 8 = 42$

17. $\int_{0}^{\pi/4} \cos x \, dx = \left[\sin x\right]_{0}^{\pi/4} = \frac{\sqrt{2}}{2} - 0 = \frac{\sqrt{2}}{2}$

18. $\int_{-1}^{1} (3x^2 - 4x + 7) \, dx = \left[x^3 - 2x^2 + 7x\right]_{-1}^{1}$
$= 6 - (-10) = 16$

19. $\int_{0}^{1} (8s^3 - 12s^2 + 5) \, ds = \left[2s^4 - 4s^3 + 5s\right]_{0}^{1} = 3 - 0 = 3$

20. $\int_{1}^{2} \frac{4}{x^2} \, dx = \left[-\frac{4}{x}\right]_{1}^{2} = -2 - (-4) = 2$

21. $\int_{1}^{27} y^{-4/3} \, dy = \left[-3y^{-1/3}\right]_{1}^{27} = -1 - (-3) = 2$

22. $\int_{1}^{4} \frac{dt}{t\sqrt{t}} = \int_{1}^{4} t^{-3/2} \, dt = \left[-2t^{-1/2}\right]_{1}^{4} = -1 - (-2) = 1$

23. $\int_{0}^{\pi/3} \sec^2 \theta \, d\theta = \left[\tan \theta\right]_{0}^{\pi/3} = \sqrt{3} - 0 = \sqrt{3}$

24. $\int_{1}^{e} \frac{1}{x} \, dx = \left[\ln |x|\right]_{1}^{e} = 1 - 0 = 1$

25. $\int_{0}^{1} \frac{36}{(2x+1)^3} \, dx = \int_{0}^{1} 36(2x+1)^{-3} \, dx$
$= \left[-9(2x+1)^{-2}\right]_{0}^{1}$
$= -1 - (-9) = 8$

26. $\int_{1}^{2} \left(x + \frac{1}{x^2}\right) dx = \int_{1}^{2} (x + x^{-2}) \, dx$
$= \left[\frac{1}{2}x^2 - x^{-1}\right]_{1}^{2}$
$= \frac{3}{2} - \left(-\frac{1}{2}\right) = 2$

27. $\int_{-\pi/3}^{0} \sec x \tan x \, dx = \left[\sec x\right]_{-\pi/3}^{0} = 1 - 2 = -1$

28. $\int_{-1}^{1} 2x \sin (1 - x^2) \, dx = \left[\cos (1 - x^2)\right]_{-1}^{1} = 1 - 1 = 0$

29. $\int_{0}^{2} \frac{2}{y+1} \, dy = \left[2 \ln |y+1|\right]_{0}^{2} = 2 \ln 3 - 0 = 2 \ln 3$

30. Graph $y = \sqrt{4 - x^2}$ on $[0, 2]$.

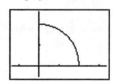

$[-1.35, 3.35]$ by $[-0.5, 2.6]$

The region under the curve is a quarter of a circle of radius 2.

$\int_{0}^{2} \sqrt{4 - x^2} \, dx = \frac{1}{4}\pi(2)^2 = \pi$

31. Graph $y = |x| \, dx$ on $[-4, 8]$.

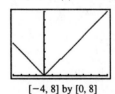

$[-4, 8]$ by $[0, 8]$

The region under the curve consists of two triangles.

$\int_{-4}^{8} |x| \, dx = \frac{1}{2}(4)(4) + \frac{1}{2}(8)(8) = 40$

32. Graph $y = \sqrt{64 - x^2}$ on $[-8, 8]$.

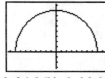

$[-9.4, 9.4]$ by $[-3.2, 9.2]$

The region under the curve $y = \sqrt{64 - x^2}$ is half a circle of radius 8.

$\int_{-8}^{8} 2\sqrt{64 - x^2}\, dx = 2\int_{-8}^{8} \sqrt{64 - x^2}\, dx = 2\left[\frac{1}{2}\pi(8)^2\right] = 64\pi$

33. (a) Note that each interval is 1 day = 24 hours
Upper estimate:
$24(0.020 + 0.021 + 0.023 + 0.025 + 0.028 + 0.031 + 0.035) = 4.392$ L
Lower estimate:
$24(0.019 + 0.020 + 0.021 + 0.023 + 0.025 + 0.028 + 0.031) = 4.008$ L

(b) $\frac{24}{2}[0.019 + 2(0.020) + 2(0.021) + \cdots + 2(0.031) + 0.035] = 4.2$ L

34. (a) Upper estimate:
$3(5.30 + 5.25 + 5.04 + \cdots + 1.11) = 103.05$ ft
Lower estimate:
$3(5.25 + 5.04 + 4.71 + \cdots + 0) = 87.15$ ft

(b) $\frac{3}{2}[5.30 + 2(5.25) + 2(5.04) + \cdots + 2(1.11) + 0]$
$= 95.1$ ft

35. One possible answer:
The dx is important because it corresponds to the actual physical quantity Δx in a Riemann sum. Without the Δx, our integral approximations would be way off.

36. $\int_{-4}^{4} f(x)\, dx = \int_{-4}^{0} f(x)\, dx + \int_{0}^{4} f(x)\, dx$
$= \int_{-4}^{0} (x - 2)\, dx + \int_{0}^{4} x^2\, dx$
$= \left[\frac{1}{2}x^2 - 2x\right]_{-4}^{0} + \left[\frac{1}{3}x^3\right]_{0}^{4}$
$= [0 - 16] + \left[\frac{64}{3} - 0\right] = \frac{16}{3}$

37. Let $f(x) = \sqrt{1 + \sin^2 x}$
$\max f = \sqrt{2}$ since $\max \sin^2 x = 1$
$\min f = 1$ since $\min \sin^2 x = 0$
$(\min f)(1 - 0) \leq \int_{0}^{1} \sqrt{1 + \sin^2 x}\, dx \leq (\max f)(1 - 0)$
$0 < 1 \leq \int_{0}^{1} \sqrt{1 + \sin^2 x}\, dx \leq \sqrt{2}$

38. (a) $av(y) = \frac{1}{4 - 0}\int_{0}^{4} \sqrt{x}\, dx = \frac{1}{4}\left[\frac{2}{3}x^{3/2}\right]_{0}^{4} = \frac{1}{4}\left(\frac{16}{3} - 0\right) = \frac{4}{3}$

(b) $av(y) = \frac{1}{a - 0}\int_{0}^{a} a\sqrt{x}\, dx = \frac{1}{a}\left[\frac{2}{3}ax^{3/2}\right]_{0}^{a} = \frac{2}{3}a^{3/2}$

39. $\frac{dy}{dx} = \sqrt{2 + \cos^3 x}$

40. $\frac{dy}{dx} = \sqrt{2 + \cos^3(7x^2)} \cdot \frac{d}{dx}(7x^2) = 14x\sqrt{2 + \cos^3(7x^2)}$

41. $\frac{dy}{dx} = \frac{d}{dx}\left(-\int_{1}^{x} \frac{6}{3 + t^4}\, dt\right) = -\frac{6}{3 + x^4}$

42. $\frac{dy}{dx} = \frac{d}{dx}\left(\int_{0}^{2x} \frac{1}{t^2 + 1}\, dt - \int_{0}^{x} \frac{1}{t^2 + 1}\, dt\right)$
$= \frac{1}{(2x)^2 + 1} \cdot 2 - \frac{1}{x^2 + 1}$
$= \frac{2}{4x^2 + 1} - \frac{1}{x^2 + 1}$

43. $c(x) = \int_{25}^{x} \frac{2}{\sqrt{t}}\, dt + 50$
$= \left[4t^{1/2}\right]_{25}^{x} + 50$
$= 4\sqrt{x} - 20 + 50$
$= 4\sqrt{x} + 30$

$c(2500) = 4\sqrt{2500} + 30 = 230$

The total cost for printing 2500 newsletters is $230.

44. $av(I) = \frac{1}{14}\int_{0}^{14} (600 + 600t)\, dt$
$= \frac{1}{14}[600t + 300t^2]_{0}^{14} = 4800$

Rich's average daily inventory is 4800 cases.

$c(t) = 0.04I(t) = 24 + 24t$

$av(c) = \frac{1}{14}\int_{0}^{14} (24 + 24t)\, dt = \frac{1}{14}\left[24t + 12t^2\right]_{0}^{14} = 192$

Rich's average daily holding cost is $192.
We could also say $(0.04)4800 = 192$.

45. $\int_{0}^{x} (t^3 - 2t + 3)\, dt = \left[\frac{1}{4}t^4 - t^2 + 3t\right]_{0}^{x}$
$= \frac{1}{4}x^4 - x^2 + 3x$

$\frac{1}{4}x^4 - x^2 + 3x = 4$

$\frac{1}{4}x^4 - x^2 + 3x - 4 = 0$

$x^4 - 4x^2 + 12x - 16 = 0$

Using a graphing calculator, $x \approx -3.09131$
or $x \approx 1.63052$.

46. (a) True, because $g'(x) = f(x)$.
  (b) True, because g is differentiable.
  (c) True, because $g'(1) = f(1) = 0$.
  (d) False, because $g''(1) = f'(1) > 0$.
  (e) True, because $g'(1) = f(1) = 0$ and $g''(1) = f'(1) > 0$.
  (f) False, because $g''(1) = f'(1) \neq 0$.
  (g) True, because $g'(x) = f(x)$, and f is an increasing function which includes the point $(1, 0)$.

47. $\int_0^1 \sqrt{1 + x^4}\, dx = F(1) - F(0)$

48. $y(x) = \int_5^x \frac{\sin t}{t}\, dt + 3$

49. $y' = 2x + \frac{1}{x}$
$y'' = 2 - \frac{1}{x^2}$
Thus, it satisfies condition **i**.
$y(1) = 1 + \int_1^1 \frac{1}{t}\, dt + 1 = 1 + 0 + 1 = 2$
$y'(1) = 2 + \frac{1}{1} = 2 + 1 = 3$
Thus, it satisfies condition **ii**.

50. Graph (b).
$y = \int_1^x 2t\, dt + 4 = \left[t^2\right]_1^x + 4 = (x^2 - 1) + 4 = x^2 + 3$

51. (a) Each interval is 5 min $= \frac{1}{12}$ h.
$\frac{1}{24}[2.5 + 2(2.4) + 2(2.3) + \cdots + 2(2.4) + 2.3]$
$= \frac{29}{12} \approx 2.42$ gal

  (b) $(60 \text{ mi/h})\left(\frac{12}{29} \text{ h/gal}\right) \approx 24.83$ mi/gal

52. (a) Using the freefall equation $s = \frac{1}{2}gt^2$ from Section 3.4, the distance A falls in 4 seconds is $\frac{1}{2}(32)(4^2) = 256$ ft. When her parachute opens, her altitude is $6400 - 256 = 6144$ ft.

  (b) The distance B falls in 13 seconds is $\frac{1}{2}(32)(13^2) = 2704$ ft. When her parachute opens, her altitude is $7000 - 2704 = 4296$ ft.

  (c) Let t represent the number of seconds after A jumps. For $t \geq 4$ sec, A's position is given by
$S_A(t) = 6144 - 16(t - 4) = 6208 - 16t$, so A lands at
$t = \frac{6208}{16} = 388$ sec. For $t \geq 45 + 13 = 58$ sec, B's position is given by
$S_B(t) = 4296 - 16(t - 58) = 5224 - 16t$,
so B lands at $t = \frac{5224}{16} = 326.5$ sec. B lands first.

53. (a) Area of the trapezoid $= \frac{1}{2}(2h)(y_1 + y_3) = h(y_1 + y_3)$
Area of the rectangle $= (2h)y_2 = 2hy_2$
$h(y_1 + y_3) + 2(2hy_2) = h(y_1 + 4y_2 + y_3)$

  (b) Let $h = \frac{b - a}{2n}$.
$S_{2n} = \frac{h}{3}[y_0 + 4y_1 + 2y_2 + 4y_3 + 2y_4 + \cdots + 2y_{2n-2} + 4y_{2n-1} + y_{2n}]$
$= \frac{1}{3}[h(y_0 + 4y_1 + y_2) + h(y_2 + 4y_3 + y_4) + \cdots + h(y_{2n-2} + 4y_{2n-1} + y_{2n})]$
Since each expression of the form $h(y_{2i-2} + 4y_{2i-1} + y_{2i})$ is equal to twice the area of the ith of n rectangles plus the area of the ith of n trapezoids, $S_{2n} = \frac{2 \cdot \text{MRAM}_n + T_n}{3}$.

54. (a) $g(1) = \int_1^1 f(t)\, dt = 0$

  (b) $g(3) = \int_1^3 f(t)\, dt = -\frac{1}{2}(2)(1) = -1$

  (c) $g(-1) = \int_1^{-1} f(t)\, dt = -\int_{-1}^1 f(t)\, dt = -\frac{1}{4}\pi(2)^2 = -\pi$

  (d) $g'(x) = f(x)$; Since $f(x) > 0$ for $-3 < x < 1$ and $f(x) < 0$ for $1 < x < 3$, $g(x)$ has a relative maximum at $x = 1$.

  (e) $g'(-1) = f(-1) = 2$
The equation of the tangent line is
$y - (-\pi) = 2(x + 1)$ or $y = 2x + 2 - \pi$

  (f) $g''(x) = f'(x), f'(x) = 0$ at $x = -1$ and $f'(x)$ is not defined at $x = 2$. The inflection points are at $x = -1$ and $x = 2$. Note that $g''(x) = f'(x)$ is undefined at $x = 1$ as well, but since $g''(x) = f'(x)$ is negative on both sides of $x = 1$, $x = 1$ is not an inflection point.

  (g) Note that the absolute maximum is $g(1) = 0$ and the absolute minimum is
$g(-3) = \int_1^{-3} f(t)\, dt = -\int_{-3}^1 f(t)\, dt = -\frac{1}{2}\pi(2)^2 = -2\pi$.
The range of g is $[-2\pi, 0]$.

55. (a) $\text{NINT}(e^{-x^2/2}, x, -10, 10) \approx 2.506628275$

$\text{NINT}(e^{-x^2/2}, x, -20, 20) \approx 2.506628275$

(b) The area is $\sqrt{2\pi}$.

56. First estimate the surface area of the swamp.

$\frac{20}{2}[146 + 2(122) + 2(76) + 2(54) + 2(40) + 2(30) + 13] = 8030 \text{ ft}^2$

$(5 \text{ ft})(8030 \text{ ft}^2) \cdot \frac{1 \text{ yd}^3}{27 \text{ ft}^3} \approx 1500 \text{ yd}^3$

57. (a) $V^2 = (V_{max})^2 \sin^2(120\pi t)$

Using NINT:

$av(V^2) = \frac{1}{1}\int_0^1 (V_{max})^2 \sin^2(120\pi t)\, dt$

$= (V_{max})^2 \int_0^1 \sin^2(120\pi t)\, dt = (V_{max})^2 \frac{1}{2} = \frac{(V_{max})^2}{2}$

$V_{rms} = \sqrt{\frac{(V_{max})^2}{2}} = \frac{V_{max}}{\sqrt{2}}$

(b) $V_{max} = 240\sqrt{2} \approx 339.41$ volts

Chapter 6
Differential Equations and Mathematical Modeling

■ Section 6.1 Antiderivatives and Slope Fields (pp. 303–315)

Exploration 1   Constructing a Slope Field

1. As i and j vary from 1 to 10, 100 ordered pairs are produced. Each ordered pair represents a distinct point in the viewing window.

2. The distance between the points with j fixed and $i = r$ and $i = r + 1$ is the distance between their x-coordinates.

$\left[\text{Xmin} + (2(r + 1) - 1)\frac{h}{2}\right] - \left[\text{Xmin} + (2r - 1)\frac{h}{2}\right]$

$= (\text{Xmin} - \text{Xmin}) + (2r + 2 - 1 - 2r + 1)\frac{h}{2} = h$

3. The distance between the points with i fixed and $j = r$ and $j = r + 1$ is the distance between their y-coordinates.

$\left[\text{Ymin} + (2(r + 1) - 1)\frac{k}{2}\right] - \left[\text{Ymin} + (2r - 1)\frac{k}{2}\right]$

$= (\text{Ymin} - \text{Ymin}) + (2r + 2 - 1 - 2r + 1)\frac{k}{2} = k$

4. Here $h = k = 1$. Each line segment in the third column has slope $\frac{4}{7}$, because the x-coordinate of the midpoint of each line segment is 2.5. The y-coordinates are $\frac{1}{2}, \frac{3}{2}, \frac{5}{2}, \ldots, \frac{19}{2}$.

The 10 graphs are graphs of the functions

$y = \left(\frac{4}{7}\right)(x - 2.5) + \frac{n}{2}, 2 \leq x \leq 3$, for $n = 1, 3, 5, \ldots, 19$.

The length of the line segment can be increased or decreased by adjusting the restriction $2 \leq x \leq 3$.

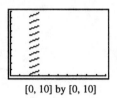

[0, 10] by [0, 10]

5. Again $h = k = 1$. The y-coordinate of the midpoint of each line segment is $\frac{7}{2}$. The x-coordinates of the midpoint of each line segment are $\frac{1}{2}, \frac{3}{2}, \frac{5}{2}, \ldots, \frac{19}{2}$. From left to right the slopes of the line segments are

$\frac{2}{\frac{1}{2} + 1}, \frac{2}{\frac{3}{2} + 1}, \frac{2}{\frac{5}{2} + 1}, \ldots, \frac{2}{\frac{19}{2} + 1}$

The 10 graphs are graphs of the functions.

$y_1 = \left(\frac{2}{\frac{1}{2} + 1}\right)\left(x - \frac{1}{2}\right) + \frac{7}{2}, 0 \leq x \leq 1,$

$y_2 = \left(\frac{3}{\frac{3}{2} + 1}\right)\left(x - \frac{3}{2}\right) + \frac{7}{2}, 1 \leq x \leq 2,$

$y_3 = \left(\frac{2}{\frac{5}{2} + 1}\right)\left(x - \frac{5}{2}\right) + \frac{7}{2}, 2 \leq x \leq 3,$

$\vdots$

$y_{10} = \left(\frac{2}{\frac{19}{2} + 1}\right)\left(x - \frac{19}{2}\right) + \frac{7}{2}, 9 \leq x \leq 10.$

[0, 10] by [0, 10]

6. For each line segment in part (5), make a column of parallel line segments as in part (4).

7. WL

Quick Review 6.1

1. $100(1.06) = \$106.00$

2. $100\left(1 + \dfrac{0.06}{4}\right)^4 \approx \$106.14$

3. $100\left(1 + \dfrac{0.06}{12}\right)^{12} \approx \$106.17$

4. $100\left(1 + \dfrac{0.06}{365}\right)^{365} \approx \$106.18$

5. $\dfrac{dy}{dx} = \dfrac{d}{dx} \sin 3x = (\cos 3x)(3) = 3 \cos 3x$

6. $\dfrac{dy}{dx} = \dfrac{d}{dx} \tan \dfrac{5}{2}x = \left(\sec^2 \dfrac{5}{2}x\right)\left(\dfrac{5}{2}\right) = \dfrac{5}{2} \sec^2 \dfrac{5}{2}x$

7. $\dfrac{dy}{dx} = \dfrac{d}{dx} Ce^{2x} = (Ce^{2x})(2) = 2Ce^{2x}$

8. $\dfrac{dy}{dx} = \dfrac{d}{dx} \ln(x+2) = \dfrac{1}{x+2}$

9.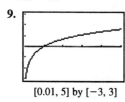

[0.01, 5] by [−3, 3]

By setting the left endpoint at $x = 0.01$ instead of $x = 0$, we avoid an error that occurs when our calculator attempts to calculate $\text{NINT}\left(\dfrac{1}{x}, x, 1, 0\right)$. The graph appears to be the same as the graph of $y = \ln x$.

10.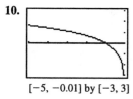

[−5, −0.01] by [−3, 3]

By setting the right endpoint at $x = -0.01$ instead of $x = 0$, we avoid an error that occurs when our calculator attempts to calculate $\text{NINT}\left(\dfrac{1}{x}, x, -1, 0\right)$. The graph appears to be the same as the graph of $y = \ln(-x)$.

Section 6.1 Exercises

1. $\displaystyle\int (x^2 - 2x + 1)\, dx = \dfrac{x^3}{3} - x^2 + x + C$

    Check:
    $\dfrac{d}{dx}\left(\dfrac{x^3}{3} - x^2 + x + C\right) = x^2 - 2x + 1$

3. $\displaystyle\int (x^2 - 4\sqrt{x})\, dx = \int (x^2 - 4x^{1/2})\, dx = \dfrac{x^3}{3} - \dfrac{8}{3}x^{3/2} + C$

    Check:
    $\dfrac{d}{dx}\left(\dfrac{x^3}{3} - \dfrac{8}{3}x^{3/2} + C\right) = x^2 - 4x^{1/2} = x^2 - \sqrt{x}$

5. $\displaystyle\int e^{4x}\, dx = \dfrac{1}{4}e^{4x} + C$

    Check:
    $\dfrac{d}{dx}\left(\dfrac{1}{4}e^{4x} + C\right) = e^{4x}$

7. $\displaystyle\int (x^5 - 6x + 3)\, dx = \dfrac{x^6}{6} - 3x^2 + 3x + C$

9. $\displaystyle\int \left(e^{t/2} - \dfrac{5}{t^2}\right) dt = \int (e^{t/2} - 5t^{-2})\, dt$
    $= 2e^{t/2} + 5t^{-1} + C$
    $= 2e^{t/2} + \dfrac{5}{t} + C$

11. $\displaystyle\int \left(x^3 - \dfrac{1}{x^3}\right) dx = \int (x^3 - x^{-3})\, dx$
    $= \dfrac{x^4}{4} + \dfrac{x^{-2}}{2} + C$
    $= \dfrac{x^4}{4} + \dfrac{1}{2x^2} + C$

13. $\displaystyle\int \dfrac{1}{3}x^{-2/3}\, dx = x^{1/3} + C$

15. $\displaystyle\int \dfrac{\pi}{2} \cos\left(\dfrac{\pi}{2}x\right) dx = \sin\left(\dfrac{\pi}{2}x\right) + C$

17. $\displaystyle\int \left(\dfrac{2}{x-1} + \dfrac{1}{x}\right) dx = 2 \ln|x-1| + \ln|x| + C$

19. $\displaystyle\int 5 \sec^2 5r\, dr = \tan 5r + C$

21. $\displaystyle\int \cos^2 x\, dx = \int \dfrac{1 + \cos 2x}{2}\, dx$
    $= \int \left(\dfrac{1}{2} + \dfrac{\cos 2x}{2}\right) dx$
    $= \dfrac{x}{2} + \dfrac{\sin 2x}{4} + C$

23. $\displaystyle\int \tan^2 \theta\, d\theta = \int (\sec^2 \theta - 1)\, d\theta = \tan \theta - \theta + C$

25. (a) Graph (b)

    (b) The slope is always positive, so (a) and (c) can be ruled out.

27. $\dfrac{dy}{dx} = 2x - 1$

    $\displaystyle\int \dfrac{dy}{dx}\, dx = \int (2x - 1)\, dx$

    $y = x^2 - x + C$

    Initial condition: $y(2) = 0$
    $0 = 2^2 - 2 + C$
    $0 = 2 + C$
    $-2 = C$
    Solution: $y = x^2 - x - 2$

    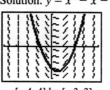

    [−4, 4] by [−3, 3]

158 Section 6.1

29. $\dfrac{dy}{dx} = \sec^2 x$

$\int \dfrac{dy}{dx}\, dx = \int \sec^2 x\, dx$

$y = \tan x + C$

Initial condition: $y\left(\dfrac{\pi}{4}\right) = -1$

$-1 = \tan \dfrac{\pi}{4} + C$

$-1 = 1 + C$

$-2 = C$

Solution: $y = \tan x - 2$

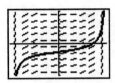

$\left[-\dfrac{\pi}{2}, \dfrac{\pi}{2}\right]$ by $[-8, 8]$

31. $\dfrac{dy}{dx} = 9x^2 - 4x + 5$

$\int \dfrac{dy}{dx}\, dx = \int (9x^2 - 4x + 5)\, dx$

$y = 3x^3 - 2x^2 + 5x + C$

Initial condition: $y(-1) = 0$

$0 = 3(-1)^3 - 2(-1)^2 + 5(-1) + C$

$0 = -10 + C$

$10 = C$

Solution: $y = 3x^3 - 2x^2 + 5x + 10$

33. $\dfrac{dy}{dt} = 2e^{-t}$

$\int \dfrac{dy}{dt}\, dt = \int 2e^{-t}\, dt$

$y = -2e^{-t} + C$

Initial condition: $y(\ln 2) = 0$

$0 = -2e^{-\ln 2} + C$

$0 = -\dfrac{2}{2} + C$

$1 = C$

Solution: $y = -2e^{-t} + 1$

35. $\dfrac{d^2 y}{d\theta^2} = \sin \theta$

$\int \dfrac{d^2 y}{d\theta^2}\, d\theta = \int \sin \theta\, d\theta$

$\dfrac{dy}{d\theta} = -\cos \theta + C_1$

Initial condition: $y'(0) = 0$

$0 = -\cos 0 + C_1$

$0 = -1 + C_1$

$1 = C_1$

First derivative: $\dfrac{dy}{d\theta} = -\cos \theta + 1$

$\int \dfrac{dy}{d\theta}\, d\theta = \int (-\cos \theta + 1)\, d\theta$

$y = -\sin \theta + \theta + C_2$

Initial condition: $y(0) = -3$

$-3 = -\sin 0 + 0 + C_2$

$-3 = C_2$

Solution: $y = -\sin \theta + \theta - 3$

37. $\dfrac{d^3 y}{dt^3} = \dfrac{1}{t^3}$

$\int \dfrac{d^3 y}{dt^3}\, dt = \int t^{-3}\, dt$

$\dfrac{d^2 y}{dt^2} = -\dfrac{1}{2} t^{-2} + C_1$

Initial condition: $y''(1) = 2$

$2 = -\dfrac{1}{2}(1)^{-2} + C_1$

$2 = -\dfrac{1}{2} + C_1$

$\dfrac{5}{2} = C_1$

Second derivative: $\dfrac{d^2 y}{dt^2} = -\dfrac{1}{2} t^{-2} + \dfrac{5}{2}$

$\int \dfrac{d^2 y}{dt^2}\, dt = \int \left(-\dfrac{1}{2} t^{-2} + \dfrac{5}{2}\right) dt$

$\dfrac{dy}{dt} = \dfrac{1}{2} t^{-1} + \dfrac{5}{2} t + C_2$

Initial condition: $y'(1) = 3$

$3 = \dfrac{1}{2}(1)^{-1} + \dfrac{5}{2}(1) + C_2$

$3 = 3 + C_2$

$0 = C_2$

First derivative: $\dfrac{dy}{dt} = \dfrac{1}{2} t^{-1} + \dfrac{5}{2} t$

$\int \dfrac{dy}{dt}\, dt = \int \left(\dfrac{1}{2} t^{-1} + \dfrac{5}{2} t\right) dt$

$y = \dfrac{1}{2} \ln |t| + \dfrac{5}{4} t^2 + C_3$

Initial condition: $y(1) = 1$

$1 = \dfrac{1}{2} \ln 1 + \dfrac{5}{4}(1)^2 + C_3$

$1 = \dfrac{5}{4} + C_3$

$-\dfrac{1}{4} = C_3$

Solution: $y = \dfrac{1}{2} \ln |t| + \dfrac{5}{4} t^2 - \dfrac{1}{4}$

39. $\frac{ds}{dt} = v = 9.8t + 5$

$\int \frac{ds}{dt} dt = \int (9.8t + 5) dt$

$s = 4.9t^2 + 5t + C$

Initial condition: $s(0) = 10$
$10 = 4.9(0)^2 + 5(0) + C$
$10 = C$
Solution: $s = 4.9t^2 + 5t + 10$

41. $\frac{dv}{dt} = a = 32$

$\int \frac{dv}{dt} dt = \int 32 \, dt$

$v = 32t + C_1$

Initial condition: $v(0) = 20$
$20 = 32(0) + C_1$
$20 = C_1$

Velocity: $\frac{ds}{dt} = v = 32t + 20$

$\int \frac{ds}{dt} dt = \int (32t + 20) dt$

$s = 16t^2 + 20t + C_2$

Initial condition: $s(0) = 0$
$0 = 16(0)^2 + 20(0) + C_2$
$0 = 0$

Solution: $s = 16t^2 + 20t$

43.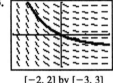

[−2, 2] by [−3, 3]

45. $\frac{d}{dx}(\tan^{-1} x + C) = \frac{1}{1 + x^2}$

47. $\frac{d}{dx}(\sec^{-1} x + C) = \frac{1}{|x|\sqrt{x^2 - 1}}$

49. (a) $\frac{dy}{dx} = x - \frac{1}{x^2}$

$\int \frac{dy}{dx} dx = \int (x - x^{-2}) dx$

$y = \frac{x^2}{2} + x^{-1} + C = \frac{x^2}{2} + \frac{1}{x} + C$

Initial condition: $y(1) = 2$
$2 = \frac{1^2}{2} + \frac{1}{1} + C$
$2 = \frac{3}{2} + C$
$\frac{1}{2} = C$

Solution: $y = \frac{x^2}{2} + \frac{1}{x} + \frac{1}{2}, x > 0$

(b) Again, $y = \frac{x^2}{2} + \frac{1}{x} + C$.

Initial condition: $y(-1) = 1$
$1 = \frac{(-1)^2}{2} + \frac{1}{(-1)} + C$
$1 = -\frac{1}{2} + C$
$\frac{3}{2} = C$

Solution: $y = \frac{x^2}{2} + \frac{1}{x} + \frac{3}{2}, x < 0$

(c) For $x < 0$, $\frac{dy}{dx} = \frac{d}{dx}\left(\frac{1}{x} + \frac{x^2}{2} + C_1\right)$

$= -\frac{1}{x^2} + x$

$= x - \frac{1}{x^2}$.

For $x > 0$, $\frac{dy}{dx} = \frac{d}{dx}\left(\frac{1}{x} + \frac{x^2}{2} + C_2\right)$

$= -\frac{1}{x^2} + x$

$= x - \frac{1}{x^2}$.

And for $x = 0$, $\frac{dy}{dx}$ is undefined.

(d) Let C_1 be the value from part (b), and let C_2 be the value from part (a). Thus, $C_1 = \frac{3}{2}$ and $C_2 = \frac{1}{2}$.

(e) $y(2) = -1$ 　　　　$y(-2) = 2$

$-1 = \frac{1}{2} + \frac{2^2}{2} + C_2$ 　　$2 = \frac{1}{(-2)} + \frac{(-2)^2}{2} + C_1$

$-1 = \frac{5}{2} + C_2$ 　　　$2 = \frac{3}{2} + C_1$

$-\frac{7}{2} = C_2$ 　　　　$\frac{1}{2} = C_1$

Thus, $C_1 = \frac{1}{2}$ and $C_2 = -\frac{7}{2}$.

51. $\int \frac{dc}{dx} dx = \int (3x^2 - 12x + 15) dx$

$c = x^3 - 6x^2 + 15x + C$

Initial condition $c(0) = 400$

$400 = 0^3 - 6(0)^2 + 15(0) + C$

$400 = C$

Solution: $c(x) = x^3 - 6x^2 + 15x + 400$

53. (a) $\int \frac{d^2s}{dt^2} dt = \int -k \, dt$

$\frac{ds}{dt} = -kt + C_1$

Initial condition: $\frac{ds}{dt} = 88$ when $t = 0$

$88 = (-k)(0) + C_1$

$88 = C_1$

Velocity: $\frac{ds}{dt} = -kt + 88$

$\int \frac{ds}{dt} dt = \int (-kt + 88) \, dt$

$s = -\frac{k}{2}t^2 + 88t + C_2$

Initial condition: $s = 0$ when $t = 0$

$0 = -\frac{k}{2}(0)^2 + 88(0) + C_2$

$0 = C_2$

Solution: $s = -\frac{kt^2}{2} + 88t$

(b) $\frac{ds}{dt} = 0$

$-kt + 88 = 0$

$t = \frac{88}{k}$

(c) $s\left(\frac{88}{k}\right) = 242$

$-\frac{k}{2}\left(\frac{88}{k}\right)^2 + 88\left(\frac{88}{k}\right) = 242$

$\frac{3872}{k} = 242$

$k = 16 \text{ ft/sec}^2$

55. $\int \frac{d^2s}{dt^2} dt = \int -5.2 \, dt$

$\frac{ds}{dt} = -5.2t + C_1$

Initial condition: $\frac{ds}{dt} = 0$ when $t = 0$

$0 = -5.2(0) + C_1$

$0 = C_1$

Velocity: $\frac{ds}{dt} = -5.2t$

$\int \frac{ds}{dt} dt = \int -5.2 \, dt$

$s = -2.6t^2 + C_2$

Initial condition: $s = 4$ when $t = 0$

$4 = -2.6(0)^2 + C_2$

$4 = C_2$

Position: $s(t) = -2.6t^2 + 4$

Solving $s(t) = 0$, we have $t^2 = \frac{4}{2.6}$, so the positive solution is $t \approx 1.240$ sec. They took about 1.240 sec to fall.

57. We use the method of Example 7.

$V = \frac{1}{3}\pi r^2 h = \frac{1}{3}\pi\left(\frac{2}{5}h\right)^2 h = \frac{4\pi}{75}h^3$

$\frac{dV}{dt} = \frac{d}{dt}\left(\frac{4\pi}{75}h^3\right)$

$-\frac{1}{6}\sqrt{h} = \frac{4\pi}{25}h^2\frac{dh}{dt}$

$-\frac{25}{24\pi} = h^{3/2}\frac{dh}{dt}$

$-\int \frac{25}{24\pi} dt = \int h^{3/2}\frac{dh}{dt} dt$

$-\frac{25}{24\pi}t = \frac{2}{5}h^{5/2} + C$

Initial condition: $h = 10$ when $t = 0$.

$-\frac{25}{24\pi}(0) = \frac{2}{5}(10)^{5/2} + C$

$C = -\frac{2}{5}(10)^{5/2}$

$-\frac{25}{24\pi}t = \frac{2}{5}h^{5/2} - \frac{2}{5}(10)^{5/2}$

$-\frac{125t}{48\pi} = h^{5/2} - 10^{5/2}$

$h^{5/2} = -\frac{125t}{48\pi} + 10^{5/2}$

$h = \left(-\frac{125t}{48\pi} + 10^{5/2}\right)^{2/5}$

The height is given by $h = \left(-\frac{125t}{48\pi} + 10^{5/2}\right)^{2/5}$ and the volume is given by

$V = \frac{4\pi}{75}h^3 = \frac{4\pi}{75}\left(-\frac{125t}{48\pi} + 10^{5/2}\right)^{6/5}$.

59. (a) $y = 1200e^{0.0625t}$

(b) $3600 = 1200e^{0.0625t}$

$3 = e^{0.0625t}$

$\ln 3 = 0.0625t$

$t = \dfrac{\ln 3}{0.0625} \approx 17.6$

It will take approximately 17.6 years.

61. (a) $\displaystyle\int xe^x \, dx = \int_0^x te^t \, dt + C$

(b) We require $\displaystyle\int_0^0 te^t \, dt + C = 1$, so $C = 1$.

The required antiderivative is $\displaystyle\int_0^x te^t \, dt + 1$.

63. Use differential equation graphing mode.
For reference, the equations of the solution curves are as follows.
$(1, 1): y = e^{(x^2-1)/2}$
$(-1, 2): y = 2e^{(x^2-1)/2}$
$(0, -2): y = -2e^{x^2/2}$
$(-2, -1): y = -e^{(x^2-4)/2}$

$[-6, 6]$ by $[-4, 4]$

The concavity of each solution curve indicates the sign of y''.

65. Use differential equation graphing mode.
For reference, the equations of the solution curves are as follows.
$(0, 1): y = -3e^{2x} + 4$
$(0, 4): y = 4$
$(0, 5): y = e^{2x} + 4$

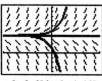

$[-3, 3]$ by $[-4, 10]$

The concavity of each solution curve indicates the sign of y''.

67. (a) $\dfrac{d}{dx}(\ln x + C) = \dfrac{1}{x}$ for $x > 0$

(b) $\dfrac{d}{dx}[\ln(-x) + C] = \dfrac{1}{-x} \dfrac{d}{dx}(-x) = \left(\dfrac{1}{-x}\right)(-1) = \dfrac{1}{x}$
for $x < 0$

(c) For $x > 0$, $\ln|x| + C = \ln x + C$, which is a solution to the differential equation, as we showed in part (a). For $x < 0$, $\ln|x| + C = \ln(-x) + C$, which is a solution to the differential equation, as we showed in part (b). Thus, $\dfrac{d}{dx} \ln|x| = \dfrac{1}{x}$ for all x except 0.

(d) For $x < 0$, we have $y = \ln(-x) + C_2$, which is a solution to the differential equation, as we showed in part (a). For $x > 0$, we have $y = \ln x + C_1$, which is a solution to the differential equation, as we showed part (b). Thus, $\dfrac{dy}{dx} = \dfrac{1}{x}$ for all x except 0.

■ Section 6.2 Integration by Substitution
(pp. 315–323)

Exploration 1 Supporting Indefinite Integrals Graphically

1. $\displaystyle\int \sqrt{1 + x^2} \cdot 2x \, dx = \int \sqrt{u} \, du$

$= \dfrac{2}{3} u^{3/2} + C$

$= \dfrac{2}{3}(1 + x^2)^{3/2} + C$

2. Their derivatives are equal: $\dfrac{dy_1}{dx} = \dfrac{dy_2}{dx} = \sqrt{1 + x^2} \cdot 2x$.

3. $y_1 = y_2 + \dfrac{2}{3}$. By Corollary 3 to the Mean Value Theorem of Section 4.2, y_1 and y_2 must differ by a constant. We find that constant by evaluating the two functions at $x = 0$.

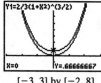

$[-3, 3]$ by $[-2, 8]$

4.

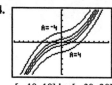

$[-10, 10]$ by $[-30, 30]$

5. The derivative with respect to x of each function graphed in part (4) is equal to $\sqrt{1 + x^2}$.

Exploration 2 Two Routes to the Integral

1. $\int_{-1}^{1} 3x^2\sqrt{x^3+1}\, dx = \int_0^2 \sqrt{u}\, du = \frac{2}{3}u^{3/2}\Big]_0^2 = \frac{4\sqrt{2}}{3}$

2. $\int 3x^2\sqrt{x^3+1}\, dx = \int \sqrt{u}\, du = \frac{2}{3}u^{3/2} = \frac{2}{3}(x^3+1)^{3/2}$ so
$\int_{-1}^{1} 3x^2\sqrt{x^3+1}\, dx = \frac{2}{3}(x^3+1)^{3/2}\Big]_{-1}^{1} = \frac{4\sqrt{2}}{3}.$

Quick Review 6.2

1. $\int_0^2 x^4\, dx = \frac{1}{5}x^5\Big]_0^2 = \frac{1}{5}(2)^5 - \frac{1}{5}(0)^5 = \frac{32}{5}$

2. $\int_1^5 \sqrt{x-1}\, dx = \int_1^5 (x-1)^{1/2}\, dx = \frac{2}{3}(x-1)^{3/2}\Big]_1^5$
$= \frac{2}{3}(4)^{3/2} - \frac{2}{3}(0)^{3/2}$
$= \frac{2}{3}(8) = \frac{16}{3}$

3. $\frac{dy}{dx} = 3^x$

4. $\frac{dy}{dx} = 3^x$

5. $\frac{dy}{dx} = 4(x^3 - 2x^2 + 3)^3(3x^2 - 4x)$

6. $\frac{dy}{dx} = 2\sin(4x-5)\cos(4x-5)\cdot 4$
$= 8\sin(4x-5)\cos(4x-5)$

7. $\frac{dy}{dx} = \frac{1}{\cos x}\cdot -\sin x = -\tan x$

8. $\frac{dy}{dx} = \frac{1}{\sin x}\cdot \cos x = \cot x$

9. $\frac{dy}{dx} = \frac{1}{\sec x + \tan x}\cdot (\sec x \tan x + \sec^2 x)$
$= \frac{\sec x \tan x + \sec^2 x}{\sec x + \tan x}$
$= \frac{\sec x(\tan x + \sec x)}{\sec x + \tan x}$
$= \sec x$

10. $\frac{dy}{dx} = \frac{1}{\csc x + \cot x}(-\csc x \cot x - \csc^2 x)$
$= -\frac{\csc x \cot x + \csc^2 x}{\csc x + \cot x}$
$= -\frac{\csc x(\cot x + \csc x)}{\csc x + \cot x}$
$= -\csc x$

Section 6.2 Exercises

1. $u = 3x$
$du = 3\, dx$
$\frac{1}{3}du = dx$
$\int \sin 3x\, dx = \frac{1}{3}\int \sin u\, du$
$= -\frac{1}{3}\cos u + C$
$= -\frac{1}{3}\cos 3x + C$

Check:
$\frac{d}{dx}\left(-\frac{1}{3}\cos 3x + C\right) = -\frac{1}{3}(-\sin 3x)(3) = \sin 3x$

3. $u = 2x$
$du = 2\, dx$
$\frac{1}{2}du = dx$
$\int \sec 2x \tan 2x\, dx = \frac{1}{2}\int \sec u \tan u\, du$
$= \frac{1}{2}\sec u + C$
$= \frac{1}{2}\sec 2x + C$

Check:
$\frac{d}{dx}\left(\frac{1}{2}\sec 2x + C\right) = \frac{1}{2}\sec 2x \tan 2x \cdot 2 = \sec 2x \tan 2x$

5. $u = \frac{x}{3}$
$du = \frac{1}{3}dx$
$3\, du = dx$
$\int \frac{dx}{x^2+9} = \int \frac{3u}{9u^2+9}$
$= \frac{3}{9}\int \frac{du}{u^2+1}$
$= \frac{1}{3}\int \frac{du}{u^2+1}$
$= \frac{1}{3}\tan^{-1} u + C$
$= \frac{1}{3}\tan^{-1}\left(\frac{x}{3}\right) + C$

Check: $\frac{d}{dx}\left(\frac{1}{3}\tan^{-1}\frac{x}{3} + C\right) = \frac{1}{3}\cdot \frac{1}{1+\left(\frac{x}{3}\right)^2}\cdot \frac{1}{3} = \frac{1}{9+x^2}$

7. $u = 1 - \cos \frac{t}{2}$

$du = \frac{1}{2}\sin \frac{t}{2}\, dt$

$2\, du = \sin \frac{t}{2}\, dt$

$\int \left(1 - \cos \frac{t}{2}\right)^2 \sin \frac{t}{2}\, dt = 2\int u^2\, du$

$\qquad = \frac{2}{3}u^3 + C$

$\qquad = \frac{2}{3}\left(1 - \cos \frac{t}{2}\right)^3 + C$

Check: $\dfrac{d}{dx}\left[\dfrac{2}{3}\left(1 - \cos \dfrac{t}{2}\right)^3 + C\right]$

$= 2\left(1 - \cos \dfrac{t}{2}\right)^2 \left(\sin \dfrac{t}{2}\right)\left(\dfrac{1}{2}\right)$

$= \left(1 - \cos \dfrac{t}{2}\right)^2 \sin \dfrac{t}{2}$

9. Let $u = 1 - x$

$du = -dx$

$\int \dfrac{dx}{(1-x)^2} = -\int \dfrac{du}{u^2}$

$\qquad = u^{-1} + C$

$\qquad = \dfrac{1}{1-x} + C$

11. Let $u = \tan x$

$du = \sec^2 x\, dx$

$\int \sqrt{\tan x}\, \sec^2 x\, dx = \int u^{1/2}\, du$

$\qquad = \dfrac{2}{3}u^{3/2} + C$

$\qquad = \dfrac{2}{3}(\tan x)^{3/2} + C$

13. Let $u = \ln x$

$du = \dfrac{1}{x}\, dx$

$\int_e^6 \dfrac{dx}{x \ln x} = \int_1^{\ln 6} \dfrac{du}{u} = \ln |u|\Big]_1^{\ln 6} = \ln(\ln 6)$

15. Let $u = 3z + 4$

$du = 3\, dz$

$\dfrac{1}{3}du = dz$

$\int \cos(3z+4)\, dz = \dfrac{1}{3}\int \cos u\, du$

$\qquad = \dfrac{1}{3}\sin u + C$

$\qquad = \dfrac{1}{3}\sin(3z+4) + C$

17. Let $u = \ln x$

$du = \dfrac{1}{x}\, dx$

$\int \dfrac{\ln^6 x}{x}\, dx = \int u^6\, du$

$\qquad = \dfrac{1}{7}u^7 + C$

$\qquad = \dfrac{1}{7}(\ln^7 x) + C$

19. Let $u = s^{4/3} - 8$

$du = \dfrac{4}{3}s^{1/3}\, ds$

$\dfrac{3}{4}du = s^{1/3}\, ds$

$\int s^{1/3} \cos(s^{4/3} - 8)\, ds = \dfrac{3}{4}\int \cos u\, du$

$\qquad = \dfrac{3}{4}\sin u + C$

$\qquad = \dfrac{3}{4}\sin(s^{4/3} - 8) + C$

21. Let $u = \cos(2t+1)$

$du = -\sin(2t+1)(2)\, dt$

$-\dfrac{1}{2}du = \sin(2t+1)\, dt$

$\int \dfrac{\sin(2t+1)}{\cos^2(2t+1)}\, dt = -\dfrac{1}{2}\int u^{-2}\, du$

$\qquad = \dfrac{1}{2}u^{-1} + C$

$\qquad = \dfrac{1}{2\cos(2t+1)} + C$

$\qquad = \dfrac{1}{2}\sec(2t+1) + C$

23. $\displaystyle\int_{\pi/4}^{3\pi/4} \cot x\, dx = \int_{\pi/4}^{3\pi/4} \dfrac{\cos x}{\sin x}\, dx$

Let $u = \sin x$

$du = \cos x\, dx$

$\displaystyle\int_{\pi/4}^{3\pi/4} \dfrac{\cos x}{\sin x}\, dx = \int_{x = \pi/4}^{x = 3\pi/4} \dfrac{1}{u}\, du$

$\qquad = \ln|u|\Big]_{x=\pi/4}^{x=3\pi/4}$

$\qquad = \ln|\sin x|\Big]_{\pi/4}^{3\pi/4}$

$\qquad = \ln\left|\dfrac{\sqrt{2}}{2}\right| - \ln\left|\dfrac{\sqrt{2}}{2}\right| = 0$

25. Let $u = x^2 + 1$

$du = 2x\, dx$

$x\, dx = \dfrac{1}{2}du$

$\displaystyle\int_{-1}^{3} \dfrac{x\, dx}{x^2 + 1} = \dfrac{1}{2}\int_2^{10} \dfrac{1}{u}\, du$

$\qquad = \dfrac{1}{2}\ln|u|\Big]_2^{10}$

$\qquad = \dfrac{1}{2}(\ln 10 - \ln 2) = \dfrac{1}{2}\ln 5 \approx 0.805$

164   Section 6.2

27. $\int \dfrac{dx}{\cot 3x} = \int \dfrac{\sin 3x}{\cos 3x}\, dx$

Let $u = \cos 3x$

$du = -3 \sin 3x\, dx$

$-\dfrac{1}{3} du = \sin 3x\, dx$

$\int \dfrac{dx}{\cot 3x} = -\dfrac{1}{3}\int \dfrac{1}{u}\, du$

$\qquad = -\dfrac{1}{3} \ln |u| + C$

$\qquad = -\dfrac{1}{3} \ln |\cos 3x| + C$

(An equivalent expression is $\dfrac{1}{3} \ln |\sec 3x| + C$.)

29. $\int \sec x\, dx = \int \sec x \cdot \left(\dfrac{\sec x + \tan x}{\sec x + \tan x}\right) dx$

$\qquad = \int \dfrac{\sec^2 x + \sec x \tan x}{\sec x + \tan x}\, dx$

Let $u = \sec x + \tan x$

$du = \sec x \tan x + \sec^2 x\, dx$

$\int \sec x\, dx = \int \dfrac{1}{u}\, du = \ln |u| + C = \ln |\sec x + \tan x| + C$

31. Let $u = y + 1$

$du = dy$

$\int_0^3 \sqrt{y+1}\, dy = \int_1^4 u^{1/2}\, du$

$\qquad = \dfrac{2}{3} u^{3/2}\Big]_1^4$

$\qquad = \dfrac{2}{3}(4)^{3/2} - \dfrac{2}{3}(1)^{3/2}$

$\qquad = \dfrac{2}{3}(8) - \dfrac{2}{3} = \dfrac{14}{3}$

33. Let $u = \tan x$

$du = \sec^2 x\, dx$

$\int_{-\pi/4}^0 \tan x \sec^2 x\, dx = \int_{-1}^0 u\, du$

$\qquad = \dfrac{1}{2} u^2 \Big]_{-1}^0$

$\qquad = \dfrac{1}{2}(0) - \dfrac{1}{2}(-1)^2$

$\qquad = -\dfrac{1}{2}$

35. Let $u = 1 + \theta^{3/2}$

$du = \dfrac{3}{2} \theta^{1/2}\, d\theta$

$\dfrac{2}{3} du = \theta^{1/2}\, d\theta$

$\int_0^1 \dfrac{10\sqrt{\theta}}{(1 + \theta^{3/2})^2}\, d\theta = \dfrac{2}{3}(10)\int_1^2 u^{-2}\, du$

$\qquad = -\dfrac{20}{3} u^{-1}\Big]_1^2$

$\qquad = -\dfrac{20}{3}\left(\dfrac{1}{2} - 1\right)$

$\qquad = -\dfrac{20}{3}\left(-\dfrac{1}{2}\right) = \dfrac{10}{3}$

37. Let $u = t^5 + 2t$

$du = (5t^4 + 2)\, dt$

$\int_0^1 \sqrt{t^5 + 2t}\,(5t^4 + 2)\, dt = \int_0^3 u^{1/2}\, du$

$\qquad = \dfrac{2}{3} u^{3/2}\Big]_0^3$

$\qquad = \dfrac{2}{3}(3)^{3/2}$

$\qquad = \dfrac{2}{3}\sqrt{27} = 2\sqrt{3}$

39. $\dfrac{dy}{dx} = (y+5)(x+2)$

$\dfrac{dy}{y+5} = (x+2)dx$

Integrate both sides.

$\int \dfrac{dy}{y+5} = \int (x+2)\, dx$

On the left,

let $u = y + 5$

$du = dy$

$\int \dfrac{1}{u}\, du = \dfrac{1}{2} x^2 + 2x + C$

$\ln |u| = \dfrac{1}{2} x^2 + 2x + C$

$\ln |y+5| = \dfrac{1}{2} x^2 + 2x + C$

$|y+5| = e^{(1/2)x^2 + 2x + C}$

$|y+5| = e^C e^{(1/2)x^2 + 2x}$

We now let $C' = e^C$ or $C' = -e^C$, depending on whether $(y+5)$ is positive or negative. Then

$y + 5 = C' e^{(1/2)x^2 + 2x}$

$y = C' e^{(1/2)x^2 + 2x} - 5$

Since C' represents an arbitrary constant (note that even the value $C' = 0$ gives a solution to the original differential equation), we may write the solution as

$y = Ce^{(1/2)x^2 + 2x} - 5$.

41. $\dfrac{dy}{dx} = (\cos x)e^{y+\sin x}$

$\dfrac{dy}{dx} = (\cos x)(e^y\, e^{\sin x})$

$\dfrac{dy}{e^y} = \cos x\, e^{\sin x}\, dx$

Integrate both sides.

$\displaystyle\int \dfrac{dy}{e^y} = \int \cos x\, e^{\sin x}\, dx$

On the right, let $u = \sin x$

$du = \cos x\, dx$

$-e^{-y} = \displaystyle\int e^u\, du$

$-e^{-y} = e^u + C$

$-e^{-y} = e^{\sin x} + C$

$e^{-y} = -e^{\sin x} + C$

(Note: technically C is now $C' = -C$.)

$-y = \ln(C - e^{\sin x})$

$y = -\ln(C - e^{\sin x})$

43. $\dfrac{dy}{dx} = -2xy^2$

$-\dfrac{dy}{y^2} = 2x\, dx$

$-\displaystyle\int \dfrac{dy}{y^2} = \int 2x\, dx$

$y^{-1} = x^2 + C$

$y = \dfrac{1}{x^2 + C}$

$y(1) = \dfrac{1}{1 + C} = 0.25$

$1 + C = 4$

$C = 3$

$y = \dfrac{1}{x^2 + 3}$

45. (a) Let $u = x + 1$

$du = dx$

$\displaystyle\int \sqrt{x+1}\, dx = \int u^{1/2}\, du$

$= \dfrac{2}{3}u^{3/2} + C$

$= \dfrac{2}{3}(x+1)^{3/2} + C$

Alternatively, $\dfrac{d}{dx}\left(\dfrac{2}{3}(x+1)^{3/2} + C\right) = \sqrt{x+1}$.

(b) By Part 1 of the Fundamental Theorem of Calculus,

$\dfrac{dy_1}{dx} = \sqrt{x+1}$ and $\dfrac{dy_2}{dx} = \sqrt{x+1}$, so both are antiderivatives of $\sqrt{x+1}$.

(c) Using NINT to find the values of y_1 and y_2, we have:

x	0	1	2	3	4
y_1	0	1.219	2.797	4.667	6.787
y_2	-4.667	-3.448	-1.869	0	2.120
$y_1 - y_2$	4.667	4.667	4.667	4.667	4.667

$C = 4\dfrac{2}{3}$

(d) $C = y_1 - y_2$

$= \displaystyle\int_0^x \sqrt{x+1}\, dx - \int_3^x \sqrt{x+1}\, dx$

$= \displaystyle\int_0^x \sqrt{x+1}\, dx + \int_x^3 \sqrt{x+1}\, dx$

$= \displaystyle\int_0^3 \sqrt{x+1}\, dx$

47. Let $u = x^4 + 9$, $du = 4x^3\, dx$.

(a) $\displaystyle\int_0^1 \dfrac{x^3\, dx}{\sqrt{x^4+9}} = \int_9^{10} \dfrac{1}{4}u^{-1/2}\, du = \dfrac{1}{2}u^{1/2}\Big]_9^{10}$

$= \dfrac{1}{2}\sqrt{10} - \dfrac{1}{2}\sqrt{9}$

$= \dfrac{1}{2}\sqrt{10} - \dfrac{3}{2} \approx 0.081$

(b) $\displaystyle\int \dfrac{x^3}{x^4+9}\, dx = \int \dfrac{1}{4}u^{-1/2}\, du$

$= \dfrac{1}{2}u^{1/2} + C$

$= \dfrac{1}{2}\sqrt{x^4+9} + C$

$\displaystyle\int_0^1 \dfrac{x^3}{x^4+9}\, dx = \dfrac{1}{2}\sqrt{x^4+9}\,\Big]_0^1$

$= \dfrac{1}{2}\sqrt{10} - \dfrac{1}{2}\sqrt{9}$

$= \dfrac{1}{2}\sqrt{10} - \dfrac{3}{2} \approx 0.081$

49. We show that $f'(x) = \tan x$ and $f(3) = 5$, where

$f(x) = \ln\left|\dfrac{\cos 3}{\cos x}\right| + 5$.

$f'(x) = \dfrac{d}{dx}\left(\ln\left|\dfrac{\cos 3}{\cos x}\right| + 5\right)$

$= \dfrac{d}{dx}(\ln|\cos 3| - \ln|\cos x| + 5)$

$= -\dfrac{d}{dx}\ln|\cos x|$

$= -\dfrac{1}{\cos x}(-\sin x) = \tan x$

$f(3) = \ln\left|\dfrac{\cos 3}{\cos 3}\right| + 5 = (\ln 1) + 5 = 5$

51. (a) $u = \sin x$, $du = \cos x\, dx$

$$\int 2 \sin x \cos x\, dx = \int 2u\, du = u^2 + C = \sin^2 x + C$$

(b) $u = \cos x$, $du = -\sin x\, dx$

$$\int 2 \sin x \cos x\, dx = \int (-2u)\, du$$
$$= -u^2 + C$$
$$= -\cos^2 x + C$$

(c) $u = 2x$, $du = 2\, dx$

$$\int 2 \sin x \cos x\, dx = \int \sin 2x\, dx$$
$$= \int \frac{1}{2} \sin u\, du$$
$$= -\frac{1}{2} \cos u + C$$
$$= -\frac{1}{2} \cos 2x + C$$

(d) $\frac{d}{dx}(\sin^2 x + C) = 2 \sin x \cos x$

$\frac{d}{dx}(-\cos^2 x + C) = (-2\cos x)(-\sin x) = 2 \sin x \cos x$

$\frac{d}{dx}\left(-\frac{1}{2}\cos 2x + C\right) = \left(\frac{1}{2} \sin 2x\right)(2)$
$= \sin 2x$
$= 2 \sin x \cos x$

■ Section 6.3 Integration by Parts
(pp. 323–329)

Exploration 1 Evaluating and Checking Integrals

1. $u = \ln x \Rightarrow du = \frac{dx}{x}$ and $dv = dx \Rightarrow v = x$. Thus,

$$\int \ln x\, dx = \int u\, dv$$
$$= uv - \int v\, du$$
$$= x \ln x - \int dx$$
$$= x \ln x - x + C$$

2. $\frac{d}{dx}(x \ln x - x) = \ln x + x\left(\frac{1}{x}\right) - 1 = \ln x$

3. The slope field of $\frac{dy}{dx} = \ln x$ shows the direction of the curve as it is graphed from left to right across the window.

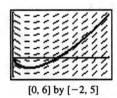
[0, 6] by [−2, 5]

4. The graph of $y_2 = x \ln x - x$ appears to be a vertical shift of the graph of $y_1 = \int_1^x \ln t\, dt$ (down 1 unit). Thus, y_2 appears to be an antiderivative of $\ln x$ which supports $x \ln x - x + C$ as the set of all antiderivatives of $\ln x$.

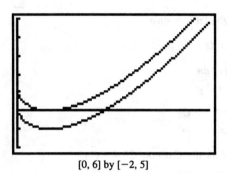
[0, 6] by [−2, 5]

Quick Review 6.3

1. $\frac{dy}{dx} = (x^3)(\cos 2x)(2) + (\sin 2x)(3x^2)$
$= 2x^3 \cos 2x + 3x^2 \sin 2x$

2. $\frac{dy}{dx} = (e^{2x})\left(\frac{3}{3x+1}\right) + \ln(3x+1)(2e^{2x})$
$= \frac{3e^{2x}}{3x+1} + 2e^{2x} \ln(3x+1)$

3. $\frac{dy}{dx} = \frac{1}{1+(2x)^2} \cdot 2$
$= \frac{2}{1+4x^2}$

4. $\frac{dy}{dx} = \frac{1}{\sqrt{1-(x+3)^2}}$

5. $y = \tan^{-1} 3x$
$\tan y = 3x$
$x = \frac{1}{3}\tan y$

6. $y = \cos^{-1}(x+1)$
$\cos y = x + 1$
$x = \cos y - 1$

7. $\int_0^1 \sin \pi x\, dx = -\frac{1}{\pi} \cos \pi x \Big]_0^1$
$= -\frac{1}{\pi} \cos \pi + \frac{1}{\pi} \cos 0$
$= -\frac{1}{\pi}(-1) + \frac{1}{\pi} = \frac{2}{\pi}$

8. $\frac{dy}{dx} = e^{2x}$
$dy = e^{2x}\, dx$

Integrate both sides.

$\int dy = \int e^{2x}\, dx$
$y = \frac{1}{2}e^{2x} + C$

9. $\frac{dy}{dx} = x + \sin x$

$dy = (x + \sin x)dx$

Integrate both sides.

$\int dy = \int (x + \sin x)\, dx$

$y = \frac{1}{2}x^2 - \cos x + C$

$y(0) = -1 + C = 2$

$\qquad C = 3$

$y = \frac{1}{2}x^2 - \cos x + 3$

10. $\frac{d}{dx}\left(\frac{1}{2}e^x(\sin x - \cos x)\right)$

$= \frac{1}{2}e^x(\cos x + \sin x) + (\sin x - \cos x)\frac{1}{2}e^x$

$= \frac{1}{2}e^x \cos x + \frac{1}{2}e^x \sin x + \frac{1}{2}e^x \sin x - \frac{1}{2}e^x \cos x$

$= e^x \sin x$

Section 6.3 Exercises

1. Let $u = x \qquad dv = \sin x\, dx$

$\quad du = dx \qquad v = -\cos x$

$\int x \sin x\, dx = -x \cos x + \int \cos x\, dx$

$\qquad = -x \cos x + \sin x + C$

Check: $\frac{d}{dx}(-x \cos x + \sin x + C)$

$\qquad = (-x)(-\sin x) + (\cos x)(-1) + \cos x$

$\qquad = x \sin x$

3. Let $u = \ln y \qquad dv = y\, dy$

$\quad du = \frac{1}{y}\, dy \qquad v = \frac{1}{2}y^2$

$\int y \ln y\, dy = \frac{1}{2}y^2 \ln y - \int \frac{1}{2}y^2 \cdot \frac{1}{y}\, dy$

$\qquad = \frac{1}{2}y^2 \ln y - \frac{1}{2}\int y\, dy$

$\qquad = \frac{1}{2}y^2 \ln y - \frac{1}{4}y^2 + C$

Check: $\frac{d}{dy}\left[\frac{1}{2}y^2 \ln y - \frac{1}{4}y^2 + C\right]$

$= \left(\frac{1}{2}y^2\right)\left(\frac{1}{y}\right) + (\ln y)(y) - \frac{1}{2}y$

$= y \ln y$

5. Let $u = x \qquad dv = \sec^2 x\, dx$

$\quad du = dx \qquad v = \tan x$

$\int x \sec^2 x\, dx = x \tan x - \int \tan x\, dx$

$\qquad = x \tan x - \int \frac{\sin x}{\cos x}\, dx$

$\qquad = x \tan x + \ln |\cos x| + C$

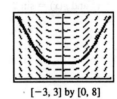

[−1.5, 1.5] by [−1, 4]

7. Let $u = t^2 \qquad dv = \sin t\, dt$

$\quad du = 2t\, dt \qquad v = -\cos t$

$\int t^2 \sin t\, dt = -t^2 \cos t + 2\int (\cos t)(t)\, dt$

Let $u = t \qquad dv = \cos t\, dt$

$\quad du = dt \qquad v = \sin t$

$-t^2 \cos t + 2\int t \cos t\, dt$

$= -t^2 \cos t + 2t \sin t - 2\int \sin t\, dt$

$= -t^2 \cos t + 2t \sin t + 2 \cos t + C$

$= (2 - t^2) \cos t + 2t \sin t + C$

[−3, 3] by [0, 8]

9. Let $u = \ln x \qquad dv = x^3\, dx$

$\quad du = \frac{1}{x}\, dx \qquad v = \frac{1}{4}x^4$

$\int x^3 \ln x\, dx = \frac{1}{4}x^4 \ln x - \frac{1}{4}\int x^4\left(\frac{1}{x}\right) dx$

$\qquad = \frac{1}{4}x^4 \ln x - \frac{1}{4}\int x^3\, dx$

$\qquad = \frac{1}{4}x^4 \ln x - \frac{1}{16}x^4 + C$

Section 6.3

11. Let $u = x^2 - 5x \qquad dv = e^x\, dx$
$\qquad du = (2x - 5)\, dx \qquad v = e^x$

$\int (x^2 - 5x)e^x\, dx = (x^2 - 5x)e^x - \int e^x(2x - 5)\, dx$

Let $\quad u = 2x - 5 \qquad dv = e^x\, dx$
$\qquad du = 2\, dx \qquad v = e^x$

$(x^2 - 5x)e^x - \int e^x(2x - 5)\, dx$
$= (x^2 - 5x)e^x - (2x - 5)e^x + \int 2e^x\, dx$
$= (x^2 - 5x)e^x - (2x - 5)e^x + 2e^x + C$
$= (x^2 - 7x + 7)e^x + C$

13. Let $u = e^y \qquad\qquad dv = \sin y\, dy$
$\qquad du = e^y\, dy \qquad\quad v = -\cos y$

$\int e^y \sin y\, dy = -e^y \cos y + \int \cos y\, e^y\, dy$

Let $\quad u = e^y \qquad\qquad dv = \cos y\, dy$
$\qquad du = e^y\, dy \qquad\quad v = \sin y$

$\int e^y \sin y\, dy = -e^y \cos y + e^y \sin y - \int \sin y\, e^y\, dy$

$2\int e^y \sin y\, dy = -e^y \cos y + e^y \sin y$

$\int e^y \sin y\, dy = \frac{1}{2}e^y(\sin y - \cos y) + C$

15. Use tabular integration with $f(x) = x^2$ and $g(x) = \sin 2x$.

$f(x)$ and its derivatives		$g(x)$ and its integrals
x^2	(+)	$\sin 2x$
$2x$	(−)	$-\dfrac{1}{2}\cos 2x$
2	(+)	$-\dfrac{1}{4}\sin 2x$
0		$\dfrac{1}{8}\cos 2x$

$\int x^2 \sin 2x\, dx = -\frac{1}{2}x^2 \cos 2x + \frac{1}{2}x \sin 2x + \frac{1}{4}\cos 2x + C$

$\qquad = \left(\dfrac{1 - 2x^2}{4}\right)\cos 2x + \dfrac{x}{2}\sin 2x + C$

$\int_0^{\pi/2} x^2 \sin 2x\, dx = \left[\left(\dfrac{1 - 2x^2}{4}\right)\cos 2x + \dfrac{x}{2}\sin 2x\right]_0^{\pi/2}$

$\qquad = \left(\dfrac{1 - 2\left(\frac{\pi}{2}\right)^2}{4}\right)(-1) + 0 - \dfrac{1}{4}(1) - 0$

$\qquad = \dfrac{\pi^2}{8} - \dfrac{1}{2} \approx 0.734$

Check: $\text{NINT}\left(x^2 \sin 2x, x, 0, \dfrac{\pi}{2}\right) \approx 0.734$

17. Let $u = e^{2x}$ $\qquad dv = \cos 3x\, dx$

$\qquad du = 2e^{2x}\, dx \qquad v = \dfrac{1}{3}\sin 3x$

$\displaystyle\int e^{2x}\cos 3x\, dx = (e^{2x})\left(\dfrac{1}{3}\sin 3x\right) - \int\left(\dfrac{1}{3}\sin 3x\right)(2e^{2x}\, dx)$

$\qquad\qquad\qquad\quad = \dfrac{1}{3}e^{2x}\sin 3x - \dfrac{2}{3}\displaystyle\int e^{2x}\sin 3x\, dx$

Let $u = e^{2x}$ $\qquad dv = \sin 3x\, dx$

$\qquad du = 2e^{2x}\, dx \qquad v = -\dfrac{1}{3}\cos 3x$

$\displaystyle\int e^{2x}\cos 3x\, dx = \dfrac{1}{3}e^{2x}\sin 3x - \dfrac{2}{3}\left[(e^{2x})\left(-\dfrac{1}{3}\cos 3x\right) - \int\left(-\dfrac{1}{3}\cos 3x\right)(2e^{2x}\, dx)\right]$

$\qquad\qquad\qquad\quad = \dfrac{1}{9}e^{2x}(3\sin 3x + 2\cos 3x) - \dfrac{4}{9}\displaystyle\int e^{2x}\cos 3x\, dx$

$\dfrac{13}{9}\displaystyle\int e^{2x}\cos 3x\, dx = \dfrac{1}{9}e^{2x}(3\sin 3x + 2\cos 3x)$

$\displaystyle\int e^{2x}\cos 3x\, dx = \dfrac{1}{13}e^{2x}(3\sin 3x + 2\cos 3x)$

$\displaystyle\int_{-2}^{3} e^{2x}\cos 3x\, dx = \left[\dfrac{1}{13}e^{2x}(3\sin 3x + 2\cos 3x)\right]_{-2}^{3}$

$\qquad\qquad\qquad\quad = \dfrac{1}{13}[e^{6}(3\sin 9 + 2\cos 9) - e^{-4}(3\sin(-6) + 2\cos(-6))]$

$\qquad\qquad\qquad\quad = \dfrac{1}{13}[e^{6}(2\cos 9 + 3\sin 9) - e^{-4}(2\cos 6 - 3\sin 6)]$

$\qquad\qquad\qquad\quad \approx -18.186$

Check: NINT($e^{2x}\cos 3x, x, -2, 3$) ≈ -18.186

19. $y = \displaystyle\int x^{2}e^{4x}\, dx$

Let $u = x^{2}$ $\qquad dv = e^{4x}\, dx$

$\qquad du = 2x\, dx \qquad v = \dfrac{1}{4}e^{4x}$

$y = (x^{2})\left(\dfrac{1}{4}e^{4x}\right) - \displaystyle\int\left(\dfrac{1}{4}e^{4x}\right)(2x\, dx)$

$\quad = \dfrac{1}{4}x^{2}e^{4x} - \dfrac{1}{2}\displaystyle\int xe^{4x}\, dx$

Let $u = x$ $\qquad dv = e^{4x}\, dx$

$\qquad du = dx \qquad v = \dfrac{1}{4}e^{4x}$

$y = \dfrac{1}{4}x^{2}e^{4x} - \dfrac{1}{2}\left[(x)\left(\dfrac{1}{4}e^{4x}\right) - \displaystyle\int\left(\dfrac{1}{4}e^{4x}\right)dx\right]$

$y = \dfrac{1}{4}x^{2}e^{4x} - \dfrac{1}{8}xe^{4x} + \dfrac{1}{32}e^{4x} + C$

$y = \left(\dfrac{x^{2}}{4} - \dfrac{x}{8} + \dfrac{1}{32}\right)e^{4x} + C$

21. $y = \int \theta \sec^{-1} \theta \, d\theta$

Let $u = \sec^{-1} \theta \qquad dv = \theta \, d\theta$

$du = \dfrac{1}{\theta\sqrt{\theta^2 - 1}} \, du \qquad v = \dfrac{1}{2}\theta^2$

Note that we are told $\theta > 1$, so no absolute value is needed in the expression for du.

$y = (\sec^{-1} \theta)\left(\dfrac{1}{2}\theta^2\right) - \int \left(\dfrac{1}{2}\theta^2\right)\left(\dfrac{1}{\theta\sqrt{\theta^2-1}}\right) d\theta$

$y = \dfrac{\theta^2}{2}\sec^{-1}\theta - \dfrac{1}{4}\int \dfrac{2|\theta| \, d\theta}{\sqrt{\theta^2-1}}$

Let $w = \theta^2 - 1$, $dw = 2\theta \, d\theta$

$y = \dfrac{\theta^2}{2}\sec^{-1}\theta - \dfrac{1}{4}\int w^{-1/2} \, dw$

$y = \dfrac{\theta^2}{2}\sec^{-1}\theta - \dfrac{1}{2}w^{1/2} + C$

$y = \dfrac{\theta^2}{2}\sec^{-1}\theta - \dfrac{1}{2}\sqrt{\theta^2 - 1} + C$

23. Let $u = x \qquad dv = \sin x \, dx$

$du = dx \qquad v = -\cos x$

$\int x \sin x \, dx = -x \cos x + \int \cos x \, dx$

$\qquad = -x \cos x + \sin x + C$

(a) $\displaystyle\int_0^\pi |x \sin x| \, dx = \int_0^\pi x \sin x \, dx$

$\qquad = \left[-x \cos x + \sin x\right]_0^\pi$

$\qquad = -\pi(-1) + 0 + 0(1) - 0$

$\qquad = \pi$

(b) $\displaystyle\int_\pi^{2\pi} |x \sin x| \, dx = -\int_\pi^{2\pi} x \sin x \, dx$

$\qquad = \left[x \cos x - \sin x\right]_\pi^{2\pi}$

$\qquad = 2\pi(1) - 0 - \pi(-1) + 0$

$\qquad = 3\pi$

(c) $\displaystyle\int_0^{2\pi} |x \sin x| \, dx = \int_0^\pi |x \sin x| \, dx + \int_\pi^{2\pi} |x \sin x| \, dx$

$\qquad = \pi + 3\pi = 4\pi$

25. First, we evaluate $\int e^{-t} \cos t \, dt$.

Let $u = e^{-t} \qquad dv = \cos t \, dt$

$du = -e^{-t} \, dt \qquad v = \sin t$

$\int e^{-t} \cos t \, dt = e^{-t} \sin t + \int \sin t \, e^{-t} \, dt$

Let $u = e^{-t} \qquad dv = \sin t \, dt$

$du = -e^{-t} \, dt \qquad v = -\cos t$

$\int e^{-t} \cos t \, dt = e^{-t} \sin t - e^{-t} \cos t - \int e^{-t} \cos t \, dt$

$2 \int e^{-t} \cos t \, dt = e^{-t}(\sin t - \cos t) + C$

$\int e^{-t} \cos t \, dt = \dfrac{1}{2} e^{-t}(\sin t - \cos t) + C$

Now we find the average value of $y = 2e^{-t} \cos t$ for $0 \le t \le 2\pi$.

Average value $= \dfrac{1}{2\pi}\displaystyle\int_0^{2\pi} 2e^{-t} \cos t \, dt$

$\qquad = \dfrac{1}{\pi}\displaystyle\int_0^{2\pi} e^{-t} \cos t \, dt$

$\qquad = \dfrac{1}{2\pi} e^{-t}(\sin t - \cos t) \Big|_0^{2\pi}$

$\qquad = \dfrac{1}{2\pi}[e^{-2\pi}(-1) - e^0(-1)]$

$\qquad = \dfrac{1 - e^{-2\pi}}{2\pi} \approx 0.159$

27. Let $w = \sqrt{x}$. Then $dw = \dfrac{dx}{2\sqrt{x}}$, so $dx = 2\sqrt{x} \, dw = 2w \, dw$.

$\int \sin \sqrt{x} \, dx = \int (\sin w)(2w \, dw) = 2\int w \sin w \, dw$

Let $u = w \qquad dv = \sin w \, dw$

$du = dw \qquad v = -\cos w$

$\int w \sin w \, dw = -w \cos w + \int \cos w \, dw$

$\qquad = -w \cos w + \sin w + C$

$\int \sin \sqrt{x} \, dx = 2\int w \sin w \, dw$

$\qquad = -2w \cos w + 2 \sin w + C$

$\qquad = -2\sqrt{x} \cos \sqrt{x} + 2 \sin \sqrt{x} + C$

29. Let $w = x^2$. Then $dw = 2x\,dx$.

$$\int x^7 e^{x^2}\,dx = \int (x^2)^3 e^{x^2}\,x\,dx = \frac{1}{2}\int w^3 e^w\,dw.$$

Use tabular integration with $f(x) = w^3$ and $g(w) = e^w$.

$f(w)$ and its derivatives		$g(w)$ and its integrals
w^3	(+)	e^w
$3w^2$	(−)	e^w
$6w$	(+)	e^w
6	(−)	e^w
0		e^w

$\int w^3 e^w\,dw = w^3 e^w - 3w^2 e^w + 6w\,e^w - 6e^w + C$

$\qquad = (w^3 - 3w^2 + 6w - 6)e^w + C$

$\int x^7 e^{x^2}\,dx = \frac{1}{2}\int w^3 e^w\,dw$

$\qquad = \frac{1}{2}(w^3 - 3w^2 + 6w - 6)e^w + C$

$\qquad = \dfrac{(x^6 - 3x^4 + 6x^2 - 6)e^{x^2}}{2} + C$

31. Let $u = x^n\qquad dv = \cos x\,dx$

$\qquad du = nx^{n-1}\,dx\qquad v = \sin x$

$\int x^n \cos x\,dx = x^n \sin x - \int (\sin x)(nx^{n-1}\,dx)$

$\qquad = x^n \sin x - n\int x^{n-1}\sin x\,dx$

33. Let $u = x^n\qquad dv = e^{ax}\,dx$

$\qquad du = nx^{n-1}\,dx\qquad v = \dfrac{1}{a}e^{ax}$

$\int x^n e^{ax}\,dx = (x^n)\left(\dfrac{1}{a}e^{ax}\right) - \int \left(\dfrac{1}{a}e^{ax}\right)(nx^{n-1}\,dx)$

$\qquad = \dfrac{x^n e^{ax}}{a} - \dfrac{n}{a}\int x^{n-1}e^{ax}\,dx,\ a \neq 0$

35. (a) Let $y = f^{-1}(x)$. Then $x = f(y)$, so $dx = f'(y)\,dy$.

Hence, $\int f^{-1}(x)\,dx = \int (y)[f'(y)\,dy] = \int y f'(y)\,dy$

(b) Let $u = y\qquad dv = f'(y)\,dy$

$\qquad du = dy\qquad v = f(y)$

$\int y f'(y)\,dy = y f(y) - \int f(y)\,dy$

$\qquad = f^{-1}(x)(x) - \int f(y)\,dy$

Hence, $\int f^{-1}(x)\,dx = \int y f'(y)\,dy$

$\qquad = x f^{-1}(x) - \int f(y)\,dy.$

37. (a) Using $y = f^{-1}(x) = \sin^{-1} x$ and $f(y) = \sin y$, $-\dfrac{\pi}{2} \le y \le \dfrac{\pi}{2}$, we have:

$\int \sin^{-1} x\,dx = x\sin^{-1} x - \int \sin y\,dy$

$\qquad = x\sin^{-1} x + \cos y + C$

$\qquad = x\sin^{-1} x + \cos(\sin^{-1} x) + C$

(b) $\int \sin^{-1} x\,dx = x\sin^{-1} x - \int x\left(\dfrac{d}{dx}\sin^{-1} x\right)dx$

$\qquad = x\sin^{-1} x - \int x\dfrac{1}{\sqrt{1-x^2}}\,dx$

$u = 1 - x^2,\ du = -2x\,dx$

$\qquad = x\sin^{-1} x + \dfrac{1}{2}\int u^{-1/2}\,du$

$\qquad = x\sin^{-1} x + u^{1/2} + C$

$\qquad = x\sin^{-1} x + \sqrt{1-x^2} + C$

(c) $\cos(\sin^{-1} x) = \sqrt{1-x^2}$

39. (a) Using $y = f^{-1}(x) = \cos^{-1} x$ and $f(y) = \cos x,\ 0 \le x \le \pi$, we have:

$\int \cos^{-1} x\,dx = x\cos^{-1} x - \int \cos y\,dy$

$\qquad = x\cos^{-1} x - \sin y + C$

$\qquad = x\cos^{-1} x - \sin(\cos^{-1} x) + C$

(b) $\int \cos^{-1} x\,dx = x\cos^{-1} x - \int x\left(\dfrac{d}{dx}\cos^{-1} x\right)dx$

$\qquad = x\cos^{-1} x - \int x\left(-\dfrac{1}{\sqrt{1-x^2}}\right)dx$

$u = 1 - x^2,\ du = -2x\,dx$

$\qquad = x\cos^{-1} x - \dfrac{1}{2}\int u^{-1/2}\,du$

$\qquad = x\cos^{-1} x - u^{1/2} + C$

$\qquad = x\cos^{-1} x - \sqrt{1-x^2} + C$

(c) $\sin(\cos^{-1} x) = \sqrt{1-x^2}$

■ Section 6.4 Exponential Growth and Decay (pp. 330–341)

Exploration 1 Slowing Down More Slowly

1. As m increases the velocity of the object represented by the graph slows down more slowly. That is, the y-coordinates of the graphs decrease to 0 more slowly as m increases.

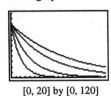

[0, 20] by [0, 120]

Section 6.4

2. As we saw in Section 5.1, $s(t) = \int_0^t v(u)\, du$ gives the distance traveled by the object over the time interval $[0, t]$. Since $s(0) = \int_0^0 v(u)\, du = 0$, the integral gives the distance traveled by the object at time t.

3. The total distance traveled is about 200 units for $m = 1$, about 400 units for $m = 2$, about 800 units for $m = 4$, and about 1200 units for $m = 6$.

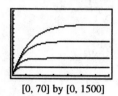

[0, 70] by [0, 1500]

Quick Review 6.4

1. $a = e^b$
2. $c = \ln d$
3. $\ln(x + 3) = 2$
   $x + 3 = e^2$
   $x = e^2 - 3$
4. $100e^{2x} = 600$
   $e^{2x} = 6$
   $2x = \ln 6$
   $x = \dfrac{1}{2}\ln 6$
5. $0.85^x = 2.5$
   $\ln 0.85^x = \ln 2.5$
   $x \ln 0.85 = \ln 2.5$
   $x = \dfrac{\ln 2.5}{\ln 0.85} \approx -5.638$
6. $2^{k+1} = 3^k$
   $\ln 2^{k+1} = \ln 3^k$
   $(k + 1)\ln 2 = k \ln 3$
   $\ln 2 = k(\ln 3 - \ln 2)$
   $k = \dfrac{\ln 2}{\ln 3 - \ln 2} \approx 1.710$
7. $1.1^t = 10$
   $\ln 1.1^t = \ln 10$
   $t \ln 1.1 = \ln 10$
   $t = \dfrac{\ln 10}{\ln 1.1} = \dfrac{1}{\log 1.1} \approx 24.159$
8. $e^{-2t} = \dfrac{1}{4}$
   $-2t = \ln\left(\dfrac{1}{4}\right)$
   $t = -\dfrac{1}{2}\ln\left(\dfrac{1}{4}\right) = \dfrac{1}{2}\ln 4 = \ln 2$

9. $\ln(y + 1) = 2x + 3$
   $y + 1 = e^{2x+3}$
   $y = -1 + e^{2x+3}$
10. $\ln|y + 2| = 3t - 1$
    $|y + 2| = e^{3t-1}$
    $y + 2 = \pm e^{3t-1}$
    $y = -2 \pm e^{3t-1}$

Section 6.4 Exercises

1. $y(t) = y_0 e^{kt}$
   $y(t) = 100e^{1.5t}$

3. $y(t) = y_0 e^{kt}$
   $y(t) = 50e^{kt}$
   $y(5) = 100 = 50e^{5k}$
   $2 = e^{5k}$
   $\ln 2 = 5k$
   $k = 0.2 \ln 2$
   Solution: $y(t) = 50e^{(0.2 \ln 2)t}$ or $y(t) = 50 \cdot 2^{0.2t}$

5. Doubling time:
   $A(t) = A_0 e^{rt}$
   $2000 = 1000 e^{0.086t}$
   $2 = e^{0.086t}$
   $\ln 2 = 0.086t$
   $t = \dfrac{\ln 2}{0.086} \approx 8.06 \text{ yr}$
   Amount in 30 years:
   $A = 1000 e^{(0.086)(30)} \approx \$13{,}197.10$

7. Initial deposit:
   $A(t) = A_0 e^{rt}$
   $2898.44 = A_0 e^{(0.0525)(30)}$
   $A_0 = \dfrac{2898.44}{e^{1.575}} \approx \$600.00$
   Doubling time:
   $A(t) = A_0 e^{rt}$
   $1200 = 600 e^{0.0525t}$
   $2 = e^{0.0525t}$
   $\ln 2 = 0.0525t$
   $t = \dfrac{\ln 2}{0.0525} \approx 13.2 \text{ years}$

9. **(a)** Annually:
$$2 = 1.0475^t$$
$$\ln 2 = t \ln 1.0475$$
$$t = \frac{\ln 2}{\ln 1.0475} \approx 14.94 \text{ years}$$

(b) Monthly:
$$2 = \left(1 + \frac{0.0475}{12}\right)^{12t}$$
$$\ln 2 = 12t \ln\left(1 + \frac{0.0475}{12}\right)$$
$$t = \frac{\ln 2}{12 \ln\left(1 + \frac{0.0475}{12}\right)} \approx 14.62 \text{ years}$$

(c) Quarterly:
$$2 = \left(1 + \frac{0.0475}{4}\right)^{4t}$$
$$\ln 2 = 4t \ln 1.011875$$
$$t = \frac{\ln 2}{4 \ln 1.011875} \approx 14.68 \text{ years}$$

(d) Continuously:
$$2 = e^{0.0475t}$$
$$\ln 2 = 0.0475t$$
$$t = \frac{\ln 2}{0.0475} \approx 14.59 \text{ years}$$

11. (a) Since there are 48 half-hour doubling times in 24 hours, there will be $2^{48} \approx 2.8 \times 10^{14}$ bacteria.

(b) The bacteria reproduce fast enough that even if many are destroyed there are still enough left to make the person sick.

13. $0.9 = e^{-0.18t}$
$$\ln 0.9 = -0.18t$$
$$t = -\frac{\ln 0.9}{0.18} \approx 0.585 \text{ days}$$

15. Since $y_0 = y(0) = 2$, we have:
$$y = 2e^{kt}$$
$$5 = 2e^{(k)(2)}$$
$$\ln 5 = \ln 2 + 2k$$
$$k = \frac{\ln 5 - \ln 2}{2} = 0.5 \ln 2.5$$
Function: $y = 2e^{(0.5 \ln 2.5)t}$ or $y \approx 2e^{0.4581t}$

17. At time $t = \frac{3}{k}$, the amount remaining is
$$y_0 e^{-kt} = y_0 e^{-k(3/k)} = y_0 e^{-3} \approx 0.0499 y_0.$$ This is less than 5% of the original amount, which means that over 95% has decayed already.

19. (a) First, we find the value of k.
$$T - T_s = (T_0 - T_s)e^{-kt}$$
$$60 - 20 = (90 - 20)e^{-(k)(10)}$$
$$\frac{4}{7} = e^{-10k}$$
$$k = -\frac{1}{10} \ln \frac{4}{7}$$

When the soup cools to 35°, we have:
$$35 - 20 = (90 - 20)e^{[(1/10) \ln (4/7)]t}$$
$$15 = 70 e^{[(1/10) \ln (4/7)]t}$$
$$\ln \frac{3}{14} = \left(\frac{1}{10} \ln \frac{4}{7}\right)t$$
$$t = \frac{10 \ln\left(\frac{3}{14}\right)}{\ln\left(\frac{4}{7}\right)} \approx 27.53 \text{ min}$$

It takes a total of about 27.53 minutes, which is an additional 17.53 minutes after the first 10 minutes.

(b) Using the same value of k as in part (a), we have:
$$T - T_s = (T_0 - T_s)e^{-kt}$$
$$35 - (-15) = [90 - (-15)]e^{[(1/10) \ln (4/7)]t}$$
$$50 = 105 e^{[(1/10) \ln (4/7)]t}$$
$$\ln \frac{10}{21} = \left(\frac{1}{10} \ln \frac{4}{7}\right)t$$
$$t = \frac{10 \ln\left(\frac{10}{21}\right)}{\ln\left(\frac{4}{7}\right)} \approx 13.26$$

It takes about 13.26 minutes.

21. Use $k = \frac{\ln 2}{5700}$ (see Example 3).
$$e^{-kt} = 0.445$$
$$-kt = \ln 0.445$$
$$t = -\frac{\ln 0.445}{k} = -\frac{5700 \ln 0.445}{\ln 2} \approx 6658 \text{ years}$$

Crater Lake is about 6658 years old.

23. Note that the total mass is $66 + 7 = 73$ kg.
$$v = v_0 e^{-(k/m)t}$$
$$v = 9e^{-3.9t/73}$$

(a) $s(t) = \int 9e^{-3.9t/73} dt = -\frac{2190}{13} e^{-3.9t/73} + C$

Since $s(0) = 0$ we have $C = \frac{2190}{13}$ and
$$\lim_{t \to \infty} s(t) = \lim_{t \to \infty} \frac{2190}{13}(1 - e^{-3.9t/73}) = \frac{2190}{13} \approx 168.5$$

The cyclist will coast about 168.5 meters.

(b) $1 = 9e^{-3.9t/73}$
$$\frac{3.9t}{73} = \ln 9$$
$$t = \frac{73 \ln 9}{3.9} \approx 41.13 \text{ sec}$$

It will take about 41.13 seconds.

25. $y = y_0 e^{-kt}$

$800 = 1000 e^{-(k)(10)}$

$0.8 = e^{-10k}$

$k = -\dfrac{\ln 0.8}{10}$

At $t = 10 + 14 = 24$ h:

$y = 1000 e^{-(-\ln 0.8/10)24}$

$= 1000 e^{2.4 \ln 0.8} \approx 585.4$ kg

About 585.4 kg will remain.

27. (a) $\dfrac{dp}{dn} = kp$

$\dfrac{dp}{p} = k\, dh$

$\displaystyle\int \dfrac{dp}{p} = \int k\, dh$

$\ln |p| = kh + C$

$e^{\ln |p|} = e^{kh+C}$

$|p| = e^C e^{kh}$

$p = A e^{kh}$

Initial condition: $p = p_0$ when $h = 0$

$p_0 = A e^0$

$A = p_0$

Solution: $p = p_0 e^{kh}$

Using the giving altitude-pressure data, we have

$p_0 = 1013$ millibars, so:

$p = 1013 e^{kh}$

$90 = 1013 e^{(k)(20)}$

$\dfrac{90}{1013} = e^{20k}$

$k = \dfrac{1}{20} \ln \dfrac{90}{1013} \approx -0.121$ km^{-1}

Thus, we have $p \approx 1013 e^{-0.121 h}$

(b) At 50 km, the pressure is

$1013 e^{((1/20)\ln(90/1013))(50)} \approx 2.383$ millibars.

(c) $900 = 1013 e^{kh}$

$\dfrac{900}{1013} = e^{kh}$

$h = \dfrac{1}{k} \ln \dfrac{900}{1013} = \dfrac{20 \ln (900/1013)}{\ln (90/1013)} \approx 0.977$ km

The pressure is 900 millibars at an altitude of about 0.977 km.

29. (a) By the Law of Exponential Change, the solution is $V = V_0 e^{-(1/40)t}$.

(b) $0.1 = e^{-(1/40)t}$

$\ln 0.1 = -\dfrac{t}{40}$

$t = -40 \ln 0.1 \approx 92.1$ sec

It will take about 92.1 seconds.

31. (a) $s(t) = \displaystyle\int v_0 e^{-(k/m)t}\, dt = -\dfrac{v_0 m}{k} e^{-(k/m)t} + C$

Initial condition: $s(0) = 0$

$0 = -\dfrac{v_0 m}{k} + C$

$\dfrac{v_0 m}{k} = C$

$s(t) = -\dfrac{v_0 m}{k} e^{-(k/m)t} + \dfrac{v_0 m}{k}$

$= \dfrac{v_0 m}{k}\left(1 - e^{-(k/m)t}\right)$

(b) $\displaystyle\lim_{t\to\infty} s(t) = \lim_{t\to\infty} \dfrac{v_0 m}{k}\left(1 - e^{-(k/m)t}\right) = \dfrac{v_0 m}{k}$

33. $\dfrac{v_0 m}{k} =$ coasting distance

$\dfrac{(0.80)(49.90)}{k} = 1.32$

$k = \dfrac{998}{33}$

We know that $\dfrac{v_0 m}{k} = 1.32$ and $\dfrac{k}{m} = \dfrac{998}{33(49.9)} = \dfrac{20}{33}$.

Using Equation 3, we have:

$s(t) = \dfrac{v_0 m}{k}(1 - e^{-(k/m)t})$

$= 1.32(1 - e^{-20t/33})$

$\approx 1.32(1 - e^{-0.606 t})$

A graph of the model is shown superimposed on a graph of the data.

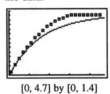

[0, 4.7] by [0, 1.4]

35. (a)
$$\frac{dT}{dt} = -k(T - T_s)$$

$$\frac{dT}{T - T_s} = -k\,dt$$

$$\int \frac{dT}{T - T_s} = -k\,dt$$

$$\ln|T - T_s| = -kt + C$$

$$|T - T_s| = e^{-kt+C}$$

$$T - T_s = \pm e^C e^{-kt}$$

$$T - T_s = Ae^{-kt}$$

Initial condition: $T = T_0$ when $t = 0$

$$T_0 - T_s = Ae^{-(k)(0)}$$

$$T_0 - T_s = A$$

Solution: $T - T_s = (T_0 - T_s)e^{-kt}$

(b) $\lim_{t \to \infty} T = \lim_{t \to \infty} [T_s + (T_0 - T_s)e^{-kt}] = T_s$

Horizontal asymptote: $T = T_s$

37. (a)

x	$\left(1 + \frac{1}{x}\right)^x$
10	2.5937
100	2.7048
1000	2.7169
10,000	2.7181
100,000	2.7183

$e \approx 2.7183$

Graphical support:

$y_1 = \left(1 + \frac{1}{x}\right)^x, y_2 = e$

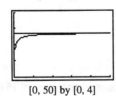

[0, 50] by [0, 4]

(b) $r = 2$

x	$\left(1 + \frac{2}{x}\right)^x$
10	6.1917
100	7.2446
1000	7.3743
10,000	7.3876
100,000	7.3889

$e^2 \approx 7.389$

Graphical support:

$y_1 = \left(1 + \frac{2}{x}\right)^x, y_2 = e^2$

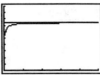

[0, 500] by [0, 10]

$r = 0.5$

x	$\left(1 + \frac{0.5}{x}\right)^x$
10	1.6289
100	1.6467
1000	1.6485
10,000	1.6487
100,000	1.6487

$e^{0.5} \approx 1.6487$

Graphical support:

$y_1 = \left(1 + \frac{0.5}{x}\right)^x, y_2 = e^{0.5}$

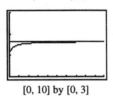

[0, 10] by [0, 3]

(c) As we compound more times, the increment of time between compounding approaches 0. Continuous compounding is based on an instantaneous rate of change which is a limit of average rates as the increment in time approaches 0.

■ Section 6.5 Population Growth
(pp. 342–349)

Quick Review 6.5

1. All real numbers

2. $\lim_{x \to \infty} f(x) = \dfrac{50}{1 + 0} = 50$

   $\lim_{x \to -\infty} f(x) = 0$

3. $y = 0, y = 50$

4. In both f' and f'', the denominator will be a power of $1 + 5e^{-0.1x}$, which is never 0. Thus, the domains of both are all real numbers.

5.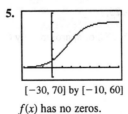

[−30, 70] by [−10, 60]

$f(x)$ has no zeros.

176 Section 6.5

6. Use NDER $f(x)$, or calculate the derivative as follows.

$$f'(x) = \frac{d}{dx} \frac{50}{1 + 5e^{-0.1x}}$$
$$= \frac{(1 + 5e^{-0.1x})(0) - (50)(5e^{-0.1x})(-0.1)}{(1 + 5e^{-0.1x})^2}$$
$$= \frac{25e^{-0.1x}}{(1 + 5e^{-0.1x})^2}$$

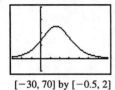

$[-30, 70]$ by $[-0.5, 2]$

(a) $(-\infty, \infty)$

(b) None

7. Use NDER(NDER $f(x)$), or calculate the second derivatives as follows.

$$f''(x) = \frac{d}{dx} \frac{25e^{-0.1x}}{(1 + 5e^{-0.1x})^2}$$
$$= \frac{(1 + 5e^{-0.1x})^2(25e^{-0.1x})(-0.1) - (25e^{-0.1x})(2)(1 + 5e^{-0.1x})(5e^{-0.1x})(-0.1)}{(1 + 5e^{-0.1x})^4}$$
$$= \frac{-2.5e^{-0.1x}[(1 + 5e^{-0.1x}) - 2(5e^{-0.1x})]}{(1 + 5e^{-0.1x})^3}$$
$$= \frac{12.5e^{-0.2x} - 2.5e^{-0.1x}}{(1 + 5e^{-0.1x})^3}$$

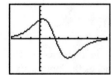

$[-30, 70]$ by $[-0.08, 0.08]$

Locate the inflection point using graphical methods, or analytically as follows.

$$f''(x) = 0$$
$$\frac{12.5e^{-0.2x} - 2.5e^{-0.1x}}{(1 + 5e^{-0.1x})^3} = 0$$
$$2.5e^{-0.1x}(5e^{-0.1x} - 1) = 0$$
$$e^{-0.1x} = \frac{1}{5}$$
$$-0.1x = -\ln 5$$
$$x = 10 \ln 5 \approx 16.094$$

(a) Since $f''(x) > 0$ for $x < 10 \ln 5$, the graph of f is concave up on the interval $(-\infty, 10 \ln 5)$, or approximately $(-\infty, 16.094)$.

(b) Since $f''(x) < 0$ for $x > 10 \ln 5$, the graph of f is concave down on the interval $(10 \ln 5, \infty)$, or approximately $(16.094, \infty)$.

8. Using the result of the previous exercise, the inflection point occurs at $x = 10 \ln 5$.

Since $f(10 \ln 5) = \dfrac{50}{1 + 5e^{-\ln 5}} = 25$,

the point of inflection is $(10 \ln 5, 25)$, or approximately $(16.094, 25)$.

9. $\dfrac{x - 12}{x^2 - 4x} = \dfrac{A}{x} + \dfrac{B}{x - 4}$

$x - 12 = A(x - 4) + Bx$

$x - 12 = (A + B)x - 4A$

Since $A + B = 1$ and $-4A = -12$, we have $A = 3$ and $B = -2$.

10. $\dfrac{2x+16}{x^2+x-6} = \dfrac{A}{x+3} + \dfrac{B}{x-2}$

$2x + 16 = A(x-2) + B(x+3)$

When $x = -3$, the equation becomes $10 = -5A$, and when $x = 2$, the equation becomes $20 = 5B$. Thus, $A = -2$ and $B = 4$.

Section 6.5 Exercises

1. (a) $\dfrac{dP}{dt} = 0.025P$

   (b) Using the Law of Exponential Change from Section 6.4, the formula is $P = 75{,}000e^{0.025t}$

   (c)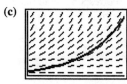

   [0, 100] by [0, 1,000,000]

3. (a) $\dfrac{dP}{dt} = \dfrac{k}{M}P(M-P)$

   $\dfrac{dP}{dt} = \dfrac{0.05}{200}P(200-P)$

   $\dfrac{dP}{dt} = 0.00025P(200-P)$

   (b) $P = \dfrac{M}{1+Ae^{-kt}}$

   $P = \dfrac{200}{1+Ae^{-0.05t}}$

   Initial condition: $P(0) = 10$

   $10 = \dfrac{200}{1+Ae^0}$

   $1 + A = \dfrac{200}{10} = 20$

   $A = 19$

   Formula: $P = \dfrac{200}{1+19e^{-0.05t}}$

   (c)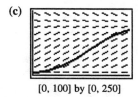

   [0, 100] by [0, 250]

5. The growth rate is -0.3 or -30%.

7. $\dfrac{dP}{dt} = 0.04P - 0.0004P^2$

   $= 0.0004P(100-P)$

   $= \dfrac{0.04}{100}P(100-P)$

   $= \dfrac{k}{M}P(M-P)$

   Thus, $k = 0.04$ and the carrying capacity is $M = 100$.

9. Choose the slope field that shows slopes that increase as y increases. (d)

11. Choose the only slope field whose slopes vary with x as well as with y. (c)

13. (a) $P(t) = \dfrac{1000}{1+e^{4.8-0.7t}}$

    $= \dfrac{1000}{1+e^{4.8}e^{-0.7t}}$

    $= \dfrac{M}{1+Ae^{-kt}}$

    This is a logistic growth model with $k = 0.7$ and $M = 1000$.

    (b) $P(0) = \dfrac{1000}{1+e^{4.8}} \approx 8$

    Initially there are 8 rabbits.

15. (a) Note that $\dfrac{dP}{dT} = \dfrac{1 \text{ person}}{14 \text{ sec}} \cdot \dfrac{365 \cdot 24 \cdot 3600 \text{ sec}}{1 \text{ yr}}$

    $\approx 2{,}252{,}571$ people per year.

    The relative growth rate is

    $\dfrac{\frac{dP}{dT}}{P} \approx \dfrac{2{,}252{,}571}{257{,}313{,}431} \approx 0.00875$ or 0.875%

    (b) The population after 8 years will be approximately $P_0 e^{rt} = 257{,}313{,}431 e^{8r}$

    $\approx 275{,}980{,}017$, where r is the unrounded rate from part (a).

17. (a) $\dfrac{dP}{dt} = 0.0015P(150-P)$

    $= \dfrac{0.225}{150}P(150-P)$

    $= \dfrac{k}{M}P(M-P)$

    Thus, $k = 0.225$ and $M = 150$.

    $P = \dfrac{M}{1+Ae^{-kt}}$

    $= \dfrac{150}{1+Ae^{-0.225t}}$

    Initial condition: $P(0) = 6$

    $6 = \dfrac{150}{1+Ae^0}$

    $1 + A = 25$

    $A = 24$

    Formula: $P = \dfrac{150}{1+24e^{-0.225t}}$

178 Section 6.5

17. continued

(b)
$$100 = \frac{150}{1 + 24e^{-0.225t}}$$
$$1 + 24e^{-0.225t} = \frac{3}{2}$$
$$24e^{-0.225t} = \frac{1}{2}$$
$$e^{-0.225t} = \frac{1}{48}$$
$$-0.225t = -\ln 48$$
$$t = \frac{\ln 48}{0.225} \approx 17.21 \text{ weeks}$$

$$125 = \frac{150}{1 + 24e^{-0.225t}}$$
$$1 + 24e^{-0.225t} = \frac{6}{5}$$
$$24e^{-0.225t} = \frac{1}{5}$$
$$e^{-0.225t} = \frac{1}{120}$$
$$-0.225t = -\ln 120$$
$$t = \frac{\ln 120}{0.225} \approx 21.28$$

It will take about 17.21 weeks to reach 100 guppies, and about 21.28 weeks to reach 125 guppies.

19. (a) $y = y_0 e^{-0.01(t/1000)} = y_0 e^{-0.00001t}$

(b) $0.9 = e^{-0.00001t}$
$\ln 0.9 = -0.00001t$
$t = -100{,}000 \ln 0.9 \approx 10{,}536$
It will take about 10,536 years.

(c) $y = y_0 e^{-(0.00001)(20{,}000)} \approx 0.819 y_0$
The tooth size will be about 81.9% of our present tooth size.

21. (a) $\frac{dx}{dt} = 1000 + 0.10x$

$$\int \frac{dx}{1000 + 0.1x} = \int dt$$
$$10 \ln |1000 + 0.1x| = t + C$$
$$\ln |1000 + 0.1x| = 0.1(t + C)$$
$$1000 + 0.1x = \pm e^{0.1(t+C)}$$
$$0.1x = -1000 \pm e^{0.1C} e^{0.1t}$$
$$x = -10{,}000 \pm 10 e^{0.1C} e^{0.1t}$$
$$x = -10{,}000 + Ae^{0.1t}$$

Initial condition: $x(0) = 1000$
$$1000 = -10{,}000 + Ae^0$$
$$11{,}000 = A$$
Solution: $x = 11{,}000 e^{0.1t} - 10{,}000$

(b) $100{,}000 = 11{,}000 e^{0.1t} - 10{,}000$
$10 = e^{0.1t}$
$\ln 10 = 0.1t$
$t = 10 \ln 10 \approx 23$ yr
It will take about 23 years.

23. (a) Note that the given years correspond to $x = 0$, $x = 20$, $x = 50$, $x = 70$, and $x = 80$.
$$y = \frac{18.70}{1 + 1.075 e^{-0.0422x}}$$

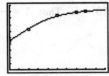

[0, 100] by [0, 20]

(b) Carrying capacity $= \lim\limits_{x \to \infty} y = 18.70$, representing 18.7 million people.

(c) Using NDER twice and solving graphically, we find that $y'' = 0$ when $x \approx 1.7$, corresponding to the year 1912. The population at this time was about $y(1.7) \approx 9.35$ million.

25.
$$\frac{dP}{dt} = \frac{k}{M} P(M - P)$$
$$\frac{M \, dP}{P(M - P)} = k \, dt$$
$$\frac{(M - P) + P}{P(M - P)} dP = k \, dt$$
$$\left(\frac{1}{P} + \frac{1}{M - P}\right) dP = k \, dt$$
$$\ln |P| - \ln |M - P| = kt + C$$
$$\ln \left|\frac{P}{M - P}\right| = kt + C$$
$$\ln \left|\frac{M - P}{P}\right| = -kt - C$$
$$\frac{M - P}{P} = \pm e^{-C} e^{-kt}$$
$$\frac{M - P}{P} = A e^{-kt}$$
$$M - P = APe^{-kt}$$
$$M = P(1 + Ae^{-kt})$$
$$P = \frac{M}{1 + Ae^{-kt}}$$

27. $\frac{dy}{dx} = (\cos x) e^{\sin x}$

$$\int dy = \int (\cos x) e^{\sin x} \, dx$$
$$\int dy = \int e^u \, du$$
$$y = e^u + C$$
$$y = e^{\sin x} + C$$

Initial value: $y(0) = 0$
$$0 = e^{\sin 0} + C$$
$$-1 = C$$
Solution: $y = e^{\sin x} - 1$

29. $\dfrac{dy}{dx} = \dfrac{x}{y}$

$\int y\, dy = \int x\, dx$

$\dfrac{y^2}{2} = \dfrac{x^2}{2} + C$

Initial condition:

$y(0) = 2$

$\dfrac{2^2}{2} = \dfrac{0^2}{2} + C$

$2 = C$

This gives $\dfrac{y^2}{2} = \dfrac{x^2}{2} + 2$, or $y^2 = x^2 + 4$.

But this equation represents two functions, $y = \pm\sqrt{x^2 + 4}$. The solution of the initial value problem is the function that satisfies the initial condition, namely $y = \sqrt{x^2 + 4}$.

31. (a) Note that $k > 0$ and $M > 0$, so the sign of $\dfrac{dP}{dt}$ is the same as the sign of $(M - P)(P - m)$. For $m < P < M$, both $M - P$ and $P - m$ are positive, so the product is positive. For $P < m$ or $P > M$, the expressions $M - P$ and $P - m$ have opposite signs, so the product is negative.

(b) $\dfrac{dP}{dt} = \dfrac{k}{M}(M - P)(P - m)$

$\dfrac{dP}{dt} = \dfrac{k}{1200}(1200 - P)(P - 100)$

$\dfrac{1200}{(1200 - P)(P - 100)} \dfrac{dP}{dt} = k$

$\dfrac{1100}{(1200 - P)(P - 100)} \dfrac{dP}{dt} = \dfrac{11}{12}k$

$\dfrac{(P - 100) + (1200 - P)}{(1200 - P)(P - 100)} \dfrac{dP}{dt} = \dfrac{11}{12}k$

$\left(\dfrac{1}{1200 - P} + \dfrac{1}{P - 100}\right) \dfrac{dP}{dt} = \dfrac{11}{12}k$

$\int \left(\dfrac{1}{1200 - P} + \dfrac{1}{P - 100}\right) dP = \dfrac{11}{12}k\, dt$

$-\ln|1200 - P| + \ln|P - 100| = \dfrac{11}{12}kt + C$

$\ln\left|\dfrac{P - 100}{1200 - P}\right| = \dfrac{11}{12}kt + C$

$\dfrac{P - 100}{1200 - P} = \pm e^C e^{11kt/12}$

$\dfrac{P - 100}{1200 - P} = Ae^{11kt/12}$

$P - 100 = 1200Ae^{11kt/12} - APe^{11kt/12}$

$P(1 + Ae^{11kt/12}) = 1200Ae^{11kt/12} + 100$

$P = \dfrac{1200Ae^{11kt/12} + 100}{1 + Ae^{11kt/12}}$

(c) $300 = \dfrac{1200Ae^0 + 100}{1 + Ae^0}$

$300(1 + A) = 1200A + 100$

$300 - 100 = 1200A - 300A$

$200 = 900A$

$A = \dfrac{2}{9}$

$P(t) = \dfrac{1200(2/9)e^{11kt/12} + 100}{1 + (2/9)e^{11kt/12}}$

$P(t) = \dfrac{1200(2)e^{11kt/12} + 100(9)}{9 + 2e^{11kt/12}}$

$P(t) = \dfrac{300(8e^{11kt/12} + 3)}{9 + 2e^{11kt/12}}$

(d)

[0, 75] by [0, 1500]

Note that the slope field is given by

$\dfrac{dP}{dt} = \dfrac{0.1}{1200}(1200 - P)(P - 100)$.

(e) $\dfrac{dP}{dt} = \dfrac{k}{M}(M - P)(P - m)$

$\dfrac{M}{(M - P)(P - m)} \dfrac{dP}{dt} = k$

$\dfrac{M}{M - m} \cdot \dfrac{M - m}{(M - P)(P - m)} \dfrac{dP}{dt} = k$

$\dfrac{(P - m) + (M - P)}{(M - P)(P - m)} \dfrac{dP}{dt} = \dfrac{M - m}{M}k$

$\left(\dfrac{1}{M - P} + \dfrac{1}{P - m}\right) \dfrac{dP}{dt} = \dfrac{M - m}{M}k$

$\int \left(\dfrac{1}{M - P} + \dfrac{1}{P - m}\right) dP = \int \dfrac{M - m}{M}k\, dt$

$-\ln|M - P| + \ln|P - m| = \dfrac{M - m}{M}kt + C$

$\ln\left|\dfrac{P - m}{M - P}\right| = \dfrac{M - m}{M}kt + C$

$\dfrac{P - m}{M - P} = \pm e^C e^{(M-m)kt/M}$

$\dfrac{P - m}{M - P} = Ae^{(M-m)kt/M}$

$P - m = (M - P)Ae^{(M-m)kt/M}$

$P(1 + Ae^{(M-m)kt/M}) = AMe^{(M-m)kt/M} + m$

$P = \dfrac{AMe^{(M-m)kt/M} + m}{1 + Ae^{(M-m)kt/M}}$

$P(0) = \dfrac{AMe^0 + m}{1 + Ae^0} = \dfrac{AM + m}{1 + A}$

$P(0)(1 + A) = AM + m$

$A(P(0) - M) = m - P(0)$

$A = \dfrac{m - P(0)}{P(0) - M} = \dfrac{P(0) - m}{M - P(0)}$

Therefore, the solution to the differential equation is

$P = \dfrac{AMe^{(M-m)kt/M} + m}{1 + Ae^{(M-m)kt/M}}$ where $A = \dfrac{P(0) - m}{M - P(0)}$.

180 Section 6.6

33. (a)
$$\frac{dP}{dt} = kP^2$$
$$\int P^{-2}\, dP = \int k\, dt$$
$$-P^{-1} = kt + C$$
$$P = -\frac{1}{kt + C}$$

Initial condition: $P(0) = P_0$
$$P_0 = -\frac{1}{C}$$
$$C = -\frac{1}{P_0}$$
Solution: $P = -\dfrac{1}{kt - (1/P_0)} = \dfrac{P_0}{1 - kP_0 t}$

(b) There is a vertical asymptote at $t = \dfrac{1}{kP_0}$

■ Section 6.6 Numerical Methods
(pp. 350–356)

Quick Review 6.6

1. $f'(x) = 3x^2 - 3$
$f'(2) = 3(2)^2 - 3 = 9$

2. $L(x) = f(2) + f'(2)(x - 2)$
$= 2 + 9(x - 2)$
$= 9x - 16$

3. $f'(x) = \sec^2 x$
$f'\left(\dfrac{\pi}{4}\right) = \sec^2 \dfrac{\pi}{4} = (\sqrt{2})^2 = 2$

4. $L(x) = f\left(\dfrac{\pi}{4}\right) + f'\left(\dfrac{\pi}{4}\right)\left(x - \dfrac{\pi}{4}\right)$
$= 1 + 2\left(x - \dfrac{\pi}{4}\right)$
$= 2x + 1 - \dfrac{\pi}{2}$

5. $f'(x) = 0.2x - 5x^{-2}$
$f'(4) = 0.2(4) - 5(4)^{-2} = 0.4875$

6. $L(x) = f(4) + f'(4)(x - 4)$
$= 2.85 + 0.4875(x - 4)$
$= 0.4875x + 0.9$

7. $L(4.1) = 0.4875(4.1) + 0.9 = 2.89875$
$f(4.1) = 0.1(4.1)^2 + \dfrac{5}{4.1} \approx 2.900512$

(a) $|L(4.1) - f(4.1)| \approx 0.001762$

(b) $\dfrac{|L(4.1) - f(4.1)|}{f(4.1)} \approx 0.00061 = 0.061\%$

8. $L(4.2) = 0.4875(4.2) + 0.9 = 2.9475$
$f(4.2) = 0.1(4.2)^2 + \dfrac{5}{4.2} \approx 2.954476$

(a) $|L(4.2) - f(4.2)| \approx 0.006976$

(b) $\dfrac{|L(4.2) - f(4.2)|}{f(4.2)} \approx 0.00236 = 0.236\%$

9. $L(4.5) = 0.4875(4.5) + 0.9 = 3.09375$
$f(4.5) = 0.1(4.5)^2 + \dfrac{5}{4.5} \approx 3.136111$

(a) $|L(4.5) - f(4.5)| \approx 0.042361$

(b) $\dfrac{|L(4.5) - f(4.5)|}{f(4.5)} \approx 0.01351 = 1.351\%$

10. $L(3.5) = 0.4875(3.5) + 0.9 = 2.60625$
$f(3.5) = 0.1(3.5)^2 + \dfrac{5}{3.5} \approx 2.653571$

(a) $|L(3.5) - f(3.5)| \approx 0.047321$

(b) $\dfrac{|L(3.5) - f(3.5)|}{f(3.5)} \approx 0.01783 = 1.783\%$

Section 6.6 Exercises

1. Check the differential equation:
$y' = \dfrac{d}{dx}(x - 1 + 2e^{-x}) = 1 + 2e^{-x}(-1) = 1 - 2e^{-x}$
$x - y = x - (x - 1 + 2e^{-x}) = 1 - 2e^{-x}$

Therefore, $y' = x - y$.

Check the initial condition:
$y(0) = 0 - 1 + 2e^{-(0)} = -1 + 2 = 1$

3. Check the differential equation:

$y' = \dfrac{d}{dx}\left(\dfrac{e^{2x} - 2\sin x - \cos x}{5}\right) = \dfrac{2e^{2x} - 2\cos x + \sin x}{5}$

$2y + \sin x = 2\left(\dfrac{e^{2x} - 2\sin x - \cos x}{5}\right) + \sin x$

$= \dfrac{2e^{2x} - 4\sin x - 2\cos x + 5\sin x}{5}$

$= \dfrac{2e^{2x} - 2\cos x + \sin x}{5}$

Therefore, $y' = 2y + \sin x$

Check the initial condition:
$y(0) = \dfrac{e^{2(0)} - 2\sin 0 - \cos 0}{5} = \dfrac{1 - 1}{5} = 0$

5. Note that we are finding an exact solution to the initial value problem discussed in Examples 1–4.

$\dfrac{dy}{dx} = 1 + y$
$\int \dfrac{dy}{1 + y} = \int dx$
$\ln|1 + y| = x + C$
$|1 + y| = e^{x+C}$
$1 + y = \pm e^{x+C}$
$y = \pm e^C e^x - 1$
$y = Ae^x - 1$

Initial condition: $y(0) = 1$
$1 = Ae^0 - 1$
$2 = A$
Solution: $y = 2e^x - 1$

7. $\dfrac{dy}{dx} = 2y(x+1)$

$\dfrac{dy}{y} = 2(x+1)\,dx$

$\int \dfrac{dy}{y} = \int (2x+2)\,dx$

$\ln|y| = x^2 + 2x + C$

$|y| = e^{x^2+2x+C}$

$y = \pm e^C\, e^{x^2+2x}$

$y = A\, e^{x^2+2x}$

Initial condition: $y(-2) = 2$
$2 = A e^{(-2)^2 + 2(-2)}$
$2 = A$
Solution: $y = 2e^{x^2+2x}$

9. To find the approximate values, set $y_1 = 2y + \sin x$ and use EULERT with initial values $x = 0$ and $y = 0$ and step size 0.1 for 10 points. The exact values are given by $y = \dfrac{1}{5}(e^{2x} - 2\sin x - \cos x)$.

x	y (Euler)	y (exact)	Error
0	0	0	0
0.1	0	0.0053	0.0053
0.2	0.0100	0.0229	0.0129
0.3	0.0318	0.0551	0.0233
0.4	0.0678	0.1051	0.0374
0.5	0.1203	0.1764	0.0561
0.6	0.1923	0.2731	0.0808
0.7	0.2872	0.4004	0.1132
0.8	0.4090	0.5643	0.1553
0.9	0.5626	0.7723	0.2097
1.0	0.7534	1.0332	0.2797

11. To find the approximate values, set $y_1 = 2y(x+1)$ and use IMPEULT with initial values $x = -2$ and $y = 2$ and step size 0.1 for 20 points. The exact values are given by $y = 2e^{x^2+2x}$.

x	y (improved Euler)	y (exact)	Error
-2	2	2	0
-1.9	1.6560	1.6539	0.0021
-1.8	1.3983	1.3954	0.0030
-1.7	1.2042	1.2010	0.0032
-1.6	1.0578	1.0546	0.0032
-1.5	0.9478	0.9447	0.0031
-1.4	0.8663	0.8634	0.0029
-1.3	0.8077	0.8050	0.0027
-1.2	0.7683	0.7658	0.0025
-1.1	0.7456	0.7432	0.0024
-1.0	0.7381	0.7358	0.0023
-0.9	0.7455	0.7432	0.0023
-0.8	0.7682	0.7658	0.0024
-0.7	0.8075	0.8050	0.0024
-0.6	0.8659	0.8634	0.0025
-0.5	0.9473	0.9447	0.0026
-0.4	1.0572	1.0546	0.0026
-0.3	1.2036	1.2010	0.0026
-0.2	1.3976	1.3954	0.0022
-0.1	1.6553	1.6539	0.0014
0	1.9996	2	0.0004

Section 6.6

13. To find the approximate values, set $y_1 = x - y$ and use EULERT and IMPEULT with initial values $x = 0$ and $y = 1$ and step size 0.1 for 20 points. The exact values are given by $y = x - 1 + 2e^{-x}$.

x	y (Euler)	y (improved Euler)	y (exact)	Error (Euler)	Error (improved Euler)
0	1	1	1	0	0
0.1	0.9000	0.9100	0.9097	0.0097	0.0003
0.2	0.8200	0.8381	0.8375	0.0175	0.0006
0.3	0.7580	0.7824	0.7816	0.0236	0.0008
0.4	0.7122	0.7416	0.7406	0.0284	0.0010
0.5	0.6810	0.7142	0.7131	0.0321	0.0011
0.6	0.6629	0.6988	0.6976	0.0347	0.0012
0.7	0.6566	0.6944	0.6932	0.0366	0.0012
0.8	0.6609	0.7000	0.6987	0.0377	0.0013
0.9	0.6748	0.7145	0.7131	0.0383	0.0013
1.0	0.6974	0.7371	0.7358	0.0384	0.0013
1.1	0.7276	0.7671	0.7657	0.0381	0.0013
1.2	0.7649	0.8037	0.8024	0.0375	0.0013
1.3	0.8084	0.8463	0.8451	0.0367	0.0013
1.4	0.8575	0.8944	0.8932	0.0357	0.0012
1.5	0.9118	0.9475	0.9463	0.0345	0.0012
1.6	0.9706	1.0050	1.0038	0.0332	0.0012
1.7	1.0335	1.0665	1.0654	0.0318	0.0011
1.8	1.1002	1.1317	1.1306	0.0304	0.0011
1.9	1.1702	1.2002	1.1991	0.0290	0.0010
2.0	1.2432	1.2716	1.2707	0.0275	0.0010

15. (a)
$$\frac{dy}{dx} = 2y^2(x - 1)$$
$$\frac{dy}{y^2} = 2(x - 1)dx$$
$$\int y^{-2} dy = \int (2x - 2) dx$$
$$-y^{-1} = x^2 - 2x + C$$

Initial value: $y(2) = -\frac{1}{2}$

$2 = 2^2 - 2(2) + C$

$2 = C$

Solution: $-y^{-1} = x^2 - 2x + 2$ or $y = -\dfrac{1}{x^2 - 2x + 2}$

$y(3) = -\dfrac{1}{3^2 - 2(3) + 2} = -\dfrac{1}{5} = -0.2$

(b) To find the approximation, set $y_1 = 2y^2(x - 1)$ and use EULERT with initial values $x = 2$ and $y = -\frac{1}{2}$ and step size 0.2 for 5 points. This gives $y(3) \approx -0.1851$; error ≈ 0.0149.

(c) Use step size 0.1 for 10 points. This gives
$y(3) \approx -0.1929$; error ≈ 0.0071.

(d) Use step size 0.05 for 20 points. This gives
$y(3) \approx -0.1965$; error ≈ 0.0035.

17. The exact solution is $y = -\dfrac{1}{x^2 - 2x + 2}$, so $y(3) = -0.2$.

(a) To find the approximation, set $y_1 = 2y^2(x - 1)$ and use IMPEULT with initial values $x = 2$ and $y = -\dfrac{1}{2}$ and step size 0.2 for 5 points. This gives $y(3) \approx -0.2024$; error ≈ 0.0024.

(b) Use step size 0.1 for 10 points. This gives $y(3) \approx -0.2005$; error ≈ 0.0005.

(c) Use step size 0.05 for 20 points. This gives $y(3) = -0.2001$; error ≈ 0.0001.

(d) As the step size decreases, the accuracy of the method increases and so the error decreases.

19. Set $y_1 = 2y + \sin x$ and use EULERG with initial values $x = 0$ and $y = 0$ and step size 0.1. The exact solution is $y = \dfrac{1}{5}(e^{2x} - 2\sin x - \cos x)$.

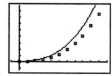

$[-0.1, 1.1]$ by $[-0.13, 0.88]$

21. Set $y_1 = 2y(x + 1)$ and use IMPEULG with initial values $x = -2$ and $y = 2$ and step size 0.1. The exact solution is $y = 2e^{x^2 + 2x}$.

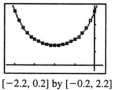

$[-2.2, 0.2]$ by $[-0.2, 2.2]$

23. To find the approximate values, set $y_1 = x + y$ and use EULERT with initial values $x = 0$ and $y = 1$ and step size -0.1 for 10 points. The exact values are given by $y = 2e^x - x - 1$.

x	y (Euler)	y (exact)	Error
0	1	1.0	0
-0.1	0.9000	0.9097	0.0097
-0.2	0.8200	0.8375	0.0175
-0.3	0.7580	0.7816	0.0236
-0.4	0.7122	0.7406	0.0284
-0.5	0.6810	0.7131	0.0321
-0.6	0.6629	0.6976	0.0347
-0.7	0.6566	0.6932	0.0366
-0.8	0.6609	0.6987	0.0377
-0.9	0.6748	0.7131	0.0383
-1.0	0.6974	0.7358	0.0384

25. Set $y_1 = y + e^x - 2$ and EULERG, with initial values $x = 0$ and $y = 2$ and step sizes 0.1 and 0.05.

(a)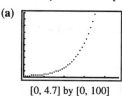

$[0, 4.7]$ by $[0, 100]$

(b)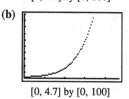

$[0, 4.7]$ by $[0, 100]$

27. Set $y_1 = y\left(\dfrac{1}{2} - \ln |y|\right)$ and use IMPEULG with initial values $x = 0$ and $y = \dfrac{1}{3}$ and step size 0.1 and 0.05.

(a)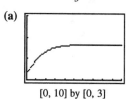

$[0, 10]$ by $[0, 3]$

(b)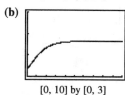

$[0, 10]$ by $[0, 3]$

29. To find the approximate values, let $y_1 = y$ and use EULERT with initial values $x = 0$ and $y = 1$ and step size 0.05 for 20 points. This gives $y(1) \approx 2.6533$.

Since the exact solution to the initial value problem is $y = e^x$, the exact value of $y(1)$ is e.

31. To find the approximate values, let $y_1 = 1 + y$ and use RUNKUTT with initial values $x = 0$ and $y = 1$ and step size 0.1 for 10 points. The exact values are given by $y = 2e^x - 1$.

x	y (Runge-Kutta)	y (exact)	Error
0	1	1	0
0.1	1.2103	1.2103	0.0000002
0.2	1.4428	1.4428	0.0000004
0.3	1.6997	1.6997	0.0000006
0.4	1.9836	1.9836	0.0000009
0.5	2.2974	2.2974	0.0000013
0.6	2.6442	2.6442	0.0000017
0.7	3.0275	3.0275	0.0000022
0.8	3.4511	3.4511	0.0000027
0.9	3.9192	3.9192	0.0000034
1.0	4.4366	4.4366	0.0000042

Chapter 6 Review Exercises
(pp. 358 – 361)

1. $\int_0^{\pi/3} \sec^2 \theta \, d\theta = \tan \theta \Big]_0^{\pi/3} = \tan \frac{\pi}{3} - \tan 0 = \sqrt{3}$

2. $\int_1^2 \left(x + \frac{1}{x^2}\right) dx = \left[\frac{1}{2}x^2 - x^{-1}\right]_1^2$

$\phantom{\int_1^2 \left(x + \frac{1}{x^2}\right) dx} = \left(\frac{1}{2}(4) - \frac{1}{2}\right) - \left(\frac{1}{2} - 1\right)$

$\phantom{\int_1^2 \left(x + \frac{1}{x^2}\right) dx} = \frac{3}{2} + \frac{1}{2}$

$\phantom{\int_1^2 \left(x + \frac{1}{x^2}\right) dx} = \frac{4}{2} = 2$

3. Let $u = 2x + 1$

$du = 2 \, dx$

$\frac{1}{2} du = dx$

$\int_0^1 \frac{36}{(2x+1)^3} dx = 18 \int_1^3 \frac{1}{u^3} du$

$\phantom{\int_0^1 \frac{36}{(2x+1)^3} dx} = 18\left(-\frac{1}{2}\right) u^{-2} \Big]_1^3$

$\phantom{\int_0^1 \frac{36}{(2x+1)^3} dx} = -9\left(\frac{1}{9} - 1\right)$

$\phantom{\int_0^1 \frac{36}{(2x+1)^3} dx} = -9\left(-\frac{8}{9}\right)$

$\phantom{\int_0^1 \frac{36}{(2x+1)^3} dx} = 8$

4. Let $u = 1 - x^2$

$du = -2x \, dx$

$-du = 2x \, dx$

$\int_{-1}^1 2x \sin(1 - x^2) \, dx = -\int_0^0 \sin u \, du = 0$

5. Let $u = \sin x$

$du = \cos x \, dx$

$\int_0^{\pi/2} 5 \sin^{3/2} x \cos x \, dx = \int_0^1 5u^{3/2} du$

$\phantom{\int_0^{\pi/2} 5 \sin^{3/2} x \cos x \, dx} = 5 \cdot \frac{2}{5} u^{5/2} \Big]_0^1$

$\phantom{\int_0^{\pi/2} 5 \sin^{3/2} x \cos x \, dx} = 2(1 - 0)$

$\phantom{\int_0^{\pi/2} 5 \sin^{3/2} x \cos x \, dx} = 2$

6. $\int_{1/2}^4 \frac{x^2 + 3x}{x} dx = \int_{1/2}^4 (x + 3) \, dx \; (x \neq 0)$

$\phantom{\int_{1/2}^4 \frac{x^2 + 3x}{x} dx} = \left(\frac{1}{2}x^2 + 3x\right)\Big]_{1/2}^4$

$\phantom{\int_{1/2}^4 \frac{x^2 + 3x}{x} dx} = \left(\frac{1}{2}(16) + 3(4)\right) - \left(\frac{1}{2}\left(\frac{1}{4}\right) + \frac{3}{2}\right)$

$\phantom{\int_{1/2}^4 \frac{x^2 + 3x}{x} dx} = 20 - \left(\frac{1}{8} + \frac{12}{8}\right)$

$\phantom{\int_{1/2}^4 \frac{x^2 + 3x}{x} dx} = 20 - \frac{13}{8}$

$\phantom{\int_{1/2}^4 \frac{x^2 + 3x}{x} dx} = \frac{147}{8}$

7. Let $u = \tan x$

$du = \sec^2 x \, dx$

$\int_0^{\pi/4} e^{\tan x} \sec^2 x \, dx = \int_0^1 e^u \, du$

$\phantom{\int_0^{\pi/4} e^{\tan x} \sec^2 x \, dx} = e^u \Big]_0^1$

$\phantom{\int_0^{\pi/4} e^{\tan x} \sec^2 x \, dx} = e^1 - e^0$

$\phantom{\int_0^{\pi/4} e^{\tan x} \sec^2 x \, dx} = e - 1$

8. Let $u = \ln r$

$du = \frac{1}{r} dr$

$\int_1^e \frac{\sqrt{\ln r}}{r} dr = \int_0^1 u^{1/2} du$

$\phantom{\int_1^e \frac{\sqrt{\ln r}}{r} dr} = \frac{2}{3} u^{3/2} \Big]_0^1$

$\phantom{\int_1^e \frac{\sqrt{\ln r}}{r} dr} = \frac{2}{3}(1 - 0)$

$\phantom{\int_1^e \frac{\sqrt{\ln r}}{r} dr} = \frac{2}{3}$

9. Let $u = 2 - \sin x$

$du = -\cos x \, dx$

$-du = \cos x \, dx$

$\int \frac{\cos x}{2 - \sin x} dx = -\int \frac{1}{u} du$

$\phantom{\int \frac{\cos x}{2 - \sin x} dx} = -\ln |u| + C$

$\phantom{\int \frac{\cos x}{2 - \sin x} dx} = -\ln |2 - \sin x| + C$

10. Let $u = 3x + 4$

$du = 3 \, dx$

$\frac{1}{3} du = dx$

$\int \frac{dx}{\sqrt[3]{3x + 4}} = \frac{1}{3} \int u^{-1/3} du$

$\phantom{\int \frac{dx}{\sqrt[3]{3x + 4}}} = \frac{1}{3} \cdot \frac{3}{2} u^{2/3} + C$

$\phantom{\int \frac{dx}{\sqrt[3]{3x + 4}}} = \frac{1}{2}(3x + 4)^{2/3} + C$

11. Let $u = t^2 + 5$

$du = 2t\, dt$

$\frac{1}{2} du = t\, dt$

$\int \frac{t\, dt}{t^2 + 5} = \frac{1}{2}\int \frac{1}{u} du = \frac{1}{2} \ln|u| + C$

$= \frac{1}{2} \ln|t^2 + 5| + C$

$= \frac{1}{2} \ln(t^2 + 5) + C$

12. Let $u = \frac{1}{\theta}$

$du = -\frac{1}{\theta^2} d\theta$

$\int \frac{1}{\theta^2} \sec \frac{1}{\theta} \tan \frac{1}{\theta} d\theta = -\int \sec u \tan u\, du$

$= -\sec u + C$

$= -\sec \frac{1}{\theta} + C$

13. Let $u = \ln y$

$du = \frac{1}{y} dy$

$\int \frac{\tan(\ln y)}{y} dy = \int \tan u\, du$

$= \int \frac{\sin u}{\cos u} du$

Let $w = \cos u$

$dw = -\sin u\, du$

$= -\int \frac{1}{w} dw$

$= \ln|w| + C$

$= -\ln|\cos u| + C$

$= -\ln|\cos(\ln y)| + C$

14. Let $u = e^x$

$du = e^x dx$

$\int e^x \sec(e^x)\, dx = \int \sec u\, du$

$= \ln|\sec u + \tan u| + C$

$= \ln|\sec(e^x) + \tan(e^x)| + C$

15. Let $u = \ln x$

$du = \frac{1}{x} dx$

$\int \frac{dx}{x \ln x} = \int \frac{1}{u} du$

$= \ln|u| + C$

$= \ln|\ln x| + C$

16. $\int \frac{dt}{t\sqrt{t}} = \int \frac{dt}{t^{3/2}}$

$= \int t^{-3/2} dt$

$= -2t^{-1/2} + C$

$= -\frac{2}{\sqrt{t}} + C$

17. Use tabular integration with $f(x) = x^3$ and $g(x) = \cos x$.

$f(x)$ and its derivatives		$g(x)$ and its integrals
x^3	(+)	$\cos x$
$3x^2$	(−)	$\sin x$
$6x$	(+)	$-\cos x$
6	(−)	$-\sin x$
0		$\cos x$

$\int x^3 \cos x\, dx$

$= x^3 \sin x + 3x^2 \cos x - 6x \sin x - 6 \cos x + C$

18. Let $u = \ln x \qquad dv = x^4 dx$

$du = \frac{1}{x} dx \qquad v = \frac{1}{5}x^5$

$\int x^4 \ln x\, dx = \frac{1}{5}x^5 \ln x - \int \frac{1}{5}x^5 \left(\frac{1}{x}\right) dx$

$= \frac{1}{5}x^5 \ln x - \frac{1}{5}\int x^4 dx$

$= \frac{1}{5}x^5 \ln x - \frac{1}{25}x^5 + C$

19. Let $u = e^{3x} \qquad dv = \sin x\, dx$

$du = 3e^{3x} dx \qquad v = -\cos x$

$\int e^{3x} \sin x\, dx = -e^{3x} \cos x + \int 3 \cos x\, e^{3x} dx$

Integrate by parts again.

Let $u = 3e^{3x} \qquad dv = \cos x\, dx$

$du = 9e^{3x} dx \qquad v = \sin x$

$\int e^{3x} \sin x\, dx = -e^{3x} \cos x + 3e^{3x} \sin x - \int 9e^{3x} \sin x\, dx$

$10\int e^{3x} \sin x\, dx = -e^{3x} \cos x + 3e^{3x} \sin x + C$

$\int e^{3x} \sin x\, dx = \frac{1}{10}[-e^{3x} \cos x + 3e^{3x} \sin x] + C$

$= \left(\frac{3 \sin x}{10} - \frac{\cos x}{10}\right) e^{3x} + C$

20. Let $u = x^2 \quad dv = e^{-3x}\,dx$

$\quad du = 2x\,dx \quad v = -\dfrac{1}{3}e^{-3x}$

$\int x^2 e^{-3x}\,dx = -\dfrac{1}{3}x^2 e^{-3x} + \dfrac{2}{3}\int e^{-3x} x\,dx$

Let $u = x \quad dv = e^{-3x}\,dx$

$\quad du = dx \quad v = -\dfrac{1}{3}e^{-3x}$

$= -\dfrac{1}{3}x^2 e^{-3x} + \dfrac{2}{3}\left[-\dfrac{1}{3}xe^{-3x} + \dfrac{1}{3}\int e^{-3x}\,dx \right]$

$= -\dfrac{1}{3}x^2 e^{-3x} - \dfrac{2}{9}xe^{-3x} + \dfrac{2}{9}\int e^{-3x}\,dx$

$= -\dfrac{1}{3}x^2 e^{-3x} - \dfrac{2}{9}xe^{-3x} - \dfrac{2}{27}e^{-3x} + C$

$= \left(-\dfrac{x^2}{3} - \dfrac{2x}{9} - \dfrac{2}{27} \right) e^{-3x} + C$

21. $\dfrac{dy}{dx} = 1 + x + \dfrac{x^2}{2}$

$dy = \left(1 + x + \dfrac{x^2}{2} \right) dx$

$\int dy = \int \left(1 + x + \dfrac{x^2}{2} \right) dx$

$y = x + \dfrac{1}{2}x^2 + \dfrac{1}{6}x^3 + C$

$y(0) = C = 1$

$y = \dfrac{x^3}{6} + \dfrac{x^2}{2} + x + 1$

Graphical support:

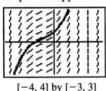

$[-4, 4]$ by $[-3, 3]$

22. $\dfrac{dy}{dx} = \left(x + \dfrac{1}{x} \right)^2$

$dy = \left(x + \dfrac{1}{x} \right)^2 dx$

$\int dy = \int \left(x + \dfrac{1}{x} \right)^2 dx$

$y = \int \left(x^2 + 2 + \dfrac{1}{x^2} \right) dx$

$y = \dfrac{1}{3}x^3 + 2x - x^{-1} + C$

$y(1) = \dfrac{1}{3} + 2 - 1 + C = 1$

$\dfrac{4}{3} + C = 1$

$C = -\dfrac{1}{3}$

$y = \dfrac{x^3}{3} + 2x - \dfrac{1}{x} - \dfrac{1}{3}$

Graphical support:

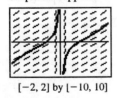

$[-2, 2]$ by $[-10, 10]$

23. $\dfrac{dy}{dt} = \dfrac{1}{t + 4}$

$dy = \dfrac{1}{t + 4}\,dt$

$\int dy = \int \dfrac{1}{t + 4}\,dt$

$y = \ln|t + 4| + C$

$y(-3) = \ln(1) + C = 2$

$C = 2$

$y = \ln(t + 4) + 2$

Graphical Support:

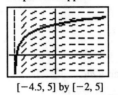

$[-4.5, 5]$ by $[-2, 5]$

24. $\dfrac{dy}{d\theta} = \csc 2\theta \cot 2\theta$

$dy = \csc 2\theta \cot 2\theta \, d\theta$

$\int dy = \int \csc 2\theta \cot 2\theta \, d\theta$

$y = -\dfrac{1}{2}\csc 2\theta + C$

$y\left(\dfrac{\pi}{4}\right) = -\dfrac{1}{2} + C = 1$

$C = \dfrac{3}{2}$

$y = -\dfrac{1}{2}\csc 2\theta + \dfrac{3}{2}$

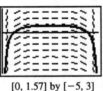

$[0, 1.57]$ by $[-5, 3]$

25.
$$\frac{d(y')}{dx} = 2x - \frac{1}{x^2}$$
$$d(y') = \left(2x - \frac{1}{x^2}\right) dx$$
$$\int d(y') = \int \left(2x - \frac{1}{x^2}\right) dx$$
$$y' = x^2 + x^{-1} + C$$
$$y'(1) = 2 + C = 1$$
$$C = -1$$
$$y' = x^2 + x^{-1} - 1$$
$$\int dy = \int (x^2 + x^{-1} - 1) \, dx$$
$$y = \frac{1}{3}x^3 + \ln x - x + C$$
$$y(1) = \frac{1}{3} + 0 - 1 + C = 0$$
$$-\frac{2}{3} + C = 0$$
$$C = \frac{2}{3}$$
$$y = \frac{x^3}{3} + \ln x - x + \frac{2}{3}$$

Graphical support:

Let $f(x) = \frac{x^3}{3} + \ln x - x + \frac{2}{3}$.

We first show the graph of $y = f'(x) = x^2 + x^{-1} - 1$, $x > 0$, along with the slope field for $y' = f''(x) = 2x - \frac{1}{x^2}$.

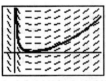

[−0.5, 4.21] by [−9, 21]

We now show the graph of $y = f(x)$ along with the slope field for $y' = f'(x) = x^2 + x^{-1} - 1$.

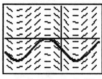

[−0.5, 4.21] by [−9, 21]

26.
$$\frac{d(r'')}{dt} = -\cos t$$
$$d(r'') = -\cos t \, dt$$
$$\int d(r'') = \int -\cos t \, dt$$
$$r'' = -\sin t + C$$
$$r''(0) = C = -1$$
$$r'' = -\sin t - 1$$
$$\int d(r') = \int (-\sin t - 1) \, dt$$
$$r' = \cos t - t + C$$
$$r'(0) = 1 + C = -1$$
$$C = -2$$
$$r' = \cos t - t - 2$$
$$\int dr = \int (\cos t - t - 2) \, dt$$
$$r = \sin t - \frac{t^2}{2} - 2t + C$$
$$r(0) = C = -1$$
$$r = \sin t - \frac{t^2}{2} - 2t - 1$$

Graphical support:

We first show the graph of $y = r'' = -\sin t - 1$ along with the slope field for $y' = r''' = -\cos t$.

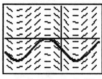

[−6, 4] by [−3, 3]

Next, we show the graph of $y = r' = \cos t - t - 2$ along with the slope field for $y' = r'' = -\sin t - 1$.

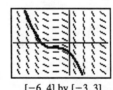

[−6, 4] by [−3, 3]

Finally we show the graph of $y = r = \sin t - \frac{t^2}{2} - 2t - 1$ along with the slope field for $y' = r' = \cos t - t - 2$.

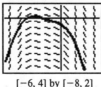

[−6, 4] by [−8, 2]

188 Chapter 6 Review

27. $\dfrac{dy}{dx} = y + 2$

$\dfrac{dy}{y+2} = dx$

$\displaystyle\int \dfrac{dy}{y+2} = \int dx$

$\ln|y+2| = x + C$

$y + 2 = Ce^x$

$y = Ce^x - 2$

$y(0) = C - 2 = 2$

$C = 4$

$y = 4e^x - 2$

Graphical support:

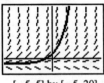

$[-5, 5]$ by $[-5, 20]$

28. $\dfrac{dy}{dx} = (2x+1)(y+1)$

$\dfrac{dy}{y+1} = (2x+1)\,dx$

$\displaystyle\int \dfrac{dy}{y+1} = \int (2x+1)\,dx$

$\ln|y+1| = x^2 + x + C$

$y + 1 = Ce^{x^2+x}$

$y = Ce^{x^2+x} - 1$

$y(-1) = C - 1 = 1$

$C = 2$

$y = 2e^{x^2+x} - 1$

Graphical support:

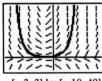

$[-3, 3]$ by $[-10, 40]$

29. $\displaystyle\int -f(x)\,dx = -\int f(x)\,dx$

$= -(1 - \sqrt{x}) + C$

$= -1 + \sqrt{x} + C$

Since $-1 + C$ is an arbitrary constant, we may write the indefinite integral as $\sqrt{x} + C$.

30. $\displaystyle\int [x + f(x)]\,dx = \int x\,dx + \int f(x)\,dx$

$= \dfrac{x^2}{2} + (1 - \sqrt{x}) + C$

$= \dfrac{x^2}{2} + 1 - \sqrt{x} + C$

Since $1 + C$ is an arbitrary constant, we may write the indefinite integral as $\dfrac{x^2}{2} - \sqrt{x} + C$.

31. $\displaystyle\int [2f(x) - g(x)]\,dx = 2\int f(x)\,dx - \int g(x)\,dx$

$= 2(1 - \sqrt{x}) - (x + 2) + C$

$= -2\sqrt{x} - x + C$

32. $\displaystyle\int [g(x) - 4]\,dx = \int g(x)\,dx - \int 4\,dx$

$= (x + 2) - 4x + C$

$= 2 - 3x + C$

Since $2 + C$ is an arbitrary constant, we may write the indefinite integral as $-3x + C$.

33. We seek the graph of a function whose derivative is $\dfrac{\sin x}{x}$. Graph (b) is increasing on $[-\pi, \pi]$, where $\dfrac{\sin x}{x}$ is positive, and oscillates slightly outside of this interval. This is the correct choice, and this can be verified by graphing $\text{NINT}\left(\dfrac{\sin x}{x}, x, 0, x\right)$.

34. We seek the graph of a function whose derivative is e^{-x^2}. Since $e^{-x^2} > 0$ for all x, the desired graph is increasing for all x. Thus, the only possibility is graph (d), and we may verify that this is correct by graphing $\text{NINT}(e^{-x^2}, x, 0, x)$.

35. (iv) The given graph looks like the graph of $y = x^2$, which satisfies $\dfrac{dy}{dx} = 2x$ and $y(1) = 1$.

36. Yes, $y = x$ is a solution.

37. (a) $\dfrac{dv}{dt} = 2 + 6t$

$\displaystyle\int dv = \int (2 + 6t)\,dt$

$v = 2t + 3t^2 + C$

Initial condition: $v = 4$ when $t = 0$

$4 = 0 + C$

$4 = C$

$v = 2t + 3t^2 + 4$

(b) $\displaystyle\int_0^1 v(t)\,dt = \int_0^1 (2t + 3t^2 + 4)\,dt$

$= \left[t^2 + t^3 + 4t\right]_0^1$

$= 6 - 0$

$= 6$

The particle moves 6 m.

38.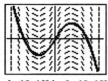
[−10, 10] by [−10, 10]

39. Set $y_1 = y + \cos x$ and use EULERT with initial values $x = 0$ and $y = 0$ and step size 0.1 for 20 points.

x	y
0	0
0.1	0.1000
0.2	0.2095
0.3	0.3285
0.4	0.4568
0.5	0.5946
0.6	0.7418
0.7	0.8986
0.8	1.0649
0.9	1.2411
1.0	1.4273
1.1	1.6241
1.2	1.8319
1.3	2.0513
1.4	2.2832
1.5	2.5285
1.6	2.7884
1.7	3.0643
1.8	3.3579
1.9	3.6709
2.0	4.0057

40. Set $y_1 = (2 - y)(2x + 3)$ and use IMPEULT with intial values $x = -3$ and $y = 1$ and step size 0.1 for 20 points.

x	y
−3	1
−2.9	0.6680
−2.8	0.2599
−2.7	−0.2294
−2.6	−0.8011
−2.5	−1.4509
−2.4	−2.1687
−2.3	−2.9374
−2.2	−3.7333
−2.1	−4.5268
−2.0	−5.2840
−1.9	−5.9686
−1.8	−6.5456
−1.7	−6.9831
−1.6	−7.2562
−1.5	−7.3488
−1.4	−7.2553
−1.3	−6.9813
−1.2	−6.5430
−1.1	−5.9655
−1.0	−5.2805

41. To estimate $y(3)$, set $y_1 = \dfrac{x - 2y}{x + 1}$ and use IMPEULT with initial values $x = 0$ and $y = 1$ and step size 0.05 for 60 points. This gives $y(3) \approx 0.9063$.

42. To estimate $y(4)$, set $y_1 = \dfrac{x^2 - 2y + 1}{x}$ and use EULERT with initial values $x = 1$ and $y = 1$ and step size 0.05 for 60 points. This gives $y(4) \approx 4.4974$.

43. Set $y_1 = e^{-(x+y+2)}$ and use EULERG with initial values $x = 0$ and $y = -2$ and step sizes 0.1 and −0.1.

(a)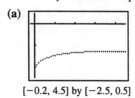
[−0.2, 4.5] by [−2.5, 0.5]

43. continued

(b) Note that we choose a small interval of x-values because the y-values decrease very rapidly and our calculator cannot handle the calculations for $x \le -1$. (This occurs because the analytic solution is $y = -2 + \ln(2 - e^{-x})$, which has an asymptote at $x = -\ln 2 \approx -0.69$. Obviously, the Euler approximations are misleading for $x \le -0.7$.)

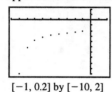

[−1, 0.2] by [−10, 2]

44. Set $y_1 = -\dfrac{x^2 + y}{e^y + x}$ and use IMPEULG with initial values $x = 0$ and $y = 0$ and step sizes 0.1 and −0.1.

(a)

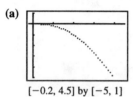

[−0.2, 4.5] by [−5, 1]

(b)

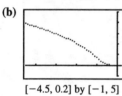

[−4.5, 0.2] by [−1, 5]

45. (a) Half-life $= \dfrac{\ln 2}{k}$

$$2.645 = \dfrac{\ln 2}{k}$$

$$k = \dfrac{\ln 2}{2.645} \approx 0.262059$$

(b) Mean life $= \dfrac{1}{k} \approx 3.81593$ years

46. $T - T_s = (T_0 - T_s)e^{-kt}$

$T - 40 = (220 - 40)e^{-kt}$

Use the fact that $T = 180$ and $t = 15$ to find k.

$180 - 40 = (220 - 40)e^{-(k)(15)}$

$$e^{15k} = \dfrac{180}{140} = \dfrac{9}{7}$$

$$k = \dfrac{1}{15} \ln \dfrac{9}{7}$$

$T - 40 = (220 - 40)e^{-((1/15)\ln(9/7))t}$

$70 - 40 = (220 - 40)e^{-((1/15)\ln(9/7))t}$

$$e^{((1/15)\ln(9/7))t} = \dfrac{180}{30} = 6$$

$$\left(\dfrac{1}{15} \ln \dfrac{9}{7}\right)t = \ln 6$$

$$t = \dfrac{15 \ln 6}{\ln(9/7)} \approx 107 \text{ min}$$

It took a total of about 107 minutes to cool from 220°F to 70°F. Therefore, the time to cool from 180°F to 70°F was about 92 minutes.

47. $T - T_s = (T_0 - T_s)e^{-kt}$

We have the system:

$$\begin{cases} 39 - T_s = (46 - T_s)e^{-10k} \\ 33 - T_s = (46 - T_s)e^{-20k} \end{cases}$$

Thus, $\dfrac{39 - T_s}{46 - T_s} = e^{-10k}$ and $\dfrac{33 - T_s}{46 - T_s} = e^{-20k}$.

Since $(e^{-10k})^2 = e^{-20k}$, this means:

$$\left(\dfrac{39 - T_s}{46 - T_s}\right)^2 = \dfrac{33 - T_s}{46 - T_s}$$

$(39 - T_s)^2 = (33 - T_s)(46 - T_s)$

$1521 - 78T_s + T_s^2 = 1518 - 79T_s + T_s^2$

$T_s = -3$

The refrigerator temperature was −3°C.

48. Use the method of Example 3 in Section 6.4.

$e^{-kt} = 0.995$

$-kt = \ln 0.995$

$$t = -\dfrac{1}{k} \ln 0.995 = -\dfrac{5700}{\ln 2} \ln 0.995 \approx 41.2$$

The painting is about 41.2 years old.

49. Use the method of Example 3 in Section 6.4.

Since 90% of the carbon-14 has decayed, 10% remains.

$e^{-kt} = 0.1$

$-kt = \ln 0.1$

$$t = -\dfrac{1}{k} \ln 0.1 = -\dfrac{5700}{\ln 2} \ln 0.1 \approx 18{,}935$$

The charcoal sample is about 18,935 years old.

50. Use $t = 1988 - 1924 = 64$ years.

$250 e^{rt} = 7500$

$e^{rt} = 30$

$rt = \ln 30$

$$r = \dfrac{\ln 30}{t} = \dfrac{\ln 30}{64} \approx 0.053$$

The rate of appreciation is about 0.053, or 5.3%.

51. Using the Law of Exponential Change in Section 6.4 with appropriate changes of variables, the solution to the differential equation is $L(x) = L_0 e^{-kx}$, where $L_0 = L(0)$ is the surface intensity. We know $0.5 = e^{-18k}$, so $k = \dfrac{\ln 0.5}{-18}$ and our equation becomes

$$L(x) = L_0 e^{(\ln 0.5)(x/18)} = L_0 \left(\dfrac{1}{2}\right)^{x/18}.$$ We now find the depth where the intensity is one-tenth of the surface value.

$$0.1 = \left(\dfrac{1}{2}\right)^{x/18}$$
$$\ln 0.1 = \dfrac{x}{18} \ln \left(\dfrac{1}{2}\right)$$
$$x = \dfrac{18 \ln 0.1}{\ln 0.5} \approx 59.8 \text{ ft}$$

You can work without artificial light to a depth of about 59.8 feet.

52. (a)
$$\dfrac{dy}{dt} = \dfrac{kA}{V}(c - y)$$
$$\int \dfrac{dy}{c - y} = \int \dfrac{kA}{V} dt$$
$$-\ln|c - y| = \dfrac{kA}{V} t + C$$
$$\ln|c - y| = -\dfrac{kA}{V} t - C$$
$$|c - y| = e^{-(kA/V)t - C}$$
$$c - y = \pm e^{-(kA/V)t - C}$$
$$y = c \pm e^{-(kA/V)t - C}$$
$$y = c + De^{-(kA/V)t}$$

Initial condition $y = y_0$ when $t = 0$
$$y_0 = c + D$$
$$y_0 - c = D$$

Solution: $y = c + (y_0 - c)e^{-(kA/V)t}$

(b) $\lim\limits_{t \to \infty} y(t) = \lim\limits_{t \to \infty} [c + (y_0 - c)e^{-(kA/V)t}] = c$

53. (a) $P(t) = \dfrac{150}{1 + e^{4.3 - t}} = \dfrac{150}{1 + e^{4.3}e^{-t}}$

This is $P = \dfrac{M}{1 + Ae^{-kt}}$ where $M = 150$, $A = e^{4.3}$, and $k = 1$. Therefore, it is a solution of the logistic differential equation.

$\dfrac{dP}{dt} = \dfrac{k}{M} P(M - P)$, or $\dfrac{dP}{dt} = \dfrac{1}{150} P(150 - P)$. The carrying capacity is 150.

(b) $P(0) = \dfrac{150}{1 + e^{4.3}} \approx 2$

Initially there were 2 infected students.

(c) $\dfrac{150}{1 + e^{4.3 - t}} = 125$
$$\dfrac{6}{5} = 1 + e^{4.3 - t}$$
$$\dfrac{1}{5} = e^{4.3 - t}$$
$$-\ln 5 = 4.3 - t$$
$$t = 4.3 + \ln 5 \approx 5.9 \text{ days}$$

It took about 6 days.

54. Use the Fundamental Theorem of Calculus.

$$y' = \dfrac{d}{dx}\left(\int_0^x \sin t^2 \, dt\right) + \dfrac{d}{dx}(x^3 + x + 2)$$
$$= (\sin x^2) + (3x^2 + 1)$$
$$y'' = \dfrac{d}{dx}(\sin x^2 + 3x^2 + 1)$$
$$= (\cos x^2)(2x) + 6x$$
$$= 2x \cos(x^2) + 6x$$

Thus, the differential equation is satisfied.

Verify the initial conditions:

$$y'(0) = (\sin 0^2) + 3(0)^2 + 1 = 1$$
$$y(0) = \int_0^0 \sin(t^2) \, dt + 0^3 + 0 + 2 = 2$$

55.
$$\dfrac{dP}{dt} = 0.002P\left(1 - \dfrac{P}{800}\right)$$
$$\dfrac{dP}{dt} = 0.002P\left(\dfrac{800 - P}{800}\right)$$
$$\dfrac{800}{P(800 - P)} dP = 0.002 \, dt$$
$$\dfrac{(800 - P) + P}{P(800 - P)} dP = 0.002 \, dt$$
$$\int \left(\dfrac{1}{P} + \dfrac{1}{800 - P}\right) dP = 0.002 \, dt$$
$$\ln|P| - \ln|800 - P| = 0.002t + C$$
$$\ln\left|\dfrac{P}{800 - P}\right| = 0.002t + C$$
$$\ln\left|\dfrac{800 - P}{P}\right| = -0.002t - C$$
$$\left|\dfrac{800 - P}{P}\right| = e^{-0.002t - C}$$
$$\dfrac{800 - P}{P} = \pm e^{-C} e^{-0.002t}$$
$$\dfrac{800}{P} - 1 = Ae^{-0.002t}$$
$$P = \dfrac{800}{1 + Ae^{-0.002t}}$$

Initial condition: $P(0) = 50$
$$50 = \dfrac{800}{1 + Ae^0}$$
$$1 + A = 16$$
$$A = 15$$

Solution: $P = \dfrac{800}{1 + 15e^{-0.002t}}$

192 Section 7.1

56. Method 1–Compare graph of $y_1 = x^2 \ln x$ with
$y_2 = \text{NDER}\left(\dfrac{x^3 \ln x}{3} - \dfrac{x^3}{9}\right)$. The graphs should be the same.
Method 2–Compare graph of $y_1 = \text{NINT}(x^2 \ln x)$ with
$y_2 = \dfrac{x^3 \ln x}{3} - \dfrac{x^3}{9}$. The graphs should be the same or differ only by a vertical translation.

57. (a) $20{,}000 = 10{,}000(1.063)^t$

$2 = 1.063^t$

$\ln 2 = t \ln 1.063$

$t = \dfrac{\ln 2}{\ln 1.063} \approx 11.345$

It will take about 11.3 years.

(b) $20{,}000 = 10{,}000 e^{0.063t}$

$2 = e^{0.063t}$

$\ln 2 = 0.063t$

$t = \dfrac{\ln 2}{0.063} \approx 11.002$

It will take about 11.0 years.

58. (a) $f'(x) = \dfrac{d}{dx}\displaystyle\int_0^x u(t)\,dt = u(x)$

$g'(x) = \dfrac{d}{dx}\displaystyle\int_3^x u(t)\,dt = u(x)$

(b) $C = f(x) - g(x)$

$= \displaystyle\int_0^x u(t)\,dt - \int_3^x u(t)\,dt$

$= \displaystyle\int_0^x u(t)\,dt + \int_x^3 u(t)\,dt$

$= \displaystyle\int_0^3 u(t)\,dt$

59. (a) $y = \dfrac{56.0716}{1 + 5.894 e^{-0.0205x}}$

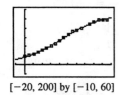

[−20, 200] by [−10, 60]

(b) The carrying capacity is about 56.0716 million people.

(c) Use NDER twice to solve $y'' = 0$. The solution is $x \approx 86.52$, representing (approximately) the year 1887. The population at this time was approximately $P(86.52) \approx 28.0$ million people.

60. (a) $T = 79.961(0.9273)^t$

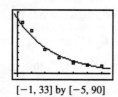

[−1, 33] by [−5, 90]

(b) Solving $T(t) = 40$ graphically, we obtain $t \approx 9.2$ sec. The temperature will reach 40° after about 9.2 seconds.

(c) When the probe was removed, the temperature was about $T(0) \approx 79.96°C$.

61. $\dfrac{v_0 m}{k} = $ coasting distance

$\dfrac{(0.86)(30.84)}{k} = 0.97$

$k \approx 27.343$

$s(t) = \dfrac{v_0 m}{k}(1 - e^{-(k/m)t})$

$s(t) = 0.97(1 - e^{-(27.343/30.84)t})$

$s(t) = 0.97(1 - e^{-0.8866t})$

A graph of the model is shown superimposed on a graph of the data.

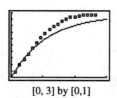

[0, 3] by [0,1]

Chapter 7
Applications of Definite Integrals

■ Section 7.1 Integral as Net Change
(pp. 363–374)

Exploration 1 Revisiting Example 2

1. $s(t) = \displaystyle\int\left(t^2 - \dfrac{8}{(t+1)^2}\right) dt = \dfrac{t^3}{3} + \dfrac{8}{t+1} + C$

$s(0) = \dfrac{0^3}{3} + \dfrac{8}{0+1} + C = 9 \Rightarrow C = 1$

Thus, $s(t) = \dfrac{t^3}{3} + \dfrac{8}{t+1} + 1$.

2. $s(1) = \dfrac{1^3}{3} + \dfrac{8}{1+1} + 1 = \dfrac{16}{3}$. This is the same as the answer we found in Example 2a.

3. $s(5) = \dfrac{5^3}{3} + \dfrac{8}{5+1} + 1 = 44$. This is the same answer we found in Example 2b.

Quick Review 7.1

1. On the interval, $\sin 2x = 0$ when $x = -\frac{\pi}{2}, 0,$ or $\frac{\pi}{2}$. Test one point on each subinterval: for $x = -\frac{3\pi}{4}$, $\sin 2x = 1$; for $x = -\frac{\pi}{4}$, $\sin 2x = -1$; for $x = \frac{\pi}{4}$, $\sin 2x = 1$; and for $x = \frac{3\pi}{4}$, $\sin 2x = -1$. The function changes sign at $-\frac{\pi}{2}, 0,$ and $\frac{\pi}{2}$. The graph is

   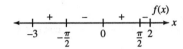

2. $x^2 - 3x + 2 = (x - 1)(x - 2) = 0$ when $x = 1$ or 2. Test one point on each subinterval: for $x = 0$, $x^2 - 3x + 2 = 2$; for $x = \frac{3}{2}$, $x^2 - 3x + 2 = -\frac{1}{4}$; and for $x = 3$, $x^2 - 3x + 2 = 2$. The function changes sign at 1 and 2. The graph is

3. $x^2 - 2x + 3 = 0$ has no real solutions, since $b^2 - 4ac = (-2)^2 - 4(1)(3) = -8 < 0$. The function is always positive. The graph is

4. $2x^3 - 3x^2 + 1 = (x - 1)^2(2x + 1) = 0$ when $x = -\frac{1}{2}$ or 1. Test one point on each subinterval: for $x = -1$, $2x^3 - 3x^2 + 1 = -4$; for $x = 0$, $2x^3 - 3x^2 + 1 = 1$; and $x = \frac{3}{2}$, $2x^3 - 3x^2 + 1 = 1$. The function changes sign at $-\frac{1}{2}$. The graph is

5. On the interval, $x \cos 2x = 0$ when $x = 0, \frac{\pi}{4}, \frac{3\pi}{4},$ or $\frac{5\pi}{4}$. Test one point on each subinterval: for $x = \frac{\pi}{8}$, $x \cos 2x = \frac{\pi\sqrt{2}}{16}$; for $x = \frac{\pi}{2}$, $x \cos 2x = -\frac{\pi}{2}$; for $x = \pi$, $x \cos 2x = \pi$; and for $x = 4$, $x \cos 2x \approx -0.58$. The function changes sign at $\frac{\pi}{4}, \frac{3\pi}{4},$ and $\frac{5\pi}{4}$. The graph is

6. $xe^{-x} = 0$ when $x = 0$. On the rest of the interval, xe^{-x} is always positive.

7. $\frac{x}{x^2 + 1} = 0$ when $x = 0$. Test one point on each subinterval: for $x = -1$, $\frac{x}{x^2 + 1} = -\frac{1}{2}$; for $x = 1$, $\frac{x}{x^2 + 1} = \frac{1}{2}$. The function changes sign at 0. The graph is

8. $\frac{x^2 - 2}{x^2 - 4} = 0$ when $x = \pm\sqrt{2}$ and is undefined when $x = \pm 2$. Test one point on each subinterval: for $x = -\frac{5}{2}$, $\frac{x^2 - 2}{x^2 - 4} = \frac{17}{9}$; for $x = -1.9$, $\frac{x^2 - 2}{x^2 - 4} \approx -4.13$; for $x = 0$, $\frac{x^2 - 2}{x^2 - 4} = \frac{1}{2}$; for $x = 1.9$, $\frac{x^2 - 2}{x^2 - 4} \approx -4.13$; and for $x = \frac{5}{2}$, $\frac{x^2 - 2}{x^2 - 4} = \frac{17}{9}$. The function changes sign at $-2, -\sqrt{2}, \sqrt{2}$ and 2. The graph is

9. $\sec(1 + \sqrt{1 - \sin^2 x}) = \frac{1}{\cos(1 + |\cos x|)}$ is undefined when $x \approx 0.9633 + k\pi$ or $2.1783 + k\pi$ for any integer k. Test for $x = 0$: $\sec(1 + \sqrt{1 - \sin^2 0}) \approx -2.4030$. Test for $x = \pm 1$: $\sec(1 + \sqrt{1 - \sin^2 1}) \approx 32.7984$. The sign alternates over successive subintervals. The function changes sign at $0.9633 + k\pi$ or $2.1783 + k\pi$, where k is an integer. The graph is

10. On the interval, $\sin\left(\frac{1}{x}\right) = 0$ when $x = \frac{1}{3\pi}$ or $\frac{1}{2\pi}$. Test one point on each subinterval: for $x = 0.1$, $\sin\left(\frac{1}{x}\right) \approx -0.54$; for $x = 0.15$, $\sin\left(\frac{1}{x}\right) \approx 0.37$; and for $x = 0.2$, $\sin\left(\frac{1}{x}\right) \approx -0.96$. The graph changes sign at $\frac{1}{3\pi}$, and $\frac{1}{2\pi}$. The graph is

Section 7.1 Exercises

1. (a) Right when $v(t) > 0$, which is when $\cos t > 0$, i.e.,

when $0 \le t < \dfrac{\pi}{2}$ or $\dfrac{3\pi}{2} < t \le 2\pi$. Left when $\cos t < 0$,

i.e., when $\dfrac{\pi}{2} < t < \dfrac{3\pi}{2}$. Stopped when $\cos t = 0$,

i.e., when $t = \dfrac{\pi}{2}$ or $\dfrac{3\pi}{2}$.

(b) Displacement =

$$\int_0^{2\pi} 5\cos t\, dt = 5\Big[\sin t\Big]_0^{2\pi} = 5[\sin 2\pi - \sin 0] = 0$$

(c) Distance $= \displaystyle\int_0^{2\pi} |5\cos t|\, dt$

$= \displaystyle\int_0^{\pi/2} 5\cos t\, dt + \int_{\pi/2}^{3\pi/2} -5\cos t\, dt + \int_{3\pi/2}^{2\pi} 5\cos t\, dt$

$= 5 + 10 + 5 = 20$

3. (a) Right when $v(t) = 49 - 9.8t > 0$, i.e., when $0 \le t < 5$.

Left when $49 - 9.8t < 0$, i.e., when $5 < t \le 10$.

Stopped when $49 - 9.8t = 0$, i.e., when $t = 5$.

(b) Displacement $= \displaystyle\int_0^{10} (49 - 9.8t)\, dt$

$= \Big[49t - 4.9t^2\Big]_0^{10} = 49[(10-10) - 0] = 0$

(c) Distance $= \displaystyle\int_0^{10} |49 - 9.8t|\, dt$

$= \displaystyle\int_0^5 (49 - 9.8t)\, dt + \int_5^{10} (-49 + 9.8t)\, dt$

$= 122.5 + 122.5 = 245$

5. (a) Right when $v(t) > 0$, which is when $\sin t \ne 0$ and

$\cos t > 0$, i.e., when $0 < t < \dfrac{\pi}{2}$ or $\dfrac{3\pi}{2} < t < 2\pi$. Left

when $\sin t \ne 0$ and $\cos t < 0$, i.e., when $\dfrac{\pi}{2} < t < \pi$ or

$\pi < t < \dfrac{3\pi}{2}$. Stopped when $\sin t = 0$ or $\cos t = 0$,

i.e., when $t = 0, \dfrac{\pi}{2}, \pi, \dfrac{3\pi}{2}$, or 2π.

(b) Displacement $= \displaystyle\int_0^{2\pi} 5\sin^2 t \cos t\, dt = 5\Big[\dfrac{1}{3}\sin^3 t\Big]_0^{2\pi}$

$= 5[0 - 0] = 0$

(c) Distance $= \displaystyle\int_0^{2\pi} |5\sin^2 t \cos t|\, dt$

$= \displaystyle\int_0^{\pi/2} 5\sin^2 t \cos t\, dt + \int_{\pi/2}^{3\pi/2} -5\sin^2 t \cos t\, dt$

$+ \displaystyle\int_{3\pi/2}^{2\pi} 5\sin^2 t \cos t\, dt$

$= \dfrac{5}{3} + \dfrac{10}{3} + \dfrac{5}{3} = \dfrac{20}{3}$

7. (a) Right when $v(t) > 0$, which is when $\cos t > 0$,

i.e., when $0 \le t < \dfrac{\pi}{2}$ or $\dfrac{3\pi}{2} < t \le 2\pi$. Left when

$\cos t < 0$, i.e., when $\dfrac{\pi}{2} < t < \dfrac{3\pi}{2}$. Stopped when

$\cos t = 0$, i.e., when $t = \dfrac{\pi}{2}$ or $\dfrac{3\pi}{2}$.

(b) Displacement $= \displaystyle\int_0^{2\pi} e^{\sin t} \cos t\, dt = \Big[e^{\sin t}\Big]_0^{2\pi}$

$= [e^0 - e^0] = 0$

(c) Distance $= \displaystyle\int_0^{2\pi} |e^{\sin t} \cos t|\, dt = \int_0^{\pi/2} e^{\sin t} \cos t\, dt +$

$\displaystyle\int_{\pi/2}^{3\pi/2} -e^{\sin t}\cos t\, dt + \int_{3\pi/2}^{2\pi} e^{\sin t} \cos t\, dt$

$= (e - 1) + \left(e - \dfrac{1}{e}\right) + \left(1 - \dfrac{1}{e}\right) = 2e - \dfrac{2}{e} \approx 4.7$

9. (a) $v(t) = \displaystyle\int a(t)\, dt = t + 2t^{3/2} + C$, and since $v(0) = 0$,

$v(t) = t + 2t^{3/2}$. Then $v(9) = 9 + 2(27) = 63$ mph.

(b) First convert units:

$t + 2t^{3/2}$ mph $= \dfrac{t}{3600} + \dfrac{t^{3/2}}{1800}$ mi/sec. Then

Distance $= \displaystyle\int_0^9 \left(\dfrac{t}{3600} + \dfrac{t^{3/2}}{1800}\right) dt$

$= \Big[\dfrac{t^2}{7200} + \dfrac{t^{5/2}}{4500}\Big]_0^9 = \left[\left(\dfrac{9}{800} + \dfrac{27}{500}\right) - 0\right] = 0.06525$ mi

$= 344.52$ ft.

11. (a) $v(t) = \displaystyle\int a(t)\, dt = \int -32\, dt = -32t + C_1$, where

$C_1 = v(0) = 90$.

Then $v(3) = -32(3) + 90 = -6$ ft/sec.

(b) $s(t) = \displaystyle\int v(t)\, dt = -16t^2 + 90t + C_2$, where

$C_2 = s(0) = 0$. Solve $s(t) = 0$:

$-16t^2 + 90t = 2t(-8t + 45) = 0$

when $t = 0$ or $t = \dfrac{45}{8} = 5.625$ sec.

The projectile hits the ground at 5.625 sec.

(c) Since starting height = ending height,

Displacement = 0.

(d) Max. Height $= s\left(\dfrac{5.625}{2}\right)$

$= -16\left(\dfrac{5.625}{2}\right)^2 + 90\left(\dfrac{5.625}{2}\right) = 126.5625$, and

Distance = 2(Max. Height) = 253.125 ft.

13. Total distance $= \displaystyle\int_0^c |v(t)|\, dt = 4 + 5 + 24 = 33$ cm

15. At $t = a$, where $\frac{dv}{dt}$ is at a maximum (the graph is steepest upward).

17. Distance = Area under curve = $4\left(\frac{1}{2} \cdot 1 \cdot 2\right) = 4$
   (a) Final position = Initial position + Distance
      $= 2 + 4 = 6$; ends at $x = 6$.
   (b) 4 meters

19. (a) Final position $= 2 + \int_0^7 v(t)\, dt$
   $= 2 - \frac{1}{2}(1)(2) + \frac{1}{2}(1)(2) + 1(2)$
   $\quad + \frac{1}{2}(2)(2) - \frac{1}{2}(2)(1)$
   $= 5$;
   ends at $x = 5$.

   (b) $\int_0^7 |v(t)|\, dt = \frac{1}{2}(1)(2) + \frac{1}{2}(1)(2) + 1(2) + \frac{1}{2}(2)(2)$
   $\quad + \frac{1}{2}(2)(1)$
   $= 7$ meters

21. $\int_0^{10} 27.08 \cdot e^{t/25}\, dt = 27.08\left[25 e^{t/25}\right]_0^{10} = 27.08[25 e^{0.4} - 25]$
   ≈ 332.965 billion barrels

23. (a) Solve $10{,}000(2 - r) = 0$: $r = 2$ miles.
   (b) Width $= \Delta r$, Length $= 2\pi r$: Area $= 2\pi r \Delta r$
   (c) Population = Population density $\times$ Area
   (d) $\int_0^2 10{,}000(2 - r)(2\pi r)\, dr = 20{,}000\pi \int_0^2 (2r - r^2)\, dr$
   $= 20{,}000\pi \left[r^2 - \frac{1}{3}r^3\right]_0^2 = 20{,}000\pi\left[\left(4 - \frac{8}{3}\right) - 0\right]$
   $= \frac{80{,}000}{3}\pi \approx 83{,}776$

25. (Answers may vary.)
Plot the speeds vs. time. Connect the points and find the area under the line graph. The definite integral also gives the area under the curve.

27. (a) $\int_{-0.5}^{10.5} (1.6x^2 + 2.3x + 5.0)\, dx$
   $= \left[\frac{1.6}{3}x^3 + \frac{2.3}{2}x^2 + 5.0x\right]_{-0.5}^{10.5} \approx 798.97$ thousand

   (b) The answer in (a) corresponds to the area of midpoint rectangles. The curve now gives a better approximation since part of each rectangle is above the curve and part is below.

29. $F(x) = kx$; $6 = k(3)$, so $k = 2$ and $F(x) = 2x$.
   (a) $F(9) = 2(9) = 18$N
   (b) $W = \int_0^9 F(x)\, dx = \int_0^9 2x\, dx = \left[x^2\right]_0^9 = 81$ N $\cdot$ cm

31. $\frac{(12-0)}{2(12)}[0.04 + 2(0.04) + 2(0.05) + 2(0.06) + 2(0.05)$
   $\qquad + 2(0.04) + 2(0.04) + 2(0.05) + 2(0.04)$
   $\qquad + 2(0.06) + 2(0.06) + 2(0.05) + 0.05] = 0.585$

The overall rate, then, is $\frac{0.585}{12} = 0.04875$.

33. (a) $\bar{x} = \frac{M_y}{M} = \frac{\sum m_k x_k}{\sum m_k}$. Taking $dm = \delta dA$ as m_k and letting $dA \to 0$, $k \to \infty$ yields $\frac{\int x\, dm}{\int dm}$.

   (b) $\bar{y} = \frac{M_y}{M} = \frac{\sum m_k y}{\sum m_k}$. Taking $dm = \delta dA$ as m_k and letting $dA \to 0$, $k \to \infty$ yields $\frac{\int y\, dm}{\int dm}$.

35. By symmetry, $y = 0$. For x, use vertical strips:

$\bar{x} = \frac{\int x\, dm}{\int dm} = \frac{\int x\delta\, dA}{\int \delta\, dA} = \frac{\int x\, dA}{\int dA}$

$= \frac{\int_0^2 x(2x)\, dx}{\int_0^2 2x\, dx}$

$= \frac{\left[\frac{2}{3}x^3\right]_0^2}{\left[x^2\right]_0^2}$

$= \frac{4}{3}$

Section 7.2 Areas in the Plane
(pp. 374–382)

Exploration 1 A Family of Butterflies

1. For $k = 1$:

$$\int_0^\pi [(2 - \sin x) - \sin x]\, dx = \int_0^\pi (2 - 2\sin x)\, dx$$
$$= 2x + 2\cos x \Big]_0^\pi$$
$$= 2\pi - 4$$

For $k = 2$:

$$\int_0^{\pi/2} [(4 - 2\sin 2x) - (2\sin 2x)]\, dx$$
$$= \int_0^{\pi/2} (4 - 4\sin 2x)\, dx$$
$$= 4x + 2\cos 2x \Big]_0^{\pi/2} = 2\pi - 4$$

2. It appears that the areas for $k \geq 3$ will continue to be $2\pi - 4$.

3. $A_k = \int_0^{\pi/k} [(2k - k\sin kx) - k\sin kx]\, dx$
$= \int_0^{\pi/k} (2k - 2k\sin kx)\, dx$

If we make the substitution $u = kx$, then $du = k\, dx$ and the u-limits become 0 to π. Thus,

$A_k = \int_0^{\pi/k} (2k - 2k\sin kx)\, dx$
$= \int_0^{\pi/k} (2 - 2\sin kx)k\, dx$
$= \int_0^\pi (2 - 2\sin u)\, du$

4. $2\pi - 4$

5. Because the amplitudes of the sine curves are k, the kth butterfly stands $2k$ units tall. The vertical edges alone have lengths $(2k)$ that increase without bound, so the perimeters are tending to infinity.

Quick Review 7.2

1. $\int_0^\pi \sin x\, dx = \Big[-\cos x\Big]_0^\pi = -[-1 - 1] = 2$

2. $\int_0^1 e^{2x}\, dx = \Big[\frac{1}{2}e^{2x}\Big]_0^1 = \frac{1}{2}(e^2 - 1) \approx 3.195$

3. $\int_{-\pi/4}^{\pi/4} \sec^2 x\, dx = \Big[\tan x\Big]_{-\pi/4}^{\pi/4} = 1 - (-1) = 2$

4. $\int_0^2 (4x - x^3)\, dx = \Big[2x^2 - \frac{1}{4}x^4\Big]_0^2 = (8 - 4) - 0 = 4$

5. $\int_{-3}^3 \sqrt{9 - x^2}\, dx = \frac{9\pi}{2}$ (This is half the area of a circle of radius 3.)

6. Solve $x^2 - 4x = x + 6$.
$x^2 - 5x - 6 = 0$
$(x - 6)(x + 1) = 0$
$x = 6$ or $x = -1$
$y = 6 + 6 = 12$ or $y = -1 + 6 = 5$
$(6, 12)$ and $(-1, 5)$

7. Solve $e^x = x + 1$. From the graphs, it appears that e^x is always greater than or equal to $x + 1$, so that if they are ever equal, this is when $e^x - (x + 1)$ is at a minimum. $\frac{d}{dx}[e^x - (x + 1)] = e^x - 1$ is zero when $e^x = 1$, i.e., when $x = 0$. Test: $e^0 = 0 + 1 = 1$. So the solution is $(0, 1)$.

8. Inspection of the graphs shows two intersection points: $(0, 0)$, and $(\pi, 0)$. Check: $0^2 - \pi \cdot 0 = \sin 0 = 0$ and $\pi^2 - \pi^2 = \sin \pi = 0$.

9. Solve $\frac{2x}{x^2 + 1} = x^3$.
$(0, 0)$ is a solution. Now divide by x.
$\frac{2}{x^2 + 1} = x^2$
$2 = x^4 + x^2$
$x^4 + x^2 - 2 = 0$
$x^2 = \frac{-1 \pm \sqrt{1 + 8}}{2} = -2$ or 1
Throw out the negative solution.
$x = \pm 1$
$y = x^3 = \pm 1$
$(0, 0), (-1, -1)$ and $(1, 1)$

10. Use the intersect function on a graphing calculator:

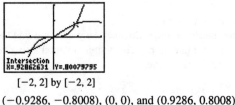

$[-2, 2]$ by $[-2, 2]$

$(-0.9286, -0.8008)$, $(0, 0)$, and $(0.9286, 0.8008)$

Section 7.2 Exercises

1. $\int_0^\pi (1 - \cos^2 x)\, dx = \Big[\frac{1}{2}x - \frac{1}{4}\sin 2x\Big]_0^\pi = \frac{\pi}{2}$

3. $\int_0^1 (y^2 - y^3)\, dy = \Big[\frac{1}{3}y^3 - \frac{1}{4}y^4\Big]_0^1 = \frac{1}{12}$

5. Use the region's symmetry:

$$2\int_0^2 [2x^2 - (x^4 - 2x^2)]\, dx = 2\int_0^2 (-x^4 + 4x^2)\, dx$$
$$= 2\left[-\frac{1}{5}x^5 + \frac{4}{3}x^3\right]_0^2$$
$$= 2\left[\left(-\frac{32}{5} + \frac{32}{3}\right) - 0\right] = \frac{128}{15}$$

7. Integrate with respect to y:

$$\int_0^1 (2\sqrt{y} - y)\, dy = \left[\frac{4}{3}y^{3/2} - \frac{1}{2}y^2\right]_0^1$$
$$= \left(\frac{4}{3} - \frac{1}{2}\right) - 0 = \frac{5}{6}$$

9. Integrate in two parts:

$$\int_{-2}^0 [(2x^3 - x^2 - 5x) - (-x^2 + 3x)]\, dx +$$
$$\int_0^2 [(-x^2 + 3x) - (2x^3 - x^2 - 5x)]\, dx$$
$$= \int_{-2}^0 (2x^3 - 8x)\, dx + \int_0^2 (-2x^3 + 8x)\, dx$$
$$= \left[\frac{1}{2}x^4 - 4x^2\right]_{-2}^0 + \left[-\frac{1}{2}x^4 + 4x^2\right]_0^2$$
$$= [0 - (8 - 16)] + [(-8 + 16) - 0] = 16$$

11. Solve $x^2 - 2 = 2$: $x^2 = 4$, so the curves intersect at $x = \pm 2$.

$$\int_{-2}^2 [2 - (x^2 - 2)]\, dx = \int_{-2}^2 (4 - x^2)\, dx$$
$$= \left[4x - \frac{1}{3}x^3\right]_{-2}^2 = \left(8 - \frac{8}{3}\right) - \left(-8 + \frac{8}{3}\right) = \frac{32}{3} = 10\frac{2}{3}$$

13. Solve $7 - 2x^2 = x^2 + 4$: $x^2 = 1$, so the curves intersect at $x = \pm 1$.

$$\int_{-1}^1 [(7 - 2x^2) - (x^2 + 4)]\, dx = \int_{-1}^1 (-3x^2 + 3)\, dx$$
$$= 3\int_{-1}^1 (1 - x^2)\, dx$$
$$= 3\left[x - \frac{1}{3}x^3\right]_{-1}^1$$
$$= 3\left[\frac{2}{3} - \left(-\frac{2}{3}\right)\right] = 4$$

15.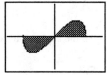

$\left[-\frac{3}{2}a, \frac{3}{2}a\right]$ by $[-a^2, a^2]$

The curves intersect at $x = 0$ and $x = \pm a$. Use the region's symmetry:

$$2\int_0^a x\sqrt{a^2 - x^2}\, dx = 2\left[-\frac{1}{3}(a^2 - x^2)^{3/2}\right]_0^a$$
$$= 2\left[0 - \left(-\frac{1}{3}a^3\right)\right]$$
$$= \frac{2}{3}a^3$$

17.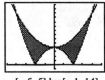

$[-5, 5]$ by $[-1, 14]$

The curves intersect at $x = 0$ and $x = \pm 4$. Because of the absolute value sign, break the integral up at $x = \pm 2$ also (where $|x^2 - 4|$ turns the corner). Use the graph's symmetry:

$$2\int_0^2 \left[\left(\frac{x^2}{2} + 4\right) - (4 - x^2)\right] dx + 2\int_2^4 \left[\left(\frac{x^2}{2} + 4\right) - (x^2 - 4)\right] dx$$
$$= 2\int_0^2 \frac{3x^2}{2}\, dx + 2\int_2^4 \left(-\frac{x^2}{2} + 8\right) dx$$
$$= 2\left[\frac{x^3}{2}\right]_0^2 + 2\left[-\frac{x^3}{6} + 8x\right]_2^4$$
$$= 2[4] + 2\left[\left(-\frac{32}{3} + 32\right) - \left(-\frac{4}{3} + 16\right)\right] = \frac{64}{3} = 21\frac{1}{3}$$

19. Solve for x: $x = \frac{y^2}{4} - 1$ and $x = \frac{y}{4} + 4$.

Now solve $\frac{y^2}{4} - 1 = \frac{y}{4} + 4$: $\frac{y^2}{4} - \frac{y}{4} - 5 = 0$,

$y^2 - y - 20 = (y - 5)(y + 4) = 0$.

The curves intersect at $y = -4$ and $y = 5$.

$$\int_{-4}^5 \left[\left(\frac{y}{4} + 4\right) - \left(\frac{y^2}{4} - 1\right)\right] dy$$
$$= \int_{-4}^5 \left(-\frac{y^2}{4} + \frac{y}{4} + 5\right) dy$$
$$= \left[-\frac{y^3}{12} + \frac{y^2}{8} + 5y\right]_{-4}^5$$
$$= \left(-\frac{125}{12} + \frac{25}{8} + 25\right) - \left(\frac{16}{3} + 2 - 20\right) = \frac{243}{8} = 30\frac{3}{8}$$

198 Section 7.2

21. Solve for x: $x = -y^2$ and $x = 2 - 3y^2$.

Now solve $-y^2 = 2 - 3y^2$: $y^2 = 1$, so the curves intersect

at $y = \pm 1$. Use the region's symmetry:

$2\int_0^1 (2 - 3y^2 + y^2)\, dy = 2\int_0^1 (2 - 2y^2)\, dy$

$= 4\int_0^1 (1 - y^2)\, dy = 4\left[y - \frac{1}{3}y^3\right]_0^1 = 4\left[\left(1 - \frac{1}{3}\right) - 0\right] = \frac{8}{3}$

23. Solve for x: $x = 3 - y^2$ and $x = -\frac{y^2}{4}$.

Now solve $3 - y^2 = -\frac{y^2}{4}$: $y^2 = 4$,

so the curves intersect at $y = \pm 2$.

Use the region's symmetry:

$2\int_0^2 \left(3 - y^2 + \frac{y^2}{4}\right) dy = 2\int_0^2 \left(3 - \frac{3y^2}{4}\right) dy$

$= 2\left[3y - \frac{y^3}{4}\right]_0^2$

$= 2(6 - 2) - 0 = 8$

25. Use the region's symmetry:

$2\int_0^{\pi/3} (8\cos x - \sec^2 x)\, dx = 2\left[8\sin x - \tan x\right]_0^{\pi/3}$

$= 2[(4\sqrt{3} - \sqrt{3}) - 0] = 6\sqrt{3}$

27.

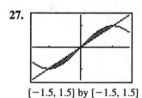

[-1.5, 1.5] by [-1.5, 1.5]

The curves intersect at $x = 0$ and $x = \pm 1$. Use the area's

symmetry:

$2\int_0^1 \left[\sin\left(\frac{\pi x}{2}\right) - x\right] dx = 2\left[-\frac{2}{\pi}\cos\left(\frac{\pi x}{2}\right) - \frac{1}{2}x^2\right]_0^1$

$= 2\left[-\frac{1}{2} - \left(-\frac{2}{\pi}\right)\right]$

$= \frac{4 - \pi}{\pi} \approx 0.273$

29. Use the region's symmetry:

$2\int_0^{\pi/4} (\tan^2 y + \tan^2 y)\, dy = 4\int_0^{\pi/4} \tan^2 y\, dy$

$= 4\left[\tan y - y\right]_0^{\pi/4}$

$= 4\left[\left(1 - \frac{\pi}{4}\right) - 0\right]$

$= 4 - \pi \approx 0.858$

31. Solve for x: $x = y^3$ and $x = y$.

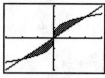

[-1.5, 1.5] by [-1.5, 1.5]

The curves intersect at $x = 0$ and $x = \pm 1$. Use the area's

symmetry: $2\int_0^1 (y - y^3)\, dy = 2\left[\frac{1}{2}y^2 - \frac{1}{4}y^4\right]_0^1 = \frac{1}{2}$

33. The curves intersect when $\sin x = \cos x$, i.e., at $x = \frac{\pi}{4}$.

$\int_0^{\pi/4} (\cos x - \sin x)\, dx = \left[\sin x + \cos x\right]_0^{\pi/4}$

$= \sqrt{2} - 1 \approx 0.414$

35. (a)

If $y = x^2 = c$, then $x = \pm\sqrt{c}$. So the points are

$(-\sqrt{c}, c)$ and $(\sqrt{c}, c)$.

(b) The two areas in Quadrant I, where $x = \sqrt{y}$, are equal:

$\int_0^c \sqrt{y}\, dy = \int_c^4 \sqrt{y}\, dy$

$\left[\frac{2}{3}y^{3/2}\right]_0^c = \left[\frac{2}{3}y^{3/2}\right]_c^4$

$\frac{2}{3}c^{3/2} = \frac{2}{3}4^{3/2} - \frac{2}{3}c^{3/2}$

$2c^{3/2} = 8$

$c^{3/2} = 4$

$c = 4^{2/3} = 2^{4/3}$

(c) Divide the upper right section into a $(4 - c)$-by-$\sqrt{c}$ rectangle and a leftover portion:

$$\int_0^{\sqrt{c}} (c - x^2)\, dx = (4 - c)\sqrt{c} + \int_{\sqrt{c}}^2 (4 - x^2)\, dx$$

$$\left[cx - \frac{1}{3}x^3\right]_0^{\sqrt{c}} = 4\sqrt{c} - c^{3/2} + \left[4x - \frac{1}{3}x^3\right]_{\sqrt{c}}^2$$

$$c^{3/2} - \frac{1}{3}c^{3/2} = 4\sqrt{c} - c^{3/2}$$
$$+ \left[\left(8 - \frac{8}{3}\right) - \left(4\sqrt{c} - \frac{1}{3}c^{3/2}\right)\right]$$

$$\frac{2}{3}c^{3/2} = 4\sqrt{c} - c^{3/2} + \frac{16}{3} - 4\sqrt{c} + \frac{1}{3}c^{3/2}$$

$$\frac{4}{3}c^{3/2} = \frac{16}{3}$$

$$c^{3/2} = 4$$

$$c = 4^{2/3} = 2^{4/3}$$

37. First find the two areas.

For the triangle, $\frac{1}{2}(2a)(a^2) = a^3$

For the parabola, $2\int_0^a (a^2 - x^2)\, dx = 2\left[a^2 x - \frac{1}{3}x^3\right]_0^a = \frac{4}{3}a^3$

The ratio, then, is $\dfrac{a^3}{\frac{4}{3}a^3} = \dfrac{3}{4}$, which remains constant as a approaches zero.

39. Neither; both integrals come out as zero because the -1-to-0 and 0-to-1 portions of the integrals cancel each other.

41.

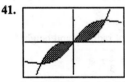

$[-1.5, 1.5]$ by $[-1.5, 1.5]$

The curves intersect at $x = 0$ and $x = \pm 1$. Use the area's symmetry:

$$2\int_0^1 \left(\frac{2x}{x^2 + 1} - x^3\right) dx = 2\left[\ln(x^2 + 1) - \frac{1}{4}x^4\right]_0^1$$
$$= 2\ln 2 - \frac{1}{2}$$
$$= \ln 4 - \frac{1}{2} \approx 0.886$$

43. First graph $y = \cos x$ and $y = x^2$.

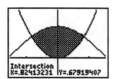

$[-1.5, 1.5]$ by $[-0.5, 1.5]$

The curves intersect at $x \approx \pm 0.8241$. Use NINT to find $2\int_0^{0.8241} (\cos x - x^2)\, dx \approx 1.0948$. Multiplying both functions by k will not change the x-value of any intersection point, so the area condition to be met is

$$2 = 2\int_0^{0.8241} (k\cos x - kx^2)\, dx$$
$$\Rightarrow 2 = k \cdot 2\int_0^{0.8241} (\cos x - x^2)\, dx$$
$$\Rightarrow 2 \approx k(1.0948)$$
$$\Rightarrow k \approx 1.8269.$$

45. By hypothesis, $f(x) - g(x)$ is the same for each region, where $f(x)$ and $g(x)$ represent the upper and lower edges. But then Area $= \int_a^b [f(x) - g(x)]\, dx$ will be the same for each.

■ Section 7.3 Volumes
(pp. 383–394)

Exploration 1 Volume by Cylindrical Shells

1. Its height is $f(x_k) = 3x_k - x_k^2$.
2. Unrolling the cylinder, the circumference becomes one dimension of a rectangle, and the height becomes the other. The thickness Δx is the third dimension of a slab with dimensions $2\pi(x_k + 1)$ by $3x_k - x_k^2$ by Δx. The volume is obtained by multiplying the dimensions together.
3. The limit is the definite integral $\int_0^3 2\pi(x + 1)(3x - x^2)\, dx$.
4. $\dfrac{45\pi}{2}$

Exploration 2 Surface Area

1. $\int_a^b 2\pi y \sqrt{1 + \left(\dfrac{dy}{dx}\right)^2}\, dx$

   The limit will exist if f and f' are continuous on the interval $[a, b]$.

2. $y = \sin x$, so $\dfrac{dy}{dx} = \cos x$ and

   $$\int_a^b 2\pi y \sqrt{1 + \left(\frac{dy}{dx}\right)^2}\, dx$$
   $$= \int_0^\pi 2\pi \sin x \sqrt{1 + \cos^2 x}\, dx \approx 14.424.$$

3. $y = \sqrt{x}$, so $\dfrac{dy}{dx} = \dfrac{1}{2\sqrt{x}}$ and

$\displaystyle\int_0^4 2\pi\sqrt{x}\sqrt{1 + \left(\dfrac{1}{2\sqrt{x}}\right)^2}\, dx \approx 36.177$.

Quick Review 7.3

1. x^2

2. $s = \dfrac{x}{\sqrt{2}}$, so Area $= s^2 = \dfrac{x^2}{2}$.

3. $\dfrac{1}{2}\pi r^2$ or $\dfrac{\pi x^2}{2}$

4. $\dfrac{1}{2}\pi\left(\dfrac{d}{2}\right)^2$ or $\dfrac{\pi x^2}{8}$

5. $b = x$ and $h = \dfrac{\sqrt{3}}{2}x$, so Area $= \dfrac{1}{2}bh = \dfrac{\sqrt{3}}{4}x^2$.

6. $b = h = x$, so Area $= \dfrac{1}{2}bh = \dfrac{x^2}{2}$.

7. $b = h = \dfrac{x}{\sqrt{2}}$, so Area $= \dfrac{1}{2}bh = \dfrac{x^2}{4}$.

8.

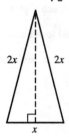

$b = x$ and $h = \sqrt{(2x)^2 - \left(\dfrac{1}{2}x\right)^2} = \dfrac{\sqrt{15}}{2}x$, so

Area $= \dfrac{1}{2}bh = \dfrac{\sqrt{15}}{4}x^2$.

9. This is a 3-4-5 right triangle. $b = 4x$, $h = 3x$, and

Area $= \dfrac{1}{2}bh = 6x^2$.

10. The hexagon contains six equilateral triangles with sides of length x, so from Exercise 5, Area $= 6\left(\dfrac{\sqrt{3}}{4}x^2\right) = \dfrac{3\sqrt{3}}{2}x^2$.

Section 7.3 Exercises

1. In each case, the width of the cross section is $w = 2\sqrt{1 - x^2}$.

(a) $A = \pi r^2$, where $r = \dfrac{w}{2}$, so $A(x) = \pi\left(\dfrac{w}{2}\right)^2 = \pi(1 - x^2)$.

(b) $A = s^2$, where $s = w$, so $A(x) = w^2 = 4(1 - x^2)$.

(c) $A = s^2$, where $s = \dfrac{w}{\sqrt{2}}$, so $A(x) = \left(\dfrac{w}{\sqrt{2}}\right)^2 = 2(1 - x^2)$.

(d) $A = \dfrac{\sqrt{3}}{4}w^2$ (see Quick Review Exercise 5), so

$A(x) = \dfrac{\sqrt{3}}{4}(2\sqrt{1-x^2})^2 = \sqrt{3}(1 - x^2)$.

3. A cross section has width $w = 2\sqrt{x}$ and area

$A(x) = s^2 = \left(\dfrac{w}{\sqrt{2}}\right)^2 = 2x$. The volume is

$\displaystyle\int_0^4 2x\, dx = \left[x^2\right]_0^4 = 16$.

5. A cross section has width $w = 2\sqrt{1 - x^2}$ and area

$A(x) = s^2 = w^2 = 4(1 - x^2)$. The volume is

$\displaystyle\int_{-1}^1 4(1 - x^2)\, dx = 4\int_{-1}^1 (1 - x^2)\, dx = 4\left[x - \dfrac{1}{3}x^3\right]_{-1}^1 = \dfrac{16}{3}$.

7. A cross section has width $w = 2\sqrt{\sin x}$.

(a) $A(x) = \dfrac{\sqrt{3}}{4}w^2 = \sqrt{3}\sin x$, and

$V = \displaystyle\int_0^\pi \sqrt{3}\sin x\, dx$

$= \sqrt{3}\displaystyle\int_0^\pi \sin x\, dx$

$= \sqrt{3}\left[-\cos x\right]_0^\pi$

$= 2\sqrt{3}$.

(b) $A(x) = s^2 = w^2 = 4\sin x$, and

$V = \displaystyle\int_0^\pi 4\sin x\, dx = 4\int_0^\pi \sin x\, dx = 4\left[-\cos x\right]_0^\pi = 8$.

9. A cross section has width $w = \sqrt{5}y^2$ and area

$\pi r^2 = \pi\left(\dfrac{w}{2}\right)^2 = \dfrac{5\pi}{4}y^4$. The volume is

$\displaystyle\int_0^2 \dfrac{5\pi}{4}y^4\, dy = \dfrac{\pi}{4}\left[y^5\right]_0^2 = 8\pi$.

11. (a) The volume is the same as if the square had moved without twisting: $V = Ah = s^2 h$.

(b) Still $s^2 h$: the lateral distribution of the square cross sections doesn't affect the volume. That's Cavalieri's Volume Theorem.

13. The solid is a right circular cone of radius 1 and height 2.

$V = \dfrac{1}{3}Bh = \dfrac{1}{3}(\pi r^2)h = \dfrac{1}{3}(\pi 1^2)2 = \dfrac{2}{3}\pi$

15. A cross section has radius $r = \tan\left(\dfrac{\pi}{4}y\right)$ and area

$A(y) = \pi r^2 = \pi\tan^2\left(\dfrac{\pi}{4}y\right)$. The volume is

$\displaystyle\int_0^1 \pi\tan^2\left(\dfrac{\pi}{4}y\right)dy = \pi\left[\dfrac{4}{\pi}\tan\left(\dfrac{\pi}{4}y\right) - y\right]_0^1$

$= \pi\left(\dfrac{4}{\pi} - 1\right)$

$= 4 - \pi$.

17.

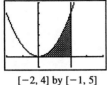

[−2, 4] by [−1, 5]

A cross section has radius $r = x^2$ and area
$A(x) = \pi r^2 = \pi x^4$. The volume is
$\int_0^2 \pi x^4 \, dx = \pi \left[\frac{1}{5}x^5\right]_0^2 = \frac{32\pi}{5}.$

19.

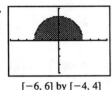

[−6, 6] by [−4, 4]

The solid is a sphere of radius $r = 3$. The volume is
$\frac{4}{3}\pi r^3 = 36\pi.$

21.

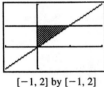

[−1, 2] by [−1, 2]

Use cylindrical shells: A shell has radius y and height y.

The volume is
$\int_0^1 2\pi(y)(y) \, dy = 2\pi\left[\frac{1}{3}y^3\right]_0^1 = \frac{2}{3}\pi.$

23.

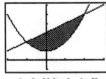

[−2, 3] by [−1, 6]

The curves intersect when $x^2 + 1 = x + 3$, which is when
$x^2 - x - 2 = (x - 2)(x + 1) = 0$, i.e., when
$x = -1$ or $x = 2$. Use washer cross sections: a washer has
inner radius $r = x^2 + 1$, outer radius $R = x + 3$, and area
$A(x) = \pi(R^2 - r^2)$
$= \pi[(x + 3)^2 - (x^2 + 1)^2]$
$= \pi(-x^4 - x^2 + 6x + 8)$. The volume is
$\int_{-1}^{2} \pi(-x^4 - x^2 + 6x + 8) \, dx$
$= \pi\left[-\frac{1}{5}x^5 - \frac{1}{3}x^3 + 3x^2 + 8x\right]_{-1}^{2}$
$= \pi\left[\left(-\frac{32}{5} - \frac{8}{3} + 12 + 16\right) - \left(\frac{1}{5} + \frac{1}{3} + 3 - 8\right)\right] = \frac{117\pi}{5}.$

25.

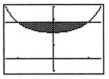

$\left[-\frac{\pi}{3}, \frac{\pi}{3}\right]$ by [−0.5, 2]

Use washer cross sections: a washer has inner radius
$r = \sec x$, outer radius $R = \sqrt{2}$, and area
$A(x) = \pi(R^2 - r^2) = \pi(2 - \sec^2 x).$

The volume is
$\int_{-\pi/4}^{\pi/4} \pi(2 - \sec^2 x) \, dx = \pi\left[2x - \tan x\right]_{-\pi/4}^{\pi/4}$
$= \pi\left[\left(\frac{\pi}{2} - 1\right) - \left(-\frac{\pi}{2} + 1\right)\right]$
$= \pi^2 - 2\pi.$

27.

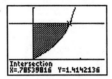

[−0.5, 1.5] by [−0.5, 2]

The curves intersect at $x \approx 0.7854$. A cross section has
radius $r = \sqrt{2} - \sec x \tan x$ and area
$A(x) = \pi r^2 = \pi(\sqrt{2} - \sec x \tan x)^2$. Use NINT to find
$\int_0^{0.7854} \pi(\sqrt{2} - \sec x \tan x)^2 \, dx \approx 2.301.$

29.

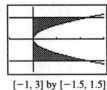

[−1, 3] by [−1.5, 1.5]

A cross section has radius $r = \sqrt{5}y^2$ and area
$A(y) = \pi r^2 = 5\pi y^4.$

The volume is $\int_{-1}^{1} 5\pi y^4 \, dy = \pi\left[y^5\right]_{-1}^{1} = 2\pi.$

31.

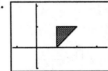

[−1.2, 3.5] by [−1, 2.1]

Use washer cross sections. A washer has inner radius $r = 1$,
outer radius $R = y + 1$, and area
$A(y) = \pi(R^2 - r^2) = \pi[(y + 1)^2 - 1] = \pi(y^2 + 2y)$. The
volume is $\int_0^1 \pi(y^2 + 2y) \, dy = \pi\left[\frac{1}{3}y^3 + y^2\right]_0^1 = \frac{4}{3}\pi.$

202 Section 7.3

33.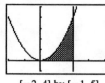

[−2, 4] by [−1, 5]

Use cylindrical shells: A shell has radius x and height x^2.

The volume is $\int_0^2 2\pi(x)(x^2)\,dx = 2\pi\left[\frac{1}{4}x^4\right]_0^2 = 8\pi$.

35.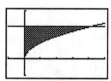

[−1, 5] by [−1, 3]

The curved and horizontal line intersect at (4, 2).

(a) Use washer cross sections: a washer has inner radius $r = \sqrt{x}$, outer radius $R = 2$, and area

$A(x) = \pi(R^2 - r^2) = \pi(4 - x)$. The volume is

$\int_0^4 \pi(4 - x)\,dx = \pi\left[4x - \frac{1}{2}x^2\right]_0^4 = 8\pi$

(b) A cross section has radius $r = y^2$ and area

$A(y) = \pi r^2 = \pi y^4$.

The volume is $\int_0^2 \pi y^4\,dy = \pi\left[\frac{1}{5}y^5\right]_0^2 = \frac{32\pi}{5}$.

(c) A cross section has radius $r = 2 - \sqrt{x}$ and area

$A(x) = \pi r^2 = \pi(2 - \sqrt{x})^2 = \pi(4 - 4\sqrt{x} + x)$.

The volume is

$\int_0^4 \pi(4 - 4\sqrt{x} + x)\,dx = \pi\left[4x - \frac{8}{3}x^{3/2} + \frac{1}{2}x^2\right]_0^4 = \frac{8\pi}{3}$.

(d) Use washer cross sections: a washer has inner radius $r = 4 - y^2$, outer radius $R = 4$, and area

$A(y) = \pi(R^2 - r^2) = \pi[16 - (4 - y^2)^2]$

$= \pi(8y^2 - y^4)$.

The volume is

$\int_0^2 \pi(8y^2 - y^4)\,dy = \pi\left[\frac{8}{3}y^3 - \frac{1}{5}y^5\right]_0^2 = \frac{224\pi}{15}$.

37.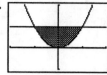

[−2, 2] by [−1, 2]

The curves intersect at (±1, 1).

(a) A cross section has radius $r = 1 - x^2$ and area

$A(x) = \pi r^2 = \pi(1 - x^2)^2 = \pi(1 - 2x^2 + x^4)$.

The volume is

$\int_{-1}^1 \pi(1 - 2x^2 + x^4)\,dx = \pi\left[x - \frac{2}{3}x^3 + \frac{1}{5}x^5\right]_{-1}^1 = \frac{16\pi}{15}$.

(b) Use cylindrical shells: a shell has radius $2 - y$ and height $2\sqrt{y}$. The volume is

$\int_0^1 2\pi(2 - y)(2\sqrt{y})\,dy = 4\pi\int_0^1 (2\sqrt{y} - y^{3/2})\,dy$

$= 4\pi\left[\frac{4}{3}y^{3/2} - \frac{2}{5}y^{5/2}\right]_0^1 = \frac{56\pi}{15}$.

(c) Use cylindrical shells: a shell has radius $y + 1$ and height $2\sqrt{y}$. The volume is

$\int_0^1 2\pi(y + 1)(2\sqrt{y})\,dy = 4\pi\int_0^1 (y^{3/2} + \sqrt{y})\,dy$

$= 4\pi\left[\frac{2}{5}y^{5/2} + \frac{2}{3}y^{3/2}\right]_0^1 = \frac{64\pi}{15}$.

39.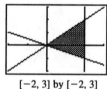

[−2, 3] by [−2, 3]

A shell has radius x and height $x - \left(-\frac{x}{2}\right) = \frac{3}{2}x$.

The volume is $\int_0^2 2\pi(x)\left(\frac{3}{2}x\right)dx = \pi\left[x^3\right]_0^2 = 8\pi$.

41.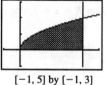

[−1, 5] by [−1, 3]

A shell has radius x and height $\sqrt{x}$. The volume is

$\int_0^4 2\pi(x)(\sqrt{x})\,dx = 2\pi\left[\frac{2}{5}x^{5/2}\right]_0^4 = \frac{128\pi}{5}$.

43. A shell has height $12(y^2 - y^3)$.

(a) A shell has radius y. The volume is
$$\int_0^1 2\pi(y)12(y^2 - y^3)\, dy = 24\pi \int_0^1 (y^3 - y^4)\, dy$$
$$= 24\pi \left[\frac{1}{4}y^4 - \frac{1}{5}y^5\right]_0^1 = \frac{6\pi}{5}.$$

(b) A shell has radius $1 - y$. The volume is
$$\int_0^1 2\pi(1-y)12(y^2 - y^3)\, dy$$
$$= 24\pi \int_0^1 (y^4 - 2y^3 + y^2)\, dy$$
$$= 24\pi \left[\frac{1}{5}y^5 - \frac{1}{2}y^4 + \frac{1}{3}y^3\right]_0^1 = \frac{4\pi}{5}.$$

(c) A shell has radius $\frac{8}{5} - y$. The volume is
$$\int_0^1 2\pi\left(\frac{8}{5} - y\right)12(y^2 - y^3)\, dy$$
$$= 24\pi \int_0^1 \left(y^4 - \frac{13}{5}y^3 + \frac{8}{5}y^2\right)\, dy$$
$$= 24\pi \left[\frac{1}{5}y^5 - \frac{13}{20}y^4 + \frac{8}{15}y^3\right]_0^1 = 2\pi.$$

(d) A shell has radius $y + \frac{2}{5}$. The volume is
$$\int_0^1 2\pi\left(y + \frac{2}{5}\right)12(y^2 - y^3)\, dy$$
$$= 24\pi \int_0^1 \left(-y^4 + \frac{3}{5}y^3 + \frac{2}{5}y^2\right)\, dx$$
$$= 24\pi \left[-\frac{1}{5}y^5 + \frac{3}{20}y^4 + \frac{2}{15}y^3\right]_0^1 = 2\pi.$$

45.

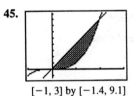

[−1, 3] by [−1.4, 9.1]

The functions intersect at $(2, 8)$.

(a) Use washer cross sections: a washer has inner radius $r = x^3$, outer radius $R = 4x$, and area
$A(x) = \pi(R^2 - r^2) = \pi(16x^2 - x^6)$. The volume is
$$\int_0^2 \pi(16x^2 - x^6)\, dx = \pi\left[\frac{16}{3}x^3 - \frac{1}{7}x^7\right]_0^2 = \frac{512\pi}{21}.$$

(b) Use cylindrical shells: a shell has a radius $8 - y$ and height $y^{1/3} - \frac{y}{4}$. The volume is
$$\int_0^8 2\pi(8 - y)\left(y^{1/3} - \frac{y}{4}\right) dy$$
$$= 2\pi \int_0^8 \left(8y^{1/3} - 2y - y^{4/3} + \frac{y^2}{4}\right) dy$$
$$= 2\pi\left[6y^{4/3} - y^2 - \frac{3}{7}y^{7/3} + \frac{1}{12}y^3\right]_0^8 = \frac{832\pi}{21}.$$

47.

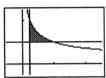

[−0.5, 2.5] by [−0.5, 2.5]

The intersection points are $\left(\frac{1}{4}, 1\right)$, $\left(\frac{1}{4}, 2\right)$, and $(1, 1)$.

(a) A washer has inner radius $r = \frac{1}{4}$, outer radius $R = \frac{1}{y^2}$, and area $\pi(R^2 - r^2) = \pi\left(\frac{1}{y^4} - \frac{1}{16}\right)$. The volume is
$$\int_1^2 \pi\left(\frac{1}{y^4} - \frac{1}{16}\right) dy = \pi\left[-\frac{1}{3y^3} - \frac{1}{16}y\right]_1^2 = \frac{11\pi}{48}.$$

(b) A shell has radius x and height $\frac{1}{\sqrt{x}} - 1$. The volume is
$$\int_{1/4}^1 2\pi(x)\left(\frac{1}{\sqrt{x}} - 1\right) dx = 2\pi\left[\frac{2}{3}x^{3/2} - \frac{1}{2}x^2\right]_{1/4}^1 = \frac{11\pi}{48}.$$

49. (a) A cross section has radius $r = \frac{x}{12}\sqrt{36 - x^2}$ and area
$A(x) = \pi r^2 = \frac{\pi}{144}(36x^2 - x^4)$. The volume is
$$\int_0^6 \frac{\pi}{144}(36x^2 - x^4)\, dx = \frac{\pi}{144}\left[12x^3 - \frac{1}{5}x^5\right]_0^6 = \frac{36\pi}{5}\ \text{cm}^3.$$

(b) $\left(\frac{36\pi}{5}\ \text{cm}^3\right)(8.5\ \text{g/cm}^3) \approx 192.3$ g

51. (a) Using $d = \frac{C}{\pi}$, and $A = \pi\left(\frac{d}{2}\right)^2 = \frac{C^2}{4\pi}$ yields the following areas (in square inches, rounded to the nearest tenth): 2.3, 1.6, 1.5, 2.1, 3.2, 4.8, 7.0, 9.3, 10.7, 10.7, 9.3, 6.4, 3.2.

(b) If $C(y)$ is the circumference as a function of y, then the area of a cross section is
$$A(y) = \pi\left(\frac{C(y)/\pi}{2}\right)^2 = \frac{C^2(y)}{4\pi},$$
and the volume is $\frac{1}{4\pi}\int_0^6 C^2(y)\, dy$.

(c) $\frac{1}{4\pi}\int_0^6 A(y)\, dy = \frac{1}{4\pi}\int_0^6 C^2(y)\, dy$
$\approx \frac{1}{4\pi}\left(\frac{6-0}{24}\right)[5.4^2 + 2(4.5^2 + 4.4^2$
$+ 5.1^2 + 6.3^2 + 7.8^2 + 9.4^2 + 10.8^2 + 11.6^2$
$+ 11.6^2 + 10.8^2 + 9.0^2) + 6.3^2] \approx 34.7\ \text{in.}^3$

204 Section 7.3

53. (a)

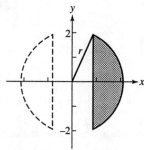

The remaining solid is that swept out by the shaded region in revolution. Use cylindrical shells: a shell has radius x and height $2\sqrt{r^2 - x^2}$. The volume is

$$\int_{\sqrt{r^2 - 2^2}}^{r} 2\pi(x)(2\sqrt{r^2 - x^2})\,dx$$

$$= 2\pi\left[-\frac{2}{3}(r^2 - x^2)^{3/2}\right]_{\sqrt{r^2 - 4}}^{r}$$

$$= -\frac{4}{3}\pi(-8) = \frac{32\pi}{3}.$$

(b) The answer is independent of r.

55. Solve $ax - x^2 = 0$: This is true at $x = 0$ and $x = a$. For revolution about the x-axis, a cross section has radius $r = ax - x^2$ and area

$A(x) = \pi r^2 = \pi(ax - x^2)^2 = \pi(a^2x^2 - 2ax^3 + x^4)$.

The volume is

$$\int_0^a \pi(a^2x^2 - 2ax^3 + x^4)\,dx = \pi\left[\frac{1}{3}a^2x^3 - \frac{1}{2}ax^4 + \frac{1}{5}x^5\right]_0^a$$

$$= \frac{1}{30}\pi a^5.$$

For revolution about the y-axis, a cylindrical shell has radius x and height $ax - x^2$. The volume is

$$\int_0^a 2\pi(x)(ax - x^2)\,dx = 2\pi\left[\frac{1}{3}ax^3 - \frac{1}{4}x^4\right]_0^a = \frac{1}{6}\pi a^4.$$

Setting the two volumes equal,

$\frac{1}{30}\pi a^5 = \frac{1}{6}\pi a^4$ yields $\frac{1}{30}a = \frac{1}{6}$, so $a = 5$.

57. $g'(y) = \frac{dx}{dy} = \frac{1}{2\sqrt{y}}$, and

$$\int_0^2 2\pi\sqrt{y}\sqrt{1 + \left(\frac{1}{2\sqrt{y}}\right)^2}\,dy = \int_0^2 \pi\sqrt{4y + 1}\,dy$$

$$= \left[\frac{\pi}{6}(4y + 1)^{3/2}\right]_0^2$$

$$= \frac{13\pi}{3} \approx 13.614$$

59. $g'(y) = \frac{dx}{dy} = \frac{1}{2}y^{-1/2}$, and

$$\int_1^3 2\pi\left[y^{1/2} - \left(\frac{1}{3}\right)^{3/2}\right]\sqrt{1 + \left[\frac{1}{2}y^{-1/2}\right]^2}\,dy$$

$$= 2\pi\int_1^3\left[y^{1/2} - \left(\frac{1}{3}\right)^{3/2}\right]\sqrt{1 + \frac{1}{4y}}\,dy.$$

Using NINT, this evaluates to ≈ 16.110

61. $f'(x) = \frac{dy}{dx} = 2x$, and

$$\int_0^2 2\pi x^2\sqrt{1 + (2x)^2}\,dx = \int_0^2 2\pi x^2\sqrt{1 + 4x^2}\,dx$$ evaluates,

using NINT, to ≈ 53.226.

63. $f'(x) = \frac{dy}{dx} = \frac{1 - x}{\sqrt{2x - x^2}}$, and

$$\int_{0.5}^{1.5} 2\pi\sqrt{2x - x^2}\sqrt{1 + \left(\frac{1 - x}{\sqrt{2x - x^2}}\right)^2}\,dx = 2\pi\int_{0.5}^{1.5} 1\,dx$$

$$= 2\pi\left[x\right]_{0.5}^{1.5}$$

$$= 2\pi \approx 6.283$$

65. Hemisphere cross sectional area: $\pi(\sqrt{R^2 - h^2})^2 = A_1$.

Right circular cylinder with cone removed cross sectional area: $\pi R^2 - \pi h^2 = A_2$

Since $A_1 = A_2$, the two volumes are equal by Cavalieri's theorem. Thus, volume of hemisphere

= volume of cylinder − volume of cone

$= \pi R^3 - \frac{1}{3}\pi R^3 = \frac{2}{3}\pi R^3$.

67. (a) Put the bottom of the bowl at $(0, -a)$. The area of a horizontal cross section is $\pi(\sqrt{a^2 - y^2})^2 = \pi(a^2 - y^2)$. The volume for height h is

$$\int_{-a}^{h-a} \pi(a^2 - y^2)\,dy = \pi\left[a^2 y - \frac{1}{3}y^3\right]_{-a}^{h-a} = \frac{\pi h^2(3a - h)}{3}.$$

(b) For $h = 4$, $y = -1$ and the area of a cross section is $\pi(5^2 - 1^2) = 24\pi$. The rate of rise is

$\frac{dh}{dt} = \frac{1}{A}\frac{dV}{dt} = \frac{1}{24\pi}(0.2) = \frac{1}{120\pi}$ m/sec.

Section 7.4 Lengths of Curves
(pp. 395–401)

Quick Review 7.4

1. $\sqrt{1 + 2x + x^2} = \sqrt{(1+x)^2}$, which, since $x \geq -1$, equals $1 + x$ or $x + 1$.

2. $\sqrt{1 - x + \frac{x^2}{4}} = \sqrt{\left(1 - \frac{x}{2}\right)^2}$, which, since $x \leq 2$, equals $1 - \frac{x}{2}$ or $\frac{2-x}{2}$.

3. $\sqrt{1 + (\tan x)^2} = \sqrt{(\sec x)^2}$, which, since $0 \leq x < \frac{\pi}{2}$, equals $\sec x$.

4. $\sqrt{1 + \left(\frac{x}{4} - \frac{1}{x}\right)^2} = \sqrt{\frac{1}{2} + \frac{1}{16}x^2 + \frac{1}{x^2}} = \frac{1}{4}\sqrt{\frac{(x^2 + 4)^2}{x^2}}$ which, since $x > 0$, equals $\frac{x^2 + 4}{4x}$.

5. $\sqrt{1 + \cos 2x} = \sqrt{2 \cos^2 x}$, which, since $-\frac{\pi}{2} \leq x \leq \frac{\pi}{2}$, equals $\sqrt{2} \cos x$.

6. $f(x)$ has a corner at $x = 4$.

7. $\frac{d}{dx}(5x^{2/3}) = \frac{10}{3\sqrt[3]{x}}$ is undefined at $x = 0$. $f(x)$ has a cusp there.

8. $\frac{d}{dx}(\sqrt[5]{x+3}) = \frac{1}{5(x+3)^{4/5}}$ is undefined for $x = -3$. $f(x)$ has a vertical tangent there.

9. $\sqrt{x^2 - 4x + 4} = |x - 2|$ has a corner at $x = 2$.

10. $\frac{d}{dx}\left(1 + \sqrt[3]{\sin x}\right) = \frac{\cos x}{3(\sin x)^{2/3}}$ is undefined for $x = k\pi$, where k is any integer. $f(x)$ has vertical tangents at these values of x.

Section 7.4 Exercises

1. (a) $y' = 2x$, so
$$\text{Length} = \int_{-1}^{2} \sqrt{1 + (2x)^2}\, dx = \int_{-1}^{2} \sqrt{1 + 4x^2}\, dx.$$

(b)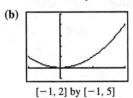

[−1, 2] by [−1, 5]

(c) Length ≈ 6.126

3. (a) $x' = \cos y$, so Length $= \int_{0}^{\pi} \sqrt{1 + \cos^2 y}\, dy$.

(b)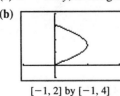

[−1, 2] by [−1, 4]

(c) Length ≈ 3.820

5. (a) $y^2 + 2y = 2x + 1$, so
$y^2 + 2y + 1 = (y + 1)^2 = 2x + 2$, and
$y = \sqrt{2x + 2} - 1$. Then $y' = \frac{1}{\sqrt{2x + 2}}$, and
$$\text{Length} = \int_{-1}^{7} \sqrt{1 + \frac{1}{2x + 2}}\, dx.$$

(b)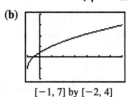

[−1, 7] by [−2, 4]

(c) Length ≈ 9.294

7. (a) $y' = \tan x$, so Length $= \int_{0}^{\pi/6} \sqrt{1 + \tan^2 x}\, dx$.

(b) $y = \int \tan x\, dx = \ln(|\sec x|)$

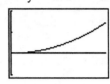

$\left[0, \frac{\pi}{6}\right]$ by [−0.1, 0.2]

(c) Length ≈ 0.549

9. (a) $y' = \sec x \tan x$, so
$$\text{Length} = \int_{-\pi/3}^{\pi/3} \sqrt{1 + \sec^2 x \tan^2 x}\, dx.$$

(b)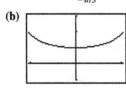

$\left[-\frac{\pi}{3}, \frac{\pi}{3}\right]$ by [−1, 3]

(c) Length ≈ 3.139

11. $y' = \frac{1}{2}(x^2 + 2)^{1/2}(2x) = x\sqrt{x^2 + 2}$, so the length is
$$\int_{0}^{3} \sqrt{1 + (x\sqrt{x^2 + 2})^2}\, dx = \int_{0}^{3} \sqrt{x^4 + 2x^2 + 1}\, dx$$
$$= \int_{0}^{3} (x^2 + 1)\, dx = \left[\frac{1}{3}x^3 + x\right]_{0}^{3} = 12.$$

13. $x' = y^2 - \frac{1}{4y^2}$, so the length is $\int_{1}^{3} \sqrt{1 + \left(y^2 - \frac{1}{4y^2}\right)^2}\, dy$
$$= \int_{1}^{3} \sqrt{\left(y^2 + \frac{1}{4y^2}\right)^2}\, dy = \left[\frac{1}{3}y^3 - \frac{1}{4y}\right]_{1}^{3} = \frac{53}{6}.$$

15. $x' = \frac{y^2}{2} - \frac{1}{2y^2}$, so the length is
$$\int_{1}^{2} \sqrt{1 + \left(\frac{y^2}{2} - \frac{1}{2y^2}\right)^2}\, dy = \int_{1}^{2} \sqrt{\left(\frac{y^2}{2} + \frac{1}{2y^2}\right)^2}\, dy$$
$$= \left[\frac{1}{6}y^3 - \frac{1}{2y}\right]_{1}^{2} = \frac{17}{12}.$$

17. $x' = \sqrt{\sec^4 y - 1}$, so the length is

$$\int_{-\pi/4}^{\pi/4} \sqrt{1 + (\sec^4 y - 1)}\, dy = \int_{-\pi/4}^{\pi/4} \sec^2 y\, dy$$
$$= \Big[\tan y\Big]_{-\pi/4}^{\pi/4} = 2.$$

19. (a) $\left(\dfrac{dy}{dx}\right)^2$ corresponds to $\dfrac{1}{4x}$ here, so take $\dfrac{dy}{dx}$ as $\dfrac{1}{2\sqrt{x}}$. Then $y = \sqrt{x} + C$, and, since $(1, 1)$ lies on the curve, $C = 0$. So $y = \sqrt{x}$.

(b) Only one. We know the derivative of the function and the value of the function at one value of x.

21. $y' = \sqrt{\cos 2x}$, so the length is

$$\int_0^{\pi/4} \sqrt{1 + \cos 2x}\, dx = \int_0^{\pi/4} \sqrt{2 \cos^2 x}\, dx$$
$$= \sqrt{2}\Big[\sin x\Big]_0^{\pi/4} = 1.$$

23. Find the length of the curve $y = \sin \dfrac{3\pi}{20} x$ for $0 \leq x \leq 20$.

$y' = \dfrac{3\pi}{20} \cos \dfrac{3\pi}{20} x$, so the length is

$$\int_0^{20} \sqrt{1 + \left(\dfrac{3\pi}{20} \cos \dfrac{3\pi}{20} x\right)^2}\, dx,$$ which evaluates, using NINT, to ≈ 21.07 inches.

25. For track 1: $y_1 = 0$ at $x = \pm 10\sqrt{5} \approx \pm 22.3607$, and

$y_1' = \dfrac{-0.2x}{\sqrt{100 - 0.2x^2}}$. NINT fails to evaluate $\int_{-10\sqrt{5}}^{10\sqrt{5}} \sqrt{1 + (y_1')^2}\, dx$ because of the undefined slope at the limits, so use the track's symmetry, and "back away" from the upper limit a little, and find

$2\int_0^{22.36} \sqrt{1 + (y_1')^2}\, dx \approx 52.548$. Then, pretending the last little stretch at each end is a straight line, add $2\sqrt{100 - 0.2(22.36)^2} \approx 0.156$ to get the total length of track 1 as ≈ 52.704. Using a similar strategy, find the length of the *right half* of track 2 to be ≈ 32.274. Now enter $Y_1 = 52.704$ and

$Y_2 = 32.274 + \text{NINT}\left(\sqrt{1 + \left(\dfrac{-0.2t}{\sqrt{150 - 0.2t^2}}\right)^2}, t, x, 0\right)$ and graph in a $[-30, 0]$ by $[0, 60]$ window to see the effect of the x-coordinate of the lane-2 starting position on the length of lane 2. (Be patient!) Solve graphically to find the intersection at $x \approx -19.909$, which leads to starting point coordinates $(-19.909, 8.410)$.

27. $f'(x) = \dfrac{(4x^2 + 1) - (8x^2 - 8x)}{(4x^2 + 1)^2} = -\dfrac{4x^2 - 8x - 1}{(4x^2 + 1)^2}$, so the length is $\int_{-1/2}^{1} \sqrt{1 + \left(\dfrac{4x^2 - 8x - 1}{(4x^2 + 1)^2}\right)^2}\, dx$ which evaluates, using NINT, to ≈ 2.1089.

29. $y = (1 - \sqrt{x})^2$, $0 \leq x \leq 1$

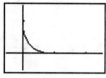

$[-0.5, 2.5]$ by $[-0.5, 1.5]$

$y' = \dfrac{\sqrt{x} - 1}{\sqrt{x}}$, but NINT may fail using y' over the entire interval because y' is undefined at $x = 0$. So, split the curve into two equal segments by solving $\sqrt{x} + \sqrt{y} = 1$ with $y = x$: $x = \dfrac{1}{4}$. The total length is $2\int_{1/4}^{1} \sqrt{1 + \left(\dfrac{\sqrt{x} - 1}{\sqrt{x}}\right)^2}\, dx$, which evaluates, using NINT, to ≈ 1.623.

31. Because the limit of the sum $\sum \Delta x_k$ as the norm of the partition goes to zero will always be the length $(b - a)$ of the interval (a, b).

33. (a) The fin is the hypotenuse of a right triangle with leg lengths Δx_k and $\dfrac{df}{dx}\Big|_{x = x_{k-1}} \Delta x_k = f'(x_{k-1})\Delta x_k$.

(b) $\displaystyle\lim_{n \to \infty} \sum_{k=1}^{n} \sqrt{(\Delta x_k)^2 + (f'(x_{k-1})\Delta x_k)^2}$
$= \displaystyle\lim_{n \to \infty} \sum_{k=1}^{n} \Delta x_k \sqrt{1 + (f'(x_{k-1}))^2}$
$= \displaystyle\int_a^b \sqrt{1 + (f'(x))^2}\, dx$

Section 7.5 Applications from Science and Statistics
(pp. 401–411)

Quick Review 7.5

1. (a) $\int_0^1 e^{-x}\, dx = \Big[-e^{-x}\Big]_0^1 = 1 - \dfrac{1}{e}$

(b) ≈ 0.632

2. (a) $\int_0^1 e^x\, dx = \Big[e^x\Big]_0^1 = e - 1$

(b) ≈ 1.718

3. (a) $\int_{\pi/4}^{\pi/2} \sin x\, dx = \Big[-\cos x\Big]_{\pi/4}^{\pi/2} = \dfrac{\sqrt{2}}{2}$

(b) ≈ 0.707

4. (a) $\int_0^3 (x^2 + 2)\, dx = \left[\frac{1}{3}x^3 + 2x\right]_0^3 = 15$

   (b) 15

5. (a) $\int_1^2 \frac{x^2}{x^3 + 1}\, dx = \left[\frac{1}{3} \ln(x^3 + 1)\right]_1^2$
   $= \frac{1}{3}[\ln 9 - \ln 2]$
   $= \frac{1}{3}\ln\left(\frac{9}{2}\right)$

   (b) ≈ 0.501

6. $\int_0^7 2\pi(x + 2) \sin x\, dx$

7. $\int_0^7 (1 - x^2)(2\pi x)\, dx$

8. $\int_0^7 \pi \cos^2 x\, dx$

9. $\int_0^7 \pi\left(\frac{y}{2}\right)^2 (10 - y)\, dy$

10. $\int_0^7 \frac{\sqrt{3}}{4} \sin^2 x\, dx$

Section 7.5 Exercises

1. $\int_0^5 xe^{-x/3}\, dx = \left[-3e^{-x/3}(3 + x)\right]_0^5 = -\frac{24}{e^{5/3}} + 9 \approx 4.4670$ J

3. $\int_0^3 x\sqrt{9 - x^2}\, dx = \left[-\frac{1}{3}(9 - x^2)^{3/2}\right]_0^3 = 9$ J

5. When the bucket is x m off the ground, the water weighs
   $F(x) = 490\left(\frac{20 - x}{20}\right) = 490\left(1 - \frac{x}{20}\right) = 490 - 24.5x$ N.

   Then
   $W = \int_0^{20} (490 - 24.5x)\, dx = \left[490x - 12.25x^2\right]_0^{20} = 4900$ J.

7. When the bag is x ft off the ground, the sand weighs
   $F(x) = 144\left(\frac{18 - x/2}{18}\right) = 144\left(1 - \frac{x}{36}\right) = 144 - 4x$ lb.

   Then
   $W = \int_0^{18} (144 - 4x)\, dx = \left[144x - 2x^2\right]_0^{18} = 1944$ ft-lb

9. (a) $F = ks$, so $21{,}714 = k(8 - 5)$ and $k = 7238$ lb/in.

   (b) $F(x) = 7238x$. $W = \int_0^{1/2} 7238x\, dx = \left[3619x^2\right]_0^{1/2}$
   $= 904.75 \approx 905$ in.-lb, and $W = \int_{1/2}^1 7238x\, dx$
   $= \left[3619x^2\right]_{1/2}^1 = 2714.25 \approx 2714$ in.-lb.

11. When the end of the rope is x m from its starting point, the $(50 - x)$ m of rope still to go weigh
    $F(x) = (0.624)(50 - x)$ N. The total work is
    $\int_0^{50} (0.624)(50 - x)\, dx = 0.624\left[50x - \frac{1}{2}x^2\right]_0^{50} = 780$ J.

13. (a) From the equation $x^2 + y^2 = 3^2$, it follows that a thin horizontal rectangle has area $2\sqrt{9 - y^2}\, \Delta y$, where y is distance from the top, and pressure $62.4y$. The total force is approximately $\sum_{k=1}^n (62.4 y_k)(2\sqrt{9 - y_k^2})\, \Delta y$
    $= \sum_{k=1}^n 124.8 y_k \sqrt{9 - y_k^2}\, \Delta y$.

    (b) $\int_0^3 124.8 y\sqrt{9 - y^2}\, dy = \left[-41.6(9 - y^2)^{3/2}\right]_0^3$
    $= 1123.2$ lb

15. (a) From the equation $x = \frac{3}{8}y$, it follows that a thin horizontal rectangle has area $\frac{3}{4}y\Delta y$, where y is the distance from the top of the triangle, the pressure is $62.4(y - 3)$. The total force is approximately
    $\sum_{k=1}^n 62.4(y_k - 3)\left(\frac{3}{4}y_k\right)\Delta y = \sum_{k=1}^n 46.8(y_k^2 - 3y_k)\Delta y$.

    (b) $\int_3^8 46.8(y^2 - 3y)\, dy = \left[15.6y^3 - 70.2y^2\right]_3^8$
    $= 3494.4 - (-210.6) - 3705$ lb

17. (a) Work to raise a thin slice $= 62.4(10 \times 12)(\Delta y)y$.
    Total work $= \int_0^{20} 62.4(120)y\, dy = 62.4\left[60y^2\right]_0^{20}$
    $= 1{,}497{,}600$ ft-lb

    (b) $(1{,}497{,}600 \text{ ft-lb}) \div (250 \text{ ft-lb/sec}) = 5990.4$ sec
    ≈ 100 min

    (c) Work to empty half the tank $= \int_0^{10} 62.4(120)y\, dy$
    $= 62.4\left[60y^2\right]_0^{10} = 374{,}400$ ft-lb, and
    $374{,}400 \div 250 = 1497.6$ sec ≈ 25 min

17. continued

(d) The weight per ft³ of water is a simple multiplicative factor in the answers. So divide by 62.4 and multiply by the appropriate weight-density

For 62.26:

$1,497,600\left(\dfrac{62.26}{62.4}\right) = 1,494,240$ ft-lb and

$5990.4\left(\dfrac{62.26}{62.4}\right) = 5976.96$ sec ≈ 100 min.

For 62.5:

$1,497,600\left(\dfrac{62.5}{62.4}\right) = 1,500,000$ ft-lb and

$5990.4\left(\dfrac{62.5}{62.4}\right) = 6000$ sec $= 100$ min.

19. Work to pump through the valve is $\pi(2)^2(62.4)(y + 15)\Delta y$ for a thin disk and

$\displaystyle\int_0^6 4\pi(62.4)(y + 15)\, dy = 249.6\pi\left[\dfrac{1}{2}y^2 + 15y\right]_0^6$

$\approx 84{,}687.3$ ft-lb

for the whole tank. Work to pump over the rim is

$\pi(2)^2(62.4)(6 + 15)\Delta y$ for a thin disk and

$\displaystyle\int_0^6 4\pi(62.4)(21)\, dy = 4\pi(62.4)(21)(6) \approx 98{,}801.8$ ft-lb for the whole tank. Through a hose attached to a valve in the bottom is faster, because it takes more time for a pump with a given power output to do more work.

21. The work is that already calculated (to pump the oil to the rim) plus the work needed to raise the entire amount 3 ft higher. The latter comes to

$\left(\dfrac{1}{3}\pi r^2 h\right)(57)(3) = 57\pi(4)^2(8) = 22{,}921.06$ ft-lb, and the total is $22{,}921.06 + 30{,}561.41 \approx 53{,}482.5$ ft-lb.

23. The work to raise a thin disk is

$\pi r^2(56)h = \pi(\sqrt{10^2 - y^2})^2(56)(10 + 2 - y)\Delta y$

$= 56\pi(12 - y)(100 - y^2)\Delta y$. The total work is

$\displaystyle\int_0^{10} 56\pi(12 - y)(100 - y^2)\, dy$, which evaluates using NINT to $\approx 967{,}611$ ft-lb. This will come to

$(967{,}611)(\$0.005) \approx \$4838$, so yes, there's enough money to hire the firm.

25. (a) The pressure at depth y is $62.4y$, and the area of a thin horizontal strip is $2\Delta y$. The depth of water is $\dfrac{11}{6}$ ft, so the total force on an end is

$\displaystyle\int_0^{11/6} (62.4y)(2\, dy) \approx 209.73$ lb.

(b) On the sides, which are twice as long as the ends, the initial total force is doubled to ≈ 419.47 lb. When the tank is upended, the depth is doubled to $\dfrac{11}{3}$ ft, and the force on a side becomes $\displaystyle\int_0^{11/3} (62.4y)(2\, dy) \approx 838.93$ lb, which means that the fluid force doubles.

27. (a) 0.5 (50%), since half of a normal distribution lies below the mean.

(b) Use NINT to find $\displaystyle\int_{63}^{65} f(x)\, dx$, where

$f(x) = \dfrac{1}{3.2\sqrt{2\pi}} e^{-(x-63.4)^2/(2 \cdot 3.2^2)}$. The result is ≈ 0.24 (24%).

(c) 6 ft = 72 in. Pick 82 in. as a conveniently high upper limit and with NINT, find $\displaystyle\int_{72}^{82} f(x)\, dx$. The result is ≈ 0.0036 (0.36%).

(d) 0 if we assume a continuous distribution. Between 59.5 in. and 60.5 in., the proportion is

$\displaystyle\int_{59.5}^{60.5} f(x)\, dx \approx 0.071$ (7.1%)

29. Integration is a good approximation to the area (which represents the probability), since the area is a kind of Riemann sum.

31. $\displaystyle\int_{6,370,000}^{35,780,000} \dfrac{1000 MG}{r^2}\, dr = 1000\, MG\left[-\dfrac{1}{r}\right]_{6,370,000}^{35,780,000}$, which for $M = 5.975 \times 10^{24}$, $G = 6.6726 \times 10^{-11}$ evaluates to

$\approx 5.1446 \times 10^{10}$ J.

33. $F = m\left(\dfrac{dv}{dt}\right) = mv\left(\dfrac{dv}{dx}\right)$, so

$W = \displaystyle\int_{x_1}^{x_2} F(x)\, dx$

$= \displaystyle\int_{x_1}^{x_2} mv\left(\dfrac{dv}{dx}\right) dx$

$= \displaystyle\int_{v_1}^{v_2} mv\, dv$

$= \dfrac{1}{2}mv_2^2 - \dfrac{1}{2}mv_1^2$

35. $0.3125 \text{ lb} = \dfrac{0.3125 \text{ lb}}{32 \text{ ft/sec}^2} = 0.009765625$ slug, and

$90 \text{ mph} = 90\left(\dfrac{5280 \text{ ft}}{1 \text{ mi}}\right)\left(\dfrac{1 \text{ hr}}{3600 \text{ sec}}\right) = 132$ ft/sec, so

Work = change in kinetic energy $= \dfrac{1}{2}(0.009765625)(132)^2$

≈ 85.1 ft-lb.

37. $2 \text{ oz} = 2 \text{ oz}\left(\dfrac{1 \text{ lb}}{16 \text{ oz}}\right)/(32 \text{ ft/sec}^2) = \dfrac{1}{256}$ slug, and

$124 \text{ mph} = 124 \text{ mph}\left(\dfrac{5280 \text{ ft}}{1 \text{ mi}}\right)\left(\dfrac{1 \text{ hr}}{3600 \text{ sec}}\right) = 181.867$ ft/sec,

so Work $= \dfrac{1}{2}\left(\dfrac{1}{256}\right)(181.867)^2 \approx 64.6$ ft-lb.

39. $6.5 \text{ oz} = 6.5 \text{ oz}\left(\dfrac{1 \text{ lb}}{16 \text{ oz}}\right)/(32 \text{ ft/sec}^2) \approx 0.01270$ slug, so

Work $= \dfrac{1}{2}(0.01270)(132)^2 \approx 110.6$ ft-lb.

■ Chapter 7 Review Exercises
(pp. 413–415)

1. $\displaystyle\int_0^5 v(t)\, dt = \int_0^5 (t^2 - 0.2t^3)\, dt$

$= \left[\dfrac{1}{3}t^3 - 0.05t^4\right]_0^5 \approx 10.417$ ft

2. $\displaystyle\int_0^7 c(t)\, dt = \int_0^7 (4 + 0.001t^4)\, dt$

$= \left[4t + 0.0002t^5\right]_0^7 \approx 31.361$ gal

3. $\displaystyle\int_0^{100} B(x)\, dx = \int_0^{100} (21 - e^{0.03x})\, dx$

$\approx \left[21x - 33.333e^{0.03x}\right]_0^{100} \approx 1464$

4. $\displaystyle\int_0^2 \rho(x)\, dx = \int_0^2 (11 - 4x)\, dx = \left[11x - 2x^2\right]_0^2 = 14$ g

5. $\displaystyle\int_0^{24} E(t)\, dt = \int_0^{24} 300\left(2 - \cos\left(\dfrac{\pi t}{12}\right)\right) dt$

$= 300\left[2t - \dfrac{12}{\pi}\sin\left(\dfrac{\pi t}{12}\right)\right]_0^{24} = 14{,}400$

6.

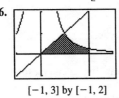

[−1, 3] by [−1, 2]

The curves intersect at $x = 1$. The area is

$\displaystyle\int_0^1 x\, dx + \int_1^2 \dfrac{1}{x^2}\, dx = \left[\dfrac{1}{2}x^2\right]_0^1 + \left[-\dfrac{1}{x}\right]_1^2$

$= \dfrac{1}{2} + \left(-\dfrac{1}{2} + 1\right) = 1.$

7.

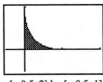

[−4, 4] by [−4, 4]

The curves intersect at $x = -2$ and $x = 1$. The area is

$\displaystyle\int_{-2}^{1} [3 - x^2 - (x+1)]\, dx = \int_{-2}^1 (-x^2 - x + 2)\, dx$

$= \left[-\dfrac{1}{3}x^3 - \dfrac{1}{2}x^2 + 2x\right]_{-2}^1$

$= \left(-\dfrac{1}{3} - \dfrac{1}{2} + 2\right) - \left(\dfrac{8}{3} - 2 - 4\right)$

$= \dfrac{9}{2}.$

8. $\sqrt{x} + \sqrt{y} = 1$ implies $y = (1 - \sqrt{x})^2 = 1 - 2\sqrt{x} + x.$

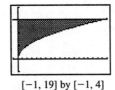

[−0.5, 2] by [−0.5, 1]

The area is $\displaystyle\int_0^1 (1 - 2\sqrt{x} + x)\, dx = \left[x - \dfrac{4}{3}x^{3/2} + \dfrac{1}{2}x^2\right]_0^1$

$= \dfrac{1}{6}.$

9. $x = 2y^2$ implies $y = \sqrt{\dfrac{x}{2}}.$

[−1, 19] by [−1, 4]

The curves intersect at $x = 18$. The area is

$\displaystyle\int_0^{18}\left(3 - \sqrt{\dfrac{x}{2}}\right) dx = \left[3x - \dfrac{4}{3}\left(\dfrac{x}{2}\right)^{3/2}\right]_0^{18} = 18,$

or $\displaystyle\int_0^3 2y^2\, dy = \left[\dfrac{2}{3}y^3\right]_0^3 = 18.$

10. $4x = y^2 - 4$ implies $x = \frac{1}{4}y^2 - 1$, and $4x = y + 16$ implies $x = \frac{1}{4}y + 4$.

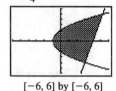

[−6, 6] by [−6, 6]

The curves intersect at $(3, -4)$ and $(5.25, 5)$. The area is

$$\int_{-4}^{5}\left[\left(\frac{1}{4}y + 4\right) - \left(\frac{1}{4}y^2 - 1\right)\right] dy$$
$$= \int_{-4}^{5}\left(-\frac{1}{4}y^2 + \frac{1}{4}y + 5\right) dy$$
$$= \left[-\frac{1}{12}y^3 + \frac{1}{8}y^2 + 5y\right]_{-4}^{5}$$
$$= \frac{425}{24} - \left(-\frac{38}{3}\right) = \frac{243}{8} = 30.375.$$

11.

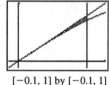

[−0.1, 1] by [−0.1, 1]

The area is $\int_{0}^{\pi/4} (x - \sin x)\, dx = \left[\frac{1}{2}x^2 + \cos x\right]_{0}^{\pi/4}$
$$= \frac{\pi^2}{32} + \frac{\sqrt{2}}{2} - 1$$
$$\approx 0.0155.$$

12.

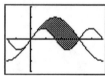

$\left[-\frac{\pi}{2}, \frac{3\pi}{2}\right]$ by [−3, 3]

The area is

$$\int_{0}^{\pi} (2\sin x - \sin 2x)\, dx = \left[-2\cos x + \frac{1}{2}\cos 2x\right]_{0}^{\pi} = 4.$$

13.

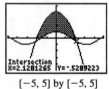

[−5, 5] by [−5, 5]

The curves intersect at $x \approx \pm 2.1281$. The area is

$$\int_{-2.1281}^{2.1281} (4 - x^2 - \cos x)\, dx,$$

which using NINT evaluates to ≈ 8.9023.

14.

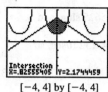

[−4, 4] by [−4, 4]

The curves intersect at $x \approx \pm 0.8256$. The area is

$$\int_{-0.8256}^{0.8256} (3 - |x| - \sec^2 x)\, dx,$$

which using NINT evaluates to ≈ 2.1043.

15. Solve $1 + \cos x = 2 - \cos x$ for the x-values at the two ends of the region: $x = 2\pi \pm \frac{\pi}{3}$, i.e., $\frac{5\pi}{3}$ or $\frac{7\pi}{3}$. Use the symmetry of the area:

$$2\int_{2\pi}^{7\pi/3} [(1 + \cos x) - (2 - \cos x)]\, dx$$
$$= 2\int_{2\pi}^{7\pi/3} (2\cos x - 1)\, dx$$
$$= 2\left[2\sin x - x\right]_{2\pi}^{7\pi/3}$$
$$= 2\sqrt{3} - \frac{2}{3}\pi \approx 1.370.$$

16. $\int_{\pi/3}^{5\pi/3} [(2 - \cos x) - (1 + \cos x)]\, dx$
$$= \int_{\pi/3}^{5\pi/3} (1 - 2\cos x)\, dx$$
$$= \left[x - 2\sin x\right]_{\pi/3}^{5\pi/3}$$
$$= 2\sqrt{3} + \frac{4}{3}\pi \approx 7.653$$

17. Solve $x^3 - x = \frac{x}{x^2 + 1}$ to find the intersection points at $x = 0$ and $x = \pm 2^{1/4}$. Then use the area's symmetry: the area is

$$2\int_{0}^{2^{1/4}}\left[\frac{x}{x^2 + 1} - (x^3 - x)\right] dx$$
$$= 2\left[\frac{1}{2}\ln(x^2 + 1) - \frac{1}{4}x^4 + \frac{1}{2}x^2\right]_{0}^{2^{1/4}}$$
$$= \ln(\sqrt{2} + 1) + \sqrt{2} - 1 \approx 1.2956.$$

18. Use the intersect function on a graphing calculator to determine that the curves intersect at $x \approx \pm 1.8933$. The area is

$$\int_{-1.8933}^{1.8933}\left(3^{1-x^2} - \frac{x^2 - 3}{10}\right) dx,$$

which using NINT evaluates to ≈ 5.7312.

19. Use the x- and y-axis symmetries of the area:

$$4\int_{0}^{\pi} x\sin x\, dx = 4\left[\sin x - x\cos x\right]_{0}^{\pi} = 4\pi.$$

20. A cross section has radius $r = 3x^4$ and area
$A(x) = \pi r^2 = 9\pi x^8$.
$V = \int_{-1}^{1} 9\pi x^8\, dx = \pi\left[x^9\right]_{-1}^{1} = 2\pi$.

21.

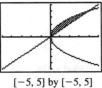

[−5, 5] by [−5, 5]

The graphs intersect at (0, 0) and (4, 4).

(a) Use cylindrical shells. A shell has radius y and height $y - \dfrac{y^2}{4}$. The total volume is

$$\int_0^4 2\pi(y)\left(y - \frac{y^2}{4}\right) dy = 2\pi \int_0^4 \left(y^2 - \frac{y^3}{4}\right) dy$$
$$= 2\pi\left[\frac{1}{3}y^3 - \frac{1}{16}y^4\right]_0^4$$
$$= \frac{32\pi}{3}.$$

(b) Use cylindrical shells. A shell has radius x and height $2\sqrt{x} - x$. The total volume is

$$\int_0^4 2\pi(x)(2\sqrt{x} - x)\, dx = 2\pi \int_0^4 (2x^{3/2} - x^2)\, dx$$
$$= 2\pi\left[\frac{4}{5}x^{5/2} - \frac{1}{3}x^3\right]_0^4$$
$$= \frac{128\pi}{15}.$$

(c) Use cylindrical shells. A shell has radius $4 - x$ and height $2\sqrt{x} - x$. The total volume is

$$\int_0^4 2\pi(4 - x)(2\sqrt{x} - x)\, dx$$
$$= 2\pi \int_0^4 (8\sqrt{x} - 4x - 2x^{3/2} + x^2)\, dx$$
$$= 2\pi\left[\frac{16}{3}x^{3/2} - 2x^2 - \frac{4}{5}x^{5/2} + \frac{1}{3}x^3\right]_0^4 = \frac{64\pi}{5}.$$

(d) Use cylindrical shells. A shell has radius $4 - y$ and height $y - \dfrac{y^2}{4}$. The total volume is

$$\int_0^4 2\pi(4 - y)\left(y - \frac{y^2}{4}\right) dy$$
$$= 2\pi \int_0^4 \left(4y - 2y^2 + \frac{y^3}{4}\right) dy$$
$$= 2\pi\left[2y^2 - \frac{2}{3}y^3 + \frac{1}{16}y^4\right]_0^4 = \frac{32\pi}{3}.$$

22. (a) Use disks. The volume is

$$\pi \int_0^2 (\sqrt{2y})^2\, dy = \pi \int_0^2 2y\, dy = \pi y^2\Big]_0^2 = 4\pi.$$

(b) $\pi \int_0^k 2y\, dy = \pi y^2\Big]_0^k = \pi k^2$

(c) Since $V = \pi k^2$, $\dfrac{dV}{dt} = 2\pi k\dfrac{dk}{dt}$.

When $k = 1$, $\dfrac{dk}{dt} = \dfrac{1}{2\pi k}\dfrac{dV}{dt} = \left(\dfrac{1}{2\pi}\right)(2) = \dfrac{1}{\pi}$,

so the depth is increasing at the rate of $\dfrac{1}{\pi}$ unit per second.

23. The football is a solid of revolution about the x-axis. A cross section has radius $\sqrt{12\left(1 - \dfrac{4x^2}{121}\right)}$ and area $\pi r^2 = 12\pi\left(1 - \dfrac{4x^2}{121}\right)$. The volume is, given the symmetry,

$$2\int_0^{11/2} 12\pi\left(1 - \frac{4x^2}{121}\right) dx = 24\pi \int_0^{11/2} \left(1 - \frac{4x^2}{121}\right) dx$$
$$= 24\pi\left[x - \left(\frac{2}{11}\right)^2\left(\frac{1}{3}\right)x^3\right]_0^{11/2}$$
$$= 24\pi\left[\frac{11}{2} - \frac{11}{6}\right]$$
$$= 88\pi \approx 276\text{ in}^3.$$

24. The width of a cross section is $2\sin x$, and the area is $\dfrac{1}{2}\pi r^2 = \dfrac{1}{2}\pi \sin^2 x$. The volume is

$$\int_0^\pi \frac{1}{2}\pi \sin^2 x\, dx = \frac{\pi}{2}\left[\frac{1}{2}x - \frac{1}{4}\sin 2x\right]_0^\pi = \frac{\pi^2}{4}.$$

25.

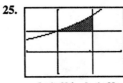

[−1, 2] by [−1, 2]

Use washer cross sections. A washer has inner radius $r = 1$, outer radius $R = e^{x/2}$, and area $\pi(R^2 - r^2) = \pi(e^x - 1)$.

The volume is

$$\int_0^{\ln 3} \pi(e^x - 1)\, dx = \pi\left[e^x - x\right]_0^{\ln 3}$$
$$= \pi(3 - \ln 3 - 1)$$
$$= \pi(2 - \ln 3).$$

26. Use cylindrical shells. Taking the hole to be vertical, a shell has radius x and height $2\sqrt{2^2 - x^2}$. The volume of the piece cut out is

$$\int_0^{\sqrt{3}} 2\pi(x)(2\sqrt{2^2 - x^2})\, dx = 2\pi \int_0^{\sqrt{3}} 2x\sqrt{4 - x^2}\, dx$$
$$= 2\pi \left[-\frac{2}{3}(4 - x^2)^{3/2} \right]_0^{\sqrt{3}}$$
$$= -\frac{4}{3}\pi(1 - 8)$$
$$= \frac{28\pi}{3} \approx 29.3215 \text{ ft}^3.$$

27. The curve crosses the x-axis at $x = \pm 3$. $y' = -2x$, so the length is $\int_{-3}^{3} \sqrt{1 + (-2x)^2}\, dx = \int_{-3}^{3} \sqrt{1 + 4x^2}\, dx$, which using NINT evaluates to ≈ 19.4942.

28.

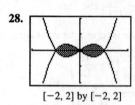

$[-2, 2]$ by $[-2, 2]$

The curves intersect at $x = 0$ and $x = \pm 1$. Use the graphs' x- and y-axis symmetry:

$\frac{d}{dx}(x^3 - x) = 3x^2 - 1$, and the total perimeter is

$4\int_0^1 \sqrt{1 + (3x^2 - 1)^2}\, dx$, which using NINT evaluates to ≈ 5.2454.

29.

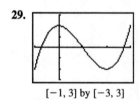

$[-1, 3]$ by $[-3, 3]$

$y' = 3x^2 - 6x$ equals zero when $x = 0$ or 2. The maximum is at $x = 0$, the minimum at $x = 2$. The distance between them along the curve is $\int_0^2 \sqrt{1 + (3x^2 - 6x)^2}\, dx$, which using NINT evaluates to ≈ 4.5920. The time taken is about $\frac{4.5920}{2} = 2.296$ sec.

30. If (b) were true, then the curve $y = k \sin x$ would have to get vanishingly short as k approached zero. Since in fact the curve's length approaches 2π instead, (b) is false and (a) is true.

31. $F'(x) = \sqrt{x^4 - 1}$, so

$$\int_2^5 \sqrt{1 + (F'(x))^2}\, dx = \int_2^5 \sqrt{x^4}\, dx$$
$$= \int_2^5 x^2\, dx$$
$$= \left[\frac{1}{3}x^3 \right]_2^5 = 39.$$

32. (a) $(100 \text{ N})(40 \text{ m}) = 4000 \text{ J}$

(b) When the end has traveled a distance y, the weight of the remaining portion is $(40 - y)(0.8) = 32 - 0.8y$. The total work to lift the rope is

$$\int_0^{40} (32 - 0.8y)\, dy = \left[32y - 0.4y^2 \right]_0^{40} = 640 \text{ J}.$$

(c) $4000 + 640 = 4640 \text{ J}$

33. The weight of the water at elevation x (starting from $x = 0$) is $(800)(8)\left(\frac{4750 - x/2}{4750}\right) = \frac{128}{95}\left(4750 - \frac{1}{2}x\right)$. The total work is $\int_0^{4750} \frac{128}{95}\left(4750 - \frac{1}{2}x\right) dx = \frac{128}{95}\left[4750x - \frac{1}{4}x^2 \right]_0^{4750}$

$= 22{,}800{,}000 \text{ ft-lb}.$

34. $F = ks$, so $k = \frac{F}{s} = \frac{80}{0.3} = \frac{800}{3}$ N/m. Then

Work $= \int_0^{0.3} \frac{800}{3}x\, dx = \left[\frac{800}{6}x^2 \right]_0^{0.3} = 12 \text{ J}.$

To stretch the additional meter,

Work $= \int_{0.3}^{1.3} \frac{800}{3}x\, dx = \left[\frac{800}{6}x^2 \right]_{0.3}^{1.3} \approx 213.3 \text{ J}.$

35. The work is positive going uphill, since the force pushes in the direction of travel. The work is negative going downhill.

36. The radius of a horizontal cross section is $\sqrt{8^2 - y^2}$, where y is distance below the rim. The area is $\pi(64 - y^2)$, the weight is $0.04\pi(64 - y^2)\, \Delta y$, and the work to lift it over the rim is $0.04\pi(64 - y^2)(y)\, \Delta y$. The total work is

$$\int_2^8 0.04\pi y(64 - y^2)\, dy = 0.04\pi \int_2^8 (64y - y^3)\, dy$$
$$= 0.04\pi \left[32y^2 - \frac{1}{4}y^4 \right]_2^8$$
$$= 36\pi \approx 113.097 \text{ in.-lb}.$$

37. The width of a thin horizontal strip is $2(2y) = 4y$, and the force against it is $80(2 - y)4y\,\Delta y$. The total force is

$$\int_0^2 320y(2 - y)\,dy = 320\int_0^2 (-y^2 + 2y)\,dy$$
$$= 320\left[-\frac{1}{3}y^3 + y^2\right]_0^2$$
$$= \frac{1280}{3} \approx 426.67 \text{ lb.}$$

38. 5.75 in. $= \frac{23}{48}$ ft, 3.5 in. $= \frac{7}{24}$ ft, and 10 in. $= \frac{5}{6}$ ft.

For the base,

Force $= 57\left(\frac{23}{48} \times \frac{7}{24} \times \frac{5}{6}\right) \approx 6.6385$ lb.

For the front and back,

Force $= \int_0^{5/6} 57\left(\frac{7}{24}\right)y\,dy = \frac{399}{24}\left[\frac{1}{2}y^2\right]_0^{5/6} \approx 5.7726$ lb.

For the sides,

Force $= \int_0^{5/6} 57\left(\frac{23}{48}\right)y\,dy = \frac{1311}{48}\left[\frac{1}{2}y^2\right]_0^{5/6} \approx 9.4835$ lb.

39. A square's height is $y = (\sqrt{6} - \sqrt{x})^2$, and its area is $y^2 = (\sqrt{6} - \sqrt{x})^4$. The volume is $\int_0^6 (\sqrt{6} - \sqrt{x})^4\,dx$, which using NINT evaluates to exactly 14.4.

40. Choose 50 cm as a conveniently large upper limit.
$\int_{-20}^{50} \frac{1}{3.4\sqrt{2\pi}} e^{-(x-17.2)^2/(2 \cdot 3.4^2)}\,dx$, evaluates, using NINT to ≈ 0.2051 (20.5%).

41. Answers will vary. Find μ, then use the fact that 68% of the class is within σ of μ to find σ, and then choose a conveniently large number b and calculate

$\int_{10}^{b} \frac{1}{\sigma\sqrt{2\pi}} e^{-(x-\mu)^2/(2\sigma^2)}\,dx$.

42. Use $f(x) = \frac{1}{\sqrt{2\pi}} e^{-x^2/2}$.

(a) $\int_{-1}^{1} f(x)\,dx$ evaluates, using NINT, to ≈ 0.6827 (68.27%).

(b) $\int_{-2}^{2} f(x)\,dx \approx 0.9545$ (95.45%).

(c) $\int_{-3}^{3} f(x)\,dx \approx 0.9973$ (99.73%).

43. Because $f(x) \geq 0$ and $\int_{-\infty}^{\infty} f(x)\,dx = 1$.

44.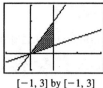

[−1, 3] by [−1, 3]

A shell has radius x and height $2x - \frac{x}{2} = \frac{3}{2}x$. The total volume is $\int_0^1 2\pi(x)\left(\frac{3}{2}x\right)\,dx = \pi\left[x^3\right]_0^1 = \pi$.

45.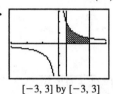

[−3, 3] by [−3, 3]

A shell has radius x and height $\frac{1}{x}$. The total volume is

$\int_{1/2}^{2} 2\pi(x)\left(\frac{1}{x}\right)\,dx = \int_{1/2}^{2} 2\pi\,dx = \left[2\pi x\right]_{1/2}^{2} = 3\pi$.

46.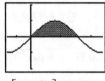

$\left[-\frac{\pi}{2}, \frac{3\pi}{2}\right]$ by [−2, 2]

A shell has radius x and height $\sin x$. The total volume is

$\int_0^{\pi} 2\pi(x)(\sin x)\,dx = 2\pi\left[\sin x - x\cos x\right]_0^{\pi} = 2\pi^2$.

47.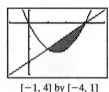

[−1, 4] by [−4, 1]

The curves intersect at $x = 1$ and $x = 3$. A shell has radius x and height $x - 3 - (x^2 - 3x) = -x^2 + 4x - 3$. The total volume is

$\int_1^3 2\pi(x)(-x^2 + 4x - 3)\,dx = 2\pi\int_1^3 (-x^3 + 4x^2 - 3x)\,dx$
$= 2\pi\left[-\frac{1}{4}x^4 + \frac{4}{3}x^3 - \frac{3}{2}x^2\right]_1^3$
$= \frac{16\pi}{3}$.

48. Use the intersect function on a graphing calculator to determine that the curves intersect at $x = \pm 1.8933$. A shell has radius x and height $3^{1-x^2} - \frac{x^2 - 3}{10}$. The volume, which is calculated using the *right half* of the area, is

$\int_0^{1.8933} 2\pi(x)\left(3^{1-x^2} - \frac{x^2 - 3}{10}\right)\,dx$, which using NINT evaluates to ≈ 9.7717.

49. (a) $y = -\frac{5}{4}(x+2)(x-2) = 5 - \frac{5}{4}x^2$

(b) Revolve about the line $x = 4$, using cylindrical shells.

A shell has radius $4 - x$ and height $5 - \frac{5}{4}x^2$. The total volume is

$$\int_{-2}^{2} 2\pi(4-x)\left(5 - \frac{5}{4}x^2\right)dx$$
$$= 10\pi \int_{-2}^{2} \left(\frac{1}{4}x^3 - x^2 - x + 4\right)dx$$
$$= 10\pi \left[\frac{1}{16}x^4 - \frac{1}{3}x^3 - \frac{1}{2}x^2 + 4x\right]_{-2}^{2}$$
$$= \frac{320}{3}\pi \approx 335.1032 \text{ in}^3.$$

50. Since $\frac{dL}{dx} = \frac{1}{x} + f'(x)$ must equal $\sqrt{1 + (f'(x))^2}$,

$1 + (f'(x))^2 = \frac{1}{x^2} + \frac{2}{x}f'(x) + (f'(x))^2$, and

$f'(x) = \frac{1}{2}x - \frac{1}{2x}$. Then $f(x) = \frac{1}{4}x^2 - \frac{1}{2}\ln x + C$, and the requirement to pass through $(1, 1)$ means that $C = \frac{3}{4}$. The function is $f(x) = \frac{1}{4}x^2 - \frac{1}{2}\ln x + \frac{3}{4} = \frac{x^2 - 2\ln x + 3}{4}$.

51. $y' = \sec^2 x$, so the area is $\int_0^{\pi/4} 2\pi(\tan x)\sqrt{1 + (\sec^2 x)^2}\, dx$, which using NINT evaluates to ≈ 3.84.

52. $x = \frac{1}{y}$ and $x' = -\frac{1}{y^2}$, so the area is

$$\int_1^2 2\pi\left(\frac{1}{y}\right)\sqrt{1 + \left(-\frac{1}{y^2}\right)^2}\, dy,$$

which using NINT evaluates to ≈ 5.02.

Chapter 8
L'Hôpital's Rule, Improper Integrals, and Partial Fractions

■ Section 8.1 L'Hôpital's Rule (pp. 417–425)

Exploration 1 Exploring L'Hôpital's Rule Graphically

1. $\lim\limits_{x\to 0}\frac{\sin x}{x} = \lim\limits_{x\to 0}\frac{\cos x}{1} = 1$

2. The two graphs suggest that $\lim\limits_{x\to 0}\frac{y_1}{y_2} = \lim\limits_{x\to 0}\frac{y_1'}{y_2'}$.

3. $y_5 = \frac{x\cos x - \sin x}{x^2}$. The graphs of y_3 and y_5 clearly show that l'Hôpital's Rule does not say that $\lim\limits_{x\to 0}\frac{y_1}{y_2}$ is equal to $\lim\limits_{x\to 0}\left(\frac{y_1}{y_2}\right)'$.

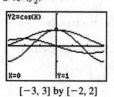

[−3, 3] by [−2, 2]

Quick Review 8.1

1.

x	$\left(1 + \frac{0.1}{x}\right)^x$
1	1.1000
10	1.1046
100	1.1051
1000	1.1052
10,000	1.1052
1,000,000	1.1052

As $x\to\infty$, $\left(1 + \frac{0.1}{x}\right)^x$ approaches 1.1052.

2.

x	$x^{1/(\ln x)}$
0.1	2.7183
0.01	2.7183
0.001	2.7183
0.0001	2.7183
0.00001	2.7183

As $x\to 0^+$, $x^{1/(\ln x)}$ approaches 2.7183.

3.

x	$\left(1 - \frac{1}{x}\right)^x$
−1	0.5
−0.1	0.78679
−0.01	0.95490
−0.001	0.99312
−0.0001	0.99908
−0.00001	0.99988
−0.000001	0.99999

As $x\to 0^-$, $\left(1 - \frac{1}{x}\right)^x$ approaches 1.

4.

x	$\left(1 + \frac{1}{x}\right)^x$
−1.1	13.981
−1.01	105.77
−1.001	1007.9
−1.0001	10010

As $x\to -1^-$, $\left(1 + \frac{1}{x}\right)^x$ goes to ∞.

5.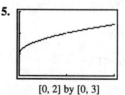

[0, 2] by [0, 3]

As $t\to 1$, $\dfrac{t-1}{\sqrt{t}-1}$ approaches 2.

6.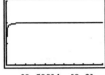

[0, 500] by [0, 3]

As $x \to \infty$, $\dfrac{\sqrt{4x^2+1}}{x+1}$ approaches 2.

7.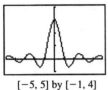

[−5, 5] by [−1, 4]

As $x \to 0$, $\dfrac{\sin 3x}{x}$ approaches 3.

8.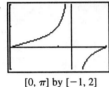

[0, π] by [−1, 2]

As $\theta \to \dfrac{\pi}{2}$, $\dfrac{\tan \theta}{2+\tan \theta}$ approaches 1.

9. $y = \dfrac{1}{h}\sin h$

10. $y = (1+h)^{1/h}$

Section 8.1 Exercises

1.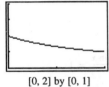

[0, 2] by [0, 1]

From the graph, the limit appears to be $\dfrac{1}{4}$.

$\displaystyle\lim_{x\to 2}\dfrac{x-2}{x^2-4} = \lim_{x\to 2}\dfrac{1}{2x} = \dfrac{1}{4}$

3.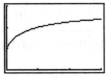

[0, 3] by [0, 3]

From the graph, the limit appears to be 1. The limit leads to the indeterminate form ∞^0.

$\ln\left(1+\dfrac{1}{x}\right)^x = x\ln\left(1+\dfrac{1}{x}\right) = \dfrac{\ln\left(1+\dfrac{1}{x}\right)}{\dfrac{1}{x}}$

$\displaystyle\lim_{x\to 0^+}\dfrac{\ln\left(1+\dfrac{1}{x}\right)}{\dfrac{1}{x}} = \lim_{x\to 0^+}\dfrac{\dfrac{1}{1+1/x}\left(-\dfrac{1}{x^2}\right)}{-\dfrac{1}{x^2}}$

$= \displaystyle\lim_{x\to 0^+}\dfrac{1}{1+\dfrac{1}{x}}$

$= \displaystyle\lim_{x\to 0^+}\dfrac{x}{x+1} = 0$

Therefore,

$\displaystyle\lim_{x\to 0^+}\left(1+\dfrac{1}{x}\right)^x = \lim_{x\to 0^+} f(x) = \lim_{x\to 0^+} e^{\ln f(x)} = e^0 = 1.$

5. $\displaystyle\lim_{x\to 1}\dfrac{x^3-1}{4x^3-x-3} = \lim_{x\to 1}\dfrac{3x^2}{12x^2-1} = \dfrac{3}{11}$

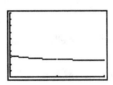

[0, 2] by [0, 1]

The graph supports the answer.

7. The limit leads to the indeterminate form 1^∞.

Let $\ln f(x) = \ln(e^x+x)^{1/x} = \dfrac{\ln(e^x+x)}{x}$.

$\displaystyle\lim_{x\to 0^+}\dfrac{\ln(e^x+x)}{x} = \lim_{x\to 0^+}\dfrac{\dfrac{e^x+1}{e^x+x}}{1} = \lim_{x\to 0^+}\dfrac{e^x+1}{e^x+x} = \dfrac{2}{1} = 2$

$\displaystyle\lim_{x\to 0^+}(e^x+x)^{1/x} = \lim_{x\to 0^+} f(x) = \lim_{x\to 0^+} e^{\ln f(x)} = e^2$

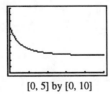

[0, 5] by [0, 10]

The graph supports the answer.

9. (a)

x	10	10^2	10^3	10^4	10^5
$f(x)$	1.1513	0.2303	0.0345	0.00461	0.00058

Estimate the limit to be 0.

(b) $\lim\limits_{x\to\infty} \dfrac{\ln x^5}{x} = \lim\limits_{x\to\infty} \dfrac{5\ln x}{x} = \lim\limits_{x\to\infty} \dfrac{5/x}{1} = \dfrac{0}{1} = 0$

11. Let $f(\theta) = \dfrac{\sin 3\theta}{\sin 4\theta}$.

θ	$\pm 10^0$	$\pm 10^{-1}$	$\pm 10^{-2}$	$\pm 10^{-3}$	$\pm 10^{-4}$
$f(\theta)$	-0.1865	0.7589	0.7501	0.7500	0.7500

Estimate the limit to be $\dfrac{3}{4}$.

$\lim\limits_{\theta\to 0} \dfrac{\sin 3\theta}{\sin 4\theta} = \lim\limits_{\theta\to 0} \dfrac{3\cos 3\theta}{4\cos 4\theta} = \dfrac{3}{4}$

13. Let $f(x) = (1+x)^{1/x}$.

x	10	10^2	10^3	10^4	10^5
$f(x)$	1.2710	1.0472	1.0069	1.0009	1.0001

Estimate the limit to be 1.

$\ln f(x) = \dfrac{\ln(1+x)}{x}$

$\lim\limits_{x\to\infty} \dfrac{\ln(1+x)}{x} = \lim\limits_{x\to\infty} \dfrac{\frac{1}{1+x}}{1} = \dfrac{0}{1} = 0$

$\lim\limits_{x\to\infty} (1+x)^{1/x} = \lim\limits_{x\to\infty} f(x) = \lim\limits_{x\to\infty} e^{\ln f(x)} = e^0 = 1$

15. $\lim\limits_{\theta\to 0} \dfrac{\sin \theta^2}{\theta} = \lim\limits_{\theta\to 0} \dfrac{2\theta \cos\theta^2}{1} = (2)(0)\cos(0)^2 = 0$

17. $\lim\limits_{t\to 0} \dfrac{\cos t - 1}{e^t - t - 1} = \lim\limits_{t\to 0} \dfrac{-\sin t}{e^t - 1} = \lim\limits_{t\to 0} \dfrac{-\cos t}{e^t} = -1$

19. $\lim\limits_{x\to\infty} \dfrac{\ln(x+1)}{\log_2 x} = \lim\limits_{x\to\infty} \dfrac{\frac{1}{x+1}}{\frac{1}{x\ln 2}}$

$= \lim\limits_{x\to\infty} \dfrac{x\ln 2}{x+1}$

$= \lim\limits_{x\to\infty} \ln 2$

$= \ln 2$

21. $\lim\limits_{y\to 0^+} \dfrac{\ln(y^2+2y)}{\ln y} = \lim\limits_{y\to 0^+} \dfrac{\frac{2y+2}{y^2+2y}}{\frac{1}{y}}$

$= \lim\limits_{y\to 0^+} \dfrac{y(2y+2)}{y^2+2y}$

$= \lim\limits_{y\to 0^+} \dfrac{2y^2+2y}{y^2+2y}$

$= \lim\limits_{y\to 0^+} \dfrac{4y+2}{2y+2}$

$= \dfrac{4(0)+2}{2(0)+2} = \dfrac{2}{2} = 1$

23. $\lim\limits_{x\to 0^+} x\ln x = \lim\limits_{x\to 0^+} \dfrac{\ln x}{\frac{1}{x}}$

$= \lim\limits_{x\to 0^+} \dfrac{\frac{1}{x}}{-\frac{1}{x^2}}$

$= \lim\limits_{x\to 0^+} \dfrac{-x^2}{x}$

$= \lim\limits_{x\to 0^+} -x = 0$

25. $\lim\limits_{x\to 0^+} (\csc x - \cot x + \cos x)$

$= \lim\limits_{x\to 0^+} \left(\dfrac{1}{\sin x} - \dfrac{\cos x}{\sin x} + \cos x\right)$

$= \lim\limits_{x\to 0^+} \dfrac{1 - \cos x + \cos x \sin x}{\sin x}$

$= \lim\limits_{x\to 0^+} \dfrac{\sin x + \cos x \cos x - \sin x \sin x}{\cos x} = 1$

27. $\lim\limits_{x\to 0^+} (\ln x - \ln \sin x) = \lim\limits_{x\to 0^+} \ln \dfrac{x}{\sin x}$

Let $f(x) = \dfrac{x}{\sin x}$.

$\lim\limits_{x\to 0^+} \dfrac{x}{\sin x} = \lim\limits_{x\to 0^+} \dfrac{1}{\cos x} = 1$

Therefore,

$\lim\limits_{x\to 0^+} (\ln x - \ln \sin x) = \lim\limits_{x\to 0^+} \ln f(x) = \ln 1 = 0$

29. The limit leads to the indeterminate form 1^∞.

Let $f(x) = (e^x + x)^{1/x}$.

$\ln(e^x+x)^{1/x} = \dfrac{\ln(e^x+x)}{x}$

$\lim\limits_{x\to 0} \dfrac{\ln(e^x+x)}{x} = \lim\limits_{x\to 0} \dfrac{\frac{e^x+1}{e^x+x}}{1} = 2$

$\lim\limits_{x\to 0} (e^x+x)^{1/x} = \lim\limits_{x\to 0} e^{\ln f(x)} = e^2$

31. $\lim\limits_{x\to\pm\infty}\dfrac{3x-5}{2x^2-x+2}=\lim\limits_{x\to\pm\infty}\dfrac{3}{4x-1}=0$

33. The limit leads to the indeterminate form ∞^0.

Let $f(x) = (\ln x)^{1/x}$.

$\ln(\ln x)^{1/x} = \dfrac{\ln(\ln x)}{x}$

$\lim\limits_{x\to\infty}\dfrac{\ln(\ln x)}{x} = \lim\limits_{x\to\infty}\dfrac{\frac{1/x}{\ln x}}{1} = \lim\limits_{x\to\infty}\dfrac{1}{x\ln x} = 0$

$\lim\limits_{x\to\infty}(\ln x)^{1/x} = \lim\limits_{x\to\infty}e^{\ln f(x)} = e^0 = 1$

35. The limit leads to the indeterminate form 0^0.

Let $f(x) = (x^2-2x+1)^{x-1}$

$\ln(x^2-2x+1)^{x-1} = (x-1)\ln(x^2-2x+1)$

$= \dfrac{\ln(x^2-2x+1)}{\frac{1}{x-1}}$

$\lim\limits_{x\to 1}\dfrac{\ln(x^2-2x+1)}{\frac{1}{x-1}} = \lim\limits_{x\to 1}\dfrac{\frac{2x-2}{x^2-2x+1}}{-\frac{1}{(x-1)^2}}$

$= \lim\limits_{x\to 1}\dfrac{\frac{2(x-1)}{(x-1)^2}}{-\frac{1}{(x-1)^2}}$

$= \lim\limits_{x\to 1}-2(x-1) = 0$

$\lim\limits_{x\to 1}(x^2-2x+1)^{x-1} = \lim\limits_{x\to 1}e^{\ln f(x)} = e^0 = 1$

37. The limit leads to the indeterminate form 1^∞.

Let $f(x) = (1+x)^{1/x}$.

$\ln(1+x)^{1/x} = \dfrac{\ln(1+x)}{x}$

$\lim\limits_{x\to 0^+}\dfrac{\ln(1+x)}{x} = \lim\limits_{x\to 0^+}\dfrac{\frac{1}{1+x}}{1} = 1$

$\lim\limits_{x\to 0^+}(1+x)^{1/x} = \lim\limits_{x\to 0^+}e^{\ln f(x)} = e^1 = e$

39. The limit leads to the indeterminate form 0^0.

Let $f(x) = (\sin x)^x$.

$\ln(\sin x)^x = x\ln(\sin x) = \dfrac{\ln(\sin x)}{\frac{1}{x}}$

$\lim\limits_{x\to 0^+}\dfrac{\ln(\sin x)}{\frac{1}{x}} = \lim\limits_{x\to 0^+}\dfrac{\frac{\cos x}{\sin x}}{-\frac{1}{x^2}}$

$= \lim\limits_{x\to 0^+}\dfrac{-x^2\cos x}{\sin x}$

$= \lim\limits_{x\to 0^+}\dfrac{x^2\sin x - 2x\cos x}{\cos x} = 0$

$\lim\limits_{x\to 0^+}(\sin x)^x = \lim\limits_{x\to 0^+}e^{\ln f(x)} = e^0 = 1$

41. The limit leads to the indeterminate form $1^{-\infty}$.

Let $f(x) = x^{1/(1-x)}$.

$\ln x^{1/(1-x)} = \dfrac{\ln x}{1-x}$

$\lim\limits_{x\to 1^+}\dfrac{\ln x}{1-x} = \lim\limits_{x\to 1^+}\dfrac{\frac{1}{x}}{-1} = -1$

$\lim\limits_{x\to 1^+}x^{1/(1-x)} = \lim\limits_{x\to 1^+}e^{\ln f(x)} = e^{-1} = \dfrac{1}{e}$

43. (a) L'Hôpital's Rule does not help because applying L'Hôpital's Rule to this quotient essentially "inverts" the problem by interchanging the numerator and denominator (see below). It is still essentially the same problem and one is no closer to a solution. Applying L'Hôpital's Rule a second time returns to the original problem.

$\lim\limits_{x\to\infty}\dfrac{\sqrt{9x+1}}{\sqrt{x+1}} = \lim\limits_{x\to\infty}\dfrac{(9/2)(9x+1)^{-1/2}}{(1/2)(x+1)^{-1/2}} = \lim\limits_{x\to\infty}\dfrac{9\sqrt{x+1}}{\sqrt{9x+1}}$

(b)

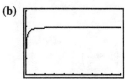

[0, 100] by [0, 4]

The limit appears to be 3.

(c) $\lim\limits_{x\to\infty}\dfrac{\sqrt{9x+1}}{\sqrt{x+1}} = \lim\limits_{x\to\infty}\dfrac{\sqrt{9+\frac{1}{x}}}{\sqrt{1+\frac{1}{x}}} = \dfrac{\sqrt{9}}{\sqrt{1}} = 3$

218 Section 8.1

45. Possible answers:

(a) $f(x) = 7(x - 3)$; $g(x) = x - 3$

$$\lim_{x \to 3} \frac{f(x)}{g(x)} = \lim_{x \to 3} \frac{7(x - 3)}{x - 3} = \lim_{x \to 3} \frac{7}{1} = 7$$

(b) $f(x) = (x - 3)^2$; $g(x) = x - 3$

$$\lim_{x \to 3} \frac{f(x)}{g(x)} = \lim_{x \to 3} \frac{(x - 3)^2}{x - 3} = \lim_{x \to 3} \frac{2(x - 3)}{1} = 0$$

(c) $f(x) = x - 3$; $g(x) = (x - 3)^3$

$$\lim_{x \to 3} \frac{f(x)}{g(x)} = \lim_{x \to 3} \frac{x - 3}{(x - 3)^3} = \lim_{x \to 3} \frac{1}{3(x - 3)^2} = \infty$$

47. Find c such that $\lim_{x \to 0} f(x) = c$.

$$\lim_{x \to 0} f(x) = \lim_{x \to 0} \frac{9x - 3 \sin 3x}{5x^3}$$
$$= \lim_{x \to 0} \frac{9 - 9 \cos 3x}{15x^2}$$
$$= \lim_{x \to 0} \frac{27 \sin 3x}{30x}$$
$$= \lim_{x \to 0} \frac{81 \cos 3x}{30} = \frac{81}{30} = \frac{27}{10}$$

Thus, $c = \frac{27}{10}$. This works since $\lim_{x \to 0} f(x) = c = f(0)$, so f is continuous.

49. (a) The limit leads to the indeterminate form 1^∞.

Let $f(k) = \left(1 + \frac{r}{k}\right)^{kt}$.

$$\ln f(k) = kt \ln\left(1 + \frac{r}{k}\right) = \frac{t \ln\left(1 + \frac{r}{k}\right)}{\frac{1}{k}}$$

$$\lim_{k \to \infty} \frac{t \ln\left(1 + \frac{r}{k}\right)}{\frac{1}{k}} = \lim_{k \to \infty} \frac{t\left(-\frac{r}{k^2}\right)\left(1 + \frac{r}{k}\right)^{-1}}{-\frac{1}{k^2}}$$

$$= \lim_{k \to \infty} \frac{rt}{1 + \frac{r}{k}} = \frac{rt}{1} = rt$$

$$\lim_{k \to \infty} A_0\left(1 + \frac{r}{k}\right)^{kt} = A_0 \lim_{k \to \infty}\left(1 + \frac{r}{k}\right)^{kt}$$
$$= A_0 \lim_{k \to \infty} e^{\ln f(k)}$$
$$= A_0 e^{rt}$$

(b) Part (a) shows that as the number of compoundings per year increases toward infinity, the limit of interest compounded k times per year is interest compounded continuously.

51. (a) $A(t) = \int_0^t e^{-x} dx = \left[-e^{-x}\right]_0^t = -e^{-t} + 1$

$$\lim_{t \to \infty} A(t) = \lim_{t \to \infty} (-e^{-t} + 1) = \lim_{t \to \infty}\left(-\frac{1}{e^t} + 1\right) = 1$$

(b) $V(t) = \pi \int_0^t (e^{-x})^2 dx$
$$= \pi \int_0^t e^{-2x} dx$$
$$= \pi \left[-\frac{1}{2}e^{-2x}\right]_0^t$$
$$= \pi\left(-\frac{1}{2}e^{-2t} + \frac{1}{2}\right)$$
$$= \frac{\pi}{2}(-e^{-2t} + 1)$$

$$\lim_{t \to \infty} \frac{V(t)}{A(t)} = \lim_{t \to \infty} \frac{\frac{\pi}{2}(-e^{-2t} + 1)}{-e^{-t} + 1} = \frac{\frac{\pi}{2}(1)}{1} = \frac{\pi}{2}$$

(c) $\lim_{t \to 0^+} \frac{V(t)}{A(t)} = \lim_{t \to 0^+} \frac{\frac{\pi}{2}(-e^{-2t} + 1)}{-e^{-t} + 1}$

$$= \lim_{t \to 0^+} \frac{\frac{\pi}{2}(2e^{-2t})}{e^{-t}}$$

$$= \frac{\frac{\pi}{2}(2)}{1} = \pi$$

53. (a) $f(x) = e^{x \ln(1 + 1/x)}$

$1 + \frac{1}{x} > 0$ when $x < -1$ or $x > 0$

Domain: $(-\infty, -1) \cup (0, \infty)$

(b) The form is 0^{-1}, so $\lim_{x \to -1^-} f(x) = \infty$

(c) $\lim_{x \to -\infty} x \ln\left(1 + \frac{1}{x}\right) = \lim_{x \to -\infty} \frac{\ln\left(1 + \frac{1}{x}\right)}{\frac{1}{x}}$

$$= \lim_{x \to -\infty} \frac{\left(-\frac{1}{x^2}\right)\left(1 + \frac{1}{x}\right)^{-1}}{-\frac{1}{x^2}}$$

$$= \lim_{x \to -\infty} \frac{1}{1 + \frac{1}{x}} = 1$$

$$\lim_{x \to -\infty} f(x) = \lim_{x \to -\infty} e^{x \ln(1 + 1/x)} = e$$

55. (a) $f'(x) = 3x^2$, $g'(x) = 2x - 1$

$$f(1) - f(-1) = 2, g(1) - g(-1) = -2$$

$$\frac{3c^2}{2c-1} = \frac{2}{-2}$$

$$3c^2 = -2c + 1$$

$$3c^2 + 2c - 1 = 0$$

$$(3c - 1)(c + 1) = 0$$

$$c = \frac{1}{3} \text{ or } c = -1$$

The value of c that satisfies the property is $c = \frac{1}{3}$.

(b) $f'(x) = -\sin x$, $g'(x) = \cos x$

$$f\left(\frac{\pi}{2}\right) - f(0) = -1, g\left(\frac{\pi}{2}\right) - g(0) = 1$$

$$\frac{-\sin c}{\cos c} = \frac{-1}{1}$$

$$\tan c = 1$$

$$c = \tan^{-1} 1 = \frac{\pi}{4} \text{ on } \left(0, \frac{\pi}{2}\right)$$

■ Section 8.2 Relative Rates of Growth
(pp. 425–433)

Exploration 1 Comparing Rates of Growth as $x \to \infty$

1. $\lim\limits_{x \to \infty} \frac{a^x}{x^2} = \lim\limits_{x \to \infty} \frac{(\ln a)(a^x)}{2x} = \lim\limits_{x \to \infty} \frac{(\ln a)^2 a^x}{2} = \infty$, so a^x grows faster than x^2 as $x \to \infty$.

2. $\lim\limits_{x \to \infty} \frac{3^x}{2^x} = \lim\limits_{x \to \infty} 1.5^x = \infty$

3. $\lim\limits_{x \to \infty} \frac{a^x}{b^x} = \lim\limits_{x \to \infty} \left(\frac{a}{b}\right)^x = \infty$ because $\frac{a}{b} > 1$.

Quick Review 8.2

1. $\lim\limits_{x \to \infty} \frac{\ln x}{e^x} = \lim\limits_{x \to \infty} \frac{\frac{1}{x}}{e^x} = 0$

2. $\lim\limits_{x \to \infty} \frac{e^x}{x^3} = \lim\limits_{x \to \infty} \frac{e^x}{3x^2} = \lim\limits_{x \to \infty} \frac{e^x}{6x} = \lim\limits_{x \to \infty} \frac{e^x}{6} = \infty$

3. $\lim\limits_{x \to -\infty} \frac{x^2}{e^{2x}} = \infty$

4. $\lim\limits_{x \to \infty} \frac{x^2}{e^{2x}} = \lim\limits_{x \to \infty} \frac{2x}{2e^{2x}} = \lim\limits_{x \to \infty} \frac{2}{4e^{2x}} = 0$

5. $-3x^4$

6. $\frac{2x^3}{x} = 2x^2$

7. $\lim\limits_{x \to \infty} \frac{f(x)}{g(x)} = \lim\limits_{x \to \infty} \frac{x + \ln x}{x} = \lim\limits_{x \to \infty} \frac{1 + \frac{1}{x}}{1} = 1$

8. $\lim\limits_{x \to \infty} \frac{f(x)}{g(x)} = \lim\limits_{x \to \infty} \frac{\sqrt{4x^2 + 5x}}{2x} = \lim\limits_{x \to \infty} \sqrt{1 + \frac{5}{4x}} = 1$

9. (a) $f(x) = \frac{e^x + x^2}{e^x} = 1 + \frac{x^2}{e^x}$

$$f'(x) = \frac{2xe^x - x^2 e^x}{e^{2x}} = \frac{2x - x^2}{e^x}$$

$$\frac{2x - x^2}{e^x} = 0$$

$$x(2 - x) = 0$$

$$x = 0 \text{ or } x = 2$$

$f'(x) < 0$ for $x < 0$ or $x > 2$

The graph decreases, increases, and then decreases.

$f(0) = 1; f(2) = 1 + \frac{4}{e^2} \approx 1.541$

f has a local maximum at $\approx (2, 1.541)$ and has a local minimum at $(0, 1)$.

(b) f is increasing on $[0, 2]$

(c) f is decreasing on $(-\infty, 0]$ and $[2, \infty)$.

10. $f(x) = \frac{x + \sin x}{x} = 1 + \frac{\sin x}{x}$, $x \neq 0$

Observe that $\left|\frac{\sin x}{x}\right| < 1$ since $|\sin x| < |x|$ for $x \neq 0$.

$\lim\limits_{x \to 0} f(x) = 1 + \lim\limits_{x \to 0} \frac{\sin x}{x} = 1 + 1 = 2$

Thus the values of f get close to 2 as x gets close to 0, so f doesn't have an absolute maximum value. f is not defined at 0.

Section 8.2 Exercises

1. $\lim\limits_{x \to \infty} \frac{x^3 - 3x + 1}{e^x} = \lim\limits_{x \to \infty} \frac{3x^2 - 3}{e^x} = \lim\limits_{x \to \infty} \frac{6x}{e^x} = \lim\limits_{x \to \infty} \frac{6}{e^x} = 0$

$x^3 - 3x + 1$ grows slower than e^x as $x \to \infty$.

3. $\lim\limits_{x \to \infty} \frac{4^x}{e^x} = \lim\limits_{x \to \infty} \left(\frac{4}{e}\right)^x = \infty$ since $\frac{4}{e} > 1$.

4^x grows faster than e^x as $x \to \infty$.

5. $\lim\limits_{x \to \infty} \frac{e^{x+1}}{e^x} = \lim\limits_{x \to \infty} e = e$

e^{x+1} grows at the same rate as e^x as $x \to \infty$.

7. $\lim\limits_{x \to \infty} \frac{e^{\cos x}}{e^x} = 0$ since $e^{\cos x} \leq e$ for all x.

$e^{\cos x}$ grows slower than e^x as $x \to \infty$.

9. $\lim\limits_{x \to \infty} \frac{x^{1000}}{e^x} = 0$ (Repeated application of L'Hôpital's Rule gets $\lim\limits_{x \to \infty} \frac{1000!}{e^x} = 0$.) x^{1000} grows slower than e^x as $x \to \infty$.

11. $\lim\limits_{x \to \infty} \frac{x^2 + 4x}{x^2} = \lim\limits_{x \to \infty} \left(1 + \frac{4}{x}\right) = 1$

$x^2 + 4x$ grows at the same rate as x^2 as $x \to \infty$.

13. $\lim\limits_{x \to \infty} \frac{15x + 3}{x^2} = \lim\limits_{x \to \infty} \left(\frac{15}{x} + \frac{3}{x^2}\right) = 0$

$15x + 3$ grows slower than x^2 as $x \to \infty$.

15. $\lim\limits_{x\to\infty} \dfrac{\ln x}{x^2} = \lim\limits_{x\to\infty} \dfrac{1/x}{2x} = \lim\limits_{x\to\infty} \dfrac{1}{2x^2} = 0$

$\ln x$ grows slower than x^2 as $x\to\infty$.

17. $\lim\limits_{x\to\infty} \dfrac{\log_2 x^2}{\ln x} = \lim\limits_{x\to\infty} \dfrac{2\log_2 x}{\ln x} = \lim\limits_{x\to\infty} \dfrac{2(\ln x)/(\ln 2)}{\ln x} = \dfrac{2}{\ln 2}$

$\log_2 x^2$ grows at the same rate as $\ln x$ as $x\to\infty$.

19. $\lim\limits_{x\to\infty} \dfrac{1/\sqrt{x}}{\ln x} = \lim\limits_{x\to\infty} \dfrac{1}{\sqrt{x}\ln x} = 0$

$\dfrac{1}{\sqrt{x}}$ grows slower than $\ln x$ as $x\to\infty$.

21. $\lim\limits_{x\to\infty} \dfrac{x - 2\ln x}{\ln x} = \lim\limits_{x\to\infty}\left(\dfrac{x}{\ln x} - 2\right) = \lim\limits_{x\to\infty}\left(\dfrac{1}{1/x} - 2\right)$
$= \lim\limits_{x\to\infty}(x - 2) = \infty$

$x - 2\ln x$ grows faster than $\ln x$ as $x\to\infty$.

23. Compare e^x to x^x.

$\lim\limits_{x\to\infty} \dfrac{e^x}{x^x} = \lim\limits_{x\to\infty}\left(\dfrac{e}{x}\right)^x = 0$

e^x grows slower than x^x.

Compare e^x to $(\ln x)^x$.

$\lim\limits_{x\to\infty} \dfrac{e^x}{(\ln x)^x} = \lim\limits_{x\to\infty}\left(\dfrac{e}{\ln x}\right)^x = 0$

e^x grows slower than $(\ln x)^x$.

Compare e^x to $e^{x/2}$.

$\lim\limits_{x\to\infty} \dfrac{e^x}{e^{x/2}} = \lim\limits_{x\to\infty} e^{x/2} = \infty$

e^x grows faster than $e^{x/2}$.

Compare x^x to $(\ln x)^x$.

$\lim\limits_{x\to\infty} \dfrac{x^x}{(\ln x)^x} = \lim\limits_{x\to\infty}\left(\dfrac{x}{\ln x}\right)^x = \infty$ since $\lim\limits_{x\to\infty}\dfrac{x}{\ln x} = \lim\limits_{x\to\infty}\dfrac{1}{1/x} = \infty$.

x^x grows faster than $(\ln x)^x$.

Thus, in order from slowest-growing to fastest-growing, we get $e^{x/2}, e^x, (\ln x)^x, x^x$.

25. Compare f_1 to f_2.

$\lim\limits_{x\to\infty} \dfrac{f_2(x)}{f_1(x)} = \lim\limits_{x\to\infty} \dfrac{\sqrt{10x+1}}{\sqrt{x}} = \lim\limits_{x\to\infty}\sqrt{10 + \dfrac{1}{x}} = \sqrt{10}$

Thus f_1 and f_2 grow at the same rate.

Compare f_1 to f_3.

$\lim\limits_{x\to\infty} \dfrac{f_3(x)}{f_1(x)} = \lim\limits_{x\to\infty} \dfrac{\sqrt{x+1}}{\sqrt{x}} = \lim\limits_{x\to\infty}\sqrt{1 + \dfrac{1}{x}} = 1$

Thus f_1 and f_3 grow at the same rate.

By transitivity, f_2 and f_3 grow at the same rate, so all three functions grow at the same rate as $x\to\infty$.

27. Compare f_1 to f_2.

$\lim\limits_{x\to\infty} \dfrac{f_2(x)}{f_1(x)} = \lim\limits_{x\to\infty} \dfrac{\sqrt{9^x + 2^x}}{3^x}$
$= \lim\limits_{x\to\infty} \dfrac{\sqrt{9^x + 2^x}}{\sqrt{9^x}}$
$= \lim\limits_{x\to\infty}\sqrt{1 + \left(\dfrac{2}{9}\right)^x} = 1$

Thus f_1 and f_2 grow at the same rate.

Compare f_1 to f_3.

$\lim\limits_{x\to\infty} \dfrac{f_3(x)}{f_1(x)} = \lim\limits_{x\to\infty} \dfrac{\sqrt{9^x - 4^x}}{3^x}$
$= \lim\limits_{x\to\infty} \dfrac{\sqrt{9^x - 4^x}}{\sqrt{9^x}}$
$= \lim\limits_{x\to\infty}\sqrt{1 - \left(\dfrac{4}{9}\right)^x} = 1$

Thus f_1 and f_3 grow at the same rate.
By transitivity, f_2 and f_3 grow at the same rate, so all three functions grow at the same rate as $x\to\infty$.

29. (a) False, since $\lim\limits_{x\to\infty} \dfrac{x}{x} = 1 \neq 0$.

(b) False, since $\lim\limits_{x\to\infty} \dfrac{x}{x+5} = 1 \neq 0$.

(c) True, since $\lim\limits_{x\to\infty} \dfrac{x}{x+5} = 1 \leq 1$.

(d) True, since $\lim\limits_{x\to\infty} \dfrac{x}{2x} = \dfrac{1}{2} \leq 1$.

(e) True, since $\lim\limits_{x\to\infty} \dfrac{e^x}{e^{2x}} = \lim\limits_{x\to\infty} \dfrac{1}{e^x} = 0$.

(f) True, since $\lim\limits_{x\to\infty} \dfrac{x + \ln x}{x} = \lim\limits_{x\to\infty} \dfrac{1 + \dfrac{1}{x}}{1} = 1 \leq 1$.

(g) False, since $\lim\limits_{x\to\infty} \dfrac{\ln x}{\ln 2x} = \lim\limits_{x\to\infty} \dfrac{1/x}{1/x} = 1 \neq 0$.

(h) True, since $\lim\limits_{x\to\infty} \dfrac{\sqrt{x^2 + 5}}{x} = \lim\limits_{x\to\infty}\sqrt{1 + \dfrac{5}{x^2}} = 1 \leq 1$.

31. From the graph, $\lim\limits_{x\to\infty} \dfrac{f(x)}{g(x)} = \infty$, so $\lim\limits_{x\to\infty} \dfrac{g(x)}{f(x)} = 0$.
Thus $g = o(f)$, so ii is true.

33. From the graph, $\lim\limits_{x\to\infty} \dfrac{f(x)}{g(x)} \leq 1$ and not equal to zero. Thus, f and g grow at the same rate, so iii is true.

35. (a) The nth derivative of x^n is $n!$, a constant. We can apply L'Hôpital's Rule n times to find $\lim_{x\to\infty} \frac{e^x}{x^n}$.

$$\lim_{x\to\infty} \frac{e^x}{x^n} = \cdots = \lim_{x\to\infty} \frac{e^x}{n!} = \infty$$

Thus e^x grows faster than x^n as $x\to\infty$ for any positive integer n.

(b) The nth derivative of a^x, $a > 1$, is $(\ln a)^n\, a^x$. We can apply L'Hôpital's Rule n times to find $\lim_{x\to\infty} \frac{a^x}{x^n}$.

$$\lim_{x\to\infty} \frac{a^x}{x^n} = \cdots = \lim_{x\to\infty} \frac{(\ln a)^n\, a^x}{n!} = \infty$$

Thus a^x grows faster than x^n as $x\to\infty$ for any positive integer n.

37. (a) $\lim_{x\to\infty} \frac{\ln x}{x^{1/n}} = \lim_{x\to\infty} \frac{\frac{1}{x}}{\frac{1}{n}x^{(1/n)-1}} = \lim_{x\to\infty} \frac{n}{x^{1/n}} = 0$

Thus $\ln x$ grows slower than $x^{1/n}$ as $x\to\infty$ for any positive integer n.

(b) $\lim_{x\to\infty} \frac{\ln x}{x^a} = \lim_{x\to\infty} \frac{\frac{1}{x}}{ax^{a-1}} = \lim_{x\to\infty} \frac{1}{ax^a} = 0$

Thus $\ln x$ grows slower than x^a as $x\to\infty$ for any number $a > 0$.

39. Compare $n \log_2 n$ to $n^{3/2}$ as $n\to\infty$.

$$\lim_{n\to\infty} \frac{n \log_2 n}{n^{3/2}} = \lim_{n\to\infty} \frac{\log_2 n}{n^{1/2}}$$

$$= \lim_{n\to\infty} \frac{\frac{(\ln n)}{(\ln 2)}}{n^{1/2}}$$

$$= \lim_{n\to\infty} \frac{\frac{1}{n \ln 2}}{\frac{1}{2n^{1/2}}}$$

$$= \lim_{n\to\infty} \frac{2}{n^{1/2}(\ln 2)} = 0$$

Thus $n \log_2 n$ grows slower than $n^{3/2}$ as $n\to\infty$.

Compare $n \log_2 n$ to $n(\log_2 n)^2$

$$\lim_{n\to\infty} \frac{n \log_2 n}{n(\log_2 n)^2} = \lim_{n\to\infty} \frac{1}{\log_2 n} = 0$$

Thus $n \log_2 n$ grows slower than $n(\log_2 n)^2$ as $n\to\infty$. The algorithm of order of $n \log_2 n$ is likely the most efficient because of the three functions, it grows the most slowly as $n\to\infty$.

41. Since f and g grow at the same rate, there exists a nonzero number L such that $\lim_{x\to\infty} \frac{f(x)}{g(x)} = L$. Then for sufficiently large x, $\frac{f(x)}{g(x)} < L + 1 \leq M$ for some integer M.

Similarly, for sufficiently large x, $\frac{g(x)}{f(x)} < \frac{1}{L} + 1 \leq N$ for some integer N.

43. (a) $\lim_{x\to\infty} \frac{x^5}{x^2} = \lim_{x\to\infty} x^3 = \infty$

x^5 grows faster than x^2.

(b) $\lim_{x\to\infty} \frac{5x^3}{2x^3} = \lim_{x\to\infty} \frac{5}{2} = \frac{5}{2}$

$5x^3$ and $2x^3$ have the same rate of growth.

(c) $m > n$ since $\lim_{x\to\infty} \frac{x^m}{x^n} = \lim_{x\to\infty} x^{m-n} = \infty$.

(d) $m = n$ since $\lim_{x\to\infty} \frac{x^m}{x^n} = \lim_{x\to\infty} x^{m-n}$ is nonzero and finite.

(e) Degree of $g >$ degree of f ($m > n$) since $\lim_{x\to\infty} \frac{g(x)}{f(x)} = \infty$.

(f) Degree of $g =$ degree of f ($m = n$) since $\lim_{x\to\infty} \frac{g(x)}{f(x)}$ is nonzero and finite.

45. (a) $\lim_{x\to\infty} \frac{|f(x)|}{|g(x)|} = \lim_{x\to\infty} \frac{-f(x)}{-g(x)} = \lim_{x\to\infty} \frac{f(x)}{g(x)} = \infty$

Thus $|f|$ grows faster than $|g|$ as $x\to\infty$ by definition.

(b) $\lim_{x\to\infty} \frac{|f(x)|}{|g(x)|} = \lim_{x\to\infty} \frac{-f(x)}{-g(x)} = \lim_{x\to\infty} \frac{f(x)}{g(x)} = L$

Thus $|f|$ grows at the same rate as $|g|$ as $x\to\infty$ by definition.

Section 8.3 Improper Integrals
(pp. 433–444)

Exploration 1 Investigating $\int_0^1 \frac{dx}{x^p}$

1. Because $\frac{1}{x^p}$ has an infinite discontinuity at $x = 0$.

2. $\int_0^1 \frac{dx}{x} = \lim_{c\to 0^+} \int_c^1 \frac{dx}{x} = \lim_{c\to 0^+} \left[\ln x\right]_c^1 = \lim_{c\to 0^+} (-\ln c) = \infty$

3. If $p > 1$, then

$$\int_0^1 \frac{dx}{x^p} = \lim_{c\to 0^+} \int_c^1 \frac{dx}{x^p}$$

$$= \lim_{c\to 0^+} \left.\frac{x^{-p+1}}{-p+1}\right]_c^1$$

$$= \lim_{c\to 0^+} \left(\frac{1 - c^{-p+1}}{-p+1}\right) = \infty \text{ because } (-p+1) < 0.$$

4. If $0 < p < 1$, then

$$\int_0^1 \frac{dx}{x^p} = \lim_{c \to 0^+} \int_c^1 \frac{dx}{x^p}$$

$$= \lim_{c \to 0^+} \frac{x^{-p+1}}{-p+1}\Big]_c^1$$

$$= \lim_{c \to 0^+} \left(\frac{1 - c^{-p+1}}{-p+1}\right) = \frac{1}{1-p}$$

Quick Review 8.3

1. $\int_0^3 \frac{dx}{x+3} = \Big[\ln|x+3|\Big]_0^3 = \ln 6 - \ln 3 = \ln 2$

2. $\int_{-1}^1 \frac{x\,dx}{x^2+1} = \Big[\frac{1}{2}\ln|x^2+1|\Big]_{-1}^1 = \frac{1}{2}\ln 2 - \frac{1}{2}\ln 2 = 0$

3. $\int \frac{dx}{x^2+4} = \frac{1}{4}\int \frac{dx}{\left(\frac{x}{2}\right)^2+1}$

$$= \frac{1}{4}\left(2\tan^{-1}\frac{x}{2}\right) + C$$

$$= \frac{1}{2}\tan^{-1}\frac{x}{2} + C$$

4. $\int \frac{dx}{x^4} = \int x^{-4}\,dx = -\frac{1}{3}x^{-3} + C$

5. $9 - x^2 > 0$ for $-3 < x < 3$
   The domain is $(-3, 3)$.

6. $x - 1 > 0$ for $x > 1$
   The domain is $(1, \infty)$.

7. $-1 \le \cos x \le 1$, so $|\cos x| \le 1$.
   $\left|\frac{\cos x}{x^2}\right| = \frac{|\cos x|}{|x^2|} \le \frac{1}{x^2}$

8. $x^2 - 1 \le x^2$ so $\sqrt{x^2-1} \le \sqrt{x^2} = x$ for $x > 1$
   $\frac{1}{\sqrt{x^2-1}} \ge \frac{1}{x}$

9. $\lim_{x \to \infty} \frac{f(x)}{g(x)} = \lim_{x \to \infty} \frac{4e^x - 5}{3e^x + 7} = \lim_{x \to \infty} \frac{4e^x}{3e^x} = \lim_{x \to \infty} \frac{4}{3} = \frac{4}{3}$
   Thus f and g grow at the same rate as $x \to \infty$.

10. $\lim_{x \to \infty} \frac{f(x)}{g(x)} = \lim_{x \to \infty} \frac{\sqrt{2x-1}}{\sqrt{x+3}}$

$$= \lim_{x \to \infty} \sqrt{\frac{2x-1}{x+3}}$$

$$= \lim_{x \to \infty} \sqrt{\frac{2 - \frac{1}{x}}{1 + \frac{3}{x}}} = \sqrt{2}$$

Section 8.3 Exercises

1. (a) The integral is improper because of an infinite limit of integration.

   (b) $\int_0^\infty \frac{dx}{x^2+1} = \lim_{b \to \infty} \int_0^b \frac{dx}{x^2+1}$

   $= \lim_{b \to \infty} \Big[\tan^{-1} x\Big]_0^b$

   $= \lim_{b \to \infty} (\tan^{-1} b - 0)$

   $= \frac{\pi}{2}$

   The integral converges.

   (c) $\frac{\pi}{2}$

3. (a) The integral involves improper integrals because the integrand has an infinite discontinuity at $x = 0$.

   (b) $\int_{-8}^1 \frac{dx}{x^{1/3}} = \int_{-8}^0 \frac{dx}{x^{1/3}} + \int_0^1 \frac{dx}{x^{1/3}}$

   $\int_{-8}^0 \frac{dx}{x^{1/3}} = \lim_{b \to 0^-} \int_{-8}^b \frac{dx}{x^{1/3}}$

   $= \lim_{b \to 0^-} \Big[\frac{3}{2}x^{2/3}\Big]_{-8}^b$

   $= \lim_{b \to 0^-} \left(\frac{3}{2}b^{2/3} - 6\right) = -6$

   $\int_0^1 \frac{dx}{x^{1/3}} = \lim_{b \to 0^+} \int_b^1 \frac{dx}{x^{1/3}}$

   $= \lim_{b \to 0^+} \Big[\frac{3}{2}x^{2/3}\Big]_b^1$

   $= \lim_{b \to 0^+} \left(\frac{3}{2} - \frac{3}{2}b^{2/3}\right)$

   $= \frac{3}{2}$

   $\int_{-8}^1 \frac{dx}{x^{1/3}} = -6 + \frac{3}{2} = -\frac{9}{2}$

   The integral converges.

   (c) $-\frac{9}{2}$

5. (a) The integral is improper because the integrand has an infinite discontinuity at 0.

   (b) $\int_0^{\ln 2} x^{-2} e^{1/x}\,dx = \lim_{b \to 0^+} \int_b^{\ln 2} x^{-2} e^{1/x}\,dx$

   $= \lim_{b \to 0^+} \Big[-e^{1/x}\Big]_b^{\ln 2}$

   $= \lim_{b \to 0^+} [-e^{1/\ln 2} + e^{1/b}] = \infty$

   The integral diverges.

   (c) No value

7. $\int_1^\infty \frac{dx}{x^{1.001}} = \lim_{b\to\infty} \int_1^b \frac{dx}{x^{1.001}}$
$= \lim_{b\to\infty} \left[-1000 x^{-0.001}\right]_1^b$
$= \lim_{b\to\infty} (-1000 b^{-0.001} + 1000) = 1000$

9. $\int_0^4 \frac{dr}{\sqrt{4-r}} = \lim_{b\to 4^-} \int_0^b \frac{dr}{\sqrt{4-r}}$
$= \lim_{b\to 4^-} \left[-2\sqrt{4-r}\right]_0^b$
$= \lim_{b\to 4^-} (-2\sqrt{4-b} + 4) = 4$

11. $\int_0^1 \frac{dx}{\sqrt{1-x^2}} = \lim_{b\to 1^-} \int_0^b \frac{dx}{\sqrt{1-x^2}}$
$= \lim_{b\to 1^-} \left[\sin^{-1} x\right]_0^b$
$= \lim_{b\to 1^-} (\sin^{-1} b - 0) = \frac{\pi}{2}$

13. $\int_{-\infty}^{-2} \frac{2\,dx}{x^2-1} = \lim_{b\to-\infty} \int_b^{-2} \frac{[(x+1)-(x-1)]\,dx}{(x+1)(x-1)}$
$= \lim_{b\to-\infty} \int_b^{-2} \left(\frac{1}{x-1} - \frac{1}{x+1}\right) dx$
$= \lim_{b\to-\infty} \left[\ln|x-1| - \ln|x+1|\right]_b^{-2}$
$= \lim_{b\to-\infty} \left[\ln\left|\frac{x-1}{x+1}\right|\right]_b^{-2}$
$= \lim_{b\to-\infty} \left(\ln 3 - \ln\left|\frac{b-1}{b+1}\right|\right)$
$= \ln 3 - \ln 1 = \ln 3$

15. $\int_0^1 \frac{\theta+1}{\sqrt{\theta^2+2\theta}} = \lim_{b\to 0^+} \int_b^1 \frac{1(2\theta+2)\,d\theta}{2\sqrt{\theta^2+2\theta}}$
$= \lim_{b\to 0^+} \left[\sqrt{\theta^2+2\theta}\right]_b^1$
$= \lim_{b\to 0^+} (\sqrt{3} - \sqrt{b^2+2b}) = \sqrt{3}$

17. First integrate $\int \frac{dx}{(1+x)\sqrt{x}}$ by letting
$u = \sqrt{x}$, so $du = \frac{1}{2\sqrt{x}}\,dx$.
$\int \frac{dx}{(1+x)\sqrt{x}} = \int \frac{2\,du}{1+u^2}$
$= 2\tan^{-1} u + C$
$= 2\tan^{-1} \sqrt{x} + C$

Now evaluate the improper integral. Note that the integrand is infinite at $x = 0$.

$\int_0^\infty \frac{dx}{(1+x)\sqrt{x}} = \int_0^1 \frac{dx}{(1+x)\sqrt{x}} + \int_1^\infty \frac{dx}{(1+x)\sqrt{x}}$
$= \lim_{b\to 0^+} \int_b^1 \frac{dx}{(1+x)\sqrt{x}} + \lim_{c\to\infty} \int_1^c \frac{dx}{(1+x)\sqrt{x}}$
$= \lim_{b\to 0^+} \left[2\tan^{-1}\sqrt{x}\right]_b^1 + \lim_{c\to\infty} \left[2\tan^{-1}\sqrt{x}\right]_1^c$
$= \lim_{b\to 0^+} (2\tan^{-1} 1 - 2\tan^{-1}\sqrt{b}) +$
$\lim_{c\to\infty} (2\tan^{-1}\sqrt{c} - 2\tan^{-1} 1)$
$= \left(\frac{\pi}{2} - 0\right) + \left(\pi - \frac{\pi}{2}\right) = \pi$

19. $\int_1^2 \frac{ds}{s\sqrt{s^2-1}} = \lim_{b\to 1^+} \int_b^2 \frac{ds}{s\sqrt{s^2-1}}$
$= \lim_{b\to 1^+} \left[\sec^{-1} s\right]_b^2$
$= \lim_{b\to 1^+} (\sec^{-1} 2 - \sec^{-1} b)$
$= \sec^{-1} 2 - \sec^{-1} 1 = \frac{\pi}{3}$

21. Integrate $\int \frac{16 \tan^{-1} x}{1+x^2}\,dx$ by letting $u = \tan^{-1} x$, so
$du = \frac{dx}{1+x^2}$.
$\int \frac{16 \tan^{-1} x}{1+x^2}\,dx = \int 16u\,du = 8u^2 + C$
$= 8(\tan^{-1} x)^2 + C$

$\int_0^\infty \frac{16 \tan^{-1} x}{1+x^2}\,dx = \lim_{b\to\infty} \int_0^b \frac{16 \tan^{-1} x}{1+x^2}\,dx$
$= \lim_{b\to\infty} \left[8(\tan^{-1} x)^2\right]_0^b$
$= \lim_{b\to\infty} \left[8(\tan^{-1} b)^2 - 0\right]$
$= 8\left(\frac{\pi}{2}\right)^2 = 2\pi^2$

23. Integrate $\int \theta e^\theta \, d\theta$ by parts.

$u = \theta \qquad dv = e^\theta \, d\theta$

$du = d\theta \qquad v = e^\theta$

$\int \theta e^\theta \, d\theta = \theta e^\theta - \int e^\theta \, d\theta = \theta e^\theta - e^\theta + C$

$\int_{-\infty}^0 \theta e^\theta \, d\theta = \lim_{b \to -\infty} \int_b^0 \theta e^\theta \, d\theta$

$= \lim_{b \to -\infty} \left[\theta e^\theta - e^\theta\right]_b^0$

$= \lim_{b \to -\infty} (-1 - be^b + e^b) = -1$

$\left(\text{Note that } \lim_{b \to -\infty} be^b = \lim_{c \to \infty} -ce^{-c} = \lim_{c \to \infty} -\frac{c}{e^c}\right.$

$\left.= \lim_{c \to \infty} -\frac{1}{e^c} = 0 \text{ and } \lim_{b \to -\infty} e^b = \lim_{c \to \infty} e^{-c} = 0.\right)$

25. $\int_{-\infty}^\infty e^{-|x|} \, dx = \int_{-\infty}^0 e^x \, dx + \int_0^\infty e^{-x} \, dx$

$\int_{-\infty}^0 e^x \, dx = \lim_{b \to -\infty} \int_b^0 e^x \, dx = \lim_{b \to -\infty} \left[e^x\right]_b^0 = \lim_{b \to -\infty} (1 - e^b) = 1$

$\int_0^\infty e^{-x} \, dx = \lim_{b \to \infty} \int_0^b e^{-x} \, dx = \lim_{b \to \infty} \left[-e^{-x}\right]_0^b$

$= \lim_{b \to \infty} (-e^{-b} + 1) = 1$

$\int_{-\infty}^\infty e^{-|x|} \, dx = 1 + 1 = 2$

27. $\int_0^{\pi/2} \tan \theta \, d\theta = \lim_{b \to \pi/2} \int_0^b \frac{\sin \theta}{\cos \theta} \, d\theta$

$= \lim_{b \to \pi/2} \left[-\ln|\cos \theta|\right]_0^b$

$= \lim_{b \to \pi/2} \left[-\ln|\cos b| + 0\right] = \infty$

The integral diverges.

29. $\int_{-\infty}^\infty 2xe^{-x^2} \, dx = \int_0^\infty 2xe^{-x^2} \, dx + \int_{-\infty}^0 2xe^{-x^2} \, dx$

$\int_0^\infty 2xe^{-x^2} \, dx = \lim_{b \to \infty} \int_0^b 2xe^{-x^2} \, dx$

$= \lim_{b \to \infty} \left[-e^{-x^2}\right]_0^b$

$= \lim_{b \to \infty} \left[-e^{-b^2} + 1\right] = 1$

$\int_{-\infty}^0 2xe^{-x^2} \, dx = \lim_{b \to -\infty} \int_b^0 2xe^{-x^2} \, dx$

$= \lim_{b \to -\infty} \left[-e^{-x^2}\right]_b^0$

$= \lim_{b \to -\infty} \left[-1 + e^{-b^2}\right] = -1$

The integral converges.

31. $0 \le \dfrac{1}{\sqrt{t} + \sin t} \le \dfrac{1}{\sqrt{t}}$ on $(0, \pi]$ since $\sin t \ge 0$ on $[0, \pi]$.

$\int_0^\pi \dfrac{dt}{\sqrt{t}} = \lim_{b \to 0^+} \int_b^\pi \dfrac{dt}{\sqrt{t}}$

$= \lim_{b \to 0^+} \left[2\sqrt{t}\right]_b^\pi$

$= \lim_{b \to 0^+} [2\sqrt{\pi} - 2\sqrt{b}]$

$= 2\sqrt{\pi}$

Since this integral converges, the given integral converges.

33. $0 \le \dfrac{1}{x^3 + 1} \le \dfrac{1}{x^3}$ on $[1, \infty)$

$\int_1^\infty \dfrac{dx}{x^3} = \lim_{b \to \infty} \int_1^b \dfrac{dx}{x^3}$

$= \lim_{b \to \infty} \left[-\dfrac{1}{2}x^{-2}\right]_1^b$

$= \lim_{b \to \infty} \left[-\dfrac{1}{2}b^{-2} + \dfrac{1}{2}\right] = \dfrac{1}{2}$

Since this integral converges, the given integral converges.

35. $\int_0^2 \dfrac{dx}{1-x} = \int_0^1 \dfrac{dx}{1-x} + \int_1^2 \dfrac{dx}{1-x}$

$\int_0^1 \dfrac{dx}{1-x} = \lim_{b \to 1^-} \int_0^b \dfrac{dx}{1-x}$

$= \lim_{b \to 1^-} \left[-\ln|1-x|\right]_0^b$

$= \lim_{b \to 1^-} (-\ln|1-b| + 0) = \infty$

Since this integral diverges, the given integral diverges.

37. $0 \le \dfrac{1}{1+e^\theta} \le \dfrac{1}{e^\theta}$ on $[1, \infty)$

$\int_1^\infty \dfrac{1}{e^\theta} \, d\theta = \lim_{b \to \infty} \int_1^b e^{-\theta} \, d\theta$

$= \lim_{b \to \infty} \left[-e^{-\theta}\right]_1^b$

$= \lim_{b \to \infty} \left[-e^{-b} + e^{-1}\right]$

$= \dfrac{1}{e}$

Since this integral converges, the given integral converges.

39. Let $f(x) = \dfrac{\sqrt{x+1}}{x^2}$ and $g(x) = \dfrac{1}{x^{3/2}}$. Both are continuous on $[1, \infty)$.

$$\lim_{x\to\infty} \frac{f(x)}{g(x)} = \lim_{x\to\infty} \frac{\sqrt{x+1}}{\sqrt{x}} = \lim_{x\to\infty} \sqrt{1 + \frac{1}{x}} = 1$$

$$\int_1^\infty \frac{1}{x^{3/2}}\,dx = \lim_{b\to\infty} \int_1^b x^{-3/2}\,dx$$

$$= \lim_{b\to\infty} \left[-2x^{-1/2}\right]_1^b$$

$$= \lim_{b\to\infty} (-2b^{-1/2} + 2) = 2$$

Since this integral converges, the given integral converges.

41. $0 \le \dfrac{1}{x} \le \dfrac{2 + \cos x}{x}$ on $[\pi, \infty)$

$$\int_\pi^\infty \frac{dx}{x} = \lim_{b\to\infty} \int_\pi^b \frac{dx}{x} = \lim_{b\to\infty} \Big[\ln x\Big]_\pi^b = \lim_{b\to\infty}(\ln b - \ln \pi) = \infty$$

Since this integral diverges, the given integral diverges.

43. First rewrite $\dfrac{1}{e^x + e^{-x}}$.

$$\frac{1}{e^x + e^{-x}} = \frac{1}{e^{-x}(e^{2x} + 1)} = \frac{e^x}{1 + (e^x)^2}$$

Integrate $\displaystyle\int \frac{e^x\,dx}{1 + (e^x)^2}$ by letting $u = e^x$ so $du = e^x\,dx$.

$$\int \frac{dx}{e^x + e^{-x}} = \int \frac{e^x\,dx}{1 + (e^x)^2}$$

$$= \int \frac{du}{1 + u^2}$$

$$= \tan^{-1} u + C$$

$$= \tan^{-1} e^x + C$$

$$\int_{-\infty}^\infty \frac{dx}{e^x + e^{-x}} = \int_{-\infty}^0 \frac{dx}{e^x + e^{-x}} + \int_0^\infty \frac{dx}{e^x + e^{-x}}$$

$$\int_{-\infty}^0 \frac{dx}{e^x + e^{-x}} = \lim_{b\to-\infty} \int_b^0 \frac{dx}{e^x + e^{-x}}$$

$$= \lim_{b\to-\infty} \Big[\tan^{-1} e^x\Big]_b^0$$

$$= \lim_{b\to-\infty} [\tan^{-1} 1 - \tan^{-1} e^b]$$

$$= \frac{\pi}{4} - 0 = \frac{\pi}{4}$$

$$\int_0^\infty \frac{dx}{e^x + e^{-x}} = \lim_{b\to\infty} \int_0^b \frac{dx}{e^x + e^{-x}}$$

$$= \lim_{b\to\infty} \Big[\tan^{-1} e^x\Big]_0^b$$

$$= \lim_{b\to\infty} [\tan^{-1} e^b - \tan^{-1} 1]$$

$$= \frac{\pi}{2} - \frac{\pi}{4} = \frac{\pi}{4}$$

Thus, the given integral converges.

45. Integrate $\displaystyle\int \frac{dy}{(1+y^2)(1 + \tan^{-1} y)}$ by letting $u = \tan^{-1} y$ so $du = \dfrac{dy}{1 + y^2}$

$$\int \frac{dy}{(1+y^2)(1 + \tan^{-1} y)} = \int \frac{du}{1 + u}$$

$$= \ln|1 + u| + C$$

$$= \ln|1 + \tan^{-1} y| + C$$

$$\int_0^\infty \frac{dy}{(1+y^2)(1 + \tan^{-1} y)} = \lim_{b\to\infty} \int_0^b \frac{dy}{(1+y^2)(1 + \tan^{-1} y)}$$

$$= \lim_{b\to\infty} \Big[\ln|1 + \tan^{-1} y|\Big]_0^b$$

$$= \lim_{b\to\infty} (\ln|1 + \tan^{-1} b| - 0)$$

$$= \ln\left(1 + \frac{\pi}{2}\right)$$

The integral converges.

47. For $x \ge 0$, $y \ge 0$ on $[1, \infty)$.

$$\text{Area} = \int_1^\infty \frac{\ln x}{x^2}\,dx = \lim_{b\to\infty} \int_1^b \frac{\ln x}{x^2}\,dx$$

Integrate $\displaystyle\int \frac{\ln x}{x^2}\,dx$ by parts.

$u = \ln x \qquad\qquad dv = \dfrac{dx}{x^2}$

$du = \dfrac{1}{x}\,dx \qquad\qquad v = -\dfrac{1}{x}$

$$\int \frac{\ln x}{x^2} = -\frac{\ln x}{x} + \int \frac{dx}{x^2} = -\frac{\ln x}{x} - \frac{1}{x} + C$$

$$\text{Area} = \lim_{b\to\infty}\left[-\frac{\ln x}{x} - \frac{1}{x}\right]_1^b = \lim_{b\to\infty}\left[-\frac{\ln b}{b} - \frac{1}{b} + 1\right] = 1$$

$\left(\text{Note that } \displaystyle\lim_{b\to\infty} \frac{\ln b}{b} = \lim_{b\to\infty} \frac{1/b}{1} = 0.\right)$

49. (a) The integral in Example 1 gives the area of region R.

$$\text{Area} = \int_1^\infty \frac{dx}{x}$$

(b) Refer to Exploration 2 of Section 7.3.

$y' = -\dfrac{1}{x^2}$

The surface area of the solid is given by the following integral.

$$\int_1^\infty 2\pi\left(\frac{1}{x}\right) \sqrt{1 + \left(-\frac{1}{x^2}\right)^2}\,dx = 2\pi \int_1^\infty \frac{1}{x}\sqrt{\frac{x^4 + 1}{x^4}}\,dx$$

$$= 2\pi \int_1^\infty \frac{\sqrt{x^4 + 1}}{x^3}\,dx$$

Since $0 \le \dfrac{1}{x} \le \dfrac{\sqrt{x^4 + 1}}{x^3}$ on $[1, \infty)$, the direct comparison test shows that the integral for the surface area diverges. The surface area is ∞.

49. continued

(c) Volume $= \int_1^\infty \pi\left(\frac{1}{x}\right)^2 dx = \pi \int_1^\infty \frac{1}{x^2} dx$

$= \pi \lim_{b \to \infty} \int_1^b \frac{1}{x^2} dx$

$= \pi \lim_{b \to \infty} \left[-\frac{1}{x}\right]_1^b$

$= \pi \lim_{b \to \infty} \left(-\frac{1}{b} + 1\right) = \pi$

(d) Gabriel's horn has finite volume so it could only hold a finite amount of paint, but it has infinite surface area so it would require an infinite amount of paint to cover itself.

51. (a) For $x \geq 6$, $x^2 \geq 6x$, so $e^{-x^2} \leq e^{-6x}$

$\int_6^\infty e^{-x^2} dx \leq \int_6^\infty e^{-6x} dx$

$= \lim_{b \to \infty} \int_6^b e^{-6x} dx$

$= \lim_{b \to \infty} \left[-\frac{1}{6} e^{-6x}\right]_6^b$

$= \lim_{b \to \infty} \left(-\frac{1}{6} e^{-6b} + \frac{1}{6} e^{-36}\right)$

$= \frac{1}{6} e^{-36} < 4 \times 10^{-17}$

(b) $\int_1^\infty e^{-x^2} dx = \int_1^6 e^{-x^2} dx + \int_6^\infty e^{-x^2} dx$

$\leq \int_1^6 e^{-x^2} dx + 4 \times 10^{-17}$

Thus, from part (a) we have shown that the error is bounded by 4×10^{-17}.

(c) $\int_1^\infty e^{-x^2} dx \approx \text{NINT}(e^{-x^2}, x, 1, 6) \approx 0.1394027926$

(This agrees with Figure 8.16.)

(d) $\int_0^\infty e^{-x^2} dx = \int_0^3 e^{-x^2} dx + \int_3^\infty e^{-x^2} dx$

$\leq \int_0^3 e^{-x^2} dx + \int_3^\infty e^{-3x} dx$

since $x^2 \geq 3x$ for $x > 3$.

$\int_3^\infty e^{-3x} dx = \lim_{b \to \infty} \int_3^b e^{-3x} dx$

$= \lim_{b \to \infty} \left[-\frac{1}{3} e^{-3x}\right]_3^b$

$= \lim_{b \to \infty} \left(-\frac{1}{3} e^{-3b} + \frac{1}{3} e^{-9}\right)$

$= \frac{1}{3} e^{-9} \approx 0.000041 < 0.000042$

53. (a) $\int_0^\infty \frac{2x \, dx}{x^2 + 1} = \lim_{b \to \infty} \int_0^b \frac{2x \, dx}{x^2 + 1}$

$= \lim_{b \to \infty} \left[\ln(x^2 + 1)\right]_0^b$

$= \lim_{b \to \infty} \ln(b^2 + 1) = \infty$

Thus the integral diverges.

(b) Both $\int_0^\infty \frac{2x \, dx}{x^2 + 1}$ and $\int_{-\infty}^0 \frac{2x \, dx}{x^2 + 1}$ must converge in order for $\int_{-\infty}^\infty \frac{2x \, dx}{x^2 + 1}$ to converge.

(c) $\lim_{b \to \infty} \int_{-b}^b \frac{2x \, dx}{x^2 + 1} = \lim_{b \to \infty} \left[\ln(x^2 + 1)\right]_{-b}^b$

$= \lim_{b \to \infty} [\ln(b^2 + 1) - \ln(b^2 + 1)]$

$= \lim_{b \to \infty} 0 = 0.$

Note that $\frac{2x}{x^2 + 1}$ is an odd function so $\int_{-b}^b \frac{2x \, dx}{x^2 + 1} = 0$.

(d) Because the determination of convergence is not made using the method in part (c). In order for the integral to converge, there must be finite areas in both directions (toward ∞ and toward $-\infty$). In this case, there are infinite areas in both directions, but when one computes the integral over an interval $[-b, b]$, there is cancellation which gives 0 as the result.

55. Suppose $0 \leq f(x) \leq g(x)$ for all $x \geq a$.

From the properties of integrals, for any $b > a$,

$\int_a^b f(x) \leq \int_a^b g(x) \, dx$.

If the infinite integral of g converges, then taking the limit in the above inequality as $b \to \infty$ shows that the infinite integral of f is bounded above by the infinite integral of g. Therefore, the infinite integral of f must be finite and it converges. If the infinite integral of f diverges, it must grow to infinity. So taking the limit in the above inequality as $b \to \infty$ shows that the infinite integral of g must also diverge to infinity.

57. (a) On a grapher, plot $\text{NINT}\left(\dfrac{\sin x}{x}, x, 0, x\right)$ or create a table of values. For large values of x, $f(x)$ appears to approach approximately 1.57.

(b) Yes, the integral appears to converge.

■ Section 8.4 Partial Fractions and Integral Tables (pp. 444–453)

Quick Review 8.4

1. Solving the first equation for B yields $B = -3A - 5$. Substitute into the second equation.
$$-2A + 3(-3A - 5) = 7$$
$$-2A - 9A - 15 = 7$$
$$-11A = 22$$
$$A = -2$$
Substituting $A = -2$ into $B = -3A - 5$ gives $B = 1$. The solution is $A = -2$, $B = 1$.

2. Solve by Gaussian elimination. Multiply first equation by -3 and add to second equation. Multiply first equation by -1 and add to third equation.
$$A + 2B - C = 0$$
$$-7B + 5C = 1$$
$$-B + 2C = 4$$
Multiply third equation by -7 and add to second equation.
$$A + 2B - C = 0$$
$$-9C = -27$$
$$-B + 2C = 4$$
Solve the second equation for C to get $C = 3$. Solve for B by substituting $C = 3$ into the third equation.
$$-B + 2(3) = 4$$
$$-B = -2$$
$$B = 2$$
Solve for A by substituting $B = 2$ and $C = 3$ into the first equation.
$$A + 2(2) - 3 = 0$$
$$A + 1 = 0$$
$$A = -1$$
The solution is $A = -1$, $B = 2$, $C = 3$.

3. $x^2 - 3x - 4 \overline{) 2x^3 - 5x^2 - 10x - 7}$ gives quotient $2x + 1$ with remainder $x - 3$:
$$2x + 1 + \dfrac{x - 3}{x^2 - 3x - 4}$$

4. $x^2 + 4x + 5 \overline{) 2x^2 + 11x + 6}$ gives quotient 2 with remainder $3x - 4$:
$$2 + \dfrac{3x - 4}{x^2 + 4x + 5}$$

5. $x^3 - 3x^2 + x - 3 = x^2(x - 3) + (x - 3)$
$$= (x - 3)(x^2 + 1)$$

6. $y^4 - 5y^2 + 4 = (y^2 - 4)(y^2 - 1)$
$$= (y - 2)(y + 2)(y - 1)(y + 1)$$

7. $\dfrac{2}{x + 3} - \dfrac{3}{x - 2} = \dfrac{2(x - 2)}{(x - 2)(x + 3)} - \dfrac{3(x + 3)}{(x - 2)(x + 3)}$
$$= \dfrac{2x - 4 - 3x - 9}{(x - 2)(x + 3)}$$
$$= \dfrac{-x - 13}{(x - 2)(x + 3)}$$
$$= -\dfrac{x + 13}{x^2 + x - 6}$$

8. $\dfrac{x - 1}{x^2 - 4x + 5} - \dfrac{2}{x + 5}$
$$= \dfrac{(x - 1)(x + 5)}{(x + 5)(x^2 - 4x + 5)} - \dfrac{2(x^2 - 4x + 5)}{(x + 5)(x^2 - 4x + 5)}$$
$$= \dfrac{x^2 + 4x - 5 - 2x^2 + 8x - 10}{(x + 5)(x^2 - 4x + 5)}$$
$$= \dfrac{-x^2 + 12x - 15}{(x + 5)(x^2 - 4x + 5)}$$

9. $\dfrac{t - 1}{t^2 + 2} - \dfrac{3t + 4}{t^2 + 1} = \dfrac{(t - 1)(t^2 + 1)}{(t^2 + 2)(t^2 + 1)} - \dfrac{(3t + 4)(t^2 + 2)}{(t^2 + 2)(t^2 + 1)}$
$$= \dfrac{t^3 - t^2 + t - 1 - 3t^3 - 4t^2 - 6t - 8}{(t^2 + 2)(t^2 + 1)}$$
$$= \dfrac{-2t^3 - 5t^2 - 5t - 9}{(t^2 + 2)(t^2 + 1)}$$
$$= -\dfrac{2t^3 + 5t^2 + 5t + 9}{(t^2 + 2)(t^2 + 1)}$$

10. $\dfrac{2}{x - 1} - \dfrac{3}{(x - 1)^2} + \dfrac{1}{(x - 1)^3}$
$$= \dfrac{2(x - 1)^2}{(x - 1)^3} - \dfrac{3(x - 1)}{(x - 1)^3} + \dfrac{1}{(x - 1)^3}$$
$$= \dfrac{2x^2 - 4x + 2 - 3x + 3 + 1}{(x - 1)^3}$$
$$= \dfrac{2x^2 - 7x + 6}{(x - 1)^3}$$

Section 8.4 Exercises

1. $x^2 - 3x + 2 = (x - 2)(x - 1)$
$$\dfrac{5x - 7}{x^2 - 3x + 2} = \dfrac{A}{x - 2} + \dfrac{B}{x - 1}$$
$$5x - 7 = A(x - 1) + B(x - 2)$$
$$= (A + B)x - (A + 2B)$$

Equating coefficients of like terms gives

$A + B = 5$ and $A + 2B = 7$.

Solving the system simultaneously yields $A = 3$, $B = 2$.
$$\dfrac{5x - 7}{x^2 - 3x + 2} = \dfrac{3}{x - 2} + \dfrac{2}{x - 1}$$

3. $\dfrac{t+1}{t^2(t-1)} = \dfrac{A}{t-1} + \dfrac{B}{t} + \dfrac{C}{t^2}$

$t + 1 = At^2 + Bt(t-1) + C(t-1)$

$\qquad = (A+B)t^2 + (-B+C)t - C$

Equating coefficients of like terms gives

$A + B = 0,\ -B + C = 1,$ and $-C = 1.$

Solving the system simultaneously yields

$A = 2,\ B = -2,\ C = -1.$

$\dfrac{t+1}{t^2(t-1)} = \dfrac{2}{t-1} - \dfrac{2}{t} - \dfrac{1}{t^2}$

5. $x^2 - 5x + 6\,\overline{\big)\,x^2 \qquad\ \ + 8}$
$\qquad\qquad\ \ \underline{x^2 - 5x + 6}$
$\qquad\qquad\qquad\quad 5x + 2$

$\dfrac{x^2+8}{x^2-5x+6} = 1 + \dfrac{5x+2}{x^2-5x+6}$

$x^2 - 5x + 6 = (x-3)(x-2)$

$\dfrac{5x+2}{x^2-5x+6} = \dfrac{A}{x-3} + \dfrac{B}{x-2}$

$5x + 2 = A(x-2) + B(x-3)$

$\qquad = (A+B)x + (-2A - 3B)$

Equating coefficients of like terms gives

$A + B = 5$ and $-2A - 3B = 2$

Solving the system simultaneously yields

$A = 17,\ B = -12.$

$\dfrac{x^2+8}{x^2-5x+6} = 1 + \dfrac{17}{x-3} - \dfrac{12}{x-2}$

7. $1 - x^2 = (1-x)(1+x)$

$\dfrac{1}{1-x^2} = \dfrac{A}{1-x} + \dfrac{B}{1+x}$

$1 = A(1+x) + B(1-x)$

$\ = (A - B)x + (A + B)$

Equating coefficients of like terms gives

$A - B = 0$ and $A + B = 1.$

Solving the system simultaneously yields

$A = \dfrac{1}{2},\ B = \dfrac{1}{2}.$

$\displaystyle\int \dfrac{dx}{1-x^2} = \int \dfrac{1/2}{1-x}\,dx + \int \dfrac{1/2}{1+x}\,dx$

$\qquad = -\dfrac{1}{2}\ln|1-x| + \dfrac{1}{2}\ln|1+x| + C$

$\qquad = \dfrac{1}{2}\ln\left|\dfrac{1+x}{1-x}\right| + C$

9. $y^2 - 2y - 3 = (y-3)(y+1)$

$\dfrac{y}{y^2-2y-3} = \dfrac{A}{y-3} + \dfrac{B}{y+1}$

$y = A(y+1) + B(y-3)$

$\ = (A+B)y + (A - 3B)$

Equating coefficients of like terms gives

$A + B = 1$ and $A - 3B = 0.$

Solving the system simultaneously yields $A = \dfrac{3}{4},\ B = \dfrac{1}{4}.$

$\displaystyle\int \dfrac{y\,dy}{y^2-2y-3} = \int \dfrac{3/4}{y-3}\,dy + \int \dfrac{1/4}{y+1}\,dy$

$\qquad = \dfrac{3}{4}\ln|y-3| + \dfrac{1}{4}\ln|y+1| + C$

11. $t^3 + t^2 - 2t = t(t^2 + t - 2) = t(t+2)(t-1)$

$\dfrac{1}{t^3+t^2-2t} = \dfrac{A}{t} + \dfrac{B}{t+2} + \dfrac{C}{t-1}$

$1 = A(t+2)(t-1) + B(t)(t-1) + C(t)(t+2)$

$\ = A(t^2 + t - 2) + B(t^2 - t) + C(t^2 + 2t)$

$\ = (A + B + C)t^2 + (A - B + 2C)t - 2A$

Equating coefficients of like terms gives

$A + B + C = 0,\ A - B + 2C = 0,$ and $-2A = 1.$

Solving the system simultaneously yields

$A = -\dfrac{1}{2},\ B = \dfrac{1}{6},\ C = \dfrac{1}{3}.$

$\displaystyle\int \dfrac{dt}{t^3+t^2-2t} = \int \dfrac{-1/2}{t}\,dt + \int \dfrac{1/6}{t+2}\,dt + \int \dfrac{1/3}{t-1}\,dt$

$\qquad = -\dfrac{1}{2}\ln|t| + \dfrac{1}{6}\ln|t+2| + \dfrac{1}{3}\ln|t-1| + C$

13. $s^2 + 4\,\overline{\big)\,s^3}$
$\qquad\quad \underline{s^3 + 4s}$
$\qquad\qquad\ \ -4s$

$\dfrac{s^3}{s^2+4} = s + \dfrac{-4s}{s^2+4}$

$\displaystyle\int \dfrac{s^3}{s^2+4}\,ds = \int s\,ds - \int \dfrac{4s}{s^2+4}\,ds$

$\qquad = \dfrac{1}{2}s^2 - 2\ln(s^2+4) + C$

15. $x^2 + x + 1\,\overline{\big)\,5x^2}$
$\qquad\qquad\ \ \underline{5x^2 + 5x + 5}$
$\qquad\qquad\qquad\ -5x - 5$

$\dfrac{5x^2}{x^2+x+1} = 5 - \dfrac{5x+5}{x^2+x+1}$

$\displaystyle\int \dfrac{5x^2\,dx}{x^2+x+1} = \int 5\,dx - 5\int \dfrac{x+1}{x^2+x+1}\,dx$

$\qquad = 5x - 5\int \dfrac{x+1}{x^2+x+1}\,dx$

To evaluate the second integral, complete the square in the denominator.

$$x^2 + x + 1 = x^2 + x + \frac{1}{4} + \frac{3}{4} = \left(x + \frac{1}{2}\right)^2 + \frac{3}{4}$$

$$\int \frac{x+1}{x^2+x+1} dx$$

$$= \int \frac{x+1}{(x+1/2)^2 + 3/4} dx$$

$$= \int \frac{x + 1/2}{(x+1/2)^2 + 3/4} dx + \int \frac{1/2}{(x+1/2)^2 + 3/4} dx$$

$$= \frac{1}{2} \ln\left[\left(x + \frac{1}{2}\right)^2 + \frac{3}{4}\right] + \frac{1}{2} \int \frac{dx}{(x+1/2)^2 + (\sqrt{3}/2)^2}$$

$$= \frac{1}{2} \ln(x^2 + x + 1) + \frac{1}{2}\left(\frac{2}{\sqrt{3}}\right) \tan^{-1}\left(\frac{x+1/2}{\sqrt{3}/2}\right)$$

$$= \frac{1}{2} \ln(x^2 + x + 1) + \frac{1}{\sqrt{3}} \tan^{-1}\left(\frac{2x+1}{\sqrt{3}}\right)$$

The second integral was evaluated by using Formula 16 from the Brief Table of Integrals.

$$\int \frac{5x^2 \, dx}{x^2 + x + 1}$$

$$= 5x - \frac{5}{2} \ln(x^2 + x + 1) - \frac{5}{\sqrt{3}} \tan^{-1}\left(\frac{2x+1}{\sqrt{3}}\right) + C$$

17. $(x^2 - 1)^2 = (x+1)^2(x-1)^2$

$$\frac{1}{(x^2-1)^2} = \frac{A}{x+1} + \frac{B}{(x+1)^2} + \frac{C}{x-1} + \frac{D}{(x-1)^2}$$

$1 = A(x+1)(x-1)^2 + B(x-1)^2 + C(x+1)^2(x-1)$
$\ + D(x+1)^2$

$= A(x^3 - x^2 - x + 1) + B(x^2 - 2x + 1)$
$\ + C(x^3 + x^2 - x - 1) + D(x^2 + 2x + 1)$

$= (A+C)x^3 + (-A+B+C+D)x^2$
$\ + (-A - 2B - C + 2D)x + (A+B-C+D)$

Equating coefficients of like terms gives

$A + C = 0, -A + B + C + D = 0,$

$-A - 2B - C + 2D = 0,$ and $A + B - C + D = 1$

Solving the system simultaneously yields

$$A = \frac{1}{4}, B = \frac{1}{4}, C = -\frac{1}{4}, D = \frac{1}{4}.$$

$$\int \frac{dx}{(x^2-1)^2}$$

$$= \int \frac{1/4}{x+1} dx + \int \frac{1/4}{(x+1)^2} dx + \int \frac{-1/4}{x-1} dx + \int \frac{1/4}{(x-1)^2} dx$$

$$= \frac{1}{4} \ln|x+1| - \frac{1}{4(x+1)} - \frac{1}{4} \ln|x-1| - \frac{1}{4(x-1)} + C$$

19. Complete the square in the denominator.

$$r^2 - 2r + 2 = r^2 - 2r + 1 + 1 = (r-1)^2 + 1$$

$$\int \frac{2 \, dr}{r^2 - 2r + 2} = \int \frac{2 \, dr}{(r-1)^2 + 1} = 2 \tan^{-1}(r-1) + C$$

21. $x^3 - 1 = (x-1)(x^2 + x + 1)$

$$\frac{x^2 - 2x - 2}{x^3 - 1} = \frac{A}{x-1} + \frac{Bx + C}{x^2 + x + 1}$$

$x^2 - 2x - 2 = A(x^2 + x + 1) + (Bx + C)(x - 1)$
$\ = (A+B)x^2 + (A - B + C)x + (A - C)$

Equating coefficients of like terms gives

$A + B = 1, A - B + C = -2,$ and $A - C = -2.$

Solving the system simultaneously yields

$A = -1, B = 2, C = 1.$

$$\int \frac{x^2 - 2x - 2}{x^3 - 1} dx = \int \frac{-1}{x-1} dx + \int \frac{2x+1}{x^2+x+1} dx$$

$$= -\ln|x-1| + \ln(x^2 + x + 1) + C$$

23. $\frac{3x^2 - 2x + 12}{(x^2+4)^2} = \frac{Ax + B}{x^2 + 4} + \frac{Cx + D}{(x^2+4)^2}$

$3x^2 - 2x + 12 = (Ax + B)(x^2 + 4) + (Cx + D)$

$\ = Ax^3 + Bx^2 + (4A + C)x + 4B + D$

Equating coefficients of like terms gives

$A = 0, B = 3, 4A + C = -2,$ and $4B + D = 12$

Solving the system simultaneously yields

$A = 0, B = 3, C = -2, D = 0.$

$$\int \frac{3x^2 - 2x + 12}{(x^2+4)^2} dx = \int \frac{3}{x^2+4} dx + \int \frac{-2x}{(x^2+4)^2} dx$$

$$= \frac{3}{2} \tan^{-1}\frac{x}{2} + \frac{1}{x^2+4} + C$$

The first integral was evaluated by using Formula 16 from the Brief Table of Integrals.

25. $\theta + 1 \overline{\smash{\big)}\ \theta}$
$\ \underline{\theta + 1}$
$\ -1$

$$\frac{\theta}{\theta + 1} = 1 - \frac{1}{\theta + 1}$$

$$\int_0^1 \frac{\theta}{\theta + 1} d\theta = \int_0^1 d\theta - \int_0^1 \frac{1}{\theta + 1} d\theta$$

$$= \left[\theta\right]_0^1 - \left[\ln|\theta + 1|\right]_0^1$$

$$= 1 - \ln 2$$

27. $\dfrac{1}{y^2 - y}\, dy = e^x\, dx$

$\displaystyle\int \dfrac{1}{y(y-1)}\, dy = \int e^x\, dx = e^x + C$

$\dfrac{1}{y(y-1)} = \dfrac{A}{y} + \dfrac{B}{y-1}$

$1 = A(y-1) + B(y)$

$ = (A+B)y - A$

Equating coefficients of like terms gives

$A + B = 0$ and $-A = 1$

Solving the system simultaneously yields $A = -1$, $B = 1$.

$\displaystyle\int \dfrac{1}{y(y-1)}\, dy = \int -\dfrac{1}{y}\, dy + \int \dfrac{1}{y-1}\, dy$

$\phantom{\int \dfrac{1}{y(y-1)}\, dy} = -\ln|y| + \ln|y-1| + C_2$

$-\ln|y| + \ln|y-1| = e^x + C$

Substitute $x = 0$, $y = 2$.

$-\ln 2 + 0 = 1 + C$ or $C = -1 - \ln 2$

The solution to the initial value problem is

$-\ln|y| + \ln|y-1| = e^x - 1 - \ln 2$.

29. $dy = \dfrac{dx}{x^2 - 3x + 2}$

$x^2 - 3x + 2 = (x-2)(x-1)$

$\dfrac{1}{x^2 - 3x + 2} = \dfrac{A}{x-2} + \dfrac{B}{x-1}$

$1 = A(x-1) + B(x-2)$

$1 = (A+B)x - A - 2B$

Equating coefficients of like terms gives

$A + B = 0$, $-A - 2B = 1$

Solving the system simultaneously yields $A = 1$, $B = -1$.

$\displaystyle\int dy = \int \dfrac{dx}{x^2 - 3x + 2} = \int \dfrac{dx}{x-2} - \int \dfrac{dx}{x-1}$

$y = \ln|x-2| - \ln|x-1| + C$

Substitute $x = 3$, $y = 0$.

$0 = 0 - \ln 2 + C$ or $C = \ln 2$

The solution to the initial value problem is

$y = \ln|x-2| - \ln|x-1| + \ln 2$

31. (a) Complete the square in the denominator.

$5 + 4x - x^2 = 5 - (x^2 - 4x)$

$ = 9 - (x^2 - 4x + 4)$

$ = 9 - (x-2)^2$

Let $u = x - 2$ so $du = dx$, and then use Formula 18 with $x = u$ and $a = 3$.

$\displaystyle\int \dfrac{dx}{5 + 4x - x^2} = \int \dfrac{du}{9 - u^2}$

$\phantom{\int \dfrac{dx}{5 + 4x - x^2}} = \dfrac{1}{6} \ln\left|\dfrac{u+3}{u-3}\right| + C$

$\phantom{\int \dfrac{dx}{5 + 4x - x^2}} = \dfrac{1}{6} \ln\left|\dfrac{x+1}{x-5}\right| + C$

(b) $\dfrac{d}{dx}\left(\dfrac{1}{2a} \ln\left|\dfrac{x+a}{x-a}\right| + C\right)$

$= \dfrac{1}{2a} \dfrac{d}{dx}\left(\ln|x+a| - \ln|x-a|\right)$

$= \dfrac{1}{2a}\left(\dfrac{1}{x+a} - \dfrac{1}{x-a}\right)$

$= \dfrac{1}{2a}\left[\dfrac{x-a}{(x+a)(x-a)} - \dfrac{x+a}{(x+a)(x-a)}\right]$

$= -\dfrac{1}{x^2 - a^2} = \dfrac{1}{a^2 - x^2}$

33. Volume $= \displaystyle\int_{0.5}^{2.5} \pi\left(\dfrac{9}{3x-x^2}\right) dx = 9\pi \int_{0.5}^{2.5} \dfrac{dx}{3x - x^2}$

$3x - x^2 = x(3-x)$

$\dfrac{1}{3x - x^2} = \dfrac{A}{x} + \dfrac{B}{3-x}$

$1 = A(3-x) + Bx$

$ = (-A+B)x + 3A$

Equating coefficients of like terms gives

$-A + B = 0$ and $3A = 1$

Solving the system simultaneously yields $A = \dfrac{1}{3}$, $B = \dfrac{1}{3}$.

$9\pi \displaystyle\int_{0.5}^{2.5} \dfrac{dx}{3x-x^2} = 3\pi\left(\int_{0.5}^{2.5} \dfrac{dx}{x} + \int_{0.5}^{2.5} \dfrac{dx}{3-x}\right)$

$\phantom{9\pi \int_{0.5}^{2.5} \dfrac{dx}{3x-x^2}} = 3\pi\left(\Big[\ln|x|\Big]_{0.5}^{2.5} + \Big[-\ln|3-x|\Big]_{0.5}^{2.5}\right)$

$\phantom{9\pi \int_{0.5}^{2.5} \dfrac{dx}{3x-x^2}} = 3\pi(\ln 2.5 - \ln 0.5 - \ln 0.5 + \ln 2.5)$

$\phantom{9\pi \int_{0.5}^{2.5} \dfrac{dx}{3x-x^2}} = 3\pi \ln 25 = 6\pi \ln 5$

35. $y = 3 \tan \theta$, $dy = 3 \sec^2 \theta \, d\theta$, $-\dfrac{\pi}{2} < \theta < \dfrac{\pi}{2}$

$9 + y^2 = 9 + 9 \tan^2 \theta = 9 \sec^2 \theta$

$$\int \dfrac{dy}{\sqrt{9 + y^2}} = \int \dfrac{3 \sec^2 \theta}{|3 \sec \theta|} d\theta$$
$$= \int \sec \theta \, d\theta$$
$$= \ln |\sec \theta + \tan \theta| + C_1$$
$$= \ln \left| \dfrac{\sqrt{9 + y^2}}{3} + \dfrac{y}{3} \right| + C_1$$
$$= \ln \left| \sqrt{9 + y^2} + y \right| - \ln 3 + C_1$$
$$= \ln \left| \sqrt{9 + y^2} + y \right| + C$$

Use Formula 88 for $\int \sec \theta \, d\theta$ with $x = \theta$ and $a = 1$. Use Figure 8.18(a) from the text with $a = 3$ to get

$\sqrt{9 + y^2} = 3 |\sec \theta|$.

37. $x = \dfrac{7}{2} \sec \theta$, $dx = \dfrac{7}{2} \sec \theta \tan \theta \, d\theta$, $0 \le \theta < \dfrac{\pi}{2}$

$4x^2 - 49 = 49 \sec^2 \theta - 49 = 49 \tan^2 \theta$

$$\int \dfrac{dx}{\sqrt{4x^2 - 49}} = \int \dfrac{7/2 \sec \theta \tan \theta \, d\theta}{|7 \tan \theta|}$$
$$= \int \dfrac{1}{2} \sec \theta \, d\theta$$
$$= \dfrac{1}{2} \ln |\sec \theta + \tan \theta| + C_1$$
$$= \dfrac{1}{2} \ln \left| \dfrac{2x}{7} + \dfrac{\sqrt{4x^2 - 49}}{7} \right| + C_1$$
$$= \dfrac{1}{2} \ln \left| 2x + \sqrt{4x^2 - 49} \right| - \dfrac{1}{2} \ln 7 + C_1$$
$$= \dfrac{1}{2} \ln \left| 2x + \sqrt{4x^2 - 49} \right| + C$$

Use Figure 8.18(c) from the text with $a = \dfrac{7}{2}$ to get

$\sqrt{x^2 - \dfrac{49}{4}} = \dfrac{7}{2} \tan \theta$.

39. $x = \sin \theta$, $dx = \cos \theta \, d\theta$, $-\dfrac{\pi}{2} < \theta < \dfrac{\pi}{2}$

$1 - x^2 = 1 - \sin^2 \theta = \cos^2 \theta$

$$\int \dfrac{x^3 \, dx}{\sqrt{1 - x^2}} = \int \dfrac{\sin^3 \theta \cos \theta \, d\theta}{|\cos \theta|}$$
$$= \int \sin^3 \theta \, d\theta$$
$$= \int (1 - \cos^2 \theta) \sin \theta \, d\theta$$
$$= -\cos \theta + \dfrac{1}{3} \cos^3 \theta + C$$
$$= \cos \theta \left(-1 + \dfrac{1}{3} \cos^2 \theta \right) + C$$
$$= \sqrt{1 - x^2} \left[-1 + \dfrac{1}{3}(1 - x^2) \right] + C$$
$$= -\dfrac{x^2 \sqrt{1 - x^2}}{3} - \dfrac{2\sqrt{1 - x^2}}{3} + C$$

Use Figure 18.8(b) from the text with $a = 1$ to get $\sqrt{1 - x^2} = \cos \theta$.

41. $z = 4 \sin \theta$, $dz = 4 \cos \theta \, d\theta$, $0 < \theta < \dfrac{\pi}{2}$

$16 - z^2 = 16 - 16 \sin^2 \theta = 16 \cos^2 \theta$

$$\int \dfrac{\sqrt{16 - z^2}}{z} dz$$
$$= \int \dfrac{|4 \cos \theta| (4 \cos \theta) \, d\theta}{4 \sin \theta}$$
$$= \int \dfrac{4 \cos^2 \theta}{\sin \theta} d\theta$$
$$= \int \dfrac{4 - 4 \sin^2 \theta}{\sin \theta} d\theta$$
$$= \int (4 \csc \theta - 4 \sin \theta) \, d\theta$$
$$= -4 \ln |\csc \theta + \cot \theta| + 4 \cos \theta + C$$
$$= -4 \ln \left| \dfrac{4}{z} + \dfrac{\sqrt{16 - z^2}}{z} \right| + 4 \left(\dfrac{\sqrt{16 - z^2}}{4} \right) + C$$
$$= -4 \ln \left| \dfrac{4 + \sqrt{16 - z^2}}{z} \right| + \sqrt{16 - z^2} + C$$

Use Formula 89 with $a = 1$ and $x = \theta$. Use Figure 8.18(b) from the text with $a = 4$ to get

$\csc \theta = \dfrac{4}{|z|}$, $\cot \theta = \dfrac{\sqrt{16 - z^2}}{|z|}$ and $\cos \theta = \dfrac{\sqrt{16 - z^2}}{4}$.

43. $dy = \dfrac{dx}{\sqrt{x^2 - 9}}$

$x = 3 \sec \theta, dx = 3 \sec \theta \tan \theta \, d\theta, 0 < \theta < \dfrac{\pi}{2}$

$x^2 - 9 = 9 \sec^2 \theta - 9 = 9 \tan^2 \theta$

$$y = \int \dfrac{dx}{\sqrt{x^2 - 9}}$$

$$= \int \dfrac{3 \sec \theta \tan \theta \, d\theta}{|3 \tan \theta|}$$

$$= \int \sec \theta \, d\theta$$

$$= \ln |\sec \theta + \tan \theta| + C$$

$$= \ln \left| \dfrac{x}{3} + \dfrac{\sqrt{x^2 - 9}}{3} \right| + C$$

Substitute $x = 5, y = \ln 3$.

$\ln 3 = \ln \left(\dfrac{5}{3} + \dfrac{4}{3} \right) + C$ or $C = 0$

The solution to the initial value problem is

$$y = \ln \left| \dfrac{x}{3} + \dfrac{\sqrt{x^2 - 9}}{3} \right|.$$

45. For $x \geq 0, y \geq 0$ on $[0, 3]$

Area $= \displaystyle\int_0^3 \dfrac{\sqrt{9 - x^2}}{3} dx$

$x = 3 \sin \theta, dx = 3 \cos \theta \, d\theta, 0 \leq \theta \leq \dfrac{\pi}{2}$

$9 - x^2 = 9 - 9 \sin^2 \theta = 9 \cos^2 \theta$

When $x = 0, \theta = 0$ and when $x = 3, \theta = \dfrac{\pi}{2}$.

$$\int_0^3 \dfrac{\sqrt{9 - x^2}}{3} dx = \int_0^{\pi/2} \dfrac{3 |\cos \theta|}{3} \cdot 3 \cos \theta \, d\theta$$

$$= \int_0^{\pi/2} 3 \cos^2 \theta \, d\theta$$

$$= 3 \int_0^{\pi/2} \dfrac{1 + \cos 2\theta}{2} d\theta$$

$$= 3 \left[\dfrac{\theta}{2} + \dfrac{\sin 2\theta}{4} \right]_0^{\pi/2}$$

$$= \dfrac{3\pi}{4} \approx 2.356$$

47. (a) $\dfrac{dx}{x(1000 - x)} = \dfrac{1}{250} dt$

$\dfrac{1}{x(1000 - x)} = \dfrac{A}{x} + \dfrac{B}{1000 - x}$

$1 = A(1000 - x) + Bx$

$= (-A + B)x + 1000A$

Equating the coefficients and solving for A and B gives

$A = \dfrac{1}{1000}, B = \dfrac{1}{1000}$

$\displaystyle\int \dfrac{dx}{x(1000 - x)} = \int \dfrac{(1/1000) \, dx}{x} + \int \dfrac{(1/1000) \, dx}{1000 - x}$

$$= \dfrac{1}{1000} \ln x - \dfrac{1}{1000} \ln (1000 - x) + C_1$$

$$= \dfrac{1}{1000} \ln \dfrac{x}{1000 - x} + C_1$$

$\dfrac{1}{1000} \ln \dfrac{x}{1000 - x} = \dfrac{t}{250} + C_2$

$\ln \dfrac{x}{1000 - x} = 4t + C$

$\dfrac{x}{1000 - x} = e^{4t + C} = Ae^{4t}$

When $t = 0, x = 2$.

$\dfrac{2}{998} = A$ or $A = \dfrac{1}{499}$

$\dfrac{x}{1000 - x} = \dfrac{1}{499} e^{4t}$

$x = \dfrac{1000}{499} e^{4t} - \dfrac{x}{499} e^{4t}$

$x \left(1 + \dfrac{e^{4t}}{499} \right) = \dfrac{1000 e^{4t}}{499}$

$x = \dfrac{1000 e^{4t}}{499 + e^{4t}}$

or $x = \dfrac{1000}{1 + 499 e^{-4t}}$

(b) $500 = \dfrac{1000}{1 + 499 e^{-4t}}$

$1 + 499 e^{-4t} = 2$

$e^{-4t} = \dfrac{1}{499}$

$t = -\dfrac{1}{4} \ln \dfrac{1}{499} \approx 1.553$

Half the population will have heard the rumor in about 1.553 days.

(c) $\dfrac{dx}{dt} = \dfrac{1}{250} x(1000 - x)$

$\dfrac{dx}{dt}$ will be greatest when $y = x(1000 - x)$ is greatest.

This occurs when $x = 500$ which occurs when

$t \approx 1.553$ as shown in part (b).

49. (a) From the figure, $\tan \dfrac{x}{2} = \dfrac{\sin x}{1 + \cos x}$.

(b) From part (a), $z = \dfrac{\sin x}{1 + \cos x}$

$z(1 + \cos x) = \sin x$

$z^2(1 + \cos x)^2 = \sin^2 x$

$z^2(1 + \cos x)^2 = 1 - \cos^2 x$

$z^2(1 + \cos x)^2 - (1 - \cos x)(1 + \cos x) = 0$

$(1 + \cos x)(z^2 + z^2 \cos x - 1 + \cos x) = 0$

$1 + \cos x = 0 \quad \text{or} \quad (z^2 + 1)\cos x = 1 - z^2$

$\cos x = -1 \qquad \cos x = \dfrac{1 - z^2}{1 + z^2}$

$\cos x = -1$ does not make sense in this case.

(c) From part (b), $\cos x = \dfrac{1 - z^2}{1 + z^2}$

$\sin^2 x = 1 - \cos^2 x$

$= 1 - \dfrac{(1 - z^2)^2}{(1 + z^2)^2}$

$= \dfrac{(1 + z^2)^2 - (1 - z^2)^2}{(1 + z^2)^2}$

$= \dfrac{1 + 2z^2 + z^4 - 1 + 2z^2 - z^4}{(1 + z^2)^2}$

$= \dfrac{4z^2}{(1 + z^2)^2}$

$\sin x = \pm \dfrac{2z}{1 + z^2}$

Only $\sin x = \dfrac{2z}{1 + z^2}$ makes sense in this case.

(d) $z = \tan \dfrac{x}{2}$

$dz = \left(\dfrac{1}{2} \sec^2 \dfrac{x}{2}\right) dx$

$dz = \dfrac{1}{2}\left(1 + \tan^2 \dfrac{x}{2}\right) dx$

$dz = \dfrac{1}{2}(1 + z^2)\, dx$

$dx = \dfrac{2\, dz}{1 + z^2}$

51. $\displaystyle\int \dfrac{dx}{1 - \cos x} = \int \dfrac{\dfrac{2\, dz}{1 + z^2}}{1 - \dfrac{1 - z^2}{1 + z^2}}$

$= \displaystyle\int \dfrac{dz}{z^2}$

$= -\dfrac{1}{z} + C = -\dfrac{1}{\tan \dfrac{x}{2}} + C$

53. $\displaystyle\int \dfrac{dt}{1 + \sin t + \cos t} = \int \dfrac{\dfrac{2\, dz}{1 + z^2}}{1 + \dfrac{2z}{1 + z^2} + \dfrac{1 - z^2}{1 + z^2}}$

$= \displaystyle\int \dfrac{dz}{z + 1}$

$= \ln |z + 1| + C$

$= \ln \left|\tan \dfrac{t}{2} + 1\right| + C$

■ Chapter 8 Review Exercises
(pp. 454–455)

1. $\displaystyle\lim_{t \to 0} \dfrac{t - \ln(1 + 2t)}{t^2} = \lim_{t \to 0} \dfrac{1 - \dfrac{2}{1 + 2t}}{2t} = \infty$ for $t \to 0^-$ and $-\infty$ for $t \to 0^+$

The limit does not exist.

2. $\displaystyle\lim_{t \to 0} \dfrac{\tan 3t}{\tan 5t} = \lim_{t \to 0} \dfrac{3 \sec^2 3t}{5 \sec^2 5t} = \dfrac{3}{5}$

3. $\displaystyle\lim_{x \to 0} \dfrac{x \sin x}{1 - \cos x} = \lim_{x \to 0} \dfrac{x \cos x + \sin x}{\sin x}$

$= \displaystyle\lim_{x \to 0} \dfrac{-x \sin x + \cos x + \cos x}{\cos x} = 2$

4. The limit leads to the indeterminate form 1^∞.

$f(x) = x^{1/(1-x)}$

$\ln f(x) = \dfrac{\ln x}{1 - x}$

$\displaystyle\lim_{x \to 1} \dfrac{\ln x}{1 - x} = \lim_{x \to 1} \dfrac{1/x}{-1} = -1$

$\displaystyle\lim_{x \to 1} x^{1/(1-x)} = \lim_{x \to 1} e^{\ln f(x)} = e^{-1} = \dfrac{1}{e}$

5. The limit leads to the indeterminate form ∞^0.

$f(x) = x^{1/x}$

$\ln f(x) = \dfrac{\ln x}{x}$

$\displaystyle\lim_{x \to \infty} \dfrac{\ln x}{x} = \lim_{x \to \infty} \dfrac{1/x}{1} = 0$

$\displaystyle\lim_{x \to \infty} x^{1/x} = \lim_{x \to \infty} e^{\ln f(x)} = e^0 = 1$

6. The limit leads to the indeterminate form 1^∞.

$$f(x) = \left(1 + \frac{3}{x}\right)^x$$

$$\ln f(x) = x \ln\left(1 + \frac{3}{x}\right) = \frac{\ln\left(1 + \frac{3}{x}\right)}{\frac{1}{x}}$$

$$\lim_{x\to\infty} \frac{\ln\left(1 + \frac{3}{x}\right)}{\frac{1}{x}} = \lim_{x\to\infty} \frac{\frac{-3/x^2}{1 + 3/x}}{-\frac{1}{x^2}} = \lim_{x\to\infty} \frac{3x}{x+3} = 3$$

$$\lim_{x\to\infty} \left(1 + \frac{3}{x}\right)^x = \lim_{x\to\infty} e^{\ln f(x)} = e^3$$

7. $\lim_{r\to\infty} \frac{\cos r}{\ln r} = 0$ since $|\cos r| \le 1$ and $\ln r \to \infty$ as $r \to \infty$.

8. $\lim_{\theta\to\pi/2} \left(\theta - \frac{\pi}{2}\right) \sec\theta = \lim_{\theta\to\pi/2} \frac{\theta - \frac{\pi}{2}}{\cos\theta} = \lim_{\theta\to\pi/2} \frac{1}{-\sin\theta} = -1$

9. $\lim_{x\to 1} \left(\frac{1}{x-1} - \frac{1}{\ln x}\right) = \lim_{x\to 1} \left[\frac{\ln x - x + 1}{(x-1)\ln x}\right]$

$$= \lim_{x\to 1} \frac{\frac{1}{x} - 1}{\frac{x-1}{x} + \ln x}$$

$$= \lim_{x\to 1} \frac{1-x}{x - 1 + x\ln x}$$

$$= \lim_{x\to 1} \frac{-1}{1 + x/x + \ln x} = -\frac{1}{2}$$

10. The limit leads to the indeterminate form ∞^0.

$$f(x) = \left(1 + \frac{1}{x}\right)^x$$

$$\ln f(x) = x\ln\left(1 + \frac{1}{x}\right) = \frac{\ln(1 + 1/x)}{1/x}$$

$$\lim_{x\to 0^+} \frac{\ln(1 + 1/x)}{1/x} = \lim_{x\to 0^+} \frac{\frac{-1/x^2}{1 + 1/x}}{-1/x^2} = \lim_{x\to 0^+} \frac{x}{x+1} = 0$$

$$\lim_{x\to 0^+} \left(1 + \frac{1}{x}\right)^x = \lim_{x\to 0^+} e^{\ln f(x)} = e^0 = 1$$

11. The limit leads to the indeterminate form 0^0.

$$f(\theta) = (\tan\theta)^\theta$$

$$\ln f(\theta) = \theta \ln(\tan\theta) = \frac{\ln(\tan\theta)}{1/\theta}$$

$$\lim_{x\to 0^+} \frac{\ln(\tan\theta)}{1/\theta} = \lim_{x\to 0^+} \frac{\frac{\sec^2\theta}{\tan\theta}}{-\frac{1}{\theta^2}}$$

$$= \lim_{x\to 0^+} -\frac{\theta^2}{\sin\theta\cos\theta}$$

$$= \lim_{x\to 0^+} \frac{-2\theta}{-\sin^2\theta + \cos^2\theta} = 0$$

$$\lim_{x\to 0^+} (\tan\theta)^\theta = \lim_{x\to 0^+} e^{\ln f(\theta)} = e^0 = 1$$

12. $\lim_{\theta\to\infty} \theta^2 \sin\left(\frac{1}{\theta}\right) = \lim_{t\to 0^+} \frac{\sin t}{t^2} = \lim_{t\to 0^+} \frac{\cos t}{2t} = \infty$

13. $\lim_{x\to\infty} \frac{x^3 - 3x^2 + 1}{2x^2 + x - 3} = \lim_{x\to\infty} \frac{3x^2 - 6x}{4x + 1} = \lim_{x\to\infty} \frac{6x - 6}{4} = \infty$

14. $\lim_{x\to\infty} \frac{3x^2 - x + 1}{x^4 - x^3 + 2} = \lim_{x\to\infty} \frac{6x - 1}{4x^3 - 3x^2} = \lim_{x\to\infty} \frac{6}{12x^2 - 6x} = 0$

15. $\lim_{x\to\infty} \frac{f(x)}{g(x)} = \lim_{x\to\infty} \frac{x}{5x} = \frac{1}{5}$

f grows at the same rate as g.

16. $\lim_{x\to\infty} \frac{f(x)}{g(x)} = \lim_{x\to\infty} \frac{\log_2 x}{\log_3 x} = \lim_{x\to\infty} \frac{(\ln x)/(\ln 2)}{(\ln x)/(\ln 3)} = \frac{\ln 3}{\ln 2}$

f grows at the same rate as g.

17. $\lim_{x\to\infty} \frac{f(x)}{g(x)} = \lim_{x\to\infty} \frac{x}{x + 1/x} = \lim_{x\to\infty} \frac{1}{1 - 1/x^2} = 1$

f grows at the same rate as g.

18. $\lim_{x\to\infty} \frac{f(x)}{g(x)} = \lim_{x\to\infty} \frac{x/100}{xe^{-x}} = \lim_{x\to\infty} \frac{e^x}{100} = \infty$

f grows faster than g.

19. $\lim_{x\to\infty} \frac{f(x)}{g(x)} = \lim_{x\to\infty} \frac{x}{\tan^{-1} x} = \infty$ since

$\lim_{x\to\infty} \tan^{-1} x = \frac{\pi}{2}$ and $\lim_{x\to\infty} x = \infty$

f grows faster than g.

20. $\lim_{x\to\infty} \frac{f(x)}{g(x)} = \lim_{x\to\infty} \frac{\csc^{-1} x}{1/x}$

$$= \lim_{x\to\infty} \frac{-\frac{1}{x\sqrt{x^2 - 1}}}{-\frac{1}{x^2}}$$

$$= \lim_{x\to\infty} \frac{x}{\sqrt{x^2 - 1}}$$

$$= \lim_{x\to\infty} \sqrt{\frac{x^2}{x^2 - 1}}$$

$$= \lim_{x\to\infty} \sqrt{\frac{1}{1 - 1/x^2}} = 1$$

f grows at the same rate as g.

21. $\displaystyle\lim_{x\to\infty}\frac{f(x)}{g(x)}=\lim_{x\to\infty}\frac{x^{\ln x}}{x^{\log_2 x}}$
$\displaystyle=\lim_{x\to\infty} x^{\ln x - \log_2 x}$
$\displaystyle=\lim_{x\to\infty} x^{\ln x - (\ln x)/\ln 2}$
$\displaystyle=\lim_{x\to\infty} x^{(\ln x)(1 - 1/\ln 2)}$
$\displaystyle=\lim_{x\to\infty}\left(\frac{1}{x}\right)^{(\ln x)(1/\ln 2 - 1)} = 0$

Note that $1 - \dfrac{1}{\ln 2} < 0$ since $\ln 2 < 1$.

f grows slower than g.

22. $\displaystyle\lim_{x\to\infty}\frac{f(x)}{g(x)}=\lim_{x\to\infty}\frac{3^{-x}}{2^{-x}}=\lim_{x\to\infty}\frac{2^x}{3^x}=\lim_{x\to\infty}\left(\frac{2}{3}\right)^x = 0$ since $\dfrac{2}{3} < 1$.

f grows slower than g.

23. $\displaystyle\lim_{x\to\infty}\frac{f(x)}{g(x)}=\lim_{x\to\infty}\frac{\ln 2x}{\ln x^2}=\lim_{x\to\infty}\frac{\ln x + \ln 2}{2\ln x}=\lim_{x\to\infty}\frac{1/x}{2/x}=\frac{1}{2}$

f grows at the same rate as g.

24. $\displaystyle\lim_{x\to\infty}\frac{f(x)}{g(x)}=\lim_{x\to\infty}\frac{10x^3 + 2x^2}{e^x}$
$\displaystyle=\lim_{x\to\infty}\frac{30x^2 + 4x}{e^x}$
$\displaystyle=\lim_{x\to\infty}\frac{60x + 4}{e^x}$
$\displaystyle=\lim_{x\to\infty}\frac{60}{e^x} = 0$

f grows slower than g.

25. $\displaystyle\lim_{x\to\infty}\frac{f(x)}{g(x)}=\lim_{x\to\infty}\frac{\tan^{-1}(1/x)}{1/x}$
$\displaystyle=\lim_{x\to\infty}\frac{\frac{1}{1+(1/x)^2}(-x^{-2})}{-x^{-2}}$
$\displaystyle=\lim_{x\to\infty}\frac{1}{1+(1/x)^2} = 1$

f grows at the same rate as g.

26. $\displaystyle\lim_{x\to\infty}\frac{f(x)}{g(x)}=\lim_{x\to\infty}\frac{\sin^{-1}(1/x)}{(1/x^2)}$
$\displaystyle=\lim_{x\to\infty}\frac{\frac{1}{\sqrt{1-(1/x^2)}}(-x^{-2})}{-2x^{-3}}$
$\displaystyle=\lim_{x\to\infty}\frac{x}{2\sqrt{1-(1/x)^2}} = \infty$

f grows faster than g.

27. (a) $\displaystyle\lim_{x\to 0} f(x) = \lim_{x\to 0}\frac{2^{\sin x} - 1}{e^x - 1} = \lim_{x\to 0}\frac{(\ln 2)(\cos x)2^{\sin x}}{e^x} = \ln 2$

(b) Define $f(0) = \ln 2$.

28. (a) $\displaystyle\lim_{x\to 0^+} f(x) = \lim_{x\to 0^+} x \ln x$
$\displaystyle= \lim_{x\to 0^+}\frac{\ln x}{1/x}$
$\displaystyle= \lim_{x\to 0^+}\frac{1/x}{-1/x^2}$
$\displaystyle= \lim_{x\to 0^+}(-x) = 0$

(b) Define $f(0) = 0$.

29. $\displaystyle\lim_{x\to\infty}\frac{\frac{1}{x^2}+\frac{1}{x^4}}{\frac{1}{x^2}}=\lim_{x\to\infty}\left(1+\frac{1}{x^2}\right) = 1 \le 1$

True

30. $\displaystyle\lim_{x\to\infty}\frac{\frac{1}{x^2}+\frac{1}{x^4}}{\frac{1}{x^4}}=\lim_{x\to\infty}(x^2 + 1) = \infty$

False

31. $\displaystyle\lim_{x\to\infty}\frac{x}{x+\ln x}=\lim_{x\to\infty}\frac{1}{1+\frac{1}{x}} = 1 \ne 0$

False

32. $\displaystyle\lim_{x\to\infty}\frac{\ln(\ln x)}{\ln x}=\lim_{x\to\infty}\frac{\frac{1}{x\ln x}}{\frac{1}{x}}=\lim_{x\to\infty}\frac{1}{\ln x} = 0$

True

33. $\displaystyle\lim_{x\to\infty}\frac{\tan^{-1} x}{1} = \frac{\pi}{2} \le \frac{\pi}{2}$

True

34. $\displaystyle\lim_{x\to\infty}\frac{\frac{1}{x^4}}{\frac{1}{x^2}+\frac{1}{x^4}}=\lim_{x\to\infty}\frac{1}{x^2+1} = 0 \le 1$

True

35. $\displaystyle\lim_{x\to\infty}\frac{\frac{1}{x^4}}{\frac{1}{x^2}+\frac{1}{x^4}}=\lim_{x\to\infty}\frac{1}{x^2+1} = 0$

True

36. $\displaystyle\lim_{x\to\infty}\frac{\ln x}{x+1}=\lim_{x\to\infty}\frac{\frac{1}{x}}{1} = 0$

True

37. $\displaystyle\lim_{x\to\infty}\frac{\ln 2x}{\ln x}=\lim_{x\to\infty}\frac{\frac{1}{x}}{\frac{1}{x}} = 1 \le 1$

True

38. $\displaystyle\lim_{x\to\infty}\frac{\sec^{-1} x}{1} = \frac{\pi}{2} \le \frac{\pi}{2}$

True

39. $x = 3\sin\theta,\ dx = 3\cos\theta\, d\theta,\ -\frac{\pi}{2} \le \theta \le \frac{\pi}{2}$

$$\int \frac{dx}{\sqrt{9-x^2}} = \int \frac{3\cos\theta\, d\theta}{|3\cos\theta|}$$

$$= \int d\theta$$

$$= \theta + C$$

$$= \sin^{-1}\frac{x}{3} + C$$

$$\int_0^3 \frac{dx}{\sqrt{9-x^2}} = \lim_{b\to 3^-} \int_0^b \frac{dx}{\sqrt{9-x^2}}$$

$$= \lim_{b\to 3^-} \left[\sin^{-1}\frac{x}{3}\right]_0^b$$

$$= \lim_{b\to 3^-} \left(\sin^{-1}\frac{b}{3}\right) = \frac{\pi}{2}$$

40. $u = \ln x \qquad dv = dx$

$du = \frac{1}{x}\, dx \qquad v = x$

$$\int \ln x\, dx = x\ln x - \int dx = x\ln x - x + C$$

$$\int_0^1 \ln x\, dx = \lim_{b\to 0^+} \int_b^1 \ln x\, dx$$

$$= \lim_{b\to 0^+} \left[x\ln x - x\right]_b^1$$

$$= \lim_{b\to 0^+} (-1 - b\ln b + b)$$

$$= -1 - \lim_{b\to 0^+} \frac{\ln b}{1/b}$$

$$= -1 - \lim_{b\to 0^+} \frac{1/b}{-1/b^2}$$

$$= -1 - \lim_{b\to 0^+} (-b) = -1$$

41. $\int_{-1}^1 \frac{dy}{y^{2/3}} = \int_{-1}^0 \frac{dy}{y^{2/3}} + \int_0^1 \frac{dy}{y^{2/3}}$

$$\int_{-1}^0 \frac{dy}{y^{2/3}} = \lim_{b\to 0^-} \int_{-1}^b y^{-2/3}\, dy$$

$$= \lim_{b\to 0^-} \left[3y^{1/3}\right]_{-1}^b$$

$$= \lim_{b\to 0^-} [3b^{1/3} + 3] = 3$$

$$\int_0^1 \frac{dy}{y^{2/3}} = \lim_{b\to 0^+} \int_b^1 y^{-2/3}\, dy$$

$$= \lim_{b\to 0^+} \left[3y^{1/3}\right]_b^1$$

$$= \lim_{b\to 0^+} [3 - 3b^{1/3}] = 3$$

$$\int_{-1}^1 \frac{dy}{y^{2/3}} = 3 + 3 = 6$$

42. $\int_{-2}^0 \frac{d\theta}{(\theta+1)^{3/5}} = \int_{-2}^{-1} \frac{d\theta}{(\theta+1)^{3/5}} + \int_{-1}^0 \frac{d\theta}{(\theta+1)^{3/5}}$

$$\int_{-2}^{-1} \frac{d\theta}{(\theta+1)^{3/5}} = \lim_{b\to -1^-} \int_{-2}^b \frac{d\theta}{(\theta+1)^{3/5}}$$

$$= \lim_{b\to -1^-} \left[\frac{5}{2}(\theta+1)^{2/5}\right]_{-2}^b$$

$$= \lim_{b\to -1^-} \left[\frac{5}{2}(b+1)^{2/5} - \frac{5}{2}\right] = -\frac{5}{2}$$

$$\int_{-1}^0 \frac{d\theta}{(\theta+1)^{3/5}} = \lim_{b\to -1^+} \int_b^0 \frac{d\theta}{(\theta+1)^{3/5}}$$

$$= \lim_{b\to -1^+} \left[\frac{5}{2}(\theta+1)^{2/5}\right]_b^0$$

$$= \lim_{b\to -1^+} \left[\frac{5}{2} - \frac{5}{2}(b+1)^{2/5}\right] = \frac{5}{2}$$

$$\int_{-2}^0 \frac{d\theta}{(\theta+1)^{3/5}} = -\frac{5}{2} + \frac{5}{2} = 0$$

43. $\int_3^\infty \frac{2\, dx}{x^2 - 2x} = \lim_{b\to\infty} \int_3^b \frac{2\, dx}{x(x-2)}$

$$\frac{2}{x(x-2)} = \frac{A}{x} + \frac{B}{x-2}$$

$$2 = A(x-2) + Bx = (A+B)x - 2A$$

where $A = -1,\ B = 1$.

$$\int_3^\infty \frac{2\, dx}{x(x-2)} = \lim_{b\to\infty} \int_3^b \left(-\frac{1}{x} + \frac{1}{x-2}\right) dx$$

$$= \lim_{b\to\infty} \left[-\ln|x| + \ln|x+2|\right]_3^b$$

$$= \lim_{b\to\infty} \left[\ln\frac{x-2}{x}\right]_3^b$$

$$= \lim_{b\to\infty} \left(\ln\frac{b+2}{b} - \ln\frac{1}{3}\right) = \ln 3$$

44. $\int_1^\infty \dfrac{3x-1}{4x^3-x^2}\,dx = \lim_{b\to\infty}\int_1^b \dfrac{3x-1}{x^2(4x-1)}\,dx$

$\dfrac{3x-1}{x^2(4x-1)} = \dfrac{A}{4x-1} + \dfrac{B}{x} + \dfrac{C}{x^2}$

$3x - 1 = Ax^2 + Bx(4x-1) + C(4x-1)$

$ = (A+4B)x^2 + (-B+4C)x - C$

where $A = -4,\ B = 1,\ C = 1$

$\int_1^\infty \dfrac{3x-1}{4x^3-x^2}\,dx = \lim_{b\to\infty}\int_1^b \left(-\dfrac{4}{4x-1} + \dfrac{1}{x} + \dfrac{1}{x^2}\right)dx$

$= \lim_{b\to\infty}\left[-\ln|4x-1| + \ln|x| - \dfrac{1}{x}\right]_1^b$

$= \lim_{b\to\infty}\left(\ln\dfrac{b}{4b-1} - \dfrac{1}{b} + \ln 3 + 1\right)$

$= \ln\dfrac{1}{4} + \ln 3 + 1 = \ln\dfrac{3}{4} + 1$

45. $u = x^2 \qquad dv = e^{-x}\,dx$

$du = 2x\,dx \qquad v = -e^{-x}$

$\int x^2 e^{-x}\,dx = -x^2 e^{-x} + \int 2x\, e^{-x}\,dx$

$u = 2x \qquad dv = e^{-x}\,dx$

$du = 2\,dx \qquad v = -e^{-x}$

$\int x^2 e^{-x}\,dx = -x^2 e^{-x} - 2x\, e^{-x} + \int 2e^{-x}\,dx$

$\phantom{\int x^2 e^{-x}\,dx} = -x^2 e^{-x} - 2x\, e^{-x} - 2e^{-x} + C$

$\int_0^\infty x^2 e^{-x}\,dx = \lim_{b\to\infty}\int_0^b x^2 e^{-x}\,dx$

$= \lim_{b\to\infty}\left[-x^2 e^{-x} - 2x\,e^{-x} - 2e^{-x}\right]_0^b$

$= \lim_{b\to\infty}\left[-\dfrac{b^2}{e^b} - \dfrac{2b}{e^b} - \dfrac{2}{e^b} + 2\right] = 2$

46. $u = x \qquad dv = e^{3x}\,dx$

$du = dx \qquad v = \dfrac{1}{3}e^{3x}$

$\int x\, e^{3x}\,dx = \dfrac{1}{3}x\, e^{3x} - \int \dfrac{1}{3}e^{3x}\,dx$

$\phantom{\int x\, e^{3x}\,dx} = \dfrac{1}{3}x\, e^{3x} - \dfrac{1}{9}e^{3x} + C$

$\int_{-\infty}^0 x\, e^{3x}\,dx = \lim_{b\to-\infty}\int_b^0 x\, e^{3x}\,dx$

$= \lim_{b\to-\infty}\left[\dfrac{1}{3}x\, e^{3x} - \dfrac{1}{9}e^{3x}\right]_b^0$

$= \lim_{b\to-\infty}\left[-\dfrac{1}{9} - \dfrac{1}{3}b\, e^{3b} + \dfrac{1}{9}e^{3b}\right] = -\dfrac{1}{9}$

47. $\dfrac{4t^3 + t - 1}{t^2(t-1)(t^2+1)} = \dfrac{A}{t} + \dfrac{B}{t^2} + \dfrac{C}{t-1} + \dfrac{Dt+E}{t^2+1}$

$4t^3 + t - 1 = At(t-1)(t^2+1) + B(t-1)(t^2+1)$

$\qquad + Ct^2(t^2+1) + (Dt+E)t^2(t-1)$

$= (A+C+D)t^4 + (-A+B-D+E)t^3$

$\qquad + (A-B+C-E)t^2 + (-A+B)t - B$

Equating coefficients of like terms gives

$A + C + D = 0,\ -A + B - D + E = 4,$

$A - B + C - E = 0,\ -A + B = 1 \text{ and } -B = -1.$

Solving the system simultaneously yields

$A = 0,\ B = 1,\ C = 2,\ D = -2,\ E = 1.$

$\int \dfrac{4t^3 + t - 1}{t^2(t-1)(t^2+1)}\,dt$

$= \int \dfrac{dt}{t^2} + \int \dfrac{2\,dt}{t-1} + \int \dfrac{-2t+1}{t^2+1}\,dt$

$= \int \dfrac{dt}{t^2} + \int \dfrac{2\,dt}{t-1} - \int \dfrac{2t\,dt}{t^2+1} + \int \dfrac{1\,dt}{t^2+1}$

$= -\dfrac{1}{t} + 2\ln|t-1| - \ln|t^2+1| + \tan^{-1} t + C$

$= -\dfrac{1}{t} + \ln\dfrac{(t-1)^2}{t^2+1} + \tan^{-1} t + C$

$\int_{-\infty}^\infty \dfrac{4t^3 + t - 1}{t^2(t-1)(t^2+1)}\,dt$

$= \int_{-\infty}^{-1} \dfrac{4t^3+t-1}{t^2(t-1)(t^2+1)}\,dt + \int_{-1}^0 \dfrac{4t^3+t-1}{t^2(t-1)(t^2+1)}\,dt$

$+ \int_0^{1/2} \dfrac{4t^3+t-1}{t^2(t-1)(t^2+1)}\,dt + \int_{1/2}^1 \dfrac{4t^3+t-1}{t^2(t-1)(t^2+1)}\,dt$

$+ \int_1^2 \dfrac{4t^3+t-1}{t^2(t-1)(t^2+1)}\,dt + \int_2^\infty \dfrac{4t^3+t-1}{t^2(t-1)(t^2+1)}\,dt$

Note that the integral must be broken up since the integrand has infinite discontinuities at $t = 0$ and $t = 1$.

$\int_{-1}^0 \dfrac{4t^3+t-1}{t^2(t-1)(t^2+1)}\,dt$

$= \lim_{b\to 0^-}\int_{-1}^b \dfrac{4t^3+t-1}{t^2(t-1)(t^2+1)}\,dt$

$= \lim_{b\to 0^-}\left[-\dfrac{1}{t} + \ln\dfrac{(t-1)^2}{t^2+1} + \tan^{-1} t\right]_{-1}^b$

$= \lim_{b\to 0^-}\left[-\dfrac{1}{b} + \ln\dfrac{(b-1)^2}{b^2+1} + \tan^{-1}\right.$

$\left. b - 1 - \ln 2 + \dfrac{\pi}{4}\right] = \infty$

Since this limit diverges, the given integral diverges.

48. $\displaystyle\int_{-\infty}^{\infty} \frac{4\,dx}{x^2+16} = \int_{-\infty}^{0} \frac{4\,dx}{x^2+16} + \int_{0}^{\infty} \frac{4\,dx}{x^2+16}$

$\displaystyle\int \frac{4\,dx}{x^2+16} = \tan^{-1}\frac{x}{4} + C$ using Formula 16 with $a=4$

$\displaystyle\int_{-\infty}^{0} \frac{4\,dx}{x^2+16} = \lim_{b\to -\infty} \int_{b}^{0} \frac{4\,dx}{x^2+16}$

$\displaystyle\qquad = \lim_{b\to -\infty}\left[\tan^{-1}\frac{x}{4}\right]_{b}^{0}$

$\displaystyle\qquad = \lim_{b\to -\infty}\left(-\tan^{-1}\frac{b}{4}\right) = \frac{\pi}{2}$

$\displaystyle\int_{0}^{\infty} \frac{4\,dx}{x^2+16} = \lim_{b\to\infty} \int_{0}^{b} \frac{4\,dx}{x^2+16}$

$\displaystyle\qquad = \lim_{b\to\infty}\left[\tan^{-1}\frac{x}{4}\right]_{0}^{b}$

$\displaystyle\qquad = \lim_{b\to\infty}\left(\tan^{-1}\frac{b}{4}\right) = \frac{\pi}{2}$

$\displaystyle\int_{-\infty}^{\infty} \frac{4\,dx}{x^2+16} = \frac{\pi}{2} + \frac{\pi}{2} = \pi$

49. Use the limit comparison test with $f(\theta) = \dfrac{1}{\sqrt{\theta^2+1}}$ and $g(\theta) = \dfrac{1}{\theta}$. Both are positive continuous functions on $[1,\infty)$.

$\displaystyle\lim_{\theta\to\infty}\frac{f(\theta)}{g(\theta)} = \lim_{\theta\to\infty}\frac{\sqrt{\theta^2+1}}{\theta} = \lim_{\theta\to\infty}\sqrt{1+\frac{1}{\theta^2}} = 1$

Since $\displaystyle\int_{1}^{\infty} g(\theta)\,d\theta = \int_{1}^{\infty}\frac{1}{\theta}\,d\theta$

$\displaystyle\qquad = \lim_{b\to\infty}\int_{1}^{b}\frac{1}{\theta}\,d\theta$

$\displaystyle\qquad = \lim_{b\to\infty}\left[\ln\theta\right]_{1}^{b}$

$\displaystyle\qquad = \lim_{b\to\infty}\ln b$

$\displaystyle\qquad = \infty,$

we know that $\displaystyle\int_{1}^{\infty} g(\theta)\,d\theta$ diverges and so $\displaystyle\int_{1}^{\infty} f(\theta)\,d\theta$ diverges. This means that the given integral diverges.

50. Evaluate $\int e^{-x}\cos x\,dx$ using integration by parts.

$u = \cos x \qquad dv = e^{-x}\,dx$
$du = -\sin x\,dx \qquad v = -e^{-x}$

$\displaystyle\int e^{-x}\cos x\,dx = -e^{-x}\cos x - \int \sin x\, e^{-x}\,dx$

Evaluate $\int \sin x\, e^{-x}\,dx$ using integration by parts.

$u = \sin x \qquad dv = e^{-x}\,dx$
$du = \cos x\,dx \qquad v = -e^{-x}$

$\displaystyle\int \sin x\, e^{-x}\,dx = -e^{-x}\sin x + \int e^{-x}\cos x\,dx$

$\displaystyle\int e^{-x}\cos x\,dx = -e^{-x}\cos x + e^{-x}\sin x - \int e^{-x}\cos x\,dx$

$\displaystyle 2\int e^{-x}\cos x\,dx = e^{-x}\sin x - e^{-x}\cos x + C_1$

$\displaystyle\int e^{-x}\cos x\,dx = \frac{e^{-x}\sin x - e^{-x}\cos x}{2} + C$

$\displaystyle\int_{0}^{\infty} e^{-u}\cos u\,du = \lim_{b\to\infty}\int_{0}^{b} e^{-x}\cos x\,dx$

$\displaystyle\qquad = \lim_{b\to\infty}\left[\frac{e^{-x}\sin x - e^{-x}\cos x}{2}\right]_{0}^{b}$

$\displaystyle\qquad = \lim_{b\to\infty}\left[\frac{e^{-b}\sin b - e^{-b}\cos b}{2} + \frac{1}{2}\right]$

$\displaystyle\qquad = \frac{1}{2}$

Note that we cannot use a comparison test since $e^{-x}\cos x < 0$ for some values on $[0,\infty)$.

51. $0 \le \dfrac{1}{z} \le \dfrac{\ln z}{z}$ on $[e,\infty)$

$\displaystyle\int_{e}^{\infty} \frac{dz}{z} = \lim_{b\to\infty}\int_{e}^{b}\frac{dz}{z} = \lim_{b\to\infty}\left[\ln|z|\right]_{e}^{b} = \lim_{b\to\infty}(\ln b - 1) = \infty$

Since this integral diverges, $\displaystyle\int_{e}^{\infty}\frac{1}{z}\,dz$ diverges, so the given integral diverges.

52. $0 \le \dfrac{e^{-t}}{\sqrt{t}} \le e^{-t}$ on $[1,\infty)$

$\displaystyle\int_{1}^{\infty} e^{-t}\,dt = \lim_{b\to\infty}\int_{1}^{b} e^{-t}\,dt$

$\displaystyle\qquad = \lim_{b\to\infty}\left[-e^{-t}\right]_{1}^{b}$

$\displaystyle\qquad = \lim_{b\to\infty}\left(-e^{-b} + \frac{1}{e}\right) = \frac{1}{e}$

Since this integral converges, the given integral converges.

53. $\int \dfrac{dx}{e^x + e^{-x}} = \int \dfrac{dx}{e^{-x}(e^{2x} + 1)} = \int \dfrac{e^x \, dx}{(e^x)^2 + 1}$

Let $u = e^x$, $du = e^x \, dx$

$\int \dfrac{dx}{e^x + e^{-x}} = \int \dfrac{du}{u^2 + 1} = \tan^{-1} u + C = \tan^{-1} e^x + C$

$\displaystyle\int_{-\infty}^{\infty} \dfrac{dx}{e^x + e^{-x}} = \int_{-\infty}^{0} \dfrac{dx}{e^x + e^{-x}} + \int_{0}^{\infty} \dfrac{dx}{e^x + e^{-x}}$

$\displaystyle\int_{-\infty}^{0} \dfrac{dx}{e^x + e^{-x}} = \lim_{b \to -\infty} \int_{b}^{0} \dfrac{dx}{e^x + e^{-x}}$

$= \displaystyle\lim_{b \to -\infty} \left[\tan^{-1} e^x\right]_{b}^{0}$

$= \displaystyle\lim_{b \to -\infty} \left(\dfrac{\pi}{4} - \tan^{-1} e^b\right) = \dfrac{\pi}{4}$

$\displaystyle\int_{0}^{\infty} \dfrac{dx}{e^x + e^{-x}} = \lim_{b \to \infty} \int_{0}^{b} \dfrac{dx}{e^x + e^{-x}}$

$= \displaystyle\lim_{b \to \infty} \left[\tan^{-1} e^x\right]_{0}^{b}$

$= \displaystyle\lim_{b \to \infty} \left(\tan^{-1} e^b - \dfrac{\pi}{4}\right)$

$= \dfrac{\pi}{2} - \dfrac{\pi}{4} = \dfrac{\pi}{4}$

Since these two integrals converge, the given integral converges.

54. The integral has an infinite discontinuity at $x = 0$.

$\displaystyle\int_{-\infty}^{\infty} \dfrac{dx}{x^2(1 + e^x)} = \int_{-\infty}^{-1} \dfrac{dx}{x^2(1 + e^x)} + \int_{-1}^{0} \dfrac{dx}{x^2(1 + e^x)}$
$\quad + \displaystyle\int_{0}^{1} \dfrac{dx}{x^2(1 + e^x)} + \int_{1}^{\infty} \dfrac{dx}{x^2(1 + e^x)}$

$0 \le \dfrac{1}{4x^2} \le \dfrac{1}{x^2(1 + e^x)}$ on $(0, 1]$ since $1 + e^x \le 4$ on $(0, 1]$.

$\displaystyle\int_{0}^{1} \dfrac{dx}{4x^2} = \lim_{b \to 0^+} \int_{b}^{1} \dfrac{dx}{4x^2} = \lim_{b \to 0^+} \left[-\dfrac{1}{4x}\right]_{b}^{1} = \lim_{b \to 0^+} \left[-\dfrac{1}{4} + \dfrac{1}{4b}\right] = \infty$

Since this integral diverges, $\displaystyle\int_{0}^{1} \dfrac{dx}{x^2(1 + e^x)}$ diverges, so the given integral diverges.

55. $x^2 - 7x + 12 = (x - 4)(x - 3)$

$\dfrac{2x + 1}{x^2 - 7x + 12} = \dfrac{A}{x - 4} + \dfrac{B}{x - 3}$

$2x + 1 = A(x - 3) + B(x - 4)$

$= (A + B)x - 3A - 4B$

Equating coefficients of like terms gives

$A + B = 2$ and $-3A - 4B = 1$.

Solving the system simultaneously yields $A = 9$, $B = -7$.

$\displaystyle\int \dfrac{2x + 1}{x^2 - 7x + 12} \, dx = \int \dfrac{9 \, dx}{x - 4} + \int \dfrac{-7 \, dx}{x - 3}$

$= 9 \ln|x - 4| - 7 \ln|x - 3| + C$

56. $\dfrac{8}{x^3(x + 2)} = \dfrac{A}{x + 2} + \dfrac{B}{x} + \dfrac{C}{x^2} + \dfrac{D}{x^3}$

$8 = Ax^3 + Bx^2(x + 2) + Cx(x + 2) + D(x + 2)$

$= (A + B)x^3 + (2B + C)x^2 + (2C + D)x + 2D$

Equating coefficients of like terms gives

$A + B = 0$, $2B + C = 0$, $2C + D = 0$, and $2D = 8$

Solving the system simultaneously yields

$A = -1$, $B = 1$, $C = -2$, $D = 4$

$\displaystyle\int \dfrac{8 \, dx}{x^3(x + 2)} = \int \dfrac{-dx}{x + 2} + \int \dfrac{dx}{x} + \int \dfrac{-2 \, dx}{x^2} + \int \dfrac{4 \, dx}{x^3}$

$= -\ln|x + 2| + \ln|x| + \dfrac{2}{x} - \dfrac{2}{x^2} + C$

57. $t^3 + t = t(t^2 + 1)$

$\dfrac{3t^2 + 4t + 4}{t^3 + t} = \dfrac{A}{t} + \dfrac{Bt + C}{t^2 + 1}$

$3t^2 + 4t + 4 = A(t^2 + 1) + (Bt + C)t$

$= (A + B)t^2 + Ct + A$

Equating coefficients of like terms gives

$A + B = 3$, $C = 4$ and $A = 4$.

Solving the system simultaneously yields

$A = 4$, $B = -1$, $C = 4$

$\displaystyle\int \dfrac{3t^2 + 4t + 4}{t^3 + t} \, dt = \int \dfrac{4 \, dt}{t} + \int \dfrac{-t + 4}{t^2 + 1} \, dt$

$= \displaystyle\int \dfrac{4 \, dt}{t} - \int \dfrac{t \, dt}{t^2 + 1} + \int \dfrac{4 \, dt}{t^2 + 1}$

$= 4 \ln|t| - \dfrac{1}{2} \ln(t^2 + 1) + 4 \tan^{-1} t + C$

58. $t^4 + 4t^2 + 3 = (t^2 + 3)(t^2 + 1)$

$\dfrac{1}{(t^2 + 1)(t^2 + 3)} = \dfrac{At + B}{t^2 + 1} + \dfrac{Ct + D}{t^2 + 3}$

$1 = (At + B)(t^2 + 3) + (Ct + D)(t^2 + 1)$

$= (A + C)t^3 + (B + D)t^2 + (3A + C)t + 3B + D$

Equating coefficients of like terms gives

$A + C = 0$, $B + D = 0$, $3A + C = 0$, and $3B + D = 1$

Solving the system simultaneously yields

$A = 0$, $B = \dfrac{1}{2}$, $C = 0$, $D = -\dfrac{1}{2}$

$\displaystyle\int \dfrac{dt}{(t^2 + 3)(t^2 + 1)} = \int \dfrac{1/2}{t^2 + 1} \, dt - \int \dfrac{1/2}{t^2 + 3} \, dt$

$= \dfrac{1}{2} \tan^{-1} t - \dfrac{1}{2\sqrt{3}} \tan^{-1} \dfrac{t}{\sqrt{3}} + C$

Evaluate the integrals using Formula 16, with $x = t$, $a = 1$ in the first integral and $a = \sqrt{3}$ in the second.

59. Long division:
$$x^3 - x \overline{\smash{)}x^3 + 1}$$
$$\underline{x^3 - x}$$
$$x + 1$$

$$\frac{x^3+1}{x^3-x} = 1 + \frac{x+1}{x^3-x} = 1 + \frac{x+1}{x(x^2-1)} = 1 + \frac{x+1}{x(x-1)(x+1)}$$

$$\frac{x+1}{x(x-1)(x+1)} = \frac{A}{x} + \frac{B}{x-1} + \frac{C}{x+1}$$

$$x + 1 = A(x-1)(x+1) + Bx(x+1) + Cx(x-1)$$

$$= (A + B + C)x^2 + (B - C)x - A$$

Equating coefficients of like terms gives

$A + B + C = 0$, $B - C = 1$, and $-A = 1$.

Solving the system simultaneously yields

$A = -1, B = 1, C = 0$

$$\int \frac{x^3+1}{x^3-x}\,dx = \int dx - \int \frac{dx}{x} + \int \frac{dx}{x-1}$$

$$= x - \ln|x| + \ln|x-1| + C$$

60. Long division:
$$x^2 + 4x + 3 \overline{\smash{)}x^3 + 4x^2}$$
$$\underline{x^3 + 4x^2 + 3x}$$
$$-3x$$

$$\frac{x^3+4x^2}{x^2+4x+3} = x + \frac{-3x}{x^2+4x+3} = x + \frac{-3x}{(x+1)(x+3)}$$

$$\frac{-3x}{(x+1)(x+3)} = \frac{A}{x+1} + \frac{B}{x+3}$$

$$-3x = A(x+3) + B(x+1)$$

$$-3x = (A+B)x + 3A + B$$

Equating coefficients of like terms gives

$A + B = -3$, $3A + B = 0$

Solving the system simultaneously yields $A = \frac{3}{2}, B = -\frac{9}{2}$.

$$\int \frac{x^3+4x^2}{x^2+4x+3}\,dx = \int x\,dx + \int \frac{3/2}{x+1}\,dx - \int \frac{9/2}{x+3}\,dx$$

$$= \frac{x^2}{2} + \frac{3}{2}\ln|x+1| - \frac{9}{2}\ln|x+3| + C$$

61. $\dfrac{dy}{y(500-y)} = 0.002\,dx$

$\dfrac{1}{y(500-y)} = \dfrac{A}{y} + \dfrac{B}{500-y}$

$1 = A(500-y) + By$

$= (B-A)y + 500A$

where $A = \dfrac{1}{500}, B = \dfrac{1}{500}$.

$$\int \frac{dy}{y(500-y)} = \int \frac{1/500}{y}\,dy + \int \frac{1/500}{500-y}\,dy$$

$$= \frac{1}{500}\ln|y| - \frac{1}{500}\ln|500-y| + C_1$$

$$= \frac{1}{500}\ln\left|\frac{y}{500-y}\right| + C_1$$

$\dfrac{1}{500}\ln\left|\dfrac{y}{500-y}\right| + C_1 = 0.002x + C_2$

$\ln\left|\dfrac{y}{500-y}\right| = x + C$

$\dfrac{y}{500-y} = ke^x$

Substitute $x = 0, y = 20$.

$\dfrac{20}{480} = ke^0$ or $k = \dfrac{1}{24}$

$\dfrac{y}{500-y} = \dfrac{1}{24}e^x$

$24y = 500e^x - ye^x$

$(e^x + 24)y = 500e^x$

$y = \dfrac{500\,e^x}{e^x + 24}$

$y = \dfrac{500}{1 + 24\,e^{-x}}$

62. $\dfrac{dy}{y^2+1} = \dfrac{dx}{x+1}$

$\int \dfrac{dy}{y^2+1} = \int \dfrac{dx}{x+1}$

$\tan^{-1} y + C_1 = \ln|x+1| + C_2$

$\tan^{-1} y = \ln|x+1| + C$

Substitute $x = 0, y = \dfrac{\pi}{4}$.

$\tan^{-1}\dfrac{\pi}{4} = C$

$\tan^{-1} y = \ln|x+1| + \tan^{-1}\dfrac{\pi}{4}$

$y = \tan\left(\ln|x+1| + \tan^{-1}\dfrac{\pi}{4}\right)$

63. $y = \frac{1}{3}\tan\theta$, $dy = \frac{1}{3}\sec^2\theta\,d\theta$, $-\frac{\pi}{2} < \theta < \frac{\pi}{2}$

$1 + 9y^2 = 1 + \tan^2\theta = \sec^2\theta$

$$\int \frac{3\,dy}{\sqrt{1+9y^2}} = \int \frac{\sec^2\theta\,d\theta}{|\sec\theta|}$$
$$= \int \sec\theta\,d\theta$$
$$= \ln|\sec\theta + \tan\theta| + C$$
$$= \ln\left|\sqrt{1+9y^2} + 3y\right| + C$$

Integrate by using Formula 88 with $a = 1$ and $x = \theta$.

Use Figure 8.18(a) from the text with $a = \frac{1}{3}$.

64. $t = \frac{1}{3}\sin\theta$, $dt = \frac{1}{3}\cos\theta\,d\theta$, $-\frac{\pi}{2} \le \theta \le \frac{\pi}{2}$

$1 - 9t^2 = 1 - \sin^2\theta = \cos^2\theta$

$$\int \sqrt{1-9t^2}\,dt = \int |\cos\theta|\left(\frac{1}{3}\cos\theta\right)d\theta$$
$$= \int \frac{1}{3}\cos^2\theta\,d\theta$$
$$= \int \frac{1+\cos 2\theta}{6}\,d\theta$$
$$= \frac{\theta}{6} + \frac{\sin 2\theta}{12} + C$$
$$= \frac{\theta}{6} + \frac{\sin\theta\cos\theta}{6} + C$$
$$= \frac{\sin^{-1} 3t}{6} + \frac{3t\sqrt{1-9t^2}}{6} + C$$
$$= \frac{1}{6}\sin^{-1} 3t + \frac{1}{2}t\sqrt{1-9t^2} + C$$

Use Figure 8.18(b) with $a = \frac{1}{3}$ and $x = t$.

65. $x = \frac{3}{5}\sec\theta$, $dx = \frac{3}{5}\sec\theta\tan\theta$, $0 \le \theta < \frac{\pi}{2}$

$25x^2 - 9 = 9\sec^2\theta - 9 = 9\tan^2\theta$

$$\int \frac{5\,dx}{\sqrt{25x^2-9}} = \int \frac{3\sec\theta\tan\theta\,d\theta}{3|\tan\theta|}$$
$$= \int \sec\theta\,d\theta$$
$$= \ln|\sec\theta + \tan\theta| + C_1$$
$$= \ln\left|\frac{5x}{3} + \frac{\sqrt{25x^2-9}}{3}\right| + C_1$$
$$= \ln(5x + \sqrt{25x^2-9}) + C$$

Integrate by using Formula 88 with $a = 1$ and $x = \theta$.

Use Figure 8.18(c) with $a = \frac{3}{5}$.

66. $x = \sin\theta$, $dx = \cos\theta\,d\theta$, $-\frac{\pi}{2} \le \theta \le \frac{\pi}{2}$

$1 - x^2 = \cos^2\theta$

$$\int \frac{4x^2\,dx}{(1-x^2)^{3/2}} = \int \frac{4\sin^2\theta\cos\theta}{|\cos^3\theta|}\,d\theta$$
$$= \int \frac{4(1-\cos^2\theta)}{\cos^2\theta}\,d\theta$$
$$= \int (4\sec^2\theta - 4)\,d\theta$$
$$= 4\tan\theta - 4\theta + C$$
$$= \frac{4x}{\sqrt{1-x^2}} - 4\sin^{-1}x + C$$

Use Figure 8.18(b) with $a = 1$.

67. For $x \ge 0$, $y \ge 0$ on $(0, 1]$.

$$\text{Volume} = \int_0^1 \pi(-\ln x)^2\,dx$$
$$= \pi\int_0^1 (\ln x)^2\,dx$$
$$= \pi\lim_{b\to 0^+}\int_b^1 (\ln x)^2\,dx$$

Evaluate $\int (\ln x)^2\,dx$ by using integration by parts.

$u = (\ln x)^2 \qquad dv = dx$

$du = \frac{2\ln x}{x}\,dx \qquad v = x$

$\int (\ln x)^2\,dx = x(\ln x)^2 - \int 2\ln x\,dx$

Evaluate $\int 2\ln x\,dx$ by using integration by parts.

$u = 2\ln x \qquad dv = dx$

$du = \frac{2}{x}\,dx \qquad v = x$

$\int 2\ln x\,dx = 2x\ln x - \int 2\,dx = 2x\ln x - 2x + C$

$\int (\ln x)^2\,dx = x(\ln x)^2 - 2x\ln x + 2x + C$

$$\text{Area} = \pi\lim_{b\to 0^+}\left[x(\ln x)^2 - 2x\ln x + 2x\right]_b^1$$
$$= \pi\lim_{b\to 0^+}[2 - b(\ln b)^2 + 2b\ln b - 2b]$$
$$= 2\pi - \lim_{b\to 0^+}\frac{\pi(\ln b)^2}{1/b} + 2\lim_{b\to 0^+}\frac{\pi\ln b}{1/b}$$
$$= 2\pi - \lim_{b\to 0^+}\frac{2\pi(\ln b)(1/b)}{-1/b^2} + 2\lim_{b\to 0^+}\frac{\pi/b}{-1/b^2}$$
$$= 2\pi - \lim_{b\to 0^+}\frac{2\pi(\ln b)}{-1/b} + 2\lim_{b\to 0^+}(-\pi b)$$
$$= 2\pi - \lim_{b\to 0^+}\frac{2\pi/b}{1/b^2} + 2\pi - \lim_{b\to 0^+} 2\pi b = 2\pi$$

68. For $x \geq 0$, $y \geq 0$ on $[0, \infty)$.

Area $= \int_0^\infty xe^{-x}\, dx = \lim_{b \to \infty} \int_0^b xe^{-x}\, dx$

Evaluate $\int xe^{-x}\, dx$ by using integration by parts.

$u = x$ $\quad\quad\quad dv = e^{-x}\, dx$

$du = dx$ $\quad\quad v = -e^{-x}$

$\int xe^{-x}\, dx = -xe^{-x} + \int e^{-x}\, dx = -xe^{-x} - e^{-x} + C$

Area $= \lim_{b \to \infty} \left[-xe^{-x} - e^{-x}\right]_0^b$

$= \lim_{b \to \infty} [-be^{-b} - e^{-b} + 1]$

$= -\lim_{b \to \infty} \dfrac{b}{e^b} + 1$

$= -\lim_{b \to \infty} \dfrac{1}{e^b} + 1 = 1$

69. (a) $\dfrac{dx}{dt} = k(a - x)^2$

$\dfrac{dx}{(a - x)^2} = k\, dt$

$\int \dfrac{dx}{(a - x)^2} = \int k\, dt = kt + C_1$

$\dfrac{1}{a - x} + C_2 = kt + C_1$

$\dfrac{1}{a - x} = kt + C$

Substitute $x = 0$, $t = 0$

$\dfrac{1}{a} = C$

$\dfrac{1}{a - x} = kt + \dfrac{1}{a}$

$\dfrac{1}{kt + 1/a} = a - x$

$x = a - \dfrac{1}{kt + 1/a}$

(b) $\dfrac{dx}{(a - x)(b - x)} = k\, dt$

$\int \dfrac{dx}{(a - x)(b - x)} = \int k\, dt = kt + C_1$

$\dfrac{1}{(a - x)(b - x)} = \dfrac{A}{a - x} + \dfrac{B}{b - x}$

$1 = A(b - x) + B(a - x)$

$= (-A - B)x + bA + aB$

Equating coefficients of like terms gives

$-A - B = 0$ and $bA + aB = 1$

Solving the system simultaneously yields

$A = -\dfrac{1}{a - b},\ B = \dfrac{1}{a - b}$

$\int \dfrac{dx}{(a - x)(b - x)} = \int \dfrac{-1/(a - b)}{a - x}\, dx + \int \dfrac{1/(a - b)}{b - x}\, dx$

$= \dfrac{\ln|a - x|}{a - b} - \dfrac{\ln|b - x|}{a - b} + C_2$

$= \dfrac{1}{a - b} \ln\left|\dfrac{a - x}{b - x}\right| + C_2$

$\dfrac{1}{a - b} \ln\left|\dfrac{a - x}{b - x}\right| + C_2 = kt + C_1$

$\ln\left|\dfrac{a - x}{b - x}\right| = (a - b)kt + C$

$\dfrac{a - x}{b - x} = De^{(a-b)kt}$

Substitute $t = 0$, $x = 0$.

$\dfrac{a}{b} = D$

$\dfrac{a - x}{b - x} = \dfrac{a}{b} e^{(a-b)kt}$

$ab - bx = abe^{(a-b)kt} - axe^{(a-b)kt}$

$x(ae^{(a-b)kt} - b) = ab(e^{(a-b)kt} - 1)$

$x = \dfrac{ab(e^{(a-b)kt} - 1)}{ae^{(a-b)kt} - b}$

Multiply the rational expression by $\dfrac{e^{bkt}}{e^{bkt}}$.

$x = \dfrac{ab(e^{akt} - e^{bkt})}{ae^{akt} - be^{bkt}}$

Chapter 9
Infinite Series

■ Section 9.1 Power Series (pp. 457–468)

Exploration 1 Power Series for Other Functions

1. $1 - x + x^2 - x^3 + \cdots + (-x)^n + \cdots$.

2. $x - x^2 + x^3 - x^4 + \cdots + (-1)^n x^{n+1} + \cdots$.

3. $1 + 2x + 4x^2 + 8x^3 + \cdots + (2x)^n + \cdots$.

4. $1 - (x - 1) + (x - 1)^2 - (x - 1)^3 + \cdots + (-1)^n(x - 1)^n + \cdots$.

5. $\dfrac{1}{3} - \dfrac{1}{3}(x - 1) + \dfrac{1}{3}(x - 1)^2 - \dfrac{1}{3}(x - 1)^3 + \cdots + \left(-\dfrac{1}{3}\right)^n (x - 1)^n + \cdots$.

This geometric series converges for $-1 < x - 1 < 1$, which is equivalent to $0 < x < 2$. The interval of convergence is $(0, 2)$.

Exploration 2 A Power Series for $\tan^{-1} x$

1. $1 - x^2 + x^4 - x^6 + \cdots + (-1)^n x^{2n} + \cdots$.

2. $\tan^{-1} x = \int_0^x \dfrac{1}{1 + t^2}\, dt$

$= \int_0^x (1 - t^2 + t^4 - t^6 + \cdots + (-1)^n t^{2n} + \cdots)\, dt$

$= \left[t - \dfrac{t^3}{3} + \dfrac{t^5}{5} - \dfrac{t^7}{7} + \cdots + (-1)^n \dfrac{t^{2n+1}}{2n + 1} + \cdots\right]_0^x$

$= x - \dfrac{x^3}{3} + \dfrac{x^5}{5} - \dfrac{x^7}{7} + \cdots + (-1)^n \dfrac{x^{2n+1}}{2n + 1} \cdots$.

3. The graphs of the first four partial sums appear to be converging on the interval $(-1, 1)$.

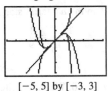

$[-5, 5]$ by $[-3, 3]$

4. When $x = 1$, the series becomes
$$1 - \frac{1}{3} + \frac{1}{5} - \frac{1}{7} + \cdots + \frac{(-1)^n}{2n+1} + \cdots.$$

This series does appear to converge. The terms are getting smaller, and because they alternate in sign they cause the partial sums to oscillate above and below a limit. The two calculator statements shown below will cause the successive partial sums to appear on the calculator each time the ENTER button is pushed. The partial sums will appear to be approaching a limit of $\pi/4$ (which is $\tan^{-1}(1)$), although very slowly.

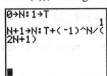

Exploration 3 A Series with a Curious Property

1. $f'(x) = 1 + x + \frac{x^2}{2!} + \frac{x^3}{3!} + \cdots + \frac{x^n}{n!} + \cdots.$

2. $f(0) = 1 + 0 + 0 + \cdots = 1.$

3. Since this function is its own derivative and takes on the value 1 at $x = 0$, we suspect that it must be e^x.

4. If $y = f(x)$, then $\frac{dy}{dx} = y$ and $y = 1$ when $x = 0$.

5. The differential equation is separable.

$\frac{dy}{y} = dx$

$\int \frac{dy}{y} = \int dx$

$\ln |y| = x + C$

$y = Ke^x$

$1 = Ke^0 \Rightarrow K = 1$

$\therefore y = e^x.$

6. The first three partial sums are shown in the graph below. It is risky to draw any conclusions about the interval of convergence from just three partial sums, but so far the convergence to the graph of $y = e^x$ only looks good on $(-1, 1)$. Your answer might differ.

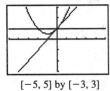

$[-5, 5]$ by $[-3, 3]$

7. The next three partial sums show that the convergence extends outside the interval $(-1, 1)$ in both directions, so $(-1, 1)$ was apparently an underestimate. Your answer in #6 might have been better, but unless you guessed "all real numbers," you still underestimated! (See Example 3 in Section 9.3.)

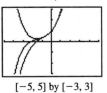

$[-5, 5]$ by $[-3, 3]$

Quick Review 9.1

1. $u_1 = \frac{4}{1+2} = \frac{4}{3}$

$u_2 = \frac{4}{2+2} = \frac{4}{4} = 1$

$u_3 = \frac{4}{3+2} = \frac{4}{5}$

$u_4 = \frac{4}{4+2} = \frac{4}{6} = \frac{2}{3}$

$u_{30} = \frac{4}{30+2} = \frac{4}{32} = \frac{1}{8}$

2. $u_1 = \frac{(-1)^1}{1} = -1$

$u_2 = \frac{(-1)^2}{2} = \frac{1}{2}$

$u_3 = \frac{(-1)^3}{3} = -\frac{1}{3}$

$u_4 = \frac{(-1)^4}{4} = \frac{1}{4}$

$u_{30} = \frac{(-1)^{30}}{30} = \frac{1}{30}$

3. (a) Since $\frac{6}{2} = \frac{18}{6} = \frac{54}{18} = 3$, the common ratio is 3.

(b) $2(3^9) = 39{,}366$

(c) $a_n = 2(3^{n-1})$

4. (a) Since $\frac{-4}{8} = \frac{2}{-4} = \frac{-1}{2} = -\frac{1}{2}$, the common ratio is $-\frac{1}{2}$.

(b) $8\left(-\frac{1}{2}\right)^9 = -\frac{1}{64}$

(c) $a_n = 8\left(-\frac{1}{2}\right)^{n-1} = 8(-0.5)^{n-1}$

5. (a) We graph the points $\left(n, \frac{1-n}{n^2}\right)$ for $n = 1, 2, 3, \ldots$.

(Note that there is a point at $(1, 0)$ that does not show in the graph.)

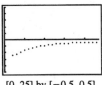

$[0, 25]$ by $[-0.5, 0.5]$

(b) $\lim_{n \to \infty} a_n = \lim_{n \to \infty} \frac{1-n}{n^2} = 0$

6. (a) We graph the points $\left(n, \left(1 + \dfrac{1}{n}\right)^n\right)$ for $n = 1, 2, 3, \ldots$.

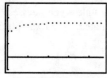

[0, 23.5] by [−1, 4]

(b) $\lim\limits_{n \to \infty} a_n = \lim\limits_{n \to \infty} \left(1 + \dfrac{1}{n}\right)^n = e$

7. (a) We graph the points $(n, (-1)^n)$ for $n = 1, 2, 3, \ldots$.

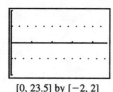

[0, 23.5] by [−2, 2]

(b) $\lim\limits_{n \to \infty} a_n$ does not exist because the values of a_n oscillate between -1 and 1.

8. (a) We graph the points $\left(n, \dfrac{1 - 2n}{1 + 2n}\right)$ for $n = 1, 2, 3, \ldots$.

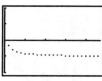

[0, 23.5] by [−2, 2]

(b) $\lim\limits_{n \to \infty} a_n = \lim\limits_{n \to \infty} \dfrac{1 - 2n}{1 + 2n} = -\dfrac{2}{2} = -1$

9. (a) We graph the points $\left(n, 2 - \dfrac{1}{n}\right)$ for $n = 1, 2, 3, \ldots$.

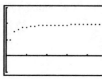

[0, 23.5] by [−1, 3]

(b) $\lim\limits_{n \to \infty} a_n = \lim\limits_{n \to \infty} \left(2 - \dfrac{1}{n}\right) = 2$

10. (a) We graph the points $\left(n, \dfrac{\ln(n + 1)}{n}\right)$ for $n = 1, 2, 3, \ldots$.

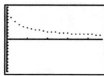

[0, 23.5] by [−1, 1]

(b) $\lim\limits_{n \to \infty} a_n = \lim\limits_{n \to \infty} \dfrac{\ln(n + 1)}{n} = \lim\limits_{n \to \infty} \dfrac{\dfrac{1}{n+1}}{1} = 0$

Section 9.1 Exercises

1. (a) Let u_n represent the value of $*$ in the nth term, starting with $n = 1$. Then $\dfrac{1}{u_1} = 1$, $-\dfrac{1}{u_2} = -\dfrac{1}{4}$, $\dfrac{1}{u_3} = \dfrac{1}{9}$, and $-\dfrac{1}{u_4} = -\dfrac{1}{16}$, so $u_1 = 1, u_2 = 4, u_3 = 9,$ and $u_4 = 16$. We may write $u_n = n^2$, or $* = n^2$.

(b) Let u_n represent the value of $*$ in the nth term, starting with $n = 0$. Then $\dfrac{1}{u_0} = 1$, $-\dfrac{1}{u_1} = -\dfrac{1}{4}$, $\dfrac{1}{u_2} = \dfrac{1}{9}$, and $-\dfrac{1}{u_3} = -\dfrac{1}{16}$, so $u_0 = 1, u_1 = 4, u_2 = 9,$ and $u_3 = 16$. We may write $u_n = (n + 1)^2$, or $* = (n + 1)^2$.

(c) If $* = 3$, the series is
$(-1)^3\left(\dfrac{-1}{1^2}\right) + (-1)^4\left(\dfrac{-1}{2^2}\right) + (-1)^5\left(\dfrac{-1}{3^2}\right) + (-1)^6\left(\dfrac{-1}{4^2}\right) + \cdots$, which is the same as the desired series. Thus let $* = 3$.

3. Different, since the terms of $\displaystyle\sum_{n=1}^{\infty} \left(-\dfrac{1}{2}\right)^{n-1}$ alternate between positive and negative, while the terms of $\displaystyle\sum_{n=1}^{\infty} -\left(\dfrac{1}{2}\right)^{n-1}$ are all negative.

5. The same, since both series can be represented as
$1 - \dfrac{1}{2} + \dfrac{1}{4} - \dfrac{1}{8} + \cdots$.

7. Converges; $\displaystyle\sum_{n=0}^{\infty} \left(\dfrac{2}{3}\right)^n = \dfrac{1}{1 - \dfrac{2}{3}} = 3$

9. Converges; $\displaystyle\sum_{n=0}^{\infty} \left(\dfrac{5}{4}\right)\left(\dfrac{2}{3}\right)^n = \dfrac{\dfrac{5}{4}}{1 - \dfrac{2}{3}} = 3\left(\dfrac{5}{4}\right) = \dfrac{15}{4}$

11. Diverges, because the terms alternate between 1 and -1 and do not approach zero.

13. Converges; $\displaystyle\sum_{n=0}^{\infty} \sin^n\left(\dfrac{\pi}{4} + n\pi\right)$

$= 1 + \left(-\dfrac{1}{\sqrt{2}}\right)^1 + \left(\dfrac{1}{\sqrt{2}}\right)^2 + \left(-\dfrac{1}{\sqrt{2}}\right)^3 + \cdots$

$= \displaystyle\sum_{n=0}^{\infty} \left(-\dfrac{1}{\sqrt{2}}\right)^n = \dfrac{1}{1 - \left(-\dfrac{1}{\sqrt{2}}\right)} = \dfrac{\sqrt{2}}{\sqrt{2} + 1}$

$= \dfrac{\sqrt{2}(\sqrt{2} - 1)}{(\sqrt{2} + 1)(\sqrt{2} - 1)} = \dfrac{2 - \sqrt{2}}{2 - 1} = 2 - \sqrt{2}$

15. Converges; since $\frac{e}{\pi} \approx 0.865 < 1$, $\sum_{n=1}^{\infty} \left(\frac{e}{\pi}\right)^n = \frac{1}{1-\left(\frac{e}{\pi}\right)}$

$= \frac{\pi}{\pi - e}$

17. Since $\sum_{n=0}^{\infty} 2^n x^n = \sum_{n=0}^{\infty} (2x)^n$, the series converges when $|2x| < 1$ and the interval of convergence is $\left(-\frac{1}{2}, \frac{1}{2}\right)$. Since the sum of the series is $\frac{1}{1-2x}$, the series represents the function $f(x) = \frac{1}{1-2x}$, $-\frac{1}{2} < x < \frac{1}{2}$.

19. Since $\sum_{n=0}^{\infty} \left(-\frac{1}{2}\right)^n (x-3)^n = \sum_{n=0}^{\infty} \left(\frac{3-x}{2}\right)^n$, the series converges when $\left|\frac{3-x}{2}\right| < 1$ and the interval of convergence is $(1, 5)$. Since the sum of the series is $\frac{1}{1-\frac{(3-x)}{2}} = \frac{2}{x-1}$, the series represents the function $f(x) = \frac{2}{x-1}$, $1 < x < 5$.

21. Since $\sum_{n=0}^{\infty} \sin^n x = \sum_{n=0}^{\infty} (\sin x)^n$, the series converges when $|\sin x| < 1$. Thus, the series converges for all values of x except odd integer multiples of $\frac{\pi}{2}$, that is, $x \neq (2k+1)\frac{\pi}{2}$ for integers k. Since the sum of the series is $\frac{1}{1-\sin x}$, the series represents the function $f(x) = \frac{1}{1-\sin x}$, $x \neq (2k+1)\frac{\pi}{2}$.

23. (a) Since the terms are all positive and do not approach zero, the partial sums tend toward infinity.

(b) The partial sums are alternately 1 and 0.

(c) The partial sums alternate between positive and negative while their magnitude increases toward infinity.

25. $\sum_{n=0}^{\infty} x^n = 20$

$\frac{1}{1-x} = 20, |x| < 1$

$1 = 20 - 20x$

$20x = 19$

$x = \frac{19}{20}$

27. Assuming the series begins at $n = 1$:

(a) $\sum_{n=1}^{\infty} 2r^{n-1} = \frac{2}{1-r} = 5, |r| < 1$

$2 = 5 - 5r$

$5r = 3$

$r = \frac{3}{5}$

Series: $\sum_{n=1}^{\infty} 2\left(\frac{3}{5}\right)^{n-1}$

(b) $\sum_{n=1}^{\infty} \frac{13}{2} r^{n-1} = \frac{\frac{13}{2}}{1-r} = 5, |r| < 1$

$\frac{13}{2} = 5 - 5r$

$5r = -\frac{3}{2}$

$r = -\frac{3}{10}$

Series: $\sum_{n=1}^{\infty} \frac{13}{2}\left(-\frac{3}{10}\right)^{n-1}$

29. Let $a = \frac{234}{1000}$ and $r = \frac{1}{1000}$, giving

$0.\overline{234} = 0.234 + 0.234(0.001) + 0.234(0.001)^2$

$+ 0.234(0.001)^3 + \ldots$

$= \sum_{n=0}^{\infty} 0.234(0.001)^n$

$= \frac{0.234}{1-0.001}$

$= \frac{0.234}{0.999}$

$= \frac{26}{111}$

31. $0.\overline{d} = \frac{d}{10}[1 + 0.1 + 0.1^2 + 0.1^3 + \cdots]$

$= \frac{d}{10} \sum_{n=0}^{\infty} (0.1)^n$

$= \frac{d}{10} \cdot \frac{1}{1-0.1}$

$= \frac{d}{10} \cdot \frac{1}{0.9}$

$= \frac{d}{9}$

33. $1.\overline{414} = 1 + 0.414 + 0.414(0.001) + 0.414(0.001)^2$

$= 1 + \sum_{n=0}^{\infty} 0.414(0.001)^n$

$= 1 + \frac{0.414}{1-0.001}$

$= 1 + \frac{46}{111}$

$= \frac{157}{111}$

Section 9.1

35. $3.\overline{142857} = 3 + 0.142857(1 + 0.000001$
$$+ 0.000001^2 + \cdots)$$
$$= 3 + 0.142857 \sum_{n=0}^{\infty} 0.000001^n$$
$$= 3 + (0.142857)\left(\frac{1}{1 - 0.000001}\right)$$
$$= 3 + \frac{0.142857}{0.999999}$$
$$= 3 + \frac{1}{7}$$
$$= \frac{22}{7}$$

37. Total time $= \sqrt{\frac{4}{4.9}} + 2\left[\sqrt{\frac{4(0.6)}{4.9}} + \sqrt{\frac{4(0.6)^2}{4.9}}\right.$
$$\left. + \sqrt{\frac{4(0.6)^3}{4.9}} + \cdots \right]$$
$$= \sqrt{\frac{4}{4.9}} + 2\sqrt{\frac{4(0.6)}{4.9}}[1 + \sqrt{0.6} + (\sqrt{0.6})^2 + \cdots]$$
$$= \sqrt{\frac{4}{4.9}} + 2\sqrt{\frac{4(0.6)}{4.9}} \cdot \frac{1}{1 - \sqrt{0.6}}$$
$$\approx 7.113 \text{ sec}$$

39. Total area $= \sum_{n=1}^{\infty} 2^n \cdot \frac{1}{2} \cdot \pi\left(\frac{1}{2^n}\right)^2$
$$= \sum_{n=1}^{\infty} \frac{\pi}{2} \cdot \left(\frac{1}{2}\right)^n$$
$$= \sum_{n=0}^{\infty} \frac{\pi}{4} \left(\frac{1}{2}\right)^n$$
$$= \frac{\frac{\pi}{4}}{1 - \left(\frac{1}{2}\right)}$$
$$= \frac{\pi}{2}$$

41. Using the notation $S_n = a + ar + ar^2 + ar^3 + \ldots + ar^{n-1}$, the formula from Exercise 40 is $S_n = \frac{a - ar^n}{1 - r}$.
If $|r| < 1$, then $\lim_{n \to \infty} r^n = 0$ and so $\sum_{n=1}^{\infty} ar^{n-1}$
$$= \lim_{n \to \infty} S_n = \lim_{n \to \infty} \frac{a - ar^n}{1 - r} = \frac{a}{1 - r}.$$
If $|r| > 1$ or $r = -1$, then r^n has no finite limit as $n \to \infty$, so the expression $\frac{a - ar^n}{1 - r}$ has no finite limit and $\sum_{n=1}^{\infty} ar^{n-1}$ diverges.
If $r = 1$, then the nth partial sum is na, which goes to $\pm\infty$.

43. Comparing $\frac{x}{1 - 2x}$ with $\frac{a}{1 - r}$, the first term is $a = x$ and the common ratio is $r = 2x$.
Series: $x + 2x^2 + 4x^3 + \cdots + 2^{n-1}x^n + \cdots$
Interval: The series converges when $|2x| < 1$, so the interval of convergence is $\left(-\frac{1}{2}, \frac{1}{2}\right)$.

45. Comparing $\frac{1}{1 + (x - 4)}$ with $\frac{a}{1 - r}$, the first term is $a = 1$ and the common ratio is $r = -(x - 4)$.
Series: $1 - (x - 4) + (x - 4)^2 - \cdots + (-1)^n(x - 4)^n + \cdots$
Interval: The series converges when $|x - 4| < 1$, so the interval of convergence is $(3, 5)$.

47. Rewriting $\frac{1}{2 - x}$ as $\frac{1}{1 - (x - 1)}$ and comparing with $\frac{a}{1 - r}$,
The first term is $a = 1$ and the common ratio is $r = x - 1$.
Series: $1 + (x - 1) + (x - 1)^2 + \cdots + (x - 1)^n + \cdots$
Interval: The series converges when $|x - 1| < 1$, so the interval of convergence is $(0, 2)$.

Alternate solution:

Rewriting $\frac{1}{2 - x}$ as $\frac{1}{2}\left(\frac{1}{1 - \left(\frac{x}{2}\right)}\right)$ and comparing with $\frac{a}{1 - r}$, the first is $a = \frac{1}{2}$ and the common ratio is $r = \frac{x}{2}$.

Series: $\frac{1}{2} + \frac{1}{4}x + \frac{1}{8}x^2 + \cdots + \frac{1}{2^{n+1}}x^n + \cdots$

Interval: The series converges when $\left|\frac{x}{2}\right| < 1$, so the interval of convergence is $(-2, 2)$.

49. (a) When $t = 1$, $S = \sum_{n=0}^{\infty} \left(\frac{1}{2}\right)^n = \frac{1}{1 - \left(\frac{1}{2}\right)} = 2.$

(b) S converges when $\left|\frac{t}{1 + t}\right| < 1$, or $|t| < |1 + t|$.

For $t < -1$, this inequality is equivalent to
$-t < -(1 + t)$, which is always false.

For $-1 \leq t < 0$, the inequality is equivalent to
$-t < 1 + t$, which is true when $t > -\frac{1}{2}$.

For $t \geq 0$, the inequality is equivalent to $t < 1 + t$,
which is always true.

Thus, S converges for all $t > -\frac{1}{2}$.

(c) For $t > -\frac{1}{2}$, we have
$$S = \sum_{n=0}^{\infty} \left(\frac{t}{1 + t}\right)^n = \frac{1}{1 - \frac{t}{1 + t}} = \frac{1 + t}{(1 + t) - t} = 1 + t, \text{ so}$$
$S > 10$ when $t > 9$.

51. Since $\frac{1}{x} = 1 - (x-1) + (x-1)^2 - (x-1)^3 + \cdots +$
$(-1)^n(x-1)^n + \cdots$, we may write $\ln x = \int_1^x \frac{1}{t} dt$
$= x - 1 - \frac{(x-1)^2}{2} + \frac{(x-1)^3}{3} - \frac{(x-1)^4}{4} + \cdots$
$+ \frac{(-1)^n(x-1)^n}{n}$

53. (a) No, because if you differentiate it again, you would have the original series for f, but by Theorem 1, that would have to converge for $-2 < x < 2$, which contradicts the assumption that the original series converges only for $-1 < x < 1$.

    (b) No, because if you integrate it again, you would have the original series for f, but by Theorem 2, that would have to converge for $-2 < x < 2$, which contradicts the assumption that the original series converges only for $-1 < x < 1$.

55. Given an $\epsilon > 0$, by definition of convergence there corresponds an N such that for all $n < N$, $|L_1 - a_n| < \epsilon$ and $|L_2 - a_n| < \epsilon$. (There is one such number for each series, and we may let N be the larger of the two numbers.) Now
$|L_2 - L_1| = |L_2 - a_n + a_n - L_1|$
$\leq |L_2 - a_n| + |a_n - L_1|$
$< \epsilon + \epsilon$
$= 2\epsilon.$

$|L_2 - L_1| < 2\epsilon$ says that the difference between two fixed values is smaller than any positive number 2ϵ. The only nonnegative number smaller than every positive number is 0, so $|L_2 - L_1| = 0$ or $L_1 = L_2$.

57. (a) $\lim_{n \to \infty} \frac{3n+1}{n+1} = 3$

    (b) The line $y = 3$ is a horizontal asymptote of the graph of the function $f(x) = \frac{3x+1}{x+1}$, which means $\lim_{x \to \infty} f(x) = 3$. Because $f(n) = a_n$ for all positive integers n, it follows that $\lim_{n \to \infty} a_n$ must also be 3.

Section 9.2 Taylor Series (pp. 469–479)

Exploration 1 Designing a Polynomial to Specifications

1. Since $P(0) = 1$, we know that the constant coefficient is 1.

   Since $P'(0) = 2$, we know that the coefficient of x is 2.

   Since $P''(0) = 3$, we know that the coefficient of x^2 is $\frac{3}{2}$.

   (The 2 in the denominator is needed to cancel the factor of 2 that results from differentiating x^2.) Similarly, we find the coefficients of x^3 and x^4 to be $\frac{4}{6}$ and $\frac{5}{24}$.

   Thus, $P(x) = 1 + 2x + \frac{3}{2}x^2 + \frac{4}{6}x^3 + \frac{5}{24}x^4$.

Exploration 2 A Power Series for the Cosine

1. $\cos(0) = 1$
   $\cos'(0) = -\sin(0) = 0$
   $\cos''(0) = -\cos(0) = -1$
   $\cos^{(3)}(0) = \sin(0) = 0$
   etc.
   The pattern 1, 0, −1, 0 will repeat forever. Therefore,
   $P_6(x) = 1 - \frac{x^2}{2} + \frac{x^4}{4!} - \frac{x^6}{6!}$, and the Taylor series is
   $1 - \frac{x^2}{2} + \frac{x^4}{4!} - \frac{x^6}{6!} + \cdots + (-1)^n \frac{x^{2n}}{(2n)!} + \cdots$.

2. A clever shortcut is simply to differentiate the previously-discovered series for $\sin x$ term-by-term!

Exploration 3 Approximating sin 13

1. 0.4201670368...

4. 20 terms.

Quick Review 9.2

1. $f(x) = e^{2x}$
   $f'(x) = 2e^{2x}$
   $f''(x) = 4e^{2x}$
   $f'''(x) = 8e^{2x}$
   $f^{(n)}(x) = 2^n e^{2x}$

2. $f(x) = \frac{1}{x-1}$
   $f'(x) = -(x-1)^{-2}$
   $f''(x) = 2(x-1)^{-3}$
   $f'''(x) = -6(x-1)^{-4}$
   $f^{(n)}(x) = (-1)^n n! (x-1)^{-(n+1)}$

3. $f(x) = 3^x$
   $f'(x) = 3^x \ln 3$
   $f''(x) = 3^x (\ln 3)^2$
   $f'''(x) = 3^x (\ln 3)^3$
   $f^{(n)}(x) = 3^x (\ln 3)^n$

4. $f(x) = \ln x$
   $f'(x) = x^{-1}$
   $f''(x) = -x^{-2}$
   $f'''(x) = 2x^{-3}$
   $f^{(4)}(x) = -6x^{-4}$
   $f^{(n)}(x) = (-1)^{n-1}(n-1)! x^{-n}$ for $n \geq 1$

5. $f(x) = x^n$
   $f'(x) = nx^{n-1}$
   $f''(x) = n(n-1)x^{n-2}$
   $f'''(x) = n(n-1)(n-2)x^{n-3}$
   $f^{(k)}(x) = \frac{n!}{(n-k)!} x^{n-k}$
   $f^{(n)}(x) = \frac{n!}{0!} x^0 = n!$

6. $\frac{dy}{dx} = \frac{d}{dx} \frac{x^n}{n!} = \frac{nx^{n-1}}{n!} = \frac{x^{n-1}}{(n-1)!}$

7. $\frac{dy}{dx} = \frac{d}{dx} \frac{2^n(x-a)^n}{n!} = \frac{2^n n(x-a)^{n-1}}{n!} = \frac{2^n(x-a)^{n-1}}{(n-1)!}$

248 Section 9.2

8. $\dfrac{dy}{dx} = \dfrac{d}{dx}\left[(-1)^n \dfrac{x^{2n+1}}{(2n+1)!}\right] = (-1)^n \dfrac{(2n+1)x^{2n}}{(2n+1)!} = \dfrac{(-1)^n x^{2n}}{(2n)!}$

9. $\dfrac{dy}{dx} = \dfrac{d}{dx}\dfrac{(x+a)^{2n}}{(2n)!} = \dfrac{2n(x+a)^{2n-1}}{(2n)!} = \dfrac{(x+a)^{2n}}{(2n-1)!}$

10. $\dfrac{dy}{dx} = \dfrac{d}{dx}\dfrac{(1-x)^n}{n!} = \dfrac{n(1-x)^{n-1}(-1)}{n!} = -\dfrac{(1-x)^{n-1}}{(n-1)!}$

Section 9.2 Exercises

1. Substitute $2x$ for x in the Maclaurin series for $\sin x$ shown at the end of Section 9.2.

$\sin 2x = 2x - \dfrac{(2x)^3}{3!} + \dfrac{(2x)^5}{5!} - \cdots + (-1)^n \dfrac{(2x)^{2n+1}}{(2n+1)!} + \cdots$

$= 2x - \dfrac{4x^3}{3} + \dfrac{4x^5}{15} - \cdots + \dfrac{(-1)^n (2x)^{2n+1}}{(2n+1)!} + \cdots$

This series converges for all real x.

3. Substitute x^2 for x in the Maclaurin series for $\tan^{-1} x$ shown at the end of Section 9.2.

$\tan^{-1} x^2 = x^2 - \dfrac{(x^2)^3}{3} + \dfrac{(x^2)^5}{5} - \cdots + (-1)^n \dfrac{(x^2)^{2n+1}}{2n+1} + \cdots$

$= x^2 - \dfrac{x^6}{3} + \dfrac{x^{10}}{5} - \cdots + \dfrac{(-1)^n x^{4n+2}}{2n+1} + \cdots$

This series converges when $|x^2| \leq 1$, so the interval of convergence is $[-1, 1]$.

5. $\cos(x+2) = (\cos 2)(\cos x) - (\sin 2)(\sin x)$

$= (\cos 2)\left[1 - \dfrac{x^2}{2!} + \dfrac{x^4}{4!} - \cdots + (-1)^n \dfrac{x^{2n}}{(2n)!} + \cdots\right]$

$- (\sin 2)\left[x - \dfrac{x^3}{3!} + \dfrac{x^5}{5!} - \cdots + (-1)^n \dfrac{x^{2n+1}}{(2n+1)!} + \cdots\right]$

$= (\cos 2) - (\sin 2)x - \dfrac{(\cos 2)x^2}{2!} + \dfrac{(\sin 2)x^3}{3!} + \dfrac{(\cos 2)x^4}{4!}$

$- \dfrac{(\sin 2)x^5}{5!} - \cdots$

We need to write an expression for the coefficient of x^k.

If k is even, the coefficient is $\dfrac{(-1)^n (\cos 2)}{(2n)!}$ where $2n = k$.

Thus the coefficient is

$\dfrac{(-1)^{k/2}(\cos 2)}{k!}$, which is the same as $\dfrac{(-1)^{\text{int}[(k+1)/2]}(\cos 2)}{k!}$.

If k is odd, the coefficient is $\dfrac{(-1)^{n+1}(\sin 2)}{(2n+1)!}$ where $2n+1 = k$. Thus the coefficient is

$\dfrac{(-1)^{(k+1)/2}(\sin 2)}{(2n+1)!}$, which is the same as $\dfrac{(-1)^{\text{int}[(k+1)/2]}(\cos 2)}{k!}$.

Hence the general term is $\dfrac{(-1)^A B x^n}{n!}$, where $A = \text{int}\left(\dfrac{n+1}{2}\right)$,

and $B = \sin 2$ if n is even and

$B = \cos 2$ if n is odd.

Another way to handle the general term is to observe that

$-\sin 2 = \cos\left(2 + \dfrac{\pi}{2}\right)$, $-\cos 2 = \cos(2 + \pi)$,

and so on, so the general term is $\left[\dfrac{1}{n!} \cos\left(2 + \dfrac{n\pi}{2}\right)\right] x^n$.

The series converges for all real x.

7. Factor out x and substitute x^3 for x in the Maclaurin series for $\dfrac{1}{1-x}$ shown at the end of Section 9.2.

$\dfrac{x}{1-x^3} = x\left(\dfrac{1}{1-x^3}\right)$

$= x[1 + x^3 + (x^3)^2 + \cdots + (x^3)^n + \cdots]$

$= x + x^4 + x^7 + \cdots + x^{3n+1} + \cdots$

The series converges for $|x^3| < 1$, so the interval of convergence is $(-1, 1)$.

9. $f(2) = \dfrac{1}{x}\bigg|_{x=2} = \dfrac{1}{2}$

$f'(2) = -x^{-2}\bigg|_{x=2} = -\dfrac{1}{4}$

$f''(2) = 2x^{-3}\bigg|_{x=2} = \dfrac{1}{4}$, so $\dfrac{f''(2)}{2!} = \dfrac{1}{8}$

$f'''(2) = -6x^{-4}\bigg|_{x=2} = -\dfrac{3}{8}$, so $\dfrac{f'''(2)}{3!} = -\dfrac{1}{16}$

$P_0(x) = \dfrac{1}{2}$

$P_1(x) = \dfrac{1}{2} - \dfrac{x-2}{4}$

$P_2(x) = \dfrac{1}{2} - \dfrac{x-2}{4} + \dfrac{(x-2)^2}{8}$

$P_3(x) = \dfrac{1}{2} - \dfrac{x-2}{4} + \dfrac{(x-2)^2}{8} - \dfrac{(x-2)^3}{16}$

11. $f\left(\dfrac{\pi}{4}\right) = \cos x\bigg|_{x=\pi/4} = \dfrac{\sqrt{2}}{2}$

$f'\left(\dfrac{\pi}{4}\right) = -\sin x\bigg|_{x=\pi/4} = -\dfrac{\sqrt{2}}{2}$

$f''\left(\dfrac{\pi}{4}\right) = -\cos x\bigg|_{x=\pi/4} = -\dfrac{\sqrt{2}}{2}$, so $\dfrac{f''\left(\dfrac{\pi}{4}\right)}{2!} = -\dfrac{\sqrt{2}}{4}$

$f'''\left(\dfrac{\pi}{4}\right) = \sin x\bigg|_{x=\pi/4} = \dfrac{\sqrt{2}}{2}$, so $\dfrac{f'''\left(\dfrac{\pi}{4}\right)}{3!} = \dfrac{\sqrt{2}}{12}$

$P_0(x) = \dfrac{\sqrt{2}}{2}$

$P_1(x) = \dfrac{\sqrt{2}}{2} - \left(\dfrac{\sqrt{2}}{2}\right)\left(x - \dfrac{\pi}{4}\right)$

$P_2(x) = \dfrac{\sqrt{2}}{2} - \left(\dfrac{\sqrt{2}}{2}\right)\left(x - \dfrac{\pi}{4}\right) - \left(\dfrac{\sqrt{2}}{4}\right)\left(x - \dfrac{\pi}{4}\right)^2$

$P_3(x) = \dfrac{\sqrt{2}}{2} - \left(\dfrac{\sqrt{2}}{2}\right)\left(x - \dfrac{\pi}{4}\right) - \left(\dfrac{\sqrt{2}}{4}\right)\left(x - \dfrac{\pi}{4}\right)^2$

$\qquad + \left(\dfrac{\sqrt{2}}{12}\right)\left(x - \dfrac{\pi}{4}\right)^3$

13. **(a)** Since f is a cubic polynomial, it is its own Taylor polynomial of order 3.
$P_3(x) = x^3 - 2x + 4$ or $4 - 2x + x^3$

(b) $f(1) = x^3 - 2x + 4\big|_{x=1} = 3$

$f'(1) = 3x^2 - 2\big|_{x=1} = 1$

$f''(1) = 6x\big|_{x=1} = 6$, so $\dfrac{f''(1)}{2!} = 3$

$f'''(1) = 6\big|_{x=1} = 6$, so $\dfrac{f'''(1)}{3!} = 1$

$P_3(x) = 3 + (x-1) + 3(x-1)^2 + (x-1)^3$

15. **(a)** Since $f(0) = f'(0) = f''(0) = f'''(0) = 0$, the Taylor polynomial of order 3 is $P_3(0) = 0$.

(b) $f(1) = x^4\big|_{x=1} = 1$

$f'(1) = 4x^3\big|_{x=1} = 4$

$f''(1) = 12x^2\big|_{x=1} = 12$, so $\dfrac{f''(1)}{2!} = 6$

$f'''(1) = 24x\big|_{x=1} = 24$, so $\dfrac{f'''(1)}{3!} = 4$

$P_3(x) = 1 + 4(x-1) + 6(x-1)^2 + 4(x-1)^3$

17. **(a)** $P_3(x) = 4 + (-1)(x-1) + \dfrac{3}{2!}(x-1)^2 + \dfrac{2}{3!}(x-1)^3$

$= 4 - (x-1) + \dfrac{3}{2}(x-1)^2 + \dfrac{1}{3}(x-1)^3$

$f(1.2) \approx P_3(1.2) \approx 3.863$

(b) Since the Taylor series of $f'(x)$ can be obtained by differentiating the terms of the Taylor series of $f(x)$, the second order Taylor polynomial of $f'(x)$ is given by $-1 + 3(x-1) + (x-1)^2$. Evaluating at $x = 1.2$, $f'(1.2) \approx -0.36$

19. **(a)** Substitute $\dfrac{x}{2}$ for x in the Maclaurin series for e^x shown at the end of Section 9.2

$e^{x/2} = 1 + \dfrac{x}{2} + \dfrac{\left(\dfrac{x}{2}\right)^2}{2} + \cdots + \dfrac{\left(\dfrac{x}{2}\right)^n}{n!} + \cdots$

$= 1 + \dfrac{x}{2} + \dfrac{x^2}{8} + \cdots + \dfrac{x^n}{2^n \cdot n!} + \cdots$

(b) $g(x) = \dfrac{e^x - 1}{x}$

$= \dfrac{1}{x}\left[\left(1 + x + \dfrac{x^2}{2!} + \dfrac{x^3}{3!} + \cdots + \dfrac{x^n}{n!} + \cdots\right) - 1\right]$

$= \dfrac{1}{x}\left(x + \dfrac{x^2}{2!} + \dfrac{x^3}{3!} + \cdots + \dfrac{x^n}{n!} + \cdots\right)$

$= 1 + \dfrac{x}{2!} + \dfrac{x^2}{3!} + \cdots + \dfrac{x^{n-1}}{n!} + \cdots$

This can also be written as

$1 + \dfrac{x}{2!} + \dfrac{x^2}{3!} + \cdots + \dfrac{x^n}{(n+1)!} + \cdots$.

(c) $g'(x) = \dfrac{d}{dx}\dfrac{e^x - 1}{x} = \dfrac{(x)(e^x) - (e^x - 1)(1)}{x^2}$

$= \dfrac{xe^x - e^x + 1}{x^2}$

$g'(1) = \dfrac{e - e + 1}{1} = 1$

From the series,

$g'(x) = \dfrac{d}{dx}\left(1 + \dfrac{x}{2!} + \dfrac{x^2}{3!} + \dfrac{x^3}{4!} + \cdots + \dfrac{x^n}{(n+1)!} + \cdots\right)$

$= \dfrac{1}{2!} + \dfrac{2x}{3!} + \dfrac{3x^2}{4!} + \cdots + \dfrac{nx^{n-1}}{(n+1)!} + \cdots$

$= \sum_{n=1}^{\infty} \dfrac{nx^{n-1}}{(n+1)!}$

Therefore, $g'(1) = \sum_{n=1}^{\infty} \dfrac{n}{(n+1)!}$, which means

$\sum_{n=1}^{\infty} \dfrac{n}{(n+1)!} = 1$.

21. **(a)** $f(0) = (1+x)^{1/2}\big|_{x=0} = 1$

$f'(0) = \dfrac{1}{2}(1+x)^{-1/2}\big|_{x=0} = \dfrac{1}{2}$

$f''(0) = -\dfrac{1}{4}(1+x)^{-3/2}\big|_{x=0} = -\dfrac{1}{4}$, so $\dfrac{f''(0)}{2!} = -\dfrac{1}{8}$

$f'''(0) = \dfrac{3}{8}(1+x)^{-5/2}\big|_{x=0} = \dfrac{3}{8}$, so $\dfrac{f'''(0)}{3!} = \dfrac{1}{16}$

$P_4(x) = 1 + \dfrac{x}{2} - \dfrac{x^2}{8} + \dfrac{x^3}{16}$

(b) Since $g(x) = f(x^2)$, the first four terms are

$1 + \dfrac{x^2}{2} - \dfrac{x^4}{8} + \dfrac{x^6}{16}$.

(c) Since $h(0) = 5$, the constant term is 5. The next three terms are obtained by integrating the first three terms of the answer to part (b). The first four terms of the series for $h(x)$ are $5 + x + \dfrac{x^3}{6} - \dfrac{x^5}{40}$.

23. First, note that $\cos 18 \approx 0.6603$.

Using $\cos x = \sum_{n=0}^{\infty}(-1)^n\dfrac{x^{2n}}{(2n)!}$, enter the following two-step commands on your home screen and continue to hit ENTER.

The sum corresponding to $N = 25$ is about 0.6582 (not withing 0.001 of exact value), and the sum corresponding to $N = 26$ is about 0.6606, which is within 0.001 of the exact value. Since we began with $N = 0$, it takes a total of 27 terms (or, up to and including the 52nd degree term).

25. (1) $\sin x$ is odd and $\cos x$ is even
(2) $\sin 0 = 0$ and $\cos 0 = 1$

250 Section 9.3

27. Since $\frac{d^3}{dx^3} \ln x = 2x^{-3}$, which is $\frac{1}{4}$ at $x = 2$, the coefficient is $\frac{\frac{1}{4}}{3!} = \frac{1}{24}$.

29. (a) Since $f'(x) = \frac{d}{dx} \frac{4x}{x^2 + 1}$

$$= \frac{(x^2 + 1)(4) - (4x)(2x)}{(x^2 + 1)^2}$$

$$= \frac{4 - 4x^2}{(x^2 + 1)^2},$$

we have $f(0) = 0, f'(0) = 4, f(\sqrt{3}) = \sqrt{3}$ and

$f'(\sqrt{3}) = -\frac{1}{2}$, so the linearizations are $L_1(x) = 4x$ and

$L_2(x) = \sqrt{3} - \frac{1}{2}(x - \sqrt{3}) = -\frac{1}{2}x + \frac{3}{2}\sqrt{3}$,

respectively.

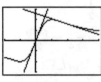

[−2, 4] by [−3, 3]

(b) $f''(a)$ must be 0 because of the inflection point, so the second degree term in the Taylor series of f at $x = a$ is zero.

31. (a) $f(x) = \frac{1}{x}(\sin x)$

$$= \frac{1}{x}\left(x - \frac{x^3}{3!} + \frac{x^5}{5!} - \cdots + (-1)^n \frac{x^{2n+1}}{(2n + 1)!} + \cdots\right)$$

$$= 1 - \frac{x^2}{3!} + \frac{x^4}{5!} - \cdots + (-1)^n \frac{x^{2n}}{(2n + 1)!} + \cdots$$

(b) Because f is undefined at $x = 0$.

(c) $k = 1$

33. (a) $f(x) = (1 + x)^m$

$f'(x) = m(1 + x)^{m-1}$

$f''(x) = m(m - 1)(1 + x)^{m-2}$

$f'''(x) = m(m - 1)(m - 2)(1 + x)^{m-3}$

(b) Differentiating $f(x)$ k times gives
$f^{(k)}(x) = m(m - 1)(m - 2) \cdots (m - k + 1)(1 + x)^{m-k}$.
Substituting 0 for x, we have
$f^{(k)}(0) = m(m - 1)(m - 2) \cdots (m - k + 1)$.

(c) The coefficient is
$\frac{f^{(k)}(0)}{k!} = \frac{m(m - 1)(m - 2) \cdots (m - k + 1)}{k!}$

(d) $f(0) = 1, f'(0) = m$, and we're done by part (c).

■ Section 9.3 Taylor's Theorem (pp. 480–487)

Exploration 1 Your Turn

1. We need to consider what happens to $R_n(x)$ as $n \to \infty$.
By Taylor's Theorem, $R_n(x) = \frac{f^{(n+1)}(c)}{(n + 1)!}(x - 0)^n$, where
$f^{(n+1)}(c)$ is the $(n + 1)$st derivative of $\cos x$ evaluated at
some c between x and 0. As with $\sin x$, we can say that
$f^{(n+1)}(c)$ lies between -1 and 1 inclusive. Therefore, no
matter what x is, we have

$$|R_n(x)| = \left|\frac{f^{(n+1)}(c)}{(n + 1)!}(x - 0)^n\right| \leq \left|\frac{1}{(n + 1)!}x^n\right| = \frac{|x|^n}{(n + 1)!}.$$

The factorial growth in the denominator, as noted in
Example 3, eventually outstrips the power growth in the
numerator, and we have $\frac{|x|^n}{(n + 1)!} \to 0$ for all x. This means
that $R_n(x) \to 0$ for all x, which completes the proof.

Exploration 2 Euler's Formula

1. $e^{ix} = 1 + ix + \frac{(ix)^2}{2!} + \frac{(ix)^3}{3!} + \cdots + \frac{(ix)^n}{n!} + \cdots$

$= 1 + ix - \frac{x^2}{2!} - i\frac{x^3}{3!} + \frac{x^4}{4!} + i\frac{x^5}{5!} - \cdots + (i)^n \frac{x^n}{n!} + \cdots$

2. If we isolate the terms in the series that have i as a factor, we get:

$e^{ix} = 1 + ix - \frac{x^2}{2!} - i\frac{x^3}{3!} + \frac{x^4}{4!} + i\frac{x^5}{5!} - \cdots + (i)^n \frac{x^n}{n!} + \cdots$

$= 1 - \frac{x^2}{2!} + \frac{x^4}{4!} - \frac{x^6}{6!} + \cdots + (-1)^n \frac{x^{2n}}{(2n)!} + \cdots$

$+ i\left(x - \frac{x^3}{3!} + \frac{x^5}{5!} - \frac{x^7}{7!} + \cdots + (-1)^n \frac{x^{2n+1}}{(2n + 1)!} + \cdots\right)$

$= \cos x + i \sin x$.

(We are assuming here that we can rearrange the terms of a convergent series without affecting the sum. It happens to be true in this case, but we will see in Section 9.5 that it is not always true.)

3. $e^{i\pi} = \cos \pi + i \sin \pi = -1 + 0 = -1$
Thus, $e^{i\pi} + 1 = 0$

Quick Review 9.3

1. Since $|f(x)| = |2 \cos (3x)| \leq 2$ on $[-2\pi, 2\pi]$ and $f(0) = 2$, $M = 2$.

2. Since $f(x)$ is increasing and positive on $[1, 2]$, $M = f(2) = 7$.

3. Since $f(x)$ is increasing and positive on $[-3, 0]$, $M = f(0) = 1$.

4. Since the minimum value of $f(x)$ is $f(-1) = -\frac{1}{2}$ and the maximum value of $f(x)$ is $f(1) = \frac{1}{2}$, $M = \frac{1}{2}$.

5. On $[-3, 1]$, the minimum value of $f(x)$ is $f(-3) = -7$ and the maximum value of $f(x)$ is $f(0) = 2$. On $(1, 3]$, f is increasing and positive, so the maximum value of f is $f(3) = 5$. Thus $|f(x)| \leq 7$ on $[-3, 3]$ and $M = 7$.

6. Yes, since each expression for an nth derivative given by the Quotient Rule will be a rational function whose denominator is a power of $x + 1$.

7. No, since the function $f(x) = |x^2 - 4|$ has a corner at $x = 2$.

8. Yes, since the derivatives of all orders for $\sin x$ and $\cos x$ are defined for all values of x.

9. Yes, since the function $f(x) = e^{-x}$ has derivatives of the form $f^{(n)}(x) = -e^{-x}$ for odd values of n and $f^{(n)}(x) = e^{-x}$ for even values of n, and both of these expressions are defined for all values of x.

10. No, since $f(x) = x^{3/2}$, we have $f'(x) = \frac{3}{2}x^{1/2}$ and $f''(x) = \frac{3}{4}x^{-1/2}$, so $f''(0)$ is undefined.

Section 9.3 Exercises

1. $f(0) = e^{-2x}\big|_{x=0} = 1$

   $f'(0) = -2e^{-2x}\big|_{x=0} = -2$

   $f''(0) = 4e^{-2x}\big|_{x=0} = 4$, so $\frac{f''(0)}{2!} = 2$

   $f'''(0) = -8e^{-2x}\big|_{x=0} = -8$, so $\frac{f'''(0)}{3!} = -\frac{4}{3}$

   $f^{(4)}(0) = 16e^{-2x}\big|_{x=0} = 16$, so $\frac{f^{(4)}(0)}{4!} = \frac{2}{3}$

   $P_4(x) = 1 - 2x + 2x^2 - \frac{4}{3}x^3 + \frac{2}{3}x^4$

   $f(0.2) \approx P_4(0.2) = 0.6704$

3. $f(0) = 5\sin(-x)\big|_{x=0} = -5\sin x\big|_{x=0} = 0$

   $f'(0) = -5\cos x\big|_{x=0} = -5$

   $f''(0) = 5\sin x\big|_{x=0} = 0$, so $\frac{f''(0)}{2!} = 0$

   $f'''(0) = 5\cos x\big|_{x=0} = 5$, so $\frac{f'''(0)}{3!} = \frac{5}{6}$

   $f^{(4)}(0) = -5\sin x\big|_{x=0} = 0$, so $\frac{f^{(4)}(0)}{4!} = 0$

   $P_4(x) = -5x + \frac{5}{6}x^3$

   $f(0.2) \approx P_4(0.2) = -\frac{149}{150} \approx -0.9933$

5. $f(0) = (1 - x)^{-2}\big|_{x=0} = 1$

   $f'(0) = 2(1 - x)^{-3}\big|_{x=0} = 2$

   $f''(0) = 6(1 - x)^{-4}\big|_{x=0} = 6$, so $\frac{f''(0)}{2!} = 3$

   $f'''(0) = 24(1 - x)^{-5}\big|_{x=0} = 24$, so $\frac{f'''(0)}{3!} = 4$

   $f^{(4)}(0) = 120(1 - x)^{-6}\big|_{x=0} = 120$, so $\frac{f^{(4)}(0)}{4!} = 5$

   $P_4(x) = 1 + 2x + 3x^2 + 4x^3 + 5x^4$

   $f(0.2) \approx P_4(0.2) = 1.56$

7. $\sin x - x + \frac{x^3}{3!}$

   $= \left(x - \frac{x^3}{3!} + \frac{x^5}{5!} - \cdots + (-1)^n \frac{x^{2n+1}}{(2n+1)!} + \cdots\right) - x + \frac{x^3}{3!}$

   $= \frac{x^5}{5!} - \frac{x^7}{7!} + \frac{x^9}{9!} - \cdots + (-1)^n \frac{x^{2n+1}}{(2n+1)!} + \cdots$

   Note: By replacing n with $n + 2$, the general term can be written as $(-1)^n \frac{x^{2n+5}}{(2n+5)!}$

9. $\sin^2 x = \frac{1}{2} - \frac{1}{2}\cos(2x)$

   $= \frac{1}{2} - \frac{1}{2}\left(1 - \frac{(2x)^2}{2!} + \frac{(2x)^4}{4!} - \frac{(2x)^6}{6!} + \cdots + (-1)^n \frac{(2x)^{2n}}{(2n)!} + \cdots\right)$

   $= \frac{4x^2}{2 \cdot 2!} - \frac{16x^4}{2 \cdot 4!} + \frac{64x^6}{2 \cdot 6!} - \cdots + (-1)^{n-1} \frac{2^{2n} x^{2n}}{2 \cdot (2n)!} + \cdots$

   $= x^2 - \frac{x^4}{3} + \frac{2x^6}{45} - \cdots + (-1)^{n-1} \frac{2^{2n-1} x^{2n}}{(2n)!} + \cdots$

   Note: By replacing n with $n + 1$, the general term can be written as $(-1)^n \frac{2^{2n+1} x^{2n+2}}{(2n+2)!}$.

11. Let $f(x) = \sin x$. Then $P_4(x) = P_3(x) = x - \frac{x^3}{6}$, so we use the Remainder Estimation Theorem with $n = 4$. Since $|f^{(5)}(x)| = |\cos x| \leq 1$ for all x, we may use $M = r = 1$, giving $|R_4(x)| \leq \frac{|x|^5}{5!}$, so we may assure that $|R_4(x)| \leq 5 \times 10^{-4}$ by requiring $\frac{|x|^5}{5!} \leq 5 \times 10^{-4}$, or $|x| \leq \sqrt[5]{0.06} \approx 0.5697$. Thus, the absolute error is no greater than 5×10^{-4} when $-0.56 < x < 0.56$ (approximately).

   Alternate method: Using graphing techniques, $\left|\sin x - \left(x - \frac{x^3}{6}\right)\right| \leq 5 \times 10^{-4}$ when $-0.57 < x < 0.57$.

252 Section 9.3

13. Let $f(x) = \sin x$. Then $P_2(x) = P_1(x) = x$, so we may use the Remainder Estimation Theorem with $n = 2$. Since $|f'''(x)| = |-\cos x| \leq 1$ for all x, we may use $M = r = 1$, giving $|R_2(x)| \leq \frac{|x|^3}{3!}$. Thus, for $|x| < 10^{-3}$, the maximum possible error is about $\frac{(10^{-3})^3}{3!} \approx 1.67 \times 10^{-10}$.

Alternate method:

Using graphing techniques, we find that when $|x| < 10^{-3}$, $|\text{error}| = |\sin x - x| \leq |\sin 10^{-3} - 10^{-3}| \approx 1.67 \times 10^{-10}$.

The inequality $x < \sin x$ is true for $x < 0$, as we may see by graphing $y = \sin x - x$.

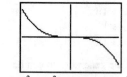

$[-10^{-3}, 10^{-3}]$ by $[-2 \times 10^{-10}, 2 \times 10^{-10}]$

15. Note that $1 + x + \frac{x^2}{2}$ is the second order Taylor polynomial for $f(x) = e^x$ at $x = 0$, so we may use the Remainder Estimation Theorem with $n = 2$. Since $|f'''(x)| = e^x$, which is less than $e^{0.1}$ when $|x| < 0.1$ and $r = 1$, giving $|R_2(x)| \leq \frac{e^{0.1}|x|^3}{3!}$. Thus, for $|x| < 0.1$, the maximum possible error is about $\frac{e^{0.1}(0.1)^3}{3!} \approx 1.842 \times 10^{-4}$.

17. All of the derivatives of $\cosh x$ are either $\cosh x$ or $\sinh x$. For any real x, $\cosh x$ and $\sinh x$ are both bounded by $e^{|x|}$. So for any real x, let $M = e^{|x|}$ and $r = 1$ in the Remainder Estimation Theorem. This gives $|R_n(x)| \leq \frac{e^{|x|}|x|^{n+1}}{(n+1)!}$. But for any fixed value of x, $\lim_{n \to \infty} \frac{e^{|x|}|x|^{n+1}}{(n+1)!} = 0$. It follows that the series converges to $\cosh x$ for all real values of x.

19. $f(0) = \ln(\cos x)\big|_{x=0} = \ln 1 = 0$

$f'(0) = \frac{1}{\cos x}(-\sin x)\big|_{x=0} = -\tan x\big|_{x=0} = 0$

$f''(0) = -\sec^2 x\big|_{x=0} = -1$ so $\frac{f''(0)}{2!} = -\frac{1}{2}$

(a) $L(x) = 0$

(b) $P_2(x) = -\frac{1}{2}x^2$

(c) The graphs of the linear and quadratic approximations fit the graph of the function near $x = 0$.

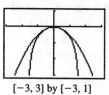

$[-3, 3]$ by $[-3, 1]$

21. $f(0) = (1 - x^2)^{-1/2}\big|_{x=0} = 1$

$f'(0) = -\frac{1}{2}(1 - x^2)^{-3/2}(-2x)\big|_{x=0} = x(1 - x^2)^{-3/2}\big|_{x=0} = 0$

$f''(0) = (x)\left[-\frac{3}{2}(1 - x^2)^{-5/2}(-2x)\right] + (1 - x^2)^{-3/2}\big|_{x=0} = 1$,

so $\frac{f''(0)}{2!} = \frac{1}{2}$

(a) $L(x) = 1$

(b) $P_2(x) = 1 + \frac{x^2}{2}$

(c) The graphs of the linear and quadratic approximations fit the graph of the function near $x = 0$.

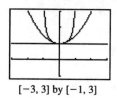

$[-3, 3]$ by $[-1, 3]$

23. $f(0) = \tan x\big|_{x=0} = 0$

$f'(0) = \sec^2 x\big|_{x=0} = 1$

$f''(0) = (2 \sec x)(\sec x \tan x)\big|_{x=0} = 0$, so $\frac{f''(0)}{2!} = 0$

(a) $L(x) = x$

(b) $P_2(x) = x$

(c) The graphs of the linear and quadratic approximations fit the graph of the function near $x = 0$.

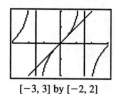

$[-3, 3]$ by $[-2, 2]$

25. Let $f(x) = e^x$. Then $P_3(x) = 1 + x + \frac{x^2}{2} + \frac{x^3}{6}$, so we may use the Remainder Estimation Theorem with $n = 3$. Since $|f^{(4)}(x)| = e^x$, which is no more than $e^{0.1}$ when $|x| \le 0.1$, we may use $M = e^{0.1}$ and $r = 1$, giving $|R_3(x)| \le \frac{e^{0.1}|x|^4}{4!}$. Thus, for $|x| \le 0.1$, the maximum possible absolute error is about $\frac{e^{0.1}(0.1)^4}{24} \approx 4.605 \times 10^{-6}$.

Alternate method:

Using graphing techniques, when $|x| \le 0.1$,

$$|\text{error}| = \left|e^x - \left(1 + x + \frac{x^2}{2} + \frac{x^3}{6}\right)\right|$$
$$\le \left|e^{0.1} - \left(1 + 0.1 + \frac{0.01}{2} + \frac{0.001}{6}\right)\right|$$
$$\approx 4.251 \times 10^{-6}.$$

27. (a) No

(b) Yes, since

$$\frac{dy}{dx} = e^{-x^2}$$
$$= 1 + (-x^2) + \frac{(-x^2)^2}{2!} + \cdots + \frac{(-x^2)^n}{n!} + \cdots$$
$$= 1 - x^2 + \frac{x^4}{2!} - \cdots + (-1)^n \frac{x^{2n}}{n!} + \cdots.$$

The constant term of y is $y(0) = 2$, and we may obtain the remaining terms of y by integrating the above series.

$$y = 2 + x - \frac{x^3}{3} + \frac{x^5}{10} - \cdots + (-1)^n \frac{x^{2n+1}}{(2n + 1)n!} + \cdots$$

By substituting $n - 1$ for n, the general term may also be written as $(-1)^{n-1} \frac{x^{2n-1}}{(2n - 1)(n - 1)!}$.

(c) The power series equals the function y for all real values of x. This is because the series for e^{-x^2} converges for all real values of x, so Theorem 2 of Section 9.1 implies that the new series also converges for all x.

29. (a)

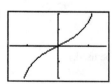

$\left[-\frac{\pi}{2}, \frac{\pi}{2}\right]$ by $[-2, 2]$
The series approximates $\tan x$.

(b)

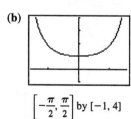

$\left[-\frac{\pi}{2}, \frac{\pi}{2}\right]$ by $[-1, 4]$
The series approximates $\sec x$.

31. (a) It works. For example, let $n = 2$. Then $P = 3.14$ and $P + \sin P \approx 3.141592653$, which is accurate to more than 6 decimal places.

(b) Let $P = \pi + x$ where x is the error in the original estimate. Then

$$P + \sin P = (\pi + x) + \sin(\pi + x) = \pi + x - \sin x$$

But by the Remainder Theorem, $|x - \sin x| < \frac{|x|^3}{6}$.

Therefore, the difference between the new estimate $P + \sin P$ and π is less than $\frac{|x|^3}{6}$.

33. $\frac{d}{dx}[e^{ax}(\cos bx + i \sin bx)]$
$= (e^{ax})(-b \sin bx + bi \cos bx) + (ae^{ax})(\cos bx + i \sin bx)$
$= (e^{ax})[(bi^2 \sin bx + bi \cos bx) + a(\cos bx + i \sin bx)]$
$= (e^{ax})[bi(\cos bx + i \sin bx) + a(\cos bx + i \sin bx)]$
$= (a + bi)(e^{ax})(\cos bx + i \sin bx)$
$= (a + bi)e^{(a+bi)x}$

■ Section 9.4 Radius of Convergence
(pp. 487–496)

Exploration 1 Finishing the Proof of the Ratio Test

1. For $\sum \frac{1}{n}$: $L = \lim_{n \to \infty} \frac{\frac{1}{n+1}}{\frac{1}{n}} = \lim_{n \to \infty} \frac{n}{n+1} = 1$.

For $\sum \frac{1}{n^2}$: $L = \lim_{n \to \infty} \frac{\frac{1}{(n+1)^2}}{\frac{1}{n^2}} = \lim_{n \to \infty} \frac{n^2}{(n+1)^2} = 1$.

2. (a) $\int_1^\infty \frac{1}{x} dx = \lim_{k \to \infty} \left(\ln x \Big|_1^k\right) = \lim_{k \to \infty} \ln k = \infty$.

(b) $\int_1^\infty \frac{1}{x^2} dx = \lim_{k \to \infty} \left(-x^{-1}\Big|_1^k\right) = \lim_{k \to \infty} \left(-\frac{1}{k} + 1\right) = 1$.

3. Figure 9.14a shows that $\sum \frac{1}{n}$ is greater than $\int_1^\infty \frac{1}{x} dx$. Since the integral diverges, so must the series.

Figure 9.14b shows that $\sum \frac{1}{n^2}$ is less than $1 + \int_1^\infty \frac{1}{x^2} dx$. Since the integral converges, so must the series.

4. These two examples prove that $L = 1$ can be true for either a divergent series or a convergent series. The Ratio Test itself is therefore inconclusive when $L = 1$.

Exploration 2 Revisiting a Maclaurin Series

1. $L = \lim_{n \to \infty} \frac{|x|^{n+1}}{n+1} \cdot \frac{n}{|x|^n} = \lim_{n \to \infty} \frac{n}{n+1}|x| = |x|$. The series converges absolutely when $|x| < 1$, so the radius of convergence is 1.

2. When $x = -1$, the series becomes
$$-1 - \frac{1}{2} - \frac{1}{3} - \cdots - \frac{1}{n} - \cdots.$$
Each term in this series is the negative of the corresponding term in the divergent series of Figure 9.14a. Just as $\sum \frac{1}{n}$ diverges to $+\infty$, this series diverges to $-\infty$.

3. Geometrically, we chart the progress of the partial sums as in the figure below:

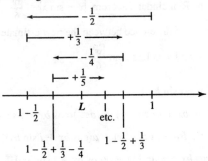

4. The series converges at the right-hand endpoint. As shown in the picture above, the partial sums are closing in on some limit L as they oscillate left and right by constantly decreasing amounts.

5. We know that the series does not converge absolutely at the right-hand endpoint, because $\sum \frac{1}{n}$ diverges (Exploration 1 of this section).

Quick Review 9.4

1. $\lim_{n\to\infty} \frac{n|x|}{n+1} = |x| \lim_{n\to\infty} \frac{n}{n+1} = |x|$

2. $\lim_{n\to\infty} \frac{n^2|x-3|}{n(n+1)} = |x-3| \lim_{n\to\infty} \frac{n^2}{n^2+n} = |x-3|$

3. $\lim_{n\to\infty} \frac{|x|^n}{n!} = 0$
   (Note: This limit is similar to the limit which is discussed at the end of Example 3 in Section 9.3.)

4. $\lim_{n\to\infty} \frac{(n+1)^4 x^2}{(2n)^4} = x^2 \lim_{n\to\infty} \frac{n^4 + 4n^3 + 6n^2 + 4n + 1}{16n^4}$
   $= x^2 \left(\frac{1}{16}\right) = \frac{x^2}{16}$

5. $\lim_{n\to\infty} \frac{|2x+1|^{n+1} 2^n}{2^{n+1}|2x+1|^n} = \lim_{n\to\infty} \frac{|2x+1|}{2} = \frac{|2x+1|}{2}$

6. Since $n^2 > 5n$ for $n \geq 6$, $a_n = n^2$, $b_n = 5n$, and $N = 6$.
7. Since $5^n > n^5$ for $n \geq 6$, $a_n = 5^n$, $b_n = n^5$ and $N = 6$.
8. Since $\sqrt{n} > \ln n$ for $n \geq 1$, $a_n = \sqrt{n}$, $b_n = \ln n$, and $N = 1$.
9. Since $10^n < n!$ (and hence $\frac{1}{10^n} > \frac{1}{n!}$) for $n \geq 25$, $a_n = \frac{1}{10^n}$, $b_n = \frac{1}{n!}$, and $N = 25$.

10. Since $n^2 < n^3$ (and hence $\frac{1}{n^2} > n^{-3}$) for $n \geq 2$, $a_n = \frac{1}{n^2}$, $b_n = n^{-3}$, and $N = 2$.

Section 9.4 Exercises

1. Diverges by the nth-Term Test, since $\lim_{n\to\infty} \frac{n}{n+1} = 1 \neq 0$.

3. Converges by the Ratio Test, since
$$\lim_{n\to\infty} \frac{a_{n+1}}{a_n} = \lim_{n\to\infty} \frac{(n+1)^2 - 1}{2^{n+1}} \cdot \frac{2^n}{n^2 - 1} = \frac{1}{2} < 1.$$

5. Converges by the Ratio Test, since
$$\lim_{n\to\infty} \frac{a_{n+1}}{a_n} = \lim_{n\to\infty} \frac{2^{n+1}}{(3^{n+1} + 1)} \cdot \frac{3^n + 1}{2^n} = \frac{2}{3} < 1.$$
Alternately, note that $\frac{2^n}{3^n + 1} < \left(\frac{2}{3}\right)^n$ for all n.
Since $\sum_{n=1}^{\infty} \left(\frac{2}{3}\right)^n$ converges, $\sum_{n=1}^{\infty} \frac{2^n}{3^n + 1}$ converges by the Direct Comparison Test.

7. Converges by the Ratio Test, since
$$\lim_{n\to\infty} \frac{a_{n+1}}{a_n} = \lim_{n\to\infty} \frac{(n+1)^2 e^{-n-1}}{n^2 e^{-n}} = e^{-1} < 1.$$

9. Converges by the Ratio Test, since
$$\lim_{n\to\infty} \frac{a_{n+1}}{a_n} = \lim_{n\to\infty} \frac{(n+4)!}{3!(n+1)!3^{n+1}} \cdot \frac{3!n! \, 3^n}{(n+3)!}$$
$$= \lim_{n\to\infty} \frac{n+4}{3(n+1)}$$
$$= \frac{1}{3} < 1.$$

11. Converges, because it is a geometric series with $r = -\frac{2}{3}$, so $|r| < 1$.

13. Diverges by the Ratio Test, since
$$\lim_{n\to\infty} \frac{a_{n+1}}{a_n} = \lim_{n\to\infty} \frac{3^{n+1}}{(n+1)^3 \, 2^{n+1}} \cdot \frac{n^3 2^n}{3^n}$$
$$= \lim_{n\to\infty} \frac{3n^3}{(n+1)^3 (2)}$$
$$= \frac{3}{2} > 1.$$
(The nth-Term Test can also be used.)

15. Converges by the Ratio Test, since
$$\lim_{n\to\infty} \frac{a_{n+1}}{a_n} = \lim_{n\to\infty} \frac{(n+1)!}{(2n+3)!} \cdot \frac{(2n+1)!}{n!}$$
$$= \lim_{n\to\infty} \frac{n+1}{(2n+3)(2n+2)}$$
$$= \lim_{n\to\infty} \frac{1}{2(2n+3)} = 0 < 1.$$

17. One possible answer:

    $\sum_{n=1}^{\infty} \frac{1}{n}$ diverges (see Exploration 1 in this section) even though $\lim_{n\to\infty} \frac{1}{n} = 0$.

19. This is a geometric series which converges only for $|x| < 1$, so the radius of convergence is 1.

21. This is a geometric series which converges only for $|-(4x + 1)| < 1$, or $\left|x + \frac{1}{4}\right| < \frac{1}{4}$, so the radius of convergence is $\frac{1}{4}$.

23. This is a geometric series which converges only for $\left|\frac{x - 2}{10}\right| < 1$, or $|x - 2| < 10$, so the radius of convergence is 10.

25. $\lim_{n\to\infty} \left|\frac{a_{n+1}}{a_n}\right| = \lim_{n\to\infty} \frac{|x|^{n+1}}{(n+1)\sqrt{n+1}\, 3^{n+1}} \cdot \frac{n\sqrt{n}\, 3^n}{|x|^n}$

    $= \lim_{n\to\infty} \frac{|x|}{3} = \frac{|x|}{3}$

    The series converges for $|x| < 3$ and diverges for $|x| > 3$, so the radius of convergence is 3.

27. $\lim_{n\to\infty} \left|\frac{a_{n+1}}{a_n}\right| = \lim_{n\to\infty} \frac{(n+1)|x+3|^{n+1}}{5^{n+1}} \cdot \frac{5^n}{n|x+3|^n}$

    $= \lim_{n\to\infty} \frac{|x+3|}{5} = \frac{|x+3|}{5}$

    The series converges for $|x + 3| < 5$ and diverges for $|x + 3| > 5$, so the radius of convergence is 5.

29. $\lim_{n\to\infty} \left|\frac{a_{n+1}}{a_n}\right| = \lim_{n\to\infty} \frac{\sqrt{n+1}\,|x|^{n+1}}{3^{n+1}} \cdot \frac{3^n}{\sqrt{n}\,|x|^n} = \lim_{n\to\infty} \frac{|x|}{3} = \frac{|x|}{3}$

    The series converges for $|x| < 3$ and diverges for $|x| > 3$, so the radius of convergence is 3.

31. $\lim_{n\to\infty} \left|\frac{a_{n+1}}{a_n}\right| = \lim_{n\to\infty} \frac{|(-2)^{n+1}|(n+2)|x-1|^{n+1}}{|-2^n|(n+1)|x-1|^n}$

    $= \lim_{n\to\infty} 2|x - 1|$

    $= 2|x - 1|$

    The series converges for $|x - 1| < \frac{1}{2}$ and diverges for $|x - 1| > \frac{1}{2}$, so the radius of convergence is $\frac{1}{2}$.

33. $\lim_{n\to\infty} \left|\frac{a_{n+1}}{a_n}\right| = \lim_{n\to\infty} \frac{|x+\pi|^{n+1}}{\sqrt{n+1}} \cdot \frac{\sqrt{n}}{|x+\pi|^n}$

    $= \lim_{n\to\infty} |x + \pi|$

    $= |x + \pi|$

    The series converges for $|x + \pi| < 1$ and diverges for $|x + \pi| > 1$, so the radius of convergence is 1.

35. This is a geometric series with first term $a = 1$ and common ratio $r = \frac{(x-1)^2}{4}$. It converges only when $\left|\frac{(x-1)^2}{4}\right| < 1$, so the interval of convergence is $-1 < x < 3$.

    $\text{Sum} = \frac{a}{1 - r} = \frac{1}{1 - \frac{(x-1)^2}{4}}$

    $= \frac{4}{4 - (x-1)^2}$

    $= \frac{4}{-x^2 + 2x + 3}$

    $= -\frac{4}{x^2 - 2x - 3}$

37. This is a geometric series with first term $a = 1$ and common ratio $r = \frac{\sqrt{x}}{2} - 1$. It converges only when $\left|\frac{\sqrt{x}}{2} - 1\right| < 1$, so the interval of convergence is $0 < x < 16$.

    $\text{Sum} = \frac{a}{1 - r} = \frac{1}{1 - \left(\frac{\sqrt{x}}{2} - 1\right)} = \frac{2}{4 - \sqrt{x}}$

39. This is a geometric series with first term $a = 1$ and common ratio $\frac{x^2 - 1}{3}$. It converges only when $\left|\frac{x^2 - 1}{3}\right| < 1$, so the interval of convergence is $-2 < x < 2$.

    $\text{Sum} = \frac{a}{1 - r} = \frac{1}{1 - \frac{x^2 - 1}{3}} = \frac{3}{3 - (x^2 - 1)} = \frac{3}{4 - x^2}$

41. Almost, but the Ratio Test won't determine whether there is convergence or divergence at the endpoints of the interval.

43. (a) For $k \leq N$, it's obvious that

    $d_1 + \cdots + d_k \leq d_1 + \cdots + d_N + \sum_{n=N+1}^{\infty} a_n$.

    For all $k > N$,

    $d_1 + \cdots + d_k = d_1 + \cdots + d_N + d_{N+1} + \cdots + d_k$

    $\leq d_1 + \cdots + d_N + a_{N+1} + \cdots + a_k$

    $\leq d_1 + \cdots + d_N + \sum_{n=N+1}^{\infty} a_n$

    (b) If $\sum a_n$ converged, that would imply that $\sum d_n$ was also convergent.

45. $\sum_{n=1}^{\infty} \dfrac{4}{(4n-3)(4n+1)} = \sum_{n=1}^{\infty}\left(\dfrac{1}{4n-3} - \dfrac{1}{4n+1}\right)$

$s_1 = 1 - \dfrac{1}{5}$

$s_2 = \left(1 - \dfrac{1}{5}\right) + \left(\dfrac{1}{5} - \dfrac{1}{9}\right) = 1 - \dfrac{1}{9}$

$s_3 = \left(1 - \dfrac{1}{5}\right) + \left(\dfrac{1}{5} - \dfrac{1}{9}\right) + \left(\dfrac{1}{9} - \dfrac{1}{13}\right) = 1 - \dfrac{1}{13}$

$s_n = 1 - \dfrac{1}{4n+1}$

$S = \lim_{n\to\infty} s_n = 1$

47. $\sum_{n=1}^{\infty} \dfrac{40n}{(2n-1)^2(2n+1)^2} = \sum_{n=1}^{\infty}\left[\dfrac{5}{(2n-1)^2} - \dfrac{5}{(2n+1)^2}\right]$

$s_1 = 5 - \dfrac{5}{9}$

$s_2 = \left(5 - \dfrac{5}{9}\right) + \left(\dfrac{5}{9} - \dfrac{5}{25}\right) = 5 - \dfrac{5}{25}$

$s_3 = \left(5 - \dfrac{5}{9}\right) + \left(\dfrac{5}{9} - \dfrac{5}{25}\right) + \left(\dfrac{5}{25} - \dfrac{5}{49}\right) = 5 - \dfrac{5}{49}$

$s_n = 5 - \dfrac{5}{(2n+1)^2}$

$S = \lim_{n\to\infty} s_n = 5$

49. $s_1 = 1 - \dfrac{1}{\sqrt{2}}$

$s_2 = \left(1 - \dfrac{1}{\sqrt{2}}\right) + \left(\dfrac{1}{\sqrt{2}} - \dfrac{1}{\sqrt{3}}\right) = 1 - \dfrac{1}{\sqrt{3}}$

$s_3 = \left(1 - \dfrac{1}{\sqrt{2}}\right) + \left(\dfrac{1}{\sqrt{2}} - \dfrac{1}{\sqrt{3}}\right) + \left(\dfrac{1}{\sqrt{3}} - \dfrac{1}{\sqrt{4}}\right) = 1 - \dfrac{1}{\sqrt{4}}$

$s_n = 1 - \dfrac{1}{\sqrt{n+1}}$

$S = \lim_{n\to\infty} s_n = 1$

51. $s_1 = \tan^{-1} 1 - \tan^{-1} 2 = \dfrac{\pi}{4} - \tan^{-1} 2$

$s_2 = (\tan^{-1} 1 - \tan^{-1} 2) + (\tan^{-1} 2 - \tan^{-1} 3)$
$= \dfrac{\pi}{4} - \tan^{-1} 3$

$s_3 = (\tan^{-1} 1 - \tan^{-1} 2) + (\tan^{-1} 2 - \tan^{-1} 3)$
$ + (\tan^{-1} 3 - \tan^{-1} 4)$
$= \dfrac{\pi}{4} - \tan^{-1} 4$

$s_n = \dfrac{\pi}{4} - \tan^{-1}(n+1)$

$S = \lim_{n\to\infty} s_n = \dfrac{\pi}{4} - \lim_{n\to\infty} \tan^{-1} n = \dfrac{\pi}{4} - \dfrac{\pi}{2} = -\dfrac{\pi}{4}$

Section 9.5 Testing Convergence at Endpoints (pp. 496–508)

Exploration 1   The p-Series Test

1. We first note that the Integral Test applies to any series of the form $\sum \dfrac{1}{n^p}$ where p is positive. This is because the function $f(x) = x^{-p}$ is continuous and positive for all $x > 0$, and $f'(x) = -p \cdot x^{p-1}$ is negative for all $x > 0$.

   If $p > 1$:
   $$\int_1^\infty \dfrac{1}{x^p}\,dx = \lim_{k\to\infty}\int_1^k \dfrac{1}{x^p}\,dx = \lim_{k\to\infty}\left(\dfrac{x^{-p+1}}{-p+1}\right]_1^k$$
   $$= \lim_{k\to\infty}\left(\dfrac{1}{1-p}\cdot\left(\dfrac{1}{k^{p-1}} - 1\right)\right)$$
   $$= 0 + \dfrac{1}{p-1}\ \text{(since } p-1 > 0\text{)}$$
   $$= \dfrac{1}{p-1} < \infty.$$
   The series converges by the Intergral Test.

2. If $0 < p < 1$:
   $$\int_1^\infty \dfrac{1}{x^p}\,dx = \lim_{k\to\infty}\int_1^k \dfrac{1}{x^p}\,dx$$
   $$= \lim_{k\to\infty}\left(\dfrac{x^{-p+1}}{-p+1}\right]_1^k$$
   $$= \lim_{k\to\infty}\left(\dfrac{1}{1-p}\cdot(k^{1-p} - 1)\right)$$
   $$= \infty\ \text{(since } 1 - p > 0\text{)}.$$
   The series diverges by the Integral Test.
   If $p \leq 0$, the series diverges by the nth Term Test. This completes the proof for $p < 1$.

3. If $p = 1$:
   $$\int_1^\infty \dfrac{1}{x^p}\,dx = \lim_{k\to\infty}\int_1^k \dfrac{1}{x}\,dx$$
   $$= \lim_{k\to\infty}\left(\ln x\right]_1^k$$
   $$= \lim_{k\to\infty} \ln k = \infty.$$
   The series diverges by the Integral Test.

Exploration 2   The Maclaurin Series of a Strange Function

1. Since $f^{(n)}(0) = 0$ for all n, the Maclaurin Series for f has all zero coefficients! The series is simply $\sum_{n=0}^{\infty} 0 \cdot x^n = 0$.

2. The series converges (to 0) for all values of x.

3. Since $f(x) = 0$ only at $x = 0$, the only place that this series actually converges to its f-value is at $x = 0$.

Quick Review 9.5

1. Converges, since it is of the form $\int_1^\infty \frac{1}{x^p}\, dx$ with $p > 1$
2. Diverges, limit comparison test with integral of $\frac{1}{x}$
3. Diverges, comparison test with integral of $\frac{1}{x}$
4. Converges, comparison test with integral of $\frac{2}{x^2}$
5. Diverges, limit comparison test with integral of $\frac{1}{\sqrt{x}}$
6. Yes, for $N = 0$
7. Yes, for $N = 2\sqrt{2}$
8. No, neither positive nor decreasing for $x > \sqrt{3}$
9. No, oscillates
10. No, not positive for $x \geq 1$

Section 9.5 Exercises

1. Diverges by the Integral Test, since $\int_1^\infty \frac{5}{x+1}\, dx$ diverges.

3. Diverges by the Direct Comparison Test, since $\frac{\ln n}{n} > \frac{1}{n}$ for $n \geq 2$ and $\sum_{n=2}^\infty \frac{1}{n}$ diverges.

5. Diverges, since it is a geometric series with
$r = \frac{1}{\ln 2} \approx 1.44$.

7. Diverges by the nth-Term Test, since $\lim_{n\to\infty} n \sin \frac{1}{n} = 1$.

9. Converges by the Direct Comparison Test, since
$\frac{\sqrt{n}}{n^2+1} < \frac{1}{n^{3/2}}$ for $n \geq 1$, and $\sum_{n=0}^\infty \frac{1}{n^{3/2}}$ converges as a p-series with $p = \frac{3}{2}$.

11. Diverges by the nth-Term Test, since
$\lim_{n\to\infty} \frac{3^{n-1}+1}{3^n} = \frac{1}{3} \neq 0$.

13. Diverges by the nth-Term Test, since $\lim_{n\to\infty} \frac{10^n}{n^{10}} = \infty$.

15. Diverges by the nth-Term Test since $\frac{\ln n}{\ln n^2} = \frac{\ln n}{2 \ln n} = \frac{1}{2}$,
which means each term is $\pm \frac{1}{2}$.

17. Converges absolutely, because, absolutely, it is a geometric series with $r = 0.1$.

19. Converges absolutely, since $\sum_{n=1}^\infty n^2 \left(\frac{2}{3}\right)^n$ converges by the Ratio Test:
$\lim_{n\to\infty} \left|\frac{a_{n+1}}{a_n}\right| = (n+1)^2 \left(\frac{2}{3}\right)^{n+1} \cdot \frac{1}{n^2}\left(\frac{3}{2}\right)^n = \frac{2}{3} < 1$.

21. Diverges by the nth-Term Test, since $\lim_{n\to\infty} \frac{n!}{2^n} = \infty$ and so the terms do not approach 0.

23. Converges conditionally:

If $u_n = \frac{1}{1+\sqrt{n}}$, then $\{u_n\}$ is a decreasing sequence of positive terms with $\lim_{n\to\infty} u_n = 0$, so $\sum_{n=1}^\infty \frac{(-1)^n}{1+\sqrt{n}}$ converges by the Alternating Series Test.

But $\sum_{n=1}^\infty \frac{1}{1+\sqrt{n}}$ diverges by direct comparison to $\sum_{n=1}^\infty \frac{1}{n^{1/2}}$, which diverges as a p-series with $p = \frac{1}{2}$.

25. Converges conditionally, since $\sum_{n=1}^\infty \frac{\cos n\pi}{n} = -\sum_{n=1}^\infty \frac{(-1)^{n+1}}{n}$.
(See Examples 2 and 4.)

27. This is a geometric series which converges only for $|x| < 1$.
    (a) $(-1, 1)$
    (b) $(-1, 1)$
    (c) None

29. This is a geometric series which converges only for
$|4x + 1| < 1$, or $-\frac{1}{2} < x < 0$.
    (a) $\left(-\frac{1}{2}, 0\right)$
    (b) $\left(-\frac{1}{2}, 0\right)$
    (c) None

31. This is a geometric series which converges only for
$\left|\frac{x-2}{10}\right| < 1$, or $-8 < x < 12$.
    (a) $(-8, 12)$
    (b) $(-8, 12)$
    (c) None

33. $\lim_{n\to\infty} \left|\frac{a_{n+1}}{a_n}\right| = \lim_{n\to\infty} \frac{|x|^{n+1}}{(n+1)\sqrt{n+1} \cdot 3^{n+1}} \cdot \frac{n\sqrt{n}\, 3^n}{|x|^n} = \frac{|x|}{3}$

The series converges absolutely for $|x| < 3$. Furthermore, when $|x| = 3$, $\sum_{n=1}^\infty \left|\frac{x^n}{n\sqrt{n}\, 3^n}\right| = \sum_{n=1}^\infty \frac{1}{n^{3/2}}$, which also converges as a p-series with $p = \frac{3}{2}$.

(a) $[-3, 3]$
(b) $[-3, 3]$
(c) None

35. $\lim\limits_{n\to\infty}\left|\dfrac{a_{n+1}}{a_n}\right| = \lim\limits_{n\to\infty}\dfrac{(n+1)|x+3|^{n+1}}{5^{n+1}} \cdot \dfrac{5^n}{n|x+3|^n} = \dfrac{|x+3|}{5}$

The series converges absolutely for $\dfrac{|x+3|}{5} < 1$, or $-8 < x < 2$. For $\dfrac{|x+3|}{5} \geq 1$, the series diverges by the nth-Term Test.

(a) $(-8, 2)$

(b) $(-8, 2)$

(c) None

37. $\lim\limits_{n\to\infty}\left|\dfrac{a_{n+1}}{a_n}\right| = \lim\limits_{n\to\infty}\dfrac{\sqrt{n+1}\,|x|^{n+1}}{3^{n+1}} \cdot \dfrac{3^n}{\sqrt{n}\,|x|^n} = \dfrac{|x|}{3}$

The series converges absolutely for $|x| < 3$, or $-3 < x < 3$.

For $|x| \geq 3$, the series diverges by the nth-Term Test.

(a) $(-3, 3)$

(b) $(-3, 3)$

(c) None

39. $\lim\limits_{n\to\infty}\left|\dfrac{a_{n+1}}{a_n}\right| = \lim\limits_{n\to\infty}\dfrac{|-2|^{n+1}(n+2)|x-1|^{n+1}}{|-2|^n(n+1)|x-1|^n} = |2(x-1)|$

The series converges absolutely for $|2(x-1)| < 1$, or $\dfrac{1}{2} < x < \dfrac{3}{2}$. For $|2(x-1)| \geq 1$, the series diverges by the nth-Term Test.

(a) $\left(\dfrac{1}{2}, \dfrac{3}{2}\right)$

(b) $\left(\dfrac{1}{2}, \dfrac{3}{2}\right)$

(c) None

41. $\lim\limits_{n\to\infty}\left|\dfrac{a_{n+1}}{a_n}\right| = \lim\limits_{n\to\infty}\dfrac{|x+\pi|^{n+1}}{\sqrt{n+1}} \cdot \dfrac{\sqrt{n}}{|x+\pi|^n} = |x+\pi|$

The series converges absolutely for $|x+\pi| < 1$, or $-\pi - 1 < x < -\pi + 1$.

Check $x = -\pi - 1$:

$\sum\limits_{n=1}^{\infty}\dfrac{(-1)^n}{\sqrt{n}}$ converges by Alternating Series Test.

Check $x = -\pi + 1$:

$\sum\limits_{n=1}^{\infty}\dfrac{1}{\sqrt{n}}$ diverges as a p-series with $p = \dfrac{1}{2}$.

(a) $[-\pi - 1, -\pi + 1)$

(b) $(-\pi - 1, -\pi + 1)$

(c) At $x = -\pi - 1$

43. $n = 13 \times 10^9 \cdot 365 \cdot 24 \cdot 3600 = 4.09968 \times 10^{17}$

$\ln(n+1) < \text{sum} < 1 + \ln n$

$\ln(4.09968 \times 10^{17} + 1) < \text{sum} < 1 + \ln(4.09968 \times 10^{17})$

$40.5548\ldots < \text{sum} < 41.5548\ldots$

$40.554 < \text{sum} < 41.555$

45.

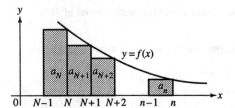

Comparing areas in the figures, we have for all $n \geq N$,

$$\int_N^{n+1} f(x)\,dx < a_N + \cdots + a_n < a_N + \int_N^n f(x)\,dx.$$

If the integral diverges, it must go to infinity, and the first inequality forces the partial sums of the series to go to infinity as well, so the series is divergent. If the integral converges, then the second inequality puts an upper bound on the partial sums of the series, and since they are a nondecreasing sequence, they must converge to a finite sum for the series. (See the explanation preceding Exercise 42 in Section 9.4.)

47. One possible answer: $\sum\limits_{n=3}^{\infty}\dfrac{1}{n\ln n}$

This series diverges by the integral test, since

$\int_3^{\infty}\dfrac{1}{x\ln x}\,dx = \lim\limits_{b\to\infty}\left[\ln|\ln x|\right]_3^b = \infty$. Its partial sums are roughly $\ln(\ln n)$, so they are much smaller than the partial sums for the harmonic series, which are about $\ln n$.

49. (a) Diverges by the Limit Comparison Test. Let

$a_n = \dfrac{n}{3n^2+1}$ and $b_n = \dfrac{1}{n}$. Then $a_n > 0$ and $b_n > 0$ for $n \geq 1$, and $\lim\limits_{n\to\infty}\dfrac{a_n}{b_n} = \lim\limits_{n\to\infty}\dfrac{n^2}{3n^2+1} = \dfrac{1}{3}$. Since $\sum\limits_{n=1}^{\infty}b_n$ diverges, $\sum\limits_{n=1}^{\infty}a_n$ diverges.

(b) $S = \sum\limits_{n=1}^{\infty}\dfrac{n}{3n^2+1} \cdot \dfrac{3}{n} = \sum\limits_{n=1}^{\infty}\dfrac{3}{3n^2+1}$.

This series converges by the Direct Comparison Test, since $\dfrac{3}{3n^2+1} < \dfrac{1}{n^2}$ and $\sum\limits_{n=1}^{\infty}\dfrac{1}{n^2}$ is convergent as a p-series with $p = 2$.

51. $\lim_{k\to\infty}\left|\frac{a_{k+1}}{a_k}\right| = \lim_{k\to\infty}\frac{2^{k+1}|x|^{k+1}}{\ln(k+3)} \cdot \frac{\ln(k+2)}{2^k|x|^k} = 2|x|$

The series converges absolutely for $|x| < \frac{1}{2}$,

or $-\frac{1}{2} < x < \frac{1}{2}$.

Check $x = -\frac{1}{2}$:

$\sum_{k=0}^{\infty} \frac{(-1)^k}{\ln(k+2)}$ converges by the Alternating Series Test.

Check $x = \frac{1}{2}$:

$\sum_{k=0}^{\infty} \frac{1}{\ln(k+2)}$ diverges by the Direct Comparison Test, since $\frac{1}{\ln(k+2)} > \frac{1}{k}$ for $k \geq 2$ and $\sum_{k=2}^{\infty} \frac{1}{k}$ diverges. The original series converges for $-\frac{1}{2} \leq x < \frac{1}{2}$.

53. $\ln(1+x) = \sum_{n=1}^{\infty}(-1)^{n+1}\frac{x^n}{n}$, so at $x = 1$, the series is $\sum_{n=1}^{\infty}\frac{(-1)^{n+1}}{n}$. This series converges by the Alternating Series Test.

55. (a) It fails to satisfy $u_n \geq u_{n+1}$ for all $n \geq N$.

(b) The sum is $\left(\sum_{n=1}^{\infty}\frac{1}{3^n}\right) - \left(\sum_{n=1}^{\infty}\frac{1}{2^n}\right) = \frac{1/3}{1-1/3} - \frac{1/2}{1-1/2}$
$= \frac{1}{2} - 1$
$= -\frac{1}{2}$.

57. (a) $\lim_{n\to\infty}\sqrt[n]{a_n} = \lim_{n\to\infty}\sqrt[n]{\frac{n^2}{2^n}} = \lim_{n\to\infty}\frac{(\sqrt[n]{n})^2}{2} = \frac{1}{2}$

The series converges.

(b) $\lim_{n\to\infty}\sqrt[n]{a_n} = \lim_{n\to\infty}\sqrt[n]{\left(\frac{n}{2n-1}\right)^n} = \lim_{n\to\infty}\frac{n}{2n-1} = \frac{1}{2}$

The series converges.

(c) $\lim_{n\to\infty,\,n\text{ odd}}\sqrt[n]{a_n} = \lim_{n\to\infty,\,n\text{ odd}}\sqrt[n]{\frac{n}{2^n}}$
$= \lim_{n\to\infty,\,n\text{ odd}}\frac{\sqrt[n]{n}}{2} = \frac{1}{2}$

$\lim_{n\to\infty,\,n\text{ even}}\sqrt[n]{a_n} = \lim_{n\to\infty,\,n\text{ even}}\sqrt[n]{\frac{1}{2^n}} = \frac{1}{2}$

Thus, $\lim_{n\to\infty}\sqrt[n]{a_n} = \frac{1}{2}$, so the series converges.

Chapter 9 Review Exercises

(pp. 509–511)

1. $\lim_{n\to\infty}\left|\frac{a_{n+1}}{a_n}\right| = \lim_{n\to\infty}\frac{|-x|^{n+1}}{(n+1)!} \cdot \frac{n!}{|-x|^n} = \lim_{n\to\infty}\frac{|x|}{n+1} = 0$

The series converges absolutely for all x.

(a) ∞

(b) All real numbers

(c) All real numbers

(d) None

2. $\lim_{n\to\infty}\left|\frac{a_{n+1}}{a_n}\right| = \lim_{n\to\infty}\frac{|x+4|^{n+1}}{(n+1)3^{n+1}} \cdot \frac{n3^n}{|x+4|^n} = \frac{|x+4|}{3}$

The series converges absolutely for $\frac{|x+4|}{3} < 1$,

or $-7 < x < -1$.

Check $x = -7$: $\sum_{n=1}^{\infty}\frac{(-1)^n}{n}$ converges

Check $x = -1$: $\sum_{n=1}^{\infty}\frac{1}{n}$ diverges.

(a) 3

(b) $[-7, -1)$

(c) $(-7, -1)$

(d) At $x = -7$

3. This is a geometric series, so it converges absolutely when $|r| < 1$ and diverges for all other values of x. Since $r = \frac{2}{3}(x-1)$, the series converges absolutely when $\left|\frac{2}{3}(x-1)\right| < 1$, or $-\frac{1}{2} < x < \frac{5}{2}$.

(a) $\frac{3}{2}$

(b) $\left(-\frac{1}{2}, \frac{5}{2}\right)$

(c) $\left(-\frac{1}{2}, \frac{5}{2}\right)$

(d) None

4. $\lim_{n\to\infty}\left|\frac{a_{n+1}}{a_n}\right| = \lim_{n\to\infty}\frac{|x-1|^{2n}}{(2n+1)!} \cdot \frac{(2n-1)!}{|x-1|^{2n-2}}$
$= \lim_{n\to\infty}\frac{|x-1|^2}{(2n+1)(2n)} = 0$

The series converges absolutely for all x.

(a) ∞

(b) All real numbers

(c) All real numbers

(d) None

5. $\lim_{n\to\infty}\left|\frac{a_{n+1}}{a_n}\right| = \lim_{n\to\infty}\frac{|3x-1|^{n+1}}{(n+1)^2} \cdot \frac{n^2}{|3x-1|^n} = |3x-1|$

The series converges absolutely for

$|3x-1| < 1$, or $0 < x < \frac{2}{3}$. Furthermore, when

$|3x-1| = 1$, we have $|a_n| = \frac{1}{n^2}$ and

$\sum_{n=1}^{\infty}\frac{1}{n^2}$ converges as a p-series with $p = 2$, so $\sum_{n=1}^{\infty} a_n$ also converges absolutely at the interval endpoints.

(a) $\frac{1}{3}$

(b) $\left[0, \frac{2}{3}\right]$

(c) $\left[0, \frac{2}{3}\right]$

(d) None

260 Chapter 9 Review

6. $\lim\limits_{n\to\infty}\left|\dfrac{a_{n+1}}{a_n}\right| = \lim\limits_{n\to\infty}\dfrac{(n+2)|x|^{3n+3}}{(n+1)|x|^{3n}} = |x|^3$

The series converges absolutely for $|x|^3 < 1$, or $-1 < x < 1$. When $|x| \geq 1$, the series diverges by the nth Term Test.

(a) 1
(b) $(-1, 1)$
(c) $(-1, 1)$
(d) None

7. $\lim\limits_{n\to\infty}\left|\dfrac{a_{n+1}}{a_n}\right| = \lim\limits_{n\to\infty}\dfrac{(n+2)|2x+1|^{n+1}}{(2n+3)2^{n+1}} \cdot \dfrac{(2n+1)2^n}{(n+1)|2x+1|^n}$

$= \dfrac{|2x+1|}{2}$

The series converges absolutely for $\dfrac{|2x+1|}{2} < 1$, or $-\dfrac{3}{2} < x < \dfrac{1}{2}$. When $\dfrac{|2x+1|}{2} \geq 1$, the series diverges by the nth-Term Test.

(a) 1
(b) $\left(-\dfrac{3}{2}, \dfrac{1}{2}\right)$
(c) $\left(-\dfrac{3}{2}, \dfrac{1}{2}\right)$
(d) None

8. $\lim\limits_{n\to\infty}\left|\dfrac{a_{n+1}}{a_n}\right| = \lim\limits_{n\to\infty}\dfrac{|x|^{n+1}}{(n+1)^{n+1}} \cdot \dfrac{n^n}{|x|^n} = |x|\lim\limits_{n\to\infty}\dfrac{n^n}{(n+1)(n+1)^n}$

$= |x|\lim\limits_{n\to\infty}\dfrac{1}{(n+1)\left(1+\dfrac{1}{n}\right)^n} = \dfrac{|x|}{e}\lim\limits_{n\to\infty}\dfrac{1}{n+1} = 0$

The series converges absolutely for all x.

Another way to see that the series must converge is to observe that for $n \geq 2x$, we have $\left|\dfrac{x^n}{n^n}\right| \leq \left(\dfrac{1}{2}\right)^n$, so the terms are (eventually) bounded by the terms of a convergent geometric series.

A third way to solve this exercise is to use the nth Root Test (see Exercises 57–58 in Section 9.5).

(a) ∞
(b) All real numbers
(c) All real numbers
(d) None

9. $\lim\limits_{n\to\infty}\left|\dfrac{a_{n+1}}{a_n}\right| = \lim\limits_{n\to\infty}\dfrac{|x|^{n+1}}{\sqrt{n+1}} \cdot \dfrac{\sqrt{n}}{|x|^n} = |x|$

The series converges absolutely for $|x| < 1$, or $-1 < x < 1$.

Check $x = -1$:
$\sum\limits_{n=1}^{\infty}\dfrac{(-1)^n}{\sqrt{n}}$ converges by the Alternating Series Test.

Check $x = 1$:
$\sum\limits_{n=1}^{\infty}\dfrac{1}{\sqrt{n}}$ diverges as a p-series with $p = \dfrac{1}{2}$.

(a) 1
(b) $[-1, 1)$
(c) $(-1, 1)$
(d) At $x = -1$

10. $\lim\limits_{n\to\infty}\left|\dfrac{a_{n+1}}{a_n}\right| = \lim\limits_{n\to\infty}\dfrac{e^{n+1}|x|^{n+1}}{(n+1)^e} \cdot \dfrac{n^e}{e^n|x|^n} = e|x|$

The series converges absolutely for $e|x| < 1$, or $-\dfrac{1}{e} < x < \dfrac{1}{e}$.

Furthermore, when $e|x| = 1$, we have $|a_n| = \dfrac{1}{n^e}$ and $\sum\limits_{n=1}^{\infty}\dfrac{1}{n^e}$ converges as a p-series with $p = e$, so $\sum\limits_{n=1}^{\infty} a_n$ also converges absolutely at the interval endpoints.

(a) $\dfrac{1}{e}$
(b) $\left[-\dfrac{1}{e}, \dfrac{1}{e}\right]$
(c) $\left[-\dfrac{1}{e}, \dfrac{1}{e}\right]$
(d) None

11. $\lim\limits_{n\to\infty}\left|\dfrac{a_{n+1}}{a_n}\right| = \lim\limits_{n\to\infty}\dfrac{(n+2)|x|^{2n+1}}{3^{n+1}} \cdot \dfrac{3^n}{(n+1)|x|^{2n-1}} = \dfrac{x^2}{3}$

The series converges absolutely when $\dfrac{x^2}{3} < 1$, or $-\sqrt{3} < x < \sqrt{3}$.

When $|x| \geq \sqrt{3}$, the series diverges by the nth Term Test.

(a) $\sqrt{3}$
(b) $(-\sqrt{3}, \sqrt{3})$
(c) $(-\sqrt{3}, \sqrt{3})$
(d) None

12. $\lim_{n\to\infty} \left|\frac{a_{n+1}}{a_n}\right| = \lim_{n\to\infty} \frac{|x-1|^{2n+3}}{2n+3} \cdot \frac{2n+1}{|x-1|^{2n+1}} = |x-1|^2$

The series converges absolutely when $|x-1|^2 < 1$, or $0 < x < 2$.

Check $x = 0$: $\sum_{n=0}^{\infty} \frac{(-1)^n(-1)^{2n-1}}{2n+1} = -\sum_{n=0}^{\infty} \frac{(-1)^n}{2n+1}$ converges conditionally by the Alternating Series Test.

Check $x = 2$: $\sum_{n=0}^{\infty} \frac{(-1)^n}{2n+1}$ converges conditionally by the Alternating Series Test.

(a) 1
(b) [0, 2]
(c) (0, 2)
(d) At $x = 0$ and $x = 2$

13. $\lim_{n\to\infty} \left|\frac{a_{n+1}}{a_n}\right| = \lim_{n\to\infty} \frac{(n+1)!|x|^{2n+2}}{2^{n+1}} \cdot \frac{2^n}{n!|x|^{2n}}$

$= \lim_{n\to\infty} \frac{(n+1)x^2}{2} = \begin{cases} 0, & x = 0 \\ \infty, & x \neq 0 \end{cases}$

The series converges only at $x = 0$.

(a) 0
(b) $x = 0$ only
(c) $x = 0$
(d) None

14. $\lim_{n\to\infty} \left|\frac{a_{n+1}}{a_n}\right| = \lim_{n\to\infty} \frac{|10x|^{n+1}}{\ln(n+1)} \cdot \frac{\ln n}{|10x|^n} = |10x|$

The series converges absolutely for $|10x| < 1$,

or $-\frac{1}{10} < x < \frac{1}{10}$.

Check $n = -\frac{1}{10}$: $\sum_{n=2}^{\infty} \frac{(-1)^n}{\ln n}$ converges by the Alternating Series Test.

Check $n = \frac{1}{10}$: $\sum_{n=2}^{\infty} \frac{1}{\ln n}$ diverges by the Direct Comparison Test, since $\frac{1}{\ln n} > \frac{1}{n}$ for $n \geq 2$ and $\sum_{n=2}^{\infty} \frac{1}{n}$ diverges.

(a) $\frac{1}{10}$
(b) $\left[-\frac{1}{10}, \frac{1}{10}\right)$
(c) $\left(-\frac{1}{10}, \frac{1}{10}\right)$
(d) At $x = -\frac{1}{10}$

15. $\lim_{n\to\infty} \left|\frac{a_{n+1}}{a_n}\right| = \lim_{n\to\infty} \frac{(n+2)!|x|^{n+1}}{(n+1)!|x|^n} = \lim_{n\to\infty} (n+2)|x| = \infty \ (x \neq 0)$

The series converges only at $x = 0$.

(a) 0
(b) $x = 0$ only
(c) $x = 0$
(d) None

16. This is a geometric series with $r = \frac{x^2-1}{2}$, so it converges absolutely when $\left|\frac{x^2-1}{2}\right| < 1$, or $-\sqrt{3} < x < \sqrt{3}$. It diverges for all other values of x.

(a) $\sqrt{3}$
(b) $(-\sqrt{3}, \sqrt{3})$
(c) $(-\sqrt{3}, \sqrt{3})$
(d) None

17. $f(x) = \frac{1}{1+x} = 1 - x + x^2 - \cdots + (-1)^n x^n + \cdots$,

evaluated at $x = \frac{1}{4}$. Sum $= \frac{1}{1+\left(\frac{1}{4}\right)} = \frac{4}{5}$.

18. $f(x) = \ln(1+x) = x - \frac{x^2}{2} + \frac{x^3}{3} - \cdots + (-1)^{n-1}\frac{x^n}{n}$,

evaluated at $x = \frac{2}{3}$. Sum $= \ln\left(1 + \frac{2}{3}\right) = \ln\left(\frac{5}{3}\right)$.

19. $f(x) = \sin x = x - \frac{x^3}{3!} + \frac{x^5}{5!} - \cdots + (-1)^n \frac{x^{2n+1}}{(2n+1)!} + \cdots$,

evaluated at $x = \pi$. Sum $= \sin \pi = 0$.

20. $f(x) = \cos x = 1 - \frac{x^2}{2!} + \frac{x^4}{4!} - \cdots + (-1)^n \frac{x^{2n}}{(2n)!} + \cdots$,

evaluated at $x = \frac{\pi}{3}$. Sum $= \cos \frac{\pi}{3} = \frac{1}{2}$.

21. $f(x) = e^x = 1 + x + \frac{x^2}{2!} + \cdots + \frac{x^n}{n!} + \cdots$, evaluated at $x = \ln 2$. Sum $= e^{\ln 2} = 2$.

22. $f(x) = \tan^{-1} x = x - \frac{x^3}{3} + \frac{x^5}{5} - \cdots + (-1)^n \frac{x^{2n+1}}{2n+1} + \cdots$,

evaluated at $x = \frac{1}{\sqrt{3}}$. Sum $= \tan^{-1}\left(\frac{1}{\sqrt{3}}\right) = \frac{\pi}{6}$. (Note that when n is replaced by $n - 1$, the general term of $\tan^{-1} x$ becomes $(-1)^{n-1}\frac{x^{2n-1}}{2n-1}$, which matches the general term given in the exercise.)

23. Replace x by $6x$ in the Maclaurin series for $\frac{1}{1-x}$ given at the end of Section 9.2.

$\frac{1}{1-6x} = 1 + (6x) + (6x)^2 + \cdots + (6x)^n + \cdots$

$= 1 + 6x + 36x^2 + \cdots + (6x)^n + \cdots$

24. Replace x by x^3 in the Maclaurin series for $\frac{1}{1+x}$ given at the end of Section 9.2.

$\frac{1}{1+x^3} = 1 - (x^3) + (x^3)^2 - \cdots + (-x^3)^n + \cdots$

$= 1 - x^3 + x^6 - \cdots + (-1)^n x^{3n} + \cdots$

25. The Maclaurin series for a polynomial is the polynomial itself: $1 - 2x^2 + x^9$.

26. $\dfrac{4x}{1-x} = 4x\left(\dfrac{1}{1-x}\right)$
$= 4x(1 + x + x^2 + \cdots + x^n + \cdots)$
$= 4x + 4x^2 + 4x^3 + \cdots + 4x^{n+1} + \cdots$

27. Replace x by πx in the Maclaurin series for $\sin x$ given at the end of Section 9.2.

$\sin \pi x = \pi x - \dfrac{(\pi x)^3}{3!} + \dfrac{(\pi x)^5}{5!} - \cdots + (-1)^n \dfrac{(\pi x)^{2n+1}}{(2n+1)!} + \cdots$

28. Replace x by $\dfrac{2x}{3}$ in the Maclaurin series for $\sin x$ given at the end of Section 9.2

$-\sin \dfrac{2x}{3} = -\left(\dfrac{2x}{3} - \dfrac{\left(\dfrac{2x}{3}\right)^3}{3!} + \dfrac{\left(\dfrac{2x}{3}\right)^5}{5!} - \cdots + (-1)^n \dfrac{\left(\dfrac{2x}{3}\right)^{2n+1}}{(2n+1)!}\right)$

$= -\dfrac{2x}{3} + \dfrac{4x^3}{81} - \dfrac{4x^5}{3645} + \cdots + \dfrac{(-1)^{n+1}\left(\dfrac{2x}{3}\right)^{2n+1}}{(2n+1)!}$

29. $-x + \sin x = -x + \left(x - \dfrac{x^3}{3!} + \dfrac{x^5}{5!} - \dfrac{x^7}{7!} + \cdots \right.$
$\left. + (-1)^n \dfrac{x^{2n+1}}{(2n+1)!} + \cdots\right)$
$= -\dfrac{x^3}{3!} + \dfrac{x^5}{5!} - \dfrac{x^7}{7!} + \cdots + (-1)^n \dfrac{x^{2n+1}}{(2n+1)!} + \cdots$

30. $\dfrac{e^x + e^{-x}}{2} = \dfrac{1}{2}\left(1 + x + \dfrac{x^2}{2!} + \cdots + \dfrac{x^n}{n!} + \cdots\right)$
$+ \dfrac{1}{2}\left(1 - x + \dfrac{x^2}{2!} + \cdots + (-1)^n\dfrac{x^n}{n!} + \cdots\right)$
$= 1 + \dfrac{x^2}{2!} + \dfrac{x^4}{4!} + \cdots + \dfrac{x^{2n}}{(2n)!} + \cdots$

31. Replace x by $\sqrt{5x}$ in the Maclaurin series for $\cos x$ given at the end of Section 9.2.

$\cos \sqrt{5x} = 1 - \dfrac{(\sqrt{5x})^2}{2!} + \dfrac{(\sqrt{5x})^4}{4!} - \cdots$
$+ (-1)^n \dfrac{(\sqrt{5x})^{2n}}{(2n)!} + \cdots$
$= 1 - \dfrac{5x}{2!} + \dfrac{(5x)^2}{4!} - \cdots + (-1)^n \dfrac{(5x)^n}{(2n)!} + \cdots$

32. Replace x by $\dfrac{\pi x}{2}$ in the Maclaurin series for e^x given at the end of Section 9.2.

$e^{\pi x/2} = 1 + \dfrac{\pi x}{2} + \dfrac{\left(\dfrac{\pi x}{2}\right)^2}{2!} + \cdots + \dfrac{\left(\dfrac{\pi x}{2}\right)^n}{n!} + \cdots$
$= 1 + \dfrac{\pi x}{2} + \dfrac{\pi^2 x^2}{8} + \cdots + \dfrac{1}{n!}\left(\dfrac{\pi x}{2}\right)^n + \cdots$

33. Use the Maclaurin series for e^x given at the end of Section 9.2.

$xe^{-x^2} = x\left[1 + (-x^2) + \dfrac{(-x^2)^2}{2!} + \cdots + \dfrac{(-x^2)^n}{n!} + \cdots\right]$
$= x - x^3 + \dfrac{x^5}{2!} - \cdots + (-1)^n \dfrac{x^{2n+1}}{n!} + \cdots$

34. Replace x by $3x$ in the Maclaurin series for $\tan^{-1} x$ given at the end of Section 9.2.

$\tan^{-1} 3x = 3x - \dfrac{(3x)^3}{3} + \dfrac{(3x)^5}{5} - \cdots + (-1)^n \dfrac{(3x)^{2n+1}}{2n+1} + \cdots$

35. Replace x by $-2x$ in the Maclaurin series for $\ln(1+x)$ given at the end of Section 9.2.

$\ln(1 - 2x) = -2x - \dfrac{(-2x)^2}{2} + \dfrac{(-2x)^3}{3} - \cdots$
$+ (-1)^{n-1}\dfrac{(-2x)^n}{n} + \cdots$
$= -2x - 2x^2 - \dfrac{8x^3}{3} - \cdots - \dfrac{(2x)^n}{n} - \cdots$

36. Use the Maclaurin series for $\ln(1+x)$ given at the end of Section 9.2.

$x \ln(1-x) = x \ln[1+(-x)]$
$= x\left[-x - \dfrac{(-x)^2}{2} + \dfrac{(-x)^3}{3} - \cdots + (-1)^{n-1}\dfrac{(-x)^n}{n} + \cdots\right]$
$= -x^2 - \dfrac{x^3}{2} - \dfrac{x^4}{3} - \cdots - \dfrac{x^{n+1}}{n} - \cdots$

37. $f(2) = (3-x)^{-1}\big|_{x=2} = 1$

$f'(2) = (3-x)^{-2}\big|_{x=2} = 1$

$f''(2) = 2(3-x)^{-3}\big|_{x=2} = 2$, so $\dfrac{f''(2)}{2!} = 1$

$f'''(2) = 6(3-x)^{-4}\big|_{x=2} = 6$, so $\dfrac{f'''(2)}{3!} = 1$

$f^{(n)}(2) = n!(3-x)^{-n-1}\big|_{x=2} = n!$, so $\dfrac{f^{(n)}(2)}{n!} = 1$

$\dfrac{1}{3-x} = 1 + (x-2) + (x-2)^2 + (x-2)^3 + \cdots$
$+ (x-2)^n + \cdots$

38. $f(-1) = (x^3 - 2x^2 + 5)\big|_{x=-1} = 2$

$f'(-1) = (3x^2 - 4x)\big|_{x=-1} = 7$

$f''(-1) = (6x - 4)\big|_{x=-1} = -10$, so $\dfrac{f''(-1)}{2!} = -5$

$f'''(-1) = 6\big|_{x=-1} = 6$, so $\dfrac{f'''(-1)}{3!} = 1$

$f^{(n)}(-1) = 0$ for $n \geq 4$.

$x^3 - 2x^2 + 5 = 2 + 7(x+1) - 5(x+1)^2 + (x+1)^3$

This is a finite series and the general term for $n \geq 4$ is 0.

39. $f(3) = \frac{1}{x}\Big|_{x=3} = \frac{1}{3}$

$f'(3) = -x^{-2}\Big|_{x=3} = -\frac{1}{9}$

$f''(3) = 2x^{-3}\Big|_{x=3} = \frac{2}{27}$, so $\frac{f''(3)}{2!} = \frac{1}{27}$

$f'''(3) = -6x^{-4}\Big|_{x=3} = -\frac{2}{27}$, so $\frac{f'''(3)}{3!} = -\frac{1}{81}$

$\frac{f^{(n)}(3)}{n!} = \frac{(-1)^n}{3^{n+1}}$

$\frac{1}{x} = \frac{1}{3} - \frac{1}{9}(x-3) + \frac{1}{27}(x-3)^2 - \frac{1}{81}(x-3)^3 + \cdots$
$+ (-1)^n \frac{(x-3)^n}{3^{n+1}}$

40. $f(\pi) = \sin x\Big|_{x=\pi} = 0$

$f'(\pi) = \cos x\Big|_{x=\pi} = -1$

$f''(\pi) = -\sin x\Big|_{x=\pi} = 0$, so $\frac{f''(\pi)}{2!} = 0$

$f'''(\pi) = -\cos x\Big|_{x=\pi} = 1$, so $\frac{f'''(\pi)}{3!} = \frac{1}{6}$

$f^{(k)}(\pi) = \begin{cases} 0, & \text{if } k \text{ is even} \\ -1, & \text{if } k = 2n+1, n \text{ even} \\ 1, & \text{if } k = 2n+1, n \text{ odd} \end{cases}$

$\sin x = -(x-\pi) + \frac{1}{3!}(x-\pi)^3 - \frac{1}{5!}(x-\pi)^5$
$+ \frac{1}{7!}(x-\pi)^7 - \cdots$
$+ (-1)^{n+1}\frac{1}{(2n+1)!}(x-\pi)^{2n+1} + \cdots$

41. Diverges, because it is -5 times the harmonic series:

$\sum_{n=1}^{\infty} \frac{-5}{n} = -5\sum_{n=1}^{\infty}\frac{1}{n} = -\infty$

42. Converges conditionally.

If $u_n = \frac{1}{\sqrt{n}}$, then $\{u_n\}$ is a decreasing sequence of positive terms with $\lim_{n\to\infty} u_n = 0$, so $\sum_{n=1}^{\infty}\frac{(-1)^n}{\sqrt{n}}$ converges by the Alternating Series Test. The convergence is conditional because $\sum_{n=1}^{\infty}\frac{1}{\sqrt{n}}$ is a divergent p-series $\left(p = \frac{1}{2}\right)$.

43. Converges absolutely by the Direct Comparison Test, since

$0 \le \frac{\ln n}{n^3} < \frac{1}{n^2}$ for $n \ge 1$ and $\sum_{n=1}^{\infty}\frac{1}{n^2}$ converges as a p-series with $p = 2$.

44. Converges absolutely by the Ratio Test, since

$\lim_{n\to\infty}\left|\frac{a_{n+1}}{a_n}\right| = \lim_{n\to\infty}\frac{n+2}{(n+1)!} \cdot \frac{n!}{n+1} = \lim_{n\to\infty}\frac{n+2}{(n+1)^2} = 0$.

45. Converges conditionally:

If $u_n = \frac{1}{\ln(n+1)}$, then $\{u_n\}$ is a decreasing sequence of positive terms with $\lim_{n\to\infty} u_n = 0$, so $\sum_{n=1}^{\infty}\frac{(-1)^n}{\ln(n+1)}$ converges by the Alternating Series Test. The convergence is conditional because $\frac{1}{\ln(n+1)} > \frac{1}{n}$ for $n \ge 1$ and $\sum_{n=1}^{\infty}\frac{1}{n}$ diverges, so $\sum_{n=1}^{\infty}\frac{1}{\ln(n+1)}$ diverges by the Direct Comparison Test.

46. Converges absolutely by the Integral Test, because

$\int_2^{\infty}\frac{1}{x(\ln x)^2}dx = \lim_{b\to\infty}\left[-\frac{1}{\ln x}\right]_2^b = \frac{1}{\ln 2}$.

47. Converges absolutely the the Ratio Test, because

$\lim_{n\to\infty}\left|\frac{a_{n+1}}{a_n}\right| = \lim_{n\to\infty}\frac{|-3|^{n+1}}{(n+1)!}\cdot\frac{n!}{|-3|^n} = \lim_{n\to\infty}\frac{3}{n+1} = 0$.

48. Converges absolutely by the Direct Comparison Test, since

$\frac{2^n 3^n}{n^n} \le \left(\frac{1}{2}\right)^n$ for $n \ge 12$ and $\sum_{n=1}^{\infty}\left(\frac{1}{2}\right)^n$ is a convergent geometric series. Alternately, we may use the Ratio Test or the nth-Root Test (see Exercise 57 and 58 in Section 9.5).

49. Diverges by the nth-Term Test, since

$\lim_{n\to\infty}\frac{(-1)^n(n^2+1)}{2n^2+n-1}$ does not exist.

50. Converges absolutely by the Direct Comparison Test, since

$\frac{1}{\sqrt{n(n+1)(n+2)}} < \frac{1}{n^{3/2}}$ and $\sum_{n=1}^{\infty}\frac{1}{n^{3/2}}$ converges as a p-series with $p = \frac{3}{2}$.

51. Converges absolutely by the Limit Comparison Test.

Let $a_n = \frac{1}{n\sqrt{n^2-1}}$ and $b_n = \frac{1}{n^2}$.

Then $\lim_{n\to\infty}\frac{a_n}{b_n} = \lim_{n\to\infty}\frac{n^2}{n\sqrt{n^2-1}} = 1$ and $\sum_{n=2}^{\infty}\frac{1}{n^2}$ converges as a p-series ($p = 2$). Therefore $\sum_{n=2}^{\infty} a_n$ converges.

52. Diverges by the nth-Term Test, since

$\lim_{n\to\infty}\left(\frac{n}{n+1}\right)^n = \lim_{n\to\infty}\left(1+\frac{1}{n}\right)^{-n} = \frac{1}{e} \ne 0$.

53. This is a telescoping series.

$$\sum_{n=3}^{\infty} \frac{1}{(2n-3)(2n-1)} = \sum_{n=3}^{\infty} \left(\frac{1}{2(2n-3)} - \frac{1}{2(2n-1)} \right)$$

$$s_1 = \frac{1}{2(2 \cdot 3 - 3)} - \frac{1}{2(2 \cdot 3 - 1)} = \frac{1}{6} - \frac{1}{10}$$

$$s_2 = \left(\frac{1}{6} - \frac{1}{10}\right) + \left(\frac{1}{10} - \frac{1}{14}\right) = \frac{1}{6} - \frac{1}{14}$$

$$s_3 = \left(\frac{1}{6} - \frac{1}{10}\right) + \left(\frac{1}{10} - \frac{1}{14}\right) + \left(\frac{1}{14} - \frac{1}{18}\right) = \frac{1}{6} - \frac{1}{18}$$

$$s_n = \frac{1}{6} - \frac{1}{2(2n-1)}$$

$$S = \lim_{n \to \infty} s_n = \frac{1}{6}$$

54. This is a telescoping series.

$$\sum_{n=2}^{\infty} \frac{-2}{n(n+1)} = \sum_{n=2}^{\infty} \left(-\frac{2}{n} + \frac{2}{n+1} \right)$$

$$s_1 = -\frac{2}{2} + \frac{2}{3} = -1 + \frac{2}{3}$$

$$s_2 = \left(-1 + \frac{2}{3}\right) + \left(-\frac{2}{3} + \frac{2}{4}\right) = -1 + \frac{2}{4}$$

$$s_3 = \left(-1 + \frac{2}{3}\right) + \left(-\frac{2}{3} + \frac{2}{4}\right) + \left(-\frac{2}{4} + \frac{2}{5}\right) = -1 + \frac{2}{5}$$

$$s_n = -1 + \frac{2}{n+2}$$

$$S = \lim_{n \to \infty} s_n = -1$$

55. (a) $P_3(x) = f(3) + f'(3)(x-3) + \frac{f''(3)}{2!}(x-3)^2 + \frac{f'''(3)}{3!}(x-3)^3$

$= 1 + 4(x-3) + 3(x-3)^2 + 2(x-3)^3$

$f(3.2) \approx P_3(3.2) = 1.936$

(b) Since the Taylor series for f' can be obtained by term-by-term differentiation of the Taylor Series for f, the second order Taylor polynomial for f' at $x = 3$ is

$4 + 6(x-3) + 6(x-3)^2$. Evaluated at $x = 2.7$,

$f'(2.7) \approx 2.74$.

(c) It underestimates the values, since $f''(3) = 6$, which means the graph of f is concave up near $x = 3$.

56. (a) Since the constant term is $f(4)$, $f(4) = 7$. Since

$-2 = \frac{f'''(4)}{3!}, f'''(4) = -12$.

(b) Note that

$P_4'(x) = -3 + 10(x-4) - 6(x-4)^2 + 24(x-4)^3$.
The second degree polynomial for f' at $x = 4$ is given by the first three terms of this expression, namely
$-3 + 10(x-4) - 6(x-4)^2$. Evaluating at $x = 4.3$,
$f'(4.3) \approx -0.54$.

(c) The fourth order Taylor polynomial for $g(x)$ at $x = 4$ is

$\int_4^x [7 - 3(t-4) + 5(t-4)^2 - 2(t-4)^3] \, dt$

$= \left[7t - \frac{3}{2}(t-4)^2 + \frac{5}{3}(t-4)^3 - \frac{1}{2}(t-4)^4 \right]_4^x$

$= 7(x-4) - \frac{3}{2}(x-4)^2 + \frac{5}{3}(x-4)^3 - \frac{1}{2}(x-4)^4$

(d) No. One would need the entire Taylor series for $f(x)$, and it would have to converge to $f(x)$ at $x = 3$.

57. (a) Use the Maclaurin series for $\sin x$ given at the end of Section 9.2.

$5 \sin\left(\frac{x}{2}\right)$

$= 5\left[\frac{x}{2} - \frac{(x/2)^3}{3!} + \frac{(x/2)^5}{5!} - \cdots + (-1)^n \frac{(x/2)^{2n+1}}{(2n+1)!} + \cdots \right]$

$= \frac{5x}{2} - \frac{5x^3}{48} + \frac{x^5}{768} - \cdots + (-1)^n \frac{5}{(2n+1)!}\left(\frac{x}{2}\right)^{2n+1} + \cdots$

(b) The series converges for all real numbers, according to the Ratio Test:

$\lim_{n \to \infty} \left| \frac{a_{n+1}}{a_n} \right| = \lim_{n \to \infty} \frac{5}{(2n+3)!} \left|\frac{x}{2}\right|^{2n+3} \cdot \frac{(2n+1)!}{5} \left|\frac{2}{x}\right|^{2n+1}$

$= \lim_{n \to \infty} \frac{|x/2|^2}{(2n+3)(2n+2)} = 0$

(c) Note that the absolute value of $f^{(n)}(x)$ is bounded by $\frac{5}{2^n}$ for all x and all $n = 1, 2, 3, \cdots$.

We may use the Remainder Estimation Theorem with $M = 5$ and $r = \frac{1}{2}$.

So if $-2 < x < 2$, the truncation error using P_n is bounded by

$\frac{5}{2^{n+1}} \cdot \frac{2^{n+1}}{(n+1)!} = \frac{5}{(n+1)!}$.

To make this less than 0.1 requires $n \geq 4$. So, two terms (up through degree 4) are needed.

58. (a) Substitute $2x$ for x in the Maclaurin series for $\frac{1}{1-x}$ given at the end of Section 9.2.

$\frac{1}{1-2x} = 1 + 2x + (2x)^2 + (2x)^3 + \cdots + (2x)^n + \cdots$

$= 1 + 2x + 4x^2 + 8x^3 + \cdots + (2x)^n + \cdots$

(b) $\left(-\frac{1}{2}, \frac{1}{2}\right)$. The series for $\frac{1}{1-t}$ is known to converge for $-1 < t < 1$, so by substituting $t = 2x$, we find the resulting series converges for $-1 < 2x < 1$.

(c) $f\left(-\dfrac{1}{4}\right) = \dfrac{2}{3}$, so one percent is approximately 0.0067. It takes 7 terms (up through degree 6). This can be found by trial and error. Also, for $x = -\dfrac{1}{4}$, the series is the alternating series $\displaystyle\sum_{n=0}^{\infty}\left(-\dfrac{1}{2}\right)^n$. If you use the Alternating Series Estimation Theorem, it shows that 8 terms (up through degree 7) are sufficient since $\left|-\dfrac{1}{2}\right|^8 < 0.0067$. It is also a geometric series, and you could use the remainder formula for a geometric series to determine the number of terms needed. (See Example 2 in Section 9.3.)

59. (a) $\displaystyle\lim_{n\to\infty}\left|\dfrac{a_{n+1}}{a_n}\right| = \lim_{n\to\infty}\dfrac{|x|^{n+1}(n+1)^{n+1}}{(n+1)!} \cdot \dfrac{n!}{|x|^n n^n}$

$= \displaystyle\lim_{n\to\infty}\dfrac{|x|(n+1)^{n+1}}{(n+1)n^n}$

$= |x|\displaystyle\lim_{n\to\infty}\left(\dfrac{n+1}{n}\right)^n = |x|e$

The series converges for $|x|e < 1$, or $|x| < \dfrac{1}{e}$, so the radius of convergence is $\dfrac{1}{e}$.

(b) $f\left(-\dfrac{1}{3}\right) \approx -\dfrac{1}{3}\cdot\dfrac{1}{1} + \left(-\dfrac{1}{3}\right)^2 \cdot \dfrac{2^2}{2!} + \left(-\dfrac{1}{3}\right)^3 \cdot \dfrac{3^3}{3!}$

$= -\dfrac{1}{3} + \dfrac{2}{9} - \dfrac{1}{6}$

$= -\dfrac{5}{18} \approx -0.278$

(c) By the Alternating Series Estimation Theorem the error is no more than the magnitude of the next term, which is $\left|\left(-\dfrac{1}{3}\right)^4 \cdot \dfrac{4^4}{4!}\right| = \dfrac{32}{243} \approx 0.132$.

60. (a) $f(3) = (x-2)^{-1}\Big|_{x=3} = 1$

$f'(3) = -(x-2)^{-2}\Big|_{x=3} = -1$

$f''(3) = 2(x-2)^{-3}\Big|_{x=3} = 2$, so $\dfrac{f''(3)}{2!} = 1$

$f'''(3) = -6(x-2)^{-4}\Big|_{x=3} = -6$, so $\dfrac{f'''(3)}{3!} = -1$

$f^{(n)}(3) = (-1)^n n!$, so $\dfrac{f^{(n)}(3)}{n!} = (-1)^n$

$f(x) = 1 - (x-3) + (x-3)^2 - (x-3)^3 + \cdots$

$\quad + (-1)^n(x-3)^n + \cdots$

(b) Integrate term by term.

$\ln|x-2| = \displaystyle\int_3^x \dfrac{1}{t-2}\,dt$

$= \left[t - \dfrac{1}{2}(t-3)^2 + \dfrac{1}{3}(t-3)^3 - \dfrac{1}{4}(t-3)^4 + \cdots\right.$

$\quad \left.+ (-1)^n\dfrac{(t-3)^{n+1}}{n+1} + \cdots\right]_3^x$

$= (x-3) - \dfrac{(x-3)^2}{2} + \dfrac{(x-3)^3}{3} - \dfrac{(x-3)^4}{4} + \cdots$

$\quad + (-1)^n\dfrac{(x-3)^{n+1}}{n+1} + \cdots$

(c) Evaluate at $x = 3.5$. This is the alternating series

$\dfrac{1}{2} - \dfrac{1}{2^2 \cdot 2} + \dfrac{1}{2^3 \cdot 3} - \cdots + (-1)^n\dfrac{1}{2^{n+1}(n+1)} + \cdots$

By the Alternating Series Estimation Theorem, since the size of the third term is $\dfrac{1}{24} < 0.05$, the first two terms will suffice. The estimate for $\ln\left(\dfrac{3}{2}\right)$ is 0.375.

61. (a) Substitute $-2x^2$ for x in the Maclaurin series for e^x given at the end of Section 9.2.

$e^{-2x^2} = 1 + (-2x^2) + \dfrac{(-2x^2)^2}{2!} + \dfrac{(-2x^2)^3}{3!}$

$\quad + \cdots + \dfrac{(-2x^2)^n}{n!} + \cdots$

$= 1 - 2x^2 + 2x^4 - \dfrac{4x^6}{3} + \cdots$

$\quad + (-1)^n\dfrac{2^n x^{2n}}{n!} + \cdots$

(b) Use the Ratio Test:

$\displaystyle\lim_{n\to\infty}\left|\dfrac{a_{n+1}}{a_n}\right| = \lim_{n\to\infty}\dfrac{2^{n+1}x^{2n+2}}{(n+1)!} \cdot \dfrac{n!}{2^n x^{2n}}$

$= \displaystyle\lim_{n\to\infty}\dfrac{2x^2}{n+1} = 0$

The series converges for all real numbers, so the interval of convergence is $(-\infty, \infty)$.

(c) This is an alternating series. The difference will be bounded by the magnitude of the fifth term, which is $\dfrac{(2x^2)^4}{4!} = \dfrac{2x^8}{3}$. Since $-0.6 \le x \le 0.6$, this term is less than $\dfrac{2(0.6)^8}{3}$ which is less than 0.02.

62. (a) $f(x) = x^2\left(\dfrac{1}{1+x}\right)$

$= x^2(1 - x + x^2 + \cdots + (-x)^n + \cdots)$

$= x^2 - x^3 + x^4 + \cdots + (-1)^n x^{n+2} + \cdots$

(b) No. At $x = 1$, the series is $\displaystyle\sum_{n=0}^{\infty}(-1)^n$ and the partial sums form the sequence 1, 0, 1, 0, 1, 0, ..., which has no limit.

266 Chapter 9 Review

63. (a) Substituting x^2 for x in the Maclaurin series for $\sin x$ given at the end of Section 9.2,
$$\sin x^2 = x^2 - \frac{x^6}{3!} + \frac{x^{10}}{5!} - \cdots + (-1)^n \frac{x^{4n+2}}{(2n+1)!}.$$
Integrating term-by-term and observing that the constant term is 0,
$$\int_0^x \sin t^2 \, dt = \frac{x^3}{3} - \frac{x^7}{7(3!)} + \frac{x^{11}}{11(5!)} - \cdots$$
$$+ (-1)^n \frac{x^{4n+3}}{(4n+3)(2n+1)!} + \cdots$$

(b) $\int_0^1 \sin x \, dx = \frac{1}{3} - \frac{1}{7(3!)} + \frac{1}{11(5!)} - \cdots$
$$+ (-1)^n \frac{1}{(4n+3)(2n+1)!} + \cdots.$$
Since the third term is $\frac{1}{11(5!)} = \frac{1}{1320} < 0.001$, it suffices to use the first two nonzero terms (through degree 7).

(c) NINT($\sin x^2, x, 0, 1$) ≈ 0.31026830

(d) $\frac{1}{3} - \frac{1}{7(3!)} + \frac{1}{11(5!)} - \frac{1}{15(7!)} = \frac{258{,}019}{831{,}600} \approx 0.31026816$
This is within 1.5×10^{-7} of the answer in (c).

64. (a) Let $f(x) = x^2 e^x \, dx$.
$$\int_0^1 x^2 e^x \, dx = \int_0^1 f(x) \, dx$$
$$\approx \frac{h}{2}\left[f(0) + 2f(0.5) + f(1)\right]$$
$$= \frac{1}{4}\left[0 + 2\frac{e^{0.5}}{4} + e\right]$$
$$= \frac{e^{0.5}}{8} + \frac{e}{4}$$
$$\approx 0.88566$$

(b) $x^2 e^x = x^2\left(1 + x + \frac{x^2}{2!} + \cdots + \frac{x^n}{n!} + \cdots\right)$
$$= x^2 + x^3 + \frac{x^4}{2!} + \cdots + \frac{x^{n+2}}{n!} + \cdots$$
$$P_4(x) = x^2 + x^3 + \frac{x^4}{2}$$
$$\int_0^1 P_4(x) = \left[\frac{x^3}{3} + \frac{x^4}{4} + \frac{x^5}{10}\right]_0^1 = \frac{41}{60} \approx 0.68333$$

(c) Since f is concave up, the trapezoids used to estimate the area lie above the curve, and the estimate is too large.

(d) Since all the derivatives are positive (and $x > 0$), the remainder, $R_n(x)$, must be positive. This means that $P_n(x)$ is smaller than $f(x)$.

(e) Let $u = x^2$, $\quad dv = e^x \, dx$
$\quad du = 2x \, dx \quad v = e^x$
$$\int x^2 e^x \, dx = x^2 e^x - \int 2x e^x \, dx$$
Let $u = 2x$, $\quad dv = e^x \, dx$
$\quad du = 2 \, dx \quad v = e^x$
$$x^2 e^x - \int 2x e^x \, dx = x^2 e^x - \left[2xe^x - \int 2e^x \, dx\right]$$
$$= x^2 e^x - 2xe^x + 2e^x + C$$
$$= (x^2 - 2x + 2)e^x + C$$
$$\int_0^1 x^2 e^x \, dx = (x^2 - 2x + 2)e^x \Big]_0^1 = e - 2 \approx 0.71828$$

65. (a) Because $[\$1000(1.08)^{-n}](1.08)^n = \$1000$ will be available after n years.

(b) Assume that the first payment goes to the charity at the end of the first year.
$$1000(1.08)^{-1} + 1000(1.08)^{-2} + 1000(1.08)^{-3} + \cdots$$

(c) This is a geometric series with sum equal to
$$\frac{1000/1.08}{1 - (1/1.08)} = \frac{1000}{0.08} = 12{,}500.$$ This means that $12,500 should be invested today in order to completely fund the perpetuity forever.

66. We again assume that the first payment occurs at the end of the year.
Present value $= 1000(1.06)^{-1} + 1000(1.06)^{-2}$
$$+ 1000(1.06)^{-3} + \cdots$$
$$= \frac{1000/1.06}{1 - (1/1.06)} = \frac{1000}{1.06 - 1} \approx 16{,}666.67$$
The present value is $16,666.67.

67. (a)

Sequence of Tosses	Payoff ($)	Probability	Term of Series
T	0	$\frac{1}{2}$	$0\left(\frac{1}{2}\right)$
HT	1	$\left(\frac{1}{2}\right)^2$	$1\left(\frac{1}{2}\right)^2$
HHT	2	$\left(\frac{1}{2}\right)^3$	$2\left(\frac{1}{2}\right)^3$
HHHT	3	$\left(\frac{1}{2}\right)^4$	$3\left(\frac{1}{2}\right)^4$
⋮	⋮	⋮	⋮

Expected payoff
$$= 0\left(\frac{1}{2}\right) + 1\left(\frac{1}{2}\right)^2 + 2\left(\frac{1}{2}\right)^3 + 3\left(\frac{1}{2}\right)^4 + \cdots$$

(b) $\dfrac{1}{(1-x)^2} = 1 + 2x + 3x^2 + \cdots + nx^{n-1} + \cdots$

(c) $\dfrac{x^2}{(1-x)^2} = x^2(1 + 2x + 3x^2 + \cdots + nx^{n-1} + \cdots)$
$\phantom{\dfrac{x^2}{(1-x)^2}} = x^2 + 2x^3 + 3x^4 + \cdots + nx^{n+1} + \cdots$

(d) If $x = \dfrac{1}{2}$, the formula in part (c) matches the nonzero terms of the series in part (a). Since $\dfrac{(1/2)^2}{[1-(1/2)]^2} = 1$, the expected payoff is \$1.

68. (a) The area of an equilateral triangle whose sides have length s is $\dfrac{1}{2}(s)\left(\dfrac{\sqrt{3}s}{2}\right) = \dfrac{s^2\sqrt{3}}{4}$. The sequence of areas removed from the original triangle is

$$\dfrac{b^2\sqrt{3}}{4} + 3\left(\dfrac{b}{2}\right)^2\dfrac{\sqrt{3}}{4} + 9\left(\dfrac{b}{4}\right)^2\dfrac{\sqrt{3}}{4} + \cdots$$
$$+ 3^n\left(\dfrac{b}{2^n}\right)^2\dfrac{\sqrt{3}}{4} + \cdots \text{ or}$$
$$\dfrac{b^2\sqrt{3}}{4} + \dfrac{3b^2\sqrt{3}}{4^2} + \dfrac{3^2b^2\sqrt{3}}{4^3} + \cdots + \dfrac{3^nb^2\sqrt{3}}{4^{n+1}} + \cdots.$$

(b) This is a geometric series with initial term $a = \dfrac{b^2\sqrt{3}}{4}$ and common ratio $r = \dfrac{3}{4}$, so the sum is $\dfrac{b^2\sqrt{3}/4}{1-(3/4)} = b^2\sqrt{3}$, which is the same as the area of the original triangle.

(c) No, not every point is removed. For example, the vertices of the original triangle are not removed. But the remaining points are "isolated" enough that there are no regions and hence no area remaining.

69. $\dfrac{1}{1-x} = 1 + x + x^2 + x^3 + \cdots$

Differentiate both sides.

$\dfrac{1}{(1-x)^2} = 1 + 2x + 3x^2 + 4x^3 + 5x^4 + \cdots$

Substitute $x = \dfrac{1}{2}$ to get the desired result.

70. (a) Note that $\sum_{n=1}^{\infty} x^{n+1}$ is a geometric series with first term $a = x^2$ and common ratio $r = x$, which explains the identity $\sum_{n=1}^{\infty} x^{n+1} = \dfrac{x^2}{1-x}$ (for $|x| < 1$).

Differentiate.

$\sum_{n=1}^{\infty} (n+1)x^n = \dfrac{(1-x)(2x) - (x^2)(-1)}{(1-x)^2} = \dfrac{2x - x^2}{(1-x)^2}$

Differentiate again.

$\sum_{n=1}^{\infty} n(n+1)x^{n-1}$
$= \dfrac{(1-x)^2(2-2x) - (2x - x^2)(2)(1-x)(-1)}{(1-x)^4}$
$= \dfrac{(1-x)(2-2x) + 2(2x - x^2)}{(1-x)^3}$
$= \dfrac{2}{(1-x)^3}$

Multiply by x.

$\sum_{n=1}^{\infty} n(n+1)x^n = \dfrac{2x}{(1-x)^3}$

Replace x by $\dfrac{1}{x}$.

$\sum_{n=1}^{\infty} \dfrac{n(n+1)}{x^n} = \dfrac{\frac{2}{x}}{\left(1-\frac{1}{x}\right)^3} = \dfrac{2x^2}{(x-1)^3}, |x| > 1$

(b) Solve $x = \dfrac{2x^2}{(x-1)^3}$ to get $x \approx 2.769$ for $x > 1$.

Chapter 10
Vectors

Section 10.1 Parametric Functions
(pp. 513–520)

Exploration 1 Investigating Cycloids

1.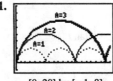

   [0, 20] by [−1, 8]

2. $x = 2na\pi$ for any integer n.

3. $a > 0$ and $1 - \cos t \geq 0$ so $y \geq 0$.

4. An arch is produced by one complete turn of the wheel. Thus, they are congruent.

5. The maximum value of y is $2a$ and occurs when $x = (2n+1)a\pi$ for any integer n.

6. The function represented by the cycloid is periodic with period $2a\pi$, and each arch represents one period of the graph. In each arch, the graph is concave down, has an absolute maximum of $2a$ at the midpoint, and an absolute minimum of 0 at the two endpoints.

Quick Review 10.1

1. $(\cos(0), \sin(0)) = (1, 0)$
2. $\left(\cos\left(\frac{3\pi}{2}\right), \sin\left(\frac{3\pi}{2}\right)\right) = (0, -1)$
3. $x^2 + y^2 = 1$ (since $\cos^2 t + \sin^2 t = 1$)
4. The portion in the first three quadrants, moving counterclockwise as t increases.
5. $x = t, y = t^2 + 1, -1 \le t \le 3$
6. The graph is a circle with radius 2 centered at $(2, 3)$. Modify the $x = \cos t, y = \sin t$ parameterization correspondingly:
   $x = 2\cos t + 2, y = 2\sin t + 3, 0 \le t \le 2\pi$.
7. $\frac{dy}{dx} = \frac{dy/dt}{dx/dt} = \frac{3\cos t}{-2\sin t}$,
   which at $t = \frac{3\pi}{4}$ equals $\frac{3(-\sqrt{2}/2)}{-2(\sqrt{2}/2)} = \frac{3}{2}$.
8. $y = \frac{3}{2}x + C$. For $t = \frac{3\pi}{4}, x = -\sqrt{2}$ and $y = \frac{3\sqrt{2}}{2}$, so
   $\frac{3\sqrt{2}}{2} = \frac{3}{2}(-\sqrt{2}) + C$ and $C = 3\sqrt{2}$.
   Thus, $y = \frac{3}{2}x + 3\sqrt{2}$.
9. $y = -\frac{2}{3}x + C$. For $t = \frac{3\pi}{4}, x = -\sqrt{2}$ and $y = \frac{3\sqrt{2}}{2}$, so
   $\frac{3\sqrt{2}}{2} = -\frac{2}{3}(-\sqrt{2}) + C$ and $C = \frac{5\sqrt{2}}{6}$.
   Thus, $y = -\frac{2}{3}x + \frac{5\sqrt{2}}{6}$.
10. $y' = \frac{3}{2}\sqrt{x}$, so
    Length $= \int_0^3 \sqrt{1 + \left(\frac{3}{2}\sqrt{x}\right)^2}\, dx$
    $= \int_0^3 \sqrt{1 + \frac{9}{4}x}\, dx$
    $= \left[\frac{8}{27}\left(1 + \frac{9}{4}x\right)^{3/2}\right]_0^3 = \frac{31^{3/2} - 8}{27}$.

Section 10.1 Exercises

1. (a) $\frac{dy}{dx} = y' = \frac{dy/dt}{dx/dt} = \frac{-2\sin t}{4\cos t} = -\frac{1}{2}\tan t$

   (b) $\frac{d^2y}{dx^2} = \frac{dy'/dt}{dx/dt} = \frac{-\frac{1}{2}\sec^2 t}{4\cos t} = -\frac{1}{8}\sec^3 t$

3. (a) $\frac{dy}{dx} = y' = \frac{dy/dt}{dx/dt}$
   $= \frac{3/(2\sqrt{3t})}{-1/(2\sqrt{t+1})}$
   $= -\frac{3\sqrt{t+1}}{\sqrt{3t}}$
   $= -\sqrt{3 + \frac{3}{t}}$

   (b) $\frac{d^2y}{dx^2} = \frac{dy'/dt}{dx/dt} = \frac{3/(2t^2\sqrt{3 + 3/t})}{-1/(2\sqrt{t+1})} = -\frac{3\sqrt{t+1}}{t^2\sqrt{3 + 3/t}}$
   $= -\frac{\sqrt{3}}{t^{3/2}}$

5. (a) $\frac{dy}{dx} = y' = \frac{dy/dt}{dx/dt} = \frac{3t^2}{2t - 3}$

   (b) $\frac{d^2y}{dx^2} = \frac{dy'/dt}{dx/dt}$
   $= \frac{[(2t-3)(6t) - (3t^2)(2)]/(2t-3)^2}{2t - 3}$
   $= \frac{12t^2 - 18t - 6t^2}{(2t-3)^3}$
   $= \frac{6t^2 - 18t}{(2t-3)^3}$

7. $\frac{dy}{dx} = \frac{dy/dt}{dx/dt} = \frac{\cos t}{-\sin t} = -\cot t$

   (a) $-\cot = 0$ when $t = \frac{\pi}{2} + k\pi$ (k any integer). Then
   $(x, y) = \left(2 + \cos\left(\frac{\pi}{2} + k\pi\right), -1 + \sin\left(\frac{\pi}{2} + k\pi\right)\right)$
   $= (2, -1 \pm 1)$. The points are $(2, 0)$ and $(2, -2)$.

   (b) $-\cot$ is undefined when $t = k\pi$ (k any integer). Then
   $(x, y) = (2 + \cos(k\pi), -1 + \sin(k\pi)) = (2 \pm 1, -1)$.
   The points are $(1, -1)$ and $(3, -1)$.

9. $\frac{dy}{dx} = \frac{dy/dt}{dx/dt} = \frac{3t^2 - 4}{-1} = 4 - 3t^2$

   (a) $4 - 3t^2 = 0$ when $t = \pm\sqrt{\frac{4}{3}} = \pm\frac{2}{\sqrt{3}}$.
   Then $(x, y) = \left(2 \mp \frac{2}{\sqrt{3}}, \pm\left(\frac{2}{\sqrt{3}}\right)^3 \mp 4\left(\frac{2}{\sqrt{3}}\right)\right)$
   $= \left(2 \mp \frac{2}{\sqrt{3}}, \pm\frac{8}{3\sqrt{3}} \mp \frac{8}{\sqrt{3}}\right)$,
   which evaluates to $\approx (0.845, -3.079)$ and $\approx (3.155, 3.079)$.

   (b) Nowhere, since $4 - 3t^2$ is never undefined.

11. $x' = -\sin t$, $y' = 1 + \cos t$, so

$$\text{length} = \int_0^\pi \sqrt{(-\sin t)^2 + (1 + \cos t)^2}\, dt$$
$$= \int_0^\pi \sqrt{2(1 + \cos t)}\, dt$$
$$= \int_0^\pi \sqrt{4 \cos^2\left(\frac{t}{2}\right)}\, dt$$
$$= \int_0^\pi 2 \cos\left(\frac{t}{2}\right) dt$$
$$= 2\left[2 \sin\left(\frac{t}{2}\right)\right]_0^\pi = 4$$

13. $x' = t^2$, $y' = t$, so

$$\text{Length} = \int_0^1 \sqrt{(t^2)^2 + t^2}\, dt$$
$$= \int_0^1 t\sqrt{t^2 + 1}\, dt$$
$$= \left[\frac{1}{3}(t^2 + 1)^{3/2}\right]_0^1$$
$$= \frac{1}{3}(2^{3/2} - 1)$$
$$= \frac{2\sqrt{2} - 1}{3} \approx 0.609$$

15. $x' = \dfrac{\sec t \tan t + \sec^2 t}{\sec t + \tan t} - \cos t = \sec t - \cos t$,

$y' = -\sin t$, so

$$\text{Length} = \int_0^{\pi/3} \sqrt{(\sec t - \cos t)^2 + (-\sin t)^2}\, dt$$
$$= \int_0^{\pi/3} \sqrt{\sec^2 t - 1}\, dt$$
$$= \int_0^{\pi/3} \tan t\, dt$$
$$= \left[\ln |\sec t|\right]_0^{\pi/3} = \ln 2$$

17. $x' = -\sin t$, $y' = \cos t$, so

$$\text{Area} = \int_0^{2\pi} 2\pi(2 + \sin t)\sqrt{(-\sin t)^2 + \cos^2 t}\, dt$$
$$= 2\pi \int_0^{2\pi} (2 + \sin t)\, dt$$
$$= 2\pi \left[2t - \cos t\right]_0^{2\pi} = 8\pi^2$$

19. $x' = 1$, $y' = 2t$, so

$$\text{Area} = \int_0^3 2\pi(t + 1)\sqrt{1 + (2t)^2}\, dt,$$

which using NINT evaluates to ≈ 178.561.

21. (a) $x(t) = 2t$, $y(t) = t + 1$, $0 \le t \le 1$

(b) $x' = 2$, $y' = 1$, so

$$\text{Area} = \int_0^1 2\pi(t + 1)\sqrt{2^2 + 1^2}\, dt$$
$$= 2\pi\sqrt{5} \int_0^1 (t + 1)\, dt$$
$$= 2\pi\sqrt{5}\left[\frac{1}{2}t^2 + t\right]_0^1$$
$$= 3\pi\sqrt{5}$$

(c) Slant height $= \sqrt{2^2 + 1^2} = \sqrt{5}$, so

Area $= \pi(1 + 2)\sqrt{5} = 3\pi\sqrt{5}$

23. (a) $x' = -2\sin 2t$, $y' = 2\cos 2t$, so

$$\text{Length} = \int_0^{\pi/2} \sqrt{(-2\sin 2t)^2 + (2\cos 2t)^2}\, dt$$
$$= \int_0^{\pi/2} 2\, dt = \pi.$$

(b) $x' = \pi \cos \pi t$, $y' = -\pi \sin \pi t$, so

$$\text{Length} = \int_{-1/2}^{1/2} \sqrt{(\pi \cos \pi t)^2 + (-\pi \sin \pi t)^2}\, dt$$
$$= \int_{-1/2}^{1/2} \pi\, dt = \pi.$$

25. In the first integral, replace t with x. Then $\dfrac{dx}{dt}$ becomes $\dfrac{dx}{dx} = 1$.

27. $x' = t$, $y' = \sqrt{2t + 1}$, so

$$\text{Total length} = \int_0^4 \sqrt{t^2 + (\sqrt{2t + 1})^2}\, dt$$
$$= \int_0^4 (t + 1)\, dt$$
$$= \left[\frac{1}{2}t^2 + t\right]_0^4 = 12.$$

Now solve $\left[\dfrac{1}{2}t^2 + t\right]_0^m = \dfrac{12}{2}$ for m:

$\dfrac{1}{2}m^2 + m = 6$, or $m^2 + 2m - 12 = 0$, and

$m = \dfrac{-2 \pm \sqrt{4 + 48}}{2} = -1 \pm \sqrt{13}$. Take the positive solution. The midpoint is at $t = \sqrt{13} - 1$, which gives

$(x, y) = \left(\dfrac{(\sqrt{13} - 1)^2}{2}, \dfrac{1}{3}(2\sqrt{13} - 1)^{3/2}\right) \approx (3.394, 5.160)$.

270   Section 10.2

29.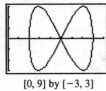
[0, 9] by [−3, 3]

Use the top half of the curve, and make use of the shape's symmetry.
$x' = 3 \cos t$, $y' = 6 \cos 2t$, so

$$\text{Area} = 2\int_0^{\pi/2} 2\pi(3 \sin 2t)\sqrt{(3 \cos t)^2 + (6 \cos 2t)^2}\, dt$$

which using NINT, evaluates to ≈ 144.513.

31. $\dfrac{dx}{dt} = a(1 - \cos t)$

(Note: integrate with respect to x from 0 to $2a\pi$; integrate with respect to t from 0 to 2π.)

$$\text{Area} = \int_0^{2a\pi} y\, dx$$
$$= \int_0^{2a\pi} a(1 - \cos t)a(1 - \cos t)\, dt$$
$$= a^2 \int_0^{2\pi} (1 - 2\cos t + \cos^2 t)\, dt$$
$$= a^2 \left[t - 2\sin t + \frac{t}{2} + \frac{1}{4}\sin 2t\right]_0^{2\pi} = 3\pi a^2$$

33. (a) $\overline{QP}$ has length t, so P can be obtained by starting at Q and moving $t \sin t$ units right and $t \cos t$ units downward.
(If either quantity is negative, the corresponding direction is reversed.) Since $Q = (\cos t, \sin t)$, the coordinates of P are
$x = \cos t + t \sin t$ and $y = \sin t - t \cos t$.

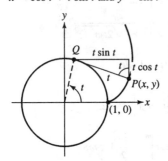

(b) $x' = t \cos t$, $y' = t \sin t$, so

$$\text{Length} = \int_0^{2\pi}\sqrt{(t\cos t)^2 + (t\sin t)^2}\, dt = \int_0^{2\pi} t\, dt$$
$$= 2\pi^2$$

For exercises 35–37, $x' = v_0 \cos\theta$ and $y' = v_0 \sin\theta - 32t$, and $y = 0$ for $t = 0$ or $t = \dfrac{v_0 \sin\theta}{16}$. The maximum height is attained in mid-flight at $t = \dfrac{v_0 \sin\theta}{32}$. To find the path length, evaluate
$\int_0^{v_0 \sin\theta/16} \sqrt{(v_0 \cos\theta)^2 + (v_0 \sin\theta - 32t)^2}\, dt$ using NINT. To find the maximum height, calculate
$y_{\max} = (v_0 \sin\theta)\left(\dfrac{v_0 \sin\theta}{32}\right) - 16\left(\dfrac{v_0 \sin\theta}{32}\right)^2$.

35. (a) The projectile hits the ground when $y = 0$.
$y = t(150 \sin 20° - 16t) = 0$
$t = 0$ or $t = \dfrac{75}{8}\sin 20° \approx 3.206$
$x' = 150 \cos 20°$, $y' = 150 \sin 20° - 32t$
$\text{Length} = \int_0^{(75 \sin 20°)/8}\sqrt{(150\cos 20°)^2 + (150\sin 20° - 32t)^2}\, dt$
which, using NINT, evaluates to ≈ 461.749 ft

(b) The maximum height of the projectile occurs when $y' = 0$,
so $t = \dfrac{75}{16}\sin 20°$, $y\left(\dfrac{75}{16}\sin 20°\right) \approx 41.125$ ft

37. (a) ≈ 840.421 ft

(b) $\dfrac{16{,}875}{64} \approx 263.672$ ft

39. In the integral $\int_a^b 2\pi y \sqrt{\left(\dfrac{dx}{dt}\right)^2 + \left(\dfrac{dy}{dt}\right)^2}\, dt$, replace t with x and y with $f(x)$.
Then $\dfrac{dx}{dt}$ becomes $\dfrac{dx}{dx} = 1$.

41. $\dfrac{dy}{dx} = -\dfrac{1}{x^2}$, so $\text{Area} = \int_1^4 2\pi\left(\dfrac{1}{x}\right)\sqrt{1 + \left(-\dfrac{1}{x^2}\right)^2}\, dx$, which, using NINT, evaluates to ≈ 9.417.

■ Section 10.2 Vectors in the Plane
(pp. 520–529)

Quick Review 10.2

1. $\sqrt{(5-1)^2 + (3-2)^2} = \sqrt{17}$

2. $\dfrac{3-2}{5-1} = \dfrac{1}{4}$

3. Solve $\dfrac{3-b}{5-3} = -4$: $b = 11$.

4. Slope of $\overline{AB}$ = Slope of $\overline{CD}$, so $\dfrac{3-0}{1-0} = \dfrac{3-0}{5-a}$ and $a = 4$.

5. Slope of $\overline{AB}$ = Slope of $\overline{CD}$, so $\dfrac{5-1}{3-1} = \dfrac{2-b}{6-8}$ and $b = 6$.

6. (a) $\theta = 120°$ **(b)** $\theta = \dfrac{2\pi}{3}$

7. (a) $\theta = -30°$ **(b)** $\theta = -\dfrac{\pi}{6}$

8. (a) $\theta = -45°$ **(b)** $\theta = -\dfrac{\pi}{4}$

9. $c^2 = 3^2 + 5^2 - 2(3)(5)\cos(30°) = 34 - 15\sqrt{3}$, so
$c = \sqrt{34 - 15\sqrt{3}} \approx 2.832$

10. $24^2 = 27^2 + 19^2 - 2(27)(19)\cos\theta$, so
$\cos\theta = \dfrac{24^2 - 27^2 - 19^2}{-2(27)(19)} = \dfrac{257}{513}$ and
$\theta = \cos^{-1}\dfrac{257}{513} \approx 1.046$ radians or $59.935°$.

Section 10.2 Exercises

1. (a) $\langle 3(3), 3(-2) \rangle = \langle 9, -6 \rangle$
   (b) $\sqrt{9^2 + (-6)^2} = \sqrt{117} = 3\sqrt{13}$

3. (a) $\langle 3 + (-2), -2 + 5 \rangle = \langle 1, 3 \rangle$
   (b) $\sqrt{1^2 + 3^2} = \sqrt{10}$

5. (a) $2\mathbf{u} = \langle 2(3), 2(-2) \rangle = \langle 6, -4 \rangle$
   $3\mathbf{v} = \langle 3(-2), 3(5) \rangle = \langle -6, 15 \rangle$
   $2\mathbf{u} - 3\mathbf{v} = \langle 6 - (-6), -4 - 15 \rangle = \langle 12, -19 \rangle$
   (b) $\sqrt{12^2 + (-19)^2} = \sqrt{505}$

7. (a) $\frac{3}{5}\mathbf{u} = \left\langle \frac{3}{5}(3), \frac{3}{5}(-2) \right\rangle = \left\langle \frac{9}{5}, -\frac{6}{5} \right\rangle$
   $\frac{4}{5}\mathbf{v} = \left\langle \frac{4}{5}(-2), \frac{4}{5}(5) \right\rangle = \left\langle -\frac{8}{5}, 4 \right\rangle$
   $\frac{3}{5}\mathbf{u} + \frac{4}{5}\mathbf{v} = \left\langle \frac{9}{5} + \left(-\frac{8}{5}\right), -\frac{6}{5} + 4 \right\rangle = \left\langle \frac{1}{5}, \frac{14}{5} \right\rangle$
   (b) $\sqrt{\left(\frac{1}{5}\right)^2 + \left(\frac{14}{5}\right)^2} = \frac{\sqrt{197}}{5}$

9. $\langle 2 - 1, -1 - 3 \rangle = \langle 1, -4 \rangle$

11. $\langle 0 - 2, 0 - 3 \rangle = \langle -2, -3 \rangle$

13. $\left\langle \cos \frac{2\pi}{3}, \sin \frac{2\pi}{3} \right\rangle = \left\langle -\frac{1}{2}, \frac{\sqrt{3}}{2} \right\rangle$

15. This is the unit vector which makes an angle of
    $120 + 90 = 210°$ with the positive x-axis;
    $\langle \cos 210°, \sin 210° \rangle = \left\langle -\frac{\sqrt{3}}{2}, -\frac{1}{2} \right\rangle$

17. The vector **v** is horizontal and 1 in. long. The vectors **u** and **w** are $\frac{11}{16}$ in. long. **w** is vertical and **u** makes a 45° angle with the horizontal. All vectors must be drawn to scale.
    (a)
    (b)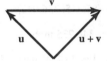
    (c)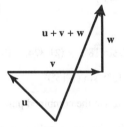

    (d)

19. $\sqrt{3^2 + 4^2} = 5; \frac{1}{5}\langle 3, 4 \rangle = \left\langle \frac{3}{5}, \frac{4}{5} \right\rangle$

21. $\sqrt{(-15)^2 + 8^2} = 17; \frac{1}{17}\langle -15, 8 \rangle = \left\langle -\frac{15}{17}, \frac{8}{17} \right\rangle$

23. $x' = \frac{1}{2\sqrt{t}}, y' = 1 + \frac{1}{\sqrt{t}}$; for $t = 1$, $x' = \frac{1}{2}$, $y' = 2$, and
    $\sqrt{(x')^2 + (y')^2} = \frac{\sqrt{17}}{2}$.
    Tangent: $\pm\frac{2}{\sqrt{17}}\left\langle \frac{1}{2}, 2 \right\rangle = \pm\left\langle \frac{1}{\sqrt{17}}, \frac{4}{\sqrt{17}} \right\rangle$,
    Normal: $\pm\frac{2}{\sqrt{17}}\left\langle 2, -\frac{1}{2} \right\rangle = \pm\left\langle \frac{4}{\sqrt{17}}, -\frac{1}{\sqrt{17}} \right\rangle$.

25. $x' = -4 \sin t, y' = 5 \cos t$; for $t = \frac{\pi}{3}$, $x' = -2\sqrt{3}$,
    $y' = \frac{5}{2}$, and $\sqrt{(x')^2 + (y')^2} = \frac{\sqrt{73}}{2}$.
    Tangent $= \pm\frac{2}{\sqrt{73}}\left\langle -2\sqrt{3}, \frac{5}{2} \right\rangle = \pm\left\langle -\frac{12}{\sqrt{219}}, \frac{5}{\sqrt{73}} \right\rangle$
    $\approx \pm\langle -0.811, 0.585 \rangle$,
    Normal $= \pm\frac{2}{\sqrt{73}}\left\langle \frac{5}{2}, 2\sqrt{3} \right\rangle = \pm\left\langle \frac{5}{\sqrt{73}}, \frac{12}{\sqrt{219}} \right\rangle$
    $\approx \pm\langle 0.585, 0.811 \rangle$.

27. $\overrightarrow{AB} = \langle 3, 1 \rangle$, $\overrightarrow{BC} = \langle -1, -3 \rangle$, and $\overrightarrow{AC} = \langle 2, -2 \rangle$.
    $\overrightarrow{BA} = \langle -3, -1 \rangle$, $\overrightarrow{CB} = \langle 1, 3 \rangle$, and $\overrightarrow{CA} = \langle -2, 2 \rangle$.
    $|\overrightarrow{AB}| = |\overrightarrow{BA}| = \sqrt{10}$, $|\overrightarrow{BC}| = |\overrightarrow{CB}| = \sqrt{10}$, and
    $|\overrightarrow{AC}| = |\overrightarrow{CA}| = 2\sqrt{2}$.

    Angle at $A = \cos^{-1}\left(\frac{\overrightarrow{AB} \cdot \overrightarrow{AC}}{|\overrightarrow{AB}||\overrightarrow{AC}|}\right)$
    $= \cos^{-1}\left(\frac{3(2) + 1(-2)}{(\sqrt{10})(2\sqrt{2})}\right)$
    $= \cos^{-1}\left(\frac{1}{\sqrt{5}}\right) \approx 63.435°$,

    Angle at $B = \cos^{-1}\left(\frac{\overrightarrow{BC} \cdot \overrightarrow{BA}}{|\overrightarrow{BC}||\overrightarrow{BA}|}\right)$
    $= \cos^{-1}\left(\frac{(-1)(-3) + (-3)(-1)}{(\sqrt{10})(\sqrt{10})}\right)$
    $= \cos^{-1}\left(\frac{3}{5}\right) \approx 53.130°$, and

    Angle at $C = \cos^{-1}\left(\frac{\overrightarrow{CB} \cdot \overrightarrow{CA}}{|\overrightarrow{CB}||\overrightarrow{CA}|}\right)$
    $= \cos^{-1}\left(\frac{1(-2) + 3(2)}{(\sqrt{10})(2\sqrt{2})}\right)$
    $= \cos^{-1}\left(\frac{1}{\sqrt{5}}\right) \approx 63.435°$.

29. (a) $\mathbf{u} \cdot (\mathbf{v} + \mathbf{w}) = u_1(v_1 + w_1) + u_2(v_2 + w_2)$
$= (u_1 v_1 + u_1 w_1) + (u_2 v_2 + u_2 w_2)$
$= (u_1 v_1 + u_2 v_2) + (u_1 w_1 + u_2 w_2)$
$= \mathbf{u} \cdot \mathbf{v} + \mathbf{u} \cdot \mathbf{w}$

(b) $(\mathbf{u} + \mathbf{v}) \cdot \mathbf{w} = (u_1 + v_1)w_1 + (u_2 + v_2)w_2$
$= (u_1 w_1 + v_1 w_1) + (u_2 w_2 + v_2 w_2)$
$= (u_1 w_1 + u_2 w_2) + (v_1 w_1 + v_2 w_2)$
$= \mathbf{u} \cdot \mathbf{w} + \mathbf{v} \cdot \mathbf{w}$

31. $(\mathbf{u} + \mathbf{v}) \cdot (\mathbf{u} - \mathbf{v})$
$= (u_1 + v_1)(u_1 - v_1) + (u_2 + v_2)(u_2 - v_2)$
$= u_1^2 - v_1^2 + u_2^2 - v_2^2$
$= (u_1^2 + u_2^2) - (v_1^2 + v_2^2)$
$= |\mathbf{u}|^2 - |\mathbf{v}|^2$

33.

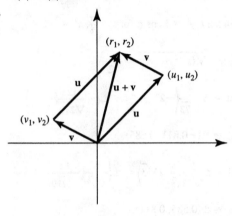

$r_1 - v_1 = u_1$ so $r_1 = u_1 + v_1$
$r_2 - v_2 = u_2$ so $r_2 = u_2 + v_2$

35. (a) Let $P = (a, b)$ and $Q = (c, d)$. Then
$\left(\frac{1}{2}\right)\overrightarrow{OP} + \left(\frac{1}{2}\right)\overrightarrow{OQ} = \left(\frac{1}{2}\right)\langle a, b\rangle + \left(\frac{1}{2}\right)\langle c, d\rangle$
$= \left\langle\frac{(a+c)}{2}, \frac{(b+d)}{2}\right\rangle = \overrightarrow{OM}$

(b) $\overrightarrow{OM} = \left(\frac{2}{3}\right)\overrightarrow{OP} + \left(\frac{1}{3}\right)\overrightarrow{OQ}$

(c) $\overrightarrow{OM} = \left(\frac{1}{3}\right)\overrightarrow{OP} + \left(\frac{2}{3}\right)\overrightarrow{OQ}$

(d) M is a fraction of the way from P to Q. Let d be this fraction. Then
$\overrightarrow{OM} = d\overrightarrow{OQ} + (1-d)\overrightarrow{OP}$.

Proof: $\overrightarrow{PM} = d\overrightarrow{PQ}$ and $\overrightarrow{MQ} = (1-d)\overrightarrow{PQ}$,
so $\overrightarrow{PQ} = \frac{1}{d}\overrightarrow{PM}$ and $\overrightarrow{PQ} = \frac{1}{1-d}\overrightarrow{MQ}$.

Therefore, $\frac{1}{d}\overrightarrow{PM} = \frac{1}{1-d}\overrightarrow{MQ}$.

But $\overrightarrow{PM} = \overrightarrow{OM} - \overrightarrow{OP}$ and $\overrightarrow{MQ} = \overrightarrow{OQ} - \overrightarrow{OM}$, so
$\frac{1}{d}\overrightarrow{OM} - \frac{1}{d}\overrightarrow{OP} = \frac{1}{1-d}\overrightarrow{OQ} - \frac{1}{1-d}\overrightarrow{OM}$.

Therefore,
$\frac{1}{d}\overrightarrow{OM} + \frac{1}{1-d}\overrightarrow{OM} = \frac{1}{d}\overrightarrow{OP} + \frac{1}{1-d}\overrightarrow{OQ}$.
$\Rightarrow \overrightarrow{OM}\left(\frac{1}{d(1-d)}\right) = \frac{1}{d}\overrightarrow{OP} + \frac{1}{1-d}\overrightarrow{OQ}$
$\Rightarrow \overrightarrow{OM} = (1-d)\overrightarrow{OP} + d\overrightarrow{OQ}$.

37. Two adjacent sides of the rhombus can be given by two vectors of the same length, $\mathbf{u}$ and $\mathbf{v}$.
Then the diagonals of the rhombus are $(\mathbf{u} + \mathbf{v})$ and $(\mathbf{u} - \mathbf{v})$. These two vectors are orthogonal since $|\mathbf{u}| = |\mathbf{v}|$ so $(\mathbf{u} + \mathbf{v}) \cdot (\mathbf{u} - \mathbf{v}) = |\mathbf{u}|^2 - |\mathbf{v}|^2 = 0$.

39. Let two adjacent sides of the parallelogram be given by two vectors $\mathbf{u}$ and $\mathbf{v}$. The diagonals are then $(\mathbf{u} + \mathbf{v})$ and $(\mathbf{u} - \mathbf{v})$. So the lengths of the diagonals satisfy
$|\mathbf{u} + \mathbf{v}|^2 = (\mathbf{u} + \mathbf{v}) \cdot (\mathbf{u} + \mathbf{v})$
$= |\mathbf{u}|^2 + 2\mathbf{u} \cdot \mathbf{v} + |\mathbf{v}|^2$
and $|\mathbf{u} - \mathbf{v}|^2 = (\mathbf{u} - \mathbf{v}) \cdot (\mathbf{u} - \mathbf{v})$
$= |\mathbf{u}|^2 - 2\mathbf{u} \cdot \mathbf{v} + |\mathbf{v}|^2$.
The two lengths will be the same if and only if $\mathbf{u} \cdot \mathbf{v} = 0$, which means that $\mathbf{u}$ and $\mathbf{v}$ are perpendicular and the parallelogram is a rectangle.

41. The slopes are the same.

43. $25°$ west of north is $90 + 25 = 115°$ north of east.
$800\langle\cos 115°, \sin 115°\rangle \approx \langle-338.095, 725.046\rangle$

45. Initial velocity is $70°$ north of east:
$325\langle\cos 70°, \sin 70°\rangle \approx \langle 111.157, 305.400\rangle$.
Wind velocity is $130°$ north of east:
$40\langle\cos 130°, \sin 130°\rangle \approx \langle-25.712, 30.642\rangle$.
Add the two vectors to get $\approx \langle 85.445, 336.042\rangle$.
The speed is the magnitude, ≈ 346.735 mph.
The direction is $\tan^{-1}\left(\frac{336.042}{85.445}\right) \approx 75.734°$ north of east, or $\approx 14.266°$ east of north.

47. Juana's pull $= 23\langle\cos 18°, \sin 18°\rangle \approx \langle 21.874, 7.107\rangle$;
Diego's pull $= 18\langle\cos(-15°), \sin(-15°)\rangle$
$\approx \langle 17.387, -4.659\rangle$. Add to get the combined pull of the children: $\approx \langle 39.261, 2.449\rangle$. The puppy pulls with an opposite force of the same magnitude:
$\sqrt{39.261^2 + 2.449^2} \approx 39.337$ lb.

49. $\overrightarrow{AB} = \langle-3-0, 4-0\rangle = \langle-3, 4\rangle = \langle 1-4, 5-1\rangle = \overrightarrow{CD}$

51. $\mathbf{u} = \langle u_1, u_2 \rangle$, $\mathbf{v} = \langle v_1, v_2 \rangle$, $\mathbf{w} = \langle w_1, w_2 \rangle$
   (i) $\mathbf{u} + \mathbf{v} = \langle u_1 + v_1, u_2 + v_2 \rangle$
        $= \langle v_1 + u_1, v_2 + u_2 \rangle = \mathbf{v} + \mathbf{u}$
   (ii) $(\mathbf{u} + \mathbf{v}) + \mathbf{w}$
        $= \langle u_1 + v_1, u_2 + v_2 \rangle + \langle w_1, w_2 \rangle$
        $= \langle (u_1 + v_1) + w_1, (u_2 + v_2) + w_2 \rangle$
        $= \langle u_1 + (v_1 + w_1), u_2 + (v_2 + w_2) \rangle$
        $= \mathbf{u} + (\mathbf{v} + \mathbf{w})$
   (iii) $\mathbf{u} + \mathbf{0} = \langle u_1, u_2 \rangle + \langle 0, 0 \rangle = \langle u_1 + 0, u_2 + 0 \rangle$
        $= \langle u_1, u_2 \rangle = \mathbf{u}$
   (iv) $\mathbf{u} + (-\mathbf{u}) = \langle u_1, u_2 \rangle + \langle -u_1, -u_2 \rangle$
        $= \langle u_1 - u_1, u_2 - u_2 \rangle = \langle 0, 0 \rangle = \mathbf{0}$
   (v) $0\mathbf{u} = 0\langle u_1, u_2 \rangle = \langle 0u_1, 0u_2 \rangle = \langle 0, 0 \rangle = \mathbf{0}$
   (vi) $1\mathbf{u} = 1\langle u_1, u_2 \rangle = \langle 1u_1, 1u_2 \rangle = \langle u_1, u_2 \rangle = \mathbf{u}$
   (vii) $a(b\mathbf{u}) = a(b\langle u_1, u_2 \rangle) = a\langle bu_1, bu_2 \rangle$
        $= \langle abu_1, abu_2 \rangle = ab\langle u_1, u_2 \rangle = (ab)\mathbf{u}$
   (viii) $a(\mathbf{u} + \mathbf{v}) = a\langle u_1 + v_1, u_2 + v_2 \rangle$
        $= \langle au_1 + av_1, au_2 + av_2 \rangle$
        $= \langle au_1, au_2 \rangle + \langle av_1, av_2 \rangle$
        $= a\langle u_1, u_2 \rangle + a\langle v_1, v_2 \rangle$
        $= a\mathbf{u} + a\mathbf{v}$
   (ix) $(a + b)\mathbf{u} = (a + b)\langle u_1, u_2 \rangle$
        $= \langle (a + b)u_1, (a + b)u_2 \rangle$
        $= \langle au_1 + bu_1, au_2 + bu_2 \rangle$
        $= \langle au_1, au_2 \rangle + \langle bu_1, bu_2 \rangle = a\mathbf{u} + b\mathbf{u}$

53. (a) Slope $= -\frac{1}{1} = -1$, so $y - y_1 = m(x - x_1)$ becomes
    $y - 1 = -(x + 2)$ or $y = -x - 1$.

    (b) Slope $= 1$, so $y - y_1 = m(x - x_1)$ becomes
    $y - 1 = x + 2$ or $y = x + 3$.

■ Section 10.3 Vector-valued Functions
(pp. 529–539)

Quick Review 10.3

1. $f'(x) = -\dfrac{x}{\sqrt{4 - x^2}}$, so for $x = 1$, $f(x) = \sqrt{3}$ and
   $f'(x) = -\dfrac{1}{\sqrt{3}}$. Then $y - \sqrt{3} = -\dfrac{1}{\sqrt{3}}(x - 1)$ or
   $y = \left(-\dfrac{1}{\sqrt{3}}\right)x + \dfrac{4}{\sqrt{3}}$.

2. $f'(x) = -\dfrac{x}{\sqrt{4 - x^2}}$, so for $x = 1$, $f(x) = \sqrt{3}$ and
   $f'(x) = -\dfrac{1}{\sqrt{3}}$. Then $y - \sqrt{3} = \sqrt{3}(x - 1)$ or $y = \sqrt{3}x$.

3. $\dfrac{dy}{dx} = \dfrac{dy/dt}{dx/dt} = \dfrac{5 \cos t}{-4 \sin t}$, which for $t = \dfrac{\pi}{2}$ equals zero.

4. $\dfrac{dy}{dx} = \dfrac{dy/dt}{dx/dt} = \dfrac{5 \cos t}{-4 \sin t}$, which for $t = \pi$ is undefined:
   the tangent line is vertical.

5. $\dfrac{dy}{dx} = \dfrac{dy/dt}{dx/dt} = \dfrac{5 \cos t}{-4 \sin t}$, which for $t = \dfrac{\pi}{6}$ equals $-\dfrac{5\sqrt{3}}{4}$.
   Also, at $t = \dfrac{\pi}{6}$, $x = 2\sqrt{3}$ and $y = \dfrac{5}{2}$. The equation for the
   tangent line is $y - \dfrac{5}{2} = -\dfrac{5\sqrt{3}}{4}(x - 2\sqrt{3})$, or
   $y = \left(-\dfrac{5\sqrt{3}}{4}\right)x + 10$.

6. $\dfrac{dy}{dx} = \dfrac{dy/dt}{dx/dt} = \dfrac{5 \cos t}{-4 \sin t}$, which for $t = \dfrac{\pi}{6}$ equals $-\dfrac{5\sqrt{3}}{4}$.
   Also, at $t = \dfrac{\pi}{6}$, $x = 2\sqrt{3}$ and $y = \dfrac{5}{2}$. The equation for the
   normal line is $y - \dfrac{5}{2} = \dfrac{4}{5\sqrt{3}}(x - 2\sqrt{3})$, or
   $y = \left(\dfrac{4\sqrt{3}}{15}\right)x + \dfrac{9}{10}$.

7. $\lim_{x \to 2} \dfrac{x - 2}{x^2 - 4} = \lim_{x \to 2} \dfrac{x - 2}{(x + 2)(x - 2)} = \lim_{x \to 2} \dfrac{1}{x + 2} = \dfrac{1}{4}$

8. $y' = 3 - 2x$, and Length $= \displaystyle\int_0^2 \sqrt{1 + (3 - 2x)^2}\, dx$, which
   using NINT evaluates to ≈ 3.400.

9. $x' = t \cos t + \sin t$, and $y' = -t \sin t + \cos t$, and
   Length $= \displaystyle\int_0^2 \sqrt{(t \cos t + \sin t)^2 + (-t \sin t + \cos t)^2}\, dt$
   $= \displaystyle\int_0^2 \sqrt{t^2 + 1}\, dt$,
   which using NINT evaluates to ≈ 2.958.

10. $y = xe^x - e^x + C$ (use integration by parts), so
    $2 = 0 - e^0 + C$ and $C = 3$:
    $y = xe^x - e^x + 3$

Section 10.3 Exercises

1. $[5 - (-1)]\mathbf{i} + (1 - 4)\mathbf{j} = 6\mathbf{i} - 3\mathbf{j}$

3. $\overrightarrow{AB} = [0 - (-3)]\mathbf{i} + (2 - 0)\mathbf{j} = 3\mathbf{i} + 2\mathbf{j}$ and
   $\overrightarrow{CD} = (0 - 4)\mathbf{i} + (-3 - 0)\mathbf{j} = -4\mathbf{i} - 3\mathbf{j}$.
   (a) $[3 + (-4)]\mathbf{i} + [2 + (-3)]\mathbf{j} = -\mathbf{i} - \mathbf{j}$
   (b) $[3 - (-4)]\mathbf{i} + [2 - (-3)]\mathbf{j} = 7\mathbf{i} + 5\mathbf{j}$

5. (a)

   [–6, 6] by [–4, 4]

   (b) $\mathbf{v}(t) = \dfrac{d}{dt}(2 \cos t)\mathbf{i} + \dfrac{d}{dt}(3 \sin t)\mathbf{j}$
       $= (-2 \sin t)\mathbf{i} + (3 \cos t)\mathbf{j}$
       $\mathbf{a}(t) = \dfrac{d}{dt}(-2 \sin t)\mathbf{i} + \dfrac{d}{dt}(3 \cos t)\mathbf{j}$
       $= (-2 \cos t)\mathbf{i} - (3 \sin t)\mathbf{j}$

   (c) $\mathbf{v}\left(\dfrac{\pi}{2}\right) = \langle -2, 0 \rangle$; speed $= \sqrt{(-2)^2 + 0^2} = 2$,
       direction $= \dfrac{1}{2}\langle -2, 0 \rangle = \langle -1, 0 \rangle$

   (d) Velocity $= 2\langle -1, 0 \rangle$

7. (a)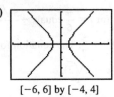
  $[-6, 6]$ by $[-4, 4]$

  (b) $\mathbf{v}(t) = \frac{d}{dt}(\sec t)\mathbf{i} + \frac{d}{dt}(\tan t)\mathbf{i} = (\sec t \tan t)\mathbf{i} + (\sec^2 t)\mathbf{j}$
  $\mathbf{a}(t) = \frac{d}{dt}(\sec t \tan t)\mathbf{i} + \frac{d}{dt}(\sec^2 t)\mathbf{j}$
  $= (\sec t \tan^2 t + \sec^3 t)\mathbf{i} + (2 \sec^2 t \tan t)\mathbf{j}$

  (c) $\mathbf{v}\left(\frac{\pi}{6}\right) = \left\langle\frac{2}{3}, \frac{4}{3}\right\rangle$; speed $= \sqrt{\left(\frac{2}{3}\right)^2 + \left(\frac{4}{3}\right)^2} = \frac{2\sqrt{5}}{3}$,
  direction $= \frac{3}{2\sqrt{5}}\left\langle\frac{2}{3}, \frac{4}{3}\right\rangle = \left\langle\frac{1}{\sqrt{5}}, \frac{2}{\sqrt{5}}\right\rangle$

  (d) Velocity $= \frac{2\sqrt{5}}{3}\left\langle\frac{1}{\sqrt{5}}, \frac{2}{\sqrt{5}}\right\rangle$

9. $\mathbf{v}(t) = (\cos t)\mathbf{i} + (2t + \sin t)\mathbf{j}$, $\mathbf{r}(0) = -\mathbf{j}$ and $\mathbf{v}(0) = \mathbf{i}$.

  So the slope is zero (the velocity vector is horizontal).

  (a) The horizontal line through $(0, -1)$: $y = -1$.
  (b) The vertical line through $(0, -1)$: $x = 0$.

11. $\left(\int_1^2 (6 - 6t)\, dt\right)\mathbf{i} + \left(\int_1^2 3\sqrt{t}\, dt\right)\mathbf{j}$
  $= \left[6t - 3t^2\right]_1^2 \mathbf{i} + \left[2t^{3/2}\right]_1^2 \mathbf{j}$
  $= -3\mathbf{i} + (4\sqrt{2} - 2)\mathbf{j}$

13. $\left(\int \sec t \tan t\, dt\right)\mathbf{i} + \left(\int \tan t\, dt\right)\mathbf{j}$
  $= (\sec t + C_1)\mathbf{i} + (\ln |\sec t| + C_2)\mathbf{j}$
  $= (\sec t)\mathbf{i} + (\ln |\sec t|)\mathbf{j} + \mathbf{C}$

15. $\mathbf{r}(t) = (t + 1)^{3/2}\mathbf{i} - e^{-t}\mathbf{j} + \mathbf{C}$, and
  $\mathbf{r}(0) = \mathbf{i} - \mathbf{j} + \mathbf{C} = \mathbf{0}$, so $\mathbf{C} = -(\mathbf{i} - \mathbf{j}) = -\mathbf{i} + \mathbf{j}$
  $\mathbf{r}(t) = ((t + 1)^{3/2} - 1)\mathbf{i} - (e^{-t} - 1)\mathbf{j}$

17. $\frac{d\mathbf{r}}{dt} = (-32t)\mathbf{j} + \mathbf{C}_1$ and $\mathbf{r}(t) = (-16t^2)\mathbf{j} + \mathbf{C}_1 t + \mathbf{C}_2$.
  $\mathbf{r}(0) = \mathbf{C}_2 = 100\mathbf{i}$, and $\left.\frac{d\mathbf{r}}{dt}\right|_{t=0} = \mathbf{C}_1 = 8\mathbf{i} + 8\mathbf{j}$. So
  $\mathbf{r}(t) = (-16t^2)\mathbf{j} + (8\mathbf{i} + 8\mathbf{j})t + 100\mathbf{i}$
  $= (8t + 100)\mathbf{i} + (-16t^2 + 8t)\mathbf{j}$.

19. $\mathbf{v}(t) = (1 - \cos t)\mathbf{i} + (\sin t)\mathbf{j}$ and $\mathbf{a}(t) = (\sin t)\mathbf{i} + (\cos t)\mathbf{j}$.
  Solve $\mathbf{v} \cdot \mathbf{a} = 0$: $(\sin t - \sin t \cos t) + (\sin t \cos t) = 0$
  implies $\sin t = 0$, which is true for $t = 0$, π, or 2π.

21. $\mathbf{v}(t) = (-3 \sin t)\mathbf{i} + (4 \cos t)\mathbf{j}$, and
  $\mathbf{a}(t) = (-3 \cos t)\mathbf{i} + (-4 \sin t)\mathbf{j}$.
  Solve $\mathbf{v} \cdot \mathbf{a} = 0$: $(9 \sin t \cos t) - (16 \sin t \cos t) = 0$, is
  true when $\sin t = 0$ or $\cos t = 0$, i.e., for
  $t = \frac{k\pi}{2}$, k any nonnegative integer.

23. $\mathbf{v}(t) = (-2 \sin t)\mathbf{i} + (\cos t)\mathbf{j}$, and
  $\mathbf{a}(t) = (-2 \cos t)\mathbf{i} + (-\sin t)\mathbf{j}$. So
  $\mathbf{v}\left(\frac{\pi}{4}\right) = (-\sqrt{2})\mathbf{i} + \left(\frac{1}{\sqrt{2}}\right)\mathbf{j}$, and
  $\mathbf{a}\left(\frac{\pi}{4}\right) = (-\sqrt{2})\mathbf{i} + \left(-\frac{1}{\sqrt{2}}\right)\mathbf{j}$.
  Then $|\mathbf{v}| = |\mathbf{a}| = \sqrt{\frac{5}{2}}$,
  $\mathbf{v} \cdot \mathbf{a} = \frac{3}{2}$, and
  $\theta = \cos^{-1}\left(\frac{\mathbf{v} \cdot \mathbf{a}}{|\mathbf{v}||\mathbf{a}|}\right) = \cos^{-1}\left(\frac{3}{5}\right) \approx 53.130°$.

25. (a) Both components are continuous at $t = 3$, so the limit is
  $3\mathbf{i} + \left(\frac{3^2 - 9}{3^2 + 3(3)}\right)\mathbf{j} = 3\mathbf{i}$.

  (b) Continuous so long as $t^2 + 3t \neq 0$, i.e., $t \neq 0, -3$
  (c) Discontinuous when $t^2 + 3t = 0$, i.e., $t = 0$ or -3

27. $\mathbf{v}(t) = (\sin t)\mathbf{i} + (1 - \cos t)\mathbf{j}$, i.e.,
  $\frac{dx}{dt} = \sin t$, and $\frac{dy}{dt} = 1 - \cos t$
  Distance $= \int_0^{2\pi/3} \sqrt{(\sin t)^2 + (1 - \cos t)^2}\, dt$
  $= \int_0^{2\pi/3} \sqrt{2 - 2 \cos t}\, dt$
  $= \int_0^{2\pi/3} 2 \sin\left(\frac{t}{2}\right) dt$
  $= \left[-4 \cos\left(\frac{t}{2}\right)\right]_0^{2\pi/3} = 2$

29. (a) $\mathbf{v}(t) = (\cos t)\mathbf{i} - (2 \sin 2t)\mathbf{j}$

  (b) $\mathbf{v}(t) = \mathbf{0}$ when both $\cos t = 0$ and $\sin 2t = 0$. $\cos t = 0$
  at $t = \frac{\pi}{2}$ and $\frac{3\pi}{2}$; $\sin 2t = 0$ at $t = 0, \frac{\pi}{2}, \pi, \frac{3\pi}{2}$, and 2π.
  So $\mathbf{v}(t) = \mathbf{0}$ at $t = \frac{\pi}{2}, \frac{3\pi}{2}$.

  (c) $x = \sin t$, $y = \cos 2t$. Relate the two using the identity
  $\cos 2u = 1 - 2 \sin^2 u$: $y = 1 - 2x^2$, where as t ranges
  over all possible values, $-1 \leq x \leq 1$. When t increases
  from 0 to 2π, the particle starts at $(0, 1)$, goes to
  $(1, -1)$, then goes to $(-1, -1)$, and then goes to $(0, 1)$,
  tracing the curve twice.

31. $\mathbf{a}(t) = 3\mathbf{i} - \mathbf{j}$, so $\mathbf{v}(t) = (3t)\mathbf{i} - t\mathbf{j} + \mathbf{C}_1$ and

$$\mathbf{r}(t) = \left(\frac{3}{2}t^2\right)\mathbf{i} - \left(\frac{1}{2}t^2\right)\mathbf{j} + \mathbf{C}_1 t + \mathbf{C}_2. \ \mathbf{r}(0) = \mathbf{C}_2 = \mathbf{i} + 2\mathbf{j},$$

and since $\mathbf{v}(0)$ must point directly from $(1, 2)$ toward $(4, 1)$ with magnitude 2,

$$\mathbf{v}(0) = \mathbf{C}_1 = 2\left(\frac{(4-1)\mathbf{i} + (1-2)\mathbf{j}}{\sqrt{(4-1)^2 + (1-2)^2}}\right)$$

$$= \frac{6}{\sqrt{10}}\mathbf{i} - \frac{2}{\sqrt{10}}\mathbf{j}$$

$$= \frac{3\sqrt{10}}{5}\mathbf{i} - \frac{\sqrt{10}}{5}\mathbf{j}$$

So $\mathbf{r}(t) = \left(\frac{3}{2}t^2 + \frac{3\sqrt{10}}{5}t + 1\right)\mathbf{i} + \left(-\frac{1}{2}t^2 - \frac{\sqrt{10}}{5}t + 2\right)\mathbf{j}$.

33. (a) The **j**-component is zero at $t = 0$ and $t = 160$: 160 seconds.

(b) $-\frac{3}{64}(40)(40 - 160) = 225$ m

(c) $\dfrac{d}{dt}\left[-\dfrac{3}{64}t(t - 160)\right] = -\dfrac{3}{32}t + \dfrac{15}{2}$, which for $t = 40$ equals $\dfrac{15}{4}$ m per second.

(d) $\mathbf{v}(t) = -\dfrac{3}{32}t + \dfrac{15}{2}$ equals 0 at $t = 80$ seconds (and is negative after that time).

35. (a) Referring to the figure, look at the circular arc from the point where $t = 0$ to the point "m". On one hand, this arc has length given by $r_0\theta$, but it also has length given by vt. Setting those two quantities equal gives the result.

(b) $\mathbf{v}(t) = \left(-v\sin\dfrac{vt}{r_0}\right)\mathbf{i} + \left(v\cos\dfrac{vt}{r_0}\right)\mathbf{j}$, and

$$\mathbf{a}(t) = \left(-\dfrac{v^2}{r_0}\cos\dfrac{vt}{r_0}\right)\mathbf{i} + \left(-\dfrac{v^2}{r_0}\sin\dfrac{vt}{r_0}\right)\mathbf{j}$$

$$= -\dfrac{v^2}{r_0}\left[\left(\cos\dfrac{vt}{r_0}\right)\mathbf{i} + \left(\sin\dfrac{vt}{r_0}\right)\mathbf{j}\right]$$

(c) From part (b) above, $\mathbf{a}(t) = -\left(\dfrac{v}{r_0}\right)^2 \mathbf{r}(t)$.

So, by Newton's second law, $\mathbf{F} = -m\left(\dfrac{v}{r_0}\right)^2 \mathbf{r}$.

Substituting for $\mathbf{F}$ in the law of gravitation gives the result.

(d) Set $\dfrac{vT}{r_0} = 2\pi$ and solve for vT.

(e) Substitute $\dfrac{2\pi r_0}{T}$ for v in $v^2 = \dfrac{GM}{r_0}$ and solve for T^2:

$$\left(\dfrac{2\pi r_0}{T}\right)^2 = \dfrac{GM}{r_0}$$

$$\dfrac{4\pi^2 r_0^2}{T^2} = \dfrac{GM}{r_0}$$

$$\dfrac{1}{T^2} = \dfrac{GM}{4\pi^2 r_0^3}$$

$$T^2 = \dfrac{4\pi^2}{GM} r_0^3$$

37. (a) Apply Corollary 3 to each component separately. If the components all differ by scalar constants, the difference vector is a constant vector.

(b) Follows immediately from (a) since any two anti-derivatives of $\mathbf{r}(t)$ must have identical derivatives, namely $\mathbf{r}(t)$.

39. Let $\mathbf{C} = \langle C_1, C_2 \rangle$. $\dfrac{d\mathbf{C}}{dt} = \left\langle \dfrac{dC_1}{dt}, \dfrac{dC_2}{dt} \right\rangle = \langle 0, 0 \rangle$.

41. $\mathbf{u} = \langle u_1, u_2 \rangle$, $\mathbf{v} = \langle v_1, v_2 \rangle$

(a) $\dfrac{d}{dt}(\mathbf{u} + \mathbf{v}) = \dfrac{d}{dt}(\langle u_1 + v_1, u_2 + v_2 \rangle)$

$$= \left\langle \dfrac{d}{dt}(u_1 + v_1), \dfrac{d}{dt}(u_2 + v_2) \right\rangle$$

$$= \langle u_1' + v_1', u_2' + v_2' \rangle$$

$$= \langle u_1', u_2' \rangle + \langle v_1', v_2' \rangle = \dfrac{d\mathbf{u}}{dt} + \dfrac{d\mathbf{v}}{dt}$$

(b) $\dfrac{d}{dt}(\mathbf{u} - \mathbf{v}) = \dfrac{d}{dt}(\langle u_1 - v_1, u_2 - v_2 \rangle)$

$$= \left\langle \dfrac{d}{dt}(u_1 - v_1), \dfrac{d}{dt}(u_2 - v_2) \right\rangle$$

$$= \langle u_1' - v_1', u_2' - v_2' \rangle$$

$$= \langle u_1', u_2' \rangle - \langle v_1', v_2' \rangle$$

$$= \dfrac{d\mathbf{u}}{dt} - \dfrac{d\mathbf{v}}{dt}$$

43. $f(t)$ and $g(t)$ differentiable at $c \Rightarrow f(t)$ and $g(t)$ continuous at $c \Rightarrow \mathbf{r}(t) = f(t)\mathbf{i} + g(t)\mathbf{j}$ is continuous at c.

45. (a) Let $\mathbf{r}(t) = f(t)\mathbf{i} + g(t)\mathbf{j}$. Then

$$\dfrac{d}{dt}\int_a^t \mathbf{r}(q)\,dq = \dfrac{d}{dt}\int_a^t [f(q)\mathbf{i} + g(q)\mathbf{j}]\,dq$$

$$= \dfrac{d}{dt}\left[\left(\int_a^t f(q)\,dq\right)\mathbf{i} + \left(\int_a^t g(q)\,dq\right)\mathbf{j}\right]$$

$$= \left(\dfrac{d}{dt}\int_a^t f(q)\,dq\right)\mathbf{i} + \left(\dfrac{d}{dt}\int_a^t g(q)\,dq\right)\mathbf{j}$$

$$= f(t)\mathbf{i} + g(t)\mathbf{j} = \mathbf{r}(t).$$

(b) Let $\mathbf{S}(t) = \int_a^t \mathbf{r}(q)\,dq$. Then part (a) shows that $\mathbf{S}(t)$ is an antiderivative of $\mathbf{r}(t)$. Let $\mathbf{R}(t)$ be any antiderivative of $\mathbf{r}(t)$. Then according to 37(b), $\mathbf{S}(t) = \mathbf{R}(t) + \mathbf{C}$.

Letting $t = a$, we have $0 = \mathbf{S}(a) = \mathbf{R}(a) + \mathbf{C}$.

Therefore, $\mathbf{C} = -\mathbf{R}(a)$ and $\mathbf{S}(t) = \mathbf{R}(t) - \mathbf{R}(a)$.

The result follows by letting $t = b$.

Section 10.4 Modeling Projectile Motion
(pp. 539–552)

Exploration 1 Hitting a Home Run

1. The graphs of the parametric equations
$x = (152 \cos 20° - 8.8)t$, $y = 3 + (152 \sin 20°)t - 16t^2$
and the fence are shown in the window [0, 450] by [−20, 60]. The fence was obtained using the line command "Line(". You can zoom in as shown in the second figure to see that the ball does just clear the fence.

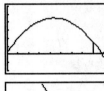

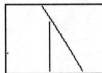

You can also use algebraic methods to show that $t \approx 2.984$ when $x = 400$, and that $y \approx 15.647$ for this value of t.

2.

angle (degrees)	25	30	45
range (ft)	≈523.707	≈588.279	≈665.629
flight time (sec)	≈4.061	≈4.789	≈6.745

3. Using the same window of part (1) we can see that the ball clears the fence.

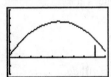

4.

angle (degrees)	25	30	45
range (ft)	≈559.444	≈630.424	≈724.988
flight time (sec)	≈4.061	≈4.789	≈6.745

Exploration 2 Hitting a Baseball

1. $x = \dfrac{152}{0.05}(1 - e^{-0.05t}) \cos 20°$

$y = 3 + \dfrac{152}{0.05}(1 - e^{-0.05t}) \sin 20°$
$+ \dfrac{32}{0.05^2}(1 - 0.05t - e^{-0.05t})$

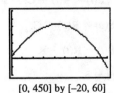

[0, 450] by [−20, 60]

2. The ball reaches a maximum height of about 43.07 ft when t is about 1.56 sec.

3. The range is about 425.47 ft and the flight time is about 3.23 sec.

Quick Review 10.4

1. $\langle 50 \cos 25°, 50 \sin 25° \rangle \approx \langle 45.315, 21.131 \rangle$

2. $\langle 80 \cos 120°, 80 \sin 120° \rangle = \langle -40, 40\sqrt{3} \rangle$

3. To find the x-intercepts, solve $2x^2 + 11x - 40 = 0$ using the quadratic formula: $x = \dfrac{-11 \pm \sqrt{11^2 - 4(2)(-40)}}{2(2)}$

$= \dfrac{5}{2}$ or -8. The x-intercepts are $\left(\dfrac{5}{2}, 0\right)$ and $(-8, 0)$. For the y-intercept, find $f(0) = 2(0)^2 + 11(0) - 40 = -40$. The y-intercept is $(0, -40)$.

4. At the vertex, $f'(x) = 4x + 11 = 0$ and $x = -\dfrac{11}{4}$. Then the vertex is $\left(-\dfrac{11}{4}, 2\left(-\dfrac{11}{4}\right)^2 + 11\left(-\dfrac{11}{4}\right) - 40\right)$
$= \left(-\dfrac{11}{4}, -\dfrac{441}{8}\right)$.

5. To find the x-intercepts, solve $20x - x^2 = 0$: $x = 0$ or 20. The x-intercepts are $(0, 0)$ and $(20, 0)$. For the y-intercept, find $y(0)$: it is already known to be 0. So the y-intercept is $(0, 0)$.

6. At the vertex, $g'(x) = 20 - 2x = 0$ and $x = 10$. Then the vertex is $(10, 20(10) - 10^2) = (10, 100)$.

7. $y = -\cos x + C$. $y\left(\dfrac{\pi}{2}\right) = -\cos\left(\dfrac{\pi}{2}\right) + C = C = 2$, so $y = -\cos x + 2$.

8. $y' = t^2 + C_1$ and $y = \dfrac{1}{3}t^3 + C_1 t + C_2$
$y'(-1) = (-1)^2 + C_1 = 1 + C_1 = 4$, so $C_1 = 3$.
$y(-1) = \dfrac{1}{3}(-1)^3 + 3(-1) + C_2 = -\dfrac{10}{3} + C_2 = 5$, so $C_2 = \dfrac{25}{3}$
$y = \dfrac{t^3}{3} + 3t + \dfrac{25}{3}$

9. $\displaystyle\int \dfrac{dy}{16 - y} = \int dt$
$-\ln|16 - y| = t + C$
$16 - y = ke^{-t}$
$y(0) = 16 - k = 20$ so $k = -4$
$y = 16 + 4e^{-t}$

10. $\displaystyle\int \dfrac{dy}{4 - 2y} = \int x \, dx$
$-\dfrac{1}{2}\ln|4 - 2y| = \dfrac{1}{2}x^2 + C$
$4 - 2y = ke^{-x^2}$
$y = 2 - \dfrac{k}{2}e^{-x^2}$
$y(0) = 2 - \dfrac{k}{2} = 1$ so $k = 2$
$y = 2 - e^{-x^2}$

Section 10.4 Exercises

1. Solve $v_x t = (840 \cos 60°)t = 21{,}000$ for t: $t = 50$ seconds.

3. (a) $t = \dfrac{2v_0 \sin \alpha}{g} = \dfrac{2(500)\sin 45°}{9.8} \approx 72.154$ seconds;

$R = \dfrac{v_0^2}{g} \sin 2\alpha = \dfrac{500^2}{9.8} \sin 90° \approx 25{,}510$ m

$= 25.510$ km downrange

(b) $y = -\left(\dfrac{g}{2v_0^2 \cos^2 \alpha}\right)x^2 + (\tan \alpha)x$

$= -\left(\dfrac{9.8}{2(500)^2 \cos^2 45°}\right)5000^2 + (\tan 45°)5000 = 4020$ m

(c) $y_{\max} = \dfrac{(v_0 \sin \alpha)^2}{2g} = \dfrac{(500 \sin 45°)^2}{2(9.8)} \approx 6377.551$ m

5. Use $y = (v_0 \sin \alpha)t - \dfrac{1}{2}gt^2 + 6.5$.

$16t^2 - 22\sqrt{2}\,t - 6.5 = 0$

$t = \dfrac{11\sqrt{2} + \sqrt{346}}{16} \approx 2.135$ seconds (by the quadratic formula). Substitute that into

$x = (v_0 \cos \alpha)t = (44 \sin 45°)t$ to obtain

$x \approx 66.4206$. 66.421 feet from the stopboard.

7. (a) Use $R = \dfrac{v_0^2}{g} \sin 2\alpha$; solve $10 = \dfrac{v_0^2}{9.8} \sin 90°$ for v_0:

$v_0 = 7\sqrt{2} \approx 9.899$ m/sec.

(b) Solve $6 = \dfrac{(7\sqrt{2})^2}{9.8} \sin 2\alpha$ for α: $\sin 2\alpha = 0.6$, so

$2\alpha = \sin^{-1} 0.6 \approx 36.870°$ and $\alpha \approx 18.435°$ or

$2\alpha = 180° - \sin^{-1} 0.6 \approx 143.130$ and $\alpha \approx 71.565°$.

9. $R = \dfrac{v_0^2}{g} \sin 2\alpha$

$(248.8 \text{ yd})(3 \text{ ft/yd}) = \dfrac{v_0^2}{32 \text{ ft/sec}^2} \sin 18°$

$v_0 \approx 278.016$ ft/sec or ≈ 189.556 mph.

11. No. For $\alpha = 30°$, $v_0 = 90$ ft/sec, and $x = 135$ ft,

$y = -\left(\dfrac{32}{2v_0^2 \cos^2 \alpha}\right)x^2 + (\tan \alpha)x$ evaluates to ≈ 29.942

feet above the ground, which is not quite high enough.

13. (a) With the origin at the launch point, use

$y = -\left(\dfrac{32}{2v_0^2 \cos^2 20°}\right)x^2 + (\tan 20°)x$. Set $x = 315$ and

$y = 37 - 3 = 34$, then solve to find

$v_0 = \dfrac{1260}{\cos 20° \sqrt{315 \tan 20° - 34}} \approx 149.307$ ft/sec.

(b) Solve $v_0(\cos 20°)t \approx 149.307(\cos 20°)t = 315$ to find

$t \approx 2.245$ seconds.

15. Use $R = \dfrac{v_0^2}{g} \sin 2\alpha$: $\sin 2\alpha = \dfrac{(9.8)(16{,}000)}{400^2} = 0.98$;

$\alpha = \dfrac{\sin^{-1} 0.98}{2} \approx 39.261°$ or $\alpha = 90 - \dfrac{\sin^{-1} 0.98}{2}$

$\approx 50.739°$.

17. With the origin at the launch point,

$y = -\left(\dfrac{32}{2v_0^2 \cos^2 40°}\right)x^2 + (\tan 40°)x$. Setting $x = 73\dfrac{5}{6}$ and

$y = -6.5$ and solving for v_0 yields $v_0 \approx 46.597$ ft/sec.

19. Integrating, $\dfrac{d}{dt}\mathbf{r}(t) = c_1\mathbf{i} + (-gt + c_2)\mathbf{j}$. The initial

condition on the velocity gives $c_1 = v_0 \cos \alpha$

and $c_2 = v_0 \sin \alpha$. Integrating again,

$\mathbf{r}(t) = ((v_0 \cos \alpha)t + c_3)\mathbf{i} + \left(\dfrac{1}{2}gt^2 + (v_0 \sin \alpha)t + c_4\right)\mathbf{j}$.

The initial condition on the position gives

$c_3 = x_0$ and $c_4 = y_0$.

21. The horizontal distance is 30 yd $-$ 6 ft $= 84$ ft. Then

$84 = (v_0 \cos \alpha)t$, where $\alpha = \tan^{-1}\left(\dfrac{68}{45}\right) \approx 56.5°$ and

$v_0 = \dfrac{16\sqrt{17}}{\sin \alpha}$ (from Exercise 20). So $t = \dfrac{84}{v_0 \cos \alpha}$

$= \dfrac{84 \tan \alpha}{16\sqrt{17}} = \dfrac{21\left(\dfrac{68}{45}\right)}{4\sqrt{17}} = \dfrac{119}{15\sqrt{17}} \approx 1.924$ seconds. Then

$y = (v_0 \sin \alpha)t - \dfrac{1}{2}gt^2$

$= (16\sqrt{17})\left(\dfrac{119}{15\sqrt{17}}\right) - \dfrac{1}{2}(32)\left(\dfrac{119}{15\sqrt{17}}\right)^2 \approx 67.698$ ft.

The height above the ground is 6 ft more than that, ≈ 73.698, and the height above the rim is about $73.698 - 70 = 3.698$ feet.

23. Angle is $\alpha \approx 62°$ (measurements may vary slightly). For

flight time $t = \dfrac{2v_0 \sin \alpha}{g} = 1$ sec, $y_{\max} = \dfrac{(v_0 \sin \alpha)^2}{2g} = \dfrac{1}{8}gt^2$

$= \dfrac{1}{8}(32)(1)^2 = 4$ ft (independent of the measured angle).

$v_0 = \dfrac{gt}{2 \sin \alpha}$, so speed of engine $= v_0 \cos \alpha = \dfrac{gt}{2 \tan \alpha}$

$\approx \dfrac{32(1)}{2 \tan(62°)} \approx 8.507$ ft/sec (changes with the angle).

25. (a)

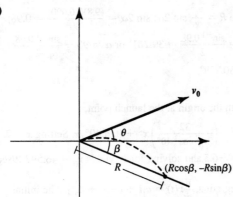

$x = (v_0 \cos\theta)t$

$y = (v_0 \sin\theta)t - \frac{1}{2}gt^2$

$x = R\cos\beta \Rightarrow R\cos\beta = (v_0 \cos\theta)t$

$\Rightarrow t = \dfrac{R\cos\beta}{v_0 \cos\theta}$. Then $y = -R\sin\beta$

$\Rightarrow -R\sin\beta = \dfrac{(v_0 \sin\theta)R\cos\beta}{v_0 \cos\theta} - \dfrac{g}{2}\dfrac{R^2\cos^2\beta}{v_0^2 \cos^2\theta}$

$\Rightarrow R = \dfrac{2v_0^2}{g\cos^2\beta}\cos\theta\sin(\theta+\beta)$.

Let $f(\theta) = \cos\theta\sin(\theta+\beta)$.

$f'(\theta) = \cos\theta\cos(\theta+\beta) - \sin\theta\sin(\theta+\beta)$

$f'(\theta) = 0 \Rightarrow \tan\theta\tan(\theta+\beta) = 1$

$\Rightarrow \tan\theta = \cot(\theta+\beta)$

$\Rightarrow \theta + \beta = 90° - \theta$. Note that $f''(\theta) < 0$, so R is maximum when $\alpha = \theta + \beta = 90° - \theta$. Thus the initial velocity bisects angle AOR.

(b)

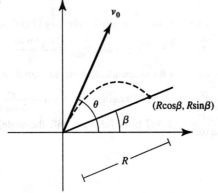

$R = \dfrac{2v_0^2}{g\cos^2\beta}\cos\theta\sin(\theta-\beta)$ is maximum when

$\tan\theta = \cot(\theta-\beta)$,

so $\theta - \beta = 90° - \theta$.

The initial velocity vector bisects the angle between the hill and the vertical for max range.

27. (a) (Assuming that "x" is zero at the point of impact.)

$\mathbf{r}(t) = (x(t))\mathbf{i} + (y(t))\mathbf{j}$, where
$x(t) = (35\cos 27°)t$ and
$y(t) = 4 + (35\sin 27°)t - 16t^2$.

(b) $y_{max} = \dfrac{(v_0 \sin\alpha)^2}{2g} + 4 = \dfrac{(35\sin 27°)^2}{64} + 4 \approx 7.945$ feet,

which is reached at $t = \dfrac{v_0 \sin\alpha}{g} = \dfrac{35\sin 27°}{32}$

≈ 0.497 seconds.

(c) For the time, solve $y = 4 + (35\sin 27°)t - 16t^2 = 0$ for t, using the quadratic formula:

$t = \dfrac{35\sin 27° + \sqrt{(-35\sin 27°)^2 + 256}}{32}$

≈ 1.201 seconds.

Then the range is about $x(1.201) = (35\cos 27°)(1.201)$

≈ 37.406 feet.

(d) For the time, solve $y = 4 + (35\sin 27°)t - 16t^2 = 7$ for t, using the quadratic formula:

$t = \dfrac{35\sin 27° \pm \sqrt{(-35\sin 27°)^2 - 192}}{32} \approx 0.254$ and

0.740 seconds. At those times the ball is about

$x(0.254) = (35\cos 27°)(0.254) \approx 7.906$ feet and

$x(0.740) = (35\cos 27°)(0.740) \approx 23.064$ feet from the impact point, or about $37.460 - 7.906 \approx 29.554$ feet and $37.460 - 23.064 \approx 14.396$ feet from the landing spot.

(e) Yes. It changes things because the ball won't clear the net ($y_{max} \approx 7.945$ ft).

29. (a) $\mathbf{r}(t) = (x(t))\mathbf{i} + (y(t))\mathbf{j}$, where

$x(t) = \left(\dfrac{1}{0.08}\right)(1 - e^{-0.08t})(152\cos 20° - 17.6)$ and

$y(t) = 3 + \left(\dfrac{152}{0.08}\right)(1 - e^{-0.08t})(\sin 20°)$
$\qquad + \left(\dfrac{32}{0.08^2}\right)(1 - 0.08t - e^{-0.08t})$

(b) Solve graphically: enter $y(t)$ for Y_1 (where X stands in for t), then use the maximum function to find that at $t \approx 1.527$ seconds the ball reaches a maximum height of about 41.893 feet.

(c) Use the zero function to find that $y = 0$ when the ball has traveled for ≈ 3.181 seconds. The range is about

$x(3.181)$

$= \left(\dfrac{1}{0.08}\right)(1 - e^{-0.08(3.181)})(152\cos 20° - 17.6)$

≈ 351.734 feet

(d) Graph $Y_2 = 35$ and use the interesect function to find that $y = 35$ for $t \approx 0.877$ and 2.190 seconds, at which times the ball is about $\approx x(0.877) \approx 106.028$ feet and $x(2.190) \approx 251.530$ feet from home plate.

(e) No; the range is less than 380 feet. To find the wind needed for a home run, first use the method of part (d) to find that $y = 20$ at $t \approx 0.376$ and 2.716 seconds. Then define

$$x(w) = \left(\frac{1}{0.08}\right)(1 - e^{-0.08(2.716)})(152 \cos 20° + w),$$

and solve $x(w) = 380$ to find $w \approx 12.846$ ft/sec. This is the speed of a wind gust needed in the direction of the hit for the ball to clear the fence for a home run.

31. The points in question are $(x, y) = \left(\dfrac{R}{2}, y_{max}\right)$. So,

$$x = \frac{v_0^2 \sin \alpha \cos \alpha}{g}, \text{ and } y = \frac{(v_0 \sin \alpha)^2}{2g}. \text{ Then}$$

$$x^2 + 4\left(y - \frac{v_0^2}{4g}\right)^2$$

$$= \left(\frac{v_0^2 \sin \alpha \cos \alpha}{g}\right)^2 + 4\left(\frac{(v_0 \sin \alpha)^2}{2g} - \frac{v_0^2}{4g}\right)^2$$

$$= \frac{v_0^4}{g^2}\left[\sin^2 \alpha \cos^2 \alpha + 4\left(\frac{\sin^2 \alpha}{2} - \frac{1}{4}\right)^2\right]$$

$$= \frac{v_0^4}{g^2}\left[\sin^2 \alpha \cos^2 \alpha + 4\left(\frac{\sin^4 \alpha}{4} - \frac{\sin^2 \alpha}{4} + \frac{1}{16}\right)\right]$$

$$= \frac{v_0^4}{g^2}\left[\sin^2 \alpha \cos^2 \alpha + (\sin^2 \alpha)(1 - \cos^2 \alpha) - \sin^2 \alpha + \frac{1}{4}\right]$$

$$= \frac{v_0^4}{g^2}\left(\frac{1}{4}\right) = \frac{v_0^4}{4g^2},$$

so the point (x, y) lies on the ellipse.

■ Section 10.5 Polar Coordinates and Polar Graphs (pp. 552–559)

Exploration 1 Investigating Polar Graphs

1. The graph is drawn in the decimal window $[-4.7, 4.7]$ by $[-3.1, 3.1]$ with $-\dfrac{\pi}{2} \leq \theta \leq \dfrac{\pi}{2}$.

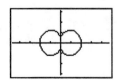

2. r_1 and r_2 are 0 for $\theta = \dfrac{\pi}{2}$.
3. π

4. If (r, θ) is a solution of $r^2 = 4 \cos \theta$, then so is $(r, -\theta)$ because $\cos(-\theta) = \cos \theta$. Thus, the graph is symmetric about the x-axis.
   If (r, θ) is a solution of $r^2 = 4 \cos \theta$, then so is $(-r, -\theta)$ because $(-r)^2 = r^2$ and $\cos(-\theta) = \cos \theta$. Thus, the graph is symmetric about the y-axis.
   The graph is symmetric about the origin because it is symmetric about both the x- and y-axes. You can also give a direct proof by showing that $(-r, \theta)$ lies on the graph if (r, θ) does.

Exploration 2 Graphing Rose Curves

All graphs are drawn in the window $[-4.7, 4.7]$ by $[-3.1, 3.1]$.

1. The graphs are rose curves with 4 petals when $n = \pm 2$, 8 petals when $n = \pm 4$, and 12 petals when $n = \pm 6$.

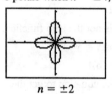

$n = \pm 2$

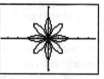

$n = \pm 4$

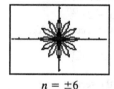

$n = \pm 6$

2. 2π
3. The graph is a rose curve with $2|n|$ petals.
4. The graphs are rose curves with 3 petals when $n = \pm 3$, 5 petals when $n = \pm 5$, and 7 petals when $n = \pm 7$.

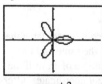

$n = \pm 3$

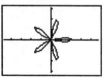

$n = \pm 5$

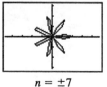

$n = \pm 7$

5. π
6. The graph is a rose cuve with $|n|$ petals.

280 Section 10.5

Quick Review 10.5

1. Slope $= \dfrac{-1-4}{3-(-2)} = -1$,
   so $y - 4 = -1[x - (-2)]$ or $y = -x + 2$.

2. $(x - 0)^2 + (y - 0)^2 = 3^2$, or $x^2 + y^2 = 9$.

3. $[x - (-2)]^2 + (y - 4)^2 = 2^2$, or $(x + 2)^2 + (y - 4)^2 = 4$.

4. (a) No; y is a function of x and is not the zero function.
   (b) No;
   $y(-x) = (-x)^3 - (-x) = -x^3 + x = -(x^3 - x) \neq y(x)$
   (c) Yes; $y(-x) = -y(x)$ (see part (b))

5. (a) No; y is a function of x and is not the zero function.
   (b) No; $y(-x) = (-x)^2 - (-x) = x^2 + x \neq y(x)$
   (c) No; $y(-x) \neq -y(x)$ (see part (b))

6. (a) No; y is a function of x and is not the zero function.
   (b) Yes; $y(-x) = \cos(-x) = \cos x = y(x)$
   (c) No; $y(-x) \neq -y(x)$ (see part (b))

7. (a) Yes; Substitute $-y$ for y in the equation to get the original equation.
   (b) Yes; Substitute $-x$ for x in the equation to get the original equation.
   (c) Yes; since the curve is symmetric with respect to both the x-axis and y-axis, it is symmetric with respect to the origin. (Also, substitute $-x$ for x and $-y$ for y in the equation to get the original equation.)

8. Solve for y: $y = (x - 2)^{1/2}$ or $-(x - 2)^{1/2}$.
   Enter the first expression for Y_1, the second for Y_2.

9. Solve for y: $y = \left(\dfrac{4 - x^2}{3}\right)^{1/2}$ or $-\left(\dfrac{4 - x^2}{3}\right)^{1/2}$.
   Enter the first expression for Y_1, the second for Y_2.

10. $(x^2 - 4x) + (y^2 + 6y + 9) = 0$
    $(x^2 - 4x + 4) + (y^2 + 6y + 9) = 4$
    $(x - 2)^2 + (x + 3)^2 = 2^2$.
    Center $= (2, -3)$, Radius $= 2$.

Section 10.5 Exercises

For Exercise 1, two pairs of polar coordinates label the same point if the r-coordinates are the same and the θ-coordinates differ by an even multiple of π, or if the r-coordinates are opposites and the θ-coordinates differ by an odd multiple of π.

1. (a) and (e) are the same.
   (b) and (g) are the same.
   (c) and (h) are the same.
   (d) and (f) are the same.

3.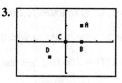
   $[-3, 3]$ by $[-2, 2]$

   (a) $\left(\sqrt{2} \cos \dfrac{\pi}{4}, \sqrt{2} \sin \dfrac{\pi}{4}\right) = (1, 1)$
   (b) $(1 \cos 0, 1 \sin 0) = (1, 0)$
   (c) $\left(0 \cos \dfrac{\pi}{2}, 0 \sin \dfrac{\pi}{2}\right) = (0, 0)$
   (d) $\left(-\sqrt{2} \cos \dfrac{\pi}{4}, -\sqrt{2} \sin \dfrac{\pi}{4}\right) = (-1, -1)$

5.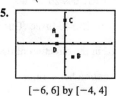
   $[-6, 6]$ by $[-4, 4]$

   (a) $r = \sqrt{(-1)^2 + 1^2} = \sqrt{2}$, $\tan \theta = \dfrac{1}{-1} = -1$ with θ in quadrant II. The coordinates are $\left(\sqrt{2}, \dfrac{3\pi}{4}\right)$. $\left(\sqrt{2}, -\dfrac{5\pi}{4}\right)$ also works, since r is the same and θ differs by 2π.

   (b) $r = \sqrt{1^2 + (-\sqrt{3})^2} = 2$, $\tan \theta = -\dfrac{\sqrt{3}}{1} = -\sqrt{3}$ with θ in quadrant IV. The coordinates are $\left(2, -\dfrac{\pi}{3}\right)$. $\left(-2, \dfrac{2\pi}{3}\right)$ also works, since r has the opposite sign and θ differs by π.

   (c) $r = \sqrt{0^2 + 3^2} = 3$, $\tan \theta = \dfrac{3}{0}$ is undefined with θ on the positive y-axis. The coordinates are $\left(3, \dfrac{\pi}{2}\right)$. $\left(3, \dfrac{5\pi}{2}\right)$ also works, since r is the same and θ differs by 2π.

   (d) $r = \sqrt{(-1)^2 + 0^2} = 1$, $\tan \theta = \dfrac{0}{-1} = 0$ with θ on the negative x-axis. The coordinates are $(1, \pi)$. $(-1, 0)$ also works, since r has the opposite sign and θ differs by π.

7.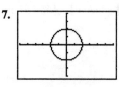
   $[-6, 6]$ by $[-4, 4]$

9.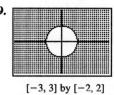
   $[-3, 3]$ by $[-2, 2]$

11.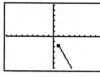
[−9, 9] by [−6, 6]

13.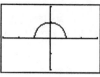
[−3, 3] by [−2, 2]

15.
[−3, 3] by [−2, 2]

17.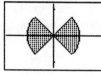
[−1.8, 1.8] by [−1.2, 1.2]

19. $y = r \sin \theta$, so the equation is $y = 0$, which is the x-axis.

21. $r = 4 \csc \theta$
$r \sin \theta = 4$
$y = r \sin \theta$, so the equation is $y = 4$, a horizontal line.

23. $x = r \cos \theta$ and $y = r \sin \theta$, so the equation is $x + y = 1$, a line (slope $= -1$, y-intercept $= 1$).

25. $x^2 + y^2 = r^2$ and $y = r \sin \theta$, so the equation is $x^2 + y^2 = 4y \Rightarrow x^2 + (y - 2)^2 = 4$, a circle (center $= (0, 2)$, radius $= 2$).

27. $r^2 \sin 2\theta = 2$
$2r^2 \sin \theta \cos \theta = 2$
$(r \sin \theta)(r \cos \theta) = 1$
$x = r \cos \theta$ and $y = r \sin \theta$, so the equation is $xy = 1$
$\left(\text{or, } y = \dfrac{1}{x} \right)$, a hyperbola.

29. $r = \csc \theta \, e^{r\cos\theta}$
$r \sin \theta = e^{r\cos\theta}$
$x = r \cos \theta$ and $y = r \sin \theta$, so the equation is $y = e^x$, the exponential curve.

31. $r \sin \theta = \ln r + \ln \cos \theta$
$r \sin \theta = \ln (r \cos \theta)$
$y = \ln x$, the logarithmic curve.

33. $r^2 = -4r \cos \theta$
$x^2 + y^2 = -4x$
$(x + 2)^2 + y^2 = 4$, a circle (center $= (-2, 0)$, radius $= 2$).

35. $r = 2 \cos \theta + 2 \sin \theta$
$r^2 = 2r \cos \theta + 2r \sin \theta$
$x^2 + y^2 = 2x + 2y$
$(x - 1)^2 + (y - 1)^2 = 2$,
a circle (center $= (1, 1)$, radius $= \sqrt{2}$).

37. $x = 7$
$r \cos \theta = 7$. The graph is a vertical line.

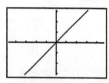
[−9.4, 9.4] by [−6.2, 6.2]

39. $x = y \Rightarrow r \cos \theta = r \sin \theta \Rightarrow \tan \theta = 1 \Rightarrow \theta = \dfrac{\pi}{4}$
More generally, $\theta = \dfrac{\pi}{4} + 2k\pi$ for any integer k.
The graph is a slanted line.

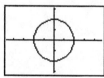
[−4.7, 4.7] by [−3.1, 3.1]

41. $x^2 + y^2 = 4$
$r^2 = 4$ or $r = 2$ (or $r = -2$)

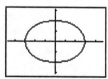
[−4.7, 4.7] by [−3.1, 3.1]

43. $\dfrac{x^2}{9} + \dfrac{y^2}{4} = 1$
$\dfrac{r^2 \cos^2 \theta}{9} + \dfrac{r^2 \sin^2 \theta}{4} = 1$
$r^2(4 \cos^2 \theta + 9 \sin^2 \theta) = 36$

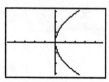
[−4.7, 4.7] by [−3.1, 3.1]

45. $y^2 = 4x$
$r^2 \sin^2 \theta = 4r \cos \theta$
$r \sin^2 \theta = 4 \cos \theta$

[−4.7, 4.7] by [−3.1, 3.1]

282 Section 10.6

47. $x^2 + (y - 2)^2 = 4$
$r^2 \cos^2 \theta + (r \sin \theta - 2)^2 = 4$
$r^2 \cos^2 \theta + r^2 \sin^2 \theta - 4r \sin \theta + 4 = 4$
$r^2 - 4r \sin \theta = 0$
$r = 4 \sin \theta$.
The graph is a circle centered at (0, 2) with radius 2.

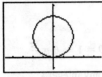

[−4.7, 4.7] by [−1.1, 5.1]

In Exercises 49–57, find the minimum θ-interval by trying different intervals on a graphing calculator.

49. (a)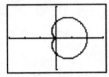

[−3, 3] by [−2, 2]
(b) Length of interval = 2π

51. (a)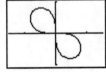

[−1.5, 1.5] by [−1, 1]
(b) Length of interval = $\dfrac{\pi}{2}$

53. (a)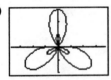

[−3.75, 3.75] by [−2, 3]
(b) Length of interval = 2π

55. (a)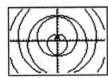

[−15, 15] by [−10, 10]
(b) Required interval = $(-\infty, \infty)$

57. (a)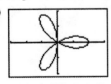

[−3, 3] by [−2, 2]
(b) Length of interval = π

59. If (r, θ) is a solution, so is $(-r, \theta)$. Therefore, the curve is symmetric about the origin. And if (r, θ) is a solution, so is $(r, -\theta)$. Therefore, the curve is symmetric about the x-axis. And since any curve with x-axis and origin symmetry also has y-axis symmetry, the curve is symmetric about the y-axis.

61. If (r, θ) is a solution, so is $(r, \pi - \theta)$. Therefore, the curve is symmetric about the y-axis. The curve does not have x-axis or origin symmetry.

63. (a) Because $r = a \sec \theta$ is equivalent to $r \cos \theta = a$, which is equivalent to the Cartesian equation $x = a$.
(b) $r = a \csc \theta$ is equivalent to $y = a$.

65. (a) We have $x = r \cos \theta$ and $y = r \sin \theta$. By taking $t = \theta$, we have $r = f(t)$, so $x = f(t) \cos t$ and $y = f(t) \sin t$.
(b) $x = 3 \cos t, y = 3 \sin t$
(c) $x = (1 - \cos t) \cos t, y = (1 - \cos t) \sin t$
(d) $x = (3 \sin 2t) \cos t, y = (3 \sin 2t) \sin t$

67. $d = \sqrt{(x_2 - x_1)^2 + (y_1 - y_2)^2}$
$= [(r_2 \cos \theta_2 - r_1 \cos \theta_1)^2 + (r_2 \sin \theta_2 - r_1 \sin \theta_1)^2]^{1/2}$
$= [r_2^2 \cos^2 \theta_2 - 2r_2 r_1 \cos \theta_2 \cos \theta_1 + r_1^2 \cos^2 \theta_1$
$\quad + r_2^2 \sin^2 \theta_2 - 2r_2 r_1 \sin \theta_2 \sin \theta_1 + r_1^2 \sin^2 \theta_1]^{1/2}$
$= \sqrt{r_1^2 + r_2^2 - 2r_1 r_2 \cos(\theta_1 - \theta_2)}$

69. (a)

[−5, 25] by [−10, 10]
The graphs are hyperbolas.
(b) Graphs for $k > 1$ are hyperbolas. As $k \to 1^+$, the right branch of the hyperbola goes to infinity and "disappears". The left branch approaches the parabola $y^2 = 4 - 4x$.

■ **Section 10.6** Calculus of Polar Curves
(pp. 559–568)

Quick Review 10.6

1. $\dfrac{dy}{dx} = \dfrac{dy/dt}{dx/dt} = \dfrac{5 \cos t}{-3 \sin t} = -\dfrac{5}{3} \cot t$

2. $-\dfrac{5}{3} \cot 2 \approx 0.763$

3. Solve $\cot t = 0$: $t = \dfrac{\pi}{2}$ or $\dfrac{3\pi}{2}$;
the corresponding points are $\left(3 \cos \dfrac{\pi}{2}, 5 \sin \dfrac{\pi}{2}\right) = (0, 5)$
and $\left(3 \cos \dfrac{3\pi}{2}, 5 \sin \dfrac{3\pi}{2}\right) = (0, -5)$

4. $-\dfrac{5}{3} \cot t$ is undefined when $t = 0, \pi,$ or 2π;
the corrresponding points are
$(3 \cos 0, 5 \sin 0) = (3 \cos 2\pi, 5 \sin 2\pi) = (3, 0)$ and
$(3 \cos \pi, 5 \sin \pi) = (-3, 0)$.

5. Length $= \int_0^\pi \sqrt{\left(\frac{dx}{dt}\right)^2 + \left(\frac{dy}{dt}\right)^2}\, dt$

$= \int_0^\pi \sqrt{9\sin^2 t + 25\cos^2 t}\, dt,$

which using NINT evaluates to ≈ 12.763.

For questions 6–8, the graph is:

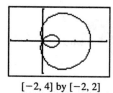

[−2, 4] by [−2, 2]

6. The upper half of the outer loop
7. The inner loop
8. The lower half of the outer loop
9. $y = 0$ for $x = 0$ or 6.

   Area $= \int_0^6 (6x - x^2)\, dx = \left[3x^2 - \frac{1}{3}x^3\right]_0^6 = 36$

10. Use a graphing calculator's intersect function to find that the curves cross at $x \approx 0.270$ and $x \approx 2.248$, then use NINT to find

    Area $= \int_{0.270}^{2.248} [2\sin x - (x^2 - 2x + 1)]\, dx \approx 2.403$.

Section 10.6 Exercises

1. $\dfrac{dy}{dx} = \dfrac{f'(\theta)\sin\theta + f(\theta)\cos\theta}{f'(\theta)\cos\theta - f(\theta)\sin\theta}$

   $= \dfrac{\cos\theta\sin\theta + (-1 + \sin\theta)\cos\theta}{\cos\theta\cos\theta - (-1 + \sin\theta)\sin\theta}$

   $= \dfrac{2\sin\theta\cos\theta - \cos\theta}{\cos^2\theta - \sin^2\theta + \sin\theta}$

   $\left.\dfrac{dy}{dx}\right|_{\theta=0} = -\dfrac{1}{1} = -1,\ \left.\dfrac{dy}{dx}\right|_{\theta=\pi} = \dfrac{1}{1} - 1$

3. $\dfrac{dy}{dx} = \dfrac{f'(\theta)\sin\theta + f(\theta)\cos\theta}{f'(\theta)\cos\theta - f(\theta)\sin\theta}$

   $= \dfrac{-3\cos\theta\sin\theta + (2 - 3\sin\theta)\cos\theta}{-3\cos\theta\cos\theta - (2 - 3\sin\theta)\sin\theta}$

   $= \dfrac{2\cos\theta - 6\sin\theta\cos\theta}{-2\sin\theta - 3(\cos^2\theta - \sin^2\theta)}$

   $\left.\dfrac{dy}{dx}\right|_{(2,0)} = \left.\dfrac{dy}{dx}\right|_{\theta=0} = \dfrac{-2}{3} = -\dfrac{2}{3},$

   $\left.\dfrac{dy}{dx}\right|_{(-1,\pi/2)} = \left.\dfrac{dy}{dx}\right|_{\theta=\pi/2} = \dfrac{0}{-1} = 0,$

   $\left.\dfrac{dy}{dx}\right|_{(2,\pi)} = \left.\dfrac{dy}{dx}\right|_{\theta=\pi} = \dfrac{2}{3},$ and

   $\left.\dfrac{dy}{dx}\right|_{(5,3\pi/2)} = \left.\dfrac{dy}{dx}\right|_{\theta=3\pi/2} = \dfrac{0}{-5} = 0.$

5.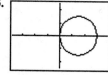

[−3.8, 3.8] by [−2.5, 2.5]

The graph passes through the pole when $r = 3\cos\theta = 0$, which occurs when $\theta = \dfrac{\pi}{2}$ and when $\theta = \dfrac{3\pi}{2}$. Since the θ-interval $0 \le \pi \le 1$ produces the entire graph, we need only consider $\theta = \dfrac{\pi}{2}$. At this point, there appears to be a vertical tangent line with equation $\theta = \dfrac{\pi}{2}$ (or $x = 0$). Confirm analytically:

$x = (3\cos\theta)\cos\theta = 3\cos^2\theta$

$y = (3\cos\theta)\sin\theta$

$\dfrac{dy}{d\theta} = (-3\sin\theta)\sin\theta + (3\cos\theta)\cos\theta = 3(\cos^2\theta - \sin^2\theta)$

and $\dfrac{dx}{d\theta} = 6\cos\theta(-\sin\theta)$.

At $\left(0, \dfrac{\pi}{2}\right),\ \left.\dfrac{dx}{d\theta}\right|_{\theta=\pi/2} = 0,$ and

$\left.\dfrac{dy}{d\theta}\right|_{\theta=\pi/2} = 3(0^2 - 1^2) = -3.$ So at $\left(0, \dfrac{\pi}{2}\right),\ \dfrac{dx}{d\theta} = 0$

and $\dfrac{dy}{d\theta} \ne 0$, so $\dfrac{dy}{dx}$ is undefined and the tangent line is vertical.

7.

[−1.5, 1.5] by [−1, 1]

The polar solutions are $\left(0, \dfrac{k\pi}{5}\right)$ for $k = 0, 1, 2, 3, 4$, and for a given k, the line $\theta = \dfrac{k\pi}{5}$ appears to be tangent to the curve at $\left(0, \dfrac{k\pi}{5}\right)$. This can be confirmed analytically by noting that the slope of the curve, $\dfrac{dy}{dx}$, equals the slope of the line, $\tan\dfrac{k\pi}{5}$. So the tangent lines are $\theta = 0\ [y = 0]$,

$\theta = \dfrac{\pi}{5}\left[y = \left(\tan\dfrac{\pi}{5}\right)x\right],\ \theta = \dfrac{2\pi}{5}\left[y = \left(\tan\dfrac{2\pi}{5}\right)x\right],$

$\theta = \dfrac{3\pi}{5}\left[y = \left(\tan\dfrac{3\pi}{5}\right)x\right],$ and $\theta = \dfrac{4\pi}{5}\left[y = \left(\tan\dfrac{4\pi}{5}\right)x\right].$

284 Section 10.6

9. $\dfrac{dy}{d\theta} = \cos\theta \sin\theta + (-1 + \sin\theta)\cos\theta$

$\qquad = \cos\theta(2\sin\theta - 1)$

$\qquad = \sin 2\theta - \cos\theta$

$\dfrac{dx}{d\theta} = \cos^2\theta - (-1 + \sin\theta)\sin\theta$

$\qquad = \cos^2\theta + \sin\theta - \sin^2\theta$

$\qquad = -2\sin^2\theta + \sin\theta + 1$

$\dfrac{dy}{d\theta} = 0$ when $\theta = \dfrac{\pi}{2}, \dfrac{3\pi}{2}$ ($\cos\theta = 0$) or when

$\theta = \dfrac{\pi}{6}, \dfrac{5\pi}{6}$ ($2\sin\theta - 1 = 0$). $\dfrac{dx}{d\theta} = 0$ when

$\sin\theta = \dfrac{-1 \pm \sqrt{9}}{-4} = -\dfrac{1}{2}$ or 1, i.e., when $\theta = \dfrac{7\pi}{6}, \dfrac{11\pi}{6}$, or

$\dfrac{\pi}{2}$. So there is a horizontal tangent line for $\theta = \dfrac{3\pi}{2}, r = -2$

$\left[\text{the line } y = -2\sin\dfrac{3\pi}{2} = 2\right]$, for $\theta = \dfrac{\pi}{6}, r = -\dfrac{1}{2}$

$\left[\text{the line } y = -\dfrac{1}{2}\sin\dfrac{\pi}{6} = -\dfrac{1}{4}\right]$ and for $\theta = \dfrac{5\pi}{6}, r = -\dfrac{1}{2}$

$\left[\text{again, the line } y = -\dfrac{1}{2}\sin\dfrac{5\pi}{6} = -\dfrac{1}{4}\right]$.

There is a vertical tangent line for $\theta = \dfrac{7\pi}{6}, r = -\dfrac{3}{2}$

$\left[\text{the line } x = -\dfrac{3}{2}\cos\dfrac{7\pi}{6} = \dfrac{3\sqrt{3}}{4}\right]$ and for

$\theta = \dfrac{11\pi}{6}, r = -\dfrac{3}{2}\left[\text{the line } x = -\dfrac{3}{2}\cos\dfrac{11\pi}{6} = -\dfrac{3\sqrt{3}}{4}\right]$.

For $\theta = \dfrac{\pi}{2}, \dfrac{dy}{d\theta} = \dfrac{dx}{d\theta} = 0$, but

$\dfrac{d}{d\theta}\left(\dfrac{dy}{d\theta}\right) = 2\cos 2\theta + \sin\theta = -1$ for $\theta = \dfrac{\pi}{2}$ and

$\dfrac{d}{d\theta}\left(\dfrac{dx}{d\theta}\right) = -4\sin\theta\cos\theta + \cos\theta = 0$ for $\theta = \dfrac{\pi}{2}$, so by

L'Hôpital's rule $\dfrac{dy}{dx}$ is undefined and the tangent line is

vertical at $\theta = \dfrac{\pi}{2}, r = 0$ [the line $x = 0$].

This information can be summarized as follows.

Horizontal at: $\left(-\dfrac{1}{2}, \dfrac{\pi}{6}\right)\ \left[y = -\dfrac{1}{4}\right]$,

$\qquad\qquad\quad \left(-\dfrac{1}{2}, \dfrac{5\pi}{6}\right)\ \left[y = -\dfrac{1}{4}\right]$,

$\qquad\qquad\quad \left(-2, \dfrac{3\pi}{2}\right)\ [y = 2]$

Vertical at: $\left(0, \dfrac{\pi}{2}\right)\ [x = 0]$,

$\qquad\qquad \left(-\dfrac{3}{2}, \dfrac{7\pi}{6}\right)\ \left[x = \dfrac{3\sqrt{3}}{4}\right]$,

$\qquad\qquad \left(-\dfrac{3}{2}, \dfrac{11\pi}{6}\right)\ \left[x = -\dfrac{3\sqrt{3}}{4}\right]$

11. $y = 2\sin^2\theta$

$\dfrac{dy}{d\theta} = 4\sin\theta\cos\theta$

$\qquad = 2\sin 2\theta$

$x = 2\sin\theta\cos\theta$

$\qquad = \sin 2\theta$

$\dfrac{dy}{d\theta} = 2\cos 2\theta$

$\dfrac{dy}{d\theta} = 0$ when $\theta = 0, \dfrac{\pi}{2}, \pi$, and $\dfrac{dx}{d\theta} = 0$ when

$\theta = \dfrac{\pi}{4}, \dfrac{3\pi}{4}$. They are never both zero.

For $\theta = 0, \dfrac{\pi}{2}, \pi$ the curve has horizontal asymptotes

at $(0, 0)$ $[y = 0 \sin 0 = 0]$, $\left(2, \dfrac{\pi}{2}\right)\left[y = 2\sin\dfrac{\pi}{2} = 2\right]$, and

$(0, \pi)$ $[y = 0 \sin\pi = 0]$. For $\theta = \dfrac{\pi}{4}, \dfrac{3\pi}{4}$ the curve has

vertical asymptotes at $\left(\sqrt{2}, \dfrac{\pi}{4}\right)\left[x = \sqrt{2}\cos\dfrac{\pi}{4} = 1\right]$ and

$\left(\sqrt{2}, \dfrac{3\pi}{4}\right)\left[x = \sqrt{2}\cos\dfrac{3\pi}{4} = -1\right]$.

This information can be summarized as follows.

Horizontal at: $(0, 0)\ [y = 0]$,

$\qquad\qquad\quad \left(2, \dfrac{\pi}{2}\right)\ [y = 2]$,

$\qquad\qquad\quad (0, \pi)\ [y = 0]$

Vertical at: $\left(\sqrt{2}, \dfrac{\pi}{4}\right)\ [x = 1]$,

$\qquad\qquad \left(\sqrt{2}, \dfrac{3\pi}{4}\right)\ [x = -1]$

13. The curve is complete for $0 \le \theta \le 2\pi$ (as can be verified by graphing). The area is

$\displaystyle\int_0^{2\pi} \dfrac{1}{2}(4 + 2\cos\theta)^2\, d\theta$

$= 2\displaystyle\int_0^{2\pi}(4 + 4\cos\theta + \cos^2\theta)\, d\theta$

$= 2\left[4\theta + 4\sin\theta + \dfrac{1}{2}\theta + \dfrac{1}{4}\sin 2\theta\right]_0^{2\pi} = 18\pi$

15. Use $r = \sqrt{2a^2\cos 2\theta}$. One lobe is complete for

$-\dfrac{\pi}{4} \le \theta \le \dfrac{\pi}{4}$. The total area is

$2\displaystyle\int_{-\pi/4}^{\pi/4}\dfrac{1}{2}(\sqrt{2a^2\cos 2\theta})^2\, d\theta = 2a^2\displaystyle\int_{-\pi/4}^{\pi/4}\cos 2\theta\, d\theta$

$\qquad\qquad\qquad\qquad\qquad\qquad\quad = 2a^2\left[\dfrac{1}{2}\sin 2\theta\right]_{-\pi/4}^{\pi/4}$

$\qquad\qquad\qquad\qquad\qquad\qquad\quad = 2a^2$

(Integrating from 0 to 2π will not work, because r is not defined over the entire interval.)

17. Use $r = \sqrt{4\sin 2\theta}$. One loop is complete for $0 \le \theta \le \dfrac{\pi}{2}$.

Its area is $\displaystyle\int_0^{\pi/2}\dfrac{1}{2}(\sqrt{4\sin 2\theta})^2\, d\theta = \displaystyle\int_0^{\pi/2} 2\sin 2\theta\, d\theta$

$\qquad\qquad\qquad\qquad\qquad\qquad = \left[-\cos 2\theta\right]_0^{\pi/2} = 2.$

19.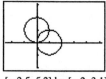

[−2.5, 5.2] by [−2, 3.1]

The circles intersect at (x, y) coordinates $(0, 0)$ and $(1, 1)$. The area shared is twice the area inside the circle $r = 2 \sin \theta$ between $\theta = 0$ and $\theta = \dfrac{\pi}{4}$.

Shared area $= 2 \int_0^{\pi/4} \dfrac{1}{2}(2 \sin \theta)^2 \, d\theta$

$= \int_0^{\pi/4} 4 \sin^2 \theta \, d\theta$

$= 4\left[\dfrac{1}{2}\theta - \dfrac{1}{4} \sin 2\theta\right]_0^{\pi/4}$

$= 4\left(\dfrac{\pi}{8} - \dfrac{1}{4}\right) = \dfrac{\pi}{2} - 1.$

21.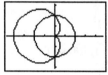

[−4.7, 4.7] by [−3.1, 3.1]

The shared area is half the circle plus two lobelike regions:

$\dfrac{1}{2}\pi(2)^2 + 2\int_0^{\pi/2} \dfrac{1}{2}[2(1 - \cos \theta)]^2 \, d\theta$

$= 2\pi + \int_0^{\pi/2} (4 - 8 \cos \theta + 4 \cos^2 \theta) \, d\theta$

$= 2\pi + 4\left[\theta - 2 \sin \theta + \dfrac{1}{2}\theta + \dfrac{1}{4} \sin 2\theta\right]_0^{\pi/2}$

$= 5\pi - 8$

23. For $a = 1$:

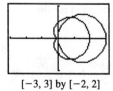

[−3, 3] by [−2, 2]

The curves intersect at the origin and when

$3a \cos \theta = a(1 + \cos \theta)$

$2 \cos \theta = 1$

$\theta = \pm \dfrac{\pi}{3}.$

Use the symmetries of the curves: the area in question is

$2\int_0^{\pi/3} \dfrac{1}{2}[(3a \cos \theta)^2 - a^2(1 + \cos \theta)^2] \, d\theta$

$= a^2 \int_0^{\pi/3} (9 \cos^2 \theta - 1 - 2 \cos \theta - \cos^2 \theta) \, d\theta$

$= a^2 \int_0^{\pi/3} (8 \cos^2 \theta - 2 \cos \theta - 1) \, d\theta$

$= a^2\left[4\theta + 2 \sin 2\theta - 2 \sin \theta - \theta\right]_0^{\pi/3} = a^2 \pi$

25.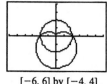

[−6, 6] by [−4, 4]

The area in question is half the circle minus two lobelike regions:

$\dfrac{1}{2}\pi(2)^2 - 2\int_0^{\pi/2} \dfrac{1}{2}[2(1 - \sin \theta)]^2 \, d\theta$

$= 2\pi - \int_0^{\pi/2} (4 - 8 \sin \theta + 4 \sin^2 \theta) \, d\theta$

$= 2\pi - 4\left[\theta + 2 \cos \theta + \dfrac{1}{2}\theta - \dfrac{1}{4} \sin 2\theta\right]_0^{\pi/2} = 8 - \pi$

27.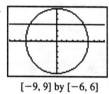

[−9, 9] by [−6, 6]

To find the integration limits, solve

$3 \csc \theta = 6$

$\theta = \dfrac{\pi}{6}, \dfrac{5\pi}{6}$. The area in question is

$\int_{\pi/6}^{5\pi/6} \dfrac{1}{2}(6^2 - 3^2 \csc^2 \theta) \, d\theta = \dfrac{1}{2}\left[36\theta + 9 \cot \theta\right]_{\pi/6}^{5\pi/6}$

$= 12\pi - 9\sqrt{3}$

29. (a) Find the area of the right half in two parts, then double the result: Right half area

$= \int_0^{\pi/4} \dfrac{1}{2} \tan^2 \theta \, d\theta + \int_{\pi/4}^{\pi/2} \dfrac{1}{2}\left(\dfrac{1}{2} \csc^2 \theta\right) d\theta$

$= \dfrac{1}{2}\left[\tan \theta - \theta\right]_0^{\pi/4} + \dfrac{1}{4}\left[-\cot \theta\right]_{\pi/4}^{\pi/2}$

$= \dfrac{1}{2}\left(1 - \dfrac{\pi}{4}\right) + \dfrac{1}{4}(0 + 1) = \dfrac{3}{4} - \dfrac{\pi}{8}.$

Total area is twice that, or $\dfrac{3}{2} - \dfrac{\pi}{4}.$

(b) Yes. $x = \tan \theta \cos \theta \Rightarrow x = \sin \theta$

$y = \tan \theta \sin \theta \Rightarrow y = \dfrac{\sin^2 \theta}{\cos \theta}$

$\lim_{\theta \to -\pi/2^+} x = -1, \quad \lim_{\theta \to -\pi/2^+} y = \infty$

$\lim_{\theta \to \pi/2^-} x = 1, \quad \lim_{\theta \to \pi/2^-} y = \infty$

31. $\frac{dr}{d\theta} = 2\theta$, so

$$\text{Length} = \int_0^{\sqrt{5}} \sqrt{(\theta^2)^2 + (2\theta)^2}\, d\theta$$
$$= \int_0^{\sqrt{5}} \theta\sqrt{\theta^2 + 4}\, d\theta$$
$$= \left[\frac{1}{3}(\theta^2 + 4)^{3/2}\right]_0^{\sqrt{5}}$$
$$= \frac{1}{3}(27 - 8) = \frac{19}{3}$$

33. $\frac{dr}{d\theta} = -\sin\theta$, so

$$\text{Length} = \int_0^{2\pi} \sqrt{(1+\cos\theta)^2 + (-\sin\theta)^2}\, d\theta$$
$$= \int_0^{2\pi} \sqrt{2 + 2\cos\theta}\, d\theta,$$
$$= \int_0^{2\pi} \sqrt{2 + 4\cos^2\left(\frac{\theta}{2}\right) - 2}\, d\theta$$
$$= \int_0^{2\pi} 2\left|\cos\left(\frac{\theta}{2}\right)\right|\, d\theta$$
$$= 4\int_0^{\pi} \cos\left(\frac{\theta}{2}\right)\, d\theta$$
$$= 8\left[\sin\left(\frac{\theta}{2}\right)\right]_0^{\pi} = 8$$

35. $\frac{dr}{d\theta} = \frac{6\sin\theta}{(1+\cos\theta)^2}$, so

$$\text{Length} = \int_0^{\pi/2} \sqrt{\frac{6^2}{(1+\cos\theta)^2} + \frac{6^2 \sin^2\theta}{(1+\cos\theta)^4}}\, d\theta, \text{ which using}$$

NINT evaluates to ≈ 6.887.

(Note: the integrand can simplify to $3\sec^3\left(\frac{\theta}{2}\right)$.)

37. $\frac{dr}{d\theta} = -\cos^2\left(\frac{\theta}{3}\right)\sin\left(\frac{\theta}{3}\right)$, so

$$\text{Length} = \int_0^{\pi/4} \sqrt{\cos^6\left(\frac{\theta}{3}\right) + \cos^4\left(\frac{\theta}{3}\right)\sin^2\left(\frac{\theta}{3}\right)}\, d\theta$$
$$= \int_0^{\pi/4} \sqrt{\cos^4\left(\frac{\theta}{3}\right)}\, d\theta$$
$$= \left[\frac{1}{2}\theta + \frac{3}{4}\sin\left(\frac{2\theta}{3}\right)\right]_0^{\pi/4} = \frac{\pi + 3}{8}.$$

39. $\frac{dr}{d\theta} = \frac{1}{2\sqrt{\cos 2\theta}}(-\sin 2\theta)(2) = -\frac{\sin 2\theta}{\sqrt{\cos 2\theta}}$, so

Surface area
$$= \int_0^{\pi/4} 2\pi\sqrt{\cos 2\theta}\cos\theta\sqrt{(\sqrt{\cos 2\theta})^2 + \left(-\frac{\sin 2\theta}{\sqrt{\cos 2\theta}}\right)^2}\, d\theta$$
$$= 2\pi\int_0^{\pi/4} \cos\theta\sqrt{\cos^2 2\theta + \sin^2 2\theta}\, d\theta$$
$$= 2\pi\int_0^{\pi/4} \cos\theta\, d\theta$$
$$= 2\pi\left[\sin\theta\right]_0^{\pi/4} = \pi\sqrt{2} \approx 4.443.$$

41.

[−1.5, 1.5] by [−1, 1]

$2r\frac{dr}{d\theta} = -2\sin 2\theta$

$\frac{dr}{d\theta} = \frac{-\sin 2\theta}{\sqrt{\cos 2\theta}}.$

Use the curve's symmetry and note that r is defined for $0 \le \theta \le \frac{\pi}{4}$: Surface area

$$= 2\int_0^{\pi/4} 2\pi\sqrt{\cos 2\theta}\sin\theta\sqrt{\cos 2\theta + \frac{\sin^2 2\theta}{\cos 2\theta}}\, d\theta$$
$$= 2\int_0^{\pi/4} 2\pi\sin\theta\, d\theta$$
$$= 4\pi\left[-\cos\theta\right]_0^{\pi/4} = (4 - 2\sqrt{2})\pi \approx 3.681$$

43. $\left(\frac{dx}{d\theta}\right)^2 + \left(\frac{dy}{d\theta}\right)^2$

$= (f'(\theta)\cos\theta - f(\theta)\sin\theta)^2$

$\quad + (f'(\theta)\sin\theta + f(\theta)\cos\theta)^2$

$= (f'(\theta)\cos\theta)^2 + (f(\theta)\sin\theta)^2 + (f'(\theta)\sin\theta)^2$

$\quad + (f(\theta)\cos\theta)^2$

$= (f(\theta))^2(\cos^2\theta + \sin^2\theta)$

$\quad + (f'(\theta))^2(\cos^2\theta + \sin^2\theta)$

$= (f(\theta))^2 + (f'(\theta))^2 = r^2 + \left(\frac{dr}{d\theta}\right)^2$

45. If $g(\theta) = 2f(\theta)$, then
$\sqrt{(g(\theta))^2 + g'(\theta)^2)} = 2\sqrt{(f(\theta)^2 + f'(\theta)^2)}$, so the length of g is 2 times the length of f.

47. (a) Let $r = 1.75 + \frac{0.06\theta}{2\pi}$.

(b) Since $\frac{dr}{d\theta} = \frac{b}{2\pi}$, this is just Equation 4 for the length of the curve.

(c) Using NINT, $\int_0^{80\pi} \sqrt{\left(1.75 + \frac{0.06\theta}{2\pi}\right)^2 + \left(\frac{0.06}{2\pi}\right)^2}\, d\theta$

evaluates to ≈ 741.420 cm ≈ 7.414 m.

(d) $\left(r^2 + \left(\frac{b}{2\pi}\right)^2\right)^{1/2} = r\left(1 + \left(\frac{b}{2\pi r}\right)^2\right)^{1/2} \approx r$

since $\left(\frac{b}{2\pi r}\right)^2$ is a very small quantity squared.

(e) $L \approx 741.420$ cm (from part (c)),

$L_a = \int_0^{80\pi} \left(1.75 + \frac{0.06\theta}{2\pi}\right)\, d\theta$

$= \left[1.75\theta + \frac{0.03\theta^2}{2\pi}\right]_0^{80\pi} = 236\pi \approx 741.416$ cm

49. $\frac{2}{3}\int_0^{2\pi} a^3(1 + \cos\theta)^3 \cos\theta \, d\theta$

$= \frac{2}{3}a^3 \int_0^{2\pi} (\cos\theta + 3\cos^2\theta + 3\cos^3\theta + \cos^4\theta) \, d\theta$

$= \frac{2}{3}a^3 \left[3\sin\theta + \frac{15}{8}\sin\theta\cos\theta + \frac{15}{8}\theta + \cos^2\theta\sin\theta \right.$

$\left. + \frac{1}{4}\cos^3\theta\sin\theta \right]_0^{2\pi}$

$= \frac{5}{2}\pi a^3$, and $\int_0^{2\pi} a^2(1 + \cos\theta)^2 \, d\theta$

$= a^2 \int_0^{2\pi} (1 + 2\cos\theta + \cos^2\theta) \, d\theta$

$= a^2 \left[\frac{3}{2}\theta + 2\sin\theta + \frac{1}{4}\sin 2\theta\right]_0^{2\pi} = 3\pi a^2.$

So $\bar{x} = \dfrac{\frac{5}{2}\pi a^3}{3\pi a^2} = \dfrac{5a}{6}.$

By symmetry, $\bar{y} = 0$, so the centroid is $\left(\dfrac{5a}{6}, 0\right)$.

Chapter 10 Review Exercises
(pp. 569–572)

1. (a) $3\langle -3, 4\rangle - 4\langle 2, -5\rangle = \langle -9 - 8, 12 + 20\rangle$
   $= \langle -17, 32\rangle$
   (b) $\sqrt{17^2 + 32^2} = \sqrt{1313}$

2. (a) $\langle -3 + 2, 4 - 5\rangle = \langle -1, -1\rangle$
   (b) $\sqrt{1^2 + 1^2} = \sqrt{2}$

3. (a) $\langle -2(-3), -2(4)\rangle = \langle 6, -8\rangle$
   (b) $\sqrt{6^2 + 8^2} = 10$

4. (a) $\langle 5(2), 5(-5)\rangle = \langle 10, -25\rangle$
   (b) $\sqrt{10^2 + 25^2} = \sqrt{725} = 5\sqrt{29}$

5. $\dfrac{\pi}{6}$ radians below the negative x-axis: $\left\langle -\dfrac{\sqrt{3}}{2}, -\dfrac{1}{2}\right\rangle$
   [assuming counterclockwise].

6. $\left\langle \dfrac{\sqrt{3}}{2}, \dfrac{1}{2}\right\rangle$

7. $2\left(\dfrac{1}{\sqrt{4^2 + 1^2}}\right)\langle 4, -1\rangle = \left\langle \dfrac{8}{\sqrt{17}}, -\dfrac{2}{\sqrt{17}}\right\rangle$

8. $-5\left(\dfrac{1}{\sqrt{(3/5)^2 + (4/5)^2}}\right)\left\langle \dfrac{3}{5}, \dfrac{4}{5}\right\rangle = \langle -3, -4\rangle$

9. (a) $\dfrac{dy}{dx} = \dfrac{dy/dt}{dx/dt} = \dfrac{(1/2)\sec t \tan t}{(1/2)\sec^2 t} = \sin t$

   For $t = \dfrac{\pi}{3}$; $x = \dfrac{\sqrt{3}}{2}$, $y = 1$, and $\dfrac{dy}{dx} = \dfrac{\sqrt{3}}{2}$. So the tangent line is $y - 1 = \dfrac{\sqrt{3}}{2}\left(x - \dfrac{\sqrt{3}}{2}\right)$ or

   $y = \dfrac{\sqrt{3}}{2}x + \dfrac{1}{4}.$

   (b) $\dfrac{d^2y}{dx^2} = \dfrac{dy'}{dx} = \dfrac{dy'/dt}{dx/dt} = \dfrac{\cos t}{(1/2)\sec^2 t} = 2\cos^3 t,$
   which for $t = \dfrac{\pi}{3}$ equals $\dfrac{1}{4}.$

10. (a) $\dfrac{dy}{dx} = \dfrac{dy/dt}{dx/dt} = \dfrac{3/t^2}{-2/t^3} = -\dfrac{3}{2}t$
    For $t = 2$: $x = \dfrac{5}{4}$, $y = -\dfrac{1}{2}$, and $\dfrac{dy}{dx} = -3$.
    So the tangent line is $y + \dfrac{1}{2} = -3\left(x - \dfrac{5}{4}\right)$ or
    $y = -3x + \dfrac{13}{4}.$

    (b) $\dfrac{d^2y}{dx^2} = \dfrac{dy'}{dx} = \dfrac{dy'/dt}{dx/dt} = \dfrac{-3/2}{-2/t^3} = \dfrac{3t^3}{4}$, which for $t = 2$ equals 6.

11. $\dfrac{dy}{dt} = \dfrac{1}{2}\tan t \sec t$ equals zero for $t = k\pi$, where k is any integer. $\dfrac{dx}{dt} = \dfrac{1}{2}\sec^2 t$ never equals zero.

    (a) Horizontal tangents at $\left(\dfrac{1}{2}\tan 0, \dfrac{1}{2}\sec 0\right) = \left(0, \dfrac{1}{2}\right)$ and
    $\left(\dfrac{1}{2}\tan\pi, \dfrac{1}{2}\sec\pi\right) = \left(0, -\dfrac{1}{2}\right).$

    (b) There are no vertical tangents, since $\dfrac{dx}{dt}$ never equals zero.

12. $\dfrac{dy}{dt} = 2\cos t$ equals zero for $\theta = \dfrac{k\pi}{2}$, where k is any odd integer. $\dfrac{dx}{dt} = 2\sin t$ equals zero for $t = k\pi$, where k is any integer.

    (a) Horizontal tangents at $\left(-2\cos\dfrac{\pi}{2}, 2\sin\dfrac{\pi}{2}\right) = (0, 2)$
    and $\left(-2\cos\dfrac{3\pi}{2}, 2\sin\dfrac{3\pi}{2}\right) = (0, -2).$

    (b) Vertical tangents at $(-2\cos 0, 2\sin 0) = (-2, 0)$ and $(-2\cos\pi, 2\sin\pi) = (2, 0).$

13. $\dfrac{dy}{dt} = -2\sin t \cos t = -\sin 2t$ equals zero for $t = \dfrac{k\pi}{2}$, where k is any integer. $\dfrac{dx}{dt} = \sin t$ equals zero for $t = k\pi$, where k is any integer. Where they are both zero, use L'Hôpital's rule:

    $\lim_{t\to k\pi}\dfrac{dy/dt}{dx/dt} = \lim_{t\to k\pi}\dfrac{-\sin 2t}{\sin t} = \lim_{t\to k\pi}\dfrac{-2\cos 2t}{\cos t} = \pm 2.$

    (a) Horizontal tangent at $\left(-\cos\dfrac{\pi}{2}, \cos^2\dfrac{\pi}{2}\right) = (0, 0).$

    (b) There are no vertical tangents.

14. $\dfrac{dy}{dt} = 9\cos t$ equals zero for $t = \dfrac{k\pi}{2}$, where k is any odd integer. $\dfrac{dx}{dt} = -4\sin t$ equals zero for $t = k\pi$, where k is any integer.

    (a) Horizontal tangents at $\left(4\cos\dfrac{\pi}{2}, 9\sin\dfrac{\pi}{2}\right) = (0, 9)$ and
    $\left(4\cos\dfrac{3\pi}{2}, 9\sin\dfrac{3\pi}{2}\right) = (0, -9).$

    (b) Vertical tangents at $(4\cos 0, 9\sin 0) = (4, 0)$ and $(4\cos\pi, 9\sin\pi) = (-4, 0).$

15.

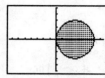

[−7.5, 7.5] by [−5, 5]

16.

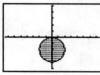

[−7.5, 7.5] by [−5, 5]

17. (a)

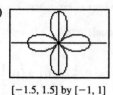

[−1.5, 1.5] by [−1, 1]

(b) 2π

18. (a)

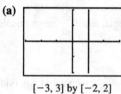

[−3, 3] by [−2, 2]

(b) π

19. (a)

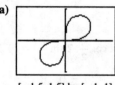

[−1.5, 1.5] by [−1, 1]

(b) $\dfrac{\pi}{2}$

20. (a)

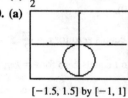

[−1.5, 1.5] by [−1, 1]

(b) π

21. $\dfrac{dy}{dx} = \dfrac{f'(\theta)\sin\theta + f(\theta)\cos\theta}{f'(\theta)\cos\theta - f(\theta)\sin\theta}$

$= \dfrac{-2\sin 2\theta \sin\theta + \cos 2\theta \cos\theta}{-2\sin 2\theta \cos\theta - \cos 2\theta \sin\theta}$

$\left(0, \dfrac{\pi}{4}\right), \left(0, \dfrac{3\pi}{4}\right), \left(0, \dfrac{5\pi}{4}\right)$ and $\left(0, \dfrac{7\pi}{4}\right)$ are polar solutions.

$\left.\dfrac{dy}{dx}\right|_{\theta=\pi/4} = \dfrac{-2/\sqrt{2}}{-2/\sqrt{2}} = 1, \left.\dfrac{dy}{dx}\right|_{\theta=3\pi/4} = \dfrac{2/\sqrt{2}}{-2/\sqrt{2}} = -1,$

$\left.\dfrac{dy}{dx}\right|_{\theta=5\pi/4} = \dfrac{2/\sqrt{2}}{2/\sqrt{2}} = 1, \left.\dfrac{dy}{dx}\right|_{\theta=7\pi/4} = \dfrac{-2/\sqrt{2}}{2/\sqrt{2}} = -1.$

The Cartesian equations are $y = \pm x$.

22. $\dfrac{dy}{dx} = \dfrac{f'(\theta)\sin\theta + f(\theta)\cos\theta}{f'(\theta)\cos\theta - f(\theta)\sin\theta}$

$= \dfrac{-2\sin 2\theta \sin\theta + (1+\cos 2\theta)\cos\theta}{-2\sin 2\theta \cos\theta - (1+\cos 2\theta)\sin\theta}$

$= \dfrac{-4\sin^2\theta\cos\theta + \cos\theta + 2\cos^3\theta - \cos\theta}{-4\cos^2\theta\sin\theta - \sin\theta - 2\cos^2\theta\sin\theta + \sin\theta}$

$= \dfrac{-4\sin^2\theta + 2\cos^2\theta}{-6\cos\theta\sin\theta}$

$= \dfrac{4\sin^2\theta - 2\cos^2\theta}{3\sin 2\theta}.$

$\left(0, \dfrac{\pi}{2}\right)$ and $\left(0, \dfrac{3\pi}{2}\right)$ are polar solutions.

$\left.\dfrac{dy}{dx}\right|_{\theta=\pi/2} = \left.\dfrac{dy}{dx}\right|_{\theta=3\pi/2} = \dfrac{4}{0}$ is undefined, so the tangent lines are vertical with equation $x = 0$.

23. $\dfrac{dy}{d\theta} = \dfrac{d}{d\theta}\left[\left(1 - \cos\left(\dfrac{\theta}{2}\right)\right)\sin\theta\right]$

$= \dfrac{1}{2}\sin\left(\dfrac{\theta}{2}\right)\sin\theta + \cos\theta - \cos\left(\dfrac{\theta}{2}\right)\cos\theta$

$\dfrac{dx}{d\theta} = \dfrac{d}{d\theta}\left[\left(1 - \cos\left(\dfrac{\theta}{2}\right)\right)\cos\theta\right]$

$= \dfrac{1}{2}\sin\left(\dfrac{\theta}{2}\right)\cos\theta - \sin\theta + \cos\left(\dfrac{\theta}{2}\right)\sin\theta$

Solve $\dfrac{dy}{d\theta} = 0$ for θ with a graphing calculator: the solutions are 0, ≈ 2.243, ≈ 4.892, ≈ 7.675, ≈ 10.323, and 4π. Using the middle four solutions to $y = r\sin\theta$ reveals horizontal tangent lines at $y \approx \pm 0.443$ and $y \approx \pm 1.739$.

Solve $\dfrac{dx}{d\theta} = 0$ for θ with a graphing calculator: the solutions are 0, ≈ 1.070, ≈ 3.531, 2π, ≈ 9.035, ≈ 11.497, and 4π.

Using the middle five solutions to find $x = r\cos\theta$ reveals vertical tangent lines at $x = 2$, $x \approx 0.067$, and $x \approx -1.104$.

Where $\dfrac{dy}{dt}$ and $\dfrac{dx}{dt}$ both equal zero ($\theta = 0, 4\pi$), close inspection of the plot shows that the tangent lines are horizontal, with equation $y = 0$. (This can be confirmed using L'Hôpital's rule.)

24. $\dfrac{dy}{d\theta} = \dfrac{d}{d\theta}[2(1 - \sin\theta)\sin\theta] = -4\sin\theta\cos\theta + 2\cos\theta$

$\dfrac{dx}{d\theta} = \dfrac{d}{d\theta}[2(1 - \sin\theta)\cos\theta]$

$= -2\cos^2\theta - 2\sin\theta + 2\sin^2\theta$

$= 4\sin^2\theta - 2\sin\theta - 2$

Solve $\dfrac{dy}{d\theta} = 0$ for θ:

the solutions are $\dfrac{\pi}{6}, \dfrac{\pi}{2}, \dfrac{5\pi}{6}$, and $\dfrac{3\pi}{2}$.

Using the first, third, and fourth solutions to find $y = r\sin\theta$ reveals horizontal tangent lines at $y = \dfrac{1}{2}$ and $y = -4$.

Solve $\dfrac{dx}{d\theta} = 0$ for θ (by first using the quadratic formula to find $\sin\theta$): the solutions are $\dfrac{\pi}{2}, \dfrac{7\pi}{6}$, and $\dfrac{11\pi}{6}$. Using the last two solutions to find $x = r\cos\theta$ reveals vertical tangent lines at $x = \pm\dfrac{3\sqrt{3}}{2} \approx \pm 2.598$. Where $\dfrac{dy}{dt}$ and $\dfrac{dx}{dt}$ both equal zero $\left(\theta = \dfrac{\pi}{2}\right)$, inspection of the plot shows that the tangent line is vertical, with equation $x = 0$. (This can be confirmed using L'Hôpital's rule.)

25.

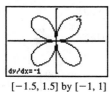

[−1.5, 1.5] by [−1, 1]

The tips have Cartesian coordinates $\left(\dfrac{1}{\sqrt{2}}, \dfrac{1}{\sqrt{2}}\right)$, $\left(-\dfrac{1}{\sqrt{2}}, \dfrac{1}{\sqrt{2}}\right), \left(-\dfrac{1}{\sqrt{2}}, -\dfrac{1}{\sqrt{2}}\right)$, and $\left(\dfrac{1}{\sqrt{2}}, -\dfrac{1}{\sqrt{2}}\right)$. From the curve's symmetries, it is evident that the tangent lines at those points have slopes of $-1, 1, -1$, and 1, respectively.

So the equations of the tangent lines are

$y - \dfrac{1}{\sqrt{2}} = -\left(x - \dfrac{1}{\sqrt{2}}\right)$ or

$y = -x + \sqrt{2}$,

$y - \dfrac{1}{\sqrt{2}} = x + \dfrac{1}{\sqrt{2}}$ or

$y = x + \sqrt{2}$,

$y + \dfrac{1}{\sqrt{2}} = -\left(x + \dfrac{1}{\sqrt{2}}\right)$ or

$y = -x - \sqrt{2}$, and

$y + \dfrac{1}{\sqrt{2}} = x - \dfrac{1}{\sqrt{2}}$ or

$y = x - \sqrt{2}$

26.

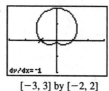

[−3, 3] by [−2, 2]

As the plot shows, the curve crosses the x-axis at (x, y)-coordinates $(-1, 0)$ and $(1, 0)$, with slope -1 and 1, respectively. (This can be confirmed analytically.) So the equations of the tangent lines are

$y - 0 = -(x + 1)$

$y = -x - 1$ and

$y - 0 = x - 1$

$y = x - 1$.

27. $r\cos\theta = r\sin\theta$

$x = y$, a line

28. $r = 3\cos\theta$

$r^2 = 3r\cos\theta$

$x^2 + y^2 = 3x$

$x^2 - 3x + \dfrac{9}{4} + y^2 = \dfrac{9}{4}$

$\left(x - \dfrac{3}{2}\right)^2 + y^2 = \left(\dfrac{3}{2}\right)^2$

a circle $\left(\text{center} = \left(\dfrac{3}{2}, 0\right), \text{radius} = \dfrac{3}{2}\right)$

29. $r = 4\tan\theta\sec\theta$

$r\cos\theta = 4\dfrac{r\sin\theta}{r\cos\theta}$

$x = 4\dfrac{y}{x}$ or $x^2 = 4y$, a parabola

30. $r\cos\left(\theta + \dfrac{\pi}{3}\right) = 2\sqrt{3}$

$r\cos\theta\cos\left(\dfrac{\pi}{3}\right) - r\sin\theta\sin\left(\dfrac{\pi}{3}\right) = 2\sqrt{3}$

$\dfrac{1}{2}r\cos\theta - \dfrac{\sqrt{3}}{2}r\sin\theta = 2\sqrt{3}$

$\dfrac{1}{2}x - \dfrac{\sqrt{3}}{2}y = 2\sqrt{3}$

$x - \sqrt{3}y = 4\sqrt{3}$ or $y = \dfrac{x}{\sqrt{3}} - 4$, a line

31. $x^2 + y^2 + 5y = 0$

$r^2 + 5r\sin\theta = 0$

$r = -5\sin\theta$

32. $x^2 + y^2 - 2y = 0$

$r^2 - 2r\sin\theta = 0$

$r = 2\sin\theta$

33. $x^2 + 4y^2 = 16$

$(r\cos\theta)^2 + 4(r\sin\theta)^2 = 16$

$r^2\cos^2\theta + 4r^2\sin^2\theta = 16$, or $r^2 = \dfrac{16}{\cos^2\theta + 4\sin^2\theta}$

34. $(x + 2)^2 + (y - 5)^2 = 16$

$(r\cos\theta + 2)^2 + (r\sin\theta - 5)^2 = 16$

35. $\frac{dx}{dt} = 2e^{2t} - \frac{1}{8}, \frac{dy}{dt} = e^t$, so

$\text{Length} = \int_0^{\ln 2} \sqrt{\left(2e^{2t} - \frac{1}{8}\right)^2 + (e^t)^2}\, dt$

$= \int_0^{\ln 2} \sqrt{4e^{4t} - \frac{1}{2}e^{2t} + \frac{1}{64} + e^{2t}}\, dt$

$= \int_0^{\ln 2} \sqrt{\left(2e^{2t} + \frac{1}{8}\right)^2}\, dt$

$= \left[e^{2t} + \frac{t}{8}\right]_0^{\ln 2}$

$= 4 + \frac{\ln 2}{8} - 1$

$= 3 + \frac{\ln 2}{8} = \frac{\ln 2 + 24}{8} \approx 3.087$.

36. $\frac{dx}{dt} = 2t, \frac{dy}{dt} = t^2 - 1$, so

$\text{Length} = \int_{-\sqrt{3}}^{\sqrt{3}} \sqrt{(2t)^2 + (t^2 - 1)^2}\, dt$

$= \int_{-\sqrt{3}}^{\sqrt{3}} \sqrt{4t^2 + t^4 - 2t^2 + 1}\, dt$

$= \int_{-\sqrt{3}}^{\sqrt{3}} \sqrt{(t^2 + 1)^2}\, dt$

$= \left[\left(\frac{t^3}{3}\right) + t\right]_{-\sqrt{3}}^{\sqrt{3}} = 4\sqrt{3}$.

37. $\frac{dr}{d\theta} = -\sin\theta$, so

$\text{Length} = \int_0^{2\pi} \sqrt{(-1 + \cos\theta)^2 + (-\sin\theta)^2}\, d\theta$

$= \int_0^{2\pi} \sqrt{2 - 2\cos\theta}\, d\theta$

$= \int_0^{2\pi} \sqrt{4\sin^2\frac{\theta}{2}}\, d\theta = \left[-4\cos\frac{\theta}{2}\right]_0^{2\pi} = 8$.

38. $\frac{dr}{d\theta} = 2\cos\theta - 2\sin\theta$, so Length

$= \int_0^{\pi/2} \sqrt{(2\sin\theta + 2\cos\theta)^2 + (2\cos\theta - 2\sin\theta)^2}\, d\theta$

$= \int_0^{\pi/2} \sqrt{8\sin^2\theta + 8\cos^2\theta}\, d\theta$

$= \int_0^{\pi/2} 2\sqrt{2}\, d\theta = \pi\sqrt{2}$.

39. $\frac{dr}{d\theta} = 8\sin^2\left(\frac{\theta}{3}\right)\cos\left(\frac{\theta}{3}\right)$, so

$\text{Length} = \int_0^{\pi/4} \sqrt{\left(8\sin^3\left(\frac{\theta}{3}\right)\right)^2 + \left(8\sin^2\left(\frac{\theta}{3}\right)\cos\left(\frac{\theta}{3}\right)\right)^2}\, d\theta$

$= \int_0^{\pi/4} 8\sin^2\left(\frac{\theta}{3}\right)\sqrt{\sin^2\left(\frac{\theta}{3}\right) + \cos^2\left(\frac{\theta}{3}\right)}\, d\theta$

$= \int_0^{\pi/4} 8\sin^2\left(\frac{\theta}{3}\right)\, d\theta$

$= 8\left[\frac{1}{2}\theta - \frac{3}{4}\sin\left(\frac{2\theta}{3}\right)\right]_0^{\pi/4} = \pi - 3$.

40. $\frac{dr}{d\theta} = \frac{-\sin 2\theta}{\sqrt{1 + \cos 2\theta}}$, so Length

$= \int_{-\pi/2}^{\pi/2} \sqrt{(\sqrt{1 + \cos 2\theta})^2 + \left(\frac{-\sin 2\theta}{\sqrt{1 + \cos 2\theta}}\right)^2}\, d\theta$

$= \int_{-\pi/2}^{\pi/2} \sqrt{\frac{(1 + \cos 2\theta)^2}{1 + \cos 2\theta} + \frac{\sin^2 2\theta}{1 + \cos 2\theta}}\, d\theta$

$= \int_{-\pi/2}^{\pi/2} \sqrt{\frac{1 + 2\cos 2\theta + \cos^2 2\theta + \sin^2 2\theta}{1 + \cos 2\theta}}\, d\theta$

$= \int_{-\pi/2}^{\pi/2} \sqrt{2}\, d\theta = \pi\sqrt{2}$.

41. $\frac{dx}{dt} = -2\sin t, \frac{dy}{dt} = 2t$, so

$\text{Length} = \int_0^{\pi/2} \sqrt{(-2\sin t)^2 + (2t)^2}\, dt$

$= \int_0^{\pi/2} 2\sqrt{t^2 + \sin^2 t}\, dt$,

which using NINT evaluates to ≈ 3.183.

42. $\frac{dx}{dt} = 3\cos t, \frac{dy}{dt} = 3\sqrt{t}$, so

$\text{Length} = \int_0^3 \sqrt{(3\cos t)^2 + (3\sqrt{t})^2}\, dt = \int_0^3 3\sqrt{t + \cos^2 t}\, dt$,

which using NINT evaluates to ≈ 12.363.

43. $\text{Area} = \int_0^{2\pi} \frac{1}{2}(2 - \cos\theta)^2\, d\theta$

$= \frac{1}{2}\int_0^{2\pi} (4 - 4\cos\theta + \cos^2\theta)\, d\theta$

$= \frac{1}{2}\left[4\theta - 4\sin\theta + \frac{1}{2}\theta + \frac{1}{4}\sin 2\theta\right]_0^{2\pi} = \frac{9\pi}{2}$

44. $\text{Area} = \int_0^{\pi/3} \frac{1}{2}\sin^2 3\theta\, d\theta = \frac{1}{2}\left[\frac{1}{2}\theta - \frac{1}{12}\sin(6\theta)\right]_0^{\pi/3} = \frac{\pi}{12}$

45.

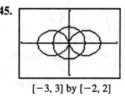

[-3, 3] by [-2, 2]

The curves cross where $\cos 2\theta = 0$, such as $\theta = \frac{\pi}{4}$. Using the curves' symmetry,

$\text{Length} = 4\int_0^{\pi/4} \frac{1}{2}[(1 + \cos 2\theta)^2 - 1]\, d\theta$

$= 2\int_0^{\pi/4} (\cos^2 2\theta + 2\cos 2\theta)\, d\theta$

$= 2\left[\frac{1}{8}\sin 4\theta + \frac{1}{2}\theta + \sin 2\theta\right]_0^{\pi/4}$

$= \frac{\pi}{4} + 2$

46.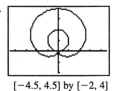

[−4.5, 4.5] by [−2, 4]

Since the two curves are covered over different θ-intervals, find the two areas separately. Then

$$\text{Area} = \int_0^{2\pi} \frac{1}{2}[2(1+\sin\theta)]^2\, d\theta - \pi r^2$$
$$= 2\int_0^{2\pi}(1 + 2\sin\theta + \sin^2\theta)\, d\theta - \pi$$
$$= 2\left[\theta - 2\cos\theta + \frac{1}{2}\theta - \frac{1}{4}\sin 2\theta\right]_0^{2\pi} - \pi = 5\pi$$

47. $\frac{dx}{dt} = t$, $\frac{dy}{dt} = 2$, so

$$\text{Area} = \int_0^{\sqrt{5}} 2\pi(2t)\sqrt{t^2 + 2^2}\, dt$$
$$= \left[\frac{4\pi}{3}(t^2+4)^{3/2}\right]_0^{\sqrt{5}} = \frac{76\pi}{3}$$

48. $\frac{dx}{dt} = 2t - \frac{1}{2t^2}$, $\frac{dy}{dt} = 4$, so

$$\text{Area} = \int_{1/\sqrt{2}}^{1} 2\pi\left(t^2 + \frac{1}{2t}\right)\sqrt{\left(2t - \frac{1}{2t^2}\right)^2 + 4^2}\, dt,$$

which using NINT evaluates to ≈ 10.110.

49. $\frac{dr}{d\theta} = \frac{-\sin 2\theta}{\sqrt{\cos 2\theta}}$, so

$$\text{Area} = \int_0^{\pi/4} 2\pi\sqrt{\cos 2\theta}\,\sin\theta\,\sqrt{\cos 2\theta + \frac{\sin^2 2\theta}{\cos 2\theta}}\, d\theta$$
$$= \int_0^{\pi/4} 2\pi\sin\theta\, d\theta$$
$$= 2\pi\Big[-\cos\theta\Big]_0^{\pi/4} = \pi(2 - \sqrt{2}) \approx 1.840$$

50.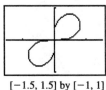

[−1.5, 1.5] by [−1, 1]

$r = \pm\sqrt{\sin 2\theta}$ and $\frac{dr}{d\theta} = \pm\frac{\cos 2\theta}{\sqrt{\sin 2\theta}}$, where $0 \leq \theta \leq \frac{\pi}{2}$, so

$$\text{Area} = 2\int_0^{\pi/2} 2\pi\sqrt{\sin 2\theta}\,\cos\theta\,\sqrt{\sin 2\theta + \frac{\cos^2 2\theta}{\sin 2\theta}}\, d\theta$$
$$= 4\pi\int_0^{\pi/2}\cos\theta\, d\theta$$
$$= 4\pi\Big[\sin\theta\Big]_0^{\pi/2} = 4\pi$$

51. (a) $\mathbf{v}(t) = \frac{d}{dt}[(4\cos t)\mathbf{i} + (\sqrt{2}\sin t)\mathbf{j}]$
$= (-4\sin t)\mathbf{i} + (\sqrt{2}\cos t)\mathbf{j}$
$\mathbf{a}(t) = \frac{d}{dt}[(-4\sin t)\mathbf{i} + (\sqrt{2}\cos t)\mathbf{j}]$
$= (-4\cos t)\mathbf{i} + (-\sqrt{2}\sin t)\mathbf{j}$

(b) $\left|\mathbf{v}\left(\frac{\pi}{4}\right)\right| = \sqrt{\left(-4\sin\frac{\pi}{4}\right)^2 + \left(\sqrt{2}\cos\frac{\pi}{4}\right)^2}$
$= \sqrt{8 + 1} = 3$

(c) At $t = \frac{\pi}{4}$, $\mathbf{v} = -2\sqrt{2}\mathbf{i} + \mathbf{j}$, $\mathbf{a} = -2\sqrt{2}\mathbf{i} - \mathbf{j}$, and
$\theta = \cos^{-1}\frac{\mathbf{v}\cdot\mathbf{a}}{|\mathbf{v}||\mathbf{a}|}$
$= \cos^{-1}\frac{8-1}{(3)(3)}$
$= \cos^{-1}\frac{7}{9} \approx 38.94°$.

52. (a) $\mathbf{v}(t) = \frac{d}{dt}[(\sqrt{3}\sec t)\mathbf{i} + (\sqrt{3}\tan t)\mathbf{j}]$
$= (\sqrt{3}\sec t\tan t)\mathbf{i} + (\sqrt{3}\sec^2 t)\mathbf{j}$
$\mathbf{a}(t) = \frac{d}{dt}[(\sqrt{3}\sec t\tan t)\mathbf{i} + (\sqrt{3}\sec^2 t)\mathbf{j}]$
$= \sqrt{3}(\sec t\tan^2 t + \sec^3 t)\mathbf{i} + (2\sqrt{3}\sec^2 t\tan t)\mathbf{j}$

(b) $|\mathbf{v}(0)| = \sqrt{3\sec^2 0\tan^2 0 + 3\sec^4 0} = \sqrt{0 + 3} = \sqrt{3}$

(c) At $t = 0$, $\mathbf{v} = \sqrt{3}\mathbf{j}$, $\mathbf{a} = \sqrt{3}\mathbf{i}$
$\theta = \cos^{-1}\frac{\mathbf{v}\cdot\mathbf{a}}{|\mathbf{v}||\mathbf{a}|} = \frac{0+0}{(\sqrt{3})(\sqrt{3})} = \cos^{-1} 0 = 90°$

53. $\mathbf{v}(t) = -\frac{t}{(1+t^2)^{3/2}}\mathbf{i} + \frac{1}{(1+t^2)^{3/2}}\mathbf{j}$

$\left|\frac{d\mathbf{r}}{dt}\right| = |\mathbf{v}(t)| = \sqrt{\left(-\frac{t}{(1+t^2)^{3/2}}\right)^2 + \left(\frac{1}{(1+t^2)^{3/2}}\right)^2} = \frac{1}{1+t^2}$,

which is at a maximum of 1 when $t = 0$.

54. $\mathbf{v}(t) = \frac{d\mathbf{r}}{dt} = (e^t\cos t - e^t\sin t)\mathbf{i} + (e^t\sin t + e^t\cos t)\mathbf{j}$

$\mathbf{a}(t) = \frac{d\mathbf{x}}{dt} = (e^t\cos t - e^t\sin t - e^t\sin t - e^t\cos t)\mathbf{i}$
$+ (e^t\sin t + e^t\cos t + e^t\cos t - e^t\sin t)\mathbf{j}$
$= (-2e^t\sin t)\mathbf{i} + (2e^t\cos t)\mathbf{j}$

$\mathbf{r}(t)\cdot\mathbf{a}(t) = (e^t\cos t)(-2e^t\sin t) + (e^t\sin t)(2e^t\cos t) = 0$

for all t. The angle between $\mathbf{r}$ and $\mathbf{a}$ is always 90°.

55. $\left(\int_0^1 (3 + 6t)\, dt\right)\mathbf{i} + \left(\int_0^1 6\pi\cos\pi t\, dt\right)\mathbf{j}$
$= \Big[3t + 3t^2\Big]_0^1 \mathbf{i} + \Big[6\sin\pi t\Big]_0^1 \mathbf{j} = 6\mathbf{i}$

56. $\left(\int_e^{e^2} \frac{2\ln t}{t} dt\right)\mathbf{i} + \left(\int_e^{e^2} \frac{1}{t \ln t} dt\right)\mathbf{j}$

$= \left[\ln^2 t\right]_e^{e^2} \mathbf{i} + \left[\ln(\ln t)\right]_e^{e^2} \mathbf{j} = 3\mathbf{i} + (\ln 2)\mathbf{j}$

57. $\mathbf{r}(t) = \int \frac{d\mathbf{r}}{dt} dt = (\cos t)\mathbf{i} + (\sin t)\mathbf{j} + \mathbf{C}$

$\mathbf{r}(0) = \mathbf{i} + \mathbf{C} = \mathbf{j}$, so $\mathbf{C} = -\mathbf{i} + \mathbf{j}$, and

$\mathbf{r}(t) = (\cos t - 1)\mathbf{i} + (\sin t + 1)\mathbf{j}$

58. $\mathbf{r}(t) = \int \frac{d\mathbf{r}}{dt} dt = (\tan^{-1} t)\mathbf{i} + \sqrt{t^2+1}\,\mathbf{j} + \mathbf{C}$

$\mathbf{r}(0) = \mathbf{j} + \mathbf{C} = \mathbf{i} + \mathbf{j}$, so $\mathbf{C} = \mathbf{i}$ and

$\mathbf{r}(t) = (\tan^{-1} t + 1)\mathbf{i} + \sqrt{t^2+1}\,\mathbf{j}$

59. $\frac{d\mathbf{r}}{dt} = \int \frac{d^2\mathbf{r}}{dt^2} dt = 2t\mathbf{j} + \mathbf{C}_1$, $\mathbf{r}(t) = \int \frac{d\mathbf{r}}{dt} dt = t^2\mathbf{j} + \mathbf{C}_1 t + \mathbf{C}_2$

$\left.\frac{d\mathbf{r}}{dt}\right|_{t=0} = \mathbf{C}_1 = \mathbf{0}$, so $\mathbf{r}(t) = t^2\mathbf{j} + \mathbf{C}_2$. And $\mathbf{r}(0) = \mathbf{C}_2 = \mathbf{i}$, so

$\mathbf{r}(t) = \mathbf{i} + t^2\mathbf{j}$

60. $\frac{d\mathbf{r}}{dt} = \int \frac{d^2\mathbf{r}}{dt^2} dt = (-2t)\mathbf{i} + (-2t)\mathbf{j} + \mathbf{C}_1$,

$\mathbf{r}(t) = \int \frac{d\mathbf{r}}{dt} dt = -t^2\mathbf{i} - t^2\mathbf{j} + \mathbf{C}_1 t + \mathbf{C}_2$

$\left.\frac{d\mathbf{r}}{dt}\right|_{t=1} = -2\mathbf{i} - 2\mathbf{j} + \mathbf{C}_1 = 4\mathbf{i}$, so $\mathbf{C}_1 = 6\mathbf{i} + 2\mathbf{j}$ and

$\mathbf{r}(t) = (-t^2 + 6t)\mathbf{i} + (-t^2 + 2t)\mathbf{j} + \mathbf{C}_2$

$\mathbf{r}(1) = 5\mathbf{i} + \mathbf{j} + \mathbf{C}_2 = 3\mathbf{i} + 3\mathbf{j}$, so $\mathbf{C}_2 = -2\mathbf{i} + 2\mathbf{j}$, and

$\mathbf{r}(t) = (-t^2 + 6t - 2)\mathbf{i} + (-t^2 + 2t + 2)\mathbf{j}$

61. (a) $\mathbf{v}(t) = \left\langle \frac{dx}{dt}, \frac{dy}{dt}\right\rangle = \left\langle -\frac{3\pi}{4}\sin\frac{\pi}{4}t, \frac{5\pi}{4}\cos\frac{\pi}{4}t\right\rangle$,

$\mathbf{v}(3) = \left\langle -\frac{3\pi}{4\sqrt{2}}, -\frac{5\pi}{4\sqrt{2}}\right\rangle$, and

$|\mathbf{v}(3)| = \sqrt{\frac{9\pi^2}{32} + \frac{25\pi^2}{32}} = \frac{\pi\sqrt{34}}{4\sqrt{2}} = \frac{\pi\sqrt{17}}{4} \approx 3.238$

(b) x-component: $\left.\frac{d^2 x}{dt^2}\right|_{t=3} = -\frac{3\pi^2}{16}\cos\left(\frac{\pi}{4}\cdot 3\right) = \frac{3\pi^2}{16\sqrt{2}}$

y-component: $\left.\frac{d^2 y}{dt^2}\right|_{t=3} = -\frac{5\pi^2}{16}\sin\left(\frac{\pi}{4}\cdot 3\right) = -\frac{5\pi^2}{16\sqrt{2}}$

(c) $\frac{x}{3} = \cos\frac{\pi}{4}t$ and $\frac{y}{5} = \sin\frac{\pi}{4}t$, so $\left(\frac{x}{3}\right)^2 + \left(\frac{y}{5}\right)^2 = 1$ or

$\frac{x^2}{9} + \frac{y^2}{25} = 1$.

62. (a) $\frac{dx}{dt} = \frac{1}{2}$ and $\frac{dy}{dt} = 5 - t$ so

Length $= \int_0^{10} \sqrt{\left(\frac{1}{2}\right)^2 + (5-t)^2}\, dt$, which using NINT

evaluates to ≈ 25.874.

(b) Volume $= \int_0^{10} \pi y^2 \frac{dx}{dt} dt$

$= \frac{\pi}{2}\int_0^{10}\left(\frac{t(10-t)}{2}\right)^2 dt$

$= \frac{\pi}{8}\int_0^{10} (100t^2 - 20t^3 + t^4)\, dt$

$= \frac{\pi}{8}\left[\frac{100}{3}t^3 - 5t^4 + \frac{1}{5}t^5\right]_0^{10} = \frac{1250\pi}{3}$

(c) Area $= \int_0^{10} 2\pi \frac{t(10-t)}{2} \sqrt{\left(\frac{1}{2}\right)^2 + (5-t)^2}\, dt$, which

using NINT evaluates to ≈ 1040.728.

63. (a) $\frac{dy}{dx} = \frac{dy/dt}{dx/dt} = \frac{e^t \sin t + e^t \cos t}{e^t \cos t - e^t \sin t} = \frac{\cos t + \sin t}{\cos t - \sin t}$

$\left.\frac{dy}{dx}\right|_{t=\pi} = \frac{-1}{-1} = 1$

(b) $\frac{dy}{dt} = e^t(\sin t + \cos t)$, $\frac{dx}{dt} = e^t(\cos t - \sin t)$

$\left(\frac{dy}{dt}\right)^2 = e^{2t}(\sin^2 t + 2\sin t \cos t + \cos^2 t)$

$= e^{2t}(1 + 2\sin t \cos t)$

$\left(\frac{dx}{dt}\right)^2 = e^{2t}(\cos^2 t - 2\cos t \sin t + \sin^2 t)$

$= e^{2t}(1 - 2\cos t \sin t)$

$|\mathbf{v}(t)| = e^t\sqrt{2}$

$|\mathbf{v}(3)| = e^3\sqrt{2}$

(c) Distance $= \int_0^3 |\mathbf{v}(t)|\, dt$

$= \int_0^3 e^t \sqrt{2}\, dt$

$= \sqrt{2}\left[e^t\right]_0^3$

$= (e^3 - 1)\sqrt{2}$

64. (a) $\mathbf{v}(t) = \left\langle \frac{dx}{dt}, \frac{dy}{dt}\right\rangle = \left\langle 2t, \frac{6}{5}t^2\right\rangle$, $\mathbf{v}(4) = \left\langle 8, \frac{96}{5}\right\rangle$, and

$|\mathbf{v}(4)| = \sqrt{8^2 + \left(\frac{96}{5}\right)^2} = \frac{104}{5}$

(b) Distance $= \int_0^4 \sqrt{(2t)^2 + \left(\frac{6}{5}t^2\right)^2}\, dt$

$= \int_0^4 \frac{2}{5}t\sqrt{25 + 9t^2}\, dt$

$= \left[\frac{2}{135}(25 + 9t^2)^{3/2}\right]_0^4 = \frac{4144}{135}$

(c) $t = \sqrt{x+2}$, so $\frac{dy}{dx} = \frac{dy/dt}{dx/dt} = \frac{6t^2/5}{2t} = \frac{3}{5}t = \frac{3}{5}\sqrt{x+2}$

65. x degrees east of north is $(90 - x)$ degrees north of east.

Add the vectors:

$\langle 540 \cos 10°, 540 \sin 10° \rangle + \langle 55 \cos(-10°), 55 \sin(-10°) \rangle$

$= \langle 595 \cos 10°, 485 \sin 10° \rangle$

$\approx \langle 585.961, 84.219 \rangle$.

Speed $\approx \sqrt{585.961^2 + 84.219^2} \approx 591.982$ mph.

Direction $\approx \tan^{-1}\left(\dfrac{585.961}{84.219}\right) \approx 81.821°$ east of north

66. Add the vectors:

$\langle 120 \cos 20°, 120 \sin 20° \rangle + \langle 300 \cos(-5°), 300 \sin(-5°) \rangle$

$\approx \langle 411.622, 14.896 \rangle$.

Direction $\approx \tan^{-1}\left(\dfrac{14.896}{411.622}\right) \approx 2.073°$

Length $\approx \sqrt{411.622^2 + 14.896^2} \approx 411.891$ lbs

67. Taking the launch point as the origin, $y = (44 \sin 45°)t - 16t^2$ equals -6.5 when $t \approx 2.135$ sec (as can be determined graphically or using the quadratic formula). Then $x \approx (44 \cos 45°)(2.135) \approx 66.421$ horizontal feet from where it left the thrower's hand. Assuming it doesn't bounce or roll, it will still be there 3 seconds after it was thrown.

68. $y_{\max} = \dfrac{(80 \sin 45°)^2}{2(32)} + 7 = 57$ feet

69. (a)

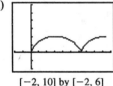

[−2, 10] by [−2, 6]

(b) $\mathbf{v}(t) = \left\langle \dfrac{dx}{dt}, \dfrac{dy}{dt} \right\rangle = \langle \pi - \pi \cos \pi t, \pi \sin \pi t \rangle$

$\mathbf{a}(t) = \left\langle \dfrac{d^2x}{dt^2}, \dfrac{d^2y}{dt^2} \right\rangle = \langle \pi^2 \sin \pi t, \pi^2 \cos \pi t \rangle$

$\mathbf{v}(0) = \langle 0, 0 \rangle$ $\mathbf{v}(1) = \langle 2\pi, 0 \rangle$
$\mathbf{a}(0) = \langle 0, \pi^2 \rangle$ $\mathbf{a}(1) = \langle 0, -\pi^2 \rangle$
$\mathbf{v}(2) = \langle 0, 0 \rangle$ $\mathbf{v}(3) = \langle 2\pi, 0 \rangle$
$\mathbf{a}(2) = \langle 0, \pi^2 \rangle$ $\mathbf{a}(3) = \langle 0, -\pi^2 \rangle$

(c) Topmost point: 2π ft/sec
center of wheel: π ft/sec
Reasons: Since the wheel rolls half a circumference, or π feet every second, the center of the wheel will move π feet every second. Since the rim of the wheel is turning at a rate of π ft/sec about the center, the velocity of the topmost point relative to the center is π ft/sec, giving it a total velocity of 2π ft/sec.

70. $v_0 = \sqrt{\dfrac{Rg}{\sin 2\alpha}}$, where $\alpha = 45°$, $g = 32$, and R = range

for 4325 yds = 12,975 ft: $v_0 \approx 644.360$ ft/sec

for 4752 yds = 14,256 ft: $v_0 \approx 675.420$ ft/sec

71. (a) $v_0 = \sqrt{\dfrac{Rg}{\sin 2\alpha}} = \sqrt{(109.5)(32)} \approx 59.195$ ft/sec

(b) The cork lands at $y = -4$, $x = 177.75$.

Solve $y = -\left(\dfrac{g}{2v_0^2 \cos^2 \alpha}\right)x^2 + (\tan \alpha)x$ for v_0, with $\alpha = 45°$: $v_0 = \sqrt{-\dfrac{gx^2}{y - x}} \approx 74.584$ ft/sec

72. (a) The javelin lands at $y = -6.5$, $x = 262\dfrac{5}{12}$.

Solve $y = -\left(\dfrac{g}{2v_0^2 \cos^2 \alpha}\right)x^2 + (\tan \alpha)x$ for v_0, with $\alpha = 40°$:

$v_0 = \sqrt{-\dfrac{gx^2}{(2 \cos^2 40°)(y - x \tan 40°)}} \approx 91.008$ ft/sec

(b) $y_{\max} = \dfrac{(v_0 \sin \alpha)^2}{2g} + 6.5$

$\approx \dfrac{(91.008 \sin 40°)^2}{64} + 6.5 \approx 59.970$ ft

73. We have $x = (v_0 t) \cos \alpha$ and

$y + \dfrac{gt^2}{2} = (v_0 t) \sin \alpha$. Squaring and adding gives

$x^2 + \left(y + \dfrac{gt^2}{2}\right)^2 = (v_0 t)^2 (\cos^2 \alpha + \sin^2 \alpha) = v_0^2 t^2$.

74. (a) $\mathbf{r}(t) = (155 \cos 18° - 11.7)t \mathbf{i} + (4 + 155 \sin 18° t - 16t^2)\mathbf{j}$
$x(t) = (155 \cos 18° - 11.7)t$
$y(t) = 4 + 155 \sin 18° t - 16t^2$

(b) $y_{\max} = \dfrac{(155 \sin 18°)^2}{2(32)} + 4 \approx 39.847$ feet, reached at

$t_{\max} = \dfrac{155 \sin 18°}{32} \approx 1.497$ sec

(c) $y(t) = 0$ when $t \approx 3.075$ sec (found using the quadratic formula), and then
$x \approx (155 \cos 18° - 11.7)(3.075) \approx 417.307$ ft.

(d) Solve $y(t) = 25$ using the quadratic formula:

$t = \dfrac{-155 \sin 18° \pm \sqrt{155^2 \sin^2 18° - 4(16)(21)}}{-32}$

≈ 0.534 and 2.460 seconds.

At those times, $x = (155 \cos 18° - 11.7)t$ equals

≈ 72.406 and ≈ 333.867 feet from home plate.

(e) Yes, the batter has hit a home run. When the ball is 380 feet from home plate (at $t \approx 2.800$ seconds), it is approximately 12.673 feet off the ground and therefore clears the fence by at least two feet.

75. (a) $\mathbf{r}(t) = \left[(155 \cos 18° - 11.7)\dfrac{1}{0.09}(1 - e^{-0.09t})\right]\mathbf{i}$
$+ \left[4 + \left(\dfrac{155 \sin 18°}{0.09}\right)(1 - e^{-0.09t})\right.$
$\left. + \dfrac{32}{0.09^2}(1 - 0.09t - e^{-0.09t})\right]\mathbf{j}$

$x(t) = (155 \cos 18° - 11.7)\dfrac{1}{0.09}(1 - e^{-0.09t})$

$y(t) = 4 + \left(\dfrac{155 \sin 18°}{0.09}\right)(1 - e^{-0.09t})$
$+ \dfrac{32}{0.09^2}(1 - 0.09t - e^{-0.09t})$

(b) Plot $y(t)$ and use the maximum function to find $y \approx 36.921$ feet at $t \approx 1.404$ seconds.

(c) Plot $y(t)$ and find that $y(t) = 0$ at $t \approx 2.959$, then plug this into the expression for $x(t)$ to find $x(2.959) \approx 352.520$ ft.

(d) Plot $y(t)$ and find that $y(t) = 30$ at $t \approx 0.753$ and 2.068 seconds. At those times, $x \approx 98.799$ and 256.138 feet (from home plate).

(e) No, the batter has not hit a home run. If the drag coefficient k is less than ≈ 0.011, the hit will be a home run. (This result can be found by trying different k-values until the parametrically plotted curve has $y \geq 10$ for $x = 380$.)

76. (a) $\overrightarrow{BD} = \overrightarrow{AD} - \overrightarrow{AB}$

(b) $\overrightarrow{AP} = \overrightarrow{AB} + \dfrac{1}{2}\overrightarrow{BD} = \dfrac{1}{2}\overrightarrow{AB} + \dfrac{1}{2}\overrightarrow{AD}$

(c) $\overrightarrow{AC} = \overrightarrow{AB} + \overrightarrow{AD}$, so by part (b), $\overrightarrow{AP} = \dfrac{1}{2}\overrightarrow{AC}$.

77. The widths between the successive turns are constant and are given by $2\pi a$.

Cumulative Review Exercises
(pp. 573–576)

1. Since the function has no discontinuity at $x = 1$, the limit is
$\dfrac{2(1)^2 - 1 - 1}{1^2 + 1 - 12} = 0$.

2. By l'Hôpital's Rule, $\lim\limits_{x \to 0} \dfrac{\sin 3x}{4x} = \lim\limits_{x \to 0} \dfrac{3 \cos 3x}{4} = \dfrac{3}{4}$.

3. By l'Hôpital's Rule, $\lim\limits_{x \to 0} \dfrac{\frac{1}{x+1} - 1}{x} = \lim\limits_{x \to 0} \dfrac{-\frac{1}{(x+1)^2}}{1} = -1$.

4. By l'Hôpital's Rule, $\lim\limits_{x \to \infty} \dfrac{x + e^x}{x - e^x} = \lim\limits_{x \to \infty} \dfrac{1 + e^x}{1 - e^x}$
$= \lim\limits_{x \to \infty} \dfrac{e^x}{-e^x} = -1$.

5. By l'Hôpital's Rule, $\lim\limits_{t \to 0} \dfrac{t(1 - \cos t)}{t - \sin t}$
$= \lim\limits_{t \to 0} \dfrac{t \sin t + (1 - \cos t)}{1 - \cos t} = \lim\limits_{t \to 0} \dfrac{t \cos t + 2 \sin t}{\sin t}$
$= \lim\limits_{t \to 0} \dfrac{-t \sin t + 3 \cos t}{\cos t} = 3$.

6. By l'Hôpital's Rule, $\lim\limits_{x \to 0^+} \dfrac{\ln(e^x - 1)}{\ln x} = \lim\limits_{x \to 0^+} \dfrac{\frac{e^x}{(e^x - 1)}}{\frac{1}{x}}$
$= \lim\limits_{x \to 0^+} \dfrac{xe^x}{e^x - 1} = \lim\limits_{x \to 0^+} \dfrac{xe^x + e^x}{e^x} = 1$

7. Use $f(x) = (e^x + x)^{1/x}$. Then $\ln f(x) = \dfrac{\ln(e^x + x)}{x}$, and
$\lim\limits_{x \to 0} \dfrac{\ln(e^x + x)}{x} = \lim\limits_{x \to 0} \dfrac{(e^x + 1)/(e^x + x)}{1} = 2$.
So $\lim\limits_{x \to 0} (e^x + x)^{1/x} = \lim\limits_{x \to 0} e^{\ln f(x)} = e^2$.

8. $\lim\limits_{x \to 0} \left(\dfrac{3x + 1}{x} - \dfrac{1}{\sin x}\right) = \lim\limits_{x \to 0} \dfrac{(3x + 1)\sin x - x}{x \sin x}$
$= \lim\limits_{x \to 0} \dfrac{(3x + 1)\cos x + 3 \sin x - 1}{x \cos x + \sin x}$
$= \lim\limits_{x \to 0} \dfrac{-(3x + 1)\sin x + 6 \cos x}{-x \sin x + 2 \cos x} = 3$

9. (a) $2(1) - 1^2 = 1$

(b) $2 - 1 = 1$

(c) 1 [from (a) and (b)]

(d) Yes, since $\lim\limits_{x \to 1} f(x) = f(1) = 1$

(e) No.
Left-hand derivative:
$\lim\limits_{h \to 0^-} \dfrac{f(1 + h) - f(1)}{h} = \lim\limits_{h \to 0^-} \dfrac{2(1 + h) - (1 + h)^2 - 1}{h}$
$= \lim\limits_{h \to 0^-} \dfrac{2 + 2h - 1 - 2h - h^2 - 1}{h}$
$= \lim\limits_{h \to 0^-} \dfrac{-h^2}{h}$
$= \lim\limits_{h \to 0^-} -h = 0$

Right-hand derivative:
$\lim\limits_{h \to 0^+} \dfrac{f(1 + h) - f(1)}{h} = \lim\limits_{h \to 0^+} \dfrac{2 - (1 + h) - 1}{h}$
$= \lim\limits_{h \to 0^+} \dfrac{-h}{h} = -1$

Since the left- and right-hand derivatives are not equal, f is not differentiable at $x = 1$.

10. Solve $4 - x^2 \leq 0$: all $x \leq -2$ and $x \geq 2$.

11. Horizontal: since as $x \to \pm\infty$, $2x^2 - x \to +\infty$ while $-1 \leq \cos x \leq 1$, the end behavior at both ends is $y = 0$.
Vertical: solve $2x^2 - x = 0$ to find $x = 0$, $x = \dfrac{1}{2}$.

12. One possible function is $y = \begin{cases} -3 + \dfrac{1}{(2 - x)}, & x < 2 \\ 3 - \dfrac{8}{x}, & x \geq 2 \end{cases}$

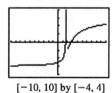

$[-10, 10]$ by $[-4, 4]$

Cumulative Review   295

13. $\dfrac{f(5) - f(0)}{5 - 0} = \dfrac{\sqrt{9} - \sqrt{4}}{5} = \dfrac{1}{5}$

14. $y' = \dfrac{(x - 2)(1) - (x + 1)(1)}{(x - 2)^2} = -\dfrac{3}{(x - 2)^2}$

15. $y' = -\sin(\sqrt{1 - 3x})\left[\dfrac{1}{2}(1 - 3x)^{-1/2}\right](-3)$

$= \dfrac{3 \sin\sqrt{1 - 3x}}{2\sqrt{1 - 3x}}$

16. $y' = \sin x \sec^2 x + \tan x \cos x = \dfrac{\sin x}{\cos^2 x} + \sin x$

$= \dfrac{(\sin x)(1 + \cos^2 x)}{\cos^2 x}$

17. $y' = \left(\dfrac{1}{x^2 + 1}\right)(2x) = \dfrac{2x}{x^2 + 1}$

18. $y' = (e^{x^2 - x})(2x - 1) = (2x - 1)e^{x^2 - x}$

19. $y' = 2x \tan^{-1} x + \dfrac{x^2}{1 + x^2}$

20. $y' = -3x^{-4}e^x + e^x x^{-3} = (x^{-3} - 3x^{-4})e^x$

21. $y' = 3\left(\dfrac{\csc x}{1 + \cos x}\right)^2 \left(\dfrac{(1 + \cos x)(-\csc x \cot x) + \csc x \sin x}{(1 + \cos x)^2}\right)$

$= \dfrac{3 \csc^2 x}{(1 + \cos x)^4}(1 - \csc x \cot x - \cos x \csc x \cot x)$

$= \dfrac{3 \csc^2 x}{(1 + \cos x)^4}(1 - \cot x \csc x - \cot^2 x)$

$= \dfrac{3 \csc^2 x}{(1 + \cos x)^4}(1 - \csc^2 x + \csc^2 x - \cot x \csc x - \cot^2 x)$

$= \dfrac{3 \csc^2 x}{(1 + \cos x)^4}(\csc^2 x - \cot x \csc x - 2\cot^2 x)$

$= \left(\dfrac{3}{(\sin^2 x)(1 + \cos x)^4}\right)\left(\dfrac{1 - \cos x - 2\cos^2 x}{\sin^2 x}\right)$

$= \left(\dfrac{3}{(\sin^2 x)(1 + \cos x)^4}\right)\left(\dfrac{(1 + \cos x)(1 - 2\cos x)}{\sin^2 x}\right)$

$= \dfrac{3(1 - 2\cos x)}{(\sin^4 x)(1 + \cos x)^3}$

22. $y' = \dfrac{d}{dx}\left(\dfrac{\pi}{2} - \sin^{-1} x\right) - \dfrac{d}{dx}\left(\dfrac{\pi}{2} - \tan^{-1} x\right)$

$= -\dfrac{1}{\sqrt{1 - x^2}} + \dfrac{1}{1 + x^2}$

23. $\dfrac{d}{dx}[\cos(xy) + y^2 - \ln x] = \dfrac{d}{dx}(0)$

$-\sin(xy)(xy' + y) + 2yy' - \dfrac{1}{x} = 0$

$y' = \dfrac{\dfrac{1}{x} + y \sin(xy)}{-x \sin(xy) + 2y} = \dfrac{1 + xy \sin(xy)}{2xy - x^2 \sin(xy)}$

24. $y' = \dfrac{1}{2}|x|^{-1/2}\dfrac{d}{dx}|x| = \dfrac{1}{2\sqrt{|x|}}\left(\dfrac{|x|}{x}\right) = \dfrac{|x|}{2x\sqrt{|x|}}$

25. $\dfrac{dy}{dx} = \dfrac{dy/dt}{dx/dt} = \dfrac{-\cos t}{-\sin t} = \cot t = \dfrac{x - 1}{1 - y}$

26. $\ln y = \ln[(\cos x)^x]$

$\ln y = x \ln(\cos x)$

$\dfrac{1}{y}\dfrac{dy}{dx} = x\left(\dfrac{1}{\cos x}\right)(-\sin x) + \ln \cos x$

$\dfrac{dy}{dx} = y \cdot \left(\ln(\cos x) - \dfrac{x \sin x}{\cos x}\right)$

$= (\cos x)^x\left(\ln(\cos x) - \dfrac{x \sin x}{\cos x}\right)$

$= (\cos x)^{x-1}[\cos x \ln(\cos x) - x \sin x]$

27. By the Fundamental theorem of Calculus,
$y' = \sqrt{1 + x^3}$.

28. $y = \Big[-\cos t\Big]_{2x}^{x^2} = -\cos(x^2) + \cos(2x)$;

$y' = 2x \sin(x^2) - 2 \sin(2x)$

29. $\dfrac{d}{dx}(y^2 + 2y) = \dfrac{d}{dx}(\sec x)$

$2yy' + 2y' = \sec x \tan x$,

$y' = \dfrac{\sec x \tan x}{2y + 2}$,

$y'' = \dfrac{(2y + 2)(\sec^3 x + \sec x \tan^2 x) - 2y' \sec x \tan x}{(2y + 2)^2}$

$= \dfrac{(2y + 2)^2(\sec^3 x + \sec x \tan^2 x) - 2 \sec^2 x \tan^2 x}{(2y + 2)^3}$

30. $\dfrac{(1 + v)u' - uv'}{(1 + v)^2}\bigg|_{x=0} = \dfrac{(1 - 3)(-1) - (2)(3)}{(1 - 3)^2} = -1$

31. (a) $v = \dfrac{dx}{dt} = 3t^2 - 12t + 9$

$a = \dfrac{dv}{dt} = 6t - 12$

(b) Solve $v = 0$ for t: $3(t - 1)(t - 3) = 0$; $t = 1$ or $t = 3$.

(c) Right: $v > 0$ for $0 \le t < 1$, $3 < t \le 5$
left: $v < 0$ for $1 < t < 3$

(d) $a = 0$ at $t = 2$, and at that instant
$v = 3(2)^2 - 12(2) + 9 = -3$ m/sec

32. For $x = 1$, $y = -1$ and

$\dfrac{dy}{dx} = 6(1)^2 - 12(1) + 4 = -2$

(a) $y + 1 = -2(x - 1)$ or $y = -2x + 1$

(b) $y + 1 = \dfrac{1}{2}(x - 1)$ or $y = \dfrac{1}{2}x - \dfrac{3}{2}$

33. For $x = \frac{\pi}{3}$, $y = \frac{\pi}{3}\cos\frac{\pi}{3} = \frac{\pi}{6}$ and
$\frac{dy}{dx} = -\frac{\pi}{3}\sin\frac{\pi}{3} + \cos\frac{\pi}{3} = -\frac{\pi\sqrt{3}}{6} + \frac{1}{2} = \frac{3 - \pi\sqrt{3}}{6}$.

(a) $y - \frac{\pi}{6} = \left(\frac{3 - \pi\sqrt{3}}{6}\right)\left(x - \frac{\pi}{3}\right)$ or
$y = \left(\frac{3 - \pi\sqrt{3}}{6}\right)\left(x - \frac{\pi}{3}\right) + \frac{\pi}{6} \approx -0.407x + 0.950$

(b) $y - \frac{\pi}{6} = \left(\frac{6}{\pi\sqrt{3} - 3}\right)\left(x - \frac{\pi}{3}\right)$ or
$y = \left(\frac{6}{\pi\sqrt{3} - 3}\right)\left(x - \frac{\pi}{3}\right) + \frac{\pi}{6} \approx 2.458x - 2.050$

34. $\frac{1}{4}(2x) + \frac{1}{9}(2yy') = 0$; $y' = -\frac{\frac{x}{2}}{\frac{2y}{9}} = -\frac{9x}{4y}$.

At $x = 1$, $y = \frac{3\sqrt{3}}{2}$, the slope is $y' = -\frac{\sqrt{3}}{2}$.

(a) $y - \frac{3\sqrt{3}}{2} = -\frac{\sqrt{3}}{2}(x - 1)$ or
$y = -\frac{\sqrt{3}}{2}x + 2\sqrt{3} \approx -0.866x + 3.464$

(b) $y - \frac{3\sqrt{3}}{2} = \frac{2}{\sqrt{3}}(x - 1)$
or $y = \frac{2}{\sqrt{3}}x + \frac{5}{2\sqrt{3}} \approx 1.155x + 1.443$

35. At $t = \frac{\pi}{3}$: $x = 1$, $y = \frac{3\sqrt{3}}{2}$, and
$\frac{dy}{dx} = \frac{dy/dt}{dx/dt} = \frac{3\cos(\pi/3)}{-2\sin(\pi/3)} = -\frac{\sqrt{3}}{2}$.

(a) $y - \frac{3\sqrt{3}}{2} = -\frac{\sqrt{3}}{2}(x - 1)$ or
$y = -\frac{\sqrt{3}}{2}x + 2\sqrt{3} \approx -0.866x + 3.464$

(b) $y - \frac{3\sqrt{3}}{2} = \frac{2}{\sqrt{3}}(x - 1)$ or
$y = \frac{2}{\sqrt{3}}x + \frac{5}{2\sqrt{3}} \approx 1.155x + 1.443$

36. At $t = \frac{\pi}{4}$: $\mathbf{r} = \sec\left(\frac{\pi}{4}\right)\mathbf{i} + \tan\left(\frac{\pi}{4}\right)\mathbf{j} = \sqrt{2}\mathbf{i} + \mathbf{j}$ and
$\mathbf{r}' = \sec\left(\frac{\pi}{4}\right)\tan\left(\frac{\pi}{4}\right)\mathbf{i} + \sec^2\left(\frac{\pi}{4}\right)\mathbf{j} = \sqrt{2}\mathbf{i} + 2\mathbf{j}$, so that
$\frac{dy}{dx} = \frac{2}{\sqrt{2}} = \sqrt{2}$.

(a) $y - 1 = \sqrt{2}(x - \sqrt{2})$ or $y = \sqrt{2}x - 1 \approx 1.414x - 1$

(b) $y - 1 = -\frac{1}{\sqrt{2}}(x - \sqrt{2})$ or
$y = -\frac{1}{\sqrt{2}}x + 2 \approx -0.707x + 2$

37. With $f(x) = \begin{cases} -x + C_1, & x < 3 \\ 2x + C_2, & x > 3 \end{cases}$, choose C_1, C_2 so that
$-3 + C_1 = 2(3) + C_2 = 1$.
$f(x) = \begin{cases} -x + 4, & x \leq 3 \\ 2x - 5, & x > 3 \end{cases}$.

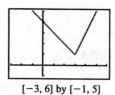

$[-3, 6]$ by $[-1, 5]$

38. (a) $x \neq 0, 2$
(b) $x = 0$
(c) $x = 2$
(d) Absolute maximum of 2 at $x = 0$:
absolute minimum of 0 at $x = -2, 2, 3$

39. According to the Mean Value Theorem the driver's speed at some time was $\frac{111}{1.5} = 74$ mph.

40. (a) Increasing in $[-0.7, 2]$ (where $f' \geq 0$), decreasing in $[-2, -0.7]$ (where $f' \leq 0$), and has a local minimum at $x \approx -0.7$.

(b) $y \approx -2x^2 + 3x + 3$

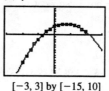

$[-3, 3]$ by $[-15, 10]$

(c) $f(x) = -\frac{2}{3}x^3 + \frac{3}{2}x^2 + 3x + C$; choose C so that
$f(0) = 1$: $f(x) = -\frac{2}{3}x^3 + \frac{3}{2}x^2 + 3x + 1$.

41. $f(x) = x^2 - 3x - \cos x + C$; choose C so that $f(0) = -2$:
$f(x) = x^2 - 3x - \cos x - 1$.

42.

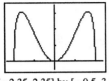

[−2.35, 2.35] by [−0.5, 3.5]

$f(x)$ is defined on $[-2, 2]$.

$f'(x) = 2x\sqrt{4 - x^2} - \dfrac{x^3}{\sqrt{4 - x^2}} = \dfrac{8x - 3x^3}{\sqrt{4 - x^2}}$; solve $f'(x) = 0$ for x to find $x = 0$, $x = \pm\dfrac{2\sqrt{6}}{3}$.

The graph of $y = f'(x)$ is shown.

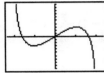

[−2.35, 2.35] by [−10, 10]

(a) $\left[-2, -\dfrac{2\sqrt{6}}{3}\right], \left[0, \dfrac{2\sqrt{6}}{3}\right]$

(b) $\left[-\dfrac{2\sqrt{6}}{3}, 0\right], \left[\dfrac{2\sqrt{6}}{3}, 2\right]$

Use NDER to plot $f''(x)$ and find that $f''(x) = 0$ for $x \approx \pm 1.042$.

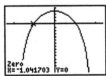

[−2.35, 2.35] by [−15, 5]

(c) Approximately $(-1.042, 1.042)$

(d) Approximately $(-2, -1.042), (1.042, 2)$

(e) Local (and absolute) maximum of approximately 3.079 at

$x = -\dfrac{2\sqrt{6}}{3}$ and $x = \dfrac{2\sqrt{6}}{3}$;

local (and absolute) minimum of 0 at $x = 0$ and at $x = \pm 2$

(f) $\approx (\pm 1.042, 1.853)$

43. (a) f has an absolute maximum at $x = 1$ and an absolute minimum at $x = 3$.

(b) f has a point of inflection at $x = 2$.

(c) The function $f(x) = \begin{cases} -\dfrac{1}{2}(x-1)^2 + 3, & -1 \le x \le 2 \\ -\dfrac{7}{2}\sqrt{x-2} + \dfrac{3}{2}, & 2 < x \le 3 \end{cases}$

is one example of a function with the given properties.

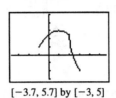

[−3.7, 5.7] by [−3, 5]

44. $y = 2\sqrt{1 - \dfrac{x^2}{16}}$, and the area of the rectangle for $x > 0$ is

$A(x) = 4x\sqrt{1 - \dfrac{x^2}{16}} = x\sqrt{16 - x^2}$.

$A'(x) = \sqrt{16 - x^2} - \dfrac{x^2}{\sqrt{16 - x^2}} = \dfrac{2(8 - x^2)}{\sqrt{16 - x^2}}$, and so

$A'(x) = 0$ when $x = \pm 2\sqrt{2}$ and $y = \sqrt{2}$. The maximum possible area is $A(2\sqrt{2}) = 8$, with dimensions $4\sqrt{2}$ by $\sqrt{2}$.

45. $f\left(\dfrac{\pi}{4}\right) = \sqrt{2}$ and $f'\left(\dfrac{\pi}{4}\right) = \sec\left(\dfrac{\pi}{4}\right)\tan\left(\dfrac{\pi}{4}\right) = \sqrt{2}$. The equation is $y - \sqrt{2} = \sqrt{2}\left(x - \dfrac{\pi}{4}\right)$ or

$y = \sqrt{2}\left(x - \dfrac{\pi}{4}\right) + \sqrt{2} \approx 1.414x + 0.303$

46. $V = s^3$

$dV = 3s^2\, ds$

Since $|ds| = 0.01s$, the error of the volume calculation is approximately $|dV| = 3s^2(0.01s) = 0.03s^3 = 0.03V$, or 3%.

47. Let s be the rope length remaining and x be the horizontal distance from the dock.

(a) $x = \sqrt{s^2 - 5^2}$, $\dfrac{ds}{dt} = -1.5$, and $\dfrac{dx}{dt} = \dfrac{s}{\sqrt{s^2 - 25}}\dfrac{ds}{dt}$,

which means that for $s = 8$ ft,

$\text{speed} = -\dfrac{dx}{dt} = -\dfrac{8}{\sqrt{64 - 25}}(-1.5) \approx 1.9$ ft/sec

(b) $\theta = \sec^{-1}\left(\dfrac{s}{5}\right)$, so $\dfrac{d\theta}{dt} = \dfrac{5}{|s|\sqrt{s^2 - 25}}\dfrac{ds}{dt}$, which for

$s = 8$ ft becomes $\dfrac{5}{8\sqrt{64 - 25}}(-1.5) = -0.15$ rad/sec.

Cumulative Review

48. (a) Let h be the level of the coffee in the pot, and let V be the volume of the coffee in the pot.

$$h = \frac{V}{16\pi}, \text{ so } \frac{dh}{dt} = \frac{dV/dt}{16\pi} = \frac{9}{16\pi} \approx 0.179 \text{ in./min.}$$

(b) Now let h be the level of the coffee in the cone, and let V be the volume of the coffee in the cone.

$$V = \frac{1}{3}\pi\left(\frac{h}{2}\right)^2 h = \frac{\pi}{12}h^3, \text{ so } \frac{dV}{dt} = \left(\frac{\pi}{4}h^2\right)\left(\frac{dh}{dt}\right) \text{ and } \frac{dh}{dt}$$

$$= \left(\frac{4}{\pi h^2}\right)\left(\frac{dV}{dt}\right) = \left(\frac{4}{25\pi}\right)(-9) = -\frac{36}{25\pi}$$

$$\approx -0.458 \text{ in./min.}$$

Since $\dfrac{dh}{dt}$ is negative, the level in the cone is falling at the rate of about 0.458 in./min.

49. (a) $(1)(0 + 1.8 + 6.4 + \cdots + 16.2) = 165$ in.
(b) $(1)(1.8 + 6.4 + 12.6 + \cdots + 0) = 165$ in.

50. $\displaystyle\int_{-2}^{1} |x|\, dx = \int_{-2}^{0} -x\, dx + \int_{0}^{1} x\, dx = \left[-\frac{1}{2}x^2\right]_{-2}^{0} + \left[\frac{1}{2}x^2\right]_{0}^{1}$
$= 2 + \dfrac{1}{2} = 2.5$

51. Using Number 29 in the Table of Integrals, with $a = 2$,

$$\int_{-2}^{2} \sqrt{4-x^2}\, dx = \left[\frac{x}{2}\sqrt{4-x^2} + 2\sin^{-1}\left(\frac{x}{2}\right)\right]_{-2}^{2}$$

$= \pi - (-\pi) = 2\pi$.

Alternately, observe that the region under the curve and above the x-axis is a semicircle of radius 2, so the area is $\dfrac{1}{2}\pi(2)^2 = 2\pi$.

52. $\displaystyle\int_{1}^{3}\left(x^2 + \frac{1}{x}\right)dx = \left[\frac{1}{3}x^3 + \ln x\right]_{1}^{3} = 9 + \ln 3 - \frac{1}{3}$
$= \ln 3 + \dfrac{26}{3} \approx 9.765$

53. $\displaystyle\int_{0}^{\pi/4} \sec^2 x\, dx = \left[\tan x\right]_{0}^{\pi/4} = 1$

54. $\displaystyle\int_{1}^{4}\frac{2+\sqrt{x}}{\sqrt{x}}dx = \int_{1}^{4}\left(\frac{2}{\sqrt{x}}+1\right)dx = \left[4\sqrt{x}+x\right]_{1}^{4} = 12 - 5$
$= 7$

55. Let $u = \ln x$, so $du = \dfrac{1}{x}dx$.

Then $\displaystyle\int \frac{dx}{x(\ln x)^2} = \int u^{-2}\,du = -\frac{1}{u} + C = -\frac{1}{\ln x} + C.$

Therefore, $\displaystyle\int_{e}^{2e}\frac{dx}{x(\ln x)^2} =$

$\left[-\dfrac{1}{\ln x}\right]_{e}^{2e} = \left[-\dfrac{1}{1+\ln 2} + \dfrac{1}{1}\right] = \dfrac{\ln 2}{1+\ln 2} \approx 0.409$

56. $\displaystyle\int \left[(3-2t)\mathbf{i} + \left(\frac{1}{t}\right)\mathbf{j}\right]dt =$

$\left[(3t - t^2)\mathbf{i} + (\ln t)\mathbf{j}\right]_{1}^{3} = (\ln 3)\mathbf{j} - 2\mathbf{i} = -2\mathbf{i} + (\ln 3)\mathbf{j}$

57. Let $u = e^x + 1$, so $du = e^x\, dx$.

Use the identity $\cot^2 u = \csc^2 u - 1$.

$\displaystyle\int e^x \cot^2(e^x + 1)\, dx = \int \cot^2 u\, du$
$= \displaystyle\int (\csc^2 u - 1)\, du$
$= -\cot u - u + C$
$= -\cot(e^x + 1) - (e^x + 1) + C$

Since $-1 + C$ is an arbitrary constant, we may redefine C and write the solution as $-\cot(e^x + 1) - e^x + C$.

58. Let $u = \dfrac{s}{2}$, so $du = \dfrac{ds}{2}$.

$\displaystyle\int \frac{ds}{s^2+4} = \int \frac{ds}{4(s/2)^2+4} = \frac{1}{2}\int \frac{ds}{2[(s/2)^2+1]} = \frac{1}{2}\int \frac{du}{u^2+1}$

$= \dfrac{1}{2}\tan^{-1} u + C = \dfrac{1}{2}\tan^{-1}\left(\dfrac{s}{2}\right) + C$

59. Let $u = \cos(x-3)$, so $du = -\sin(x-3)\, dx$.

$\displaystyle\int \frac{\sin(x-3)}{\cos^3(x-3)}dx = \int (-u^{-3})\, du = \frac{1}{2}u^{-2} + C$

$= \dfrac{1}{2\cos^2(x-3)} + C$

60. Use integration by parts.

$u = e^{-x}$ $\qquad dv = \cos 2x\, dx$
$du = -e^{-x}dx$ $\qquad v = \dfrac{1}{2}\sin 2x$

$\displaystyle\int e^{-x}\cos 2x\, dx = \frac{1}{2}e^{-x}\sin 2x + \int \frac{1}{2}e^{-x}\sin 2x\, dx$

Now let

$u = e^{-x}$ $\qquad dv = \dfrac{1}{2}\sin 2x\, dx$
$du = -e^{-x}dx$ $\qquad v = -\dfrac{1}{4}\cos 2x$

Then

$\displaystyle\int e^{-x}\cos 2x\, dx$
$= \dfrac{1}{2}e^{-x}\sin 2x - \dfrac{1}{4}e^{-x}\cos 2x - \dfrac{1}{4}\int e^{-x}\cos 2x\, dx$

so

$\displaystyle\int e^{-x}\cos 2x\, dx = \frac{e^{-x}}{5}(2\sin 2x - \cos 2x) + C$

61. $\dfrac{x+2}{x^2-5x-6} = \dfrac{x+2}{(x+1)(x-6)} = \dfrac{A}{x+1} + \dfrac{B}{x-6}$

$x + 2 = A(x - 6) + B(x + 1) = (A + B)x + (B - 6A)$

Solving $A + B = 1$, $B - 6A = 2$ yields $A = -\dfrac{1}{7}$, $B = \dfrac{8}{7}$ so

$\dfrac{x+2}{x^2-5x-6} = \dfrac{8}{7(x-6)} - \dfrac{1}{7(x+1)}$. Then

$\displaystyle\int \dfrac{x+2}{x^2-5x-6}\,dx = \int\left(\dfrac{8}{7(x-6)} - \dfrac{1}{7(x+1)}\right)dx$

$= \dfrac{8}{7}\ln|x-6| - \dfrac{1}{7}\ln|x+1| + C = \dfrac{1}{7}\ln\dfrac{(x-6)^8}{|x+1|} + C$

62. Area $\approx \dfrac{5}{2}[3 + 2(8.3) + 2(9.9) + \cdots + 2(8.3) + 3] = 359$;

Volume $\approx 25 \times 359 = 8975$ ft^3

63. $y = -(t+1)^{-1} - \dfrac{1}{2}e^{-2t} + C$; $y(0) = -1 - \dfrac{1}{2} + C = 2$, so

$C = \dfrac{7}{2}$ and $y = -\dfrac{1}{t+1} - \dfrac{1}{2}e^{-2t} + \dfrac{7}{2}$.

64. $y' = -\dfrac{1}{2}\cos 2\theta - \sin\theta + C_1$, and $y'\left(\dfrac{\pi}{2}\right) = 0 \Rightarrow$

$y' = -\dfrac{1}{2}\cos 2\theta - \sin\theta + \dfrac{1}{2}$.

$y = -\dfrac{1}{4}\sin 2\theta + \cos\theta + \dfrac{1}{2}\theta + C_2$, and $y\left(\dfrac{\pi}{2}\right) = 0$

$\Rightarrow y = -\dfrac{1}{4}\sin 2\theta + \cos\theta + \dfrac{1}{2}\theta - \dfrac{\pi}{4}$

65. Use integration by parts.

$u = x^2 \qquad dv = \sin x\,dx$

$du = 2x\,dx \qquad v = -\cos x$

$\displaystyle\int x^2 \sin x\,dx = -x^2 \cos x + \int 2x \cos x\,dx$

Now let

$u = x \qquad dv = 2\cos x\,dx$

$du = dx \qquad v = 2\sin x$

$\displaystyle\int x^2 \sin x\,dx = -x^2 \cos x + 2x \sin x - \int 2 \sin x\,dx$

$= -x^2 \cos x + 2x \sin x + 2 \cos x + C$

$= (2 - x^2)\cos x + 2x \sin x + C$

The graph of the slope field of the differential equation $\dfrac{dy}{dx} = x^2 \sin x$ and the antiderivative $y = (2 - x^2)\cos x + 2x \sin x$ is shown below.

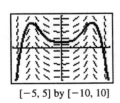

$[-5, 5]$ by $[-10, 10]$

66. Use integration by parts.

$u = x \qquad\qquad dv = e^x\,dx$

$du = dx \qquad\qquad v = e^x$

$\displaystyle\int x e^x\,dx = xe^x - \int e^x\,dx = xe^x - e^x + C = e^x(x - 1) + C$

Confirm by differentiation:

$\dfrac{d}{dx}[e^x(x-1) + C] = e^x + (x-1)e^x = xe^x$

67. (a) $y = Ce^{kt}$, with $6{,}000 = Ce^{k(2)}$ and $10{,}000 = Ce^{k(5)}$.

Then $\dfrac{10{,}000}{6{,}000} = e^{k(5-2)}$, so $\dfrac{5}{3} = e^{3k}$ and therefore

$k = \dfrac{\ln\left(\dfrac{5}{3}\right)}{3} \approx 0.170$.

Furthermore, $C = \dfrac{6{,}000}{e^{2k}} \approx 4268$. The approximate number of bacteria is given by $y = 4268e^{0.170t}$.

(b) About 4268

68. Let t be the time in minutes where $t = 0$ represents right now, and let $T(t)$ be the number of degrees above room temperature. Then we may write $T(t) = T_0 e^{-kt}$ where

$T(0) = 50$ and $T(-15) = 65$, giving $T_0 = 50$ and

$k = \dfrac{1}{15}\ln\dfrac{13}{10} \approx 0.0175$.

(a) $50e^{-k(120)} \approx 6.13°C$ above room temperature.

(b) Solving $5 = 50e^{-kt}$ gives $t = \dfrac{\ln 0.1}{-k} \approx 131.6$ minutes, or about 2 hours and 12 minutes from now.

69. $\dfrac{dy}{dx} = 0.08y\left(1 - \dfrac{y}{500}\right)$

$\dfrac{500\,dy}{y(500-y)} = 0.08\,dx$

$\dfrac{(500-y)+y}{y(500-y)}\,dy = 0.08\,dx$

$\left(\dfrac{1}{y} + \dfrac{1}{500-y}\right)dy = 0.08\,dx$

Integrate both sides.

$\ln|y| - \ln|500 - y| = 0.08x + C_1$

$\dfrac{y}{500-y} = C_2 e^{0.08x}$

$y \cdot (1 + C_2 e^{0.08x}) = 500C_2 e^{0.08x}$

$y = \dfrac{500}{1 + Ce^{-0.08x}}$

300 Cumulative Review

70. $\dfrac{dy}{dx} = (y - 4)(x + 3)$

$\dfrac{dy}{y - 4} = (x + 3)\, dx$

$\displaystyle\int \dfrac{dy}{y - 4} = \int (x + 3)\, dx$

$\ln |y - 4| = \dfrac{x^2}{2} + 3x + C_1$

$y - 4 = e^{C_1} e^{(x^2/2) + 3x} + 4$

$y = Ce^{(x^2/2) + 3x} + 4$

71. Use EULERT.

x	y
0	0
0.1	0.1
0.2	0.2095
0.3	0.3285
0.4	0.4568
0.5	0.5946
0.6	0.7418
0.7	0.8986
0.8	1.0649
0.9	1.2411
1.0	1.4273

72. The region has four congruent portions, so

$\text{Area} = 4\displaystyle\int_0^{\pi/2} \sin 2x\, dx = 4\left[-\dfrac{1}{2}\cos 2x\right]_0^{\pi/2} = 4$

73. Solve $5 - x^2 = x^2 - 3$ to find the integration limits:

$2x^2 = 8 \Rightarrow x = \pm 2$. Then

$\text{Area} = \displaystyle\int_{-2}^{2} [(5 - x^2) - (x^2 - 3)]\, dx = \int_{-2}^{2} (8 - 2x^2)\, dx$

$= \left[8x - \dfrac{2}{3}x^3\right]_{-2}^{2} = \dfrac{64}{3}$

74. Solve $y^2 - 3 = y + 2$ to find the integration limits:

$y^2 - y - 5 = 0 \Rightarrow y = \dfrac{1 \pm \sqrt{21}}{2}$. Then

$\text{Area} = \displaystyle\int_{(1 - \sqrt{21})/2}^{(1 + \sqrt{21})/2} [(y + 2) - (y^2 - 3)]\, dy \approx 16.039$.

75. $\text{Area} = \dfrac{1}{2}\displaystyle\int_0^{2\pi} r^2\, d\theta = \dfrac{1}{2}\int_0^{2\pi} 9(1 + \cos\theta)^2\, d\theta$

$= \dfrac{9}{2}\displaystyle\int_0^{2\pi} (1 + 2\cos\theta + \cos^2\theta)\, d\theta$

$= \dfrac{9}{2}\displaystyle\int_0^{2\pi}\left(1 + 2\cos\theta + \dfrac{1 + \cos 2\theta}{2}\right) d\theta$

$= \dfrac{9}{2}\displaystyle\int_0^{2\pi}\left(\dfrac{3}{2} + \cos\theta + \dfrac{1}{2}\cos 2\theta\right) d\theta$

$= \dfrac{9}{2}\left[\dfrac{3}{2}\theta + 2\sin\theta - \dfrac{1}{4}\sin 2\theta\right]_0^{2\pi}$

$= \dfrac{9}{2}(3\pi - 0) = \dfrac{27\pi}{2} \approx 42.412$

76. $\text{Volume} = \displaystyle\int_{-1}^{1} \pi \left(\dfrac{x^3}{2}\right)^2 dx = \dfrac{\pi}{4}\left[\dfrac{1}{7}x^7\right]_{-1}^{1} = \dfrac{\pi}{14} \approx 0.224$

77. Solve $4x - x^2 = 0$ to find the limit of integration:

$x = 0$ or $x = 4$. By the cylindrical shell method,

$\text{Volume} = \displaystyle\int_0^4 2\pi x(4x - x^2)\, dx = 2\pi\int_0^4 (4x^2 - x^3)\, dx$

$= 2\pi\left[\dfrac{4}{3}x^3 - \dfrac{1}{4}x^4\right]_0^4 = \dfrac{128\pi}{3}$

≈ 134.041.

78. The average value is the integral divided by the interval length. Using NINT,

$\text{average value} = \dfrac{1}{\pi}\displaystyle\int_0^{\pi} \sqrt{\sin x}\, dx \approx 0.763$

79. $y' = \sec^2 x$, so we may use NINT to obtain

$\text{Length} = \displaystyle\int_{-\pi/4}^{\pi/4} \sqrt{1 + (\sec^2 x)^2}\, dx \approx 2.556$.

80. $\dfrac{dx}{dt} = \cos t$ and $\dfrac{dy}{dt} = 1 - \sin t$, so we may use NINT to obtain

$\text{Length} = \displaystyle\int_{-\pi/2}^{\pi/2} \sqrt{\cos^2 t + (1 - \sin t)^2}\, dt$

$= \displaystyle\int_{-\pi/2}^{\pi/2} \sqrt{2 - 2\sin t}\, dt = 4$.

81. Using NINT,

$\text{Length} = \displaystyle\int_0^{\pi} \sqrt{\theta^2 + \left(\dfrac{dr}{d\theta}\right)^2}\, d\theta = \int_0^{\pi} \sqrt{1 + \theta^2}\, d\theta \approx 6.110$.

82. $\dfrac{dy}{dx} = -\dfrac{1}{2}e^{-x/2}$, so we may use NINT to obtain

$\text{Area} = \displaystyle\int_0^2 2\pi e^{-x/2}\sqrt{1 + \left(\dfrac{-e^{-x/2}}{2}\right)^2}\, dx$

$= \displaystyle\int_0^2 \pi e^{-x/2}\sqrt{4 + e^{-x}}\, dx \approx 8.423$.

83. $\dfrac{dx}{dt} = \cos t$ and $\dfrac{dy}{dt} = 1 - \sin t$, so

$\text{Area} = \displaystyle\int_0^{\pi/2} 2\pi \sin t\sqrt{\cos^2 t + (1 - \sin t)^2}\, dt$

$= \displaystyle\int_0^{\pi/2} 2\pi \sin t\sqrt{2 - 2\sin t}\, dt \approx 3.470$.

84. $\dfrac{dr}{d\theta} = 1$, so we may use NINT to obtain

$\text{Area} = \displaystyle\int_{\pi/2}^{\pi} 2\pi\theta \sin\theta\sqrt{\theta^2 + 1}\, d\theta \approx 32.683$.

85. $\text{Volume} = \displaystyle\int_0^1 \pi\left(\dfrac{\sqrt{x} - x^2}{2}\right)^2 dx$

$= \displaystyle\int_0^1 \dfrac{\pi}{4}(x - 2x^{5/2} + x^4)\, dx$

$= \dfrac{\pi}{4}\left[\dfrac{1}{2}x^2 - \dfrac{4}{7}x^{7/2} + \dfrac{1}{5}x^5\right]_0^1$

$= \dfrac{9\pi}{280} \approx 0.101$.

86. Use the region's symmetry:

$$\text{Volume} = 2\int_0^{\pi/4} \pi(2\tan x)^2\, dx$$

$$= 8\pi \int_0^{\pi/4} \tan^2 x\, dx$$

$$= 8\pi \int_0^{\pi/4} (\sec^2 x - 1)\, dx$$

$$= 8\pi \Big[\tan x - x\Big]_0^{\pi/4}$$

$$= 8\pi\left(1 - \frac{\pi}{4}\right)$$

$$= 8\pi - 2\pi^2 \approx 5.394.$$

87. (a) $F = kx \Rightarrow 200 = k(0.8) \Rightarrow k = 250$ N/m, so for

$F = 300$ N, $x = \dfrac{300}{250} = 1.2$ m.

(b) Work $= \displaystyle\int_0^{1.2} 250x\, dx = \Big[125x^2\Big]_0^{1.2} = 180$ J

88. (a) The work required to raise a thin disk at height y from the bottom is

$$(\text{weight})(\text{distance}) = \left[60\pi\left(\frac{y}{2}\right)^2 dy\right](12 - y).$$

Total work $= \displaystyle\int_0^{10} 15\pi y^2(12 - y)\, dy$

$= 15\pi \displaystyle\int_0^{10} (-y^3 + 12y^2)\, dy = 15\pi\left[-\frac{1}{4}y^4 + 4y^3\right]_0^{10}$

$= 22{,}500\pi \approx 70{,}686$ ft-lb.

(b) $\dfrac{22{,}500\pi}{275} \approx 257$ sec $= 4$ min, 17 sec

89. The sideways force exerted by a thin disk at depth y is its edge area times the pressure, or

$(2\pi\, dy)(849y) = 1698\pi y\, dy.$

Total force $= \displaystyle\int_0^H 1698\pi y\, dy = 849\pi H^2$, where H is depth.

Solve: $40{,}000 = 849\pi H^2$

$\Rightarrow H = \sqrt{\dfrac{40{,}000}{849\pi}}$ and $V = \pi H \approx 12.166$ ft^3.

90. Use l'Hôpital's Rule: $\lim\limits_{x\to\infty} \dfrac{\ln x}{\sqrt{x}} = \lim\limits_{x\to\infty} \dfrac{\frac{1}{x}}{\frac{x^{-1/2}}{2}} = \lim\limits_{x\to\infty} \dfrac{2}{\sqrt{x}} = 0.$

$f(x) = \ln x$ grows slower than $g(x) = \sqrt{x}.$

91. Use the limit comparison test with $f(t) = \dfrac{1}{t^2 - 4}$ and

$g(t) = \dfrac{1}{t^2}$. Since f and g are both continuous on $[3, \infty)$,

$\lim\limits_{t\to\infty} \dfrac{f(t)}{g(t)} = 1$, and $\displaystyle\int_3^\infty g(t)\, dt$ converges, we conclude that

$\displaystyle\int_3^\infty f(t)\, dt = \int_3^\infty \dfrac{dt}{t^2 - 4}$ converges.

92. Use the comparison test: for $x \geq 2$, $\dfrac{1}{\ln x} > \dfrac{1}{x}$, and

$\lim\limits_{b\to\infty} \displaystyle\int_2^b \dfrac{dx}{x} = \lim\limits_{b\to\infty} (\ln b - \ln 2) = \infty$. Both integrals diverge.

93. $\displaystyle\int_{-\infty}^\infty e^{-|x|}\, dx = \int_{-\infty}^0 e^x\, dx + \int_0^\infty e^{-x}\, dx = 2\int_0^\infty e^{-x}\, dx$

Since $\displaystyle\int_0^\infty e^{-x}\, dx = \lim\limits_{b\to\infty} \int_0^b e^{-x}\, dx = \lim\limits_{b\to\infty} \Big[-e^{-x}\Big]_0^b = 1$, the original integral converges.

94. $\displaystyle\int_0^1 \dfrac{4r\, dr}{\sqrt{1 - r^2}} = \lim\limits_{b\to 1^-} \int_0^b \dfrac{4r\, dr}{\sqrt{1 - r^2}} = \lim\limits_{b\to 1^-} \Big[-4\sqrt{1 - r^2}\Big]_0^b$

$= \lim\limits_{b\to 1^-} (-4\sqrt{1 - b^2} + 4) = 4.$ The integral converges.

95. $\displaystyle\int_0^{10} \dfrac{dx}{1 - x} = \int_0^1 \dfrac{dx}{1 - x} + \int_1^{10} \dfrac{dx}{1 - x}$

Since $\displaystyle\int_0^1 \dfrac{dx}{1 - x} = \lim\limits_{b\to 1^-} \int_0^b \dfrac{dx}{1 - x}$

$= \lim\limits_{b\to 1^-} \Big[-\ln(1 - x)\Big]_0^b = \infty$, the original integral diverges.

96. $\displaystyle\int_0^2 \dfrac{dx}{\sqrt[3]{x - 1}} = \int_0^1 \dfrac{dx}{\sqrt[3]{x - 1}} + \int_1^2 \dfrac{dx}{\sqrt[3]{x - 1}}$

$= \lim\limits_{b\to 1^-} \displaystyle\int_0^b \dfrac{dx}{\sqrt[3]{x - 1}} + \lim\limits_{a\to 1^+} \int_a^2 \dfrac{dx}{\sqrt[3]{x - 1}}$

$= \lim\limits_{b\to 1^-} \Big[\tfrac{3}{2}(x - 1)^{2/3}\Big]_0^b + \lim\limits_{a\to 1^+} \Big[\tfrac{3}{2}(x - 1)^{2/3}\Big]_a^2 = 0.$

The whole integral converges.

97. We know that

$\dfrac{1}{1 + x} = 1 - x + x^2 - x^3 + \cdots + (-1)^n x^n + \cdots$

for $-1 < x < 1$. Substituting $2x$ for x yields

$\dfrac{1}{1 + 2x} = 1 - 2x + 4x^2 - 8x^3 + \cdots + (-1)^n 2^n x^n + \cdots$

for $-1 < 2x < 1$, so the interval of convergence is

$-\dfrac{1}{2} < x < \dfrac{1}{2}.$

98. (a)

$$\cos t^2 = 1 - \frac{(t^2)^2}{2!} + \frac{(t^2)^4}{4!} - \frac{(t^2)^6}{6!} + \cdots + (-1)^n \frac{(t^2)^{2n}}{(2n)!} + \cdots$$

$$= 1 - \frac{t^4}{2!} + \frac{t^8}{4!} - \frac{t^{12}}{6!} + \cdots + (-1)^n \frac{t^{4n}}{(2n)!}$$

Integrating each term with respect to t from 0 to x yields

$$x - \frac{x^5}{5(2!)} + \frac{x^9}{9(4!)} - \frac{x^{13}}{13(6!)} + \cdots + (-1)^n \frac{x^{4n+1}}{(4n+1)(2n)!} + \cdots.$$

(b) $-\infty < x < \infty$; Since the cosine series converges for all real numbers, so does the integrated series, by the term-by-term integration theorem (Section 9.1, Theorem 2).

99. $\ln(2 + 2x) = \ln[2(x+1)] = \ln 2 + \ln(x+1)$

$$= \ln 2 + x - \frac{x^2}{2} + \frac{x^3}{3} - \frac{x^4}{4} + \cdots + (-1)^{n-1} \frac{x^n}{n} + \cdots$$

Since by the Ratio Test $\lim_{n\to\infty} \left|\frac{a_{n+1}}{a_n}\right| = \lim_{n\to\infty} \left|\frac{n}{n+1} x\right| = |x|$, the series converges for $-1 < x \leq 1$.

$\left(\sum_{n=1}^{\infty} \frac{(-1)^n}{n}\right.$ converges, but $\left.\sum_{n=1}^{\infty} -\frac{1}{n}\right.$ does not.$\left.\right)$

100. Let $f(x) = \sin x$. Then $f'(x) = \cos x$, $f''(x) = -\sin x$, $f'''(x) = -\cos x$, $f^{(4)}(x) = \sin x$, and so on. At $x = 2\pi$ the sine terms are zero and the cosine terms alternate between 1 and -1, so the Taylor series is

$$(x - 2\pi) - \frac{(x - 2\pi)^3}{3!} + \frac{(x - 2\pi)^5}{5!} - \cdots$$
$$+ (-1)^n \frac{(x - 2\pi)^{2n+1}}{(2n+1)!} + \cdots.$$

101. The first six terms of the Maclaurin series are

$1 - x + \frac{x^2}{2!} - \frac{x^3}{3!} + \frac{x^4}{4!} - \frac{x^5}{5!} + \frac{x^6}{6!}$. By the Alternating

Series Estimation Theorem, $|\text{error}| \leq \left|\frac{x^7}{7!}\right| \leq \frac{1}{7!} < 0.001$.

102. $f(0) = 1, f'(0) = \frac{1}{3(0+1)^{2/3}} = \frac{1}{3}$,

$f''(0) = -\frac{2}{9(0+1)^{2/3}} = -\frac{2}{9}$,

$\cdots f^{(n)}(0) = (-1)^{n-1} \frac{2 \cdot 5 \cdot \cdots \cdot (3n-4)}{3^n}$, so the

Taylor series is

$$1 + \frac{1}{3}x - \frac{2}{2! \cdot 3^2}x^2 + \frac{2 \cdot 5}{3! \cdot 3^3}x^3 - \frac{2 \cdot 5 \cdot 8}{4! \cdot 3^4}x^4 + \cdots$$
$$+ (-1)^{n-1} \frac{2 \cdot 5 \cdot \cdots \cdot (3n-4)}{n! \cdot 3^n} x^n + \cdots.$$

Since by the Ratio Test

$$\lim_{n\to\infty} \left|\frac{a_{n+1}}{a_n}\right|$$

$$= \lim_{n\to\infty} \left|\frac{2 \cdot 5 \cdot \cdots \cdot (3n-4)(3n-1)x^{n+1}}{(n+1)! 3^{n+1}} \cdot \frac{n! 3^n}{2 \cdot 5 \cdot \cdots \cdot (3n-4)x^n}\right|$$

$$= \lim_{n\to\infty} \left|\frac{(3n-1)x}{(n+1)(3)}\right| = |x|, \text{ the radius of convergence is 1.}$$

103. Using the Ratio Test, $\lim_{n\to\infty} \left|\frac{a_{n+1}}{a_n}\right| = \lim_{n\to\infty} \left|\frac{2}{3^{n+1}} \cdot \frac{3^n}{2}\right| = \frac{1}{3}$, so the series converges.

104. Note that $a_n > \frac{1}{n}$ for every n. By the Direct Comparison Test, since $\sum_{n=1}^{\infty} \frac{1}{n}$ diverges, so does $\sum_{n=1}^{\infty} \frac{2}{\sqrt{n}}$.

105. Use the alternating series test.

Note that $\sum_{n=0}^{\infty} \frac{(-1)^n}{n+1} = \sum_{n=0}^{\infty} (-1)^n u_n$, where $u_n = \frac{1}{n+1}$.

Since each u_n is positive, $u_n > u_{n+1}$ for all n, and $\lim_{n\to\infty} u_n = 0$, the original series converges.

106. Using the Ratio Test,

$$\lim_{n\to\infty} \left|\frac{a_{n+1}}{a_n}\right| = \lim_{n\to\infty} \left|\frac{3^{n+1}}{(n+1)!} \cdot \frac{n!}{3^n}\right| = \lim_{n\to\infty} \frac{3}{n+1} = 0, \text{ and the}$$

series converges.

107. (a) Using the Ratio Test,

$$\lim_{n\to\infty} \left|\frac{a_{n+1}}{a_n}\right| = \lim_{n\to\infty} \left|\frac{(x+2)^{n+1}}{n+1} \cdot \frac{n}{(x+2)^n}\right| = |x+2|, \text{ which}$$

means that the series converges for $-1 < x + 2 < 1$, or $-3 < x < -1$. Furthermore, at $x = -3$, the series is $\sum_{n=1}^{\infty} \frac{1}{n}$, which diverges, and at $x = -1$, the series is $\sum_{n=1}^{\infty} \frac{(-1)^n}{n}$, which converges. The interval of convergence is $-3 < x \leq -1$ and the radius of convergence is 1.

(b) $-3 < x < -1$

(c) At $x = -1$

108. (a) Using the Ratio Test, $\lim_{n\to\infty} \left|\dfrac{a_{n+1}}{a_n}\right|$

$= \lim_{n\to\infty} \left|\dfrac{x^{n+1}}{(n+1)\ln^2(n+1)} \cdot \dfrac{n \ln^2 n}{x^n}\right|$

$= \lim_{n\to\infty} \left|\dfrac{nx \ln^2(n)}{(n+1)\ln^2(n+1)}\right|$

$= |x| \left(\lim_{n\to\infty} \dfrac{n}{n+1}\right)\left(\lim_{n\to\infty} \dfrac{\ln n}{\ln(n+1)}\right)^2 = |x|$, which means

that the series converges for $-1 < x < 1$. At $x = \pm 1$,

the series converges by the Integral Test:

$\int_2^\infty \dfrac{1}{x(\ln x)^2} dx = \lim_{b\to\infty} \int_2^b \dfrac{1}{x(\ln x)^2} dx = \left[-\dfrac{1}{\ln x}\right]_2^b$

$= \lim_{b\to\infty}\left(-\dfrac{1}{\ln b} + \dfrac{1}{\ln 2}\right) = \dfrac{1}{\ln 2}$. So the

convergence interval is $-1 \le x \le 1$ and the radius of

convergence is 1.

(b) $-1 \le x \le 1$

(c) Nowhere

109. $\dfrac{1}{\sqrt{2^2+(-3)^2}}\langle 2, -3\rangle = \left\langle\dfrac{2}{\sqrt{13}}, -\dfrac{3}{\sqrt{13}}\right\rangle$

110. $\left\langle 1\cos\dfrac{\pi}{3}, 1\sin\dfrac{\pi}{3}\right\rangle = \left\langle\dfrac{1}{2}, \dfrac{\sqrt{3}}{2}\right\rangle$

111. $\dfrac{dy}{dx}\bigg|_{t=3\pi/4} = \dfrac{dy/dt}{dx/dt}\bigg|_{t=3\pi/4} = \dfrac{-3\sin t}{4\cos t}\bigg|_{t=3\pi/4} = \dfrac{3}{4}$. The

tangent vectors are $\dfrac{1}{\sqrt{3^2+4^2}}\langle 4, 3\rangle = \left\langle\dfrac{4}{5}, \dfrac{3}{5}\right\rangle$ and

$\left\langle -\dfrac{4}{5}, -\dfrac{3}{5}\right\rangle$. The normal vectors are $\left\langle\dfrac{3}{5}, -\dfrac{4}{5}\right\rangle$ and $\left\langle -\dfrac{3}{5}, \dfrac{4}{5}\right\rangle$.

112. (a) $\mathbf{v}(t) = \dfrac{d\mathbf{r}}{dt} = (-\cos t)\mathbf{i} + (1+\sin t)\mathbf{j}$

$\mathbf{a}(t) = \dfrac{d\mathbf{v}}{dt} = (\sin t)\mathbf{i} + (\cos t)\mathbf{j}$

(b) Using NINT, the distance traveled is

$\int_{\pi/2}^{3\pi/2} |\mathbf{v}(t)|dt = \int_{\pi/2}^{3\pi/2}\sqrt{(-\cos t)^2 + (1+\sin t)^2}\, dt$

$= \int_{\pi/2}^{3\pi/2}\sqrt{2+2\sin t}\, dt = 4$.

113. Yes. The path of the ball is given by

$x = 100(\cos 45°)t = 50\sqrt{2}t$ and

$y = -16t^2 + 100(\sin 45°)t = -16t^2 + 50\sqrt{2}t$.

When $x = 130$, we have $t = \dfrac{13}{5\sqrt{2}}$ and so

$y = -16\left(\dfrac{13}{5\sqrt{2}}\right)^2 + 50\sqrt{2}\left(\dfrac{13}{5\sqrt{2}}\right) = 75.92$ ft, high enough

to easily clear the 35-ft tree.

114. Since $r\cos\theta = x$, $r\sin\theta = y$, the Cartesian equation is $x - y = 2$. The graph is a line with slope 1 and y-intercept -2.

115.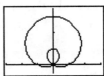

$[-3, 3]$ by $[-0.5, 3.5]$

The shortest possible θ-interval has length 2π.

116. $x = r\cos\theta = \cos\theta - \cos^2\theta$,

$y = r\sin\theta = \sin\theta - \sin\theta\cos\theta$

$\dfrac{dy}{dx} = \dfrac{dy/d\theta}{dx/d\theta} = \dfrac{\cos\theta + \sin^2\theta - \cos^2\theta}{-\sin\theta + 2\sin\theta\cos\theta}$

Zeros of $\dfrac{dy}{d\theta}$:

$\cos\theta + \sin^2\theta - \cos^2\theta = 0$

$\cos\theta + 1 - 2\cos^2\theta = 0$

$(2\cos\theta + 1)(\cos\theta - 1) = 0$

$\theta = 0, \theta = \dfrac{2\pi}{3}, \theta = \dfrac{4\pi}{3}, \text{ or } \theta = 2\pi$

Zeros of $\dfrac{dx}{d\theta}$:

$-\sin\theta + 2\sin\theta\cos\theta = 0$

$\sin\theta = 0$ or $\cos\theta = \dfrac{1}{2}$

$\theta = 0, \theta = \pi, \theta = \dfrac{\pi}{3}$ or $\theta = \dfrac{5\pi}{3}, \theta = 2\pi$

There are horizontal tangents $\left(\dfrac{dy}{d\theta} = 0, \dfrac{dx}{d\theta} \ne 0\right)$ at

$\theta = \dfrac{2\pi}{3}$ and at $\theta = \dfrac{4\pi}{3}$, and vertical tangents

$\left(\dfrac{dx}{d\theta} = 0, \dfrac{dy}{d\theta} \ne 0\right)$ at $\theta = \dfrac{\pi}{3}$, at $\theta = \pi$, and at $\theta = \dfrac{5\pi}{3}$.

For $\theta = 0$ (or 2π), $\dfrac{dy}{dx}$ becomes $\dfrac{0}{0}$ and l'Hôpital's Rule

leads to $\dfrac{dy}{dx}\bigg|_{\theta=0} = \dfrac{-\sin(0) + 4\sin(0)\cos(0)}{-\cos(0) + 2\cos^2(0) - 2\sin^2(0)} = 0$, so

this is another horizontal tangent line.

Horizontal tangents:

At $\theta = 0$ or $\theta = 2\pi$, we have $r = 0$ and the Cartesian coordinates are $(0, 0)$, so the tangent is $y = 0$.

At $\theta = \frac{2\pi}{3}$, we have $r = \frac{3}{2}$ and the Cartesian coordinates are $\left(\frac{3}{2}\cos\frac{2\pi}{3}, \frac{3}{2}\sin\frac{2\pi}{3}\right) = \left(-\frac{3}{4}, \frac{3\sqrt{3}}{4}\right)$, so the tangent is $y = \frac{3\sqrt{3}}{4}$.

At $\theta = \frac{4\pi}{3}$, we again have $r = \frac{3}{2}$ and the Cartesian coordinates are $\left(\frac{3}{2}\cos\frac{4\pi}{3}, \frac{3}{2}\sin\frac{4\pi}{3}\right) = \left(-\frac{3}{4}, -\frac{3\sqrt{3}}{4}\right)$, so the tangent is $y = -\frac{3\sqrt{3}}{4}$.

Vertical tangents:

At $\theta = \frac{\pi}{3}$, we have $r = \frac{1}{2}$ and the Cartesian coordinates are

$\left(\frac{1}{2}\cos\frac{\pi}{3}, \frac{1}{2}\sin\frac{\pi}{3}\right) = \left(\frac{1}{4}, \frac{\sqrt{3}}{4}\right)$, so the tangent is $x = \frac{1}{4}$.

At $\theta = \pi$, we have $r = 2$ and the Cartesian coordinates are $(2\cos\pi, 2\sin\pi) = (-2, 0)$, so the tangent is $x = -2$.

At $\theta = \frac{5\pi}{3}$, we have $r = \frac{1}{2}$ and the Cartesian coordinates are $\left(\frac{1}{2}\cos\frac{5\pi}{3}, \frac{1}{2}\sin\frac{5\pi}{3}\right) = \left(\frac{1}{4}, -\frac{\sqrt{3}}{4}\right)$, so the tangent is $x = \frac{1}{4}$. In summary, the horizontal tangents are $y = 0$, $y = -\frac{3\sqrt{3}}{4}$, and $y = \frac{3\sqrt{3}}{4}$, and the vertical tangents are $x = -2$ and $x = \frac{1}{4}$.

■ Appendix A2
(pp. 581–584)

1. Step 1: The formula holds for $n = 1$, because $|x_1| = |x_1|$.
Step 2: Suppose $|x_1 + x_2 + \cdots + x_k| \leq |x_1| + |x_2| + \cdots + |x_k|$.
Then, $|x_1 + x_2 + \cdots + x_{k+1}|$
$\leq |x_1 + x_2 + \cdots + x_k| + |x_{k+1}|$ by the triangle inequality.
So, by the transitivity of $\leq$,
$|x_1 + x_2 + \cdots + x_{k+1}| \leq |x_1| + |x_2| + \cdots + |x_{k+1}|$.
The mathematical induction principle now guarantees the original formula for all n.

3. Step 1: The formula holds for $n = 1$, because $\frac{d}{dx}(x) = 1$.
Step 2: Suppose $\frac{d}{dx}(x^k) = kx^{k-1}$. Then
$\frac{d}{dx}(x^{k+1}) = \frac{d}{dx}(x \cdot x^k) = x \cdot \frac{d}{dx}(x^k) + x^k \cdot \frac{d}{dx}(x)$
$= x \cdot kx^{k-1} + x^k = kx^k + x^k = (k+1)x^k$.
The mathematical induction principle now guarantees the original formula for any positive integer n.

5. Step 1: The formula holds for $n = 1$, because $\frac{2}{3} = 1 - \frac{1}{3}$.
Step 2: Suppose $\frac{2}{3^1} + \frac{2}{3^2} + \cdots + \frac{2}{3^k} = 1 - \frac{1}{3^k}$. Then
$\frac{2}{3^1} + \frac{2}{3^2} + \cdots + \frac{2}{3^{k+1}} = 1 - \frac{1}{3^k} + \frac{2}{3^{k+1}}$
$= 1 - \frac{3-2}{3^{k+1}} = 1 - \frac{1}{3^{k+1}}$.
The mathematical induction principle now guarantees the original formula for all positive integers n.

7. Experiment:

n	1	2	3	4	5	6
2^n	2	4	8	16	32	64
n^2	1	4	9	16	25	36

Step 1: The inequality holds for $n = 5$, because $32 > 25$.
Step 2: Suppose $2^k > k^2$. For $k \geq 3$, $k > 2 + \frac{1}{k}$ $\left(\text{since } \frac{1}{k} < 1\right)$, and so $k^2 > 2k + 1$. Then by the transitivity of $>$, $2^k > 2k + 1$. And thus
$2^{k+1} = 2^k + 2^k > k^2 + 2k + 1 = (k+1)^2$.
The mathematical induction principle now guarantees the original inequality for all $n \geq 5$.

9. Step 1: The formula holds for $n = 1$, because
$1^2 = \frac{1(1 + 1/2)(1 + 1)}{3} = 1$.
Step 2: Suppose $1^2 + 2^2 + \cdots + k^2 = \frac{k(k + 1/2)(k + 1)}{3}$.
Then $1^2 + 2^2 + \cdots + (k+1)^2$
$= \frac{k(k + 1/2)(k + 1)}{3} + (k+1)^2$
$= \frac{k(k + 1/2)(k + 1) + 3(k + 1)^2}{3}$
$= \frac{k^3 + (9/2)k^2 + (13/2)k + 3}{3}$
$= \frac{(k + 1)(k + 3/2)(k + 2)}{3}$
$= \frac{(k + 1)[(k + 1) + 1/2][(k + 1) + 1]}{3}$.

The mathematical induction principle now guarantees the original formula for all positive integers n.

11. (a) Step 1: The formula holds for $n = 1$, because
$$\sum_{k=1}^{1}(a_k + a_k) = \sum_{k=1}^{1}a_k + \sum_{k=1}^{1}b_k = a_1 + b_1.$$
Step 2: Suppose $\sum_{k=1}^{i}(a_k + b_k) = \sum_{k=1}^{i}a_k + \sum_{k=1}^{i}b_k$. Then
$$\sum_{k=1}^{i+1}(a_k + b_k) = \left[\sum_{k=1}^{i}(a_k + b_k)\right] + (a_{i+1} + b_{i+1})$$
$$= \left[\sum_{k=1}^{i}a_k\right] + \left[\sum_{k=1}^{i}b_k\right] + a_{i+1} + b_{i+1}$$
$$= \sum_{k=1}^{i+1}a_k + \sum_{k=1}^{i+1}b_k.$$

The mathematical induction principle now guarantees the original formula for every positive integer n.

(b) Step 1: The formula holds for $n = 1$, because
$$\sum_{k=1}^{1}(a_k - b_k) = \sum_{k=1}^{1}a_k - \sum_{k=1}^{1}b_k = a_1 - b_1.$$
Step 2: Suppose $\sum_{k=1}^{i}(a_k - b_k) = \sum_{k=1}^{i}a_k - \sum_{k=1}^{i}b_k$. Then
$$\sum_{k=1}^{i+1}(a_k - b_k) = \left[\sum_{k=1}^{i}(a_k - b_k)\right] + (a_{i+1} - b_{i+1})$$
$$= \left[\sum_{k=1}^{i}a_k\right] - \left[\sum_{k=1}^{i}b_k\right] + a_{i+1} - b_{i+1}$$
$$= \sum_{k=1}^{i+1}a_k - \sum_{k=1}^{i+1}b_k.$$

The mathematical induction principle now guarantees the original formula for every positive integer n.

(c) Step 1: The formula holds for $n = 1$, because
$$\sum_{k=1}^{1}ca_k = c \cdot \sum_{k=1}^{1}a_k = ca_1.$$
Step 2: Suppose $\sum_{k=1}^{i}ca_k = c \cdot \sum_{k=1}^{i}a_k$. Then $\sum_{k=1}^{i+1}ca_{k+1}$
$$= \left[\sum_{k=1}^{i}ca_k\right] + ca_{k+1} = \left[c \cdot \sum_{k=1}^{i}a_k\right] + ca_{k+1}$$
$$= c\left[\left(\sum_{k=1}^{i}a_k\right) + a_{k+1}\right] = c \cdot \sum_{k=1}^{i+1}a_{k+1}.$$

The mathematical induction principle how guarantees the original formula for every positive integer n.

(d) Step 1: The formula $\sum_{k=1}^{n}c = n \cdot c$ holds for $n = 1$, because
$$\sum_{k=1}^{1}c = 1 \cdot c = c.$$
Step 2: Suppose $\sum_{k=1}^{i}c = i \cdot c$. Then
$$\sum_{k=1}^{i+1}c = i \cdot c + c = (i + 1) \cdot c.$$

The mathematical induction principle now guarantees the original formula for every positive integer n.

Appendix A3
(pp. 584–592)

1.
```
    a        c        b
<---+--------+--------+--->  x
    4/9     1/2      4/7
```

Step 1: $\left|x - \frac{1}{2}\right| < \delta \Rightarrow -\delta < x - \frac{1}{2} < \delta$
$$\Rightarrow -\delta + \frac{1}{2} < x < \delta + \frac{1}{2}$$
Step 2: $-\delta + \frac{1}{2} = \frac{4}{9} \Rightarrow \delta = \frac{1}{18}$, or $\delta + \frac{1}{2} = \frac{4}{7} \Rightarrow \delta = \frac{1}{14}$.
The value of δ which assures $\left|x - \frac{1}{2}\right| < \delta$
$\Rightarrow \frac{4}{9} < x < \frac{4}{7}$ is the smaller value, $\delta = \frac{1}{18}$.

3. Step 1: $|x - 3| < \delta \Rightarrow -\delta < x - 3 < \delta$
$$\Rightarrow -\delta + 3 < x < \delta + 3$$
Step 2: From the graph, $-\delta + 3 = 2.61 \Rightarrow \delta = 0.39$, or $\delta + 3 = 3.41 \Rightarrow \delta = 0.41$; thus $\delta = 0.39$.

5. Step 1: $|(2x - 2) - (-6)| < 0.02 \Rightarrow |2x + 4| < 0.02$
$$\Rightarrow -0.02 < 2x + 4 < 0.02$$
$$\Rightarrow -4.02 < 2x < -3.98 \Rightarrow -2.01 < x < -1.99$$
Step 2: $|x - (-2)| < \delta \Rightarrow -\delta < x + 2 < \delta$
$$\Rightarrow -\delta - 2 < x < \delta - 2 \Rightarrow \delta = 0.01.$$

7. Step 1: $|\sqrt{19 - x} - 3| < 1 \Rightarrow -1 < \sqrt{19 - x} - 3 < 1$
$$\Rightarrow 2 < \sqrt{19 - x} < 4 \Rightarrow 4 < 19 - x < 16$$
$$\Rightarrow -4 > x - 19 > -16 \Rightarrow 15 > x > 3$$
or $3 < x < 15$
Step 2: $|x - 10| < \delta \Rightarrow -\delta < x - 10 < \delta$
$$\Rightarrow -\delta + 10 < x < \delta + 10.$$
Then $-\delta + 10 = 3 \Rightarrow \delta = 7$, or $\delta + 10 = 15$
$\Rightarrow \delta = 5$; thus $\delta = 5$.

9. (a) $\lim_{x \to -5} \frac{x^2 + 6x + 5}{x + 5} = \lim_{x \to -5} \frac{(x + 5)(x + 1)}{x + 5} = \lim_{x \to -5}(x + 1)$
$= -4, x \neq -5$.

(b) Step 1: $\left|\left(\frac{x^2 + 6x + 5}{x + 5}\right) - (-4)\right| < 0.05$
$$\Rightarrow -0.05 < \frac{(x + 5)(x + 1)}{x + 5} + 4 < 0.05$$
$$\Rightarrow -4.05 < x + 1 < -3.95, x \neq -5$$
$$\Rightarrow -5.05 < x < -4.95, x \neq -5.$$
Step 2: $|x - (-5)| < \delta \Rightarrow -\delta < x + 5 < \delta$
$$\Rightarrow -\delta - 5 < x < \delta - 5.$$
Then $-\delta - 5 = -5.05 \Rightarrow \delta = 0.05$, or
$\delta - 5 = -4.95 \Rightarrow \delta = 0.05$; thus $\delta = 0.05$.

11. (a) $\lim_{x \to 1} (\sin x) = \sin 1 \approx 0.841$

(b) Step 1: $|\sin x - \sin 1| < 0.01$

$\Rightarrow -0.01 < \sin x - \sin 1 < 0.01$

$\Rightarrow \sin 1 - 0.01 < \sin x < \sin 1 + 0.01$

$\Rightarrow \sin^{-1}(\sin 1 - 0.01) < x$
$< \sin^{-1}(\sin 1 + 0.01)$

Step 2: $|x - 1| < \delta \Rightarrow -\delta < x - 1 < \delta$

$\Rightarrow 1 - \delta < x < 1 + \delta.$

Then $1 - \delta = \sin^{-1}(\sin 1 - 0.01)$

$\Rightarrow \delta = 1 - \sin^{-1}(\sin 1 - 0.01) \approx 0.0182$, or

$1 + \delta = \sin^{-1}(\sin 1 + 0.01)$

$\Rightarrow \delta = \sin^{-1}(\sin 1 + 0.01) - 1 \approx 0.0188.$

Choose $\delta = 0.018.$

Alternately, graph $y_1 = \sin x$, $y_2 = \sin 1 - 1$, and $y_3 = \sin 1 + 1$. The curve intersects the lines at

$x \approx 0.98175 = 1 - 0.01825$ and at

$x \approx 1.01878 = 1 + 0.01878$. We may choose

$\delta = 0.018.$

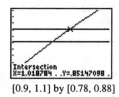

[0.9, 1.1] by [0.78, 0.88]

13. Step 1: For $x \neq 1$, $|x^2 - 1| < \epsilon \Rightarrow -\epsilon < x^2 - 1 < \epsilon$

$\Rightarrow 1 - \epsilon < x^2 < 1 + \epsilon$

$\Rightarrow \sqrt{1 - \epsilon} < |x| < \sqrt{1 + \epsilon}$

$\Rightarrow \sqrt{1 - \epsilon} < x < \sqrt{1 + \epsilon}$ near $x = 1.$

Step 2: $|x - 1| < \delta \Rightarrow -\delta < x - 1 < \delta$

$\Rightarrow -\delta + 1 < x < \delta + 1.$

Then $-\delta + 1 = \sqrt{1 - \epsilon} \Rightarrow \delta = 1 - \sqrt{1 - \epsilon},$

or $\delta + 1 = \sqrt{1 + \epsilon} \Rightarrow \delta = \sqrt{1 + \epsilon} - 1.$

Choose $\delta = \min\{1 - \sqrt{1 - \epsilon}, \sqrt{1 + \epsilon} - 1\},$

that is, the smaller of the two distances.

15. (a) $\sqrt{(5 + \delta) - 5} = \epsilon \Rightarrow \sqrt{\delta} = \epsilon \Rightarrow \delta = \epsilon^2$

$\Rightarrow I = (5, 5 + \epsilon^2)$

(b) $\lim_{x \to 5^+} \sqrt{x - 5} = 0$

17. If L, c, and k are real numbers and $\lim_{x \to c} f(x) = L$, show that for any $\epsilon > 0$, there is a $\delta > 0$ such that $0 < |x - c| < \delta$ $\Rightarrow |k \cdot f(x) - k \cdot L| < \epsilon.$

Proof: For any $\epsilon > 0$, let $\epsilon' = \frac{\epsilon}{|k|}$. Since $\lim_{x \to c} f(x) = L$, there is a $\delta > 0$ such that $0 < |x - c| < \delta \Rightarrow |f(x) - L| < \epsilon'$
$= \frac{\epsilon}{|k|}.$ Therefore, $0 < |x - c| < \delta \Rightarrow |k \cdot f(x) - k \cdot L| < \epsilon.$

19. $\lim_{x \to c} [f_1(x) + f_2(x) + f_3(x)] = \lim_{x \to c} [f_1(x) + f_2(x)] + L_3$
$= L_1 + L_2 + L_3$, by two applications of the Sum Rule.

To generalize:

Step 1 ($n = 1$): $\lim_{x \to c} f_1(x) = L_1$ as given.

Step 2: Suppose $\lim_{x \to c} [f_1(x) + f_2(x) + \cdots + f_k(x)]$
$= L_1 + L_2 + \cdots + L_k.$ Then

$\lim_{x \to c} [f_1(x) + f_2(x) + \cdots + f_{k+1}(x)]$

$= \lim[f_1(x) + f_2(x) + \cdots + f_k(x)] + L_{k+1}$

$= L_1 + L_2 + \cdots + L_{k+1}$, by the Sum Rule.

21. $\lim_{x \to c} x^n = \lim_{x \to c} \underbrace{(x \cdot x \cdot \cdots \cdot x)}_{n \text{ factors}} = \underbrace{c \cdot c \cdot \cdots \cdot c}_{n \text{ factors}} = c^n$

23. By the Quotient Rule, $\lim_{x \to c} \frac{f(x)}{g(x)} = \frac{\lim_{x \to c} f(x)}{\lim_{x \to c} g(x)} = \frac{f(c)}{g(c)}.$

■ Appendix A5.1
(pp. 593–606)

1. $(x - h)^2 + (y - k)^2 = a^2$
$(x - 0)^2 + (y - 2)^2 = 2^2$
$x^2 + (y - 2)^2 = 4$

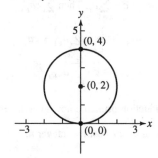

3. Complete the squares.
$x^2 + y^2 + 4x - 4y + 4 = 0$
$x^2 + 4x + 4 + y^2 - 4y + 4 = 4$
$(x + 2)^2 + (y - 2)^2 = 2^2$
Center $= (-2, 2)$; radius $= 2$

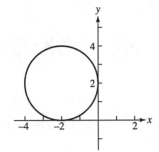

5. The circle with center at $(1, 0)$ and radius 2 plus its interior.

7. $y^2 = 8x \Rightarrow 4p = 8 \Rightarrow p = 2$; focus is $(2, 0)$, directrix is $x = -2$

9. $x^2 = -6y \Rightarrow 4p = 6 \Rightarrow p = \frac{3}{2}$; focus is $\left(0, -\frac{3}{2}\right)$, directrix is $y = \frac{3}{2}$

11. $\frac{x^2}{4} - \frac{y^2}{9} = 1 \Rightarrow c = \sqrt{4 + 9} = \sqrt{13}$
$\Rightarrow$ foci are $(\pm\sqrt{13}, 0)$; vertices are $(\pm 2, 0)$;
asymptotes are $y = \pm\frac{3}{2}x$

13. $\frac{x^2}{2} + y^2 = 1 \Rightarrow c = \sqrt{2 - 1} = 1 \Rightarrow$ foci are $(\pm 1, 0)$; vertices are $(\pm\sqrt{2}, 0)$

15. $y^2 = 12x \Rightarrow 4p = 12 \Rightarrow p = 3$; focus is $(3, 0)$, directrix is $x = -3$

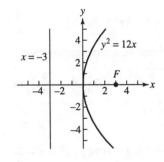

17. $16x^2 + 25y^2 = 400 \Rightarrow \frac{x^2}{25} + \frac{y^2}{16} = 1$
$\Rightarrow c = \sqrt{a^2 - b^2} = \sqrt{25 - 16} = 3$

foci are $(\pm 3, 0)$

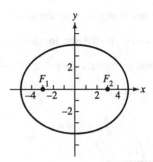

19. Foci: $(\pm\sqrt{2}, 0)$, Vertices: $(\pm 2, 0) \Rightarrow a = 2, c = \sqrt{2}$
$\Rightarrow b^2 = a^2 - c^2 = 4 - (\sqrt{2})^2 = 2 \Rightarrow \frac{x^2}{4} + \frac{y^2}{2} = 1$

21. $x^2 - y^2 = 1 \Rightarrow c = \sqrt{a^2 + b^2} = \sqrt{1 + 1} = \sqrt{2}$;
asymptotes are $y = \pm x$,
foci are $(\pm\sqrt{2}, 0)$

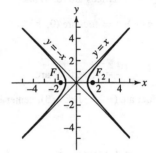

23. Foci: $(0, \pm\sqrt{2})$, Asymptotes: $y = \pm x \Rightarrow c = \sqrt{2}$
and $\frac{b}{a} = 1 \Rightarrow a = b \Rightarrow c^2 = a^2 + b^2 = 2a^2 \Rightarrow 2 = 2a^2$
$\Rightarrow a = 1 \Rightarrow b = 1 \Rightarrow y^2 - x^2 = 1$

25. (a) $y^2 = 8x \Rightarrow 4p = 8 \Rightarrow p = 2 \Rightarrow$ directrix is $x = -2$, focus is $(2, 0)$, and vertex is $(0, 0)$; therefore the new directrix is $x = -1$, the new focus is $(3, -2)$, and the new vertex is $(1, -2)$.

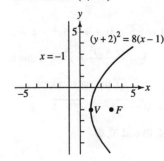

27. (a) $\frac{x^2}{16} - \frac{y^2}{9} = 1 \Rightarrow$ center is $(0, 0)$, vertices are $(-4, 0)$ and $(4, 0)$, and the asymptotes are $y = \pm\frac{3x}{4}$, $c = \sqrt{a^2 + b^2} = \sqrt{25} = 5 \Rightarrow$ foci are $(-5, 0)$ and $(5, 0)$; therefore the new center is $(2, 0)$, the new vertices are $(-2, 0)$ and $(6, 0)$, the new foci are $(-3, 0)$ and $(7, 0)$, and the new asymptotes are $y = \pm\frac{3(x-2)}{4}$.

(b)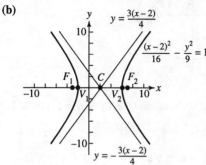

29. Original ellipse: $\frac{x^2}{6} + \frac{y^2}{9} = 1$; vertices are $(0, 3)$ and $(0, -3)$; $c^2 = 9 - 6 \Rightarrow c = \sqrt{3} \Rightarrow$ foci are $(0, \pm\sqrt{3})$; center is $(0, 0)$.
New ellipse: $\frac{(x+2)^2}{6} + \frac{(y+1)^2}{9} = 1$; vertices are $(-2, 2)$ and $(-2, -4)$; foci are $(-2, -1 \pm \sqrt{3})$; center is $(-2, -1)$.

31. Original hyperbola: $y^2 - x^2 = 1$; vertices are $(0, 1)$ and $(0, -1)$; $c^2 = 1 + 1 \Rightarrow c = \sqrt{2} \Rightarrow$ foci are $(0, \pm\sqrt{2})$; center is $(0, 0)$; asymptotes are $y = \pm x$.
New hyperbola: $(y - 1)^2 - (x + 1)^2 = 1$; vertices are $(-1, 2)$ and $(-1, 0)$; foci are $(-1, 1 \pm \sqrt{2})$; center is $(-1, 1)$; asymptotes are $y = \pm(x + 1) + 1$.

33. $2x^2 + 2y^2 - 28x + 12y + 114 = 0$
$\Rightarrow x^2 - 14x + 49 + y^2 + 6y + 9 = -57 + 49 + 9$
$\Rightarrow (x - 7)^2 + (y + 3)^2 = 1$; this is a circle: center at $C(7, -3)$, $a = 1$

35. $x^2 + 5y^2 + 4x = 1 \Rightarrow x^2 + 4x + 4 + 5y^2 = 4 + 1$
$\Rightarrow (x + 2)^2 + 5y^2 = 5 \Rightarrow \frac{(x+2)^2}{5} + y^2 = 1$; this is an ellipse: the center is $(-2, 0)$, the vertices are $(-2 \pm \sqrt{5}, 0)$; $c = \sqrt{a^2 - b^2} = \sqrt{5 - 1} = 2$
$\Rightarrow$ the foci are $(-4, 0)$ and $(0, 0)$

37. Volume of the Parabolic Solid:
$$V_1 = \int_0^{b/2} 2\pi x\left(h - \frac{4h}{b^2}x^2\right)dx = 2\pi h \int_0^{b/2}\left(x - \frac{4x^3}{b^2}\right)dx$$
$$= 2\pi h\left[\frac{x^2}{2} - \frac{x^4}{b^2}\right]_0^{b/2} = \frac{\pi h b^2}{8};$$
Volume of the Cone: $V_2 = \frac{1}{3}\pi\left(\frac{b}{2}\right)^2 h = \frac{1}{3}\pi\left(\frac{b^2}{4}\right)h = \frac{\pi h b^2}{12}$;
therefore $V_1 = \frac{3}{2}V_2$

39. Let $P_1(-p, y_1)$ be any point on $x = -p$, and let $P(x, y)$ be a point where a tangent intersects $y^2 = 4px$. Now
$y^2 = 4px \Rightarrow 2y\frac{dy}{dx} = 4p \Rightarrow \frac{dy}{dx} = \frac{2p}{y}$; then the slope of a tangent line from P_1 is $\frac{y - y_1}{x - (-p)} = \frac{dy}{dx} = \frac{2p}{y}$
$\Rightarrow y^2 - yy_1 = 2px + 2p^2$. Since $x = \frac{y^2}{4p}$, we have
$y^2 - yy_1 = 2p\left(\frac{y^2}{4p}\right) + 2p^2 \Rightarrow y^2 - yy_1 = \frac{1}{2}y^2 + 2p^2$
$\Rightarrow \frac{1}{2}y^2 - yy_1 - 2p^2 = 0$
$\Rightarrow y = \frac{2y_1 \pm \sqrt{4y_1^2 + 16p^2}}{2} = y_1 \pm \sqrt{y_1^2 + 4p^2}$. Therefore the slopes of the two tangents from P_1 are
$m_1 = \frac{2p}{y_1 + \sqrt{y_1^2 + 4p^2}}$ and $m_2 = \frac{2p}{y_1 - \sqrt{y_1^2 + 4p^2}}$
$\Rightarrow m_1 m_2 = \frac{4p^2}{y_1^2 - (y_1^2 + 4p^2)} = -1 \Rightarrow$ the lines are perpendicular.

41. (a) Around the x-axis: $9x^2 + 4y^2 = 36 \Rightarrow y^2 = 9 - \frac{9}{4}x^2$
$\Rightarrow y = \pm\sqrt{9 - \frac{9}{4}x^2}$ and we use the positive root
$\Rightarrow V = 2\int_0^2 \pi\left(\sqrt{9 - \frac{9}{4}x^2}\right)^2 dx = 2\int_0^2 \pi\left(9 - \frac{9}{4}x^2\right) dx$
$= 2\pi\left[9x - \frac{3}{4}x^3\right]_0^2 = 24\pi$

(b) Around the y-axis: $9x^2 + 4y^2 = 36 \Rightarrow x^2 = 4 - \frac{4}{9}y^2$
$\Rightarrow x = \pm\sqrt{4 - \frac{4}{9}y^2}$ and we use the positive root
$\Rightarrow V = 2\int_0^3 \pi\left(\sqrt{4 - \frac{4}{9}y^2}\right)^2 dy = 2\int_0^3 \pi\left(4 - \frac{4}{9}y^2\right) dy$
$= 2\pi\left[4y - \frac{4}{27}y^3\right]_0^3 = 16\pi$

43. $x^2 - y^2 = 1 \Rightarrow x = \pm\sqrt{1 + y^2}$ on the interval $-3 \leq y \leq 3$
$\Rightarrow V = \int_{-3}^{3} \pi(\sqrt{1 + y^2})^2 \, dy = 2\int_{0}^{3} \pi(\sqrt{1 + y^2})^2 \, dy$
$= 2\pi\int_{0}^{3} (1 + y^2) \, dy = 2\pi\left[y + \frac{y^3}{3}\right]_{0}^{3} = 24\pi$

45. $\frac{dr_A}{dt} = \frac{dr_B}{dt} \Rightarrow \frac{d}{dt}(r_A - r_B) = 0$
$\Rightarrow r_A - r_B = $ a constant

■ Appendix A5.2
(pp. 606–611)

1. $16x^2 + 25y^2 = 400 \Rightarrow \frac{x^2}{25} + \frac{y^2}{16} = 1$
$\Rightarrow c = \sqrt{a^2 - b^2} = \sqrt{25 - 16} = 3$
$\Rightarrow e = \frac{c}{a} = \frac{3}{5}$; $F(\pm 3, 0)$; directrices are
$x = 0 \pm \frac{a}{e} = \pm\frac{5}{\left(\frac{3}{5}\right)} = \pm\frac{25}{3}$

3. $3x^2 + 2y^2 = 6 \Rightarrow \frac{x^2}{2} + \frac{y^2}{3} = 1$
$\Rightarrow c = \sqrt{a^2 - b^2} = \sqrt{3 - 2} = 1 \Rightarrow e = \frac{c}{a} = \frac{1}{\sqrt{3}}$;
$F(0, \pm 1)$; directrices are $y = 0 \pm \frac{a}{e} = \pm\frac{\sqrt{3}}{\left(\frac{1}{\sqrt{3}}\right)} = \pm 3$

5. Foci: $(0, \pm 3)$, $e = 0.5 \Rightarrow c = 3$ and $a = \frac{c}{e} = \frac{3}{0.5} = 6$
$\Rightarrow b^2 = 36 - 9 = 27 \Rightarrow \frac{x^2}{27} + \frac{y^2}{36} = 1$

7. Vertices: $(\pm 10, 0)$, $e = 0.24 \Rightarrow a = 10$ and
$c = ae = 10(0.24) = 2.4 \Rightarrow b^2 = 100 - 5.76 = 94.24$
$\Rightarrow \frac{x^2}{100} + \frac{y^2}{94.24} = 1$

9. Focus: $(\sqrt{5}, 0)$, Directrix: $x = \frac{9}{\sqrt{5}} \Rightarrow c = ae = \sqrt{5}$ and
$\frac{a}{e} = \frac{9}{\sqrt{5}} \Rightarrow \frac{ae}{e^2} = \frac{9}{\sqrt{5}} \Rightarrow \frac{\sqrt{5}}{e^2} = \frac{9}{\sqrt{5}} \Rightarrow e^2 = \frac{5}{9} \Rightarrow e = \frac{\sqrt{5}}{3}$.
Then $PF = \frac{\sqrt{5}}{3}PD$
$\Rightarrow \sqrt{(x - \sqrt{5})^2 + (y - 0)^2} = \frac{\sqrt{5}}{3}\left|x - \frac{9}{\sqrt{5}}\right|$
$\Rightarrow (x - \sqrt{5})^2 + y^2 = \frac{5}{9}\left(x - \frac{9}{\sqrt{5}}\right)^2$
$\Rightarrow x^2 - 2\sqrt{5}x + 5 + y^2 = \frac{5}{9}\left(x^2 - \frac{18}{\sqrt{5}}x + \frac{81}{5}\right)$
$\Rightarrow \frac{4}{9}x^2 + y^2 = 4 \Rightarrow \frac{x^2}{9} + \frac{y^2}{4} = 1$

11. $e = \frac{4}{5} \Rightarrow$ take $c = 4$ and $a = 5$; $c^2 = a^2 - b^2$
$\Rightarrow 16 = 25 - b^2 \Rightarrow b^2 = 9 \Rightarrow b = 3$;
therefore $\frac{x^2}{25} + \frac{y^2}{9} = 1$

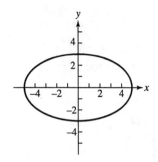

13. One axis is from $(1, 1)$ to $(1, 7)$ and is 6 units long; the other axis is from $(3, 4)$ to $(-1, 4)$ and is 4 units long. Therefore, $a = 3$, $b = 2$ and the major axis is vertical. The center is the point $C(1, 4)$ and the ellipse is given by
$\frac{(x - 1)^2}{4} + \frac{(y - 4)^2}{9} = 1$; $c^2 = a^2 - b^2 = 3^2 - 2^2 = 5$
$\Rightarrow c = \sqrt{5}$; therefore the foci are $F(1, 4 \pm \sqrt{5})$, the
eccentricity is $e = \frac{c}{a} = \frac{\sqrt{5}}{3}$, and the directrices are
$y = 4 \pm \frac{a}{e} = 4 \pm \frac{3}{\left(\frac{\sqrt{5}}{3}\right)} = 4 \pm \frac{9\sqrt{5}}{5}$.

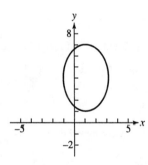

15. $9x^2 - 16y^2 = 144 \Rightarrow \frac{x^2}{16} - \frac{y^2}{9} = 1$
$\Rightarrow c = \sqrt{a^2 + b^2} = \sqrt{16 + 9} = 5 \Rightarrow e = \frac{c}{a} = \frac{5}{4}$;
asymptotes are $y = \pm\frac{3}{4}x$; $F(\pm 5, 0)$; directrices are
$x = 0 \pm \frac{a}{e} = \pm\frac{16}{5}$.

17. $8x^2 - 2y^2 = 16 \Rightarrow \frac{x^2}{2} - \frac{y^2}{8} = 1$
$\Rightarrow c = \sqrt{a^2 + b^2} = \sqrt{2 + 8} = \sqrt{10}$
$\Rightarrow e = \frac{c}{a} = \frac{\sqrt{10}}{\sqrt{2}} = \sqrt{5}$; asymptotes are $y = \pm 2x$;
$F(\pm\sqrt{10}, 0)$; directrices are $x = 0 \pm \frac{a}{e} = \pm\frac{\sqrt{2}}{\sqrt{5}} = \pm\frac{2}{\sqrt{10}}$

19. Vertices $(0, \pm 1)$ and $e = 3 \Rightarrow a = 1$ and $e = \dfrac{c}{a} = 3$
$\Rightarrow c = 3a = 3 \Rightarrow b^2 = c^2 - a^2 = 9 - 1 = 8$
$\Rightarrow y^2 - \dfrac{x^2}{8} = 1$

21. Focus $(4, 0)$ and directrix $x = 2 \Rightarrow c = ae = 4$ and $\dfrac{a}{e} = 2$
$\Rightarrow \dfrac{ae}{e^2} = 2 \Rightarrow \dfrac{4}{e^2} = 2 \Rightarrow e^2 = 2 \Rightarrow e = \sqrt{2}$.
Then $PF = \sqrt{2}PD \Rightarrow \sqrt{(x-4)^2 + (y-0)^2} = \sqrt{2}|x - 2|$
$\Rightarrow (x - 4)^2 + y^2 = 2(x - 2)^2$
$\Rightarrow x^2 - 8x + 16 + y^2 = 2(x^2 - 4x + 4)$
$\Rightarrow -x^2 + y^2 = -8 \Rightarrow \dfrac{x^2}{8} - \dfrac{y^2}{8} = 1$

23. $\sqrt{(x-1)^2 + (y+3)^2} = \dfrac{3}{2}|y - 2|$
$\Rightarrow x^2 - 2x + 1 + y^2 + 6y + 9 = \dfrac{9}{4}(y^2 - 4y + 4)$
$\Rightarrow 4x^2 - 5y^2 - 8x + 60y + 4 = 0$
$\Rightarrow 4(x^2 - 2x + 1) - 5(y^2 - 12y + 36) = -4 + 4 - 180$
$\Rightarrow \dfrac{(y-6)^2}{36} - \dfrac{(x-1)^2}{45} = 1$

25. The ellipse must pass through $(0, 0) \Rightarrow c = 0$; the point $(-1, 2)$ lies on the ellipse $\Rightarrow -a + 2b = -8$. The ellipse is tangent to the x-axis $\Rightarrow$ its center is on the y-axis, so $a = 0$ and $b = -4 \Rightarrow$ the equation is $4x^2 + y^2 - 4y = 0$. Next, $4x^2 + y^2 - 4y + 4 = 4 \Rightarrow 4x^2 + (y - 2)^2 = 4$
$\Rightarrow x^2 + \dfrac{(y-2)^2}{4} = 1 \Rightarrow a = 2$ and $b = 1$ (now using the standard symbols) $\Rightarrow c^2 = a^2 - b^2 = 4 - 1 = 3$
$\Rightarrow c = \sqrt{3} \Rightarrow e = \dfrac{c}{a} = \dfrac{\sqrt{3}}{2} = \dfrac{\sqrt{3}}{2}$.

27. To prove the reflective property for hyperbolas:
$b^2 x^2 - a^2 y^2 = a^2 b^2$
$2b^2 x - 2a^2 y y' = 0$
$y' = \dfrac{b^2 x}{a^2 y}$
Let $P(x_0, y_0)$ be a point of tangency (see the accompanying figure). The slope from P to $F(-c, 0)$ is $\dfrac{y_0}{x_0 + c}$ and from P to $F_2(c, 0)$ it is $\dfrac{y_0}{x_0 - c}$. Let the tangent through P meet the x-axis in point A, and define the angles $\angle F_1 PA = \alpha$ and $\angle F_2 PA = \beta$. We will show that $\tan \alpha = \tan \beta$.

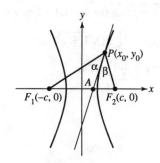

From the preliminary result in Exercise 26,

$$\tan \alpha = \dfrac{\left(\dfrac{x_0 b^2}{y_0 a^2} - \dfrac{y_0}{x_0 + c}\right)}{1 + \left(\dfrac{x_0 b^2}{y_0 a^2}\right)\left(\dfrac{y_0}{x_0 + c}\right)}$$

$$= \dfrac{x_0^2 b^2 + x_0 b^2 c - y_0^2 a^2}{x_0 y_0 a^2 + y_0 a^2 c + x_0 y_0 b^2}$$

$$= \dfrac{a^2 b^2 + x_0 b^2 c}{x_0 y_0 c^2 + y_0 a^2 c} = \dfrac{b^2}{y_0 c}.$$ In a similar manner,

$$\tan \beta = \dfrac{\left(\dfrac{y_0}{x_0 - c} - \dfrac{x_0 b^2}{y_0 a^2}\right)}{1 + \left(\dfrac{y_0}{x_0 - c}\right)\left(\dfrac{x_0 b^2}{y_0 a^2}\right)} = \dfrac{b^2}{y_0 c}.$$ Since $\tan \alpha = \tan \beta$, and α and β are acute angles, we have $\alpha = \beta$.

■ Appendix A5.3
(pp. 612–618)

1. $x^2 - 3xy + y^2 - x = 0$
$\Rightarrow B^2 - 4AC = (-3)^2 - 4(1)(1) = 5 > 0 \Rightarrow$ Hyperbola

3. $3x^2 - 7xy + \sqrt{17}y^2 = 1$
$\Rightarrow B^2 - 4AC = (-7)^2 - 4(3)\sqrt{17} \approx -0.477 < 0$
$\Rightarrow$ Ellipse

5. $x^2 + 2xy + y^2 + 2x - y + 2 = 0$
$\Rightarrow B^2 - 4AC = 2^2 - 4(1)(1) = 0 \Rightarrow$ Parabola

7. $x^2 + 4xy + 4y^2 - 3x = 6$
$\Rightarrow B^2 - 4AC = 4^2 - 4(1)(4) = 0 \Rightarrow$ Parabola

9. $xy + y^2 - 3x = 5 \Rightarrow B^2 - 4AC = 1^2 - 4(0)(1) = 1 > 0$
$\Rightarrow$ Hyperbola

11. $3x^2 - 5xy + 2y^2 - 7x - 14y = -1$
$\Rightarrow B^2 - 4AC = (-5)^2 - 4(3)(2) = 1 > 0 \Rightarrow$ Hyperbola

13. $x^2 - 3xy + 3y^2 + 6y = 7$
$\Rightarrow B^2 - 4AC = (-3)^2 - 4(1)(3) = -3 < 0 \Rightarrow$ Ellipse

15. $6x^2 + 3xy + 2y^2 + 17y + 2 = 0$
$\Rightarrow B^2 - 4AC = 3^2 - 4(6)(2) = -39 < 0 \Rightarrow$ Ellipse

17. $\cot 2\alpha = \dfrac{A-C}{B} = \dfrac{0}{1} = 0 \Rightarrow 2\alpha = \dfrac{\pi}{2} \Rightarrow \alpha = \dfrac{\pi}{4}$;

therefore $x = x' \cos \alpha - y' \sin \alpha, y = x' \sin \alpha + y' \cos \alpha$

$\Rightarrow x = \dfrac{\sqrt{2}}{2}x' - \dfrac{\sqrt{2}}{2}y', y = \dfrac{\sqrt{2}}{2}x' + \dfrac{\sqrt{2}}{2}y'$

$\Rightarrow \left(\dfrac{\sqrt{2}}{2}x' - \dfrac{\sqrt{2}}{2}y'\right)\left(\dfrac{\sqrt{2}}{2}x' + \dfrac{\sqrt{2}}{2}y'\right) = 2$

$\Rightarrow \dfrac{1}{2}x'^2 - \dfrac{1}{2}y'^2 = 2 \Rightarrow x'^2 - y'^2 = 4 \Rightarrow$ Hyperbola

19. $\cot 2\alpha = \dfrac{A-C}{B} = \dfrac{3-1}{2\sqrt{3}} = \dfrac{1}{\sqrt{3}} \Rightarrow 2\alpha = \dfrac{\pi}{3} \Rightarrow \alpha = \dfrac{\pi}{6}$;

therefore $x = x' \cos \alpha - y' \sin \alpha, y = x' \sin \alpha + y' \cos \alpha$

$\Rightarrow x = \dfrac{\sqrt{3}}{2}x' - \dfrac{1}{2}y', y = \dfrac{1}{2}x' + \dfrac{\sqrt{3}}{2}y'$

$\Rightarrow 3\left(\dfrac{\sqrt{3}}{2}x' - \dfrac{1}{2}y'\right)^2$
$\quad + 2\sqrt{3}\left(\dfrac{\sqrt{3}}{2}x' - \dfrac{1}{2}y'\right)\left(\dfrac{1}{2}x' + \dfrac{\sqrt{3}}{2}y'\right)$
$\quad + \left(\dfrac{1}{2}x' + \dfrac{\sqrt{3}}{2}y'\right)^2 - 8\left(\dfrac{\sqrt{3}}{2}x' - \dfrac{1}{2}y'\right)$
$\quad + 8\sqrt{3}\left(\dfrac{1}{2}x' + \dfrac{\sqrt{3}}{2}y'\right) = 0$

$\Rightarrow 4x'^2 + 16y' = 0 \Rightarrow$ Parabola

21. $\cot 2\alpha = \dfrac{A-C}{B} = \dfrac{1-1}{-2} = 0 \Rightarrow 2\alpha = \dfrac{\pi}{2} \Rightarrow \alpha = \dfrac{\pi}{4}$;

therefore $x = x' \cos \alpha - y' \sin \alpha, y = x' \sin \alpha + y' \cos \alpha$

$\Rightarrow x = \dfrac{\sqrt{2}}{2}x' - \dfrac{\sqrt{2}}{2}y', y = \dfrac{\sqrt{2}}{2}x' + \dfrac{\sqrt{2}}{2}y'$

$\Rightarrow \left(\dfrac{\sqrt{2}}{2}x' - \dfrac{\sqrt{2}}{2}y'\right)^2$
$\quad - 2\left(\dfrac{\sqrt{2}}{2}x' - \dfrac{\sqrt{2}}{2}y'\right)\left(\dfrac{\sqrt{2}}{2}x' + \dfrac{\sqrt{2}}{2}y'\right)$
$\quad + \left(\dfrac{\sqrt{2}}{2}x' + \dfrac{\sqrt{2}}{2}y'\right)^2 = 2$

$\Rightarrow 2y'^2 = 2 \Rightarrow y'^2 = 1 \Rightarrow$ Parallel horizontal lines

23. $\cot 2\alpha = \dfrac{A-C}{B} = \dfrac{\sqrt{2}-\sqrt{2}}{2\sqrt{2}} = 0 \Rightarrow 2\alpha = \dfrac{\pi}{2} \Rightarrow \alpha = \dfrac{\pi}{4}$;

therefore $x = x' \cos \alpha - y' \sin \alpha, y = x' \sin \alpha + y' \cos \alpha$

$\Rightarrow x = \dfrac{\sqrt{2}}{2}x' - \dfrac{\sqrt{2}}{2}y', y = \dfrac{\sqrt{2}}{2}x' + \dfrac{\sqrt{2}}{2}y'$

$\Rightarrow \sqrt{2}\left(\dfrac{\sqrt{2}}{2}x' - \dfrac{\sqrt{2}}{2}y'\right)^2$
$\quad + 2\sqrt{2}\left(\dfrac{\sqrt{2}}{2}x' - \dfrac{\sqrt{2}}{2}y'\right)\left(\dfrac{\sqrt{2}}{2}x' + \dfrac{\sqrt{2}}{2}y'\right)$
$\quad + \sqrt{2}\left(\dfrac{\sqrt{2}}{2}x' + \dfrac{\sqrt{2}}{2}y'\right)^2 - 8\left(\dfrac{\sqrt{2}}{2}x' - \dfrac{\sqrt{2}}{2}y'\right)$
$\quad + 8\left(\dfrac{\sqrt{2}}{2}x' + \dfrac{\sqrt{2}}{2}y'\right) = 0$

$\Rightarrow 2\sqrt{2}x'^2 + 8\sqrt{2}y' = 0 \Rightarrow x'^2 + 4y' = 0 \Rightarrow$ Parabola

25. $\cot 2\alpha = \dfrac{A-C}{B} = \dfrac{3-3}{2} = 0 \Rightarrow 2\alpha = \dfrac{\pi}{2} \Rightarrow \alpha = \dfrac{\pi}{4}$;

therefore $x = x' \cos \alpha - y' \sin \alpha,$

$y = x' \sin \alpha + y' \cos \alpha \Rightarrow x = \dfrac{\sqrt{2}}{2}x' - \dfrac{\sqrt{2}}{2}y'$,

$y = \dfrac{\sqrt{2}}{2}x' + \dfrac{\sqrt{2}}{2}y'$

$\Rightarrow 3\left(\dfrac{\sqrt{2}}{2}x' - \dfrac{\sqrt{2}}{2}y'\right)^2$
$\quad + 2\left(\dfrac{\sqrt{2}}{2}x' - \dfrac{\sqrt{2}}{2}y'\right)\left(\dfrac{\sqrt{2}}{2}x' + \dfrac{\sqrt{2}}{2}y'\right)$
$\quad + 3\left(\dfrac{\sqrt{2}}{2}x' + \dfrac{\sqrt{2}}{2}y'\right)^2 = 19$

$\Rightarrow 4x'^2 + 2y'^2 = 19 \Rightarrow$ Ellipse

27. $\cot 2\alpha = \dfrac{A-C}{B} = \dfrac{14-2}{16} = \dfrac{3}{4} \Rightarrow \cos 2\alpha = \dfrac{3}{5}$ (if we choose 2α in Quadrant I);

thus $\sin \alpha = \sqrt{\dfrac{1 - \cos 2\alpha}{2}} = \sqrt{\dfrac{1 - (3/5)}{2}} = \dfrac{1}{\sqrt{5}}$ and

$\cos \alpha = \sqrt{\dfrac{1 + \cos 2\alpha}{2}} = \sqrt{\dfrac{1 + (3/5)}{2}} = \dfrac{2}{\sqrt{5}}$

$\left(\text{or } \sin \alpha = -\dfrac{2}{\sqrt{5}} \text{ and } \cos \alpha = \dfrac{1}{\sqrt{5}}\right)$

29. $\tan 2\alpha = \dfrac{-1}{1-3} = \dfrac{1}{2} \Rightarrow 2\alpha \approx 26.57° \Rightarrow \alpha \approx 13.28°$

$\Rightarrow \sin \alpha \approx 0.23, \cos \alpha \approx 0.97;$ then $A' \approx 0.88, B' \approx 0.00,$

$C' \approx 3.12, D' \approx 0.74, E' \approx -1.20,$ and $F' = -3$

$\Rightarrow 0.88x'^2 + 3.12y'^2 + 0.74x' - 1.20y' - 3 = 0,$ an

ellipse

31. $\tan 2\alpha = \dfrac{-4}{1-4} = \dfrac{4}{3} \Rightarrow 2\alpha \approx 53.13° \Rightarrow \alpha \approx 26.57°$

$\Rightarrow \sin \alpha \approx 0.45, \cos \alpha \approx 0.89;$ then $A' \approx 0.00, B' \approx 0.00,$

$C' \approx 5.00, D' = 0, E' = 0,$ and $F' = -5$

$\Rightarrow 5.00y'^2 - 5 = 0$ or $y' = \pm 1.00,$ parallel lines

33. $\tan 2\alpha = \dfrac{5}{3-2} = 5 \Rightarrow 2\alpha \approx 78.69° \Rightarrow \alpha \approx 39.35°$

$\Rightarrow \sin \alpha \approx 0.63, \cos \alpha \approx 0.77;$ then $A' \approx 5.05, B' \approx 0.00,$

$C' \approx -0.05, D' \approx -5.07, E' \approx -6.19,$ and $F' = -1$

$\Rightarrow 5.05x'^2 - 0.05y'^2 - 5.07x' - 6.19y' - 1 = 0,$

a hyperbola

35. (a) $A' = \cos 45° \sin 45° = \left(\dfrac{\sqrt{2}}{2}\right)\left(\dfrac{\sqrt{2}}{2}\right) = \dfrac{1}{2}, B' = 0,$

$C' = -\sin 45° \cos 45° = -\dfrac{1}{2}, F' = -1$

$\Rightarrow \dfrac{1}{2}x'^2 - \dfrac{1}{2}y'^2 = 1 \Rightarrow x'^2 - y'^2 = 2$

(b) $A' = \dfrac{1}{2}, C' = -\dfrac{1}{2}$ (see part (a) above),

$D' = E' = B' = 0, F' = -a \Rightarrow \dfrac{1}{2}x'^2 - \dfrac{1}{2}y'^2 = a$

$\Rightarrow x'^2 - y'^2 = 2a$

37. The one curve that meets all three of the stated criteria is the ellipse $x^2 + 4xy + 5y^2 - 1 = 0$. The reasoning: The symmetry about the origin means that $(-x, -y)$ lies on the graph whenever (x, y) does. Adding $Ax^2 + Bxy + Cy^2 + Dx + Ey + F = 0$ and $A(-x)^2 + B(-x)(-y) + C(-y)^2 + D(-x) + E(-y) + F = 0$ and dividing by the result by 2 produces the equivalent equation $Ax^2 + Bxy + Cy^2 + F = 0$. Substituting $x = 1$, $y = 0$ (because the point $(1, 0)$ lies on the curve) shows further that $A = -F$. Then $-Fx^2 + Bxy + Cy^2 + F = 0$. By implicit differentiation, $-2Fx + By + Bxy' + 2Cyy' = 0$, so substituting $x = -2$, $y = 1$, and $y' = 0$ (from Property 3) gives $4F + B = 0$ $\Rightarrow B = -4F \Rightarrow$ the conic is $-Fx^2 - 4Fxy + Cy^2 + F = 0$. Now substituting $x = -2$ and $y = 1$ again gives $-4F + 8F + C + F = 0 \Rightarrow C = -5F \Rightarrow$ the equation is now $-Fx^2 - 4Fxy - 5Fy^2 + F = 0$. Finally, dividing through by $-F$ gives the equation $x^2 + 4xy + 5y^2 - 1 = 0$.

39. $\alpha = 90° \Rightarrow x = x' \cos 90° - y' \sin 90° = -y'$ and $y = x' \sin 90° + y' \cos 90° = x'$

(a) $\dfrac{x'^2}{b^2} + \dfrac{y'^2}{a^2} = 1$

(b) $\dfrac{y'^2}{a^2} - \dfrac{x'^2}{b^2} = 1$

(c) $x'^2 + y'^2 = a^2$

(d) $y = mx \Rightarrow x' = m(-y') \Rightarrow y' = -\dfrac{1}{m}x'$

(e) $y = mx + b \Rightarrow x' = m(-y') + b \Rightarrow y' = -\dfrac{1}{m}x' + \dfrac{b}{m}$

41. (a) $B^2 - 4AC = 1 - 4(0)(0) = 1 \Rightarrow$ hyperbola

(b) $xy + 2x - y = 0 \Rightarrow y(x - 1) = -2x \Rightarrow y = -\dfrac{2x}{x - 1}$

(c) $y = -\dfrac{2x}{x - 1} \Rightarrow \dfrac{dy}{dx} = \dfrac{2}{(x - 1)^2}$ and we want

$\dfrac{-1}{\left(\frac{dy}{dx}\right)} = -2$, the slope of $y = -2x$

$\Rightarrow -2 = -\dfrac{(x - 1)^2}{2} \Rightarrow (x - 1)^2 = 4$

$\Rightarrow x = 3$ or $x = -1$; $x = 3 \Rightarrow y = -3 \Rightarrow (3, -3)$ is a point on the hyperbola where the line with slope $m = -2$ is normal $\Rightarrow$ the line is $y + 3 = -2(x - 3)$ or $y = -2x + 3$;

$x = -1 \Rightarrow y = -1 \Rightarrow (-1, -1)$ is a point on the hyperbola where the line with slope $m = -2$ is normal $\Rightarrow$ the line is $y + 1 = -2(x + 1)$ or $y = -2x - 3$

[−9.4, 9.4] by [−6.1, 6.1]

43. (a) $B^2 - 4AC = 4^2 - 4(1)(4) = 0$, so the discriminant indicates that this conic is a parabola.

(b) The left-hand side of $x^2 + 4xy + 4y^2 + 6x + 12y + 9 = 0$ factors as a perfect square: $(x + 2y + 3)^2 = 0 \Rightarrow x + 2y + 3 = 0$ $\Rightarrow 2y = -x - 3$; thus the curve is a degenerate parabola (i.e., a straight line).

45. Assume the ellipse has been rotated to eliminate the xy-term $\Rightarrow$ the new equation is $A'x'^2 + C'y'^2 = 1$

$\Rightarrow$ the semi-axes are $\sqrt{\dfrac{1}{A'}}$ and $\sqrt{\dfrac{1}{C'}}$ $\Rightarrow$ the area is

$\pi\left(\sqrt{\dfrac{1}{A'}}\right)\left(\sqrt{\dfrac{1}{C'}}\right) = \dfrac{\pi}{\sqrt{A'C'}} = \dfrac{2\pi}{\sqrt{4A'C'}}$.

Since $B^2 - 4AC = B'^2 - 4A'C' = -4A'C'$

(because $B' = 0$) we find that the area is $\dfrac{2\pi}{\sqrt{4AC - B^2}}$ as claimed.

Appendix A6
(pp. 618–627)

1. $\sinh x = -\dfrac{3}{4} \Rightarrow \cosh x = \sqrt{1 + \sinh^2 x} = \sqrt{1 + \left(-\dfrac{3}{4}\right)^2}$

$= \sqrt{1 + \dfrac{9}{16}} = \sqrt{\dfrac{25}{16}} = \dfrac{5}{4}$, $\tanh x = \dfrac{\sinh x}{\cosh x}$

$= \dfrac{\left(-\frac{3}{4}\right)}{\left(\frac{5}{4}\right)} = -\dfrac{3}{5}$, $\coth x = \dfrac{1}{\tanh x} = -\dfrac{5}{3}$, $\operatorname{sech} x = \dfrac{1}{\cosh x}$

$= \dfrac{4}{5}$, and $\operatorname{csch} x = \dfrac{1}{\sinh x} = -\dfrac{4}{3}$

3. $\cosh x = \dfrac{17}{15}$, $x > 0 \Rightarrow \sinh x = \sqrt{\cosh^2 x - 1}$

$= \sqrt{\left(\dfrac{17}{15}\right)^2 - 1} = \sqrt{\dfrac{289}{225} - 1} = \sqrt{\dfrac{64}{225}} = \dfrac{8}{15}$,

$\tanh x = \dfrac{\sinh x}{\cosh x} = \dfrac{\left(\frac{8}{15}\right)}{\left(\frac{17}{15}\right)} = \dfrac{8}{17}$, $\coth x = \dfrac{1}{\tanh x}$

$= \dfrac{17}{8}$, $\operatorname{sech} x = \dfrac{1}{\cosh x} = \dfrac{15}{17}$, and $\operatorname{csch} x = \dfrac{1}{\sinh x} = \dfrac{15}{8}$

In Exercises 5–9, graphical support may consist of showing that the graph of the original expression minus the simplified one is the line $y = 0$.

5. $2 \cosh (\ln x) = 2\left(\dfrac{e^{\ln x} + e^{-\ln x}}{2}\right) = e^{\ln x} + \dfrac{1}{e^{\ln x}} = x + \dfrac{1}{x}$

7. $\cosh 5x + \sinh 5x = \dfrac{e^{5x} + e^{-5x}}{2} + \dfrac{e^{5x} - e^{-5x}}{2} = e^{5x}$

9. $(\sinh x + \cosh x)^4 = \left(\dfrac{e^x - e^{-x}}{2} + \dfrac{e^x + e^{-x}}{2}\right)^4 = (e^x)^4 = e^{4x}$

11. (a) $\sinh 2x = \sinh (x + x) = \sinh x \cosh x + \cosh x \sinh x$
$= 2 \sinh x \cosh x$

(b) $\cosh 2x = \cosh (x + x) = \cosh x \cosh x + \sinh x \sinh x$
$= \cosh^2 x + \sinh^2 x$

13. $y = 6 \sinh \frac{x}{3} \Rightarrow \frac{dy}{dx} = 6\left(\cosh \frac{x}{3}\right)\left(\frac{1}{3}\right) = 2 \cosh \frac{x}{3}$

15. $y = 2\sqrt{t} \tanh \sqrt{t} = 2t^{1/2} \tanh t^{1/2}$
$\Rightarrow \frac{dy}{dt} = [\text{sech}^2(t^{1/2})]\left(\frac{1}{2}t^{-1/2}\right)(2t^{1/2}) + (\tanh t^{1/2})(t^{-1/2})$
$= \text{sech}^2 \sqrt{t} + \frac{\tanh \sqrt{t}}{\sqrt{t}}$

17. $y = \ln (\sinh z) \Rightarrow \frac{dy}{dz} = \frac{\cosh z}{\sinh z} = \coth z$

19. $y = (\text{sech } \theta)(1 - \ln \text{sech } \theta)$
$\Rightarrow \frac{dy}{d\theta} = \left(\frac{-\text{sech } \theta \tanh \theta}{\text{sech } \theta}\right)(\text{sech } \theta)$
$\qquad + (-\text{sech } \theta \tanh \theta)(1 - \ln \text{sech } \theta)$
$= \text{sech } \theta \tanh \theta - (\text{sech } \theta \tanh \theta)(1 - \ln \text{sech } \theta)$
$= (\text{sech } \theta \tanh \theta)[1 - (1 - \ln \text{sech } \theta)]$
$= (\text{sech } \theta \tanh \theta)(\ln \text{sech } \theta)$

21. $y = \ln \cosh x - \frac{1}{2} \tanh^2 x$
$\Rightarrow \frac{dy}{dx} = \frac{\sinh x}{\cosh x} - \left(\frac{1}{2}\right)(2 \tanh x)(\text{sech}^2 x)$
$= \tanh x - (\tanh x)(\text{sech}^2 x) = (\tanh x)(1 - \text{sech}^2 x)$
$= (\tanh x)(\tanh^2 x) = \tanh^3 x$

23. $y = (x^2 + 1) \text{sech } (\ln x) = (x^2 + 1)\left(\frac{2}{e^{\ln x} + e^{-\ln x}}\right)$
$= (x^2 + 1)\left(\frac{2}{x + x^{-1}}\right) = (x^2 + 1)\left(\frac{2x}{x^2 + 1}\right) = 2x \Rightarrow \frac{dy}{dx} = 2$

25. $y = \sinh^{-1} \sqrt{x} = \sinh^{-1} (x^{1/2})$
$\Rightarrow \frac{dy}{dx} = \frac{\left(\frac{1}{2}\right)x^{-1/2}}{\sqrt{1 + (x^{1/2})^2}} = \frac{1}{2\sqrt{x}\sqrt{1 + x}} = \frac{1}{2\sqrt{x(1 + x)}}$

27. $y = (1 - \theta) \tanh^{-1} \theta$
$\Rightarrow \frac{dy}{d\theta} = (1 - \theta)\left(\frac{1}{1 - \theta^2}\right) + (-1) \tanh^{-1} \theta$
$= \frac{1}{1 + \theta} - \tanh^{-1} \theta$

29. $y = (1 - t) \coth^{-1} \sqrt{t} = (1 - t) \coth^{-1} (t^{1/2})$
$\Rightarrow \frac{dy}{dt} = (1 - t)\left[\frac{(1/2)t^{-1/2}}{1 - (t^{1/2})^2}\right] + (-1) \coth^{-1} (t^{1/2})$
$= \frac{1}{2\sqrt{t}} - \coth^{-1} \sqrt{t}$

31. $y = \cos^{-1} x - x \text{ sech}^{-1} x$
$\Rightarrow \frac{dy}{dx} = \frac{-1}{\sqrt{1 - x^2}} - \left[x\left(\frac{-1}{x\sqrt{1 - x^2}}\right) + (1) \text{sech}^{-1} x\right]$
$= \frac{-1}{\sqrt{1 - x^2}} + \frac{1}{\sqrt{1 - x^2}} - \text{sech}^{-1} x = -\text{sech}^{-1} x$

33. $y = \text{csch}^{-1}\left(\frac{1}{2}\right)^\theta$
$\Rightarrow \frac{dy}{d\theta} = -\frac{\left[\ln\left(\frac{1}{2}\right)\right]\left(\frac{1}{2}\right)^\theta}{\left(\frac{1}{2}\right)^\theta \sqrt{1 + \left[\left(\frac{1}{2}\right)^\theta\right]^2}}$
$= -\frac{\ln (1) - \ln (2)}{\sqrt{1 + \left(\frac{1}{2}\right)^{2\theta}}} = \frac{\ln 2}{\sqrt{1 + \left(\frac{1}{2}\right)^{2\theta}}}$

35. $y = \sinh^{-1} (\tan x) \Rightarrow \frac{dy}{dx} = \frac{\sec^2 x}{\sqrt{1 + (\tan x)^2}}$
$= \frac{\sec^2 x}{\sqrt{\sec^2 x}} = \frac{\sec^2 x}{|\sec x|} = \frac{|\sec x||\sec x|}{|\sec x|} = |\sec x|$

37. (a) If $y = \tan^{-1} (\sinh x) + C$, then $\frac{dy}{dx}$
$= \frac{\cosh x}{1 + \sinh^2 x} = \frac{\cosh x}{\cosh^2 x} = \text{sech } x$, which verifies the formula.

(b) If $y = \sin^{-1} (\tanh x) + C$, then $\frac{dy}{dx}$
$= \frac{\text{sech}^2 x}{\sqrt{1 - \tanh^2 x}} = \frac{\text{sech}^2 x}{\text{sech } x} = \text{sech } x$, which verifies the formula.

39. If $y = \frac{x^2 - 1}{2} \coth^{-1} x + \frac{x}{2} + C$, then
$\frac{dy}{dx} = x \coth^{-1} x + \left(\frac{x^2 - 1}{2}\right)\left(\frac{1}{1 - x^2}\right) + \frac{1}{2} = x \coth^{-1} x$,
which verifies the formula.

41. Let $u = 2x$ and $du = 2 \, dx$.
$\int \sinh 2x \, dx = \frac{1}{2} \int \sinh u \, du = \frac{\cosh u}{2} + C = \frac{\cosh 2x}{2} + C$

43. Let $u = \frac{x}{2} - \ln 3$ and $du = \frac{1}{2} dx$.
$\int 6 \cosh \left(\frac{x}{2} - \ln 3\right) dx = 12 \int \cosh u \, du = 12 \sinh u + C$
$= 12 \sinh \left(\frac{x}{2} - \ln 3\right) + C$

45. Let $u = \frac{x}{7}$ and $du = \frac{1}{7} dx$.
$\int \tanh \frac{x}{7} dx = 7 \int \frac{\sinh u}{\cosh u} du = 7 \ln |\cosh u| + C_1$
$= 7 \ln \left|\cosh \frac{x}{7}\right| + C_1$
$= 7 \ln \left|\frac{e^{x/7} + e^{-x/7}}{2}\right| + C_1$
$= 7 \ln \left|e^{x/7} + e^{-x/7}\right| - 7 \ln 2 + C_1$
$= 7 \ln \left|e^{x/7} + e^{-x/7}\right| + C$

47. Let $u = \left(x - \frac{1}{2}\right)$ and $du = dx$.
$\int \text{sech}^2 \left(x - \frac{1}{2}\right) dx = \int \text{sech}^2 u \, du = \tanh u + C$
$= \tanh \left(x - \frac{1}{2}\right) + C$

49. Let $u = \sqrt{t} = t^{1/2}$ and $du = \dfrac{dt}{2\sqrt{t}}$.

$\int \dfrac{\text{sech}\sqrt{t}\,\tanh\sqrt{t}}{\sqrt{t}}\,dt = 2\int \text{sech}\,u\,\tanh u\,du$
$= 2(-\text{sech}\,u) + C = -2\,\text{sech}\sqrt{t} + C$

51. Let $u = \sinh x$, $du = \cosh x\,dx$, the lower limit is

$\sinh(\ln 2) = \dfrac{e^{\ln 2} - e^{-\ln 2}}{2} = \dfrac{2 - \left(\frac{1}{2}\right)}{2} = \dfrac{3}{4}$ and the upper

limit is $\sinh(\ln 4) = \dfrac{e^{\ln 4} - e^{-\ln 4}}{2} = \dfrac{4 - \left(\frac{1}{4}\right)}{2} = \dfrac{15}{8}$.

$\int_{\ln 2}^{\ln 4} \coth x\,dx = \int_{\ln 2}^{\ln 4} \dfrac{\cosh x}{\sinh x}\,dx = \int_{3/4}^{15/8} \dfrac{1}{u}\,du$

$= \Big[\ln|u|\Big]_{3/4}^{15/8} = \ln\left|\dfrac{15}{8}\right| - \ln\left|\dfrac{3}{4}\right| = \ln\left|\dfrac{15}{8}\cdot\dfrac{4}{3}\right| = \ln\dfrac{5}{2} \approx 0.916$

53. $\displaystyle\int_{-\ln 4}^{-\ln 2} 2e^{\theta}\cosh\theta\,d\theta = \int_{-\ln 4}^{-\ln 2} 2e^{\theta}\!\left(\dfrac{e^{\theta} + e^{-\theta}}{2}\right)d\theta$

$= \displaystyle\int_{-\ln 4}^{-\ln 2} (e^{2\theta} + 1)\,d\theta = \left[\dfrac{e^{2\theta}}{2} + \theta\right]_{-\ln 4}^{-\ln 2}$

$= \left(\dfrac{e^{-2\ln 2}}{2} - \ln 2\right) - \left(\dfrac{e^{-2\ln 4}}{2} - \ln 4\right)$

$= \left(\dfrac{1}{8} - \ln 2\right) - \left(\dfrac{1}{32} - \ln 4\right) = \dfrac{3}{32} - \ln 2 + 2\ln 2$

$= \dfrac{3}{32} + \ln 2 \approx 0.787$

55. $\displaystyle\int_{-\pi/4}^{\pi/4} \cosh(\tan\theta)\,\sec^2\theta\,d\theta = \int_{-1}^{1} \cosh u\,du$

$= \Big[\sinh u\Big]_{-1}^{1} = \sinh(1) - \sinh(-1)$

$= \left(\dfrac{e^1 - e^{-1}}{2}\right) - \left(\dfrac{e^{-1} - e^1}{2}\right) = \dfrac{e - e^{-1} - e^{-1} + e}{2}$

$= e - e^{-1} \approx 2.350$, where $u = \tan\theta$, $du = \sec^2\theta\,d\theta$, the

lower limit is $\tan\left(-\dfrac{\pi}{4}\right) = -1$ and the upper limit is

$\tan\left(\dfrac{\pi}{4}\right) = 1$.

57. $\displaystyle\int_1^2 \dfrac{\cosh(\ln t)}{t}\,dt = \int_0^{\ln 2}\cosh u\,du = \Big[\sinh u\Big]_0^{\ln 2}$

$= \sinh(\ln 2) - \sinh(0) = \dfrac{e^{\ln 2} - e^{-\ln 2}}{2} - 0$

$= \dfrac{2 - \frac{1}{2}}{2} = \dfrac{3}{4}$, where $u = \ln t$, $du = \dfrac{1}{t}\,dt$, the lower limit is

$\ln 1 = 0$ and the upper limit is $\ln 2$

59. $\displaystyle\int_{-\ln 2}^{0}\cosh^2\!\left(\dfrac{x}{2}\right)dx = \int_{-\ln 2}^{0}\dfrac{\cosh x + 1}{2}\,dx$

$= \dfrac{1}{2}\displaystyle\int_{-\ln 2}^{0}(\cosh x + 1)\,dx = \dfrac{1}{2}\Big[\sinh x + x\Big]_{-\ln 2}^{0}$

$= \dfrac{1}{2}[(\sinh 0 + 0) - (\sinh(-\ln 2) - \ln 2)]$

$= \dfrac{1}{2}\left[(0 + 0) - \left(\dfrac{e^{-\ln 2} - e^{\ln 2}}{2} + \ln 2\right)\right]$

$= \dfrac{1}{2}\left[-\dfrac{(1/2) - 2}{2} + \ln 2\right]$

$= \dfrac{1}{2}\left(1 - \dfrac{1}{4} + \ln 2\right) = \dfrac{3}{8} + \dfrac{1}{2}\ln 2 = \dfrac{3}{8} + \ln\sqrt{2} \approx 0.722$

61. $\cosh^2 x - \sinh^2 x = 1$, so $\displaystyle\int_0^2 \pi(\cosh^2 x - \sinh^2 x)\,dx$

$= \pi\displaystyle\int_0^2 1\,dx = 2\pi$

63. $\displaystyle\int_0^{\ln\sqrt{199}} \pi(1 - \tanh x)^2\,dx$

$= \pi\displaystyle\int_0^{\ln\sqrt{199}} (1 - 2\tanh x + \tanh^2 x)\,dx$

$= \pi\displaystyle\int_0^{\ln\sqrt{199}} (2 - 2\tanh x - \text{sech}^2 x)\,dx$

$= \pi\Big[2x - 2\ln(\cosh x) - \tanh x\Big]_0^{\ln\sqrt{199}}$

$= \pi\Big[2\ln\sqrt{199} - 2\ln[\cosh(\ln\sqrt{199})] - \tanh(\ln\sqrt{199})\Big]$

$= \pi\left[2\ln\sqrt{199} - 2\ln\!\left(\dfrac{e^{\ln\sqrt{199}} + e^{-\ln\sqrt{199}}}{2}\right)\right.$

$\left.\quad - \dfrac{e^{\ln\sqrt{199}} - e^{-\ln\sqrt{199}}}{e^{\ln\sqrt{199}} + e^{-\ln\sqrt{199}}}\right]$

$= \pi\left[\ln 199 - \ln\!\left(\dfrac{(\sqrt{199} + 1/\sqrt{199})^2}{4}\right) - \dfrac{\sqrt{199} - 1/\sqrt{199}}{\sqrt{199} + 1/\sqrt{199}}\right]$

$= \pi\left[\ln 199 - \ln\!\left(\dfrac{199 + 2 + 1/199}{4}\right) - \dfrac{199 - 1}{199 + 1}\right]$

$= \pi\left[\ln 199 - \ln\!\left(\dfrac{10{,}000}{199}\right) - \dfrac{99}{100}\right] = \left(2\ln\dfrac{199}{100} - \dfrac{99}{100}\right)\pi$

≈ 1.214

65. (a) Let $E(x) = \dfrac{f(x) + f(-x)}{2}$ and $O(x) = \dfrac{f(x) - f(-x)}{2}$. Then

$E(x) + O(x) = \dfrac{f(x) + f(-x)}{2} + \dfrac{f(x) - f(-x)}{2} = \dfrac{2f(x)}{2} =$

$f(x)$. Also, $E(-x) = \dfrac{f(-x) + f(-(-x))}{2} = \dfrac{f(x) + f(-x)}{2} =$

$E(x) \Rightarrow E(x)$ is even, and $O(-x) = \dfrac{f(-x) - f(-(-x))}{2}$

$= -\dfrac{f(x) - f(-x)}{2} = -O(x) \Rightarrow O(x)$ is odd.

Consequently, $f(x)$ can be written as a sum of an even and an odd function.

(b) Even part: $\dfrac{e^x + e^{-x}}{2} = \cosh x$

odd part: $\dfrac{e^x - e^{-x}}{2} = \sinh x$

67. Note that $\dfrac{dv}{dt} = \sqrt{\dfrac{mg}{k}} \operatorname{sech}^2\left(\sqrt{\dfrac{gk}{m}}\, t\right)\left(\sqrt{\dfrac{gk}{m}}\right)$

$= g \operatorname{sech}^2\left(\sqrt{\dfrac{gk}{m}}\, t\right).$

Then $m\dfrac{dv}{dt} = mg \operatorname{sech}^2\left(\sqrt{\dfrac{gk}{m}}\, t\right)$ and

$mg - kv^2 = mg - k\left[\sqrt{\dfrac{mg}{k}} \tanh\left(\sqrt{\dfrac{gk}{m}}\, t\right)\right]^2$

$= mg\left[1 - \tanh^2\left(\sqrt{\dfrac{gk}{m}}\, t\right)\right]$

$= mg \operatorname{sech}^2\left(\sqrt{\dfrac{gk}{m}}\, t\right).$

Thus, $m\dfrac{dv}{dt}$ and $mg - kv^2$ are equal to the same quantity, so the differential equation is satisfied. Furthermore, the initial condition is satisfied because $v(0) = \sqrt{\dfrac{mg}{k}} \tanh 0 = 0$.

69. $\dfrac{dy}{dx} = \dfrac{-1}{x\sqrt{1-x^2}} + \dfrac{x}{\sqrt{1-x^2}}$

$\Rightarrow y = \int \dfrac{-1}{x\sqrt{1-x^2}}\, dx + \int \dfrac{x}{\sqrt{1-x^2}}\, dx$

$\Rightarrow y = \operatorname{sech}^{-1}(x) - \sqrt{1-x^2} + C;\ x = 1$ and

$y = 0 \Rightarrow C = 0 \Rightarrow y = \operatorname{sech}^{-1}(x) - \sqrt{1-x^2}$

71. $y = a \cosh(x/a) \Rightarrow y' = \sinh(x/a)$

$\Rightarrow y'' = (1/a)\cosh(x/a) = (1/a)\sqrt{\cosh^2(x/a)}$

$= (1/a)\sqrt{1 + \sinh^2(x/a)} = (1/a)\sqrt{1 + (y')^2}.$

Also, $y'(0) = \sinh(0) = 0$ and $y(0) = a\cosh(0) = a$.